P9-EDM-812

DHEA *in* HUMAN HEALTH *and* AGING

DHEA *in* HUMAN HEALTH *and* AGING

Edited by

RONALD ROSS WATSON

CRC Press is an imprint of the
Taylor & Francis Group, an **informa** business

CRC Press
Taylor & Francis Group
6000 Broken Sound Parkway NW, Suite 300
Boca Raton, FL 33487-2742

Printed in the United States of America on acid-free paper
Version Date: 20110601

International Standard Book Number: 978-1-4398-3883-9 (Hardback)

Visit the Taylor & Francis Web site at
http://www.taylorandfrancis.com

and the CRC Press Web site at
http://www.crcpress.com

Contents

Preface

Dehydroepiandrosterone (DHEA) and its sulfate ester are secretory products of adrenal glands. They are the most abundant hormones in the systemic circulation of humans, convertible into androgens and estrogens in the periphery with the potential to act alone or through their estrogen end products. Based on their abundance and reduced production during the frequent diseases of aging, they should be critical to many aspects of health. Therefore, DHEA is commonly used in the United States and some other countries as a nutritional supplement for antiaging, metabolic support, and other uses. Yet, substantial mystery exists about the role of low levels of DHEA in seniors when chronic disease states are prevalent. While animal studies clearly show substantial benefits to DHEA/DHEAS supplementation in reducing cancer growth, AIDS progression, and other physiological dysfunctions, these are only partially defined for humans. A major goal of this book is to document the role or lack thereof for DHEA and its reduced presence as well as supplementation on *human* physiological dysfunction.

DHEA and DHEAS are steroids synthesized in human adrenals and in the brain, further suggesting a role of these hormones in brain function and development. Despite intensifying research, many questions concerning their mechanisms of action and their potential involvement in illnesses remain unanswered. The endocrine system, specifically the hypothalamus–pituitary–adrenal axis, plays an important role in modulating immune function. With aging, an imbalance occurs between two adrenal hormones, cortisol and DHEA, that have opposing actions on immune function. There is continuing controversy on whether DHEA treatment benefits adrenal-deficient and elderly people with an age-related decline in DHEA and its sulfate ester. Available studies have demonstrated beneficial effects of DHEA on health perception, vitality, fatigue, and (in women) sexuality. DHEA restores low circulating androgens to the normal range in women, and side effects are mostly mild. DHEA may affect production of Th1- and Th2-associated cytokines, suggesting their significance in diseases in which imbalanced lymphocyte activity plays the essential role. DHEA's decline with age makes it an interesting marker for many diseases of aging, and it has been tested extensively as a supplement to restore DHEA levels, thus changing disease and disease risks. However, how many therapies and benefits can be supported by research? How successful are DHEA levels in predicting risk of disease? Individual authors review such questions as they describe DHEA's role and changes in humans.

This edition has a unique focus on the *role of DHEA in human health*. To accomplish this goal, the editor selected authors who are researchers with experience in studying the role of DHEA deficiency and supplementation in various diseases and physiological states in *humans*.

This book is a comprehensive set of reviews on the biology of DHEA relevant to *human health*. The origin of circulating DHEA and adrenal-derived androgens in humans and nonhuman primates is largely distinct from other mammalian species. However, there is a section of potential future uses of DHEA, focusing on model studies that have not or cannot yet be done in humans. There are numerous areas with clinical potential importance and varying levels of study that will be related to low DHEA with age or physiological changes and their impacts on health.

In addition, DHEA is widely available as a dietary supplement in the United States and some other countries. Thus, the sections focus on prevention as well as treatment of various human disease states by changing DHEA levels. Often, the role of DHEA is still controversial in certain human health conditions, and the researchers help define what is known for the various conditions being treated. The book will also have some overview chapters on age-induced deficiency, supplementation for aged women, and the role of therapy for aging. Prevention of disease with DHEA includes diabetes, fitness, infectious disease, cancer, AIDS, bone health, and cardiovascular diseases. Treatment covers some similar areas, with autism and mental health being important additions. Animal model information is included as needed to help understand studies done in humans. An important area will be reviews of the effects of loss of adrenal gland function and the subsequent reduction in DHEA production and of its replacement as a therapy. DHEA is one of two hormones sold over the counter as a dietary supplement, and it is thus readily available to consumers in many countries.

Finally, a group of chapters reviews mechanisms of action of DHEA in human diseases including prostate and ovarian health, vascular modification, and adverse effects in DHEA-supplemented women, and finally, stress, memory, aggression, and Alzheimer's disease. The book should be of extensive interest to gerontologists, physicians, physiologists, biologists, public health workers, and biomedical researchers.

Acknowledgments

An especial thanks is extended to Bethany L. Stevens, the editorial assistant to Dr. Ronald Ross Watson. She spent many hours working with the publisher and with the authors of this book. She made it possible for Dr. Watson to function as the editor by lightening his load and taking responsibility for routine questions, format, and style. Support for editorial assistance was provided by Southwest Scientific Editing & Consulting, LLC. The help of Mari Stoddard, Arizona Health Sciences Librarian, was crucial to finding all of the authors and topics that appear in the book.

Editor

Ronald R. Watson, PhD, attended the University of Idaho but graduated from Brigham Young University in Provo, Utah, with a degree in chemistry in 1966. He obtained his PhD in biochemistry from Michigan State University in 1971. He completed his postdoctoral schooling in nutrition and microbiology at the Harvard School of Public Health, where he gained two years of postdoctoral research experience in immunology and nutrition.

From 1973 to 1974, Dr. Watson was an assistant professor of immunology and performed research at the University of Mississippi Medical Center in Jackson. He was an assistant professor of microbiology and immunology at the Indiana University Medical School from 1974 to 1978 and an associate professor at Purdue University in the Department of Food and Nutrition from 1978 to 1982. In 1982, Dr. Watson joined the faculty at the University of Arizona Health Sciences Center in the Department of Family and Community Medicine of the School of Medicine. He is currently a professor of health promotion sciences in the Mel and Enid Zuckerman Arizona College of Public Health.

Dr. Watson is a member of several national and international nutrition, immunology, cancer, and alcoholism research societies. Among his patents is one on a dietary supplement, passion fruit peel extract, with more pending. He did research on DHEA's effects on mouse AIDS and immune function for 20 years. He edited a book on melatonin (*Health Promotion and Aging: The Role of Dehydroepiandrosterone (DHEA)*, Harwood Academic Publishers, 1999, 164 pages). For 30 years, he was funded by the Wallace Research Foundation to study dietary supplements in health promotion. Dr. Watson has edited more than 35 books on nutrition, dietary supplements, and over-the-counter agents, and 53 other scientific books. He has published more than 500 research and review articles.

Contributors

Clarence N. Ahlem
Product Development
Harbor Biosciences
San Diego, California

Ryuichi Ajisaka
Division of Sports Medicine
Comprehensive Human Sciences
University of Tsukuba
Tennodai, Tsukuba, Japan

Hulya Aksoy
Department of Biochemistry
School of Medicine
Ataturk University
Erzurum, Turkey

H. Elliott Albers
Neuroscience Institute
Center for Behavioral Neuroscience
Georgia State University
Atlanta, Georgia

Toshihiro Ansai
Division of Community Oral Health Science
Kyushu Dental College
Kitakyushu, Japan

Julia T. Arnold
National Center for Complementary and Alternative Medicine
National Institutes of Health
Bethesda, Maryland

Dominick L. Auci
Allergy, Autoimmunity, and Inflammation
Harbor Biosciences
San Diego, California

Adriane Belló-Klein
Universidade Federal do Rio Grande do Sul
Instituto de Ciencias Básicas da Saúde
Porto Alegre, Brazil

David Ben-Nathan
Laboratory for Vaccine Control
Kimron Veterinary Institute
Beit Dagan, Israel

Sébastien Bonnet
Centre de Recherche de l'Hotel-Dieu de Quebec
Quebec City, Quebec, Canada

Viviana Castilla
Laboratorio de Virología
Departamento de Química Biológica Facultad de Ciencias Exactas y Naturales
Universidad de Buenos Aires Ciudad Universitaria
Buenos Aires, Argentina

Mei-Ling Cheng
Department of Medical Biotechnology and Laboratory Science
Chang Gung University
Taoyuan, Taiwan

Inseon S. Choi
Department of Allergy
Chonnam National University Medical School
Gwangju, Republic of Korea

Jan Croonenberghs
University Center of Child and Adolescent Psychiatry
A.Z. Middelheim Faculty of Medicine
University of Antwerp
Antwerp, Belgium

Dirk Deboutte
University Center of Child and Adolescent Psychiatry
A.Z. Middelheim Faculty of Medicine
University of Antwerp
Antwerp, Belgium

Gregory E. Demas
Department of Biology
Program in Neuroscience and Center for the Integrative Study of Animal Behavior
Indiana University
Bloomington, Indiana

Joseph S. Dillon
Division of Endocrinology
Department of Internal Medicine
University of Iowa
Iowa City, Iowa

Eric Dumas De La Roque
Réanimation neonatale
Hôpital des Enfants
CHU Bordeaux
Bordeaux, France

Laïla El Kihel
University of Caen
UFR Sc. Pharmaceutiques
Centre d'Etudes et de Recherche sur le Médicament de Normandie
Caen, France

Ahmed I. El-Sakka
Department of Urology
Faculty of Medicine
Suez Canal University
Ismailia, Egypt

James M. Frincke
Harbor Biosciences
San Diego, California

Jonathan P. Fry
Department of Neuroscience, Physiology and Pharmacology
University College, London
London, United Kingdom

Norma Galindo-Sevilla
Departamento de Infectologia e Inmunologia
Instituto Nacional de Perinatología
Mexico City, Mexico

Marta Garaulet
Department of Physiology
University of Murcia
Murcia, Spain

Alessandro D. Genazzani
Department of Obstetrics and Gynecology
Gynecological Endocrinology Center
University of Modena and Reggio Emilia
Modena, Italy

Melanie Hamann
Department of Veterinary Medicine
Institute of Pharmacology and Toxicology
Freie Universitat Berlin
Berlin, Germany

Juan Jose Hernández-Morante
Facultad de Enfermeria
Universidad Catolica San Antonio de Murcia
Murcia, Spain

Hung-yao Ho
Department of Medical Biotechnology and Laboratory Science
Graduate Institute of Medical Biotechnology
Chang Gung University
Kwei-san, Taoyuan, Taiwan

Masaaki Ii
Department of Pharmacology
Osaka Medical College
Takatsuki, Osaka, Japan

Maria Helena Vianna Metello Jacob
Universidade Luterana do Brazil
Moradas da Colina, Guaiba, Brazil

Catherine M. Jankowski
Division of Geriatric Medicine
University of Colorado Denver
Aurora, Colorado

Ippei Kanazawa
Department of Internal Medicine
Shimane University Faculty of Medicine
Izumo-shi, Shimane, Japan

Rajesh Kannangai
Department of Clinical Virology
Christian Medical College
Vellore, India

Alicja Kasperska-Zajac
Clinical Department of Internal Diseases, Allergology and Clinical Immunology
Medical University of Silesia
Katowice, Poland

Surendra S. Katyare
Department of Biochemistry
Faculty of Science
The Maharaja Sayajirao University of Baroda
Vadodara, India

Wendy M. Kohrt
Division of Geriatric Medicine
University of Colorado, Denver
Aurora, Colorado

Alexander W. Krug
Division of Endocrinology, Diabetes and Hypertension
Harvard Medical School
Brigham and Women's Hospital
Boston, Massachusetts

Chia-Hua Kuo
Laboratory of Exercise Biochemistry
Taipei Sports University
Taipei City, Taiwan

Fernand Labrie
Research Center in Molecular Endocrinology, Oncology and Human Genomics
Laval University
Quebec City, Quebec, Canada

Chiara Lanzoni
Department of Obstetrics and Gynecology
Gynecological Endocrinology Center
University of Modena and Reggio Emilia
Modena, Italy

Bill L. Lasley
Center for Health and the Environment
University of California
Davis, California

Emilia P. Liao
Division of Endocrinology and Metabolism
Beth Israel Medical Center
Albert Einstein College of Medicine
New York, New York

Rebeca López-Marure
Departamento de Biología Celular
Instituto Nacional de Cardiologia Ignacio Chávez
Mexico City, Mexico

Roger M. Loria
Department of Microbiology, Immunology, Pathology and Emergency Medicine
Virginia Commonwealth University School of Medicine
Richmond, Virginia

Michael Maes
Clinical Research Center for Mental Health
Antwerp, Belgium

Javier Mancilla-Ramírez
Instituto Nacional de Perinatologia
Mexico City, Mexico

Radmila Mileusnic
Department of Life Sciences
Faculty of Science
The Open University
Milton Keynes, United Kingdom

Hiren R. Modi
Brain Physiology and Metabolism Section
National Institute on Aging
National Institutes of Health
Bethesda, Maryland

Zeina Nahleh
Department of Internal Medicine
Division of Hematology-Oncology
TTUHSC-Paul L. Foster School of Medicine
El Paso, Texas

Brianne O'Leary
Department of Internal Medicine
Division of Endocrinology
University of Iowa
Iowa City, Iowa

Roxane Paulin
Centre de Recherche de l'Hôtel-Dieu de Québec
Quebec City, Quebec, Canada

Fátima Pérez-de Heredia
Immunonutrition Group
Department of Metabolism and Nutrition
ICTAN-Instituto del Frío (CSIC)
Madrid, Spain

Iván Pérez-Neri
Department of Neurochemistry
National Institute of Neurology and Neurosurgery
Mexico City, Mexico

Leonid Poretsky
Division of Endocrinology and Metabolism
Gerald J. Friedman Diabetes Institute
Beth Israel Medical Center
Albert Einstein College of Medicine
New York, New York

Christopher L. Reading
Harbor Biosciences
San Diego, California

Aled Rees
Centre for Endocrine and Diabetes Sciences
Cardiff University School of Medicine
Heath Park, Cardiff, United Kingdom

Maria Flavia M. Ribeiro
Universidade Federal do Rio Grande do Sul
Instituto de Ciencias Básicas da Saúde
Porto Alegre, Brazil

Sam Rice
Diabetes Centre
Prince Philip Hospital
Llanelli, Wales, United Kingdom

Angelika Richter
Department of Veterinary Medicine
Institute of Pharmacology and Toxicology
Freie Universität Berlin
Berlin, Germany

Camilo Ríos
Department of Neurochemistry
National Institute of Neurology and Neurosurgery
Mexico City, Mexico

Carla M. Romero
Department of Medicine
Beth Israel Medical Center
Albert Einstein College of Medicine
New York, New York

Alex Sander da Rosa Araújo
Universidade Federal do Rio Grande do Sul
Instituto de Ciencias Básicas da Saúde
Porto Alegre, Brazil

Koji Sato
Faculty of Sport and Health Science
Ritsumeikan University
Kusatsu, Shiga, Japan

Kiran K. Soma
Graduate Program in Neuroscience
Departments of Psychology and Zoology
University of British Columbia
Vancouver, British Columbia, Canada

Gopalan Sridharan
Sri Narayani Hospital and Research Center
Vellore, India

Elizabeth Sujkovic
Faculty of Medicine
Imperial College
Hammersmith Hospital Campus
London, United Kingdom

Nishant Tageja
Department of Internal Medicine
Wayne State University
Detroit, Michigan

Katelijne Van Praet
University Center of Child and Adolescent Psychiatry
A.Z. Middelheim Faculty of Medicine
University of Antwerp
Antwerp, Belgium

Mónica B. Wachsman
Laboratorio de Virología
Departamento de Química Biológica
Facultad de Ciencias Exactas y Naturales
Universidad de Buenos Aires Ciudad Universitaria
Buenos Aires, Argentina

Toru Yamaguchi
Department of Internal Medicine
Shimane University Faculty of Medicine
Izumo-shi, Shimane, Japan

Akihiro Yoshida
Division of Community Oral Health Science
Kyushu Dental College
Kitakyushu, Japan

Barnett Zumoff
Department of Medicine
Division of Endocrinology
Beth Israel Medical Center
Albert Einstein College of Medicine
New York, New York

Section I

Overviews of Key DHEA Modified Conditions

1 DHEA versus Androstenediol in Middle-Aged Women

Bill L. Lasley

CONTENTS

INTRODUCTION

Higher circulating levels of dehydroepiandrosterone (DHEA) are associated with superior health and vitality in older adults, and this benefit is largely attributed directly to either DHEA or to its peripheral conversion to steroids with specific biological activities. Such speculations have led to widespread over-the-counter access and use of DHEA supplements particularly for both middle-aged men and women. However, there is now information to suggest that some of the benefits that have been previously associated with higher endogenous DHEA in middle-aged women are at least partially attributable to the adrenal secretion of androstenediol (Adiol), which is secreted in parallel with DHEA and is the next steroid in the delta-five steroidogenic pathway. Adiol, which is structurally a C-19 androgen, has inherent estrogenic bioactivity because of the 3–17 diols, reaches effective circulating levels, and contributes to a potential positive endocrine effect in some women.

Recent observations indicate that there is a little-recognized rise in DHEA/dehydroepiandrosterone sulfate (DHEAS) that occurs during the menopausal transition (Lasley et al. 2002; Crawford et al. 2009), and this rise is accompanied by a parallel rise in Adiol (Lasley et al. 2011). The circulating concentration of Adiol during the menopausal transition in some women can exceed the concentrations that would not likely be achieved by conversion of exogenous DHEA, indicating that there is a direct secretion of Adiol rather than the metabolism of DHEA to Adiol. This chapter examines these and other data to explore the possibility that DHEA alone may not responsible for all the benefits that have been observed when endogenous circulating DHEA is elevated in older age groups and provides evidence that increased circulating Adiol may provide significant hormonal support for many women.

BACKGROUND

The prevailing dogma up until 2002 was that both men and women have a gradual and continuous decline in circulating DHEA and DHEAS starting in the fourth decade of life. At least one earlier report has suggested a gender difference in circulating DHEAS (Sulcová et al. 1997), but the accepted gradual decline in DHEA in both genders starting in the fourth decade of life has been so well established that DHEA is used in many research studies as a biomarker for somatic aging. However, this concept began to change when the longitudinal data from the Study of Women's Health Across the Nation (SWAN) was able to show that a large majority of middle-aged women exhibited a discernible positive inflection of DHEAS (Lasley et al. 2002) that was not observable when annual measurements of the same DHEAS data were plotted or analyzed according to chronological age (Crawford et al. 2009). In fact, when the same annual measurements of DHEAS are plotted by chronological age, a clear, continuous decline in DHEAS is observed through the fifth decade of life and onward (Crawford et al. 2009). This dichotomy explains completely why this phenomenon had been so long overlooked. The reason that this relatively large hormone dynamic is not detectable based on chronological age is that the individual stages of ovarian function during the menopausal transition are found to occur over many different years of age in individual women. Thus, the alignment of longitudinal hormone data by chronological age results in the combining of three to five different ovarian stages at each year of age between the ages of 45 and 55. Since the rise in DHEAS is specific to the early and late perimenopause stages of ovarian function, the rise in circulating DHEAS is obscured when the ovarian-stage rise is diluted by data from women of the same age who are outside these two specific ovarian stages. Until the longitudinal SWAN data was available for analysis, there simply was not enough hormone data to provide a comprehensive evaluation of the sequence of hormone change with respect to changes in ovarian function.

A clear increase in DHEAS is observed in 85% of all women studied but only when the annual DHEAS measurements of women are aligned by the stage of ovarian function according to the Stages of Reproductive Aging Workshop (STRAW) convention (Crawford et al. 2009; Soules et al. 2001). DHEAS begins to rise in the early perimenopause, plateaus in the early postmenopause, and returns to premenopausal concentrations in the late postmenopause. It is likely that more than 85% of all women experience a positive inflection of DHEAS production because some of the smaller increases will be nullified by the ongoing age-related decline. Although the mean circulating levels of DHEAS differ among women from different ethnicities as they enter the menopausal transition and the subsequent age-related mean rate of decline of DHEAS varies between ethnicities, the relative increase and variation between individuals in the rise of DHEAS seen during the menopausal transition has a similar pattern in all of the women who expressed it. This observation suggests that while the secretion of DHEA and production of DHEAS in premenopausal women have a substantial genetic foundation, the increase in the circulating concentrations of these steroids during the menopausal transition is less genetically controlled and more likely to be induced by some common mechanism or event.

Since DHEA can be produced by both the ovary and the adrenal and then sulfated by the adrenal, the increase in circulating DHEAS is closely linked to the ovarian stage of the menopausal transition, and it is important to understand the source of the increase in DHEAS before attempting to investigate the mechanism driving it. Data from the SWAN study was also able to show that women who had undergone bilateral salpingo-oophorectomy in the early perimenopause exhibit a similar rise in DHEAS in annual samples during the next 3 to 4 years (Lasley et al. 2010). This observation indicates that while the significant changes in circulating sex steroid levels in the perimenopausal transition are triggered by the initial decline in ovarian function during the early perimenopause, the presence of the ovaries are not required to sustain that rise. Furthermore, the observation indicates that some, if not most, of the rise in DHEAS observed at this time is not attributable to ovarian steroidogenesis, but rather to a change in adrenal weak androgen production by the adrenal cortex.

A relatively large body of information can be found in the literature that demonstrates adrenal steroid production can be induced in animal models through an array of endocrine manipulations. However, because the murine animal models that have most often been used for these studies do not synthesize the same weak androgens in the adrenal gland as higher primates, these studies have not provided the direct evidence required to direct research into the potential induction of delta-five androgens in humans. However, the report of the presence of luteinizing hormone (LH) receptors in some human adrenal cells (Pabon et al. 1996) along with the rise in gonadotropins that occurs at the time of the DHEAS rise provides enough parallels to the rodent experimental data and therefore allows us to begin to formulate a reasonable hypothesis on the mechanism by which adrenal pathways are modulated during the menopausal transition. The testing of these hypotheses will hopefully allow us to address some of the perplexing issues of the menopausal transition.

SIGNIFICANCE

Despite clear evidence that higher endogenous concentrations of DHEA has substantial benefit in middle-aged women (Davis et al. 2008), of the numerous DHEA intervention studies conducted to date, most of them have failed to provide strong or convincing positive evidence (Baulieu et al. 2000; Percheron et al. 2003) that DHEA or its downstream metabolites are responsible for these benefits. Certainly, a partial explanation for these failures is that it has not been clear who in the population could or would benefit from the intervention with DHEA. A second possibility is that these interventions were not of sufficient length of time to generate a full effect. However, metabolic studies following DHEA supplementation now indicate that exogenous DHEA is not efficiently converted to the estrogenic compounds that were originally anticipated. In fact, in women, DHEA is converted to bioactive androgens rather than classical bioactive estrogens. The simplest explanation for this paradox is that some of the beneficial effects associated with higher DHEA levels are in some part due to the effects of Adiol, which is secreted in parallel with DHEA and can provide additional estrogenic support. The dichotomy between the benefits of higher endogenous DHEA and marginal effects of DHEA intervention seems to be that the former is associated with higher Adiol in women during the menopausal transition, while supplemental exogenous DHEA is not efficiently converted to strong bioactive estrogens.

The current literature is not convincing in that oral DHEA supplements should be expected to have significant positive effect particularly in middle-aged women (Bird et al. 1978). A 1-year daily oral DHEA (50 mg/day) regimen only modestly decreased bone turnover, slightly increased libido, and improved skin hydration in women over 70 years old, indicating that DHEA supplementation normalized some effects of aging without dramatic improvements in general health (Baulieu et al. 2000). Similarly, a 1-year study with 50 mg/day oral DHEA revealed no positive effect on muscle status (Percheron et al. 2003). More recently, a larger study using 50 mg daily, oral DHEA, revealed only modest effect of bone mineral density (BMD) and bone resorption in women but not in men (Von Muhlen et al. 2008). However, an intravaginal suppository of DHEA provided a highly efficient treatment of age-related vaginal atrophy (Labrie et al. 2009a) but no significant change in circulating sex steroid concentrations (Labrie et al. 2009b). More revealing, a transdermal delivery of DHEA to women led to a fivefold increase in circulating DHEAS, but less than a two and one-half-fold increase in circulating Adiol glucuronide (Labrie et al. 2007) and a 25 mg oral dose for 3 months resulted in a doubling of circulating concentrations of Adiol (0.32–0.66 ng/mL; Stanczyk et al. 2009) suggesting that peripheral conversion of exogenous DHEA to Adiol is relatively modest. In contrast, the rise in circulating Adiol can increase fivefold in some women (<0.3 to >1.5 ng/mL) during the menopausal transition.

A consistent finding in SWAN has been the finding of only a modest association between circulating estradiol (E2) and individual phenotypes during the menopausal transition (Randolph et al. 2003). While there are several plausible explanations, including the difficulty in establishing mean circulating estrogen levels in women with cyclic ovarian activity, the simplest may be that classical estrogens are not the only estrogenic hormone contributing to total estrogenicity during the menopausal transition. When the total circulating estrogen receptor alpha ligand load (ERLL) was measured using a cell-based bioassay, E2 was closely correlated to ERLL while circulating Adiol was significantly correlated to ERLL only when E2 concentrations were in the lowest quartile (Lasley et al. 2011). This observation suggests that when E2 levels are reasonably high, E2 concentrations alone are sufficient to maintain an "estrogenized" condition. However, when E2 concentrations are low, then the contribution of non-E2 compounds may be important for an optimal estrogenized condition to exist. When both circulating E2 and Adiol concentrations are low, then a poorly estrogenized condition would more likely exist. Thus, the measurement of circulating Adiol, either alone or in combination with E2, may more accurately predict the phenotypes observed during the MT than measurements of E2 alone.

ESTROGENIC EFFECTS OF ANDROSTENEDIOL

Adiol has been long recognized as having estrogenic bioactivity (Adams et al. 1990; Adams 1998) of 5%–10% of the estrogenic bioactivity of E2 but is found in circulating concentrations that are 10–100 times greater than E2 during the menopausal transition (Lasley et al. 2011). Thus, Adiol has the potential to contribute as much, if not more, to estrogenize women as does E2 when E2 levels are low and Adiol concentrations are high. If Adiol adds significant estrogenization to middle-aged women, then it could positively affect all estrogen target organs and possibly health outcomes. Although SWAN did not measure Adiol concentrations in all women, paired measurements of DHEAS and Adiol for approximately 200 subjects revealed a strong, positive, and linear relationship between circulating DHEAS and Adiol concentrations (Lasley et al. 2011). Therefore, it is possible to assume that circulating Adiol and the estrogenicity associated with it are high when DHEAS and/or DHEA are high and Adiol would have its greatest positive effects when DHEAS/DHEA circulating concentrations are the highest. If Adiol behaves as predicted as a 17-hydroxylated, weak estrogen receptor (ER) agonist, then it would bind to sex hormone binding globulin (SHBG) and the ERs, but less strongly than E2. It would likely have pure agonistic effects when endogenous true estrogen concentrations are extremely low but would act as a weak antagonist when the classical estrogen levels were normal or high. Since E2 levels vary widely across the ovarian cycle and Adiol would be expected to remain relatively constant, then the positive and negative effects of Adiol could also be transient. The stronger positive effects may vary on a day-to-day basis and may primarily affect acute-onset symptoms. Long-term effects of relatively high Adiol could be to supplement classical estrogen action on all target cells although some antagonisms at extremely high concentrations cannot be ruled out.

INTEGRATION OF THE CURRENT DATA

There are at least three endocrine conundrums of the menopausal transition: First, individual women have a wide range of symptoms and health outcome trajectories that begin to appear during the early perimenopause when all women exhibit a similar rise in the follicle-stimulating hormone and little, if any, clinical change in ovarian function. Second, estrogen intervention during this time is highly efficacious despite there being little clinical agreement that measuring classical estrogen production during the early perimenopausal can identify women who will benefit from estrogen augmentation. Third, during the menopausal transition, higher circulating endogenous DHEA is shown to be beneficial but DHEA intervention has little, if any, positive effect. Each of these apparent paradoxes is logically explained based on the following three assumptions: (1) E2 is not the only estrogen responsible for regulating and maintaining an appropriate estrogenized condition. Thus, the wide range of symptoms and health outcomes that are observed among individual women during the menopausal transition may be a result of the wide range of adrenal response in terms of the increased production of steroids during the menopausal transition. (2) Intervention with classical estrogens, which is highly effective, does not prove that the hormone deficit that the intervention is resolving is simply a deficit in classical estrogens. Other steroids such as Adiol can transduce the same intercellular signal transduction pathways and would not be individually recognized. (3) The association of higher circulating DHEAS with superior health outcomes in middle-aged women is not proof that DHEAS is the causal agent for healthier aging. In fact, there is declining acceptance that there is substantial benefit. Bioactive hormones that have a similar profile, such as Adiol, should be considered for their contribution.

POSSIBLE REGULATORY MECHANISM

There is vast literature that demonstrates the ability of LH to induce its own receptors in the adrenal of mice (Kero et al. 2000; Bielinska et al. 2004). These receptors then become functional when the ovary is removed, indicating a two-step process in the stimulation by LH to drive adrenal steroids. Similarly, numerous reports indicate LH or human chorionic gonadotropin (hCG) can induce adrenal tumors or drive steroid production in humans (Carlson 2007). In addition, LH receptors have been reported in human adrenal cells (Rao, Zhou, and Lei 2004). Taken together, these data suggest that there is a basic mechanism by which LH can regulate adrenal steroids and that this can occur in humans if the conditions are correct. The difference would be that the murine models would not respond by producing androgens because the steroidogenic machinery for synthesizing adrenal androgens is a unique primate trait. Since we now observe the induction of a highly variable amount of adrenal steroids in a majority of women (but not in men), and this event occurs when the ovarian function begins to lose its long-loop inhibition of gonadotropins, we can speculate how the ability to increase adrenal delta-five steroids might occur. The most logical explanation, and the one that is consistent with observations in mice, is that the hCG produced in pregnancy over a woman's lifetime induces LH receptors in the adrenals. These receptors then become functional as ovarian failure removes the inhibitory factor that has been demonstrated in mice by ovariectomy.

CONCLUSION

The identification of an increase in weak androgen production by the adrenal in middle-aged women may have profound implications relating to the endocrinology of the menopausal transition. Although peripheral conversion of increased DHEA to more bioactive steroids may be amplified as a potential contributor to the total steroidal milieu for middle-aged women, the direct, parallel secretion of Adiol, which has inherent estrogenic activity, is equally, if not more, important. In either case, the inclusion of the adrenal in the study of the endocrinology of the menopausal transition will likely be a productive research avenue in understanding the very wide array and degree of severity of symptoms that women incur in the face of a relatively similar and subtle decline in ovarian function during the menopausal transition. Although the mechanism of this phenomenon is yet to be determined, its occurrence at the time of a subtle change in gonadal function provides an important clue in light of the experiments in murine models in which induced adrenal LH receptors become functional only following ovariectomy. We can speculate at this time that in women hCG of pregnancy may act to induce LH receptors in the zona reticularis and the initial decline in ovarian function during the early perimenopause activates these receptors to promote increased synthesis of the delta-five steroids. In addition, there is the possibility that exogenous steroid hormone intervention as hormone replacement therapies may further modulate adrenal steroid production and act peripherally to provide additional support to target organs.

REFERENCES

Adams, J. B. 1998. Adrenal androgens and human breast cancer: A new appraisal. *Breast Cancer Res Treat* 51(2):183–8.

Adams, J. B., P. Martyn, F. T. Lee, N. S. Phillips, and D. L. Smith. 1990. Metabolism of 17 beta-estradiol and the adrenal-derived estrogen 5-androstene-3 beta, 17 beta-diol (hermaphrodiol) in human mammary cell lines. *Ann N Y Acad Sci* 595:93–105.

Baulieu, E. E., G. Thomas, S. Legrain, N. Lahlou, M. Roger, B. Debuire, G. Francounau et al. 2000. DHEA, DHEA sulfate, and aging: Contribution of the DHEAge study to a sociobiomedical issue. *Proc Natl Acad Sci* 97:4279–84.

Bielinska, M., E. Genova, I. Boime, H. Parviainen, S. Kiiveri, N. Rahman, J. Leppaluoto, M. Heikinheimo, and D. B. Wilson. 2004. Nude mice as a model for gonadotropin-induced adrenocortical neoplasia. *Endocr Res* 30:913–7.

Bird, C. E., J. Murphy, K. Boroomand, W. Finnis, D. Dressel, and A. F. Clark. 1978. Dehydroepiandrosterone: Kinetics of metabolism in normal men and women. *J Clin Endocrinol Metab* 47:818–22.

Carlson, H. E. 2007. Human adrenal cortex hyperfunction due to LH/hCG. *Mol Cell Endocrinol* 269:46–50.

Crawford, S., N. Santoro, G. A. Laughlin, M. F. Sowers, D. McConnell, K. Sutton-Tyrrell, G. Weiss, M. Vuga, J. Randolph, and B. L. Lasley. 2009. Circulating dehydroepiandrosterone sulfate concentrations during the menopausal transition. *J Clin Endocrinol Metab* 94(8):2945–51.

Davis, S. R., S. M. Shah, D. P. McKenzie, J. Kulkarni, S. L. Davison, and R. J. Bell. 2008. Dehydroepiandrosterone sulfate levels are associated with more favorable cognitive function in women. *J Clin Endocrinol Metab* 93:801–8.

Kero, J., M. Poutanen, F. P. Zhang, N. Rahman, A. M. McNicol, J. H. Nilson, R. A. Keri, I. T. Huhtaniemi. 2000. Elevated luteinizing hormone induces expression of its receptor and promotes steroidogenesis in the adrenal cortex. *J Clin Invest* 105:633–41.

Labrie, F., D. Archer, C. Bouchard, M. Fortier, L. Cusan, J. L. Gomez, G. Girard et al. 2009a. Intravaginal dehydroepiandrosterone (Prasterone), a physiological and highly efficient treatment of vaginal atrophy. *Menopause* 16(5):907–22.

Labrie, F., D. Archer, C. Bouchard, M. Fortier, L. Cusan, J. L. Gomez, G. Girard et al. 2009b. Serum steroid levels during 12 week intravaginal dehydroepiandrosterone administration. *Menopause* 16(5):897–906.

Labrie, F., A. Berlanger, P. Berlanger, R. Berube, C. Martel, L. Cusan, J. L. Gomez et al. 2007. Metabolism of DHEA in postmenopausal women following percutaneous administration. *J Steroid Biochem Mol Biol* 103(2):178–88.

Lasley, B. L., N. Santoro, J. F. Randolf, E. B. Gold, S. Crawford, G. Weiss, D. S. McConnell, and M. F. Sowers. 2002. The relationship of circulating dehydroepiandrosterone, testosterone, and estradiol to stages of the menopausal transition and ethnicity. *J Clin Endocrinol Metab* 87:3760–7.

Lasley, B. L. et al. 2010. Circulating dehydroepiandrosterone sulfate levels in women with bilateral salpingo-oophorectomy during the menopausal transition. *Menopause*, doi:10.1097/gme.0b13e318fb53fc.

Lasley, B. L. et al. 2011. Circulating dehydroepiandrosterone sulfate is associated with an increase in circulating androstenediol. Unpublished manuscript.

Pabon, J. E., X. Li, Z. M. Lei, J. S. Sanfilippo, M. A. Yussman, and C. V. Rao. 1996. Novel presence of luteinizing hormone/chorionic gonadotropin receptors in human adrenal glands. *J Clin Endocrinol Metab* 81:2397–400.

Percheron, G., J.-Y. Hogrel, S. Denot-Ledunois, G. Fayet, F. Forette, E. E. Baulieu, M. Fardeau, and J.-F. Marini. 2003. Effect of a 1-year oral administration of dehydroepiandrosterone to 60-to-80 year-old individuals on muscle function and cross-sectional area. *Arch Intern Med* 163:720–7.

Randolph Jr., J. F., M. Sowers, E. B. Gold, B. A. Mohr, J. Luborsky, N. Santoro, D. S. McConnell et al. 2003. Reproductive hormones in the early menopausal transition: Relationship to ethnicity, body size, and menopausal status. *J Clin Endocrinol Metab* 88:1516–22.

Rao, C., X. L. Zhou, and Z. M. Lei. 2004. Functional luteinizing hormone/chorionic gonadotropin receptors in human adrenal cortical H295R cells. *Biol Reprod* 71:579–87.

Soules, M. R., S. Sherman, E. Parrott, R. Rebar, N. Santoro, W. Utian, and N. Woods. 2001. Executive summary: Stages of reproductive aging workshop (STRAW). *Fertil Steril* 76(5):874–8.

Stanczyk, F. Z., C. C. Slater, D. E. Ramos, C. Azen, G. Cherala, C. Hakala, G. Abraham, and S. Roy. 2009. Pharmacokinetics of dehydroepiandrosterone and its metabolites after long-term oral dehydroepiandrosterone treatment in postmenopausal women. *Menopause* 16(2):272–8.

Sulcová, J., M. Hill, R. Hampl, and L. Stárka. 1997. Age and sex related differences in serum levels of unconjugated dehydroepiandrosterone and its sulphate in normal subjects. *J Endocrinol* 154(1):57–62.

Von Muhlen, D., G. A. Laughlin, D. Kritz-Silverstein, J. Bergstrom, and R. Bettencourt. 2008. Effect of dehydroepiandrosterone supplementation on bone mineral density, bone markers and body composition in older adults. *Osteoporos Int* 19(5):699–707.

2 DHEA as a Putative Replacement Therapy in the Elderly

Alessandro D. Genazzani and Chiara Lanzoni

CONTENTS

INTRODUCTION

Dehydroepiandrosterone (DHEA) is typically secreted by the adrenal gland, and its secretory rate changes throughout human life. When human development is completed and adulthood is reached, DHEA and dehydroepiandrostenedione sulfate (DHEAS) levels start to decline so that at 70–80 years of age, peak concentrations are only 10%–20% of those in young adults. This age-associated decrease is indicated as "adrenopause," and because many age-related disturbances have been observed to correlate with the decline of DHEA(S) levels, the possibility of using DHEA as replacement therapy in the aged should be considered.

The use of DHEA as treatment for aging men and women has been proposed, and this chapter aims at showing the beneficial effects reported in humans. Since a lot of interesting results have been produced on experimental animals, suggesting that DHEA has a positive modulation on most of the age-related disturbances, recently a new interest is growing on the use of DHEA supplementation in humans. Indeed, recent studies suggest that DHEA seems to be beneficial in hypoandrogenic males as well as in postmenopausal and aging women. Since menopause is the stage that induces a dramatic change in the steroid milieu in a woman's life, the use of DHEA as "hormone replacement treatment" is reported to reduce most of the symptoms and to restore the stability in both the androgenic and estrogenic environment. Since menopause is the beginning of the biological transition toward senescence in women, it is of great interest to better understand what DHEA might help and do to solve and/or overcome the problems in this peculiar part of human life.

Though most data are suggestive of the use of DHEA as a hormonal treatment, more defined and specific clinical trials are needed to reveal the features and any hidden risks of this steroid before it is used as a safe and standard treatment. In addition, DHEA is perceived in different ways all over the world: it is considered just a "dietary supplement" in the United States, a "true hormone" in many European countries, and illegal for use as "a hormonal treatment" by the European Health System. This chapter examines various viewpoints regarding the use of DHEA as an experimental hormonal replacement therapy.

DHEA SECRETION

DHEA and its sulfate ester DHEAS are the major secretory products of the human adrenal cortex. Serum DHEA(S) are distinct from the other major adrenal steroids—cortisol and aldosterone—in displaying a significant decrease with age. Indeed, secretion of DHEA in humans and in several nonhuman primates follows a characteristic age-related pattern (Orentreich et al. 1984). In fact, DHEA is the major secretory product of human fetal adrenals, leading to high circulating DHEAS levels at birth. Later, during postnatal involution of the fetal zone of the adrenals, DHEAS serum concentrations decrease to almost undetectable levels during the first year of human life. DHEAS levels remain low until they gradually increase again between the sixth and tenth years of age, owing to increasing DHEA production in the adrenal zona reticularis, a phenomenon termed "adrenarche" (Sizonenko and Paunier 1975; Reiter, Fuldauer, and Root 1977; Sklar, Kaplan, and Grumbach 1980; Palmert et al. 2001). Peak DHEAS concentrations are reached in early adulthood, followed by a steady decline throughout adult life, so that at 70–80 years of age, peak concentrations are only 10%–20% of those in young adults (Orentreich et al. 1984, 1992; Sizonenko and Paunier 1975; Reiter, Fuldauer, and Root 1977; Sklar, Kaplan, and Grumbach 1980; Palmert et al. 2001; Figure 2.1). This age-associated decrease is indicated as "adrenopause" in spite of the continued secretion of adrenal glucocorticoids and mineral corticoids throughout life. The age-related decline in DHEA(S) concentrations shows high interindividual variability (Palmert et al. 2001) and is accompanied by a reduction in the size of the zona reticularis (Parker et al. 1997). By contrast,

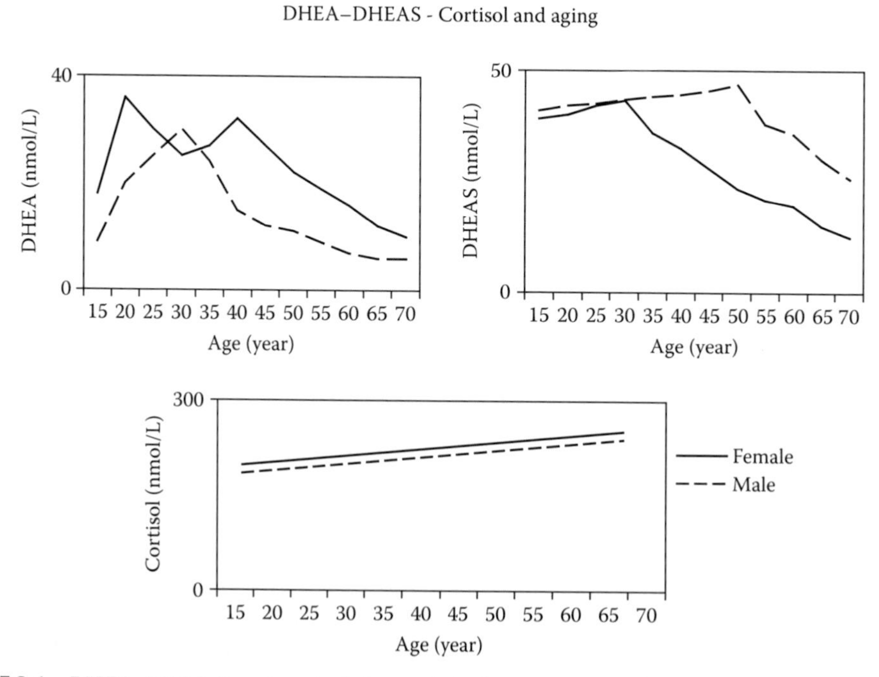

FIGURE 2.1 DHEA, DHEA-S, and cortisol change their plasma concentrations in men and women throughout life. Little differences were observed in women who showed a more rapid decay of DHEA-S concentrations close to the 40s. Cortisol plasma levels increase slowly but constantly in both men and women.

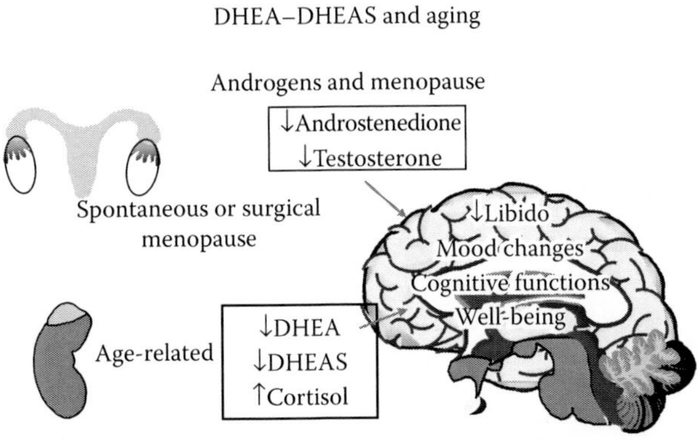

FIGURE 2.2 In general when perimenopause starts and later menopause takes place, consistent changes of steroid plasma levels occur in women. Most of the androgens are produced by the ovaries but with the age-related decline in DHEA(S) synthesis and secretion, a significant reduction of androgen plasma levels take place affecting the CNS function.

serum cortisol concentrations can even increase with aging (Laughlin and Barrett-Connor 2000), and for this reason, it has been hypothesized that the increase in the cortisol:DHEAS ratio is associated with cognitive impairment in elderly people (Kalmijn et al. 1998; Figure 2.1). Increased cortisol plasma levels seem to be related to a dysfunction of some central nervous system (CNS) areas; thus, a neurotoxic role for cortisol has been supposed (Figure 2.2). Several studies have shown a positive correlation between cortisol plasma levels and memory disturbances, a finding that is supported by the fact that patients affected by Alzheimer's disease (AD) show high cortisol plasma levels. Furthermore, circulating concentrations of progesterone, 17-hydroxyprogesterone, cortisol, testosterone, androstenedione, and DHEAS during the follicular phase in constipated young women were lower than the concentrations in controls during the follicular phase of the menstrual cycle (Donald et al. 1985; Heymen, Wexner, and Gulledge 1993; Sonnenberg, Tsou, and Müller 1994; Chami et al. 1995; Meiring and Joubert 1998; Caranasos and Busby 1985; Devroede et al. 1989; Brook 1991; Whitehead 2007; Blandizzi 2007).

During the luteal phase of the menstrual cycle, reductions were identified in estriol, cortisol, and testosterone in the constipated group. Likewise, circulating concentrations of DHEAS were found to be lower in depressed patients than in comparably healthy controls. DHEAS:cortisol ratios in morning serum and salivary samples were lower than those retrieved during other times of the day in depressed patients. Constipation is a prominent symptom among patients with depression, anorexia nervosa, weight loss, sleep disorders, fatigue, and decreased sexual interest, particularly with menarche or after a stressful emotional experience or surgical operation. Recognizing major depression in constipated patients by measuring DHEAS:cortisol ratios in saliva and serum may be plausible, but this possibility needs to be confirmed in well-designed studies (Donald et al. 1985; Heymen, Wexner, and Gulledge 1993; Sonnenberg, Tsou, and Müller 1994; Chami et al. 1995; Meiring and Joubert 1998; Caranasos and Busby 1985; Devroede et al. 1989; Brook 1991; Whitehead 2007; Blandizzi 2007).

DHEA "INTRACRINOLOGY"

DHEA exerts its action either indirectly in peripheral target tissues of sex steroid action (through its conversion to androgens, estrogens, or both) or directly as a neurosteroid (acting on neurotransmitter receptors in the brain). Since the human steroidogenic enzyme P450c17 converts virtually no 17α-hydroxyprogesterone to androstenedione, the biosynthesis of all sex steroids in humans derives from DHEA. DHEA can be converted to androstenedione by the activity of 3β-hydroxysteroid dehydrogenase (3β-HSD) and then further converted to testosterone and estradiol by isoenzymes of 17β-hydroxysteroid dehydrogenase (17β-HSD) and by P450 aromatase, respectively. Although DHEAS is the hydrophilic storage form that circulates in the blood, only lipophilic DHEA can be converted intracellularly to androgens and estrogens. Thus, the tissue-specific synthesis of DHEA sulfotransferase and steroid sulfates determines the ratio of DHEA activation (by conversion to sex steroids) to transient DHEA inactivation (by secretion of the sulfate ester back into the bloodstream).

In addition, almost ubiquitous production of 3β-HSD, 17β-HSD, 5α-reductase, and P450 aromatase results in a widespread peripheral conversion of DHEA to sex steroids (Martel et al. 1992; Martel et al. 1994; Jakob et al. 1997; Krazeisen et al. 1999; English et al. 2000). It has been estimated that 30%–50% of androgen synthesis in men and 50%–100% of estrogen synthesis in pre- and postmenopausal women occur in peripheral target cells (Labrie 1991). Studies of the pharmacokinetics and bioconversion of DHEA in humans with low serum DHEAS reveal that DHEA has a specific sexually dimorphic pattern of conversion: in fact, DHEA administration induces significant increases in circulating androgens in women (Arlt et al. 1998) and in circulating estrogens in men (Arlt et al. 1999). In men with adrenal insufficiency and hypogonadism without androgen replacement, and thus total androgen depletion, DHEA administration results in a significant increase in circulating androgens, although this increase is still far from achieving normal male serum concentrations (English et al. 2000). These findings suggest that DHEA administration causes androgenic and estrogenic effects in both women and men, respectively. However, circulating sex steroids do not correctly reflect the intracrine, tissue-specific action of DHEA. Men with an age-associated decline of DHEAS secretion and normal gonad function do not show an increase in circulating androgens after DHEA ingestion but do show a significant increase in circulating 5α-androstane-3α, 17β-diolglucuronide (ADG; Arlt et al. 1999). This reflects the increased peripheral androgen synthesis in DHEA-treated men because ADG is the main metabolite of dihydrotestosterone (DHT; Young et al. 1997; Moghissi, Ablan, and Horton 1984), and its synthesis occurs only in peripheral androgen target tissues.

All the enzymes required in transforming DHEA into androgens and/or estrogens are expressed in a large series of peripheral target tissues, thus permitting all androgen-sensitive and estrogen-sensitive tissues to produce and control the intracellular levels of sex steroids according to the local needs. This new field of endocrinology is called intracrinology. In women after menopause, all estrogens and almost all androgens are produced locally in peripheral tissues from circulating DHEA, which indirectly affects, among others, bone formation, adiposity, muscle, insulin and glucose metabolism, skin, libido, and well-being. In men, whose androgen secretion from testis are lifelong, the contribution of DHEA to androgens has been best evaluated in the prostate, where about 50% of androgens are produced locally and derived from DHEA. Such findings led to the development of a combined androgen blockade (CAB) to treat prostate cancer. In fact, this treatment adds a pure antiandrogen to medical (GnRH agonist) or surgical castration in order to block both the androgens produced locally and their receptors. Indeed, CAB has been the first treatment demonstrated to cure or prolong life in advanced prostate cancer.

In addition to indirect endocrine and intracrine effects after peripheral conversion to androgens and estrogens, DHEA directly acts as a neurosteroid. Baulieu et al. (2000) were the first to provide compelling evidence for DHEA synthesis in the CNS, demonstrating steady DHEAS levels in brain tissue from adrenalectomized and gonadectomized rats (Giagulli et al. 1989). In fact, recent studies have demonstrated the synthesis of P450c17 and other steroidogenic enzymes in the brain (Corpechot et al. 1981; Compagnone et al. 1995; Zwain and Yen 1999a,b). Compagnone and Mellon (Zwain and Yen 1999a,b) showed that DHEA and DHEAS have direct and differential effects on neuronal growth and development. They also showed that DHEA interacts with the N-methyl-d-aspartate (NMDA) receptor, thus supporting earlier findings describing DHEA as a modulator of the NMDA response via the sigma receptor (Compagnone and Mellon 1998) and as an allosteric antagonist of the γ-aminobutyric acid ($GABA_A$) receptor (Debonnel, Bergeron, and de Montigny 1996; Majewska et al. 1990). Allopregnanolone is a neurosteroid with a number of properties that may be relevant to the pathophysiology and treatment of AD and other neurodegenerative disorders, demonstrating pronounced neuroprotective actions in the setting of excitotoxicity, traumatic brain injury, and neurodegeneration. It also increases myelination, enhances neurogenesis, decreases inflammation, and reduces apoptosis. (Ciriza, Azcoitia, and Garcia-Segura 2004; Lockhart et al. 2002; Djebaili et al. 2005; Djebaili, Hoffman, and Stein 2004; He 2004a,b; Ahmad et al. 2005; Griffin et al. 2004; Mellon, Gong, and Schonemann 2008; Azcoitia et al. 2003; Ghoumari et al. 2003; Liao et al. 2009; Wang et al. 2005; VanLandingham et al. 2007; Charalampopoulos et al. 2004, 2006; Xilouri and Papazafiri 2006; Francis 2005; Hynd, Scott, and Dodd 2004; Lipton 2005; Wenk, Parsons, and Danysz 2006; Blennow, de Leon, and Zetterberg 2006; Hardy and Selkoe 2002; Ariza et al. 2006). Since excitotoxicity, neurodegeneration, and traumatic brain injury, as well as dysregulation in myelination, neurogenesis, apoptosis, and inflammation, have been implicated in the pathogenesis and clinical course of AD, deficits in allopregnanolone could represent a critical component of AD pathophysiology. In addition to allopregnanolone, other neurosteroids such as DHEAS and pregnenolone may be candidate modulators of AD pathophysiology. For example, DHEA appears to be elevated in postmortem brain tissue and in the CNS of AD patients when compared to control subjects, and DHEA is positively correlated with the Braak-to-Braak neuropathological disease stage (Ariza et al. 2006; Jellinger 2004; Marx et al. 2006; Naylor et al. 2008; Steckelbroeck et al. 2004; Brown et al. 2003). Like allopregnanolone, DHEAS demonstrates a number of neuroprotective effects. For example, DHEAS is protective against amyloid-beta protein toxicity and a number of other insults involving oxidative stress, including anoxia, glucocorticoid-induced toxicity, and NMDA-induced excitotoxicity (Kim et al. 2003; Cardounel, Regelson, and Kalimi 1999; Tamagno et al. 2003; Marx et al. 2000; Karishma and Herbert 2002; Kimonides et al. 1999).

In addition, DHEA enhances neurogenesis in rodent models and augments cell proliferation of human neural stem cells (Kurata et al. 2004; Azizi et al. 2009; Suzuki et al. 2004). Naylor et al. (2008) studied allopregnanolone, DHEA, and pregnenolone in postmortem brain tissues of AD patients and demonstrated that neurosteroids are altered in the temporal cortex of these patients, which is related to a neuropathological disease stage. Moreover, they demonstrated that the presence of allele APOE4, which is a lipoprotein, is associated with reduced allopregnanolone levels in the temporal cortex and can be considered a risk factor for developing late-onset AD.

In spite of continuing efforts, the search for a specific DHEA receptor has not been fruitful. High-affinity binding sites for DHEA have been described in murine (Demirgoren et al. 1991) and human (Meikle et al. 1992) T cells, but these sites also effectively bind DHT. High-affinity binding sites for DHEA were identified on plasma membranes derived from bovine aortic cells, presenting evidence for the activation of endothelial nitric oxide synthase by DHEA via G proteins (Okabe et al. 1995). However, potential competition of binding by DHT was not tested. Similarly, Liu and Dillon (2002) showed that DHEA activates phosphorylation of extracellular signal-regulated kinase 1 in human vascular smooth muscle cells, independent of androgen receptor (AR) and estrogen receptor (ER). However, whether inhibition of downstream conversion of DHEA (e.g., by the 3β-HSD inhibitor trilostane) might alter these effects was not investigated. Thus, it seems possible that, particularly outside the CNS, effects of DHEA are only indirectly mediated through its bioconversion to other steroids. Nevertheless, since the expectancy of life is longer, the age-related decrease in DHEA(S) secretion raised the question of whether aging is, in part, a consequence of DHEA deficiency and potentially reversible by DHEA administration. This idea has been strengthened by a number of animal experiments suggesting that DHEA is a multifunctional hormone with possible anticancer, immune-enhancing, neurotropic, and general antiaging effects. To better understand what makes DHEA a putative "antiaging" compound, it is well worth mentioning the most relevant data and results observed in experimental animals and humans.

DEHYDROEPIANDROSTERONE AND PROSTATE CANCER

An important example of DHEA "intracrinology," as coined by Fernand Labrie, is its metabolism in the prostate tissue. Prostate cells control the level of intracellular, active sex steroids using catalyzing enzymes 17β-HSD, 3β-HSD, and 5α-reductase (Labrie et al. 1993; Gingras and Simard 1999). The adrenal steroid, DHEA, is an important source of androgens, which, when metabolized by prostate cells, contribute up to one-sixth of DHT present in the prostate (Geller 1985). This very large pool of DHEA(S) is at its peak when men are young, yet we do not observe high rates of cancer in young men. Can DHEA or DHEAS perhaps play a protective role in normal prostates but contribute to prostate cancer progression in prostatic tissues at advanced ages, at least in the context of the reactive or senescent stromal microenvironment or tumor environment represented? When one considers the latency period for cancers, the early high exposure to DHEA in young men hypothetically can be confounded by risk factors such as smoking, inflammation, or diet, providing a tissue microenvironment that alters DHEA metabolism and thus an altered androgen/estrogen balance (Carruba 2006). Similar hypotheses of early hormonal exposure are proposed for increasing breast cancer risk resulting from early menarche or late, full-term pregnancy (Pike et al. 1983; Martin and Boyd 2008). Alternatively, DHEA has been shown to be a cancer preventive against carcinogen-induced rodent prostate cancers in *in vivo* and *in vitro* studies (Green et al. 2001; Rao et al. 1999; Lubet et al. 1998; Perkins et al. 1997; Ciolino et al. 2003), whether by inhibition of glucose-6-phosphate dehydrogenase (Schwartz and Pashko 1995) or other carcinogen-metabolizing enzymes. The relevance of these studies to human biology is uncertain; however, as the amounts of DHEA and DHEAS are much lower in rodents than in humans, the physiological importance of these adrenal steroids are unknown in rodents.

DHEA can directly activate AR or ERβ in the prostatic epithelium or the AR or ERα in the prostatic stroma. DHEA can be a direct ligand for the AR in mutant prostate epithelial cells such as LNCaP and induce weak androgenic effects, potentially promoting prostate cancer growth. This relationship is demonstrated by its stimulation of prostate cancer LNCaP cell proliferation and modulation of cellular PSA, AR, ERβ, and insulin-like growth factor (IGF) axis gene and protein expression, in a pattern similar to DHT and testosterone, although on a lesser scale and delayed in time (Arnold et al. 2005). ERβ is an important target in the prostate (Weihua, Warner, and Gustafsson 2002) for endogenous and exogenous estrogens and phytoestrogens and may play a role in modulating androgen activity. DHEA has been shown to exert direct agonist effects on ERβ, as observed in competitive receptor binding assays in which DHEA displayed a higher affinity for ERβ than for AR or ERα, with ERβ being the preferred target for the transcriptional effects of DHEA (Chen et al. 2005). Indirect effects refer to DHEA metabolism into androgenic ligands (including androstenedione, testosterone, and DHT) or estrogenic ligands (including 7-hydroxy-DHEA [7-OH-DHEA], 3β-adiol, or 17β-estradiol). Receptors for DHEA or DHEAS have not been definitively isolated (Widstrom and Dillon 2004). DHEAS is present in high levels in the prostate, as is the sulfatase that converts DHEAS to DHEA. Prostate stromal and epithelial cells possess the enzymatic machinery to metabolize DHEA (intracrine) to more active androgenic and/or estrogenic steroids (Labrie et al. 1998; Klein et al. 1988; Voigt and Bartsch 1986) and express secondary mediators (paracrine) for epithelial growth and differentiation. Alternatively, DHEA metabolites may act on ER in the prostate, potentially antagonizing androgenic effects on prostate cancer growth, such as metabolism to 7-OH-DHEA, a known ligand for ER (Martin et al. 2004). The complexity of the balance between androgenic and estrogenic effects whether as direct ligands or metabolites of DHEA on the prostate is matched by the complexity of estrogen action through the ERα sv ERβ. The balance between ERα sv ERβ, including the temporal and spatial expression, determines the response of prostate to estrogen. The balance of androgen and estrogen levels is most important for prostatic development and differentiation. Estrogens have long been used in prostate cancer therapy, and the role of estrogens in the prostate has been elegantly studied and reviewed (Carruba 2007; Risbridger, Ellem, and McPherson 2007). Estrogens have beneficial effects that support normal growth of the prostate but can also be detrimental to prostate growth and differentiation (McPherson et al. 2007). Estrogens acting through the prostate stromal ERα may be growth promoting, whereas estrogens acting through the epithelial ERβ may be antagonistic to ERα or AR-activated pathways (Chang and Prins 1999; Signoretti and Loda 2001). Excessive estrogen induces squamous metaplasia and can act synergistically with androgens to induce glandular hyperplasia (Isaacs 1984). To the contrary, estrogens can inhibit prostate cancer xenograft growth in female intact and ovariectomized mice, in the absence of androgens (Corey et al. 2002). These inhibitory effects were postulated to occur by direct actions via the ER or by E2 effects on other cells secreting secondary factors, which influence cancer cell growth. In addition, ERβ knockout mice exhibit an increased epithelial proliferation compared with that observed in wild-type mice (Weihua et al. 2001), suggesting that ERβ may inhibit prostate growth. What regulates the direction of estrogen action? What are downstream signal transduction pathways or gene effects of ER ligand/receptor complexes in either stromal cells or epithelial cells? How do nongenomic effects of estrogen influence prostate functioning? Paracrine functions become important when considering that stromal cells possess aromatase allowing conversion of testosterone to estrogens (Ellem and Risbridger 2007). Ellem and Risbridger proposed a positive-feedback cycle in which increased stromal aromatase production may increase local estrogens, which then promote inflammation. The inflammation may further stimulate aromatase expression leading to progression of prostate cancer. In the context of reactive stroma, the relationship between reactive stroma and aromatase expression has not been validated. A final possibility is that DHEA remains as a prohormone, not metabolized, acting as a sort of "hormonal buffer" (Regelson, Loria, and Kalimi 1988) to be used or metabolized in case of any excess of endogenous androgen or estrogen levels.

DHEA ADMINISTRATION IN EXPERIMENTAL MODELS

It should be made clear that the use of experimental models (i.e., animals) is far from being perfectly superimposable on human biology, but at least such experimental designs permit researchers to better define the range of biological action of DHEA and its metabolites.

DHEA cardioprotective properties were first identified during a study by Eich et al. It was observed that in a hypercholesterolemic rabbit model, chronic DHEA administration produced a 45% reduction in the number of significantly stenosed vessels in the transplanted hearts ($p < .05$), as compared with controls, and a 62% reduction in nontransplanted hearts ($p < .05$), yielding an overall 50% reduction in the number of significantly stenosed vessels in both the transplanted and nontransplanted hearts. This reduction in luminal stenosis was observed in the absence of any significant alterations in lipid profiles (Williams et al. 2002; Eich et al. 1993).

These observations were confirmed by other studies. Indeed, Gordon et al. claimed that atherogenic insult resulted in severe atherosclerosis in animals not treated with DHEA, whereas those receiving DHEA experienced a 50% reduction in plaque size ($p = .006$), inversely related to the serum level of DHEA attained (Gordon, Bush, and Weisman 1988). These beneficial actions were not attributable to differences in body weight gain, food intake, total plasma cholesterol, or distribution of cholesterol among the VLDL, LDL, or HDL fractions. The results show that high levels of plasma DHEA inhibit the development of atherosclerosis and provide an important experimental link to the epidemiologic studies correlating low DHEAS plasma levels with an enhanced risk of cardiovascular mortality (Eich et al. 1993).

Other studies reported that DHEA had a sort of anticancer effect. Schwartz and Pashko (1995, 2004) observed that DHEA administration to mice and rats inhibited the development of experimentally induced tumors of the breast, lung, colon, liver, skin, and lymphatic tissue. In the two-stage skin-tumor genesis model of mice, DHEA treatment inhibited tumor initiation as well as epidermal hyperplasia tumor promotion of papillomas. Much evidence suggests that DHEA produces its antiproliferative and tumor-preventive effects by inhibiting glucose-6-phosphate dehydrogenase and the pentose phosphate pathways. These pathways are an important source of NADPH, a critical reductant for many biochemical reactions that generate oxygen-free radicals, which may act as second messengers in stimulating hyperplasia.

In addition, long-term DHEA treatment in mice has also reduced the amount of weight gain (apparently by enhancing thermogenesis) and seems to induce many of the beneficial effects of food restriction, which have been shown to inhibit the development of many age-associated diseases, including cancer. Indeed, Schwartz (1995, 2004) demonstrated that an adrenalectomy completely reverses the antihyperplastic and antitumor-promoting effects of food restriction. This data supports the claim that food restriction enhances levels of adrenocortical steroids, such as cortisol and DHEA, which in turn mediate the tumor-inhibitory effect of underfeeding (Gordon, Bush, and Weisman 1988; Schwartz and Pashko 2004). Moreover, DHEA administration has been supposed to have antiobesity and insulin sensitizing effects in both rodents and dogs (Melchior and Ritzmann 1992; Morales et al. 1998), since chronic DHEA administration affects body composition by inducing an increase in lean mass and a decrease in fat mass.

Last but not least, DHEA has been demonstrated to exert a neurotropic action at the GABA receptor, enhancing maze performance and memory in mice. When administered *in vivo*, DHEAS blocked the anxiolytic effect of ethanol, and this supported the hypothesis that neurosteroids could be involved in the termination of the stress response.

Though such positive effects on experimental animals might suggest a putative role of DHEA treatment in humans, we should consider that in most animal studies, pharmacological doses of DHEA have been used, yielding DHEA levels far beyond the physiological ones. Even more importantly, experiments have been performed mainly in rodents, which belong to a different species from human beings.

DHEA SUPPLEMENTATION IN THE ELDERLY

Though the use of DHEA as a treatment has not yet been properly defined (DHEA is considered a dietary supplement in the United States, whereas in Europe it is considered a hormone), a relatively limited amount of objective data has been produced and published on humans. The first studies of humans were published many years ago, and most of them focused on metabolic effects and symptoms usually associated with aging, such as hyperlipidemia, decreased insulin sensitivity, increased fat mass, reduced muscle mass, and decreased bone mineral density (BMD). When DHEA was administered in physiological (25–50 mg) or near to physiological daily doses (100 mg), a significant decrease in apolipoprotein A1 (Casson et al. 1998; Morales et al. 1994) and HDL-C was observed in women (Casson et al. 1998; Morales et al. 1994; Diamond et al. 1996; Lasco et al. 2001; Villareal, Holloszy, and Kohrt 2000) but not in men (Casson et al. 1998; Morales et al. 1994; Diamond et al. 1996). This was probably related to an increase in circulating androgen concentrations in women but not in men. Fasting glucose and insulin levels, as well as insulin responses to oral and intravenous glucose loads, were found to be unchanged by DHEA administration (Casson et al. 1998; Diamond et al. 1996; Lasco et al. 2001; Villareal, Holloszy, and Kohrt 2000; Dhatariya, Bigelow, and Nair 2005). Yen, Morales, and Khorram (1995) observed an increase in lean body mass and muscular strength and a decrease in fat mass in age-advanced men receiving DHEA, though this was not confirmed by all studies. However, a recent study demonstrated a significant decrease of insulinemia and glucagon under DHEA administration in menopausal women, thus supporting a putative effect of DHEA on the metabolic control of glucose and on insulin sensitivity (Dhatariya, Bigelow, and Nair 2005).

It is well known that circulating interleukin 6 (IL-6) increases with age, and several epidemiological studies have reported a negative correlation of serum DHEA and DHEAS with IL-6 (Ghoumari et al. 2003; Liao et al. 2009). Though *in vitro* evidence has supported a DHEA-induced inhibition of IL-6 production by human peripheral mononuclear blood cells (Liao et al. 2009), and a potential link between endocrinosenescence and immunosenescence, not all studies are confirmatory (Wang et al. 2005; He 2004a), and several effects have been described on the immune system (VanLandingham et al. 2007; Ariza et al. 2006). Daynes et al. (1995) suggested that steroids may be regulators of the mammalian immune response, and data drawn from cellular and animal models suggest that nonpharmacological doses of DHEA have a positive immunoregulatory action (Jellinger 2004; Steckelbroeck et al. 2004). In systemic lupus erythematosus, characterized by immune deficits and an unbalanced cytokine secretion, several randomized clinical studies demonstrated a beneficial effect of DHEA of 200 mg/day, but not of 100 mg/day, on disease course, with a concomitant decrease in corticosteroid administration (Brown et al. 2003; Kim et al. 2003; Cardounel, Regelson, and Kalimi 1999; Tamagno et al. 2003). Multiple mechanisms of action mediate these effects, including enhanced secretion of IL-2 and inhibited release of the inflammatory cytokine IL-6 (Young et al. 1999; Straub et al. 1998; Delpedro et al. 1998; Young et al. 2001; McLachlan, Serkin, and Bakouche 1996; Casson et al. 1993; Solerte et al. 1999; Daynes, Dudley, and Araneo 1990; Schmidt et al. 2000; Sansoni et al. 1993; Daynes et al. 1993; Ershler et al. 1993; Krishnaraj and Blandford 1987; Murasko et al. 1986; Carson et al. 2000; Daynes et al. 1995; Araneo, Woods, and Daynes 1993; Loria et al. 1988; Loria, Regelson, and Padgett 1990; Chang et al. 2002; van Vollenhoven, Engelman, and Mc Guire 1994; van Vollenhoven, Engleman, and Mc Guire 1995; van Vollenhoven et al. 1999).

An important target of DHEA action seems to be the skin and skin components. As in adrenal insufficiency, sebum secretion and skin hydratation increased after DHEA administration in elderly people (Baulieu et al. 2000), suggesting an androgenic activity of DHEA. In addition to emphasizing the importance of tissue-specific bioconversion of DHEA, Labrie et al. (1997) demonstrated the estrogenic effects on the vaginal epithelium. From the very beginning of DHEA administration in healthy elderly volunteers, an increase in self-perception of well-being was often reported, although this was not assessed by validated psychometric questionnaires nor confirmed by other studies (Labrie et al. 1997). Conversely, a significant effect of DHEA replacement on well-being, mood, and sexuality was observed in women with adrenal insufficiency. From these data, it was clearly evident that in elderly subjects with low DHEA plasma levels, DHEA supplementation could potentially improve impaired well-being, mood, and sexuality, especially in androgen-deficient male subjects (Arlt et al. 2001). It is, however, important to state that not all studies reported positive or significant improvements during DHEA administration (Muller et al. 2006), and though many studies were able to demonstrate changes and improvements during DHEA administration in human biological tissues and organs, a greater amount of data are needed to transform "observations" on DHEA efficacy into definitive data for the use of DHEA as a standard substitutive treatment.

DHEA SUPPLEMENTATION IN WOMEN

Of greater relevance is to focus on the putative role of DHEA supplementation in postmenopausal or aging women. In fact, the almost exclusive focus on the importance of ovarian estrogens in women's reproductive physiology, mainly during the menopausal transition, removed attention from the dramatic 70% fall in circulating DHEA that starts to occur between the ages of 20 and 30 and increases up to the threshold of 50 years (Genazzani et al. 2004; Migeon et al. 1957; Vermeulen and Verdonck 1976; Vermeulen et al. 1982; Orentreich et al. 1984; Bélanger et al. 1994). Since DHEA is converted to androgens and estrogens in peripheral tissues, such a fall in serum DHEA and DHEAS explains why women at menopause are not only lacking estrogens but also androgens, as demonstrated by the 50%–60% decrease in serum androsterone glucuronide (Labrie et al. 1997d). From our discussion to this point, it is clear why DHEA was thought to be the putative solution: DHEA administration in postmenopausal women might counteract aging changes in the female organism. In fact, DHEA can restore both the androgen and estrogen milieus due to the different concentrations of steroidogenical enzymes expressed in the target tissue (Figure 2.3).

The beneficial effects of DHEA treatment have been recently evaluated in postmenopausal women, and the most relevant data demonstrates a clinical use for osteopenia or osteoporosis in elderly women. Indeed, one of the most relevant consequences of menopause is osteoporosis, which

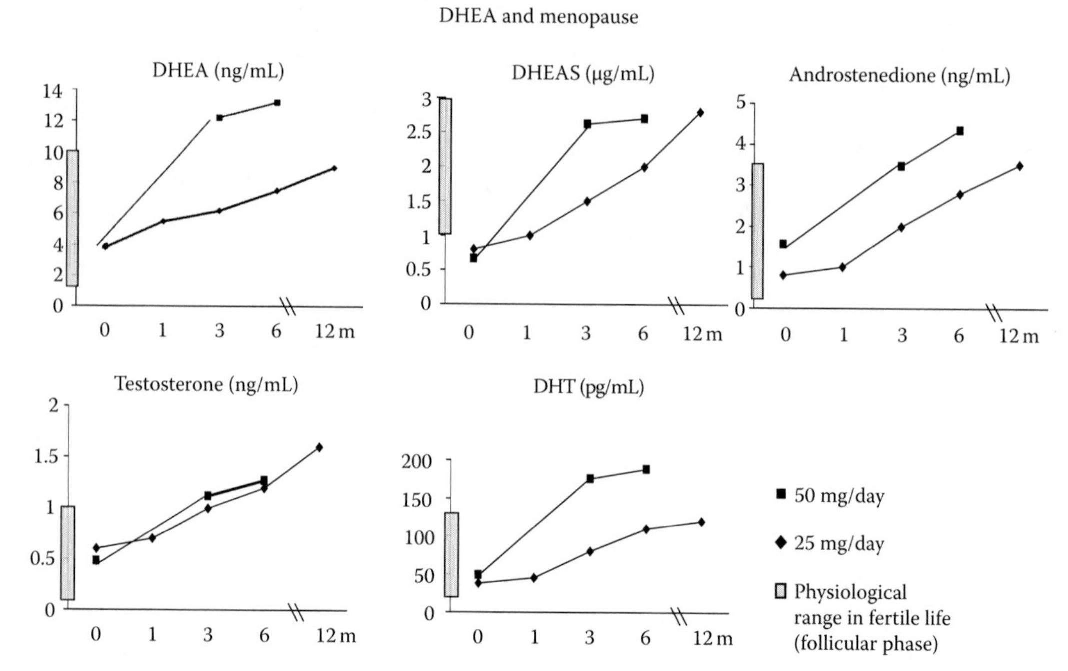

FIGURE 2.3 DHEA administration of 25 mg every day induces a smaller increase of androgens, especially of the active metabolites such as androstenedione and DHT. See various references of Genazzani et al.

is often counteracted by DHEA supplementation and transformation in both estradiol and androgens. A key role of androgens in bone physiology is well documented (Labrie et al. 1997c; Chesnut et al. 1983; Need et al. 1987; Savvas et al. 1988; Labrie et al. 1997b; Martel et al. 1998; Miller et al. 2002). In fact, both testosterone and DHT, also derived from DHEA, increased the transcription of α(I) procollagen mRNA in osteoblast-like osteosarcoma cells (Benz et al. 1991; Table 2.1). BMD of the lumbar spine, femoral trochanter, and total body was increased more by estrogen and testosterone implants than by E2 alone over a 24-month treatment period in postmenopausal women (Benz et al. 1991; Genazzani and Pluchino 2010).

Administration of 50 mg/day of oral DHEA, when coadministered with vitamin D and calcium supplements, induced large and clinically important improvements in lumbar spine BMD in older women. Spine BMD increased by 2% during DHEA treatment for a total increase of 4% from baseline. Similar increases in spine BMD induced by pharmacotherapy are associated with a 30%–50% reduction in vertebral fracture risk (Cefalu 2004). Furthermore, 2-year DHEA therapy produces an increase similar to or larger than that which results from administration of 2-year oral estrogen (5%–12%), bisphosphonates (14%), and selective estrogen receptor modulators (2%–11%; Villareal et al. 2001; The Writing Group for the PEPI Trial 1996; Gallagher, Kable, and Goldgar 1991; Hosking et al. 1998; Fogelman et al. 2000; Välimäki et al. 2007; Delmas et al. 1997; Eastell et al. 2006). The robustness of these findings is supported by the 2% increase in spine BMD that occurred when women in the placebo group crossed over to DHEA supplementation.

The significant improvements in spine BMD in women were accompanied by increases in IGF-1, testosterone, and estrogen. Because these are bone active hormones (Tracz et al. 2006; Prestwood et al. 2003; Giustina, Mazziotti, and Canalis 2008), it is possible that some or all of these increases might have mediated the improvements in spine BMD. Furthermore, circulating DHEA may have a direct effect on bone through a yet-to-be identified DHEA receptor (Wang et al. 2007) or by conversion to androgens or estrogens within bone cells (i.e., an intracrine system; Labrie et al. 2001).

No effect of DHEA supplementation on BMD in men was evident in the Weiss study (Weiss et al. 2009). Although spine BMD increased by 1%–2%, it did so in both the DHEA and placebo groups,

TABLE 2.1
Biological Effects of DHEA Administration in Experimental Models and in Clinical Trials in Humans

DHEA Effects	Studies on Experimental Animals	Studies on Humans
Atherosclerosis inhibition	Williams et al. (2002) Eich et al. (1993)	
Anticancer property	Gordon, Bush, and Weisman (1988) Schwartz and Pashko (2004)	Arnold et al. (2005)
Lean mass increase and fat mass reduction		Morales et al. (1998) Lasco et al. (2001)
Insulin sensitivity increase	Schwartz and Pashko (1995)	Morales et al. (1998) Lasco et al. (2001) Dhatariya, Bigelow, and Nair (2005)
Positive immunoregulatory action		Carson et al. (2000) Loria et al. (1988)
Well-being, mood, sexuality improvement		Arlt et al. (2001)
Bone mineral density increase		Weiss et al. (2009) Jankowski et al. 2006
Menopausal genitourinary disturbances reduction		Labrie et al. (2009a,b,c)
Cortisol decrease		Stomati et al. (2000) Genazzani et al. (2003)
Neuroprotective role		Charalampopoulos et al. (2004), (2006)
Anxiolytic effects		Stomati et al. (2000) Genazzani et al. (2003)
Menopausal termoregolatory disturbances decrease		Pye, Mansel, and Hughes (1985) Sherwin and Gelfand (1984) Stomati et al. (2000) Genazzani et al. (2003)

which suggests that the effect might have been mediated by vitamin D and calcium supplementation that was provided to all participants. Interestingly, Jankowski et al. (2006), who provided vitamin D and calcium supplements to participants with apparent deficiencies, reported similar results, that is, an increasing tendency of BMD in men in the placebo and DHEA groups.

Hip BMD did not change in either men or women in response to DHEA supplementation. Greater adaptive responses in the spine than in the hip have also been reported in response to estrogen replacement therapy (ERT; Duan et al. 1997) and exercise training (Villareal et al. 2003); however, the reason for this is not clear. A possible explanation is that the spine contains more trabecular bone, which has a greater rate of turnover than does cortical bone (Seeman et al. 1982), therefore making trabecular bone more responsive to therapeutic interventions. However, this explanation should be interpreted with caution, because little data is available on the proportion of trabecular bone in the vertebrae (Nottestad et al. 1987). Furthermore, because anteroposterior DXA scans of the spine include the cortical posterior vertebral elements (i.e., spinous processes), anteroposterior spine BMD contains a substantial proportion of cortical bone.

Other studies have assessed the effect of DHEA replacement therapy on BMD in older men and women. Shorter-term trials have yielded mixed results (Morales et al. 1998), whereas longer-term trials have reported beneficial effects of DHEA supplementation on bone, with more consistent benefits being shown in women.

It is conceivable that inadequate dietary intakes of vitamin D and calcium, as is common in older adults (Lips 2001; Ervin and Kennedy-Stephenson 2002), may attenuate the beneficial effects of DHEA supplementation on BMD, because these nutrients are important for optimal bone health (Gennari 2001). Indeed, studies by Weiss (Weiss et al. 2009) and Jankowski (Jankowski et al. 2006) showed larger 1-year increases (2%) in BMD than did other trials, and these were the only two trials that administered vitamin D and calcium supplements.

No adverse effects of DHEA supplementation were observed by Weiss et al. (2009); however, this trial was not designed to detect rare or subtle side effects. It should also be noted that because DHEA supplementation resulted in small but significant increases in circulating concentrations of estrogen, testosterone, and IGF-1, all of which may promote tumorigenesis, individuals taking DHEA supplements over a long term may need to be monitored regularly for hormone-sensitive cancer. The results of this study suggest that long-term (1 and 2 years) DHEA supplementation (50 mg/day) in combination with dietary vitamin D and calcium supplementation in older women has a substantial beneficial effect on spine BMD.

In light of the possibility that DHEA replacement therapy has other physiological benefits, such as improvements in glucoregulatory function, immunoregulation, and psychological state, and has no known major side effects, DHEA supplementation may be an attractive option for improving or preserving bone health in older women. In contrast to our findings in women, we found no evidence of a beneficial effect of DHEA supplementation in men above and beyond the effect of vitamin D and calcium supplementation.

No less important is the point that androgens are known to play a role in women's arousability and sensual pleasure as well as in the intensity and ease of orgasm. Androgens are also involved in the neurovascular smooth muscle response of swelling and increased lubrication (Diamond et al. 1996), being DHEA metabolized to both androgens and estrogens at the vaginal level. Estrogens, on the other hand, affect the vulval and vaginal congestive responses, and being capable of affecting mood, they have an influence on the whole set of sexual interests (Diamond et al. 1996). It is well known that loss of libido and/or sexual satisfaction are common in early postmenopause and that the addition of androgens to hormone replacement therapy (HRT) has been proved to be beneficial in treating these problems (Basson 2004; Greenblatt et al. 1950; Sherwin and Gelfand 1987). On such basis, DHEA administration might be suggested as a good controlled source of androgens, avoiding the use of testosterone administration, as reported in earlier studies (Sherwin 1988; Shifren et al. 2000). In fact, the benefits of androgens added to ERT or HRT on general well-being, energy, mood and general quality of life have been described (Sherwin and Gelfand 1987; Sherwin 1988; Shifren et al. 2000; Goldstat et al. 2003) and also improvements in the major psychological and psychosomatic symptoms, namely irritability, nervousness, memory and insomnia have been observed following addition of androgens to ERT (Sherwin and Gelfand 1985; Notelovitz et al. 1991).

Among the effects of DHEA, there is also the consistent reduction of hot flashes. In fact, DHEA therapy is successful in reducing hot flashes in hypogonadal men (Pye, Mansel, and Hughes 1985); similarly, the addition of androgens is effective in relieving hot flashes in women who have experienced unsatisfactory results when taking estrogen alone (De Fazio et al. 1984). Hot flashes are a primary menopausal symptom and one of the main reasons that menopausal women start HRT. Estrogen treatment is very effective in reducing or eliminating this symptom, and when DHEA is used as replacement therapy for "menopausal" women, it has been reported to have beneficial effects on hot flashes (Sherwin and Gelfand 1984; Baulieu 1999). Researchers have demonstrated the efficacy of low-dose DHEA administration in endocrine and psychoneuroendocrine parameters in early and late menopause, confirming that a daily low-dose DHEA (25 mg) supplementation increases adrenal androgen plasma levels (i.e., DHEA and DHEAS), which are significantly impaired during menopause (Stomati et al. 2000).

At present, hormone therapy is the easiest way to reverse the effects of hypoestrogenism in most (if not all) organs and tissues, but it has been demonstrated that though beneficial ERT and HRT were able to affect adrenal steroid synthesis and secretion, inducing a slight decrease in

DHEA(S) production and secretion as well as in cortisol secretion (Bernardi et al. 2003; Pluchino et al. 2005), only tibolone administration did not affect DHEA(S) secretion (Pluchino et al. 2005). This observation further supports the hypothesis that using DHEA as replacement therapy for menopausal women should be a precursor to using other sexual steroids (Genazzani and Pluchino 2010; Figure 2.4).

Until now, most studies have been conducted using a daily dosage of 50 mg or higher of DHEA, which also resulted in higher levels of some endocrine parameters (androstenedione, testosterone, 17-hydroxyprogesterone, and DHT; Stomati et al. 2000). However, when a lower daily DHEA dosage (25 or 10 mg; Pluchino et al. 2008) was administered, more positive hormonal effects than those that were observed using 50 mg but with a lower androgenic milieu were reported (Pluchino et al. 2008; Figures 2.2 and 2.4). With such a low dosage, both estrogen and androgen concentrations increased but at a lower rate than with the 50 mg/day therapy (Stomati et al. 2000; Genazzani et al. 2003). In addition, the use of a lower DHEA dose was as effective on β-endorphin, gonadotropins, the somatotropic axis (GH-IGF-1), and subjective symptoms as the 50 mg/day dosage. Interestingly, these positive effects were present though the increase in DHEAS, E2, and testosterone concentrations was lower and less rapid than during the 50-mg schedule, thus confirming that a lower dose of DHEA is enough for the metabolization or synthesis of adequate DHEA-derived steroids (Genazzani et al. 2003).

Most typical neuroendocrine changes during menopause are primarily due to the hypoestrogenic condition and in part due to the reduction of neurosteroid (i.e., allopregnanolone) synthesis and concentrations. Allopregnanolone, the most potent endogenous anxiolytic steroid, showed increasing plasma concentration throughout the treatment interval, thus suggesting that DHEA administration positively affects psychoneuroendocrine parameters through specific neuromodulatory effects on the CNS (Figure 2.4). Such effects are greatly modulated by the GABAergic activity of allopregnanolone as well as by the increase in β-endorphin plasma levels.

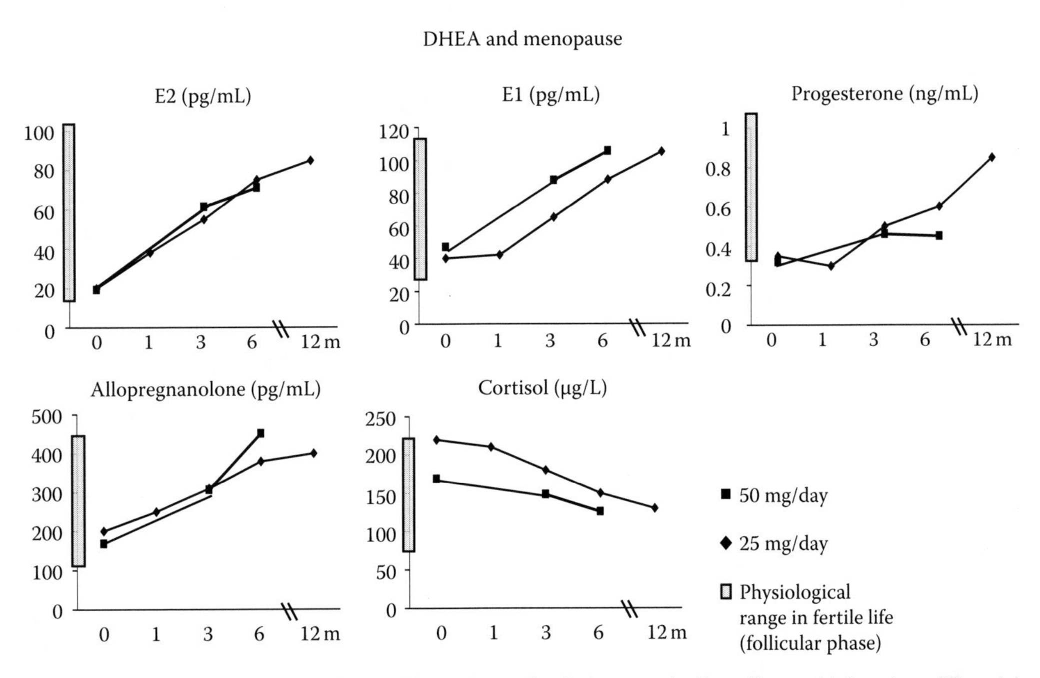

FIGURE 2.4 DHEA administration at 25 mg every day induces a similar effect to higher dose (50 mg) in terms of cortisol decrease and allopregnanolone, progesterone, estradiol (E2), and estriol (E1) production. See various references of Genazzani et al.

Of great relevance is the observation that during DHEA administration, cortisol plasma levels decrease, thus confirming that DHEA administration blunts the activity of the hypothalamic–pituitary–adrenal axes and suggesting the hypothesis of a sort of neuroprotective role against age-induced hypercortisolemia (Stomati et al. 2000; Genazzani et al. 2003; Figure 2.4).

Interestingly, DHEA supplementation was not reported to induce changes in endometrial thickness (Sherwin and Gelfand 1984; Baulieu 1999). This is probably due to the absence in the endometrial tissue of the specific enzymes responsible for the conversion of DHEA to estrogens or to the apparent equilibrium between the main DHEA metabolites, that is, estradiol and progesterone. This effect is comparable to that of a continuous combined treatment using estrogens plus progestagens.

All these data seem to indicate that DHEA is a putative hormonal compound and a drug for hormonal replacement in postmenopausal women and suggests that the beneficial effects of DHEA administration in postmenopausal women are exerted through the transformation of DHEA into androgens and/or estrogens in specific intracrine target tissues, thus limiting the possibility of side effects. An example of this is probably the absence of any stimulation on the endometrium (Gingras and Simard 1999; Rao et al. 1999). This effect should eliminate the need for progestin replacement therapy, thus reducing the fear of progestin-induced breast cancer in postmenopausal women, as stated by the Women's Health Initiative 2002 (Sherwin and Gelfand 1984; Baulieu 1999).

Panjari and Davis (2010) recently reviewed the published literature on the effects of DHEA treatment on postmenopausal women. The authors included only randomized, controlled trials that compared DHEA therapy with placebos in postmenopausal women not receiving other hormonal treatments. The end points analyzed were measures of sexual function, well-being, and safety such as lipids and carbohydrate profiles. However, only nine well-designed studies that analyze sexual function are available in the literature, and only seven trials address the issue of well-being, and all studies differ in dose and treatment time, age of women, and measured functions. More findings are available on the effects on lipid levels and insulin sensitivity, but studies still lack definitive evidence. The authors concluded that there is little convincing data to support the use of oral DHEA in healthy, postmenopausal women to improve conditions related to the aging process, such as reduced sexual function and reduced well-being.

The study by Panjari and Davis (2010), however, renews the attention to and debate about one of the most attractive and controversial issues in the physiology of the aging process—an issue that is still far from being clearly defined by the scientific community.

To be sure, the marked age-related decline in serum DHEA and DHEAS has suggested that a deficiency of these steroids may be causally related to the development of a series of diseases that are generally associated with aging.

The postulated consequences of low DHEA levels include insulin resistance, obesity, cardiovascular diseases, cancer, reduction of the immune defense system, and psychosocial problems such as depression and a general deterioration of a sense of well-being and cognitive function. As a consequence, DHEA replacement may seem an attractive treatment opportunity. Nevertheless, the analyses of clinical outcomes are far from being conclusive, and many issues should still be addressed.

Although DHEA preparations have been available in the market since the 1990s, there are very few definitive reports on the biological functions of DHEA. It is known that this steroid serves as a precursor of estrogens and androgens, and many believe that DHEA is merely an inactive precursor pool for the formation of bioactive steroid hormones.

In addition, there is increasing evidence for DHEA acting in its own right through a dedicated, although as yet unidentified, receptor. The existence of such a receptor for DHEA has been particularly investigated in brain tissue and vascular cells. In the brain, DHEA is a neurosteroid that acts as a modulator of neurotransmitter receptors, such as gamma aminobutyric acid type A, N methyl-D-aspartate, and sigma-1 receptors. In the vessels, DHEA binds with high affinity to the endothelial cell membrane, and it is not displaced by structurally related steroids. Binding of DHEA to the cell membrane is coupled to recruitment of G proteins such as G{alpha}i2 and G{alpha}i3 that mediate the rapid activation of intracellular signaling cascades.

Although debate still surrounds DHEA receptors, these findings corroborate the evidence that DHEA is not just a prehormone of the adrenals, but rather a hormone in its own right and that it modulates a series of biological processes with a remarkable tropism for the CNS.

Clinically, the range of women who would benefit from DHEA therapy is not clearly defined, nor is the dosage of hormone treatment. Whether DHEA therapy could be prescribed as a general anti-aging therapy or could be an alternative treatment for women suffering from androgen deficiency syndrome remains uncertain across studies (Genazzani et al. 2003; Panjari and Davis 2010; Labrie et al. 2005; Simoncini et al. 2003).

DHEA LOCAL THERAPY

Vaginal dryness is found in 75% of postmenopausal women. However, only 20%–25% of symptomatic women suffering from vaginal atrophy seek medical treatment for a variety of reasons, commonly the fear of estrogen-related side effects. Thus, there is a clear medical need and a major opportunity to remove the fear of breast cancer associated with today's estrogen-based therapies while improving the quality of life of the vast majority of women (75%–80%) who are presently left with the problem of vaginal atrophy for a large proportion of their lifetime.

Although intravaginal formulations are developed to avoid systemic exposure to estrogens, a series of studies has unanimously demonstrated that all intravaginal estrogen formulations lead to a significant increase in serum-estrogen levels measured directly by radioimmunoassay or through their systemic effects. Serum levels of estradiol are increased fivefold following administration of a 25-mg estradiol pill (Vagifem, Novo Nordisk, Princeton, NJ) or 1 g of 0.625-mg conjugated estrogen cream (Premarin, Wyeth Laboratories, Collegeville, PA). These findings were obtained using mass spectrometry, the most accurate and precise technology, thus indicating that the effects of estrogens applied locally in the vagina are unlikely to be limited to the vagina and that systemic activity should be expected.

Since serum DHEA is the exclusive source of sex steroids after menopause, the 60% decrease in circulating DHEA already found at the time of menopause is accompanied by a similar 60% decrease in the total androgen pool. Among the androgen target tissues, recent preclinical data obtained in experimented animals has clearly shown the beneficial effects of androgens made locally from DHEA in the vagina, not only in the superficial epithelial layer but also, and most importantly, on collagen formation of the lamina propria and muscularis. This data clearly indicates the importance of androgens for normal vaginal physiology, a role that cannot be achieved by treatment with estrogens alone.

A recent prospective, randomized, placebo-controlled phase III, 12-week clinical trial, performed on 218 postmenopausal women with moderate to severe symptoms of vaginal atrophy, demonstrated rapid and highly beneficial effects on all symptoms and signs of vaginal atrophy as well as on sexual dysfunction without significant changes in serum-estrogen and serum-androgen levels (N.A.M.S 2007; Rioux et al. 2000; Kendall et al. 2006; Labrie et al. 2009a,b,c; Labrie et al. 2005; Labrie 2007; Labrie 2010; Labrie et al. 2006; Berger et al. 2005; Labrie et al. 2008).

DHEA AND FERTILITY

Casson et al. (1998) were the first to suggest that DHEA supplementation might improve selected aspects of ovarian function in women with diminished ovarian reserve. However, because the authors reported fairly small benefits from a short-term supplementation protocol, their observations failed to attract follow-up. This, however, changed when a woman of advanced reproductive age, after self-medication with DHEA, experienced surprising gains in ovarian function. That experience led to a series of studies, investigating DHEA supplementation in infertile women with significant degrees of diminished ovarian reserve.

Those studies suggested that DHEA supplementation improves the response to ovarian stimulation with gonadotropins by increasing oocyte yield and embryo numbers (Barad and Gleicher, 2005). Explaining the rather small benefits initially observed by Casson et al. (1998) after only short-term use, DHEA effects increased over time, reaching peaks after approximately 4 to 5 months of supplementation. DHEA, however, also increased oocyte and embryo quality, spontaneous pregnancy rates in prognostically otherwise highly unfavorable patients on no further active treatments, pregnancy rates with *in vitro* fertilization (IVF), time to pregnancy, and cumulative pregnancy rates.

The reason why DHEA would positively affect ovarian function parameters and pregnancy chances in women with diminished ovarian reserve is still unknown, but it was suggested that the effect may be mediated by IGF-1 (Casson et al. 1998). Because DHEA effects peak at 4 to 5 months, a time period similar to the complete follicular recruitment cycle, it has been speculated about a DHEA effect on follicular recruitment, possibly mediated via suppressive effects on apoptosis. Following a small pilot study of insufficient statistical power, the possibility has been noted that DHEA may reduce aneuploidy in embryos.

Since approximately 80% of spontaneous pregnancy loss is the consequence of chromosomal abnormalities, reduced aneuploidy should also reduce miscarriage rates. As women get older and ovarian function progressively declines, miscarriage rates rise because of increasing aneuploidy. If DHEA beneficially affected ploidy, DHEA supplementation as an additional benefit in older women with severely diminished ovarian reserve should result in reduced miscarriage rates.

Since women with diminished ovarian reserve produce only small oocyte and embryo numbers with IVF, preimplantation genetic diagnosis in association with IVF is only rarely indicated, and, indeed, may be detrimental (Casson et al. 2000; Barad and Gleicher 2006; Barad, Brill, and Gleicher 2007; Casson et al. 1998; Gleicher, Weghofer, and Barad 2007; Morales et al. 2008; te Velde and Pearson 2002; Pal and Santoro 2003; Gleicher, Weghofer, and Barad 2008).

CONCLUSION

Recent studies have focused on the adrenal changes during aging and have mainly evaluated the possible use of DHEA as a precursor of both androgens and estrogens. Up to now, no double-blind, randomized, controlled trials have been designed on DHEA administration, probably because of the widely different perspectives about this compound across the world, but mainly because it is a "natural hormone," not protected by any copyright, so no pharmaceutical company would probably invest a great amount of money to support studies on a compound over which the company would never have any kind of exclusivity. Probably for this reason, most published studies are spontaneous and based on a relatively large number of patients. Nevertheless, several studies have confirmed the positive effects of DHEA administration in healthy elderly people, mostly in subjects older than 70 years, focusing on skin, bone density, muscle strength, and several neuropsychological symptoms. The fact that DHEA supplementation positively affects libido and sexual satisfaction in addition to promoting an increased sense of well-being more consistently in elderly women than in men demonstrated that DHEA might restore a great part, if not all, of the compromised steroid milieu typical of elder women. The recommended daily dosage for postmenopausal women is probably around 25 mg or lower (15–20 mg). Using such dosages, researchers have found the androgenic side effects to be minimal and reversible, but obviously more studies and controlled clinical trials are needed to disclose any hidden risks and features of "natural DHEA" in order to definitively determine whether DHEA can be used as a hormone replacement treatment or just as a "dietary supplementation."

REFERENCES

Ahmad, I., S. Lope-Piedrafita, X. Bi, C. Hicks, Y. Yao, C. Yu, E. Chaitkin et al. 2005. Allopregnanolone treatment, both as a single injection or repetitively, delays demyelination and enhances survival of Niemann-Pick C mice. *J Neurosci Res* 82:811–21.

Araneo, B. A., M. L. Woods, and R. A. Daynes. 1993. Reversal of the immunosenescent phenotype by dehydroepiandrosterone: Hormone treatment provides an adjuvant effect on the immunization of aged mice with recombinant hepatitis B surface antigen. *J Infect Dis* 167:830–40.

Ariza, M., R. Pueyo, M. Matarin Mdel, C. Junque, M. Mataro, I. Clemente, P. Moral, M. A. Poca, A. Garnacho, and J. Sahuquillo. 2006. Influence of APOE polymorphism on cognitive and behavioural outcome in moderate and severe traumatic brain injury. *J Neurol Neurosurg Psychiatry* 77:1191–3.

Arlt, W., F. Callies, I. Koehler, J. C. van Vlijmen, M. Fassnacht, C. J. Strasburger, M. J. Seibel et al. 2001. Dehydroepiandrosterone supplementation in healthy men with an age-related decline of dehydroepiandrosterone secretion. *J Clin Endocrinol Metab* 86:4686–92.

Arlt, W., J. Haas, F. Callies, M. Reincke, D. Hubler, M. Oettel, M. Ernst, H. M. Schulte, and B. Allolio. 1999. Biotransformation of oral dehydroepiandrosterone in elderly men: Significant increase in circulating estrogens. *J Clin Endocrinol Metab* 84:2170–6.

Arlt, W., H. G. Justl, F. Callies, M. Reincke, D. Hubler, M. Oettel, M. Ernst, H. M. Schulte, and B. Allolio. 1998. Oral dehydroepiandrosterone for adrenal androgen replacement: Pharmacokinetics and peripheral conversion to androgens and estrogens in young healthy females after dexamethasone suppression. *J Clin Endocrinol Metab* 83:1928–34.

Arnold, J. T., H. Le, K. K. McFann, and M. R. Blackman. 2005. Comparative effects of DHEA vs. testosterone, dihydrotestosterone, and estradiol on proliferation and gene expression in human LNCaP prostate cancer cells. *Am J Physiol Endocrinol Metab* 288:E573–84.

Azcoitia, I., E. Leonelli, V. Magnaghi, S. Veiga, L. M. Garcia-Segura, and R. C. Melcangi. 2003. Progesterone and its derivatives dihydroprogesterone and tetrahydroprogesterone reduce myelin fiber morphological abnormalities and myelin fiber loss in the sciatic nerve of aged rats. *Neurobiol Aging* 24:853–60.

Azizi, H., N. Z. Mehrjardi, E. Shahbazi, K. Hemmesi, M. K. Bahmani, and H. Baharvand. 2009. Dehydroepiandrosterone stimulates neurogenesis in mouse embryonal carcinoma cell- and human embryonic stem cell-derived neural progenitors and induces dopaminergic neurons. *Stem Cells Dev* 19(6):809–18.

Barad, D., H. Brill, and N. Gleicher. 2007. Update on the use of dehydroepiandrosterone supplementation among women with diminished ovarian function. *J Assist Reprod Genet* 24:629–34.

Barad, D. H., and N. Gleicher. 2005. Increased oocyte production after treatment with dehydroepiandrosterone. *Fertil Steril* 84:756.e1–3.

Barad, D., and N. Gleicher. 2006. Effect of dehydroepiandrosterone on oocyte and embryo yields, embryo grade and cell number in IVF. *Hum Reprod* 21:2845–949.

Basson, R. 2004. A new model of female sexual desire. *Endocrinol News* 29:22.

Baulieu, E. E. 1999. Neuroactive neurosteroids: Dehydroepiandrosterone (DHEA) and DHEA sulphate. *Acta Paediatr* 88(Suppl):78–80.

Baulieu, E. E., G. Thomas, S. Legrain, N. Lahlou, M. Roger, B. Debuire, V. Faucounau et al. 2000. Dehydroepiandrosterone (DHEA), DHEA sulfate, and aging: Contribution of the DHEAge Study to a sociobiomedical issue. *Proc Natl Acad Sci U S A* 97:4279–84.

Bélanger, A., B. Candas, A. Dupont, L. Cusan, P. Diamond, J. L. Gomez, and F. Labrie. 1994. Changes in serum concentrations of conjugated and unconjugated steroids in 40- to 80-year-old men. *J Clin Endocrinol Metab* 79:1086–90.

Benz, D. J., M. R. Haussler, M. A. Thomas, B. Speelman, and B. S. Komm. 1991. High-affinity androgen binding and androgenic regulation of $\alpha 1$(I)-procollagen and transforming growth factor-β steady state messenger ribonucleic acid levels in human osteoblast-like osteosarcoma cells. *Endocrinology* 128:2723–30.

Berger, L., M. El-Alfy, C. Martel, and F. Labrie. 2005. Effects of dehydroepiandrosterone, premarin and acolbifene on histomorphology and sex steroid receptors in the rat vagina. *J Steroid Biochem Mol Biol* 96:201–15.

Bernardi, F., M. Pieri, M. Stomati, S. Luisi, M. Palumbo, N. Pluchino, C. Ceccarelli, and A. R. Genazzani. 2003. Effect of different hormonal replacement therapies on circulating allopregnanolone and dehydroepiandrosterone levels in postmenopausal women. *Gynecol Endocrinol* 17:65–77.

Blandizzi, C. 2007. Enteric alpha-2 adrenoceptors: Pathophysiological implications in functional and inflammatory bowel disorders. *Neurochem Int* 51:282–8.

Blennow, K., M. J. de Leon, and H. Zetterberg. 2006. Alzheimer's disease. *Lancet* 368:387–403.

Brook, A. 1991. Bowel distress and emotional conflict. *J R Soc Med* 84:39–42.

Brown, R. C., Z. Han, C. Cascio, and V. Papadopoulos. 2003. Oxidative stress-mediated DHEA formation in Alzheimer's disease pathology. *Neurobiol Aging* 24:57–65.

Caranasos, G., and J. Busby. 1985. Sleep disorders, constipation, and pain: George Caranasos and Jan Busby. Interview by Richard L. Peck. *Geriatrics* 40:87–8, 90.

Cardounel, A., W. Regelson, and M. Kalimi. 1999. Dehydroepiandrosterone protects hippocampal neurons against neurotoxin-induced cell death: Mechanism of action. *Proc Soc Exp Biol Med* 222:145–9.

Carruba, G. 2006. Estrogens and mechanisms of prostate cancer progression. *Ann N Y Acad Sci* 1089:201–17.

Carruba, G. 2007. Estrogen and prostate cancer: An eclipsed truth in an androgen dominated scenario. *J Cell Biochem* 102:899–911.

Carson, P. J., K. L. Nichol, J. O'Brien, P. Hilo, and E. N. Janoff. 2000. Immune function and vaccine responses in healthy advanced elderly patients. *Arch Intern Med* 160(13):2017–24.

Casson, P. R., R. N. Andersen, H. G. Herrod, F. B. Stentz, A. B. Straughn, G. E. Abraham, and J. E. Buster. 1993. Oral dehydroepiandrosterone in physiologic doses modulates immune function in postmenopausal women. *Am J Obstet Gynecol* 169:1536–9.

Casson, P. R., M. S. Linday, M. D. Pisarska, S. A. Carson, and J. E. Buster. 2000. Dehydroepiandrosterone supplementation augments ovarian stimulation in poor responders: A case series. *Hum Reprod* 15:2129–32.

Casson, P. R., N. Santoro, K. Elkind-Hirsch, S. A. Carson, P. J. Hornsby, G. Abraham, and J. E. Buster. 1998. Postmenopausal dehydroepiandrosterone administration increases free insulin-like growth factor-I and decreases high-density lipoprotein: A six-month trial. *Fertil Steril* 70:107–10.

Cefalu, C. A. 2004. Is bone mineral density predictive of fracture risk reduction? *Curr Med Res Opin* 20:341–9.

Chami, T. N., A. E. Andersen, M. D. Crowell, M. M. Schuster, and W. E. Whitehead. 1995. Gastrointestinal symptoms in bulimia nervosa: Effects of treatment. *Am J Gastroenterol* 90:88–92.

Chang, D. M., J. L. Lan, H. Y. Lin, and S. F. Luo. 2002. Dehydroepiandrosterone treatment of women with mild-to-moderate systemic lupus erythematosus: A multicenter randomized, double-blind, placebo-controlled trial. *Arthritis Rheum* 46(11):2924–7.

Chang, W. Y., and G. S. Prins. 1999. Estrogen receptor-beta: Implications for the prostate gland. *Prostate* 40:115–24.

Charalampopoulos, I., V. I. Alexaki, C. Tsatsanis, V. Minas, E. Dermitzaki, I. Lasaridis, L. Vardouli et al. 2006. Neurosteroids as endogenous inhibitors of neuronal cell apoptosis in aging. *Ann N Y Acad Sci* 1088:139–52.

Charalampopoulos, I., C. Tsatsanis, E. Dermitzaki, V. I. Alexaki, E. Castanas, A. N. Margioris, and A. Gravanis. 2004. Dehydroepiandrosterone and allopregnanolone protect sympathoadrenal medulla cells against apoptosis via antiapoptotic Bcl-2 proteins. *Proc Natl Acad Sci U S A* 101:8209–14.

Chen, F., K. Knecht, E. Birzin, J. Fisher, H. Wilkinson, M. Mojena, C. T. Moreno et al. 2005. Direct agonist/antagonist functions of dehydroepiandrosterone. *Endocrinology* 146:4568–76.

Chesnut, C. H., J. L. Ivey, H. E. Gruber, M. Matthews, W. B. Nelp, K. Sisom, and D. J. Baylink. 1983. Stanozolol in postmenopausal osteoporosis: Therapeutic efficacy and possible mechanisms of action. *Metabolism* 32:571–80.

Ciolino, H., C. MacDonald, O. Memon, M. Dankwah, and G. C. Yeh. 2003. Dehydroepiandrosterone inhibits the expression of carcinogen-activating enzymes in vivo. *Int J Canc* 105:321–5.

Ciriza, I., I. Azcoitia, and L. M. Garcia-Segura. 2004. Reduced progesterone metabolites protect rat hippocampal neurones from kainic acid excitotoxicity in vivo. *J Neuroendocrinol* 16:58–63.

Compagnone, N. A., A. Bulfone, J. L. Rubenstein, and S. H. Mellon. 1995. Steroidogenic enzyme P450c17 is expressed in the embryonic central nervous system. *Endocrinology* 136:5212–23.

Compagnone, N. A., and S. H. Mellon. 1998. Dehydroepiandrosterone: A potential signalling molecule for neocortical organization during development. *Proc Natl Acad Sci U S A* 95:4678–83.

Corey, E., J. E. Quinn, M. J. Emond, K. R. Buhler, L. G. Brown, and R. L. Vessella. 2002. Inhibition of androgen-independent growth of prostate cancer xenografts by 17beta-estradiol. *Clin Cancer Res* 8:1003–7.

Corpechot, C., P. Robel, M. Axelson, J. Sjovall, and E. E. Baulieu. 1981. Characterization and measurement of dehydroepiandrosterone sulfate in rat brain. *Proc Natl Acad Sci U S A* 78:4704–7.

Daynes, R. A., B. A. Araneo, W. B. Ershler, C. Maloney, G. Z. Li, and S. Y. Ryu. 1993. Altered regulation of IL6 production with normal aging: Probable linkage to the age-associated decline in dehydroepiandrosterone and its sulfate derivative. *Immunology* 150:S219–30.

Daynes, R. A., B. A. Araneo, J. Hennebold, E. Enioutina, and H. H. Mu. 1995. Steroids as regulators of the mammalian immune response. *J Invest Dermatol* 105:14S–9S.

Daynes, R. A., D. J. Dudley, and B. A. Araneo. 1990. Regulation of murine lymphokine production in vivo. II. Dehydroepiandrosterone is a natural enhancer of interleukin 2 synthesis by helper T cells. *Eur J Immunol* 20:793–802.

De Fazio, J., D. R. Meldrum, J. H. Winer, and H. L. Judd. 1984. Direct action of androgen on hot flushes in the human male. *Maturitas* 6:3–8.

Debonnel, G., R. Bergeron, and C. de Montigny. 1996. Potentiation of neuronal NMDA response induced by dehydroepiandrosterone and its suppression by progesterone: Effects mediated via sigma receptors. *J Neurosci* 16:1193–202.

Delmas, P. D., N. H. Bjarnason, B. H. Mitlak, A. C. Ravoux, A. S. Shah, W. J. Huster, M. Draper, and C. Christiansen. 1997. Effects of raloxifene on bone mineral density, serum cholesterol concentrations, and uterine endometrium in postmenopausal women. *N Engl J Med* 337:1641–7.

Delpedro, A. D., M. J. Barjavel, Z. Mamdouh, and O. Bakouche. 1998. Activation of human monocytes by LPS and DHEA. *J Interferon Cytokine Res* 18:125–35.

Demirgoren, S., M. D. Majewska, C. E. Spivak, and E. D. London. 1991. Receptor binding and electrophysiological effects of dehydroepiandrosterone sulfate, an antagonist of the GABAA receptor. *Neuroscience* 45:127–35.

Devroede, G., G. Girard, M. Bouchoucha, T. Roy, R. Black, M. Camerlain, G. Pinard, J. C. Schang, and P. Arhan. 1989. Idiopathic constipation by colonic dysfunction. Relationship with personality and anxiety. *Dig Dis Sci* 34:1428–33.

Dhatariya, K., M. L. Bigelow, and K. S. Nair. 2005. Effects of dehydroepiandrosterone replacement on insulin sensitivity and lipids hypoadrenal women. *Diabetes* 54:765–9.

Diamond, P., L. Cusan, J. L. Gomez, A. Bélanger, and F. Labrie. 1996. Metabolic effects of 12-month percutaneous dehydroepiandrosterone replacement therapy in postmenopausal women. *J Endocrinol* 150(Suppl.):S43–50.

Djebaili, M., Q. Guo, E. H. Pettus, S. W. Hoffman, and D. G. Stein. 2005. The neurosteroids progesterone and allopregnanolone reduce cell death, gliosis, and functional deficits after traumatic brain injury in rats. *J Neurotrauma* 22:106–18.

Djebaili, M., S. W. Hoffman, and D. G. Stein. 2004. Allopregnanolone and progesterone decrease cell death and cognitive deficits after a contusion of the rat pre-frontal cortex. *Neuroscience* 123:349–59.

Donald, I. P., R. G. Smith, J. G. Cruikshank, R. A. Elton, and M. E. Stoddart. 1985. A study of constipation in the elderly living at home. *Gerontology* 31:112–8.

Duan, Y., A. Tabensky, V. DeLuca, and E. Seeman. 1997. The benefit of hormone replacement therapy on bone mass is greater at the vertebral body than posterior processes or proximal femur. *Bone* 21:447–51.

Eastell, R., R. A. Hannon, J. Cuzick, M. Dowsett, G. Clack, and J. E. Adams. 2006. Effect of an aromatase inhibitor on BMD and bone turnover markers: 2-year results of the Anastrozole, Tamoxifen, Alone or in Combination (ATAC) trial (18233230). *J Bone Miner Res* 21:1215–23.

Eich, D. M., J. E. Nestler, D. E. Johnson et al. 1993. Inhibition of accelerated coronary atherosclerosis with dehydroepiandrosterone in heterotopic rabbit model of cardiac transplantation. *Circulation* 87:261–9.

Ellem, S. J., and G. P. Risbridger. 2007. Treating prostate cancer: A rationale for targeting local oestrogens. *Nat Rev Cancer* 7:621–7.

English, M. A., S. V. Hughes, K. F. Kane, M. J. Langman, P. M. Stewart, and M. Hewison. 2000. Oestrogen inactivation in the colon: Analysis of the expression and regulation of 17β-hydroxysteroid dehydrogenase isozymes in normal colon and colonic cancer. *Br J Cancer* 83:550–8.

Ershler, W. B., W. H. Sun, N. Binkley et al. 1993. Interleukin 6 and aging: Blood levels and mononuclear cell production increase with advancing age and in vitro production is modifiable by dietary restriction. *Lymphokine Cytokine Res* 12:225–30.

Ervin, R. B., and J. Kennedy-Stephenson. 2002. Mineral intakes of elderly adult supplement and non-supplement users in the third national health and nutrition examination survey. *J Nutr* 132:3422–7.

Fogelman, I., C. Ribot, R. Smith, D. Ethgen, E. Sod, and J. Y. Reginster. 2000. Risedronate reverses bone loss in postmenopausal women with low bone mass: Results from a multinational, double-blind, placebo-controlled trial. BMD-MN Study Group. *J Clin Endocrinol Metab* 85:1895–900.

Francis, P. T. 2005. The interplay of neurotransmitters in Alzheimer's disease. *CNS Spectr* 10:6–9.

Gallagher, J. C., W. T. Kable, and D. Goldgar. 1991. Effect of progestin therapy on cortical and trabecular bone: Comparison with estrogen. *Am J Med* 90:171–8.

Geller, J. 1985. Rationale for blockade of adrenal as well as testicular androgens in the treatment of advanced prostate cancer. *Semin Oncol* 12:28–35.

Genazzani, A. R., S. Inglese, I. Lombardi, M. Pieri, F. Bernardi, A. D. Genazzani, L. Rovati, and M. Luisi. 2004. Long-term low dose dehydroepiandrosterone replacement therapy in aging males with partial androgen deficiency. *Aging Male* 7:133–43.

Genazzani, A. R., and N. Pluchino. 2010. DHEA therapy in postmenopausal women: The need to move forward beyond the lack of evidence. *Climacteric* 13:314–6.

Genazzani, A. D., M. Stomati, F. Bernardi, M. Pieri, L. Rovati, and A. R. Genazzani. 2003. Long term low-dose dehydroepiandrosterone in early and late postmenopausal women modulates endocrine parameters and synthesis of neuoractive steroids. *Fertil Steril* 80:1495–501.

Gennari, C. 2001. Calcium and vitamin D nutrition and bone disease of the elderly. *Public Health Nutr* 4:547–59.

Ghoumari, A. M., C. Ibanez, M. El-Etr, P. Leclerc, B. Eychenne, B. W. O'Malley, E. E. Baulieu, and M. Schumacher. 2003. Progesterone and its metabolites increase myelin basic protein expression in organotypic slice cultures of rat cerebellum. *J Neurochem* 86:848–59.

Giagulli, V. A., L. Verdonck, R. Giorgino, and A. Vermeulen. 1989. Precursors of plasma androstanediol- and androgen-glucuronides in women. *J Steroid Biochem* 33:935–40.

Gingras, S., and J. Simard. 1999. Induction of 3beta-hydroxysteroid dehydrogenase/isomerase type 1 expression by interleukin-4 in human normal prostate epithelial cells, immortalized keratinocytes, colon, and cervix cancer cell lines. *Endocrinology* 140:4573–84.

Giustina, A., G. Mazziotti, and E. Canalis. 2008. Growth hormone, insulin-like growth factors, and the skeleton. *Endocr Rev* 29:535–59.

Gleicher, N., A. Weghofer, and D. Barad. 2007. Increased euploid embryos after supplementation with dehydroepiandrosterone (DHEA) in women with premature ovarian aging. *Fertil Steril* 88(Suppl 1):S232.

Gleicher, N., A. Weghofer, and D. Barad. 2008. Preimplantation genetic screening (PGS), "established" and ready for prime time? *Fertil Steril* 89:780–8.

Goldstat, R., E. Briganti, J. Tran, R. Wolfe, and S. R. Davis. 2003. Transdermal testosterone therapy improves well-being, mood, and sexual function in premenopausal women. *Menopause* 10:390–8.

Gordon, G. B., D. E. Bush, and H. F. Weisman. 1988. Reduction of atherosclerosis by administration of dehydroepiandrosterone. A study in the hypercholesterolemic New Zealand white rabbit with aortic intimal injury. *J Clin Invest* 82:712–20.

Green, J. E., M. A. Shibata, E. Shibata, R. C. Moon, M. R. Anver, G. Kelloff, and R. Lubet. 2001. 2-difluoromethylornithine and dehydroepiandrosterone inhibit mammary tumor progression but not mammary or prostate tumor initiation in C3(1)/SV40 T/t-antigen transgenic mice. *Cancer Res* 61:7449–55.

Greenblatt, R. B., W. E. Barfield, J. F. Garner, G. L. Calk, and J. P. Harrod Jr. 1950. Evaluation of an estrogen, androgen, estrogen–androgen combination, and a placebo in the treatment of the menopause. *J Clin Endocrinol Metab* 10:1547–58.

Griffin, L. D., W. Gong, L. Verot, and S. H. Mellon. 2004. Niemann-Pick type C disease involves disrupted neurosteroidogenesis and responds to allopregnanolone. *Nat Med* 10:704–11.

Hardy, J., and D. J. Selkoe. 2002. The amyloid hypothesis of Alzheimer's disease: Progress and problems on the road to therapeutics. *Science* 297:353–6.

He, J., C. O. Evans, S. W. Hoffman, N. M. Oyesiku, and D. G. Stein. 2004a. Progesterone and allopregnanolone reduce inflammatory cytokines after traumatic brain injury. *Exp Neurol* 189:404–12.

He, J., S. W. Hoffman, and D. G. Stein. 2004b. Allopregnanolone, a progesterone metabolite, enhances behavioral recovery and decreases neuronal loss after traumatic brain injury. *Restor Neurol Neurosci* 22:19–31.

Heymen, S., S. D. Wexner, and A. D. Gulledge. 1993. MMPI assessment of patients with functional bowel disorders. *Dis Colon Rectum* 36:593–6.

Hosking, D., C. E. Chilvers, C. Christiansen et al. 1998. Prevention of bone loss with alendronate in postmenopausal women under 60 years of age. Early Postmenopausal Intervention Cohort Study Group. *N Engl J Med* 338:485–92.

Hynd, M. R., H. L. Scott, and P. R. Dodd. 2004. Glutamate-mediated excitotoxicity and neurodegeneration in Alzheimer's disease. *Neurochem Int* 45:583–95.

Isaacs, J. T. 1984. The aging ACI/Seg versus copenhagen male rat as a model system for the study of prostatic carcinogenesis. *Cancer Res* 44:5785–96.

Jakob, F., H. Siggelkow, D. Homann, J. Kohrle, J. Adamski, and N. Schutze. 1997. Local estradiol metabolism in osteoblast- and osteoclast-like cells. *J Steroid Biochem Mol Biol* 61:167–74.

Jankowski, C. M., W. S. Gozansky, R. S. Schwartz et al. 2006. Effects of dehydroepiandrosterone replacement therapy on bone mineral density in older adults: A randomized, controlled trial. *J Clin Endocrinol Metab* 91:2986–93.

Jellinger, K. A. 2004. Traumatic brain injury as a risk factor for Alzheimer's disease. *J Neurol Neurosurg Psychiatry* 75:511–2.

Kalmijn, S., L. J. Launer, R. P. Stolk, F. H. de Jong, H. A. Pols, A. Hofman, M. M. Breteler, and S. W. Lamberts. 1998. A prospective study on cortisol, dehydroepiandrosterone sulfate, and cognitive function in the elderly. *J Clin Endocrinol Metab* 83:3487–92.

Karishma, K. K., and J. Herbert. 2002. Dehydroepiandrosterone (DHEA) stimulates neurogenesis in the hippocampus of the rat, promotes survival of newly formed neurons and prevents corticosterone-induced suppression. *Eur J Neurosci* 16:445–53.

Kendall, A., M. Dowsett, E. Folkerd, and I. Smith. 2006. Caution: Vaginal estradiol appears to be contraindicated in postmenopausal women on adjuvant aromatase inhibitors. *Ann Oncol* 17:584–7.

Kim, S. B., M. Hill, Y. T. Kwak, R. Hampl, D. H. Jo, and R. Morfin. 2003. Neurosteroids: Cerebrospinal fluid levels for Alzheimer's disease and vascular dementia diagnostics. *J Clin Endocrinol Metab* 88:5199–206.

Kimonides, V. G., M. G. Spillantini, M. V. Sofroniew, J. W. Fawcett, and J. Herbert. 1999. Dehydroepiandrosterone antagonizes the neurotoxic effects of corticosterone and translocation of stress-activated protein kinase 3 in hippocampal primary cultures. *Neuroscience* 89:429–36.

Klein, H., M. Bressel, H. Kastendieck, and K. D. Voigt. 1988. Quantitative assessment of endogenous testicular and adrenal sex steroids and of steroid metabolizing enzymes in untreated human prostatic cancerous tissue. *J Steroid Biochem* 30:119–30.

Krazeisen, A., R. Breitling, K. Imai, S. Fritz, G. Moller, and J. Adamski. 1999. Determination of cDNA, gene structure and chromosomal localization of the novel human 17β-hydroxysteroid dehydrogenase type 7(1). *FEBS Lett* 460:373–9.

Krishnaraj, R., and G. Blandford. 1987. Age-associated alterations in human natural killer cells. *Clin Immunol Immunopathol* 45:268–85.

Kurata, K., M. Takebayashi, S. Morinobu, and S. Yamawaki. 2004. beta-estradiol, dehydroepiandrosterone, and dehydroepiandrosterone sulfate protect against N-methyl-D-aspartate-induced neurotoxicity in rat hippocampal neurons by different mechanisms. *J Pharmacol Exp Ther* 311:237–45.

Labrie, F. 1991. Intracrinology. *Mol Cell Endocrinol* 78:C113–8. Abstract.

Labrie, F. 2007. Drug insight: Breast cancer prevention and tissuetargeted hormone replacement therapy. *Nat Clin Pract Endocrinol Metab* 3:584–93.

Labrie, F. 2010. DHEA after Menopause - Sole source of sex steroids and potential sex steroid deficiency treatment. *Menopause Manag* 19:14–24.

Labrie, F., D. Archer, C. Bouchard, M. Fortier, L. Cusan, J. L. Gomez, G. Girard et al. 2009a. Effect of intravaginal dehydroepiandrosterone (Prasterone) on libido and sexual dysfunction in postmenopausal women. *Menopause* 16:923–31.

Labrie, F., D. Archer, C. Bouchard, M. Fortier, L. Cusan, J. L. Gomez, G. Girard et al. 2009b. Intravaginal dehydroepiandrosterone (Prasterone), a physiological and highly efficient treatment of vaginal atrophy. *Menopause* 16:907–22.

Labrie, F., D. Archer, C. Bouchard, M. Fortier, L. Cusan, J. L. Gomez, G. Girard et al. 2009c. Serum steroid levels during 12-week intravaginal dehydroepiandrosterone administration. *Menopause* 16:897–906.

Labrie, F., A. Bélanger, P. Bélanger, R. Bérubé, C. Martel, L. Cusan, J. Gomez et al. 2006. Androgen glucuronides, instead of testosterone, as the new markers of androgenic activity in women. *J Steroid Biochem Mol Biol* 99:182–8.

Labrie, F., A. Bélanger, L. Cusan, J. L. Gomez, and B. Candas. 1997c. Marked decline in serum concentrations of adrenal C19 sex steroid precursors and conjugated androgen metabolites during aging. *J Clin Endocrinol Metab* 82:2396–402.

Labrie, F., A. Belanger, V. Luu-The, C. Labrie, J. Simard, L. Cusan, J. L. Gomez, and B. Candas. 1998. DHEA and the intracrine formation of androgens and estrogens in peripheral target tissues: Its role during aging. *Steroids* 63:322–8.

Labrie, F., L. Cusan, J. L. Gomez, I. Côté, R. Bérubé, P. Bélanger, C. Martel, and C. Labrie. 2008. Effect of intravaginal DHEA on serum DHEA and eleven of its metabolites in postmenopausal women. *J Steroid Biochem Mol Biol* 111:178–94.

Labrie, F., L. Cusan, J. L. Gomez, I. Côté, R. Bérubé, P. Bélanger, C. Martel, and C. Labrie. 2009. Effect of one-week treatment with vaginal estrogen preparations on serum estrogen levels in postmenopausal women. *Menopause* 16:30–6.

Labrie, F., P. Diamond, L. Cusan, J. L. Gomez, and A. Bélanger. 1997b. Effect of 12-month DHEA replacement therapy on bone, vagina, and endometrium in postmenopausal women. *J Clin Endocrinol Metab* 82:3498–505.

Labrie, F., P. Diamond, L. Cusan, J. L. Gomez, A. Belanger, and B. Candas. 1997. Effect of 12-month dehydroepiandrosterone replacement therapy on bone, vagina, and endometrium in postmenopausal women. *J Clin Endocrinol Metab* 82:3498–505.

Labrie, F., A. Dupont, J. Simard, V. Luu-The, and A. Belanger. 1993. Intracrinology: The basis for the rational design of endocrine therapy at all stages of prostate cancer. *Eur Urol* 24(Suppl. 2):94–105.

Labrie, F., V. Luu-The, A. Bélanger, S.-X. Lin, J. Simard, and C. Labrie. 2005. Is DHEA a hormone? Starling Review. *J Endocrinol* 187:169–96.

Labrie, F., V. Luu-The, C. Labrie, and J. Simard. 2001. DHEA and its transformation into androgens and estrogens in peripheral target tissues: Intracrinology. *Front Neuroendocrinol* 22:185–212.

Labrie, F., V. Luu-The, S. X. Lin, C. Labrie, J. Simard, R. Breton, and A. Bélanger. 1997d. The key role of 17β-HSDs in sex steroid biology. *Steroids* 62:148–58.

Lasco, A., N. Frisina, N. Morabito, A. Gaudio, E. Morini, A. Trifiletti, G. Basile, V. Nicita-Mauro, and D. Cucinotta. 2001. Metabolic effects of dehydroepiandrosterone replacement therapy in postmenopausal women. *Eur J Endocrinol* 145:457–61.

Laughlin, G. A., and E. Barrett-Connor. 2000. Sexual dimorphism in the influence of advanced aging on adrenal hormone levels: The Rancho Bernardo Study. *J Clin Endocrinol Metab* 85:3561–8.

Liao, G., S. Cheung, J. Galeano, A. X. Ji, Q. Qin, and X. Bi. 2009. Allopregnanolone treatment delays cholesterol accumulation and reduces autophagic/lysosomal dysfunction and inflammation in Npc1-/- mouse brain. *Brain Res* 1270:140–51.

Lips, P. 2001. Vitamin D deficiency and secondary hyperparathyroidism in the elderly: Consequences for bone loss and fractures and therapeutic implications. *Endocr Rev* 22:477–501.

Lipton, S. A. 2005. The molecular basis of memantine action in Alzheimer's disease and other neurologic disorders: Low-affinity, uncompetitive antagonism. *Curr Alzheimer Res* 2:155–65.

Liu, D., and J. S. Dillon. 2002. Dehydroepiandrosterone activates endothelial cell nitric oxide synthase by a specific plasma membrane receptor coupled to Gα i2,3. *J Biol Chem* 277:21379–88.

Lockhart, E. M., D. S. Warner, R. D. Pearlstein, D. H. Penning, S. Mehrabani, and R. M. Boustany. 2002. Allopregnanolone attenuates N-methyl-D-aspartate-induced excitotoxicity and apoptosis in the human NT2 cell line in culture. *Neurosci Lett* 328:33–6.

Loria, R. M., T. H. Inge, S. S. Cook, A. K. Szakal, and W. Regelson. 1988. Protection against acute lethal viral infection with the native dehydroepiandrosterone (DHEA). *J Med Virol* 26:301–14.

Loria, R. M., W. Regelson, and D. A. Padgett. 1990. Immune response facilitation to virus and bacterial infections with dehydroepiandrosterone (DHEA). In *The Biologic Role of Dehydroepiandrosterone (DHEA)*, ed. M. Kalimi and W. Regelson, 101–26. New York: Walter de Gruyter & Co.

Lubet, R. A., G. B. Gordon, R. A. Prough, X. D. Lei, M. You, Y. Wang, C. J. Grubbs et al. 1998. Modulation of methylnitrosourea-induced breast cancer in Sprague–Dawley rats by dehydroepiandrosterone: Dose-dependent inhibition, effects of limited exposure, effects on peroxisomal enzymes, and lack of effects on levels of Ha-Ras mutations. *Cancer Res* 58:921–6.

Majewska, M. D., S. Demirgoren, C. E. Spivak, and E. D. London. 1990. The neurosteroid dehydroepiandrosterone sulfate is an allosteric antagonist of the GABAA receptor. *Brain Res* 526:143–6.

Martel, C., M. H. Melner, D. Gagne, J. Simard, and F. Labrie. 1994. Widespread tissue distribution of steroid sulfatase, 3β-hydroxysteroid dehydrogenase/Δ5-Δ4 isomerase (3β-HSD), 17β-HSD, 5α-reductase and aromatase activities in the rhesus monkey. *Mol Cell Endocrinol* 104:103–111.

Martel, C., E. Rheaume, M. Takahashi, C. Trudel, J. Couet, V. Luu-The, J. Simard, and F. Labrie. 1992. Distribution of 17β-hydroxysteroid dehydrogenase gene expression and activity in rat and human tissues. *J Steroid Biochem Mol Biol* 41:597–603. Abstract.

Martel, C., A. Sourla, G. Pelletier, C. Labrie, M. Fournier, S. Picard, S. Li, M. Stojanovic, and F. Labrie. 1998. Predominant androgenic component in the stimulatory effect of dehydroepiandrosterone on bone mineral density in the rat. *J Endocrinol* 157:433–42.

Martin, L. J., and N. F. Boyd. 2008. Mammographic density. Potential mechanisms of breast cancer risk associated with mammographic density: Hypotheses based on epidemiological evidence. *Breast Cancer Res* 10:201.

Martin, C., M. Ross, K. E. Chapman, R. Andrew, P. Bollina, J. R. Seckl, and F. K. Habib. 2004. CYP7B generates a selective estrogen receptor beta agonist in human prostate. *J Clin Endocrinol Metab* 89:2928–35.

Marx, C. E., L. F. Jarskog, J. M. Lauder, J. H. Gilmore, J. A. Lieberman, and A. L. Morrow. 2000. Neurosteroid modulation of embryonic neuronal survival in vitro following anoxia. *Brain Res* 871:104–12.

Marx, C. E., W. T. Trost, L. J. Shampine, R. D. Stevens, C. M. Hulette, D. C. Steffens, J. F. Ervin et al. 2006. The neurosteroid allopregnanolone is reduced in prefrontal cortex in Alzheimer's disease. *Biol Psychiatry* 60:1287–94.

McLachlan, J. A., C. D. Serkin, and O. Bakouche. 1996. Dehydroepiandrosterone modulation of lipopolysaccharide-stimulated monocyte cytotoxicity. *J Immunol* 156:328–35.

McPherson, S. J., S. J. Ellem, E. R. Simpson, V. Patchev, K. H. Fritzemeier, and G. P. Risbridger. 2007. Essential role for estrogen receptor beta in stromal–epithelial regulation of prostatic hyperplasia. *Endocrinology* 148:566–74.

Meikle, A. W., R. W. Dorchuck, B. A. Araneo, J. D. Stringham, T. G. Evans, S. L. Spruance, and R. A. Daynes. 1992. The presence of a dehydroepiandrosterone-specific receptor binding complex in murine T cells. *J Steroid Biochem Mol Biol* 42:293–304.

Meiring, P. J., and G. Joubert. 1998. Constipation in elderly patients attending a polyclinic. *S Afr Med J* 88:888–90.

Melchior, C. L., and R. F. Ritzmann. 1992. Dehydroepiandrosterone enhances the hypnotic and hypothermic effects of ethanol and pentobarbital. *Pharmacol Biochem Behav* 43:223–7.

Mellon, S. H., W. Gong, and M. D. Schonemann. 2008. Endogenous and synthetic neurosteroids in treatment of Niemann-Pick Type C disease. *Brain Res Rev* 57:410–20.

Migeon, C. J., A. R. Keller, B. Lawrence, and T. H. Shepart II. 1957. Dehydroepiandrosterone and androsterone levels in human plasma. Effect of age and sex: day-to-day and diurnal variations. *J Clin Endocrinol Metab* 17:1051–62.

Miller, K. K., B. M. Biller, J. Hier, E. Arena, and A. Klibanski. 2002. Androgens and bone density in women with hypopituitarism. *J Clin Endocrinol Metab* 87:2770–6.

Moghissi, E., F. Ablan, and R. Horton. 1984. Origin of plasma androstanediol glucuronide in men. *J Clin Endocrinol Metab* 59:417–21.

Morales, A. J., R. H. Haubrich, J. Y. Hwang, H. Asakura, and S. S. Yen. 1998. The effect of six months treatment with a 100 mg daily dose of dehydroepiandrosterone (DHEA) on circulating sex steroids, body composition and muscle strength in age-advanced men and women. *Clin Endocrinol (Oxf)* 49:421–32.

Morales, A. J., J. J. Nolan, J. C. Nelson, and S. S. Yen. 1994. Effects of replacement dose of dehydroepiandrosterone in men and women of advancing age. *J Clin Endocrinol Metab* 78:1360–7.

Morales, C., A. Sánchez, J. Bruguera, E. Margarit, A. Borrell, V. Borobio, and A. Soler. 2008. Cytogenetic study of spontaneous abortions using semi-direct analysis of chorionic villi samples detects the broadest spectrum of chromosome abnormalities. *Am J Med Genet* 146A:66–70.

Muller, M., A. W. Van Den Bend, Y. T. Van Der Schouw, D. E. Grobbee, and S. W. J. Lamberts. 2006. Effects of dehydroepiandrosterone and atamestane supplementation on frailty in elderly men. *J Clin Endocrinol Metab* 91(10):3988–91.

Murasko, D. M., B. J. Nelson, R. Silver, D. Matour, and D. Kaye. 1986. Immunologic response in an elderly population with a mean age of 85. *Am J Med* 81:612–8.

N.A.M.S. 2007. The role of local vaginal estrogen for treatment of vaginal atrophy in postmenopausal women: 2007 position statement of the North American Menopause Society. *Menopause* 14:357–69.

Naylor, J. C., C. M. Hulette, D. C. Steffens, L. J. Shampine, J. F. Ervin, V. M. Payne, M. W. Massing et al. 2008. Cerebrospinal fluid dehydroepiandrosterone levels are correlated with brain dehydroepiandrosterone levels, elevated in Alzheimer's disease, and related to neuropathological disease stage. *J Clin Endocrinol Metab* 93:3173–8.

Need, A. G., M. Horowitz, H. A. Morris, C. J. Walker, and B. E. Nordin. 1987. Effects of nandrolone therapy on forearm bone mineral content in osteoporosis. *Clin Orthop Relat Res* 225:273–8.

Notelovitz, M., N. Watts, C. Timmons, A. Addison, B. Wiita, and L. Downey. 1991. Effects of estrogen plus low dose androgen vs estrogen alone on menopausal symptoms in oophorectomized/hysterectomized women. In *Proceedings of the North American Menopause Society*, 101. Montreal, Canada.

Nottestad, S. Y., J. J. Baumel, D. B. Kimmel, R. R. Recker, and R. P. Heaney. 1987. The proportion of trabecular bone in human vertebrae. *J Bone Miner Res* 2:221–9.

Okabe, T., M. Haji, R. Takayanagi, M. Adachi, K. Imasaki, F. Kurimoto, T. Watanabe, and H. Nawata. 1995. Up-regulation of high-affinity dehydroepiandrosterone binding activity by dehydroepiandrosterone in activated human T lymphocytes. *J Clin Endocrinol Metab* 80:2993–6.

Orentreich, N., J. L. Brind, R. L. Rizer, and J. H. Vogelman. 1984. Age changes and sex differences in serum dehydroepiandrosterone sulfate concentrations throughout adulthood. *J Clin Endocrinol Metab* 59:551–5.

Orentreich, N., J. L. Brind, J. H. Vogelman, R. Andres, and H. Baldwin. 1992. Long-term longitudinal measurements of plasma dehydroepiandrosterone sulfate in normal men. *J Clin Endocrinol Metab* 75:1002–4.

Pal, L., and N. Santoro. 2003. Age-related decline in fertility. *Endocrinol Metab Clin North Am* 32:669–88.

Palmert, M. R., D. L. Hayden, M. J. Mansfield, J. F. Crigler Jr, W. F. Crowley Jr, D. W. Chandler, and P. A. Boepple. 2001. The longitudinal study of adrenal maturation during gonadal suppression: Evidence that adrenarche is a gradual process. *J Clin Endocrinol Metab* 86:4536–42.

Panjari, M., and S. R. Davis. 2010. DHEA for postmenopausal women: A review of the evidence. *Maturitas* 66(2):172–9. Epub, 2010 Jan 20.

Parker Jr, C. R., R. L. Mixon, R. M. Brissie, and W. E. Grizzle. 1997. Aging alters zonation in the adrenal cortex of men. *J Clin Endocrinol Metab* 82:3898–901.

Perkins, S. N., S. D. Hursting, D. C. Haines, S. J. James, B. J. Miller, and J. M. Phang. 1997. Chemoprevention of spontaneous tumorigenesis in nullizygous p53-deficient mice by dehydroepiandrosterone and its analog 16alpha-fluoro-5-androsten-17-one. *Carcinogenesis* 18:989–94.

Pike, M. C., M. D. Krailo, B. E. Henderson, J. T. Casagrande, and D. G. Hoel. 1983. 'Hormonal' risk factors, 'breast tissue age' and the age-incidence of breast cancer. *Nature* 303:767–70.

Pluchino, N., A. D. Genazzani, F. Bernardi, E. Casarosa, M. Pieri, M. Palumbo, G. Picciarelli, M. Gabbanini, M. Luisi, and A. R. Genazzani. 2005. Tibolone, transdermal estradiol or oral estrogen–progestin therapies: Effects on circulating allopregnanolone, cortisol and dehydroepiandrosterone levels. *Gynecol Endocrinol* 20:144–9.

Pluchino, N., F. Ninni, M. Stomati, L. Freschi, E. Casarosa, V. Valentino, S. Luisi, A. D. Genazzani, E. Poti, and A. R. Genazzani. 2008. One-year therapy with 10 mg/day DHEA alone or in combination with HRT in postmenopausal women: Effects on hormonal milieu. *Maturitas* 59:293–303.

Prestwood, K. M., A. M. Kenny, A. Kleppinger, and M. Kulldorff. 2003. Ultralow-dose micronized 17beta-estradiol and bone density and bone metabolism in older women: A randomized controlled trial. *JAMA* 290:1042–8.

Pye, J. K., R. E. Mansel, and L. E. Hughes. 1985. Clinical experience of drug treatments for mastalgia. *Lancet* ii:373–7.

Rao, K. V., W. D. Johnson, M. C. Bosland, R. A. Lubet, V. E. Steele, G. J. Kelloff, and D. L. McCormick. 1999. Chemoprevention of rat prostate carcinogenesis by early and delayed administration of dehydroepiandrosterone. *Cancer Res* 59:3084–9.

Regelson, W., R. Loria, and M. Kalimi. 1988. Hormonal intervention: "buffer hormones" or "state dependency." The role of dehydroepiandrosterone (DHEA), thyroid hormone, estrogen and hypophysectomy in aging. *Ann N Y Acad Sci* 521:260–73.

Reiter, E. O., V. G. Fuldauer, and A. W. Root. 1977. Secretion of the adrenal androgen, dehydroepiandrosterone sulfate, during normal infancy, childhood, and adolescence, in sick infants, and in children with endocrinologic abnormalities. *J Pediatr* 90:766–70.

Rioux, J. E., C. Devlin, M. M. Gelfand, W. M. Steinberg, and D. S. Hepburn. 2000. 17beta-estradiol vaginal tablet versus conjugated equine estrogen vaginal cream to relieve menopausal atrophic vaginitis. *Menopause* 7:156–61.

Risbridger, G. P., S. J. Ellem, and S. J. McPherson. 2007. Estrogen action on the prostate gland: A critical mix of endocrine and paracrine signaling. *J Mol Endocrinol* 39:183–8.

Sansoni, P., A. Cossarizza, V. Brianti et al. 1993. Lymphocytes subsets and natural killer cell activity in healthy old people and centenarians. *Blood* 82(9):2767–73.

Savvas, M., J. W. W. Studd, I. Fogelman, M. Dooley, J. Montgomery, and B. Murby. 1988. Skeletal effects of oral oestrogen compared with subcutaneous oestrogen and testosterone in postmenopausal women. *Br Med J* 297:331–3.

Schmidt, M., M. Kreutz, G. Loffler, J. Scholmerich, and R. H. Straub. 2000. Conversion of dehydroepiandrosterone to downstream steroid hormones in macrophages. *J Endocrinol* 164:161–9.

Schwartz, A. G., and L. L. Pashko. 1995. Cancer prevention with dehydroepiandrosterone and non-androgenic structural analogs. *J Cell Biochem* 22(Suppl.):210–7.

Schwartz, A. G., and L. L. Pashko. 2004. Dehydroepiandrosterone, glucose-6-phosphate dehydrogenase, and longevity. *Ageing Res Rev* 3:171–87.

Seeman, E., H. W. Wahner, K. P. Offord, R. Kumar, W. J. Johnson, and B. L. Riggs. 1982. Differential effects of endocrine dysfunction on the axial and the appendicular skeleton. *J Clin Invest* 69:1302–9.

Sherwin, B. B. 1988. Affective changes with estrogen and androgen replacement therapy in surgically menopausal women. *J Affect Disord* 14:177–87.

Sherwin, B. B., and M. M. Gelfand. 1984. Effects of parenteral administration of estrogen and androgen on plasma hormone levels and hot flushes in the surgical menopause. *Am J Obstet Gynecol* 148:552–7.

Sherwin, B. B., and M. M. Gelfand. 1985. Differential symptom response to parenteral estrogen and/or androgen administration in the surgical menopause. *Am J Obstet Gynecol* 151:153–60.

Sherwin, B. B., and M. M. Gelfand. 1987. The role of androgen in the maintenance of sexual functioning in oophorectomized women. *Psychosom Med* 49:397–409.

Shifren, J. L., G. D. Braunstein, J. A. Simon, P. R. Casson, J. E. Buster, G. P. Redmond, R. E. Burki et al. 2000. Transdermal testosterone treatment in women with impaired sexual function after oophorectomy. *N Engl J Med* 343:682–8.

Signoretti, S., and M. Loda. 2001. Estrogen receptor beta in prostate cancer: Brake pedal or accelerator? *Am J Pathol* 159:13–6.

Simoncini, T., P. Mannella, L. Fornari et al. 2003. Dehydroepiandrosterone modulates endothelial nitric oxide synthesis via direct genomic and nongenomic mechanisms. *Endocrinology* 144:3449–55.

Sizonenko, P. C., and L. Paunier. 1975. Hormonal changes in puberty III: Correlation of plasma dehydroepiandrosterone, testosterone, FSH, and LH with stages of puberty and bone age in normal boys and girls and in patients with Addison's disease or hypogonadism or with premature or late adrenarche. *J Clin Endocrinol Metab* 41:894–904.

Sklar, C. A., S. L. Kaplan, and M. M. Grumbach. 1980. Evidence for dissociation between adrenarche and gonadarche: Studies in patients with idiopathic precocious puberty, gonadal dysgenesis, isolated gonadotropin deficiency, and constitutionally delayed growth and adolescence. *J Clin Endocrinol Metab* 51:548–56.

Solerte, S. B., M. Fioravanti, G. Vignati, A. Giustina, L. Cravello, and E. Ferrari. 1999. Dehydroepiandrosterone sulfate enhances natural killer cell cytotoxicity in humans via locally generated immunoreactive insulin-like growth factor I. *J Clin Endocrinol Metab* 84:3260–7.

Sonnenberg, A., V. T. Tsou, and A. D. Müller. 1994. The "institutional colon": A frequent colonic dysmotility in psychiatric and neurologic disease. *Am J Gastroenterol* 89:62–6.

Steckelbroeck, S., Y. Jin, S. Gopishetty, B. Oyesanmi, and T. M. Penning. 2004. Human cytosolic 3alpha-hydroxysteroid dehydrogenases of the aldo-keto reductase superfamily display significant 3beta-hydroxysteroid dehydrogenase activity: Implications for steroid hormone metabolism and action. *J Biol Chem* 279:10784–95.

Stomati, M., P. Monteleone, E. Casarosa, B. Quirici, S. Puccetti, F. Bernardi, A. D. Genazzani, L. Rovati, M. Luisi, and A. R. Genazzani. 2000. Six-months oral dehydroepiandrosterone supplementation in early and late postmenopause. *Gynecol Endocrinol* 14:342–63.

Straub, R. H., L. Konecna, S. Hrach, G. Rothe, M. Kreutz, J. Scholmerich, W. Falk, and B. Lang. 1998. Serum dehydroepiandrosterone (DHEA) and DHEA sulfate are negatively correlated with serum interleukin-6 (IL-6), and DHEA inhibits IL-6 secretion from mononuclear cells in man in vitro: Possible link between endocrinosenescence and immunosenescence. *J Clin Endocrinol Metab* 83:2012–7.

Suzuki, M., L. S. Wright, P. Marwah, H. A. Lardy, and C. N. Svendsen. 2004. Mitotic and neurogenic effects of dehydroepiandrosterone (DHEA) on human neural stem cell cultures derived from the fetal cortex. *Proc Natl Acad Sci U S A* 101:3202–7.

Tamagno, E., M. Guglielmotto, P. Bardini, G. Santoro, A. Davit, D. Di Simone, O. Danni, and M. Tabaton. 2003. Dehydroepiandrosterone reduces expression and activity of BACE in NT2 neurons exposed to oxidative stress. *Neurobiol Dis* 14:291–301.

te Velde, E. R., and P. L. Pearson. 2002. The variability of female reproductive ageing. *Hum Reprod Update* 8:141–54.

The Writing Group for the PEPI Trial. 1996. Effects of hormone replacement therapy on endometrial histology in postmenopausal women. The Postmenopausal Estrogen/Progestin Interventions (PEPI) Trial. *JAMA* 275:370–5.

Tracz, M. J., K. Sideras, E. R. Bolona et al. 2006. Testosterone use in men and its effects on bone health. A systematic review and meta-analysis of randomized placebo-controlled trials. *J Clin Endocrinol Metab* 91:2011–6.

Välimäki, M. J., J. Farrerons-Minguella, J. Halse et al. 2007. Effects of risedronate 5 mg/d on bone mineral density and bone turnover markers in late-postmenopausal women with osteopenia: A multinational, 24-month, randomized, double-blind, placebo-controlled, parallel-group, phase III trial. *Clin Ther* 29:1937–49.

van Vollenhoven, R. F., E. G. Engelman, and J. L. Mc Guire. 1994. An open study of dehydroepiandrosterone in systemic lupus erythematosus. *Arthritis Rheum* 37(9):1305–10.

van Vollenhoven, R. F., E. G. Engleman, and J. L. Mc Guire. 1995. Dehydroepiandrosterone in systemic lupus erythematosus: Results of a double-blind, placebo-controlled, randomized clinical trial. *Arthritis Rheum* 38:1826–31.

van Vollenhoven, R. F., J. L. Park, M. C. Genovese, J. P. West, and J. L. McGuire. 1999. A double-blind, placebo-controlled, clinical trial of dehydroepiandrosterone in severe systemic lupus erythematosus. *Lupus* 8(3):181–7.

VanLandingham, J. W., M. Cekic, S. Cutler, S. W. Hoffman, and D. G. Stein. 2007. Neurosteroids reduce inflammation after TBI through CD55 induction. *Neurosci Lett* 425:94–8.

Vermeulen, A., J. P. Deslypene, W. Schelfhout, L. Verdonck, and R. Rubens. 1982. Adrenocortical function in old age: Response to acute adrenocorticotropin stimulation. *J Clin Endocrinol Metab* 54:187–91.

Vermeulen, A., and L. Verdonck. 1976. Radioimmunoassays of 17β-hydroxy-5α-androstan-3-one, 4-androstene-3,17-dione, dehydroepiandrosterone, 17β-hydroxyprogesterone and progesterone and its application to human male plasma. *J Steroid Biochem* 7:1–10.

Villareal, D. T., E. F. Binder, D. B. Williams, K. B. Schechtman, K. E. Yarasheski, and W. M. Kohrt. 2001. Bone mineral density response to estrogen replacement in frail elderly women: A randomized controlled trial. *JAMA* 286:815–20.

Villareal, D. T., E. F. Binder, K. E. Yarasheski et al. 2003. Effects of exercise training added to ongoing hormone replacement therapy on bone mineral density in frail elderly women. *J Am Geriatr Soc* 51:985–90.

Villareal, D. T., J. O. Holloszy, and W. M. Kohrt. 2000. Effects of DHEA replacement on bone mineral density and body composition in elderly women and men. *Clin Endocrinol (Oxf.)* 53:561–8.

Voigt, K. D., and W. Bartsch. 1986. Intratissular androgens in benign prostatic hyperplasia and prostatic cancer. *J Steroid Biochem* 25:749–57.

Wang, J. M., P. B. Johnston, B. G. Ball, and R. D. Brinton. 2005. The neurosteroid allopregnanolone promotes proliferation of rodent and human neural progenitor cells and regulates cell-cycle gene and protein expression. *J Neurosci* 25:4706–18.

Wang, L., Y. D. Wang, W. J. Wang, Y. Zhu, and D. J. Li. 2007. Dehydroepiandrosterone improves murine osteoblast growth and bone tissue morphometry via mitogen-activated protein kinase signaling pathway independent of either androgen receptor or estrogen receptor. *J Mol Endocrinol* 38:467–79.

Weihua, Z., S. Makela, L. C. Andersson, S. Salmi, S. Saji, J. I. Webster, E. V. Jensen, S. Nilsson, M. Warner, and J. A. Gustafsson. 2001. A role for estrogen receptor beta in the regulation of growth of the ventral prostate. *Proc Natl Acad Sci U S A* 98:6330–5.

Weihua, Z., M. Warner, and J. A. Gustafsson. 2002. Estrogen receptor beta in the prostate. *Mol Cell Endocrinol* 193:1–5.

Weiss, P., K. Shah, L. Fontana, C. P. Lambert, J. Holloszy, and D. T. Villareal. 2009. Dehydroepiandrosterone replacement therapy in older adults: 1- and 2-Y effects on bone. *Am J Clin Nutr* 89:1459–67.

Wenk, G. L., C. G. Parsons, and W. Danysz. 2006. Potential role of N-methyl-D-aspartate receptors as executors of neurodegeneration resulting from diverse insults: Focus on memantine. *Behav Pharmacol* 17:411–24.

Whitehead, W. E. 2007. Twin studies used to prove that the comorbidity of major depressive disorder with IBS is NOT influenced by heredity. *Am J Gastroenterol* 102:2230–1.

Widstrom, R. L., and J. S. Dillon. 2004. Is there a receptor for dehydroepiandrosterone or dehydroepiandrosterone sulfate? *Semin Reprod Med* 22:289–98.

Williams, M. R., S. Ling, T. Dawood, K. Hashimura, A. Dai, H. Li, J. P. Liu, J. W. Funder, K. Sudhir, and P. A. Komesaroff. 2002. Dehydroepiandrosterone inhibits human vascular smooth muscle cell proliferation independent of ARs and ERs. *J Clin Endocrinol Metab* 87:176–81.

Xilouri, M., and P. Papazafiri. 2006. Anti-apoptotic effects of allopregnanolone on P19 neurons. *Eur J Neurosci* 23:43–54.

Yen, S. S., A. J. Morales, and O. Khorram. 1995. Replacement of DHEA in aging men and women. Potential remedial effects. *Ann N Y Acad Sci* 774:128–42.

Young, J., B. Couzinet, K. Nahoul, S. Brailly, P. Chanson, E. E. Baulieu, and G. Schaison. 1997. Panhypopituitarism as a model to study the metabolism of dehydroepiandrosterone (DHEA) in humans. *J Clin Endocrinol Metab* 82:2578–85.

Young, D. G., G. Skibinski, J. I. Mason, and K. James. 1999. The influence of age and gender on serum dehydroepiandrosterone sulphate (DHEA-S), IL-6, IL-6 soluble receptor (IL-6 sR) and transforming growth factor β1 (TGF-β1) levels in normal healthy blood donors. *Clin Exp Immunol* 117:476–81.

Young, D. G., G. Skibinski, A. Skibinska, J. I. Mason, and K. James. 2001. Preliminary studies on the effect of dehydroepiandrosterone (DHEA) on both constitutive and phytohaemagglutinin (PHA)-inducible IL-6 and IL-2 mRNA expression and cytokine production in human spleen mononuclear cell suspensions in vitro. *Clin Exp Immunol* 123:28–35.

Zwain, I. H., and S.S. Yen. 1999a. Dehydroepiandrosterone: Biosynthesis and metabolism in the brain. *Endocrinology* 140:880–7.

Zwain, I. H., and S.S. Yen. 1999b. Neurosteroidogenesis in astrocytes, oligodendrocytes, and neurons of cerebral cortex of rat brain. *Endocrinology* 140:3843–52.

Section II

Prevention of Disease by DHEA

3 Dehydroepiandrosterone *Its Metabolites and Resistance to Infections*

Roger M. Loria and David Ben-Nathan

CONTENTS

INTRODUCTION

This chapter focuses on the effects of dehydroepiandrosterone (DHEA) and its derivatives Δ5-androstene-3β, 17β-diol (AED) and Δ5-androstene-3β, 7β, 17β-triol (AET) during infections. The goal is to provide an overview of the advances but not a complete and exhaustive literature review. Any publication related to immune deficiency syndrome and human immunodeficiency virus (HIV) is excluded, as this topic is being considered in another chapter of this book.

DHEA was first reported to increase host resistance against lethal infections by human viruses (e.g., coxsackievirus B4 and herpes simplex type 2) and against bacterial infection (e.g., *Enterococcus faecalis*) in a mouse model (Loria et al. 1988). This finding was independently confirmed and extended by Ben-Nathan et al. (1991), who reported that a single, subcutaneous (SC) injection of DHEA protected mice from lethal infection of three different arthropod-borne agents: (1) West Nile virus (WNV), (2) Sindbis virus (SVNI), and (3) Semliki Forest virus (SFV). *In vivo*, the DHEA protective effects were found to require the host to have an intact immune system (Loria et al. 1988; Loria and Padgett 1992), indicating that DHEA mediated the increase in host resistance by upregulating host immunity. Subsequent studies showed that two metabolic derivatives of DHEA, AED, and AET, are markedly more effective in upregulating host immunity than the precursor DHEA (Loria and Padgett 1992; Padgett, Loria, and Sheridan 2000). These agents, when injected at significantly lower concentrations than the precursor, provided host protection against much higher infectious doses. AED effectively protected the host from coxsackie virus-induced tissue pathology and also myocardiopathy and pancreopathy. Both DHEA and AED protected the host from a lethal infection with *Enterococcus faecalis* and *Pseudomonas aeruginosa* and from a lethal dose of lipopolysaccharide of 800 μg/mouse (E. coli O55:B5 enterotoxin). *In vivo*, there was no significant difference in the protective effects of AED and DHEA against a lethal challenge with *E. faecalis*, but AED was significantly more effective in protecting animals from a lethal challenge with *P. aeruginosa*. Two hours prior to a lethal Pseudomonas challenge (LD100), treatment with a dose of 20-mg DHEA protected 43% of animals, whereas 2 mg of AED protected 67% of Pseudomonas-infected animals. Similarly, AED was markedly more effective in protecting female CD-1 mice from a lipopolysaccharide (LPS) challenge. However, DHEA reduced tumor necrosis factor alpha (TNF-α) levels by 50%, whereas AED was ineffective. These results show that the protective effects of DHEA and AED against bacterial and endotoxin challenges are distinct from each other and the enhanced AED protective effect is not mediated by the reduction of TNF-α. These observations establish that these androstene steroids upregulate host resistance and serve as effective immune modulators against viral and bacterial infections, illustrating their immunomodulatory properties (Ben-Nathan, Padgett, and Loria 1999; Loria and Padgett 1992; Padgett, Loria, and Sheridan 2000).

DHEA AND POLYMICROBIAL INFECTIONS—CECAL LIGATION AND PUNCTURE

Oberbeck et al. (2001) provided significant additional information on the protective effects of DHEA in models of polymicrobial infection and sepsis following cecal ligation and puncture (CLIP). This group reported that 8–9-week-old male mice treated with 30 mg/kg DHEA SC before and 24 hours after CLIP showed a significant increase in survival rates compared with untreated CLIP animals. Forty-eight hours after CLIP, DHEA treatment reduced mortality from 47% to 13.4% and by 72 hours from 66.7% to 26.8% in untreated animals, respectively ($p < .001$). Also in this model, DHEA treatment reduced the levels of TNF-α to 20.7 ± 1.4 pg/mL as compared to 32.4 ± 6.6 pg/mL in untreated animals ($p < .01$).

In subsequent studies, Oberbeck et al. (2007) reported an increased HSP-70 concentration in the lungs and spleens of DHEA-treated septic animals, but no such increase was evident in untreated septic animals. This study confirmed the previous observation by showing that in DHEA-treated CLIP animals, mortality was reduced to 22% (7/32) compared to 50% (16/32) of CLIP-saline-treated animals ($p < .05$). van Griensven et al. (2002) measured delayed-type hypersensitivity (DTH) ratio following primary and secondary challenges 48 hours after CLIP. DHEA treatment had a significant effect on delayed hypersensitivity response and reduced perturbation of CD8+ and CD56+ cells.

The anti-inflammatory phase following CLIP-induced severe sepsis is associated with an elevation in interleukin 10 (IL-10), and mediates a significant downregulation of the splenic macrophage toll-like receptors, TLR2 and TLR4. These receptors function to bind gram positive and gram negative bacterial enterotoxins in sepsis. Treatment with DHEA significantly suppresses the levels of plasma IL-10, whereas it restores TLR2 and TLR4 mRNA expression on splenic macrophages and the production of TNF-α (Matsuda et al. 2005).

Barkhausen et al. (2009) investigated the protective function of DHEA in a trauma model consisting of femur fracture, hemorrhagic shock, and CLIP-induced polymicrobial sepsis. Morbidity was primarily modulated by TNF-α expression, which was both time and organ dependent. DHEA treatment induced the restoration of TNF-α levels in the liver and lung at 48 hours; TNF-α levels in these tissues were depressed in the untreated trauma hemorrhage-CLIP group. After 96 hours, an attenuation of the DHEA effect was evident.

Contrary to the previous reports, Hildebrand et al. (2003) showed a 100% survival of DHEA-treated CLIP-TNF-RI (-/-) knockout mice compared to 50% survival of CLIP-TNF-RI (-/-) untreated mice. There was no significant difference of DHEA treatment in CLIP-IL-6 (-/-) knockout mice, indicating that the effects of DHEA were independent of TNF-α. This contradicts the previous reports that showed that the protective effect of DHEA is associated with the reduction of the systemic inflammation, that is, TNF-α, IL-1β, IL-6, and IL-10. As stated by these authors and suggested by their findings, whether the *in vivo* effects of DHEA are independent of TNF-α remains to be confirmed.

In their seminal work, Gianotti et al. (1996) reported that DHEA and prednisone exert opposite effects during gut-derived sepsis following a challenge by C^{14} Escherichia coli in female BALB/c mice with 20% thermal injury. DHEA treatment at 25 mg/kg/day increased survival to 72% compared to 16% in untreated controls. In contrast, a dose of 10 mg/kg/day of prednisone markedly reduced survival. Furthermore, prednisone increased the E.coli translocation to the mesenteric lymph nodes, resulting in a significant increase in the number of viable E. coli in this tissue. In contrast, DHEA treatment at a dose of 25 mg/kg/day significantly reduced E.coli translocation and the number of viable E.coli in the mesenteric lymph node. The number of viable E.coli as measured by the number of colony-forming units per gram tissue in DHEA-treated animals was reported to be 398 ± 255 compared to 6066 ± 3222 for the untreated group ($p < .05$). This observation clearly illustrates the immune-modulating properties of this hormone and its ability to reduce the number of viable microorganisms in the target tissue.

Additional support for the immune-modulating properties of DHEA treatment is derived from the study in which C3H/HeN mice underwent trauma and hemorrhage followed by CLIP procedure. Fourty-eight hours after the procedure, a dose of 4 mg/kg DHEA was administered. The treatment mediated a significant increase in survival compared to untreated controls. Splenocyte and macrophage counts were restored to normal level after 24 hours. Trauma and hemorrhage per se mediated a 54.4% reduction in IL-2 levels, which was restored to control levels following DHEA treatment (Angele et al. 1998).

PROTECTIVE EFFICACY OF DHEA AGAINST LETHAL VIRAL ENCEPHALITIS

WNV, a mosquito-transmitted flavivirus, was first isolated from a febrile adult woman in the West Nile District of Uganda in 1937 (Smithburn et al. 1940). WNV is a single-stranded RNA virus and is a member of the Japanese encephalitis antigenic complex of the genus *Flavivirus* (Burke and Monath 2001; Scherret et al. 2001). Until 1999, WNV was found in Africa, the Middle East, parts of Asia, Southern Europe, and Australia. It then suddenly emerged in New York, rapidly spread throughout the United States, and has since then caused considerable acute mortality and morbidity (Nash et al. 2001). The clinical manifestations of WNV in humans range from asymptomatic seroconversion to fatal meningoencephalitis, with symptoms including cognitive dysfunction, muscle weakness, and flaccid paralysis (Carson et al. 2006; Debiasi and Tyler 2006; Hayes and Gubler 2006; Hayes et al. 2005). Depressed immunity, age, and genetic factors (Glass et al. 2006; Yakub et al. 2005) are correlated with a greater risk for neurological diseases. There is no effective WNV vaccine to protect populations at risk. Currently, the only effective way to provide immediate resistance to WNV is by the passive administration of WNV-specific antibodies (Ben-Nathan et al. 2003; Ben-Nathan et al. 2009; Wang and Fikrig 2004; Mehlhop and Diamond 2008).

In our study, we used the murine model of WNV to determine the protective efficacy of DHEA against lethal viral encephalitis. In this model, WNV caused a systemic infection and invaded the central nervous system (CNS), resulting in death within 1–2 weeks (Ben-Nathan et al. 1992; Ben-Nathan et al. 1996). The *in vivo* activity of DHEA was tested by either intraperitoneal injection with dimethyl sulfoxide (DMSO) and paraffin oil as the solvent or soybean oil for SC injection. A single injection of DHEA (20 mg/kg), administered subcutaneously before or after virus inoculation (500 PFU/mouse), protected 70% of the mice against lethal WNV infection. The drug was more effective against WNV when injected prior to infection but was still effective when injected 1 day after infection (Table 3.1). The protective efficacy of DHEA (20 mg/kg) against different infectious doses of WNV 10, 100, or 1000 PFU/mouse is given in Table 3.2. Again, DHEA treatment protected 70% of the mice against lethal WNV infection compared to no protection in the control WNV-infected DHEA-untreated group. DHEA treatment reduced viremia and death rate, delayed the

TABLE 3.1
DHEA Treatment Increases Survival of Mice Infected with West Nile Virus

Day of DHEA Treatment	Mortality D/T	Percent Survival
−1	3/10	70*
0	5/10	50*
1	5/10	50*
2	6/10	40
3	8/10	20
Control	9/10	10

Note: Mice were injected once SC with 1 g/kg of DHEA on days −1, 0, 1, 2, and 3 after virus infection; WNV: 100 PFU/mouse were injected intraperitoneally.
D/T = dead/total; PFU = plaque-forming units.
*$p < .05$ compared to control nontreated DHEA group.

TABLE 3.2
Protective Efficacy of DHEA in Mice Infected with West Nile Virus

	10 PFU		100 PFU		1000 PFU	
Treatment Group	D/T	Percent Survival	D/T	Percent Survival	D/T	Percent Survival
WNV	8/18	56	17/18	6	10/10	0
WNV + DHEA	4/18	78	5/18	72*	3/10	70*

Note: DHEA at a dose of 20 mg/mouse was injected intraperitoneally on days −1, 0, 2, 4, and 6 after WNV infection.
PFU = plaque-forming units; D/T = dead/total.
*$p < .05$ compared to control nontreated DHEA group.

onset of the disease, and increased the ability of the host to control virus replication and neuroinvasiveness through various immune mechanisms (Ben-Nathan et al. 1991). In addition, DHEA treatment reduced viremia and WNV level in the spleen by 2 log10 PFU and by 2–3 log10 PFU in the brain of infected mice compared to nontreated mice. Administration of DHEA caused an increase in thymus and spleen weight in control mice and in WNV-infected mice. This effect was more pronounced in the immunosuppressed mice with the DHEA-treated virus-infected animals thymus and spleen tissues exceeding weight of the same tissues of DHEA-treated control–uninfected mice (Ben-Nathan et al. 1992). This observation extended to the *in vivo* action of AED on spleen size as illustrated in Figure 3.1. A dose of 32 g/kg AED to C57BL/6J mice did not have a significant effect on the spleen. However, a dose of 0.08 g/kg clearly protected the spleen from coxsackievirus B4 pathology, but a higher dose of 0.32 g/kg resulted in significant proliferation of the spleen, but only in infected animals (Loria and Padgett 1992).

Similar results were reported against infection with Venezuelan equine encephalomyelitis virus (VEE) in mice following vaccination with the TC-83 VEE virus (Negrette et al. 2001). Researchers administered DHEA in a single dose (10 mg/kg), 4 hours before vaccination. This treatment increased antibody titers against TC-83 VEE virus at 14 days after immunization. When vaccinated animals were challenged with live VEE virus 21 days after immunization and treated with DHEA,

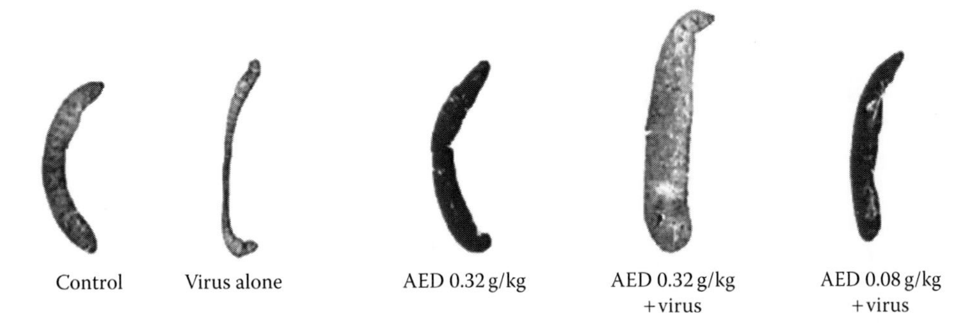

FIGURE 3.1 Protective action of androstenediol on the spleen following infection with coxsackievirus B4. Treatment of virus-infected animals with a dose of 0.32 g/kg resulted in a marked proliferation of the organ. Treatment with the same dose of 5 androsterone 3β, 17β diol without infection did not result in such enlargement. Treatment with a lower dose of 0.08 g/kg protected the spleen from virus-mediated destruction. Each section is a sagittal cross section. (From *Arch Virol*, Androstenediol regulates systemic resistances against lethal infections in mice, 127, 103–115, R. M. Loria and D. A. Padgett. With permission.)

there was evidence of reduced viremia levels and a reduction in brain virus levels. It is suggested that DHEA treatment could enhance the efficiency of immunization against VEE virus in mice (Negrette et al. 2001).

DHEA AND ITS DERIVATIVES ANDROSTENEDIOL AND ANDROSTENETRIOL

DHEA is converted to AED in the skin and in the brain (Toth and Faredin 1985a, 1985b). Consequently, we tested the effects of AED on the course and outcome of influenza A infection of C57BL/6J male mice. A single SC injection of 320 mg/kg dose of AED to animals infected with influenza A/Puerto Rico/8/34 (PR8) protected 80% of the animals from mortality, whereas only 30% of untreated animals survived (Figure 3.2). The type of immune response following AED treatment and influenza virus infection was determined by the *in vitro* cytokine secretion from isolated T cells from the infected target tissues—that is, ex vivo the lung, the lymph node, and spleen, which were exposed again to influenza virus antigen. AED treatment mediated a significant increase in the levels of IFN-γ from the lung draining lymph node and the spleen at 3, 5, and 7 days post-influenza virus infection. In contrast, AED treatment reduced the secretion of IL-10 from the lung draining lymph node at days 3 and 5 postinfection. AED did not affect splenocyte levels of IL-10 production, which indicated that the cytokine response was specific to the influenza virus–infected target tissue.

Viral infection, as well as influenza infection, is associated with an elevation in corticosterone levels. Two corticosterone peaks, at day 3 and at day 7 postinfection, were observed and coincided with the elevation of IL-1 and IL-6 cytokines during infection. The second peak of corticosterone coincided with the high levels of mononuclear cells in the lungs. As illustrated in Figure 3.3, pretreatment with AED counteracted the infection-mediated elevation in corticosterone levels (Padgett, Loria, and Sheridan 1997).

Herpes virus-1 ocular infection is 80% lethal in female ICR mice if untreated. Treatment with AED-S, (referred to as AED in this report), at doses of 10 and 50 μg/mL, administered with the drinking water, significantly reduced HSV-1 mortality to 38% and 30%, respectively. In this model, AED-S did not mediate a reduction in virus titers in the trigeminal ganglion or the eye and did not influence the level of serum corticosterone (Daigle and Carr 1998). A type 1 cytokine response with an increase in IL-2, IFN-γ, and natural killer cell activity was observed following subcutaneous administration of 320/mg/kg AED to HSV-1 infected animals (Carr 1998).

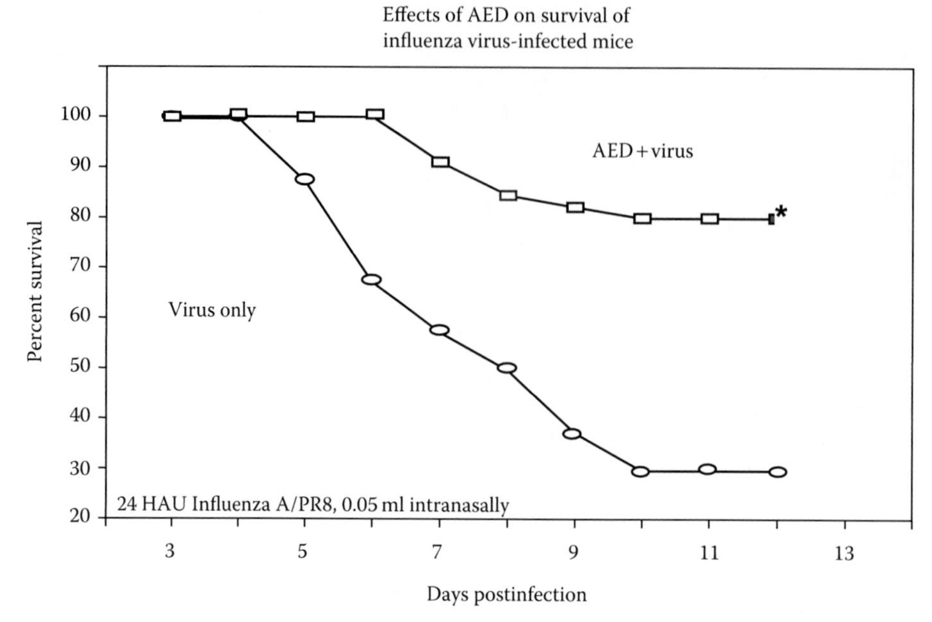

FIGURE 3.2 AED treatment improved survival of influenza-infected mice. Male C57BLr6 mice were treated with 320 mg/kg AED (filled squares), $n = 45$ or control vehicle (open circles), $n = 40$ four hours prior to infection with 24 HAU influenza A/PR8 virus, $p \leq .005$. (Reprinted from *J Neuroimmunol*, 78, Padgett, D. A., R. M. Loria, and J. F. Sheridan, Endocrine regulation of the immune response to influenza virus infection with a metabolite of DHEA-androstenediol, 203–11, Copyright (1997), with permission from Elsevier.)

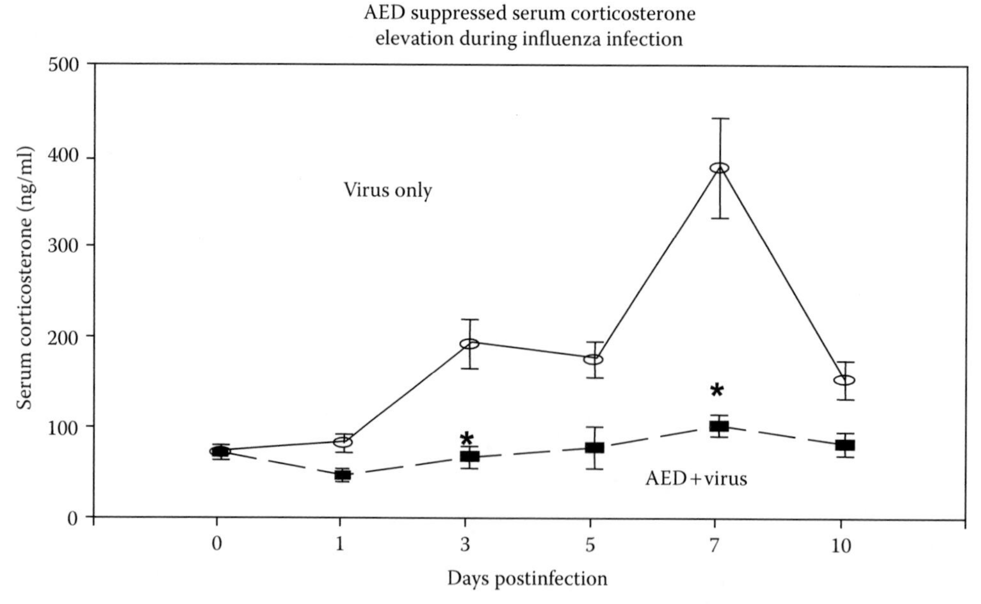

FIGURE 3.3 AED suppressed serum corticosterone elevation during influenza infection. Pretreatment of influenza A/PR8 virus-infected mice with 320 mg/kg AED (filled squares) prevented viral-induced elevation of serum corticosterone as shown in vehicle-treated animals (open circles). Data represent the mean ± S.D. of 5–10 animals at each time point. $*p \leq .05$. (Reprinted from *J Neuroimmunol*, 78, Padgett, D. A., R. M. Loria, and J. F. Sheridan, Endocrine regulation of the immune response to influenza virus infection with a metabolite of DHEA-androstenediol, 203–11, Copyright (1997), with permission from Elsevier.)

PROTECTION FROM RADIATION INJURY AND INFECTION

Androstenes upregulate immunity and counteract corticosteroid immune suppression (Padgett and Loria 1998). These properties are essential for recovery following exposure to whole body lethal radiation because the suppression of host immunity is a major factor in morbidity and mortality from radiation sickness. We reported that a single SC injection of 0.75 mg of 5 androstene 3β, 7β, 17β triol (AET) or 8-mg AED protected 25 g C57BL/6J male mice from whole body gamma radiation of 8 Gy. The whole body radiation dose used in this experiment, 8 Gy was considerably higher than the Hiroshima LD100 for humans of 6 Gy. Treatment with either AET or AED resulted in 70%–60% survival, respectively ($p < .001$; Loria et al. 2000). Whole body gamma radiation exposure with a 6-Gy dose resulted in a loss of 80%–90% of the total spleen cell counts. The data show that both AET and AED function to restore the spleen cell counts within 14–21 days and restore bone marrow CD11b, B220, CD4+, and CD8+ over this period. Whole body radiation exposure at a high dose mediates a suppression of host resistance with the resulting infections being a major cause for the morbidity and mortality. We tested the level of immune suppression by exposing male C57BL/6J mice to a dose of 6-Gy gamma radiation and the effectiveness of AED or AET to restore host resistance following viral infection (Figure 3.4; Loria et al. 2000). When normal C57BL/6J mice are infected with 100 PFU of coxsackievirus B4, the infection is inconsequential and not associated with mortality or other symptoms. However, the exposure of such infected animals to 6 Gy whole body gamma radiation results in 56% mortality, unmasking the levels of immune suppression caused by radiation and demonstrating the level of increased susceptibility to infection. A single SC injection of 8 mg AED or 0.75 mg AET to a 25 g mouse protected up to 78%–88% of irradiated mice from mortality resulting from coxsackievirus B4 infection. Similar findings were reported on the effects of AED administered 2 hours after exposure to 3 Gy whole body irradiated B6D2Fl/J female mice infected with 5.6×10^6 CFU Klebsiella pneumonia. A dose of 160 mg/kg AED protected 50% of infected irradiated animals, whereas DHEA at the same dose did not

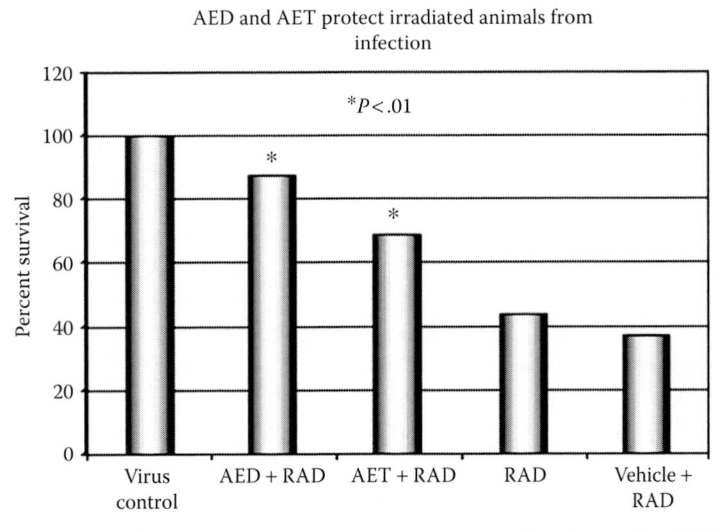

FIGURE 3.4 AED and AET protected irradiated mice from virus-mediated mortality. C57BL/6J mice were exposed to whole body irradiation of 6 Gy and challenged with 100 plaque-forming units of coxsackievirus B4. Unexposed control animals survived this infectious challenge; however, only 44% of irradiated infected mice survived. A single SC injection of AED 8 mg/25 g mouse or AET 0.75 mg/25 g mouse protected the irradiated mice from mortality resulting from coxsackievirus B4 challenge. The vehicle (DMSO:ethanol 1:1) did not have any protective effect against virus infection. (Adapted from Loria, R. M., D. H. Conrad, T. Huff, H. Carter, and D. Ben-Nathan. 2000. *Ann NY Acad Sci* 917:860–7.)

have a protective effect. A higher dose of AED, 320 mg/kg, protected 90% of the exposed animals (Whitnall et al. 2000).

EFFECTS OF DHEA AND ITS DERIVATIVES ON TUBERCULOSIS

Rook and Hernandez-Pando (1997) made the seminal observation that the ratio of cortisol to DHEA may be an important factor influencing the immunopathology in patients with tuberculosis (TB). Because immunity to TB is dependent on a TH1 response, the finding is highly relevant.

Using a model of pulmonary TB in the BALB/c mouse, these authors reported that there was a change from a protective TH1 pattern to a nonprotective TH2 pattern during TB-mediated adrenal hyperplasia with progressive loss of TNF-α and progression of disease. Administration of DHEA and AED was protective, and AED mediated a reduction in bacterial counts and increased survival. This treatment was associated with elevation in TH1 cytokines and increased cell-mediated immunity. The therapeutic effects of DHEA and AED were obtained by administering the agent in the first 3 weeks of infection, which coincides with the TB-mediated adrenal hyperplasia phase. The results underline the significance of a role played by glucocorticoids in the pathogenesis of the disease (Hernandez-Pando et al. 1998; Hernandez-Pando et al. 1998b).

Bottasso et al. (2009) extended these findings to human studies and reported that measurement of cytokine levels in the blood from TB patients showed an elevation in IFN-γ, IL-10, and IL-6 levels. This was associated with a marked increase in the level of growth hormone but only mildly elevated levels of cortisol, prolactin, and thyroid hormones. However, both DHEA and testosterone levels were markedly decreased. Santucci et al. reported a clinical correlation of the dysregulation of the immune endocrine response in human TB. They showed a significant increase in IL-18 in patients with severe TB and a reduced body mass index, which was concordant with the elevation in cortisol levels and decreased DHEA concentration. The immunoendocrine imbalance in patients with TB with low DHEA and increased cortisol levels may be a focus for therapeutic considerations.

EFFECTS OF DHEA AND ITS DERIVATIVES ON HOST RESISTANCE IN AGING

Degelau, Guay, and Hallgren (1997) tested the simultaneous administration of 7.5-mg dehydroepiandrosterone sulfate (DHEAS) and influenza vaccine to a human geriatric group. Hemagglutination inhibition (HI) antibody levels were enhanced against the 1993–94 H3N2 Beijing influenza antigen, but no increase was evident in the response to the H1N1 or the B antigen. No conclusion could be reached from this study. In contrast, DHEA treatment was found to be effective in augmenting the response to immunization of old mice challenged with intranasal infection (Danenberg et al. 1995). However, in a second human study, these investigators could not confirm the previous observation on the benefit of DHEA during influenza vaccination. They report a negative effect of DHEA administration on the response to A/Texas antigen. (Ben-Yehuda et al. 1998). Subjects who were never immunized before had significantly higher antibody titers against influenza A strains than those who were immunized before. Additional experiments to determine the ability of AED-sulfate (AED-S) to mitigate the age-associated immune senescence and increased susceptibility to influenza infection were performed. 10-month-old animals were treated with AED-S during vaccination with a commercially available influenza vaccine preparation. This resulted in an elevation in circulating anti-influenza immunoglobulin G to levels comparable to those that were observed in the younger 3-month-old mice. AED-S administered with the drinking water protected 10-month-old animals from an intranasal challenge with a lethal dose of influenza virus 21 days after secondary vaccination. Treatment with AED-S alone without vaccination protected 40% of 10-month-old animals, as shown in Figure 3.5. However, the combination of vaccination

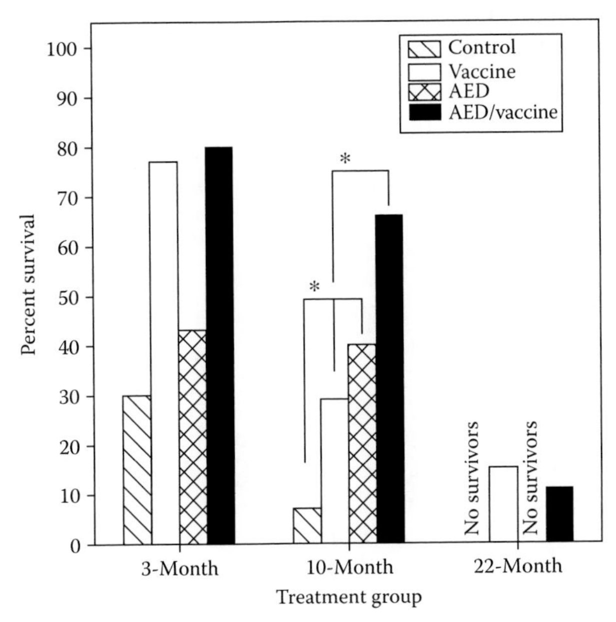

FIGURE 3.5 Influence of androstenediol-sulfate (AED-S) on vaccination efficacy. AED-S was added to the daily water supply (25 mg/mL) beginning 10 days prior to vaccination and continued until 7 days prior to viral challenge. Fluzone trivalent vaccine was diluted 1/200 in phosphate-buffered saline, and 200 mL was injected intramuscularly into 3-, 10-, and 22-month-old mice. Animals were boosted with the same dosage 21 days after primary vaccination. To determine the efficacy of the vaccination dose, mice were inoculated with 40 HAU of influenza A/PR8 virus 21 days after secondary boost, and percent survival was determined (n = 30 for 3- and 10-month-old controls, n = 29 for 22-month-old controls, n = 33 for 22-month-old a + vaccine, and n = 35 for all other groups). *Analysis of frequency of survival showed overall significant effects ($x^2 = 136.15$, $df = 11$, $p = .001$). Post hoc comparisons revealed that treatment with AED-S produced significantly improved survival rates for 10-month-old mice, whether or not those mice were vaccinated. (From Padgett, D. A., R. C. MacCullum, R. M. Loria, and J. F. Sheridan, *J Gerontol*, 2000, by permission of Oxford University Press.)

and AED-S treatment increased survival to 66%. These findings show that AED-S mediated an increase in antibody titers but also increased host resistance to lethal influenza virus infection. This effect was not observed in young 3- or 22-month-old animals.

EFFECTS OF DHEA ON PARASITIC INFECTIONS

Major strides have taken place in the studies of the roles of DHEA and DHEAS in the treatment of parasitic infections, including among many *Taenia crassiceps, Trypanosoma cruzi, Criptosporidium parvum,* and *Schistosoma mansoni.*

Administration of DHEA either *in vivo* or *in vitro* markedly reduced the tapeworm growth and reproduction of *T. crassiceps* during its larval stage (Vargas-Villavicencio, Larralde, and Morales-Montor 2008). Administration of DHEA at a dose of 200 μg/25 g BALB/c female or male mice 1 week prior to infection and every other day for 8 weeks resulted in a 50% reduction of parasite load compared to untreated, infected animals. The protective effect was independent of the host immune response because DHEA did not affect the levels of IL-1, IFN-γ, IL-4, or IL-10 mRNA. *In vitro* evidence showed a dose-dependent effect of DHEA treatment on the reduction of motility and viability of *T. crassiceps.* These findings may indicate a metabolic effect of the hormone on parasitic infection independent of the immune upregulation evident in other infections.

However, in the case of experimental Chagas' disease in the Wistar rat, DHEA treatment was shown to modulate the immune response during the acute and chronic phases of the disease. Results show that SC administration of 40 mg/kg DHEA was associated with ex vivo elevation in IL-12 and NO levels during the acute phase and with an increased spleen cell proliferation during the chronic phase of the disease (Brazao et al. 2009). Similar effects were observed in a study that combined treatment with zinc and DHEA to animals infected with *T. cruzi*. The findings showed an increased macrophage count and an increased level of IFN-γ and NO (Brazao et al. 2009). Dos Santos, Toldo, and do Prado Junior (2005) reported that DHEAS treatment was effective in reducing the mortality rate of animals infected with *T. cruzi* (Bolivia strain). DHEAS treatment was superior to treatment with benznidazole alone or to the combined treatment of DHEAS and benznidazole. DHEAS administration to rats infected with *T. cruzi* also enhanced the levels of peritoneal macrophages, IFN-γ, IL-2, and NO production (Domingues Santos et al.).

Cryptosporidiosis is a life-threatening parasitic disease in the immune-compromised host. Ten golden Syrian hamsters were treated with DHEA for 7 days prior to infection with 1×10^6 *C. parvum* oocysts. DHEA was shown to be an effective prophylactic agent in this model (Rasmussen and Healey 1992). This experiment was reproduced in mice with similar findings showing a significant reduction in intestinal and stool oocyts counts. DHEA was more effective if administered prior to infection. The relationship between DHEAS and the severity of *S. mansoni* infection was examined in 135 Ethiopians. Using enzyme-linked immunosorbent assay (ELISA) measurements, Abebe et al. (2003) reported a significant negative correlation between serum levels of DHEAS and intensity of *S. mansoni* infection.

FUTURE CONSIDERATIONS

There is a paucity of studies on the effects of androstenetriol (AET), and considerable additional studies are needed to establish the differences between DHEA, AED, and AET in their mode of action, range of protection, counteracting cortisone immune suppression, and immune-regulating activity. The great specificity of these agents is further demonstrated by the fact that the epimer of 17β AED (17α AED, Δ5-androstene-3β, 17α-diol) induces apoptosis in myeloid cells and autophagy in human gliomas (Huynh, Carter, and Loria 2000; Loria 2002; Graf et al. 2009b). Graf et al. (2009b) reported that the antitumor activity of 17α AED was strictly dependent on the hydroxyl group at position 17 in the alpha configuration. Furthermore, the same agent, 17α AED, induced apoptosis in U937 lymphoma cells but autophagy in human T98 glioblastoma, indicating a unique response of different target tissues (Graf et al. 2009a). Similar considerations exist with regard to 16α-bromoepiandrosterone and other derivatives of DHEA. Significant additional research will be needed before we have a more complete understanding of the function and mechanisms of the androstenes hormones. Recent studies have expanded the possible scope of the action of DHEA by demonstrating its ability to suppress Japanese encephalitis virus (JEV) replication and virus cytotoxicity via a nonsteroid hormone action in neuroblastoma cells. Chang et al. (2005) suggested that the mechanism of this antiviral effect is mediated by modulation of the MAPK signaling pathway. In addition, two recent studies reported that DHEA and epiandrosterone effectively inhibited the replication of JUNV, a member of the arenavirus genus. This effect was independent on virus adsorption or cell uptake but in part due to inhibition of viral glycoprotein G1 at the cell surface. Acosta et al. (2008) stated that the inhibitory activity of DHEA, epiandrosterone, and several other derivatives was greater than that of ribavirin. These findings are further extended to include inhibition of adenovirus replication by reducing the synthesis of an unspecified adenovirus protein (Romanutti et al. 2009). Finally, the recent report by Vargas-Villavicencio, Larralde, and Morales-Montor (2008) demonstrates that DHEA may inhibit parasite proliferation by a metabolic pathway independent of the host immunity.

The reported direct inhibition of viral replication *in vitro* by DHEA and its derivatives has not been extended to demonstrate its *in vivo* effects, where immunological mechanisms may be a significant factor in reducing infections. Further studies are needed to resolve these differences.

SUMMARY

It is evident from the advances in the last 20 years that the protective properties of the androstene hormones DHEA and AED have been documented as highly effective against a wide range of infectious diseases including bacterial, viral, and parasitic infections. The data clearly illustrate that, in experimental models, these agents are highly effective to increase survival and to modulate host immunity to favor the TH1 response. Table 3.3 clearly shows that DHEA treatment results in a significant increase in survival in various different animal models of polymicrobial sepsis following CLIP and hemmorhagic trauma. In general, the different reports show that DHEA did not have a significant influence on the levels of TNF-α in sham-treated animals. However, following CLIP–sepsis procedures, DHEA treatment reduced elevated plasma TNF-α levels. In the event that plasma TNF-α levels are significantly suppressed, DHEA treatment will mediate an increase in plasma TNF-α levels. These observations indicate that DHEA may also function to restore TNF-α equilibrium under different conditions. Similarly, DHEA and AED are highly effective in protecting the host from lethal viral infections, including human enteroviruses, herpes virus-1 and 2, influenza virus, the arthropod-borne viruses, and other RNA and DNA viruses. Restoration of the waning host immunity in the aging host and after immune suppression by whole body radiation exposure and stress further attests to the role of these agents in upregulating host resistance as evident by the protections to infectious challenges. The data indicate that DHEA and its derivatives, AED and AET, with their ability to counteract corticosterone immune suppression *in vivo* (see Figure 3.3) constitute a new subclass of steroid hormones with specific physiological properties (Loria 2002). The advances of the last two decades confirm and significantly extend the *in vivo* protective effects induced by DHEA (Loria et al. 1988; Ben-Nathan et al. 1991, 1992; Padgett, Loria, and Sheridan 2000) and provide a clear rationale for extending the research into the clinical use of these agents for the treatment of infectious diseases.

TABLE 3.3
Effects of DHEA Treatment in Models of Sepsis

Reference	Animal Model	Increased Percent Survival with DHEA Treatment[a]	Effects of DHEA Treatment on Plasma TNF-α levels
Ben-Nathan, Padgett, and Loria (1999)	Lethal Pseudomonas aeruginosa challenge	43%	Reduced by 50%
Oberbeck et al. (2001, 2007)	CLIP–sepsis	39.9% at 72 hours	Reduced by 37%
Van Griensven et al. (2002)	CLIP–sepsis	30% at 96 hours	NA
Matsuda et al. (2005)	CLIP–sepsis	Delayed mortality	Increased levels above vehicle-treated CLIP[b]
Barkhausen et al. (2009)	Trauma (femur fracture) hemorrhage + CLIP–sepsis	38% at 96 hours	Reduced by 44% compared to CLIP vehicle (96 hours)

[a] Calculated as increased survival over untreated animals.

[b] This study measures TNF-α levels at 2 and 6 hours after CLIP procedures, which were significantly lower than sham vehicle or DHEA-treated animals.

REFERENCES

Abebe, F., K. I. Birkeland, P. I. Gaarder, B. Petros, and S. G. Gundersen. 2003. The relationships between dehydroepiandrosterone sulphate (DHEAS), the intensity of Schistosoma mansoni infection and parasite-specific antibody responses. A cross-sectional study in residents of endemic communities in north-east Ethiopia. *Apmis* 111:319–28.

Acosta, E. G., A. C. Bruttomesso, J. A. Bisceglia, M. B. Wachsman, L. R. Galagovsky, and V. Castilla. 2008. Dehydroepiandrosterone, epiandrosterone and synthetic derivatives inhibit Junin virus replication in vitro. *Virus Res* 135:203–12.

Angele, M. K., R. A. Catania, A. Ayala, W. G. Cioffi, K. I. Bland, and I. H. Chaudry. 1998. Dehydroepiandrosterone: An inexpensive steroid hormone that decreases the mortality due to sepsis following trauma-induced hemorrhage. *Arch Surg* 133:1281–8.

Barkhausen, T., F. Hildebrand, C. Krettek, and M. Van Griensven. 2009. DHEA-dependent and organ-specific regulation of TNF-alpha mRNA expression in a murine polymicrobial sepsis and trauma model. *Crit Care* 13:R114.

Ben-Nathan, D., O. Gershoni-Yahalom, I. Samina, Y. Khinich, I. Nur, O. Laub, A. Gottreich, M. Simanov, A. Porgador, B. Rager-Zisman, and N. Orr. 2009. Using high titer West Nile intravenous immunoglobulin from selected Israeli donors for treatment of West Nile virus infection. *BMC Infect Dis* 9:18.

Ben-Nathan, D., I. Huitinga, S. Lustig, N. Van Rooijen, and D. Kobiler. 1996. West Nile virus neuroinvasion and encephalitis induced by macrophage depletion in mice. *Arch Virol* 141:459–69.

Ben-Nathan, D., B. Lachmi, S. Lustig, and G. Feuerstein. 1991. Protection by dehydroepiandrosterone in mice infected with viral encephalitis. *Arch Virol* 120:263–71.

Ben-Nathan, D., S. Lustig, D. Kobiler, H. D. Danenberg, E. Lupu, and G. Feuerstein. 1992. Dehydroepiandrosterone protects mice inoculated with West Nile virus and exposed to cold stress. *J Med Virol* 38:159–66.

Ben-Nathan, D., S. Lustig, G. Tam, S. Robinzon, S. Segal, and B. Rager-Zisman. 2003. Prophylactic and therapeutic efficacy of human intravenous immunoglobulin in treating West Nile virus infection in mice. *J Infect Dis* 188:5–12.

Ben-Nathan, D., D. A. Padgett, and R. M. Loria. 1999. Androstenediol and dehydroepiandrosterone protect mice against lethal bacterial infections and lipopolysaccharide toxicity. *J Med Microbiol* 48:425–31.

Ben-Yehuda, A., H. D. Danenberg, Z. Zakay-Rones, D. J. Gross, and G. Friedman. 1998. The influence of sequential annual vaccination and of DHEA administration on the efficacy of the immune response to influenza vaccine in the elderly. *Mech Ageing Dev* 102:299–306.

Bottasso, O., M. L. Bay, H. Besedovsky, and A. Del Rey. 2009. Immunoendocrine alterations during human tuberculosis as an integrated view of disease pathology. *Neuroimmunomodulation* 16:68–77.

Brazao, V., F. H. Santello, L. C. Caetano, M. Del Vecchio Filipin, M. Paula Alonso Toldo, and J. C. Do Prado Jr. 2009. Immunomodulatory effects of zinc and DHEA on the Th-1 immune response in rats infected with Trypanosoma cruzi. *Immunobiology* 215(5):427–34.

Burke, S. D., and T. P. Monath. 2001. Flaviviruses. In *Fields Virology*, ed. B. N. Field, D. M. Knipe, and P. M. Howley, 1043–125. Philadelphia: Lippincott-Raven.

Carr, D. J. 1998. Increased levels of IFN-gamma in the trigeminal ganglion correlate with protection against HSV-1-induced encephalitis following subcutaneous administration with androstenediol. *J Neuroimmunol* 89:160–7.

Carson, P. J., P. Konewko, K. S. Wold, P. Mariani, S. Goli, P. Bergloff, and R. D. Crosby. 2006. Long-term clinical and neuropsychological outcomes of West Nile virus infection. *Clin Infect Dis* 43:723–30.

Chang, C. C., Y. C. Ou, S. L. Raung, and C. J. Chen. 2005. Antiviral effect of dehydroepiandrosterone on Japanese encephalitis virus infection. *J Gen Virol* 86:2513–23.

Daigle, J., and D. J. Carr. 1998. Androstenediol antagonizes herpes simplex virus type 1-induced encephalitis through the augmentation of type I IFN production. *J Immunol* 160:3060–6.

Danenberg, H. D., A. Ben-Yehuda, Z. Zakay-Rones, and G. Friedman. 1995. Dehydroepiandrosterone (DHEA) treatment reverses the impaired immune response of old mice to influenza vaccination and protects from influenza infection. *Vaccine* 13:1445–8.

Debiasi, R. L., and K. L. Tyler. 2006. West Nile virus meningoencephalitis. *Nat Clin Pract Neurol* 2:264–75.

Degelau, J., D. Guay, and H. Hallgren. 1997. The effect of DHEAS on influenza vaccination in aging adults. *J Am Geriatr Soc* 45:747–51.

Domingues Santos, C., R. M. Loria, L. G. Rodrigues Oliveira, C. Collins Kuehn, M. P. Alonso Toldo, S. Albuquerque, and J. C. do Prado, Jr. Effects of dehydroepiandrosterone-sulfate (DHEA-S) and benznidazole treatments during acute infection of two different Trypanosoma cruzi strains. *Immunobiology* 215(12):980–6.

Dos Santos, C. D., M. P. Toldo, and J. C. Do Prado Junior. 2005. Trypanosoma cruzi: The effects of dehydroepiandrosterone (DHEA) treatment during experimental infection. *Acta Trop* 95:109–15.

Gianotti, L., J. W. Alexander, R. Fukushima, and T. Pyles. 1996. Steroid therapy can modulate gut barrier function, host defense, and survival in thermally injured mice. *J Surg Res* 62:53–8.

Glass, W. G., D. H. Mcdermott, J. K. Lim, S. Lekhong, S. F. Yu, W. A. Frank, J. Pape, R. C. Cheshier, and P. M. Murphy. 2006. CCR5 deficiency increases risk of symptomatic West Nile virus infection. *J Exp Med* 203:35–40.

Graf, M. R., W. Jia, R. S. Johnson, P. Dent, C. Mitchell, and R. M. Loria. 2009a. Autophagy and the functional roles of Atg5 and beclin-1 in the anti-tumor effects of 3beta androstene 17alpha diol neuro-steroid on malignant glioma cells. *J Steroid Biochem Mol Biol* 115:137–45.

Graf, M. R., W. Jia, M. L. Lewbart, and R. M. Loria. 2009b. The anti-tumor effects of androstene steroids exhibit a strict structure-activity relationship dependent upon the orientation of the hydroxyl group on carbon-17. *Chem Biol Drug Des* 74:625–9.

Hayes, E. B., and D. J. Gubler. 2006. West Nile virus: Epidemiology and clinical features of an emerging epidemic in the United States. *Annu Rev Med* 57:181–94.

Hayes, E. B., J. J. Sejvar, S. R. Zaki, R. S. Lanciotti, A. V. Bode, and G. L. Campbell. 2005. Virology, pathology, and clinical manifestations of West Nile virus disease. *Emerg Infect Dis* 11:1174–9.

Hernandez-Pando, R., M. De La Luz Streber, H. Orozco, K. Arriaga, L. Pavon, S. A. Al-Nakhli, and G. A. Rook. 1998a. The effects of androstenediol and dehydroepiandrosterone on the course and cytokine profile of tuberculosis in BALB/c mice. *Immunology* 95:234–41.

Hernandez-Pando, R., M. De La Luz Streber, H. Orozco, K. Arriaga, L. Pavon, O. Marti, S. L. Lightman, and G. A. Rook. 1998b. Emergent immunoregulatory properties of combined glucocorticoid and anti-glucocorticoid steroids in a model of tuberculosis. *Qjm* 91:755–66.

Hildebrand, F., H. C. Pape, P. Hoevel, C. Krettek, and M. Van Griensven. 2003. The importance of systemic cytokines in the pathogenesis of polymicrobial sepsis and dehydroepiandrosterone treatment in a rodent model. *Shock* 20:338–46.

Huynh, P. N., W. H. Carter Jr, and R. M. Loria. 2000. 17 alpha androstenediol inhibition of breast tumor cell proliferation in estrogen receptor-positive and -negative cell lines. *Cancer Detect Prev* 24:435–44.

Loria, R. M. 2002. Immune up-regulation and tumor apoptosis by androstene steroids. *Steroids* 67:953–66.

Loria, R. M., and D. A. Padgett. 1992. Androstenediol regulates systemic resistance against lethal infections in mice. *Arch Virol* 127:103–15.

Loria, R. M., D. H. Conrad, T. Huff, H. Carter, and D. Ben-Nathan. 2000. Androstenetriol and androstenediol. Protection against lethal radiation and restoration of immunity after radiation injury. *Ann N Y Acad Sci* 917:860–7.

Loria, R. M., T. H. Inge, S. S. Cook, A. K. Szakal, and W. Regelson. 1988. Protection against acute lethal viral infections with the native steroid dehydroepiandrosterone (DHEA). *J Med Virol* 26:301–14.

Matsuda, A., K. Furukawa, H. Suzuki, T. Matsutani, T. Tajiri, and I. H. Chaudry. 2005. Dehydroepiandrosterone modulates toll-like receptor expression on splenic macrophages of mice after severe polymicrobial sepsis. *Shock* 24:364–9.

Mehlhop, E., and M. S. Diamond. 2008. The molecular basis of antibody protection against West Nile virus. *Curr Top Microbiol Immunol* 317:125–53.

Nash, D., F. Mostashari, A. Fine, J. Miller, D. O'Leary, K. Murray, A. Huang et al. 2001. The outbreak of West Nile virus infection in the New York City area in 1999. *N Engl J Med* 344:1807–14.

Negrette, B., E. Bonilla, N. Valero, D. Giraldoth, S. Medina-Leendertz, and F. Anez. 2001. In mice the efficiency of immunization with Venezuelan Equine Encephalomyelitis virus TC-83 is transiently increased by dehydroepiandrosterone. *Invest Clin* 42:235–40.

Oberbeck, R., M. Dahlweid, R. Koch, M. Van Griensven, A. Emmendorfer, H. Tscherne, and H. C. Pape. 2001. Dehydroepiandrosterone decreases mortality rate and improves cellular immune function during polymicrobial sepsis. *Crit Care Med* 29:380–4.

Oberbeck, R., H. Deckert, J. Bangen, P. Kobbe, and D. Schmitz. 2007. Dehydroepiandrosterone: A modulator of cellular immunity and heat shock protein 70 production during polymicrobial sepsis. *Intensive Care Med* 33:2207–13.

Padgett, D. A., and R. M. Loria. 1998. Endocrine regulation of murine macrophage function: Effects of dehydroepiandrosterone, androstenediol, and androstenetriol. *J Neuroimmunol* 84:61–8.

Padgett, D. A., R. M. Loria, and J. F. Sheridan. 1997. Endocrine regulation of the immune response to influenza virus infection with a metabolite of DHEA-androstenediol. *J Neuroimmunol* 78:203–11.

Padgett, D. A., R. M. Loria, and J. F. Sheridan. 2000. Steroid hormone regulation of antiviral immunity. *Ann NY Acad Sci* 917:935–43.

Padgett D. A., R. C. MacCallum, R. M. Loria, and J. F. Sheridan. 2000. Adrostenediol-induced restoration of responsiveness to Influenza vaccination in mice. *J Gerontology* 55A(9):B418–24.

Rasmussen, K. R., and M. C. Healey. 1992. Dehydroepiandrosterone-induced reduction of Cryptosporidium parvum infections in aged Syrian golden hamsters. *J Parasitol* 78:554–7.

Romanutti, C., A. C. Bruttomesso, V. Castilla, J. A. Bisceglia, L. R. Galagovsky, and M. B. Wachsman. 2009. In vitro antiviral activity of dehydroepiandrosterone and its synthetic derivatives against vesicular stomatitis virus. *Vet J* 182:327–35.

Rook, G. A., and R. Hernandez-Pando. 1997. Pathogenetic role, in human and murine tuberculosis, of changes in the peripheral metabolism of glucocorticoids and antiglucocorticoids. *Psychoneuroendocrinology* 22 Suppl 1:S109–13.

Santucci, N., L. D'Attilio, H. Besedovsky, A. Del Rey, M. L. Bay, and O. A Bottasso. 2010. Clinical correlate of the dysregulated immunoendocrine response in human tuberculosis. *Neuroimmunomodulation* 17:184–7.

Scherret, J. H., M. Poidinger, J. S. Mackenzie, A. K. Broom, V. Deubel, W. I. Lipkin, T. Briese, E. A. Gould, and R. A. Hall. 2001. The relationships between West Nile and Kunjin viruses. *Emerg Infect Dis* 7:697–705.

Smithburn K. C., T. P. Hughes, A. W. Burke, and J. H. Paul. 1940. A neutrotropic virus isolated from the blood of a native of Uganda. *Am J Trop Med* 20:471–92.

Toth, I., and I. Faredin. 1985a. Concentrations of androgens and C19-steroid sulphates in abdominal skin of healthy women and men. *Acta Med Hung* 42:13–20.

Toth, I., and I. Faredin. 1985b. Steroids excreted by human skin. II. C19-steroid sulphates in human axillary sweat. *Acta Med Hung* 42:21–8.

Van Griensven, M., M. Kuzu, M. Breddin, F. Bottcher, C. Krettek, H. C. Pape, and T. Tschernig. 2002. Polymicrobial sepsis induces organ changes due to granulocyte adhesion in a murine two hit model of trauma. *Exp Toxicol Pathol* 54:203–9.

Vargas-Villavicencio, J. A., C. Larralde, and J. Morales-Montor. 2008. Treatment with dehydroepiandrosterone in vivo and in vitro inhibits reproduction, growth and viability of Taenia crassiceps metacestodes. *Int J Parasitol* 38:775–81.

Wang, T., and E. Fikrig. 2004. Immunity to West Nile virus. *Curr Opin Immunol* 16:519–23.

Whitnall, M. H., T. B. Elliott, R. A. Harding, C. E. Inal, M. R. Landauer, C. L. Wilhelmsen, L. Mckinney et al. 2000. Androstenediol stimulates myelopoiesis and enhances resistance to infection in gamma-irradiated mice. *Int J Immunopharmacol* 22:1–14.

Yakub, I., K. M. Lillibridge, A. Moran, O. Y. Gonzalez, J. Belmont, R. A. Gibbs, and D. J. Tweardy. 2005. Single nucleotide polymorphisms in genes for 2′-5′-oligoadenylate synthetase and RNase L inpatients hospitalized with West Nile virus infection. *J Infect Dis* 192:1741–48.

4 DHEA and Glucose Metabolism in Skeletal Muscle

Koji Sato and Ryuichi Ajisaka

CONTENTS

INTRODUCTION

Dehydroepiandrosterone (DHEA) is a precursor of sex steroid hormones and is converted to testosterone by 17β-hydroxysteroid dehydrogenase (HSD) and 3β-HSD enzymes (Labrie et al. 2005; Figure 4.1). We found that 3β-HSD, 17β-HSD, and aromatase cytochrome P450 (P450arom) existed in cultured skeletal muscle cells, and steroid hormones including testosterone were locally synthesized from DHEA in skeletal muscle (Aizawa et al. 2007); therefore, skeletal muscle is presumed to be capable of synthesizing and metabolizing sex steroid hormones. In a skeletal muscle specimen testosterone administration accelerated protein synthesis and anabolism (Urban et al. 1995), resulting in muscle growth and/or increased strength (Ihemelandu 1980). Thus, sex steroid hormones are important for muscle configuration and function. Sex steroid hormones generally decrease with aging, and it has been reported that a deficiency of sex steroid hormones in both men and women is a risk factor for developing obesity and type 2 diabetes (Ding et al. 2007). Yamaguchi et al. (1998) reported that patients with hyperinsulinemia had low DHEA levels. Therefore, lower sex steroid hormone levels may be related to the pathogenesis of obesity and type 2 diabetes.

THE IMPORTANCE OF SEX STEROID HORMONES FOR DIABETES

Sex Steroid Hormones and Exercise

Chronic exercise increases muscle size, strength, energy metabolism, antioxidant capacity, and changes in muscle fiber type (Staron et al. 1991). Sex steroid hormones partly contribute to these exercise-induced muscular adaptations. A single bout of exercise increases testosterone levels in plasma in men but not in women, even though there is no sex difference in chronic exercise-induced adaptations in skeletal muscle. The skeletal muscle can synthesize sex steroid hormones from circulating dehydroepiandrosterone sulfate (DHEAS) or testosterone, and they act as an autocrine and local paracrine muscle growth factor. Although it is unclear whether the muscular steroid genetic system is changed by a single bout of exercise stimulation and hormonal response

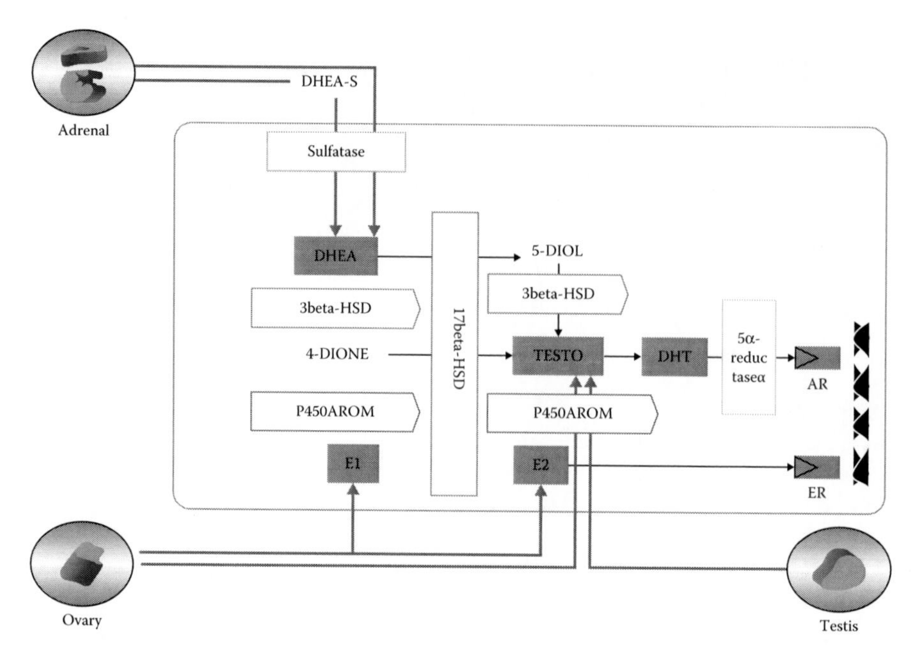

FIGURE 4.1 The schema of pathway in sex steroid hormones metabolism. The expression of 3β, 17β-HSD, and P450 (steroidogenesis-related enzymes) in skeletal muscle has been established, however, the expression of 5α-reductase in skeletal muscle has not been established.

differs between men and women, Aizawa et al. (2008) reported that testosterone levels and the expression of 17β-HSD and 3β-HSD proteins in skeletal muscle increased by acute exercise in both sexes. By comparison, muscular estradiol levels increased in men following exercise but remained unchanged in women. Moreover, after acute exercise, the expression of P450arom increased in men but decreased in women. In addition, 5α-reductase expression increased significantly in both sexes even though the expression of 5α-reductase at rest (pre-exercise) was lower in women. Therefore, a single session of exercise can influence the steroid genetic system in skeletal muscle, but these alterations differ between sexes.

Sex Steroid Hormones and Glucose Metabolism

In the 1980s, Coleman, Leiter, and Schwizer (1982) reported that dietary administration of DHEA to db/db mice induced the remission of hyperglycemia and was strongly correlated with improved insulin sensitivity in these animals. A more recent study reported that DHEA was shown to protect against the development of visceral obesity and muscular insulin resistance in rats fed with a high-fat diet (Hansen et al. 1997). Another study demonstrated that DHEA increased glucose uptake rates in human fibroblasts and suggested that this increased uptake might be mediated by activation of PI3-kinase (Nakashima et al. 1995). According to Kang et al. (2004), the principal steroidal androgens, testosterone and dihydrotestosterone (DHT), are thought to predominantly mediate the biological effects of DHEA. Additionally, DHT activated the Akt/PKB pathway through stimulation of PI3-kinase in osteoblast cells (Kang et al. 2004). However, the mechanism by which sex steroid hormones stimulate the glucose–signaling pathway is still unclear.

Effects of Sex Steroid Hormones on Skeletal Muscle

Skeletal muscles are sex steroid hormone–sensitive tissues, and they express receptors for both androgen and estrogen (Glenmark et al. 2004). Both testosterone and estradiol affect the growth, strength, and antioxidant action of skeletal muscles (Ihemelandu 1980). The administration of testosterone has been reported to induce increased strength through an acceleration of protein synthesis in the muscle (Urban et al. 1995). Similarly, a long-term administration of DHEA promotes muscle strength and mass in humans, whereas a deficiency of sex steroid hormones leads to retardation in skeletal muscle development (Morales et al. 1998). In a recent study, we demonstrated from circulating DHEA in the skeletal muscle cells the presence of a local and active metabolic machinery or pathway (mRNA and protein expressions of 3β-HSD, 17β-HSD, P450arom, and 5α-reductase) for the synthesis of sex steroid hormones (testosterone and estrogen; Aizawa et al. 2007). Moreover, Han et al. (1998) reported that 2 weeks of DHEA supplementation and/or 2 weeks of wheel-running exercise training improved PI3-kinase activity and insulin infusion rates in old rats, but no additive effect was observed from the combination of DHEA supplementation and exercise training. It has been fully elucidated that exercise improves the glucose-metabolism signaling pathway in patients with insulin resistance; in addition, acute DHEA administration enhances glucose metabolism in liver and adipose tissues (Mohan et al. 1990; Ishizuka et al. 2007). Therefore, DHEA may induce more prominent effects on insulin resistance and can mitigate the strain of each single therapy.

Effect of DHEA on Diabetes, *In Vitro* Study

According to Kang et al., the principal steroidal androgens synthesized from DHEA—testosterone and DHT—are thought to predominantly mediate the biological effects of DHEA. Additionally, DHT was shown to activate the Akt pathway through stimulation of PI3-kinase in osteoblast cells (Kang et al. 2004). In rat adipocytes, DHEA addition stimulated glucose uptake via enhancement of PI3-kinase and PKC activities (Ishizuka et al. 2007). However, the effect of DHEA on the glucose metabolism in skeletal muscle has remained unclear, as has whether DHT can be synthesized in skeletal muscle.

In our earlier studies (Aizawa et al. 2007; Sato et al. 2008, 2009), steroid genesis-related enzymes such as 17β-HSD, 3β-HSD, P450arom, and 5α-reductase mRNA were detected in skeletal muscle cells, and muscular testosterone, estrogen, and DHT concentrations increased with the addition of DHEA (Figure 4.2). Furthermore, Akt and PKCζ/λ phosphorylations and glucose transporter type 4 (GLUT4) translocation were significantly increased by DHEA or testosterone addition. Moreover, DHEA- and testosterone-induced increases in both Akt and PKCζ/λ phosphorylations and GLUT4 expression were blocked by a DHT inhibitor. Thus, DHT activates the glucose metabolism-related signaling pathway in skeletal muscle cells (Sato et al. 2008; Figure 4.3).

Effect of DHEA Administration on Diabetes, *In Vivo* Study

According to Coleman, Leiter, and Schwizer (1982), acute DHEA administration decreased blood glucose levels in type 2 diabetes mouse models. However, the mechanism underlying this decrease has been unclear. In our study (Sato et al. 2009), acute DHEA injection (2 mg/body weight kg) improved the glucose-metabolism signaling pathway (Akt/PKCζ/λ-GLUT4) in skeletal muscles, resulting in decreased hyperglycemia in streptozotocin (STZ)-induced, type 1 diabetes rats (Figure 4.4). We found that acute DHEA injection increased GLUT4 content in skeletal muscle approximately 2.7-fold compared to STZ control rats and decreased blood glucose levels from 22.8 mmol/L at baseline to 16.7 mmol/L at 90 minutes after the injection. Although DHEA reduced hyperglycemia as insulin did, acute DHEA injection (2 mg/body weight kg) had a weaker effect on blood glucose levels than on insulin levels. However, the decrease in hyperglycemia with DHEA injection was achieved without changing serum insulin levels. Consequently, a single DHEA injection has an independent effect on decreasing hyperglycemia in rats with type 1 diabetes mellitus. In addition, the DHEA-induced Akt/PKCζ/λ-GLUT4 activations were blocked by a DHT inhibitor (Figure 4.4). Thus, the

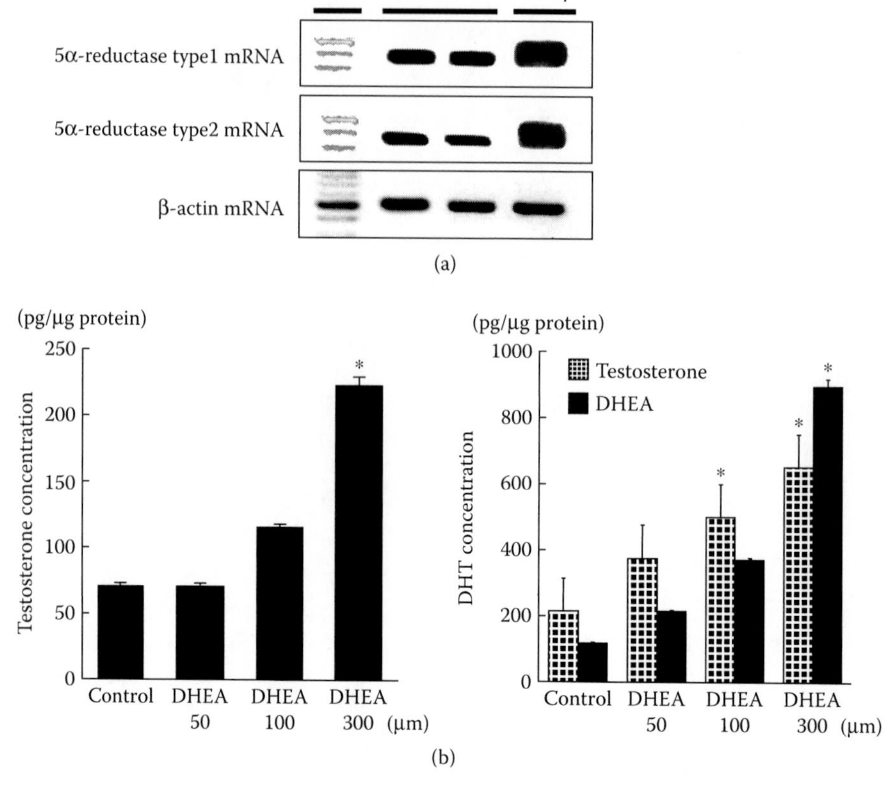

FIGURE 4.2 5α-reductase type 1 and type 2 mRNA expressions (a) and changes in DHT concentration with the addition of testosterone or DHEA to cultured skeletal muscle cells (b). (a) Representative images of expressions of 5α-reductase type 1 and type 2 mRNA revealed by RT-PCR. The mRNA of the rat ovary was used as a positive control, and β-actin mRNA was used as an internal control. (b) Changes in intramuscular concentrations of DHT with testosterone or DHEA addition.

DHEA effects might be achieved partly by DHT converted from DHEA in the skeletal muscle (Sato et al. 2009).

Patients with both obesity and type 2 diabetes have lower levels of serum DHEAS. Moreover, many reports have described that a 2-week administration of DHEA decreased serum insulin levels in old-age rats and improved hepatic glucose metabolism-related enzyme activities in diet-induced, or Zucker, obese rats (Leighton, Tagliaferro, and Newsholme 1987; Lea-Currie, Wen, and McIntosh 1997). Moreover, Han et al. (1998) demonstrated the enhancement in PI3-kinase activity and insulin sensitivity by either 2 weeks of wheel-running exercise or DHEA administration in 25-month-old rats. However, it remains unclear whether the combination of chronic DHEA administration and exercise training improves insulin resistance more effectively in rats with obesity and hyperglycemia. However, in our study, the combination of long-term DHEA administration and exercise training induced a more significant decrease in the fasting glucose level than in the DHEA administration alone; in addition, this combination induced a more significant decrease in abdominal fat weight compared with exercise training alone. Moreover, this combination more clearly improved the insulin-resistance index and decreased body weight than did either DHEA administration or exercise training alone (Figure 4.5). These results suggest that the long-term combination treatment improved obesity and insulin-resistant conditions more effectively than either treatment did alone (Sato, in press).

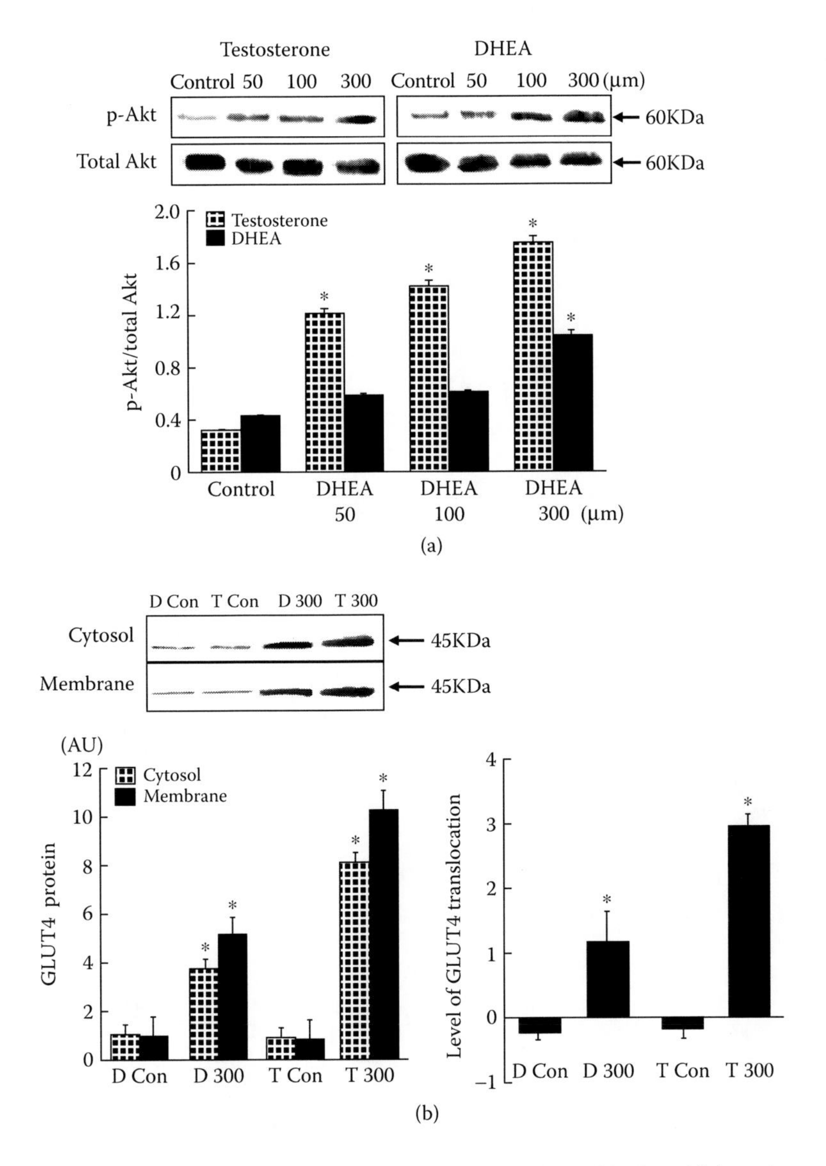

FIGURE 4.3 (a) Glucose transporter type 4 (GLUT4) translocation with the addition of testosterone or DHEA to cultured skeletal muscle cells. Representative images of immunoblotting for GLUT4 proteins in cytosol and the membrane are shown in the upper figures. Arrows indicate immunoblot bands for GLUT4 protein. (b) The left lower figure shows the results of statistical analyses for levels of GLUT4 protein expression by densitometer and the right lower figure shows the difference in GLUT4 protein levels in cytosol and membrane fractions as the level of GLUT4 translocation. Data are the means ± standard error (SE) of three independent measurements. D Con: DHEA control; T Con: testosterone control; D 300: DHEA 300 μM addition. T 300: testosterone 300 μM addition. $*p < .01$ compared with the control.

Future Use of DHEA for Diabetes Patients

According to our earlier study (Sato et al. 2009), acute DHEA injection for type 1 diabetes-model rats had less of a decreasing effect on blood glucose than insulin did. However, DHEA might be a candidate for improving and controlling blood glucose levels and for restoring impaired insulin-signal transduction in the skeletal muscle of type 1 diabetes mellitus patients. Moreover, the combination of exercise training and DHEA administration improved insulin

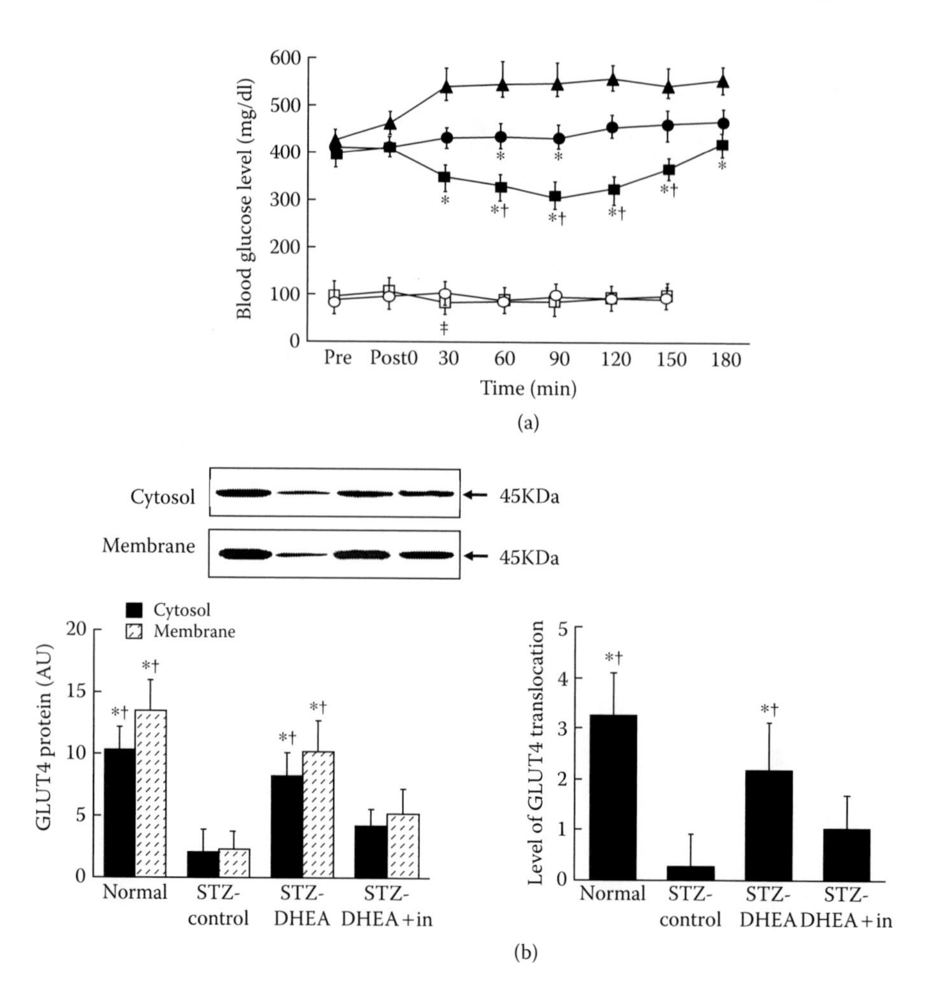

FIGURE 4.4 Effects of 2-mg DHEA injection on blood glucose levels (a) and glucose transporter type 4 (GLUT4) translocation in skeletal muscle (b) in normal control rats and rat models of type 1 diabetes mellitus. (a) *White circle*: normal control (n = 10). *White square*: normal control with DHEA injection (n = 10). *Black triangle*: streptozotocin (STZ) control (n = 10). *Black square*: STZ-DHEA injection (n = 10). *Black circle*: STZ-DHEA and DHT inhibitor injections (n = 10). Data are means ± SE. *$p < .01$ compared with the STZ-control; †$p < .05$ compared with DHT inhibitor injection; ‡$p < .05$ compared with normal control. (b) Representative images of immunoblotting for GLUT4 proteins in cytosol and the membrane are depicted in the upper panels. The arrows indicate immunoblot bands for the GLUT4 protein. The left lower panel presents the results of statistical analyses for levels of GLUT4 protein expression by densitometry. The right lower panel shows the difference in GLUT4 protein levels between cytosol and membrane fractions as the level of GLUT4 translocation. Data are means ± SE. *$p < .01$ compared with the control; †$p < .05$ compared with DHT inhibitor injection.

resistance more effectively in high-sucrose diet-induced obese rats. In addition, long-term DHEA administration of obesity in hyperglycemic rats increased the glucose metabolism-related signaling pathway at the same level as exercise training did. Therefore, the combination of exercise training and DHEA administration can mitigate the strain of each single therapy. Furthermore, DHEA administration can mitigate the strain on obese patients with type 2 diabetes who cannot participate in exercise training. It might provide new therapeutic and preventive candidates for restoring impaired insulin-signal transduction in skeletal muscle for patients with obesity and hyperglycemia.

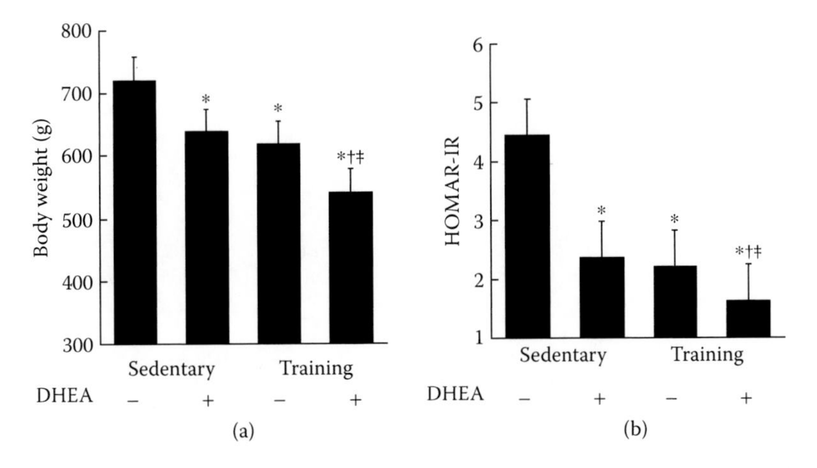

FIGURE 4.5 Effects of DHEA treatment and/or exercise training on (a) the body weight and (b) HOMA-IR index. Data are the means ± SE for 10 animals. $*p < .01$ compared with the sedentary control group; $\dagger p < .01$ compared with the DHEA treatment group; $\ddagger p < .01$ compared with the exercise training group.

SUMMARY

- Steroidogenesis-related enzymes such as 17β-HSD, 3β-HSD, P450arom, and 5α-reductase were expressed in skeletal muscle cells, and skeletal muscle is capable of synthesizing and metabolizing sex steroid hormones.
- DHT synthesized from DHEA affected the activation of the glucose metabolism-related signaling pathway in skeletal muscle cells.
- DHT synthesized from DHEA enhanced the glucose metabolism-related signaling pathway, and it led to the improvement of hyperglycemia in STZ-induced, type 1 diabetes-model rats.
- Exercise training increased protein expression in steroidogenesis-related enzymes and sex steroid hormone concentrations in skeletal muscle.
- The combination of DHEA administration and exercise training improved insulin resistance more effectively than DHEA administration or exercise training alone.

REFERENCES

Aizawa, K., M. Iemitsu, S. Maeda, S. Jesmin, T. Otsuki, C. N. Mowa, T. Miyauchi, and N. Mesaki. 2007. Expression of steroidogenic enzymes and synthesis of sex steroid hormones from DHEA in skeletal muscle. *Am J Physiol* 292:E577–84.

Aizawa, K., M. Iemitsu, T. Otsuki, S. Maeda, T. Miyauchi, and N. Mesaki. 2008. Sex differences in steroidogenesis in skeletal muscle following a single bout of exercise in rats. *J Appl Physiol* 104:67–74.

Coleman, D. L., E. H. Leiter, and R. W. Schwizer. 1982. Therapeutic effects of dehydroepiandrosterone (DHEA) in diabetic mice. *Diabetes* 31:830–3.

Ding, E. L., Y. Dong, J. E. Manson, N. Rifai, J. E. Buring, and S. Liu. 2007. Plasma sex steroid hormones and risk of developing type 2 diabetes in women: A prospective study. *Diabetologia* 50:2076–84.

Glenmark, B., M. Nilsson, H. Gao, J. A. Gustafsson, K. Dahlman-Wright, and H. Westerblad. 2004. Difference in skeletal muscle function in males vs. females: Role of estrogen receptor-beta. *Am J Physiol* 287:E1125–31.

Han, D. H., P. A. Hansen, M. M. Chen, and J. O. Holloszy. 1998. DHEA treatment reduces fat accumulation and protects against insulin resistance in male rats. *J Gerontol A Biol Sci Med Sci* 53:19–24.

Hansen, P. A., D. H. Han, L. A. Nolte, M. Chen, and J. O. Holloszy. 1997. DHEA protects against visceral obesity and muscle insulin resistance in rats fed a high-fat diet. *Am J Physiol* 273:R1704–8.

Ihemelandu, E. C. 1980. Effect of oestrogen on muscle development of female rabbits. *Acta Anat* 108:310–5.

Ishizuka, T., A. Miura, K. Kajita, M. Matsumoto, C. Sugiyama, K. Matsubara, T. Ikeda et al. 2007. Effect of dehydroepiandrosterone on insulin sensitivity in Otsuka Long-Evans Tokushima-fatty rats. *Acta Diabetol* 44:219–26.

Kang, H. Y., C. L. Cho, K. L. Huang, J. C. Wang, Y. C. Hu, H. K. Lin, C. Chang, and K. E. Huang. 2004. Nongenomic androgen activation of phosphatidylinositol 3-kinase/Akt signalling pathway in MC3T3-E1 osteoblasts. *J Bone Miner Res* 19:1181–90.

Labrie, F., V. Luu-The, A. Belanger, S. X. Lin, J. Simard, G. Pelletier, and C. Labrie. 2005. Is dehydroepiandrosterone a hormone? *J Endocrinol* 187:169–96.

Lea-Currie, Y. R., P. Wen, and M. K. McIntosh. 1997. Dehydroepiandrosterone-sulfate (DHEAS) reduces adipocyte hyperplasia associated with feeding rats a high-fat diet. *Int J Obes Relat Metab Disord* 21:1058–64.

Leighton, B., A. R. Tagliaferro, and E. A. Newsholme. 1987. The effect of DHEA acetate on liver peroxisomal enzyme activities of male and female rats. *J Nutr* 117:1287–90.

Mohan, P. F., J. S. Ihnen, B. E. Levin, and M. P. Cleary. 1990. Effects of dehydroepiandrosterone treatment in rats with diet-induced obesity. *J Nutr* 120:1103–14.

Morales, A. J., R. H. Haubrich, J. Y. Hwang, H. Asakura, and S. S. Yen. 1998. The effect of six months treatment with a 100mg daily dose of dehydroepiandrosterone (DHEA) on circulating sex steroids, body composition and muscle strength in age-advanced men and women. *Clin Endocrinol (Oxf)* 49:421–32.

Nakashima, N., M. Haji, F. Umeda, and H. Nawata. 1995. Effect of dehydroepiandrosterone on glucose uptake in cultured rat myoblasts. *Horm Metab Res* 27:491–4.

Sato, K., M. Iemitsu, K. Aizawa, and R. Ajisaka. 2008. Testosterone and DHEA activate the glucose metabolism-related signaling pathway in skeletal muscle. *Am J Physiol* 294:E961–8.

Sato, K., M. Iemitsu, K. Aizawa, and R. Ajisaka. 2009. DHEA improves impaired activation of Akt and PKC ζ/λ pathway in skeletal muscle and improves hyperglycemia in streptozotocin-induced diabetes rats. *Acta Physiol* 197:217–25.

Staron, R. S., M. J. Leonardi, D. L. Karapondo, E. S. Malicky, J. E. Falkel, F. C. Hagerman, and R. S. Hikida. 1991. Strength and skeletal muscle adaptations in heavy-resistance-trained women after detraining and retraining. *J Appl Physiol* 70:631–40.

Urban, R. J., Y. H. Bodenburg, C. Gilkison, J. Foxworth, A. R. Coggan, R. R. Wolfe, and A. Ferrando. 1995. Testosterone administration to elderly men increases skeletal muscle strength and protein synthesis. *Am J Physiol* 269:H820–6.

Yamaguchi, Y., S. Tanaka, T. Yamakawa, M. Kimura, K. Ukawa, Y. Yamada, M. Ishihara, and H. Sekihara. 1998. Reduced serum dehydroepiandrosterone levels in diabetic patients with hyperinsulinaemia. *Clin Endocrinol* 49:377–83.

5 DHEA and Dyskinesias

Melanie Hamann and Angelika Richter

CONTENTS

SUMMARY

The term "dyskinesia" is particularly used for hereditary paroxysmal movement disorders such as dystonia, choreoathetosis, and ballism, and for drug-induced complications such as levodopa-induced dyskinesia (LID) or tardive dyskinesia (TD). Dyskinesias represent common and severe movement disorders and are regarded as a basal ganglia disorder, that is, they are based within brain structures that are crucial for the initiation and processing of voluntary and automatic movements. However, the limited knowledge about the pathophysiology of various types of dyskinesias hampers the development of effective therapeutics. Dyskinesias are difficult to treat and often lead to severe disabilities in patients. Although the etiology of different forms (e.g., hereditary and drug-induced) of this disabling condition is heterogeneous, it cannot be excluded that they share, at least in part, common pathophysiological mechanisms and have identical targets for new strategies in prevention and therapy. For example, clinical and epidemiological studies in humans revealed that the severity of dyskinesias and the frequency of paroxysmal forms are altered by factors such as the onset of puberty, pregnancy, cyclical changes, and stress, indicating an underlying hormonal component (Kranick et al. 2010; Soorani-Lunsing et al. 1994; Weber and Lerche 2009). Therefore, this chapter focuses on the possible pathophysiological, preventive, and therapeutic role of the "neuroactive steroid" dehydroepiandrosterone (DHEA) in drug-induced and hereditary paroxysmal dyskinesias.

INTRODUCTION

Steroids that are active in the central nervous system, also referred to as neurosteroids or neuroactive steroids, are synthesized de novo in the brain from cholesterol or peripheral steroid precursors (Mellon and Griffin 2002). DHEA is a neurosteroid, and there is evidence that it is produced in the central nervous system independent of its peripheral origin. Also, the enzymes required for its biosynthesis are found in astrocytes and neurons (Pérez-Neri et al. 2008). DHEA concentrations in the rodent brain exceed its low peripheral concentrations and are independent of adrenal or gonadal synthesis (Corpechot et al. 1981; Wolf and Kirschbaum 1999). Like other steroids, DHEA and its metabolites act in the central nervous system by genomic actions mediated through classic nuclear steroid receptors, commonly known for steroids. These genomic actions are slow and occur after several hours or days (Lambert et al. 2003). In addition, a fast-acting paracrine, nongenomic mechanism of DHEA and its sulfated metabolite dehydroepiandrosterone sulfate (DHEAS) is known (Mellon and Griffin 2002; Pérez-Neri et al. 2008). This type of action has been shown to be mediated through neurotransmitter receptors. It has been reported that DHEA and its metabolites act as $GABA_A$, *N*-methyl-D-aspartic acid (NMDA), and sigma receptors (Pérez-Neri et al. 2008). Furthermore, effects of DHEA on glutamate and acetylcholine release were evidenced (Zheng 2009). These effects on different neurotransmitter systems suggest that DHEA could represent an interesting candidate for the prevention and therapy of different neurological disorders such as movement disorders.

Movement disorders are a clinically, pathophysiologically, and genetically heterogeneous group of neurological conditions. Despite their high variability in clinical signs, genetic background, and possibilities of treatment, they all share the common features of impaired planning, control, or execution of movement (Klein 2005). The term "dyskinesia" is used particularly for paroxysmal movement disorders such as dystonia, choreoathethosis, and ballism, and for drug-induced complications such as LID in patients with Parkinson's disease (PD) or TD in antipsychotic-treated patients with schizophrenia (Damier 2009; van Rootselaar et al. 2009). Although it is presumed that dyskinesias are based on biochemical dysfunctions within the basal ganglia, that is, subcortical nuclei in the central nervous system, which play a crucial role in the initiation and control of movement, their pathophysiology remains unknown so far. Furthermore, the heterogeneity in clinical signs, genetic findings, and the variability in treatment efforts hamper the elucidation of the underlying mechanisms, although the different types of the disorder share common underlying mechanisms. Based on the unknown pathophysiology, in general, the efficacy of the empirically conducted pharmacotherapy of this disorder is often disappointing (Jankovic 2009). Thus, dyskinesias often lead to severe disabilities in affected patients. Their pathophysiology needs to be elucidated further. Furthermore, new and innovative preventive and therapeutic strategies are necessary to ameliorate the condition of patients with dyskinesias. The neurosteroid DHEA with its modulating effects on different neurotransmitter systems within the central nervous system could represent a useful "new" compound for the prevention and therapy of drug-induced and/or hereditary paroxysmal dyskinesias, which are discussed in the following section.

DHEA AND DRUG-INDUCED DYSKINESIAS

Definition, Clinical Features, and Pathophysiology of Drug-Induced Dyskinesias

Levodopa-Induced Dyskinesias

Levodopa remains the most effective compound in replacing the deficiency of dopamine caused by the degeneration of dopaminergic neurons in patients suffering from PD (Cenci 2007). After a period of stable response to dopaminergic medication, patients gradually develop motor fluctuations and dyskinesias that hamper the effective treatment with levodopa, particularly in younger patients who seem to be more susceptible to these side effects (Halkias et al. 2007). LID, representing an important source of disability, are abnormal, involuntary movements, including dystonic and choreatic movements, which are heterogeneous with respect to affected body parts and the time course after levodopa intake (Jankovic 2005). Fabbrini et al. (2007) estimated that 11% of patients with PD who were treated for less than 5 years, 32% of those who were treated for 6–9 years, and 89% of those who were treated for more than 10 years were at risk of developing LID. Some factors such as severity of parkinsonism (which reflects the degree of striatal denervation), the dose of levodopa, and the duration of treatment as well as earlier age at PD onset are constantly correlated with an earlier onset and higher prevalence of LID (Cenci and Lundblad 2006; Cenci 2007).

Once established, LID is difficult to treat. The only currently available drug with an evidence-based recommendation on efficacy for dyskinesia is amantadine, a glutamate receptor antagonist (Fabbrini et al. 2007). However, its use is limited by the development of tolerance, psychomimetic adverse effects, and a rebound after its withdrawal (Thomas et al. 2004). Thus, it is essential to develop new therapeutic strategies for the treatment of LID, and neurosteroids like DHEA could represent interesting candidates.

The underlying mechanisms of LID, including possible hormonal influences, are unclear, but it is suggested that an increased activity of GABAergic striatal projection neurons within the basal ganglia, together with a nonphysiological, pulsatile stimulation of dopamine receptors, represent the pathophysiological basis of LID (Calabresi et al. 2008; Del Sorbo and Albanese 2008; Deogaonkar and Subramanian 2005). This suggestion is in coincidence with the general hypothesis about the pathophysiology of hyperkinetic movement disorders (Figure 5.1).

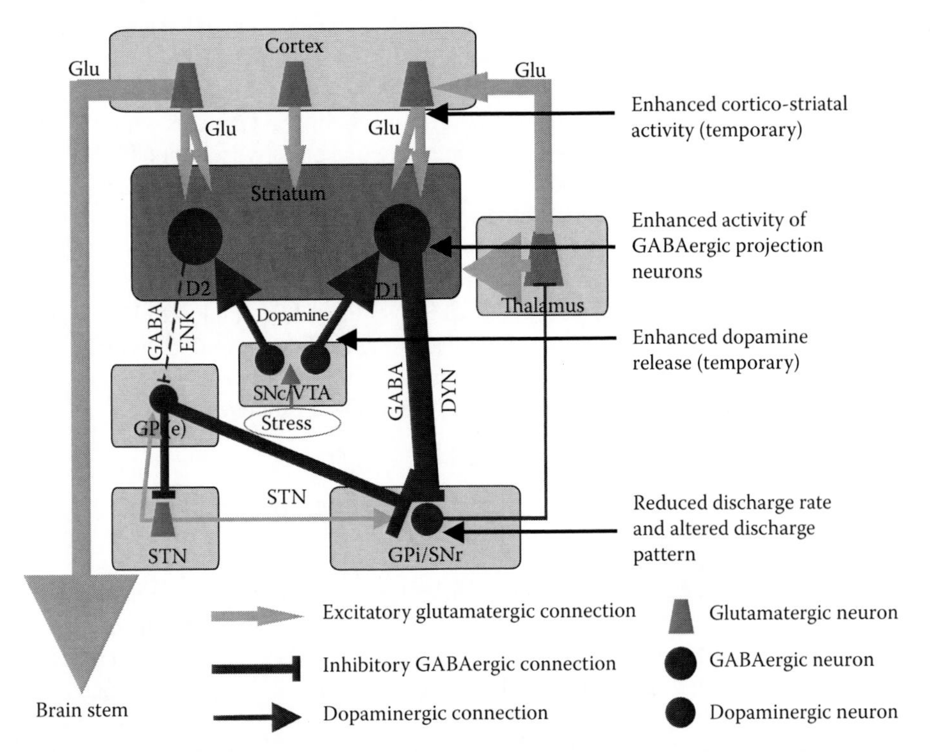

FIGURE 5.1 Hypothesis about the pathophysiology of hyperkinetic movement disorders like dyskinesias. Changes in neuronal activities are indicated by the thickness of the arrows. An enhanced activity of striatal GABAergic projection neurons of the direct pathway, which coexpress dynorphin (DYN), leads to a reduced discharge rate and an altered firing pattern of neurons of the globus pallidus pars internus (GPi) and the substantia nigra pars reticulata (SNr). In addition, the GABAergic projection neurons of the indirect pathway, which coexpress enkephalin (ENK), are underactive, leading to overinhibition of the subthalamic nucleus (STN) and intensifying the reduced neuronal activity of GPi/SNr neurons. In hereditary dyskinesias, both effects are often induced or triggered by stress, which causes a temporary enhanced dopamine release from neurons of the substantia nigra pars compacta (SNc) and the ventral tegmental area (VTA). In drug-induced dyskinesias, this effect is caused by the drug application itself or by enhanced sensitivity of dopamine receptors. The overinhibition of GPi/SNr neurons in turn causes enhanced activity of thalamocortical connections. This results in an overexcitation of the brain stem and therefore leads to hyperkinetic movement disorders. GABA = γ-aminobutyric acid; Glu = glutamate; D1 = dopamine receptor type 1; D2 = dopamine receptor type 2.

Tardive Dyskinesia

TD typically starts after months or years of treatment with antipsychotics that act primarily as dopamine receptor–blocking agents (Soares-Weiser and Fernandez 2007). It consists of involuntary movements that usually start orofacially involving the muscles of the tongue, lips, mouth, or face (Jankovic 1995). The disorder is progressive during antipsychotic treatments; its severity can increase and affect any part of the body. Thus, a wide range of movements including myoclonic jerks, tics, chorea, and dystonic postures can occur, and it is known that stress and anxiety can trigger the involuntary movements (Haddad and Dursun 2008). In contrast to LID, older patients are at an increased risk of developing TD (Saltz et al. 1991).

The ideal management of TD is to discontinue the antipsychotic medication, which is of course not possible in all cases (Haddad and Dursun 2008). A switch to an antipsychotic, such as clozapine or quetiapine, that lowers the incidence of TD is proposed (Emsley et al. 2004; Liebermann et al. 1991). Tetrabenazine, a compound that inhibits the vesicular monoamine transporter and thereby leads to a decrease in dopamine resources within the central nervous system, is the only licensed agent for the treatment of TD until now, but unfortunately, it is also known to cause depression (Kenney, Hunter, and Jankovic 2007). A combination of tetrabenazine and anticholinergics is suggested as the most effective treatment of TD (Haddad and Dursun 2008). The efficacy of this combination clearly points to the proposed hypothesis that dyskinesias, in general, and TDs, in particular, are caused by imbalances between the dopaminergic and cholinergic systems within the basal ganglia (Pisani et al. 2007). There is evidence that an increased sensitivity of dopamine receptors by the chronic blockade through antipsychotics plays an important pathophysiological role in TD (Hitri et al. 1978; Soares-Weiser and Fernandez 2007). Thus, during a 2-week administration of dopamine receptor–blocking agents to rodent models of TD, the affinity and number of dopamine D2 receptors increased, but it remains unclear how this leads to the development of TD and why only a part of antipsychotic-treated patients are affected (Soares-Weiser and Fernandez 2007). Apart from the alterations of the dopamine D2 receptors, candidate gene studies showed an association between a serine-to-glycine polymorphism in exon 1 of the dopamine D3 receptor gene and TD, but the functional role of this defect for the development of the disorder is unclear so far (Steen et al. 1997).

Possible Pathophysiological, Preventive, and Therapeutic Role of DHEA in Drug-Induced Dyskinesias

Several findings clearly indicate that hormonal influences are involved in the pathophysiology of drug-induced dyskinesias. Although the incidence of PD is higher in men than in women with a ratio of 49:1 (Wooten et al. 2004), women are more frequently affected by LID (Zappia et al. 2005). Acute drug-induced dystonic reactions occurred in association with polycystic ovarian syndrome, a condition with inadequate luteal phase function and unopposed continuous estrogen influence without normal mitigating effects of progesterone (Bonuccelli et al. 1991). It is commonly stated that women are more vulnerable to develop TD than men (Haddad and Dursun 2008), although one study found the converse, with a higher risk in men (van Os et al. 1999).

DHEA and its sulfated form DHEAS represent negative modulators of the $GABA_A$ receptors (Mellon and Griffin 2002). Thus, DHEA could have antidyskinetic properties by reducing the suggested increased GABAergic activity in the striatal output pathways. It was found that chronic administration of levodopa to parkinsonian 6-hydroxydopamine (OHDA)-lesioned rats leads to changes in the subunit composition of $GABA_A$ receptors within the basal ganglia (Katz, Nielsen, and Soghomonian 2005). It is also known that the subunit composition of $GABA_A$ receptors potently influences the pharmacological properties of neurosteroids at these receptors (Belelli et al. 2002; Belelli and Lambert 2005). Therefore, it can be speculated that the effects of DHEA on dyskinesias can change with the progress of LID and may be highly selective across different regions of the brain and different types of neurons.

DHEA is known to be a precursor of estradiol and can therefore acutely increase the dopaminergic activity. Furthermore, DHEA and estradiol are known to protect dopaminergic neurons (Charalampopoulos et al. 2008; D'Astous et al. 2003; Morissette et al. 2008). Interestingly, it was found that serum concentrations of DHEA and DHEAS decrease extensively in men and women during aging (Bélanger, Candas, and Dupont 1994; Morales et al. 1994). Taken together, these findings point to a possible usefulness of DHEA in the prevention of LID in PD because it is known that the severity of LID is similar to that of parkinsonism (Cenci 2007). On the other hand, one should consider that the increase of dopaminergic activity by DHEA via its "metabolite" estradiol may also aggravate the pulsatile stimulation of dopamine receptors, suggested as a pathophysiological mechanism for the development of LID. The proposed enhancement of dopaminergic activity by DHEA could also worsen TD because it would aggravate the enhanced sensitivity of the dopamine receptors. Nevertheless, other mechanisms of DHEA in the central nervous system could ameliorate LID and TD. Imbalances between the dopaminergic and the cholinergic neurotransmitter systems within the basal ganglia are proposed to represent an underlying mechanism of dyskinesia (Pisani et al. 2007). Because DHEA is capable of modifying the release of acetylcholine, it is thinkable that the enhanced dopaminergic activity in LID and TD may be counteracted by changes in the acetylcholine release by DHEA.

In accordance with the contradictory suggestions about the possible effects of DHEA on LID and TD, previous studies have illustrated a controverse modulatory efficacy of DHEA on drug-induced dyskinesias. Although DHEA worsened haloperidol-induced vacuous chewing movements in a rat model of TD, it did not influence LID in MPTP-treated parkinsonian monkeys and the 6-OHDA rat model of PD (Bélanger et al. 2006; Bishnoi, Chopra, and Kulkarni 2008; Paquette et al. 2009). Regarding these controversial findings, it has to be considered that the complexity of the brain suggests that DHEA may induce effects in several brain regions, which may counteract each other. In addition, with regard to the influence of the subunit composition of $GABA_A$ receptors on the pharmacological properties of neurosteroids such as DHEA, it can be suggested that the effects of these substances seem to be highly selective across different brain regions and different types of neurons and maybe also different types of receptors that cannot be excluded by the systemic administration of DHEA in the animal models of LID and TD. Further studies on these models with local applications of DHEA directly into distinct brain regions can clarify this suggestion.

DHEA AND HEREDITARY PAROXYSMAL DYSKINESIAS

Definition, Clinical Features, and Pathophysiology of Hereditary Paroxysmal Dyskinesias

Paroxysmal dyskinesias are a heterogeneous group of hyperkinetic movement disorders characterized by recurrent episodes of involuntary movements that are mostly a combination of dystonia, chorea, athetosis, and ballism. In contrast to epilepsy, consciousness is never lost in patients with paroxysmal dyskinesia. The disease can occur symptomatically after encephalitis, trauma, stroke, and others with lesions predominantly found in the basal ganglia (Weber and Lerche 2009). However, an underlying cerebral lesion is not present. Thus, paroxysmal dyskinesias are commonly idiopathic forms with a hereditary background (Bhatia 2001). The disorder is classified primarily according to the duration and the trigger of the episodes of involuntary movements as described in the following paragraphs.

Paroxysmal kinesigenic dyskinesia (PKD) is induced by sudden voluntary movements and lasts seconds to minutes. The brief attacks of involuntary movements in patients with PKD may include dystonic postures, chorea, athetosis, or ballism (Bhatia 1999). PKD is a rare neurological condition, but 27% of cases were reported to have a hereditary background with an autosomal dominant inheritance (Weber and Lerche 2009). The age of onset is usually during childhood or early adulthood with a range from 6 months to 40 years (Bhatia 1999; Fahn 1994). Furthermore, it is known that the frequency of the attacks diminishes with age (Bhatia 1999). The best therapeutic results can be achieved by the application of anticonvulsant drugs such as phenytoin or carbamazepine (van Rootselaar et al. 2009).

Paroxysmal nonkinesigenic dyskinesia (PND) is characterized by prolonged episodes (2 minutes until 4 hours) of involuntary movements, which also comprise a combination of dystonic posturing with chorea and ballism. In contrast to PKD, this form of paroxysmal dyskinesia is not induced by sudden movements but precipitated by factors such as intake of alcohol, coffee, or tea; fatigue, stress, and anxiety can also induce dyskinetic episodes (Bhatia 1999, 2001). In most cases, the onset could be observed in childhood, and interestingly, the episodes tend to diminish with age. Anticonvulsants are not effective in PND, and the most benefit could be observed after treatment with GABA-mimetic drugs such as benzodiazepines.

Paroxysmal exercise-induced dyskinesia (PED) is an intermediate form with dystonic episodes that last for 5–30 minutes, and which are induced by continued and prolonged exercise. They cannot be precipitated by stress, anxiety, alcohol, or coffee like PND and they are not inducible by sudden movements like PKD. The attacks are usually dystonic and appear in the body parts involved in the exercise, most commonly the legs after prolonged walking or running (Bhatia 2001). PED is a rare disorder, and only a few families have been described with an autosomal dominant pattern of inheritance. The most benefit could be reached with acetazolamide, whereas anticonvulsants were not found to be useful in the treatment of this condition (van Rootselaar et al. 2009; Weber and Lerche 2009).

In *paroxysmal hypnogenic dyskinesia* (PHD), patients typically awake with a cry and have involuntary dystonic and ballistic movements lasting up to 45 seconds (Bhatia 2001). Several attacks could occur each night. It is found that this disorder does not represent a typical dyskinesia. Instead, this nocturnal dyskinesia is due to mesial frontal lobe seizures, that is, a type of epilepsy (Bhatia 2001). Therefore, the term "autosomal dominant nocturnal frontal lobe epilepsy" is now used to describe this condition with the intent to distinguish it from other types of aforementioned paroxysmal dyskinesias (Scheffer et al. 1995). The antiepileptic drug carbamazepine is effective in most cases (Scheffer et al. 1994).

Recently, the first genes have been identified for PND (myofibrillogenesis regulator-1 = MR-1) and PED (SLC2A1; Ghezzi et al. 2009; Suls et al. 2008). Only the mutation in SLC2A1, which encodes the glucose transporter type 1 (GLUT1), is known to lead to a reduced transport of glucose across the blood-brain barrier (Weber and Lerche 2009). This finding can explain the occurrence of dyskinesia after prolonged exercise as a sign of energy deficit causing episodic dysfunction in the basal ganglia (van Rootselaar et al. 2009). In PED and PKD, genetic loci are found near ion channel genes that lead to the hypothesis that a channelopathy might be involved in the pathophysiology of paroxysmal dyskinesias, comparable to episodic ataxias and epilepsy. In contrast, the function of the MR-1 protein and the pathophysiology of PND are still poorly understood, although it is presumed that the MR-1 gene leads to a defect in the stress response pathway and not to a channelopathy (van Rootselaar et al. 2009). Despite these findings, the pathophysiology of paroxysmal dyskinesias remains largely unknown. Furthermore, the heterogeneity in the clinical signs in the response to medication and different gene defects allows us to presume that the pathophysiology of different types of paroxysmal dyskinesias differs. Nevertheless, it can be presumed that they may share, at least partly, common underlying mechanisms. These mechanisms may be consistent with the favored hypothesis for the occurrence of hyperkinetic movement disorders, which suggests dysfunctions of and imbalances between different neurotransmitter systems within the basal ganglia-thalamo-cortical loop, as illustrated in Figure 5.1. Similar to drug-induced dyskinesias, it is presumed that (temporary) dopaminergic and/or glutamatergic overactivity may induce an enhanced activity of striatal GABAergic projection neurons, resulting in a disinhibition of the thalamus and thereby to an overexcitation of the cortex.

Possible Pathophysiological, Preventive, and Therapeutic Role of DHEA for Hereditary Dyskinesias

Comparable to drug-induced dyskinesias, different facts point to underlying hormonal components in the pathophysiology of hereditary paroxysmal dyskinesias. Houser et al. (1999) found a clear male predominance in 26 patients with idiopathic PKD (23 men, 3 women), and twice as many men exhibited writer's cramp, a special type focal dystonia (Soland, Bhatia, and Marsden 1996). The gender predominance in PND seems to be unclear; on the one hand, a female predominance in PND has been reported (Bressman, Fahn, and Burke 1988), but otherwise it was reported that more men are affected than women (1.4:1; Fahn 1994). Measurements of DHEA levels in patients with hereditary paroxysmal dyskinesias are missing, and effects of the application of exogenous DHEA or metabolites are still not known. Therefore, the special role of DHEA in the pathophysiology, prevention, or therapy of inborn paroxysmal dyskinesias remains elusive so far.

Animal models can be helpful to further clarify the underlying mechanisms of these disorders and the possible role of steroids and thereby DHEA in prevention and therapy of these movement disorders. Animal models of inborn paroxysmal dyskinesias are rare and restricted to the dt^{sz} mutant hamster, which represents one of the most extensively investigated and unique animal models of movement disorders (for review see Raike, Jinnah, and Hess 2005; Richter 2005). The dt^{sz} hamsters show all clinical and pharmacological characteristics of PND in human patients (Nardocci et al. 2002; Raike, Jinnah, and Hess 2005; Richter 2005). A clear difference in the severity and frequency of stress-inducible paroxysmal nonkinesigenic dyskinetic episodes between men and women does not exist in this animal model (Löscher et al. 1995). Nevertheless, an age-dependent spontaneous remission of PND, coinciding with puberty and the occurrence of a triphasic relapse occurring in women associated with pregnancy, parturition, and nursing, indicated the involvement of steroids in the pathophysiology of PND in the dt^{sz} hamster (Khalifa and Iturrian 1993; Richter and Löscher 1998). An influence of peripheral sex hormones was excluded because castrations did not change the severity or frequency of dyskinetic episodes and the typical age-dependent time course (Löscher et al. 1995). Therefore, it is concluded that neurosteroids such as DHEA as endogenous modulators of different neurotransmitter systems may be involved in the pathophysiology of dyskinesia in this animal model. Different neurochemical, immunohistochemical, pharmacological, and electrophysiological studies on the dt^{sz} hamster clearly indicated that a reduced GABAergic inhibition, a temporary dopaminergic and glutamatergic overactivity in the striatum, that is, an important basal ganglia nucleus for the initiation and processing of movements, is involved in the pathophysiology (for review see Richter and Löscher 1998; Richter 2005). With regard to the nongenomic effects of DHEA as a negative modulator of the $GABA_A$, a positive modulator of the glutamatergic NMDA receptor, and as a possible enhancer of dopaminergic activity by estradiol, it was presumed that DHEA would worsen the severity of dyskinesia in this animal model. However, the systemic administration of DHEA in this animal model failed to exert significant effects (Hamann, Richter, and Richter 2007). Nevertheless, this lack of effects does not obligatorily stand for an unimportant role of DHEA in the pathophysiology of dyskinesias. It can be presumed that it can be based on the influence of $GABA_A$ receptor subunit composition on DHEA impact, mentioned earlier, or on opposed effects of DHEA in different brain regions or on different neurotransmitter systems, which could not be excluded by the systemic administration of DHEA. Furthermore, it can be speculated that DHEA and its metabolites might have many more effects on neuronal excitability than what is currently known. For example, in addition to the negative modulatory effects of DHEA on $GABA_A$ and the positive modulatory effects on NMDA receptors, both of which lead to an enhanced neuronal excitability, it was evidenced that DHEAS depressed calcium voltage-gated currents (Ffrench-Mullen and Spence 1991). To our knowledge, this is the only demonstration of a DHEA metabolite action that leads to reduced neuronal excitability, which is obviously in contrast to the effects on $GABA_A$ and NMDA receptors.

CONCLUSIONS

It can be stated that our knowledge about the possible pathophysiological, preventive, and therapeutic role of DHEA in the different types of dyskinesias is only marginal until now. The available data were mainly obtained from a small number of studies in animal models, and there is a lack of studies in dyskinetic patients until now. DHEA has widespread genomic and nongenomic effects in the central nervous system and is capable of modulating different neurotransmitter systems and different brain regions. Furthermore, its effects may be conflictive in the different forms of dyskinesias. It is not clear from the present data whether DHEA may worsen or ameliorate dyskinesias because of its negative modulation of $GABA_A$, positive modulation of glutamatergic NMDA receptors, changes in acetylcholine and dopamine release by reducing neuronal excitability via depression of voltage-gated calcium channels, or even other mechanisms that are not yet known. Therefore, a final statement about the suitability of DHEA for the prevention or therapy of dyskinesias cannot be given at the moment, and patients who suffer from dyskinesia should be very cautious with the intake of DHEA. Further preclinical studies in animal models and epidemiological and clinical studies in patients should clarify the role of DHEA and other neurosteroids in the pathophysiology, prevention, and therapy of dyskinesias and other movement disorders.

REFERENCES

Bélanger, A., B. Candas, A. Dupont, L. Cusan, P. Diamond, J. L. Gomez, and F. Labrie. 1994. Changes in serum concentrations of conjugated and unconjugated steroids in 40- to 80-year-old men. *J Clin Endocrinol Metab* 79:1086–90.

Bélanger, N., L. Grégoire, P. J. Bédard, and T. Di Paolo. 2006. DHEA improves symptomatic treatment of moderately and severely impaired MPTP monkeys. *Neurobiol Aging* 27:1684–93.

Belelli, D., A. Casula, A. Ling, and J. J. Lambert. 2002. The influence of subunit composition on the interaction of neurosteroids with GABA(A) receptors. *Neuropharmacology* 43(4):651–61.

Belelli, D., and J. J. Lambert. 2005. Neurosteroids: Endogenous regulators of the GABA(A) receptor. *Nat Rev Neurosci* 6(7):565–75.

Bhatia, K. P. 1999. The paroxysmal dyskinesias. *J Neurol* 246:149–55.

Bhatia, K. P. 2001. Familial (idiopathic) paroxysmal dyskinesias: An update. *Semin Neurol* 21(3):69–74.

Bishnoi, M., K. Chopra, and S. K. Kulkarni. 2008. Modulatory effect of neurosteroids in haloperidol-induced vacuous chewing movements and related behaviors. *Psychopharmacol (Berl)* 196:243–54.

Bonuccelli, U., A. Nocchiero, A. Napolitano et al. 1991. Domperidone-induced acute dystonia and polycystic ovary syndrome. *Mov Disord* 6:79–81.

Bressman, S. B., S. Fahn, and R. E. Burke. 1988. Paroxysmal non-kinesigenic dyskinesia. *Adv Neurol* 50:403–13.

Calabresi, P., M. Di Filippo, V. Ghiglieri, and B. Picconi. 2008. Molecular mechanisms underlying levodopa-induced dyskinesia. *Mov Disord* 23(Suppl. 3):S570–9.

Cenci, M. A. 2007. L-DOPA-induced dyskinesia: Cellular mechanisms and approaches to treatment. *Parkinsonism Relat Disord* 13:S263–7.

Cenci, M. A., and M. Lundblad. 2006. Post- versus presynaptic plasticity in L-DOPA-induced dyskinesia. *J Neurochem* 99:381–92.

Charalampopoulos, I., E. Remboutsika, A. N. Margioris, and A. Gravanis. 2008. Neurosteroids as modulators of neurogenesis and neuronal survival. *Trends Endocrinol Metab* 19(8):300–7.

Corpechot, C., P. Robel, M. Axelson, J. Sjovall, and E. E. Baulieu. 1981. Characterization and measurement of dehydroepiandrosterone sulphate in rat brain. *Proc Natl Acad Sci USA* 78:4704–7.

Damier, P. 2009. Drug-induced dyskinesias. *Curr Opin Neurol* 22(4):394–9.

D'Astous, M., M. Morissette, B. Tanguay, S. Callier, and T. Di Paolo. 2003. Dehydroepiandrosterone (DHEA) such as 17beta-estradiol prevents MPTP-induced dopamine depletion in mice. *Synapse* 47:10–4.

Del Sorbo, F., and A. Albanese. 2008. Levodopa-induced dyskinesias and their management. *J Neurol* 255(Suppl. 4):32–41.

Deogaonkar, M., and T. Subramanian. 2005. Pathophysiological basis of drug-induced dyskinesias in Parkinson's disease. *Brain Res Brain Res Rev* 50(1):156–68.

Emsley, R., H. J. Turner, J. Schronen, K. Botha, R. Smit, and P. P. Oosthuizen. 2004. A single-blind, randomized trial comparing quetiapine and haloperidol in the treatment of tardive dyskinesia. *J Clin Psychiatry* 65(5):696–701.

Fabbrini, G., J. M. Brotchie, F. Grandas, M. Nomoto, and C. C. Goetz. 2007. Levodopa-induced dyskinesias. *Mov Disord* 22:1379–89.
Fahn, S. 1994. The paroxysmal dyskinesias. In *Movement Disorders*, ed. C. D. Marsden and S. Fahn, 310–45. Oxford, England: Butterworth-Heinemann.
Ffrench-Mullen, J. M., and K. T. Spence. 1991. Neurosteroids block Ca2+ channel current in freshly isolated hippocampal CA1 neurons. *Eur J Pharmacol* 202(2):269–72.
Ghezzi, D., C. Viscorni, A. Ferlini et al. 2009. Paroxysmal non-kinesigenic dyskinesia is caused by mutations of the MR-1 mitochondrial targeting sequence. *Hum Mol Genet* 18(6):1058–64.
Haddad, P. M., and S. M. Dursun. 2008. Neurological complications of psychiatric drugs: Clinical features and management. *Hum Psychopharmacol Clin Exp* 23:15–26.
Halkias, I. A., I. Haq, Z. Huang, and H. H. Fernandez. 2007. When should levodopa therapy be initiated in patients with Parkinson's disease? *Drugs Aging* 24:261–73.
Hamann, M., F. Richter, and A. Richter. 2007. Acute effects of neurosteroids in a rodent model of primary paroxysmal dystonia. *Horm Behav* 52:220–7.
Hitri, A., W. J. Weiner, R. L. Borison, B. I. Diamond, P. A. Nausieda, and H. L. Klawans. 1978. Dopamine binding following prolonged haloperidol treatment. *Ann Neurol* 3:134–40.
Houser, M. K., V. L. Soland, K. P. Bhatia, N. P. Quinn, and C. D. Marsden. 1999. Paroxysmal kinesigenic choreoathethosis: A report of 26 patients. *J Neurol* 246(2):120–6.
Jankovic, J. 1995. Tardive syndromes and other drug-induced movement disorders. *Clin Neuropharmacol* 18:197–214.
Jankovic, J. 2005. Motor fluctuations and dyskinesias in Parkinson's disease: Clinical manifestations. *Mov Disord* 20(11):S11–16.
Jankovic, J. 2009. Treatment of hyperkinetic movement disorders. *Lancet Neurol* 8(9): 844–56.
Katz, J., K. M. Nielsen, and J. J. Soghomonian. 2005. Comparative effects of acute and chronic administration of levodopa to 6-hydroxydopamine-lesioned rats on the expression of glutamic acid decarboxylase in the neostriatum and GABAA receptor subunits in the substantia nigra pars reticulata. *Neuroscience* 132(3):833–42.
Kenney, C., C. Hunter, and J. Jankovic. 2007. Long-term tolerability of tetrabenazine in the treatment of hyperkinetic movement disorders. *Mov Disord* 22(2):193–7.
Khalifa, A. E., and W. B. Iturrian. 1993. Triphasic relapse in dystonic hamsters during pregnancy as rigorous neurochemical model for movement disorder. *Soc Neurosci Abstr* 19:1626.
Klein, C. 2005. Movement disorders: Classifications. *J Inherit Metab Dis* 28:425–39.
Kranick, S. M., E. M. Mowry, A. Colcher, S. Horn, and L. I. Golbe. 2010. Movement disorders and pregnancy: A review of the literature. *Mov Disord* 23(6):665–71.
Lambert, J. J., D. Belelli, D. R. Peden, A. W. Vardy, and J. A. Peters. 2003. Neurosteroid modulation of GABAA receptors. *Prog Neurobiol* 71(1):67–80.
Liebermann, J., B. Saltz, C. John, S. Pollak, M. Borenstein, and J. Kane. 1991. The effect of clozapine in tardive dyskinesia. *Br J Psychiatry* 154:503–10.
Löscher, W., T. Blanke, A. Richter, and H. O. Hoppen. 1995. Gonadal sex hormones and dystonia: Experimental studies in genetically dystonic hamsters. *Mov Disord* 10(1):92–102.
Mellon, S. H., and L. D. Griffin. 2002. Neurosteroids: Biochemistry and physiological significance. *Trends Endocrinol Metab* 13(1):35–43.
Morales, A. J., J. J. Nolan, J. C. Nelson, and S. S. Yen. 1994. Effects of replacement dose of dehydroepiandrosterone in men and women of advancing age. *J Clin Endocrinol Metab* 78:1360–7.
Morissette, M., S. Al Sweidi, S. Callier, and T. Di Paolo. 2008. Estrogen and SERM neuroprotection in animal models of Parkinson's disease. *Mol Cell Endocrinol* 290:60–9.
Nardocci, N., E. Fernandez-Alvarez, N. W. Wood, S. D. Spacy, and A. Richter. 2002. The paroxysmal dyskinesias. In *Epilepsy and Movement Disorders*, ed. R. Guerrini, J. Aicardi, F. Andermann, and M. Hallett, 125–39. Cambridge, MA: Cambridge University Press.
Paquette, M. A., K. Foley, E. G. Brudney, C. K. Meshul, S. W. Johnson, and S. P. Berger. 2009. The sigma-1 antagonist BMY-14802 inhibits L-DOPA-induced abnormal involuntary movements by a WAY-100635-sensitive mechanism. *Psychopharmacol* 204:743–54.
Pérez-Neri, I., S. Montes, C. Ojeda-López, J. Ramirez-Bermúdez, and C. Rios. 2008. Modulation of neurotransmitter systems by dehydroepiandrosterone and dehydroepiandrosterone sulphate: Mechanism of action and relevance to psychiatric disorders. *Prog Neuropsychopharmacol Biol Psychiatry* 32:1118–30.
Pisani, A., G. Bernardi, J. Ding, and D. J. Surmeier. 2007. Re-emergence of striatal cholinergic interneurons in movement disorders. *Trends Neurosci* 30(10):545–53.
Raike, R. S., H. A. Jinnah, and E. J. Hess. 2005. Animal models of generalized dystonia. *NeuroRx* 2(3):504–12.

Richter, A. 2005. The genetically dystonic hamster: An animal model of paroxysmal dystonia. In *Animal Models of Movement Disorders*, ed. M. LeDoux, 459–66. Amsterdam, New York: Elsevier Academic Press.

Richter, A., and W. Löscher. 1998. Pathophysiology of idiopathic dystonia: Findings from genetic animal models. *Prog Neurobiol* 54(6):633–77.

Saltz, B. L., M. G. Woener, J. M. Kane et al. 1991. A prospective study of tardive dyskinesia incidence in the elderly. *JAMA* 66:2402–6.

Scheffer, I. E., K. P. Bhatia, I. Lopes-Cendes et al. 1994. Autosomal dominant frontal lobe epilepsy misdiagnosed as a sleep disorder. *Lancet* 43:515–7.

Scheffer, I. E., K. P. Bhatia, I. Lopes-Cendes et al. 1995. Autosomal dominant nocturnal frontal lobe epilepsy: A distinctive clinical disorder. *Brain* 118:61–73.

Soares-Weiser, K., and H. H. Fernandez. 2007. Tardive dyskinesia. *Semin Neurol* 27(2):159–69.

Soland, V. L., K. P. Bhatia, and C. D. Marsden. 1996. Sex prevalence of focal dystonias. *J Neurol Neurosurg Psychiatry* 60(2):204–5.

Soorani-Lunsing, R. J., M. Hadders-Algra, H. J. Huisjes, and B. C. Touwen. 1994. Neurobehavioural relationships after the onset of puberty. *Dev Med Child Neurol* 36(4):334–43.

Steen, V. M., R. Lovlie, T. MacEwan, and R. G. McCreadie. 1997. Dopamine D3-receptor gene variant and susceptibility to tardive dyskinesia in schizophrenic patients. *Mol Psychiatry* 2:139–45.

Suls, A., P. Dedeken, K. Goffin et al. 2008. Paroxysmal exercise-induced dyskinesia and epilepsy is due to mutations in SLC2A1, encoding the glucose transporter GLUT1. *Brain* 131(7):1831–44.

Thomas, A., D. Iacono, A. L. Luciano, K. Armellino, A. Di Iorio, and M. Onofrj. 2004. Duration of amantadine benefit on dyskinesia of severe Parkinson's disease. *J Neurol Neurosurg Psychiatry* 75:141–3.

van Os, J., E. Walsh, E. van Horn, T. Tattan, R. Bale, and S. G. Thompson. 1999. Tardive dyskinesia in psychosis: Are women really more at risk? UK700 group. *Acta Psychiatr Scand* 99(4):288–93.

van Rootselaar, A.-F., S. S. van Westrum, D. N. Velis, and M. A. J. Tijssen. 2009. The paroxysmal dyskinesias. *Pract Neurol* 9:102–9.

Weber, Y. G., and H. Lerche. 2009. Genetics of paroxysmal dyskinesias. *Curr Neurol Neurosci Rep* 9(3):206–11.

Wolf, O. T., and C. Kirschbaum. 1999. Actions of dehydroepiandrosterone and its sulfate in the central nervous system: Effects on cognition and emotion in animals and humans. *Brain Res Rev* 30:264–88.

Wooten, G. F., L. J. Currie, V. E. Bovbjerg, J. K. Lee, and J. Patrie. 2004. Are men at greater risk for Parkinson's disease than women? *J Neurol Neurosurg Psychiatry* 75:637–9.

Zappia, M., G. Annesi, G. Nicoletti et al. 2005. Sex differences in clinical and genetic determinants of levodopa peak-dose dyskinesias in Parkinson disease: An exploratory study. *Arch Neurol* 62:601–5.

Zheng, P. 2009. Neuroactive steroid regulation of neurotransmitter release in the CNS: Action, mechanism and possible significance. 89:134–52.

6 Dehydroepiandrosterone
Its Effects on Cell Proliferation and Cancer

Rebeca López-Marure

CONTENTS

INTRODUCTION

Dehydroepiandrosterone (DHEA) is a steroid mainly synthesized in the adrenal cortex and gonads (Matsuzaki and Honda 2006). DHEA and its sulfated form (DHEAS) are the most abundant circulating hormones in humans, and their plasma levels decline with age and vary with sex, ethnicity, and environmental factors (Orentreich et al. 1984). Although DHEA's effects are very controversial, numerous animal studies have demonstrated the beneficial effects of DHEA in preventing obesity, diabetes, cancer, and heart disease; in enhancing the immune system; and even in prolonging life spans.

In addition, both clinical observations and laboratory experiments have shown that DHEA levels possibly play a role in determining the risk of developing cancer. For example, because DHEA levels in humans are known to decline with age, it has been suggested that low levels of this steroid or its metabolites are associated with the presence and increased risk of developing mammary cancer and possibly other cancers. Unfortunately, not all studies have observed that DHEA has anticarcinogenic effects.

While some studies have indeed demonstrated that DHEA has a protective effect against cancer, others have concluded that DHEA can actually induce cancer. This contrast highlights the fact that the administration of DHEA results in a wide range of physiological and pharmacological effects in laboratory animals and their cells. Some of these effects include (1) protection against the development of spontaneous and carcinogen-induced tumors, (2) inhibition of carcinogen binding to DNA, (3) inhibition of DNA synthesis, (4) proliferation of animal cells under basal conditions and when stimulated by various mitogens, (5) inhibition of lipogenesis, (6) suppression of weight gain in rodents without significant reduction in food intake, (7) amelioration of genetic diabetes in mice, and (8) retardation of autoimmune diseases associated with aging (Gordon, Shantz, and Talalay 1987). However, the actions of exogenous DHEA depend on the concentrations used. Physiological doses induce neurological and immunological functions and cell proliferation, whereas pharmacological doses affect cardiological and metabolic functions, antineoplastic effects, and inhibition of cell proliferation (Matsuzaki and Honda 2006).

DHEA has been shown to inhibit breast, prostate, lung, liver, skin, and thyroid carcinogenesis in animal model systems. The chemopreventive and antiproliferative effects of DHEA are explained by at least four mechanisms: (1) depletion of nicotinamide adenine dinucleotide phosphate hydrogen (NADPH) and ribose-5-phosphate by inhibiting glucose-6-phosphate dehydrogenase (G6PD) activity, (2) suppression of cholesterol biosynthetic pathways by inhibiting 3-hydroxy-3-methyl-glutaryl coenzyme-A (HMG-CoA) reductase, (3) interference with cell proliferation signaling pathways, and (4) suppression of nitric oxide generation through downregulation of nitric oxide synthase II (Matsuzaki and Honda 2006).

This chapter summarizes some of these seemingly contradictory effects and outlines the conditions under which each effect was observed. The discussion that follows focuses on the effects of DHEA on the proliferation of normal and tumoral cells in *in vitro* and *in vivo* studies.

EFFECTS OF DHEA *IN VIVO*

The physiological functions of DHEA in preventing human carcinogenesis are still controversial, but many reports have shown that pharmacological doses of DHEA have chemopreventive and antiproliferative effects and inhibit tumor growth in rats and mice. DHEA has been shown to inhibit breast, prostate, lung, liver, skin, and thyroid carcinogenesis in animal model systems. Furthermore, DHEA has been shown to inhibit the development of not only spontaneous tumors (Tannenbaum and Silverstone 1953; Shimokawa, Yu, and Masoro 1991), but also chemically induced (Tannenbaum and Silverstone 1953; Klurfeld, Weber, and Kritchevsky 1987) and radiation-induced (Gross and Dreyfuss 1984, 1986) tumors.

As mentioned, despite the fact that the administration of DHEA in the diet of mice and rats has generally been found to reduce tumor growth, some studies indicate that DHEA can, on the contrary, induce tumor formation in mice following long-term oral DHEA treatment, perhaps due in part to the direct action of the steroid on cells. Several studies on the effects of DHEA induced in models *in vivo* are described in the next section and are shown in Tables 6.1 and 6.2.

DHEA and Its Anticarcinogenic Action

Mice and Rats

Skin

The topical application of DHEA to the skin of CD-1 mice encouraged the inhibition of papilloma formation induced by 7,12-dimethylbenz(a)anthracene (DMBA) and tetradecanoylphorbol-13-acetate at both initiation and promotion stages (Pashko et al. 1984). In addition, DHEA inhibited

TABLE 6.1
Effects of Dehydroepiandrosterone (DHEA) Treatment on Normal Tissues *In Vivo*

Species	Effect of DHEA	Reference
Mice		
Thymus	Inhibition of the expression of proliferating cell nuclear antigen; induction of apoptosis in combination with calorie restriction	Wang et al. (1997)
Bone marrow	Inhibition of bone marrow growth and lymphopoiesis	Catalina et al. (2003); Risdon, Cope, and Bennett (1990)
Rats		
Vagina	Induction of a typical androgenic effect of epithelial mucification	Berger, El-Alfy, and Labrie (2008)
Ovary	Induction of proliferation of ovarian mast cells	Parshad and Batth (1999)
Liver	Inhibition of the growth of physiologically proliferating liver tissue	Mayer et al. (2003)
Brain	Increase in the number of newly formed neurons in hippocampus	Charalampopoulos et al. (2008)
Other organs	Increase in liver, colon, and small intestine weights, and induction of thymic atrophy	Pelissier et al. (2004)
Dogs		
Brain	Reduction in DNA damage induced by H_2O_2	Shen et al. (2001)
Humans		
Blood	Induction of an increase in serum levels of DHEA and decrease in cholesterol in menopausal women	Williams et al. (2004)
Bone	Induction of an increase in the bone mineral density	Labrie et al. (1998)

DMBA-induced papilloma and carcinoma development in the complete carcinogenesis model (Pashko et al. 1985). Specifically in mice, oral administration in the diet at 0.6% for 2 weeks inhibited the rate of binding of 3H-DMBA to skin DNA (Pashko and Schwartz 1983). Furthermore, in the two-stage skin tumorigenesis model in mice, DHEA treatment inhibited tumor initiation, tumor promoter–induced epidermal hyperplasia, and papilloma promotion (Schwartz and Pashko 1995).

Breast

In C3H-Av/A and C3HA/A mice, DHEA treatment inhibited the development of spontaneous breast cancer (Schwartz et al. 1981). Treatment with increasing doses of DHEA delivered constantly by silastic implants of increasing length and number caused a progressive inhibition of tumor development of mammary carcinoma induced by DMBA in rats (Li et al. 1993). It is noteworthy that tumor size in the group of animals treated with the highest dose (663.0-cm-long implants) of DHEA was similar to that found in ovariectomized animals, thus showing a complete blockage of estrogen action by DHEA. Such data clearly demonstrate that circulating levels of DHEA comparable to those observed in normal adult premenopausal women exert a potent inhibitory effect on the development of mammary carcinoma induced by DMBA in rats (Li et al. 1993). Another interesting finding is that DHEA seems to be an effective chemopreventive agent in the N-methyl-N-nitrosourea (MNU)–induced rat mammary tumor model. In studies with this model, the highest dose of DHEA (600 ppm) significantly decreased tumor incidence from 95% to 45%. Additionally, it increased tumor latency and decreased tumor multiplicity from 4.1 to 0.5 tumors per rat. Thus, it can be concluded that DHEA has striking chemopreventive efficacy in this model at doses far below its maximally tolerated dose (Lubet et al. 1995).

TABLE 6.2
Effects of Dehydroepiandrosterone (DHEA) Treatment on Tumors *In Vivo*

Species	Effect of DHEA	Reference
Mice		
Skin	Inhibition of papilloma formation induced by chemicals and inhibition of carcinoma development	Pashko and Schwartz (1983); Pashko et al. (1984, 1985); Schwartz and Pashko (1995)
Breast	Inhibition of spontaneous cancer development	Schwartz et al. (1981); Green et al. (2001)
Pancreas	Inhibition of growth and tumor size and weight of pancreatic cancer cell lines inoculated; inhibition of pancreatic cancer xenograft growth	Melvin et al. (1997); Muscarella et al. (1998)
Leukemia	Inhibition of leukemia cell growth	Catalina et al. (2003)
Myeloma	Inhibition of multiple myeloma U-266 cells inoculated	Liu et al. (2005)
Ehrlich's ascites	Inhibition of tumor cell proliferation and arrest of cells in the G0/G1 phases of the cell cycle	Boros et al. (1997); Raïs et al. (1999)
Rats		
Breast	Inhibition of tumor development induced by chemicals and inhibition of cell proliferation and induction of apoptosis	Li et al. (1993); Lubet et al. (1995); Shilkaitis et al. (2005)
Liver	Induction of hepatocarcinogenesis	Rao et al. (1992); Prough et al. (1995); Metzger et al. (1995)
Dogs and Cats		
Mastosarcoma	Induction of regression in spontaneous mastosarcomas	Regelson, Loria, and Kalimi (1988)
Hepatoma	Inhibition of hepatoma cell proliferation	Regelson, Loria, and Kalimi (1988)

DHEA was found to reduce invasive carcinoma growth in transgenic mice that spontaneously developed multifocal mammary lesions that evolved into invasive, hormone-independent carcinomas. DHEA was observed to reduce tumor burden by 50% at 2000 mg/kg (Green et al. 2001). DHEA induced a similar decrease in tumor multiplicity and tumor burden in rats treated with N-nitroso-N-methylurea to develop mammary hyperplastic and premalignant lesions. DHEA was also observed to induce a senescent phenotype in tumor cells, inhibit cell proliferation, and increase the number of apoptotic cells with the involvement of p16 and p21 proteins to mediate these effects (Shilkaitis et al. 2005).

Thymus

Investigations have shown that DHEA induces histological changes in the thymus, which may be due to modulation of thymocyte apoptosis and proliferation. In such studies, calorie restriction (CR) treatment did not affect thymic histology. Treatment with DHEA and CR in combination resulted in a decreased expression of the proliferating cell nuclear antigen (PCNA) proliferation marker in the thymus in p53-deficient ($p53^{-/-}$) mice (Wang et al. 1997). DHEA decreased Bcl-2 but not Bax mRNA levels in the thymus; therefore, the effect of DHEA on thymocytes may have involved perturbation of the apoptotic pathway and an increase of apoptosis in the thymus, resulting in thymic atrophy (Wang et al. 1997). Previous studies on p53-deficient ($p53^{-/-}$) mice showed that treatment with DHEA decreased lymphoma development (Hursting, Perkins, and Phang 1994; Hursting et al. 1995). In conclusion, it is interesting that treatment with both DHEA and CR decreased expression of the PCNA proliferation marker in the thymus of mice.

Pancreas

In a study by Melvin et al. (1997), two human pancreatic cancer cell lines, MiaPaCa-2 and Panc-1, were inoculated into the flanks of nude athymic mice. After 2 weeks, it was observed that DHEA inhibited tumor growth, size, and weight. In addition, DHEA administered by intraperitoneal injection in athymic mice inhibited the growth of pancreatic cancer xenografts in nude mice. In mice fed with a diet supplemented with 0.6% DHEA ad libitum, the mean weight of the pancreatic tumor was reduced by DHEA, which was in turn associated with an increased plasma DHEAS concentration (Muscarella et al. 1998).

Bone Marrow and Leukemia

Dietary DHEA inhibited the proliferation of syngeneic bone marrow cells infused into lethally irradiated mice. DHEA reduced food intake, and this was related to the inhibition of bone marrow and leukemia cell growth (Catalina et al. 2003).

Myeloma

DHEA injection induced a dramatic inhibition of multiple myeloma U-266 cells implanted intraperitoneally in severe combined immunodeficiency-hIL-6 transgenic mice (Liu et al. 2005).

Lymphopoiesis

Some studies showed that feeding DHEA to mice resulted in the inhibition of lymphopoiesis but not myelopoiesis. Thus, DHEA treatment may exert a specific effect on thymocytes (Risdon, Cope, and Bennett 1990).

Ehrlich's Ascites Tumor

DHEAS inhibited Ehrlich's ascites tumor cell proliferation by 46% in C57BL/6 mice (Boros et al. 1997). Increasing doses of DHEA were administered by daily intraperitoneal injections to Ehrlich's ascites tumor-hosting mice for 4 days. In this experiment, tumors showed a dose-dependent increase in their G0-G1 cell populations after DHEA treatment and a simultaneous decrease in cells advancing to the S and G2-M cell cycle phases (Raïs et al. 1999). DHEA decreased both the cell volume and cell number of tumors compared with the controls. Pentose cycle (PC) inhibitors and DHEA efficiently regulated the cell cycle and tumor proliferation processes, suggesting that DHEA induces a G1 phase cycle arrest in Ehrlich's tumor cells through inhibition of the PC (Raïs et al. 1999).

Vagina

The intravaginal application of DHEA at daily doses of 0.33 mg in ovariectomized rats for 2 weeks induced a typical androgenic effect of epithelial mucification, indicating that DHEA can exert beneficial effects in the vagina (Berger, El-Alfy, and Labrie 2008).

Ovary

Ovarian sections of immature rats treated with DHEA (6.0 mg/100 g body weight) showed a majority of antral follicles either undergoing atresia or in early stages of cystogenesis. Treatment for 8 days with DHEA resulted in an increased number of alcian blue–positive ovarian mast cells; in addition, a significant rise in ovarian mast cell counts was recorded after 24 days (Parshad and Batth 1999).

Brain

DHEA increased the number of newly formed neurons in the rat dentate gyrus of the hippocampus and antagonized the suppressive effect of corticosterone on both neurogenesis and neuronal precursor proliferation (Charalampopoulos et al. 2008).

Other Organs

DHEA treatment administered by intraperitoneal injections significantly increased liver, colon, and small intestine cell weights. After 7 days, DHEA exerted an antioxidant effect in all organs studied. In the colon, oxidative damage protection was accompanied by goblet cell proliferation and increased acidic mucus production. After 2 days, the antioxidant effect of DHEA was mainly observed in the liver in Wistar rats (Pelissier et al. 2004).

Dogs and Cats

DHEA inhibited cell replication and uridine incorporation into proliferating hepatoma cells and hepatocytes, and at 10 mg/kg per day induced regression in spontaneous mastosarcomas in dogs and cats (Regelson, Loria, and Kalimi 1988).

Older male dogs receiving daily DHEA treatment for 7 months had significantly less DNA damage detectable in their brains compared with age-matched control dogs. After 7 months of treatment, DHEA-treated dogs also had a significant reduction in DNA damage in peripheral blood lymphocytes (PBLs) compared with pretreatment levels. The PBLs of dogs treated with DHEA were more resistant to H_2O_2-induced DNA damage than PBLs of untreated dogs (Shen et al. 2001). These results suggest that DHEA supplementation can significantly reduce steady-state levels of DNA damage in the mammalian brain.

Humans

Studies of DHEA in humans are scarce. Some of the studies have shown that the administration of DHEA for 12 months to postmenopausal women induced an increase in bone mineral density. This increase was associated with increased plasma osteocalcin, a marker of bone formation, indicating that DHEA could have a beneficial use as hormone replacement therapy in women (Labrie et al. 1998). In another study, DHEA administered to healthy menopausal women, a group with consistently low circulating estrogen and androgen levels, significantly increased serum levels of DHEA, and total plasma cholesterol concentrations decreased after 12 weeks of DHEA administration. DHEA also increased endothelium-dependent cutaneous blood flow in response to iontophoresed acetylcholine, suggesting that the changes in large vessel and microvascular endothelial function observed after oral administration of DHEA are potentially beneficial (Williams et al. 2004). It was shown that DHEA inhibited interleukin (IL)-6 production in bone marrow mononuclear cells from patients with myeloma (Liu et al. 2005).

DHEA and Its Carcinogenic Action

Contrary to its anticarcinogenic effects, it has been shown that long-term administration of DHEA causes hepatocarcinogenesis. DHEA-induced carcinogenesis is most likely dose dependent. For example, rats eliciting hepatocarcinoma were fed diets high in DHEA (5,000 or 10,000 mg DHEA/kg diet ad libitum). In contrast, the highest concentration of DHEA used to blunt breast carcinogenesis, 600 mg/kg diet, was probably lower than that required for high incidences of hepatocarcinogenesis. At this lower dosage, MNU measured as induced fatty acyl CoA oxidase activity or levels of W 4 A mRNAs was only modestly increased (10%–20% of maximum) compared with higher concentrations of DHEA (>2000 mg DHEA/kg diet) used in studies in which maximal peroxisome proliferation was observed. At doses of 25–120 mg DHEA/kg diet, the cancer chemopreventive action of DHEA was still striking, whereas no significant increase in the markers of peroxisome proliferation was noted (Prough et al. 1995).

In another study, DHEA was administered in the diet at a concentration of 0.45% to F-344 rats for up to 84 weeks. At the termination of the experiment, 14 of 16 rats had developed hepatocellular carcinomas. DHEA was also shown to markedly inhibit liver cell [^{3}H]thymidine labeling indices, suggesting that cell proliferation is not a critical feature in liver tumor development with this agent (Rao et al. 1992).

These results demonstrate that DHEA acts as a peroxisome proliferator and hepatocarcinogen in rats upon long-term treatment with high doses (0.6%) in the diet. Ultrastructural studies of tumors revealed a marked proliferation of mitochondria and a moderate proliferation of peroxisomes in all lesions (Metzger et al. 1995). DHEA also enhanced hepatocarcinogenesis induced by N-nitrosomorpholine (NNM); nevertheless, hepatocellular carcinomas under DHEA treatment seemed to have a less malignant phenotype compared with tumors induced by NNM only. Thus, DHEA resulted in a growth stimulation of the late basophilic lesion type and in a growth inhibition of early preneoplastic lesions. DHEA also inhibited the growth of physiologically proliferating liver tissue (Mayer, Forstner, and Kopplow 2003). Despite the fact that DHEA exerts anticarcinogenic effects in a variety of tissues, the peroxisome-proliferative property makes it a hepatocarcinogen (Rao et al. 1992).

EFFECTS OF DHEA *IN VITRO*

The growth of some mammalian cell lines in culture was inhibited by moderate concentrations of DHEA over a range of 10^{-4} to 10^{-5} M, and there appeared to be a wide variation in sensitivity among cell lines. This sensitivity to growth inhibition correlates well with the susceptibility of G6PD of these cells to inhibition by DHEA. Some studies have suggested that G6PD inhibition by DHEA accounts for its antiproliferative activity; however, other studies have shown the contrary. G6PD is a key protective enzyme responsible for maintaining adequate levels of the major cellular reducing agent NADPH, which is needed for the reduction of folic acid to tetrahydrofolic acid, is required for ribonucleotide and thymidylate synthesis, and is also required as a cofactor for ribonucleotide reductase. Therefore, it has been suggested that another probable mechanism inhibiting DHEA is a reduced NADPH pool size (Whitcomb and Schwartz 1985). NADPH generates o_2^-, and there is evidence that o_2^- may play an important role in tumor promotion (Babior 1982). Although DHEA has an inhibitory effect on the proliferation of several cell types, some studies have reported a proliferative effect. All these effects induced *in vitro* by DHEA in normal and tumor cells are shown in Tables 6.3 and 6.4.

NORMAL CELLS

Lymphocytes and Leukocytes

DHEA inhibited the [^{3}H]thymidine uptake of mitogen-stimulated, peripheral blood mononuclear cells obtained from individuals with normal G6PD and with Mediterranean-type G6PD deficiency (G6PD$^-$). The data suggest that the inhibition of [^{3}H]thymidine uptake induced on lymphocytes by DHEA probably does not depend on the inhibition of G6PD (Ennas et al. 1987).

TABLE 6.3
Effects of Dehydroepiandrosterone Treatment on Normal Cells *In Vitro*

Cell	Effect on Proliferation	Concentration	Reference
Blood mononuclear cells	Inhibition	20–80 μM	Ennas et al. (1987)
Lymphocytes	Stimulation	15 ng/ml	Corsini et al. (2005)
T cells	Inhibition	10 μM	Solano et al. (2008)
Neurons	Promotion of neurite length	10^{-10} to 10^{-12} M	Compagnone and Mellon (1998)
Endothelial cells	Stimulation	1–100 nM; 0.1–10 nM	Williams et al. (2004); Altman et al. (2008); Mohan and Benghuzzi (1997)
	Inhibition	25–50 ng/ml; 10 μM; 100 μM; 10–100 μM	Hinson and Khan (2004); Zapata et al. (2005); Varet, Vincent, and Akwa (2004)
Adipocytes	Inhibition	10^{-8} M	Rice et al. (2010)
Epithelial cells	Stimulation	100 nM	Sun et al. (2010)

TABLE 6.4
Effects of Dehydroepiandrosterone Treatment on Tumor Cells *In Vitro*

Cell	Effect on Proliferation	Concentration	Reference
Breast			
ZR-75-1	Stimulation	10 μM	Poulin and Labrie (1986)
MCF-7SH	Stimulation	100 nM	Maggiolini et al. (1999)
MCF-7	Stimulation	500 nM	Boccuzzi et al. (1992); Najid and Habrioux (1990)
		<10 μM; 10–100 nM; 0.1–10 nM	Di Monaco et al. (1997); Schmitt et al. (2001); Gayosso, Montano, and López-Marure (2006)
	Inhibition	10 μM; 10^{-6} M; 100 μM	Di Monaco et al. (1997); Andò et al. (2002); Gayosso, Montano, and López-Marure (2006)
T47D	Stimulation	—	Liberato, Sonohara, and Brentani (1993)
Prostate			
LNCaP	Stimulation	100 nM	Arnold et al. (2005)
Colon			
Caco-2	Inhibition	100 μM	Yoshida et al. (2003)
HT-29	Inhibition	100 μM	Yoshida et al. (2003); Jiang et al. (2005)
Hepatoma			
HepG2	Inhibition	10–100 mM	Yoshida et al. (2003); Ho et al. (2008)
		50 μM	Jiang et al. (2005)
Melanoma			
B16	Inhibition	1 nM	Kawai et al. (1995)
A375-SM	Stimulation	—	Richardson et al. (1999)
Myeloma			
U-266, NOP-2, IL-KM3	Inhibition	—	Liu et al. (2005)
Neuroblastoma			
	Inhibition	1 nM–1 μM	Gil-ad et al. (2001)
Cervix			
C33A, CASKI, HeLa	Inhibition	50–70 μM	Girón et al. (2009)

One work showed that peripheral blood leukocytes obtained from elderly donors (≥65 years) had a reduced expression of receptors for activated C kinase (RACK-1). The treatment of these cells with DHEA resulted in increased RACK-1 in leukocytes and in lymphocyte proliferation (Corsini et al. 2005). Another study showed that DHEA diminished proliferation of T cells isolated from lymph nodes of prepuberal BALB/c mice, and this effect was associated with decreased levels of the antioxidant molecule glutathione (Solano et al. 2008).

Neurons

In the early 1980s, it was reported that neurons had the ability to synthesize steroids (Baulieu, Robel, and Schumacher 2001). It was found that the brain synthesizes DHEA, and this steroid was dubbed a "neurosteroid" to differentiate it from steroids produced in the periphery. The DHEA produced in the brain has important effects on cortical neuronal growth in the rodent fetus. DHEA at nanomolar concentrations promoted lengthening of neuronal axons, the principal structures that transmit messages downstream from the neuronal soma (Compagnone and Mellon 1998). DHEA and DHEAS are involved in protection from neural apoptosis within the central nervous system (CNS) (Charalampopoulos et al. 2006). Besides its neuroprotective and prosurvival effects, DHEA also appears to affect neurogenesis.

DHEA produced locally has important effects on mouse embryonic neocortical neurons. Interestingly, sulfation of DHEA to DHEAS completely abolished its axon-promoting effect. It is important to note that the effects of DHEA appeared to be focused on the neuronal axons (i.e., the transmitting part of a neuron); DHEA did not appear to affect the receiving mechanism because it exerted a limited effect on dendrite growth (Brinton 1994). DHEA also induced spine synapse formation in hippocampal neurons (MacLusky, Hajszan, and Leranth 2004; Hajszan, MacLusky, and Leranth 2004).

DHEA is able to promote neurogenesis and neuronal survival in human neural stem cell cultures in both an epidermal growth factor– and leukemia inhibitory factor–dependent manner. More specifically, it was shown that DHEA, and not its derivatives, increased proliferation of long-term neural stem cells acting via N-methyl-D-aspartate (NMDA) and sigma-1 receptors, leading to increased neurogenesis (Suzuki et al. 2004).

In cultures of neural progenitor cells isolated from rat embryonic forebrains, DHEA upregulated and DHEAS downregulated the activity of Akt, a prosurvival serine/threonine protein kinase (Zhang et al. 2002; Mirescu, Peters, and Gould 2004). These findings suggest that DHEA and DHEAS constitute opposing forces in CNS.

Another study showed that DHEA (1 nM–10 μM) decreased the viability of primary neuronal cells but not of whole-brain cultured cells. In a human neuroblastoma cell line (type SK-N-SH), DHEA decreased [^{3}H]thymidine uptake and cell viability, whereas DHEAS did not significantly modify, or only slightly stimulated, cell viability and uptake of [^{3}H]thymidine. The combination of DHEA and DHEAS neutralized the toxic effect of DHEA, suggesting that DHEA and DHEAS have differing roles: DHEA possesses both a neurotoxic (expressed only in isolated neurons) and antiproliferative effect and DHEAS demonstrates only a slight neuroprotective effect (Gil-ad et al. 2001).

DHEA induced an increase of Tubulin-III and tau-positive cells in neuronal-component bone marrow mesenchymal stem cells (MSCs), indicating neuronal differentiation. Western blot analysis revealed that the Tubulin-III protein was more strongly induced by DHEA than Tau protein. The expression of neuronal-specific genes such as Isl-1, Tubulin-III, Pax6, and Nestin was also detected by reverse transcriptase–polymerase chain reaction and BrdU incorporation and was found to have increased significantly after DHEA induction, suggesting that DHEA can affect neuronal-competent MSCs inducing the expression of a comprehensive set of genes and proteins that define neuronal cells. DHEA was also able to induce the division of neuronal-competent MSCs, thereby increasing the number of cells with major neuronal characteristics (Shiri et al. 2009).

On the other hand, DHEA was able to induce neurogenesis in mouse P19 embryonal carcinoma cell neural progenitors (ECC-NPs) and in human embryonic stem cell–derived neural progenitors. In ECC-NPs, DHEA induced an increase of Nestin and Tuj1 markers. The expression of neuronal-specific genes such as Mash1, Pax6, Tuj1, and TH was also detected by RT-PCR. BrdU incorporation and estrogen receptors (EsR) were increased after DHEA induction. Moreover, apoptosis was significantly decreased after DHEA treatment (Azizi et al. 2010).

Endothelial Cells

Several works have shown that the effect of DHEA on the proliferation of endothelial cells (EC) depends on the concentration used. Some of these works are described below.

Incubation of DHEA with EC for 24 hours had a minimal effect on cell proliferation. However, at 48 hours of incubation, 25 and 50 ng of DHEA inhibited EC proliferation by 50%. EC proliferation was stimulated by 40% in the presence of 5 ng of DHEA and low-density lipoproteins (LDL) at 24-hour incubation, suggesting that DHEA protected EC against LDL-induced cytotoxic effects (Mohan and Benghuzzi 1997). In another study, it was found that DHEAS at physiological concentrations (10 μM) inhibited human vascular endothelial cell growth, which was reversible following removal of the steroid (Hinson and Khan 2004).

DHEA at a concentration of 1–100 nM increased the proliferation of bovine aortic endothelial cells (BAEC) and human umbilical vein endothelial cells (HUVEC) by mechanism(s) independent of either androgen receptors (AR) or EsR (Williams et al. 2004).

My work group has been studying the effects of DHEA on EC. Our results have shown that DHEA inhibits the proliferation of EC derived from human umbilical cords (HUVEC). This inhibitory effect is associated with an arrest in the G1 phase of the cell cycle, with decreased levels of phosphorylated retinoblastoma protein and increased expression of p53 and p21 mRNAs (Zapata et al. 2005). We also have shown that DHEA has an anti-inflammatory and antioxidant effect in EC since DHEA inhibited the tumor necrosis factor (TNF)-α–and oxidized LDL–induced expression of intracellular adhesion molecule (ICAM)-1, E-selectin, reactive oxygen species production, and U937 cells' adhesion to HUVEC, and interfered with nuclear factor (NF)-κB translocation and I-κB-α degradation. DHEA also decreased the expression of monocyte chemoattractant protein-1 and IL-8 mRNAs and delayed the kinetics of LDL oxidation *in vitro* (Gutiérrez et al. 2007; López-Marure et al. 2007).

Due to the antiproliferative effect of DHEA on EC, its effect on angiogenesis is significant. Varet, Vincent, and Akwa (2004) showed that at physiological concentrations found in human plasma following DHEA therapy (1–50 nM), DHEA had no action on angiogenesis *in vitro*. In contrast, higher concentrations of DHEA (10–100 μM), which can be found in tissues after local administration or storage, inhibited *in vitro* EC proliferation (blockage in G2/M), migration, capillary tube formation, and *in vivo* angiogenesis in the Matrigel plug assay. Liu et al. (2008) showed that DHEA increased endothelial migration. DHEA also increased the formation of primitive capillary tubes *in vitro* in solubilized basement membranes and enhanced angiogenesis *in vivo* in a chick embryo chorioallantoic membrane assay.

Other studies showed that DHEA significantly increased cell viability, reduced caspase-3 activity, protected both bovine and human vascular EC against serum deprivation–induced apoptosis, and enhanced antiapoptotic Bcl-2 protein expression. DHEA acted as a survival factor for these EC by triggering the Galphai-PI3K/Akt-Bcl-2 pathway to protect cells against apoptosis (Liu et al. 2007). DHEA also stimulated phosphorylation of FoxO1 via phosphoinositide-3-kinase– and protein kinase A–dependent pathways in EC (BAEC) that negatively regulates ET-1 promoter activity and secretion (Chen et al. 2008).

In aortic EC, TNF-α–induced upregulation of the expression of inflammatory genes IL-8 and ICAM-1 was attenuated by incubation with DHEAS. Treatment of EC with DHEAS dramatically inhibited the TNF-α–induced activation of NF-κB, an inflammatory transcription factor, and increased protein levels of the NF-κB inhibitor, I-κB-α. These results signified the ability of DHEAS to directly inhibit the inflammatory process (Altman et al. 2008). In these cells, DHEA at physiological concentrations (0.1–10 nm) increased proliferation by 30% at 24 hours.

These findings indicate that exposure to DHEA at concentrations found in human blood causes vascular endothelial proliferation by a plasma membrane–initiated activity that is Gi/o and extracellular signal–regulated kinase (ERK1/2) dependent (Liu et al. 2008).

DHEA inhibited interferon (IFN)-γ–induced expression of CD40 and CD40L in HUVEC in a dose-dependent manner. Moreover, DHEA inhibited IFN-γ–induced activation of ERK1/2, suggesting that DHEA can inhibit the expression of molecules involved in the inflammatory process in EC activated with IFN-γ (Li, Xia, and Wang 2009).

Adipocytes

DHEA (from 10^{-8} M) caused concentration-dependent proliferation inhibition of white (3T3-L1) and brown (PAZ6) preadipocyte cell lines. Cell cycle analysis demonstrated unaltered apoptosis but indicated blockades at G1/S and G2/M in 3T3-L1 and PAZ6, respectively. In human primary subcutaneous and omental preadipocytes, DHEA significantly inhibited proliferation (Rice et al. 2010).

Epithelial Cells

DHEA stimulated normal prostatic epithelial cell proliferation, and the AR was involved in DHEA-induced prostate-specific antigen (PSA) expression in normal prostatic epithelial cells. This stimulation effect induced by DHEA was mediated by the activation of NF-κB via the PI3K/AKT pathway (Sun et al. 2010).

Chromaffin Cells

DHEA by itself did not affect the proliferation of chromaffin cells from young bovine. In contrast, DHEA significantly decreased the cell proliferation induced by insulin-like growth factor II (IGF-II). Basic fibroblast growth factor (bFGF)–induced proliferation was not affected by DHEA, suggesting that DHEA inhibits proliferation induced by IGF-II by interacting with any signaling pathways that differ from the bFGF-mediated signal transduction, and that DHEA exerts a paracrine function in the control of growth of chromaffin cells (Sicard et al. 2006).

Tumor Cells

Breast

The effect of DHEA on the proliferation of breast cancer has been widely studied. In this section, *in vitro* studies showing the effects of DHEA on breast cancer cell lines will be described.

In ZR-75-1 cells, DHEA and DHEAS at a concentration of 10 μM stimulated the proliferation of ZR-75-1 cells (Poulin and Labrie 1986).

In MCF-7 cells, DHEA at a concentration of 500 nM stimulated MCF-7 cell growth in a steroid-free medium, whereas in a medium supplemented with 1 nM E2, DHEA partly antagonized the stimulatory effect of the estrogen (Boccuzzi et al. 1992). DHEA induced the proliferation, and it had a more potent action than DHEAS (Najid and Habrioux 1990). It has been proposed that the noncompetitive inhibition of G6PD contributes to DHEA's antitumor action. In human breast cancer cell lines MCF-7 and MDA-MB-231, DHEA inhibited both G6PD activity and cell growth only at concentrations above 10 μM; however, at lower concentrations, DHEA increased the proliferation but did not affect G6PD activity, suggesting that DHEA plays a role in the growth of breast cancer independently of the G6PD modulation (Di Monaco et al. 1997). DHEA stimulated the proliferation of MCF-7SH cells, an estrogen-independent MCF-7 variant (Maggiolini et al. 1999). The enhanced aromatization of DHEA in aromatase-transfected MCF-7 cells conferred biological advantages such as proliferative stimulation similar to that induced by estradiol (Maggiolini et al. 2001). In MCF-7 cells, the conversion of DHEA to estrogens, particularly estradiol, was required to exert a mitogenic response (Schmitt et al. 2001). DHEA inhibited MCF-7 cell proliferation in either the presence or the absence of E2, but this effect was abrogated by hydroxyflutamide, suggesting a specific role of the AR in inhibiting MCF-7 cell proliferation (Andò et al. 2002). DHEA had a proliferative effect at physiological concentrations and an antiproliferative effect at supraphysiologic concentrations, and when it was used at physiologic concentrations, it increased the proliferation of MCF-7 cells (Gayosso, Montano, and López-Marure 2006).

In T47D cells, a significant enhancement of growth and progesterone receptor expression was observed after treatment with DHEA and the induction of transforming growth factor-α mRNA in the breast cancer cell lines (T47D) expressing estrogen, androgen, and progesterone receptors (Liberato, Sonohara, and Brentani 1993). Other studies showed increased proliferation of T47D cells treated with DHEAS (Morris et al. 2001).

It has been shown that DHEA effects differ depending on the presence of EsR. For example, DHEAS induced growth of 43.4% in ER-positive/AR-positive cells but inhibited growth of ER-negative/AR-positive cells by 22%. Stimulation with DHEAS induced proliferation through the ER but inhibited cells via the AR (Toth-Fejel et al. 2004). Also, DHEA did not stimulate a proliferation of the EsR-negative BT-20 cell line (Najid and Habrioux 1990).

Prostate

It has been shown that DHEA does not have an effect on prostate growth and on transplanted human prostatic cancer in nude mice (Regelson, Loria, and Kalimi 1988; Van Weerden et al. 1992). DHEAS at concentrations of 10^{-6} to 10^{-9} M did not affect *in vitro* proliferation of AXC/SSh prostate cancer cells; however, 10^{-5} M DHEAS decreased proliferation of selected clonal lines of AXC/SSh prostate cancer cells (Huot and Shain 1988).

Most recently, it was shown that DHEA induced the proliferation of human LNCaP prostate cancer cells and increased gene and/or protein expression of PSA, IGF receptor, IGF-I, and IGF-binding protein 2 (Arnold et al. 2005, 2007).

Colon

DHEA at a concentration of 50 μM induced the incorporation of products of [^{3}H]mevalonate metabolism into several size classes of cellular proteins in HT-29 SF human colonic adenocarcinoma cells. Post-translational processing and membrane association of p21ras were both found to be inhibited by DHEA. Therefore, it is possible that the inhibition of isoprenylation of p21ras and other cellular proteins by DHEA may contribute to its anticancer effects (Schulz and Nyce 1991).

DHEA inhibited the growth of human colonic adenocarcinoma cells Caco-2 and HT-29, and this effect was not correlated with the inhibition of G6PD (Yoshida et al. 2003).

Hepatoma

DHEA induced growth arrest of hepatoma cells HepG2 (Yoshida et al. 2003; Ho et al. 2008). Growth inhibition was associated with increased G6PD activity and insensitivity to reversal by mevalonate. Thus, DHEA did not act via inhibition of G6PD and 3-hydroxy-3-methylglutamyl CoA. Growth stagnation was accompanied by reduced expression of nucleus-encoded mitochondrial genes, morphological and functional alterations of mitochondria, and depletion of intracellular adenosine triphosphate, suggesting that DHEA suppresses cell growth by altering mitochondrial gene expression, morphology, and functions (Ho et al. 2008).

Proliferation of HepG2 and HT-29 cells was significantly inhibited after 24 hours of incubation with 100 μM DHEA treatment; the inhibitory effect was stronger on HepG2 cells than on HT-29 cells. The effect of DHEAS on cell proliferation in both cell lines was weaker than that of the DHEA. DHEA treatment induced cell arrest in the G0/G1 phase in both cell lines. Apoptosis of HepG2 cells was significantly triggered (18.6% ± 2.2%) by treatment with 100 μM DHEA for 24 hours, but not by DHEAS. DHEA treatment for 24 hours markedly inhibited phosphorylation of Akt (Thr308 and Ser473) in HepG2 cells, suggesting that the induction of apoptosis through the inhibition of the Akt signaling pathway is one of the antiproliferative mechanisms of DHEA in certain tumors (Jiang et al. 2005, 2007).

Melanoma

DHEA dose dependently inhibited the growth of B16 mouse melanoma cells and enhanced melanin production, which implies the induction of differentiation. Although DHEA did not promote the translocation of protein kinase C from the cytosolic to the membrane fraction, the total protein kinase C activity was upregulated by treatment with DHEA (Kawai et al. 1995).

DHEA at a concentration of 1 nM enhanced the invasion of A375-SM melanoma cells through fibronectin; DHEA also induced proliferation when cells were cultured on plastic (Richardson et al. 1999).

Myeloma

DHEAS and DHEA suppressed IL-6 production from a bone marrow stromal cell line, KM-102, and in bone marrow mononuclear cells from patients with myeloma. DHEA and DHEAS inhibited the proliferation of several kinds of human myeloma cell lines *in vitro*: U-266, an IL-6–responsive but independent and IL-6–producing cell line; NOP-2, an IL-6–independent cell line; and IL-KM3, an IL-6–dependent cell line. DHEA upregulated the expression of peroxisome proliferator–activated receptor (PPAR), PPAR-β, but not PPAR-γ or PPAR-α, and the expression of the I-κB-α gene in myeloma cells and bone marrow stromal cells, which could explain the suppressive effect of DHEA on IL-6 production through the downregulation of NF-κB activity (Liu et al. 2005).

Pancreatic Adenocarcinoma

DHEAS treatment produced a dose-dependent inhibition of cell proliferation in human pancreatic adenocarcinoma cell lines MiaPaCa-2, Capan-1, Capan-2, CAV, and Panc-1 (Melvin et al. 1997). DHEAS inhibited cell proliferation by 22% at dose 5×10^{-7} µM, but increasing DHEAS did not further increase this inhibitory effect (Boros et al. 1997). The hepatic anticarcinogenic effect of DHEA in the liver has been associated with peroxisome and catalase induction and by effects inhibiting carcinogen binding or activation by DHEA (Prough, Wu, and Milewich 1990; Mayer, Weber, and Bannasch 1990).

Neuroblastoma

In a human neuroblastoma cell line, DHEA (1 nM–1 µM) decreased [^{3}H]thymidine uptake and cell viability after 24 hours. DHEAS did not significantly modify, or only slightly stimulated, cell viability and uptake of [^{3}H]thymidine. The combination of DHEA and DHEAS neutralized the toxic effect of DHEA in both primary neuronal culture and neuroblastoma cell line. Flow cytometric analysis of DNA fragmentation in neuroblastoma cells treated with 100 nM DHEA/DHEAS for 24 hours showed increased apoptotic events, suggesting a neuroprotective role for DHEA and that roles of DHEA and DHEAS are different: DHEA possesses a neurotoxic (expressed only in isolated neurons) and antiproliferative effect and DHEAS possesses only a slight neuroprotective effect (Gil-ad et al. 2001).

DHEA enhanced the differentiating effect of retinoic acid (RA) on a human neuroblastoma cell line (SK-N-BE) via a signaling that is not RA receptor mediated. DHEA also upregulated MMP-9 expression (Silvagno et al. 2002).

Cervical Cancer

DHEA inhibited cell proliferation of human uterine cervical cancers either not infected or infected with human papilloma virus (HPV) in a dose-dependent manner. The antiproliferative effect was not abrogated by inhibitors of AR and EsR or by an inhibitor of the conversion of testosterone to estradiol, and this effect was associated with increased necrotic cell death in HPV-negative cells and apoptosis in HPV-positive cells (Girón et al. 2009).

CONCLUSION

The majority of *in vitro* and *in vivo* studies indicated that DHEA has a protective effect against cancer, although the physiological functions of DHEA in preventing human tumorigenesis are still controversial. This protection has been associated mainly with its potent antiproliferative effect. Apparently, the effect induced by DHEA is most likely dose dependent. DHEA has shown a proliferative effect at physiological concentrations and an antiproliferative and chemoprotective effect at pharmacological concentrations. All studies performed on *in vitro* and *in vivo* models have suggested that DHEA at high concentrations is potentially beneficial in chemoprevention and as an antiproliferative, antitumor agent; therefore, DHEA could be a useful alternative in the treatment of cancer.

REFERENCES

Altman, R., D. D. Motton, R. S. Kota, and J. C. Rutledge. 2008. Inhibition of vascular inflammation by dehydroepiandrosterone sulfate in human aortic endothelial cells: Roles of PPARalpha and NF-kappaB. *Vascul Pharmacol* 48:76–84.

Andò, S., F. De Amicis, V. Rago et al. 2002. Breast cancer: From estrogen to androgen receptor. *Mol Cell Endocrinol* 193:121–8.

Arnold, J. T., H. Le, and K. K. McFann et al. 2005. Comparative effects of DHEA vs. testosterone, dihydrotestosterone, and estradiol on proliferation and gene expression in human LNCaP prostate cancer cells. *Am J Physiol Endocrinol Metab* 288:E573–84.

Arnold, J. T., X. Liu, J. D. Allen et al. 2007. Androgen receptor or estrogen receptor-beta blockade alters DHEA-, DHT-, and E(2)-induced proliferation and PSA production in human prostate cancer cells. *Prostate* 67:1152–62.

Azizi, H., N. Z. Mehrjardi, E. Shahbazi et al. 2010. Dehydroepiandrosterone stimulates neurogenesis in mouse embryonal carcinoma cell- and human embryonic stem cell-derived neural progenitors and induces dopaminergic neurons. *Stem Cells Dev* 19:809–18.

Babior, B. M. 1982. The enzymatic basis for 0-.2 production by human neutrophils. *Can J Physiol Pharmacol* 60:1353–58.

Baulieu, E. E., P. Robel, and M. Schumacher. 2001. Neurosteroids: Beginning of the story. *Int Rev Neurobiol.* 46:1–32.

Berger, L., M. El-Alfy, and F. Labrie. 2008. Effects of intravaginal dehydroepiandrosterone on vaginal histomorphology, sex steroid receptor expression and cell proliferation in the rat. *J Steroid Biochem Mol Biol* 109:67–80.

Boccuzzi, G., E. Brignardello, and M. di Monaco et al. 1992. Influence of dehydroepiandrosterone and 5-en-androstene-3 beta, 17 beta-diol on the growth of MCF-7 human breast cancer cells induced by 17 beta-estradiol. *Anticancer Res* 12:799–803.

Boros, L. G., J. Puigjaner, M. Cascante et al. 1997. Oxythiamine and dehydroepiandrosterone inhibit the non-oxidative synthesis of ribose and tumor cell proliferation. *Cancer Res* 57:4242–48.

Brinton, R. D. 1994. The neurosteroid 3 alpha-hydroxy-5 alphapregnan-20-one induces cytoarchitectural regression in cultured fetal hippocampal neurons. *J Neurosci* 14:2763–74.

Catalina, F., L. Milewich, V. Kumar et al. 2003. Dietary dehydroepiandrosterone inhibits bone marrow and leukemia cell transplants: Role of food restriction. *Exp Biol Med (Maywood)* 228:1303–20.

Charalampopoulos, I., V. I. Alexaki, C. Tsatsanis et al. 2006. Neurosteroids as endogenous inhibitors of neuronal cell apoptosis in aging. *Ann N Y Acad Sci* 1088:139–52.

Charalampopoulos, I., E. Remboutsika, A. N. Margioris et al. 2008. Neurosteroids as modulators of neurogenesis and neuronal survival. *Trends Endocrinol Metab* 19:300–7.

Chen, H., A. S. Lin, Y. Li et al. 2008. Dehydroepiandrosterone stimulates phosphorylation of FoxO1 in vascular endothelial cells via phosphatidylinositol 3-kinase- and protein kinase A-dependent signaling pathways to regulate ET-1 synthesis and secretion. *J Biol Chem* 283:29228–38.

Compagnone, N. A., and S. H. Mellon. 1998. Dehydroepiandrosterone: A potential signaling molecule for neocortical organization during development. *Proc Natl Acad Sci USA* 95:4678–83.

Corsini, E., M. Racchi, E. Sinforiani et al. 2005. Age-related decline in RACK-1 expression in human leukocytes is correlated to plasma levels of dehydroepiandrosterone. *J Leukoc Biol* 77:247–56.

Di Monaco, M., A. Pizzini, V. Gatto et al. 1997. Role of glucose-6-phosphate dehydrogenase inhibition in the antiproliferative effects of dehydroepiandrosterone on human breast cancer cells. *Br J Cancer* 75:589–92.

Ennas, M. G., S. Laconi, S. Dessí et al. 1987. Influence of dehydroepiandrosterone on G-6-PD activity and 3H-thymidine uptake of human lymphocytes in vitro. *Toxicol Pathol* 15:241–4.

Gayosso, V., L. F. Montano, and R. López-Marure. 2006. DHEA-induced antiproliferative effect in MCF-7 cells is androgen- and estrogen receptor-independent. *Cancer J* 12:160–5.

Gil-ad, I., B. Shtaif, R. Eshet et al. 2001. Effect of dehydroepiandrosterone and its sulfate metabolite on neuronal cell viability in culture. *Isr Med Assoc J* 3:639–43.

Girón, R. A., L. F. Montaño, M. L. Escobar et al. 2009. Dehydroepiandrosterone inhibits the proliferation and induces the death of HPV-positive and HPV-negative cervical cancer cells through an androgen- and estrogen-receptor independent mechanism. *FEBS J* 276:5598–609.

Gordon, G. B., L. M. Shantz, and P. Talalay. 1987. Modulation of growth, differentiation and carcinogenesis by dehydroepiandrosterone. *Adv Enzyme Regul* 26:355–82.

Green, J. E., M. A. Shibata, E. Shibata et al. 2001. 2-difluoromethylornithine and dehydroepiandrosterone inhibit mammary tumor progression but not mammary or prostate tumor initiation in C3(1)/SV40 T/t-antigen transgenic mice. *Cancer Res* 61:7449–55.

Gross, L., and Y. Dreyfuss. 1984. Reduction in the incidence of radiation-induced tumors in rats after restriction of food intake. *Proc Natl Acad Sci USA* 81:7596–8.

Gross, L., and Y. Dreyfuss. 1986. Inhibition of the development of radiation-induced leukemia in mice by reduction of food intake. *Proc Natl Acad Sci USA* 83:7928–31.

Gutiérrez, G., C. Mendoza, E. Zapata et al. 2007. Dehydroepiandrosterone inhibits the TNF-alpha-induced inflammatory response in human umbilical vein endothelial cells. *Atherosclerosis* 190:90–9.

Hajszan, T., N. J. MacLusky, and C. Leranth. 2004. Dehydroepiandrosterone increases hippocampal spine synapse density in ovariectomized female rats. *Endocrinology* 145:1042–5.

Hinson, J. P., and M. Khan. 2004. Dehydroepiandrosterone sulphate (DHEAS) inhibits growth of human vascular endothelial cells. *Endocr Res* 30:667–71.

Ho, H. Y., M. L. Cheng, H. Y. Chiu et al. 2008. Dehydroepiandrosterone induces growth arrest of hepatoma cells via alteration of mitochondrial gene expression and function. *Int J Oncol* 33:969–77.

Huot, R. I., and S. A. Shain. 1988. Differential metabolism of dehydroepiandrosterone sulfate and estrogen conjugates by normal or malignant AXC/SSh rat prostate cells and effects of these steroid conjugates on cancer cell proliferation in vitro. *J Steroid Biochem* 29:617–21.

Hursting, S. D., S. N. Perkins, D. C. Haines et al. 1995. Chemoprevention of spontaneous tumorigenesis in p53-knockout mice. *Cancer Res* 55:3949–53.

Hursting, S. D., S. N. Perkins, and J. M. Phang. 1994. Calorie restriction delays spontaneous tumorigenesis in p53-knockout transgenic mice. *Proc Natl Acad Sci* 91:7036–40.

Jiang, Y., T. Miyazaki, A. Honda et al. 2005. Apoptosis and inhibition of the phosphatidylinositol 3-kinase/Akt signaling pathway in the anti-proliferative actions of dehydroepiandrosterone. *J Gastroenterol* 40:490–7.

Jiang, Y. F., P. W. Zhao, Y. Tan et al. 2007. Molecular mechanisms of DHEA and DHEAs on apoptosis and cell cycle arrest via Akt pathway in hepatoma cell lines. *Zhonghua Gan Zang Bing Za Zhi* 15:441–4.

Kawai, S., N. Yahata, S. Nishida et al. 1995. Dehydroepiandrosterone inhibits B16 mouse melanoma cell growth by induction of differentiation. *Anticancer Res* 15:427–31.

Klurfeld, D. M., M. M. Weber, and D. Kritchevsky. 1987. Inhibition of chemically induced mammary and colon tumor promotion by caloric restriction in rats fed increased dietary fat. *Cancer Res* 47:2759–62.

Labrie, F., A. Bélanger, V. Luu-The et al. 1998. DHEA and the intracrine formation of androgens and estrogens in peripheral target tissues: Its role during aging. *Steroids* 63:322–8.

Li, Y., Z. Xia, and M. Wang. 2009. Dehydroepiandrosterone inhibits CD40/CD40L expression on human umbilical vein endothelial cells induced by interferon gamma. *Int Immunopharmacol* 9:168–72.

Li, S., X. Yan, A. Bélanger et al. 1993. Prevention by dehydroepiandrosterone of the development of mammary carcinoma induced by 7,12-dimethylbenz(a)anthracene (DMBA) in the rat. *Breast Cancer Res Treat* 29:203–17.

Liberato, M. H., S. Sonohara, and M. M. Brentani. 1993. Effects of androgens on proliferation and progesterone receptor levels in T47D human breast cancer cells. *Tumour Biol* 14:38–45.

Liu, D., M. Iruthayanathan, L. L. Homan et al. 2008. Dehydroepiandrosterone stimulates endothelial proliferation and angiogenesis through extracellular signal-regulated kinase 1/2-mediated mechanisms. *Endocrinology* 149:889–98.

Liu, S., H. Ishikawa, F. J. Li et al. 2005. Dehydroepiandrosterone can inhibit the proliferation of myeloma cells and the interleukin-6 production of bone marrow mononuclear cells from patients with myeloma. *Cancer Res* 65:2269–76.

Liu, D., H. Si, K. A. Reynolds et al. 2007. Dehydroepiandrosterone protects vascular endothelial cells against apoptosis through a Galphai protein-dependent activation of phosphatidylinositol 3-kinase/Akt and regulation of antiapoptotic Bcl-2 expression. *Endocrinology* 148:3068–76.

López-Marure, R., C. Huesca-Gómez, M. J. Ibarra-Sánchez et al. 2007. Dehydroepiandrosterone delays LDL oxidation in vitro and attenuates several oxLDL-induced inflammatory responses in endothelial cells. *Inflamm Allergy Drug Targets* 6:174–82.

Lubet, R. A., D. M. McCormick, G. M. Gordon et al. 1995. Effects of dehydroepiandrosterone on MNU-induced breast cancer in Sprague-Dawley rats. *Ann N Y Acad Sci* 774:340–1.

MacLusky, N. J., T. Hajszan, and C. Leranth. 2004. Effects of dehydroepiandrosterone and flutamide on hippocampal CA1 spine synapse density in male and female rats: Implications for the role of androgens in maintenance of hippocampal structure. *Endocrinology* 145:4154–61.

Maggiolini, M., A. Carpino, D. Bonofiglio, V. Pezzi, V. Rago, S. Marsico, D. Picard, and S. Andò. 2001. The direct proliferative stimulus of dehydroepiandrosterone on MCF7 breast cancer cells is potentiated by overexpression of aromatase. *Mol Cell Endocrinol* 184:163–71.

Maggiolini, M., O. Donzé, E. Jeannin et al. 1999. Adrenal androgens stimulate the proliferation of breast cancer cells as direct activators of estrogen receptor alpha. *Cancer Res* 59:4864–9.

Matsuzaki, Y., and A. Honda. 2006. Dehydroepiandrosterone and its derivatives: Potentially novel anti-proliferative and chemopreventive agents. *Curr Pharm Des* 12:3411–21.

Mayer, D., K. Forstner, and K. Kopplow. 2003. Induction and modulation of hepatic preneoplasia and neoplasia in the rat by dehydroepiandrosterone. *Toxicol Pathol* 31:103–12.

Mayer, D., E. Weber, and P. Bannasch. 1990. Modulation of liver carcinogenesis by dehydroepiandrosterone. In *The Biologic Role of Dehydroepiandrosterone (DHEA)*, ed. M. Kalimi and W. Regelson, 361–85. New York: Walter de Gruyter.

Melvin, W. S., L. G. Boros, P. Muscarella et al. 1997. Dehydroepiandrosterone-sulfate inhibits pancreatic carcinoma cell proliferation in vitro and in vivo. *Surgery* 121:392–7.

Metzger, C., D. Mayer, H. Hoffmann et al. 1995. Sequential appearance and ultrastructure of amphophilic cell foci, adenomas, and carcinomas in the liver of male and female rats treated with dehydroepiandrosterone. *Toxicol Pathol* 23:591–605.

Mirescu, C., J. D. Peters, and E. Gould. 2004. Early life experience alters response of adult neurogenesis to stress. *Nat Neurosci* 7:841–6.

Mohan, P. F., and H. Benghuzzi. 1997. Effect of dehydroepiandrosterone on endothelial cell proliferation. *Biomed Sci Instrum* 33:550–5.

Morris, K. T., S. Toth-Fejel, J. Schmidt et al. 2001. High dehydroepiandrosterone-sulfate predicts breast cancer progression during new aromatase inhibitor therapy and stimulates breast cancer cell growth in tissue culture: A renewed role for adrenalectomy. *Surgery* 130:947–53.

Muscarella, P., L. G. Boros, W. E. Fisher et al. 1998. Oral dehydroepiandrosterone inhibits the growth of human pancreatic cancer in nude mice. *J Surg Res* 79:154–7.

Najid, A., and G. Habrioux. 1990. Biological effects of adrenal androgens on MCF-7 and BT-20 human breast cancer cells. *Oncology* 47:269–2.

Orentreich, N., J. L. Brind, R. L. Rizer et al. 1984. Age changes and sex differences in serum dehydroepiandrosterone sulfate concentrations throughout adulthood. *J Clin Endocrinol Metab* 59:551–5.

Parshad, R. K., and B. K. Batth. 1999. Cystogenesis of antral follicles induced by dehydroepiandrosterone (DHEA) stimulates mast cell proliferation and maturation in the house rat (Rattus rattus) ovary. *Indian J Exp Biol* 37:933–5.

Pashko, L. L., G. C. Hard, R. J. Rovito et al. 1985. Inhibition of 7,12-dimethylbenz(a)anthncene-induced skin papillomas and carcinomas by 3p-methylandrost-5-en-17-one in mice. *Cancer Res* 45:164–6.

Pashko, L. L., R. J. Rovito, J. R. Williams et al. 1984. Dehydroepiandrosterone (DHEA) and 3p-methylandrost-5-en-17-one: Inhibitors of 7,12-dimethylbenz(a)anthracene (DMBA)-initiated and 12-O-tetradecanoylphorbol-1 3-acetate (TPA)-promoted skin papilloma formation in mice. *Carcinogenesis* 5:463–6.

Pashko, L. L., and A. G. Schwartz. 1983. Effect of food restriction, dehydroepiandrosterone, or obesity on the binding of 3H-7,12-dimethylbenz(a)anthracene to mouse skin DNA. *J Gerontol* 38:8–12.

Pelissier, M. A., C. Trap, M. I. Malewiak et al. 2004. Antioxidant effects of dehydroepiandrosterone and 7alpha-hydroxy-dehydroepiandrosterone in the rat colon, intestine and liver. *Steroids* 69:137–44.

Poulin, R., and F. Labrie. 1986. Stimulation of cell proliferation and estrogenic response by adrenal C19-delta 5-steroids in the ZR-75-1 human breast cancer cell line. *Cancer Res* 46:4933–7.

Prough, R. A., X. D. Lei, G. H. Xiao et al. 1995. Regulation of cytochromes P450 by DHEA and its anticarcinogenic action. *Ann N Y Acad Sci* 774:187–99.

Prough, A., H. G. Wu, and C. Milewich. 1990. Effect of DHEA on rodent liver microsomal mitochondrial and peroxisomal proteins. In *The Biologic Role of Dehydroepiandrosterone (DHEA)*, ed. M. Kalimi and W. Regelson, 253–79. New York: Walter de Gruyter.

Raïs, B., B. Comin, J. Puigjaner et al. 1999. Oxythiamine and dehydroepiandrosterone induce a G1 phase cycle arrest in Ehrlich's tumor cells through inhibition of the pentose cycle. *FEBS Lett* 456:113–8.

Rao, M. S., V. Subbarao, A. V. Yeldandi et al. 1992. Hepatocarcinogenicity of dehydroepiandrosterone in the rat. *Cancer Res* 52:2977–9.

Regelson, W., R. Loria, and M. Kalimi. 1988. Hormonal intervention: "Buffer hormones" or "state dependency." The role of dehydroepiandrosterone (DHEA), thyroid hormone, estrogen and hypophysectomy in aging. *Ann N Y Acad Sci* 521:260–73.

Rice, S. P., L. Zhang, F. Grennan-Jones et al. 2010. Dehydroepiandrosterone (DHEA) treatment in vitro inhibits adipogenesis in human omental but not subcutaneous adipose tissue. *Mol Cell Endocrinol* 320:51–7.

Richardson, B., A. Price, M. Wagner et al. 1999. Investigation of female survival benefit in metastatic melanoma. *Br J Cancer* 80:2025–33.

Risdon, G., J. Cope, and M. Bennett. 1990. Mechanism of chemoprevention by dietary dehydroepiandrosterone. *Am J Pathol* 136:759–69.

Schmitt, M., K. Klinga, B. Schnarr et al. 2001. Dehydroepiandrosterone stimulates proliferation and gene expression in MCF-7 cells after conversion to estradiol. *Mol Cell Endocrinol* 173:1–13.

Schulz, S., and J. W. Nyce. 1991. Inhibition of protein isoprenylation and p21ras membrane association by dehydroepiandrosterone in human colonic adenocarcinoma cells in vitro. *Cancer Res* 51:6563–7.

Schwartz, A., G. Hard, L. Pashko et al. 1981. Dehydroepiandrosterone: An anti-obesity and anti-carcinogenic agent. *Nutr Cancer* 3:46–53.

Schwartz, A. G., and L. L. Pashko. 1995. Cancer prevention with dehydroepiandrosterone and non-androgenic structural analogs. *J Cell Biochem* 22:210–17.

Shen, S., D. M. Cooley, L. T. Glickman et al. 2001. Reduction in DNA damage in brain and peripheral blood lymphocytes of elderly dogs after treatment with dehydroepiandrosterone (DHEA). *Mutat Res* 481:153–62.

Shilkaitis, A., A. Green, V. Punj et al. 2005. Dehydroepiandrosterone inhibits the progression phase of mammary carcinogenesis by inducing cellular senescence via a p16-dependent but p53-independent mechanism. *Breast Cancer Res* 7:R1132–40.

Shimokawa, I., B. P. Yu, and E. J. Masoro. 1991. Influence of diet on fatal neoplastic disease in male Fischer 344 rats. *J Gerontol* 46:228–32.

Shiri, E. H., N. Z. Mehrjardi, M. Tavallaei et al. 2009. Neurogenic and mitotic effects of dehydroepiandrosterone on neuronal-competent marrow mesenchymal stem cells. *Int J Dev Biol* 53:579–84.

Sicard, F., A. W. Krug, C. G. Ziegler et al. 2006. Role of DHEA and growth factors in chromaffin cell proliferation. *Ann N Y Acad Sci* 1073:312–6.

Silvagno, F., V. Guarnieri, A. Capizzi et al. 2002. Synergistic effect of retinoic acid and dehydroepiandrosterone on differentiation of human neuroblastoma cells. *FEBS Lett* 532:153–8.

Solano, M. E., V. Sander, M. R. Wald et al. 2008. Dehydroepiandrosterone and metformin regulate proliferation of murine T lymphocytes. *Clin Exp Immunol* 153:289–96.

Sun, H. Z., T. W. Yang, W. J. Zang et al. 2010. Dehydroepiandrosterone-induced proliferation of prostatic epithelial cell is mediated by NFKB via PI3K/AKT signaling pathway. *J Endocrinol* 204:311–8.

Suzuki, M., L. S. Wright, P. Marwah et al. 2004. Mitotic and neurogenic effects of dehydroepiandrosterone (DHEA) on human neural stem cell cultures derived from the fetal cortex. *Proc Natl Acad Sci USA* 101:3202–7.

Tannenbaum, A., and H. Silverstone. 1953. Nutrition in relation to cancer. *Adv Cancer Res* 1:451–501.

Toth-Fejel, S., J. Cheek, K. Calhoun et al. 2004. Estrogen and androgen receptors as comediators of breast cancer cell proliferation: Providing a new therapeutic tool. *Arch Surg* 139:50–4.

Van Weerden, W. M., A. Van Kreuningenn, N. M. Elissen et al. 1992. Effects of adrenal androgens on the transplantable human prostate tumor. *Endocrinology* 131:2909–13.

Varet, J., L. Vincent, and Y. Akwa. 2004. Dose-dependent effect of dehydroepiandrosterone, but not of its sulphate ester, on angiogenesis. *Eur J Pharmacol* 502:21–30.

Wang, T. T., S. D. Hursting, S. N. Perkins et al. 1997. Effects of dehydroepiandrosterone and calorie restriction on the Bcl-2/Bax-mediated apoptotic pathway in p53-deficient mice. *Cancer Lett* 116:61–9.

Whitcomb, J. M., and A. G. Schwartz. 1985. Dehydroepiandrosterone and 16a-Br-epiandrosterone inhibit 12-O-tetradecanoylphorbol-1 3-acetate stimulation of superoxide radical production by human polymorphonuclear leukocytes. *Carcinogenesis* 6:333–5.

Williams, M. R., T. Dawood, S. Ling et al. 2004. Dehydroepiandrosterone increases endothelial cell proliferation in vitro and improves endothelial function in vivo by mechanisms independent of androgen and estrogen receptors. *J Clin Endocrinol Metab* 89:4708–15.

Yoshida, S., A. Honda, Y. Matsuzaki et al. 2003. Anti-proliferative action of endogenous dehydroepiandrosterone metabolites on human cancer cell lines. *Steroids* 68:73–83.

Zapata, E., J. L. Ventura, K. De la Cruz et al. 2005. Dehydroepiandrosterone inhibits the proliferation of human umbilical vein endothelial cells by enhancing the expression of p53 and p21, restricting the phosphorylation of retinoblastoma protein, and is androgen- and estrogen-receptor independent. *FEBS J* 272:1343–53.

Zhang, L., B. Li, W. Ma et al. 2002. Dehydroepiandrosterone (DHEA) and its sulfated derivative (DHEAS) regulate apoptosis during neurogenesis by triggering the Akt signaling pathway in opposing ways. *Brain Res Mol Brain Res* 98:58–66.

7 Dehydroepiandrosterone and Energy Metabolism

Surendra S. Katyare and Hiren R. Modi

CONTENTS

INTRODUCTION

GENERAL

Dehydroepiandrosterone (DHEA) and its sulfate ester (DHEAS) are synthesized in the highest quantities by the young adult human adrenals and are the most abundant circulating steroids in the human organism (Milgrom 1990). The plasma levels of DHEA show a characteristic age-dependent pattern. The levels are low in infancy, start increasing in adolescence, and reach maximum values around the age of 20–30 years (Hinson and Raven 1999; Buvat 2003; Celec and Starka 2003). The levels decline thereafter, and in elderly individuals, the levels fall to about 10% of the young adult values (Hinson and Raven 1999; Parker 1999). Based on this characteristic pattern, DHEA is considered to be "a youth hormone" or "a fountain of youth" (Hinson and Raven 1999; Buvat 2003; Celec and Starka 2003; Kim and Morley 2005). However, the concept of a "hormonal fountain of youth" seems to be debatable (Kim and Morley 2005). Any hope of a fountain of youth to stop people from getting older, however, may perhaps be a long way off since we have just begun to understand the complex causes of aging: genetic, physical, and hormonal. One wonders whether the age-related decline in hormone systems is physiological, perhaps conveying a benefit, or if the changes are pathological, causing harm.

The present review will focus on the effects of DHEA, DHEAS, and derivatives of DHEA (Robinzon et al. 2003) on energy metabolism. This will include the survey of effects of DHEA under *in vitro* and *in vivo* conditions. Newer data, which show beneficial effects of DHEA in the developmental and functional maturation of liver and brain mitochondria and in improvement of mitochondrial energy transduction functions using a rat model, are presented. It is anticipated that improved energy metabolism will certainly have a bearing on health, diseases, aging, and improvement of neurotransmitter and memory functions (Racchi, Balduzzi, and Corsini 2003).

BIOSYNTHESIS

The Mitochondrial Connection

Biosynthesis of DHEA from cholesterol (CHL) is only a two-step process involving two cytochrome P450s. In the first step, CHL is converted to pregnenolone. This step is mediated by CHL side chain cleavage enzyme CYP11A1. Interestingly, this cleavage enzyme system is present in the mitochondria. The conversion of pregnenolone to DHEA requires both 17-alpha-hydroxylase and 17,20-lyase activities of CYP17, which is present in the endoplasmic reticulum. Additionally, the steroidogenic acute regulatory protein regulates the flux of CHL into the biosynthetic pathway and represents the mechanism of acute regulation. It is important that in addition to possessing CYP11A1 and CYP17, a steroidogenic cell should not contain other enzymes that will interfere with the flux of pregnenolone to DHEA. The fetal adrenal cortex and the zona reticularis that exclusively synthesize DHEA and DHEAS possess and represent these characteristics (Auchus 2004; Miller 2002). Quantitative regulation of steroidogenesis occurs at the first step, the conversion of CHL to pregnenolone. Chronic quantitative regulation is principally at the level of transcription of the CYP11A1 gene encoding P450scc, which is the enzymatically rate-limiting step (Miller 2009).

In the liver, DHEA is first converted to 7-alpha-OH-DHEA, which is subsequently oxidized to 7-oxo-DHEA in both the liver and kidney. In the liver, interconversion of 7-oxo-DHEA and 7-OH-DHEA isomers is largely catalyzed by 11-beta-hydroxysteroid dehydrogenase (HSD1), whereas in the kidney the unidirectional oxidation of 7-alpha-OH-DHEA to 7-oxo-DHEA is catalyzed by nicotinamide adenine dinucleotide–dependent 11-beta-HSD2 and nicotinamide adenine dinucleotide phosphate ($NADP^+$)–dependent 11-beta-HSD3. These distinct species-specific routes of metabolism of DHEA and the interconversion of its metabolites make it difficult to extrapolate the results of animal studies to humans (Robinzon et al. 2003). Interestingly, however, 7-alpha-OH-DHEA and 7-oxo-DHEA have been shown to be more effective in the process of thermogenesis (Bobyleva et al. 1993; Lardy et al. 1998).

DHEA Biosynthesis in the Brain

Significant quantities of DHEA are also present in the human and rat brain, although the levels are lower in the rat (Corpechot et al. 1981; Vallee et al. 2000; Racchi, Balduzzi, and Corsini 2003;

Steckelbroeck et al. 2002; Ren and Hou 2005). Interestingly, it has been reported that the content of DHEA in the brain declines in an age-dependent manner (Weill-Engerer et al. 2002; Kazihnitkova et al. 2004). The brain derives its DHEA partly from the adrenals. However, the brain itself is capable of synthesizing DHEA, DHEAS, and pregnenolone (Racchi, Balduzzi, and Corsini 2003). Hence, DHEA, DHEAS, and pregnenolone are considered to be neurosteroids (Racchi, Balduzzi, and Corsini 2003). Although DHEA and DHEAS are synthesized in the highest quantities, there are no known receptors for DHEA (Milgrom 1990).

FUNCTIONAL ASPECTS

Alteration in Energy Metabolism

The role of DHEA is generally unclear except that it is a precursor to sex steroids (Milgrom 1990). Interestingly, however, it has been shown that when the experimental animals are fed diets supplemented with DHEA, this results in proliferation of mitochondria in the liver and stimulation of mitochondrial functional parameters (Bellei et al. 1992; Song et al. 1989). Besides, importance of DHEA in energy metabolism has also been indicated (Chiu et al. 1997). It has been suggested that alteration in energy metabolism in postmenopausal women might be related to the reduction of DHEAS. DHEAS may modulate L-carnitine level and carnitine acetyltransferase activity in estrogen-deficient rats. The potential role of DHEAS in the regulation of fatty acid oxidation in postmenopausal women may be worthy of investigation (Chiu et al. 1997).

Neurotransmitter Role

Since DHEA, DHEAS, and pregnenolone are known to be synthesized by the brain, they are considered neurosteroids (Racchi, Balduzzi, and Corsini 2003). The concentration of DHEA in the brain is in the range of a nanogram per gram of tissue, with the highest concentration seen in the anterior pituitary (Racchi, Balduzzi, and Corsini 2003; Ren and Hou 2005; Vallee et al. 2000; Weill-Engerer et al. 2002), and the concentration of DHEAS is about 10 times higher (Racchi, Balduzzi, and Corsini 2003). An age-dependent decline in the content of DHEAS in the brain has been reported (Racchi, Balduzzi, and Corsini 2003). Presence of DHEA and DHEAS in human brain and decreased contents of DHEA and DHEAS in brain in Alzheimer's disease and dementia have been reported (Weill-Engerer et al. 2002). Studies indicate that the exogenous supplementation with DHEA helps improve memory and behavior in elderly population (Buvat 2003).

DHEA in Health and Disease

While DHEA is freely available in the United States and needs no prescription (Hinson and Raven 1999), the claims for its beneficial effects in health, disease, and improvement of memory and behavior in elderly persons have not been clearly established and seem to be equivocal (Hinson and Raven 1999; Oliver and Clemens 1999). DHEA and DHEAS have been implicated as one of the powerful modulators of memory processes and sleep states in young and aged subjects with memory impairment. As these processes depend on the integrity of cholinergic systems, a specific effect of DHEA and DHEAS on these systems may account for their effects on sleep and memory. The specific modulation of basal forebrain and brain stem cholinergic systems by DHEA and DHEAS may account for the effects of these compounds on sleep and memory processes (George et al. 2006).

DHEA and Aging

DHEA supplementation to counteract its gradual decrease over age is beneficial. Positive effects on the cardiovascular system, body composition, bone mineral density, skin, central nervous system, and immune system have been reported. Improvement of sexual function by DHEA has been demonstrated (Saad et al. 2005).

DHEA AND MITOCHONDRIAL FUNCTION

In Vitro Effects

DHEA inhibited mitochondrial respiratory rates with substrates of tricarboxylic acid cycle, with the exception of succinate in mitochondria isolated from adrenals, heart, kidneys, brain, and brown adipose tissue. The aspartate–malate shuttle was inhibited, but the α-glycerophosphate shuttle was totally insensitive to DHEA. High-amplitude swelling of mitochondria in hypotonic media was also inhibited by DHEA and a few other steroids (Mohan and Cleary 1989b). DHEA and related nonsulfated steroids decreased state 3 mitochondrial respiration, respiratory control, and oxidative phosphorylation capacity with malate + pyruvate as substrate in a dose-dependent manner (McIntosh, Pan, and Berdanier 1993). Morin et al. (2002) examined the effects of DHEA, DHEAS, α-estradiol, and β-estradiol on functional attributes of purified rat brain mitochondria submitted to various stresses including anoxia–reoxygenation, uncoupling, and apoptosis. DHEA and α-estradiol partly preserved the mitochondrial functions altered by anoxia–reoxygenation in a concentration-dependent manner. Also, DHEA and α-estradiol (1 μM) limited the release of cytochrome *c* from mitochondria subjected to the anoxia–reoxygenation by about 50% and inhibited the oxidation of NADH-cytochrome *c* system and uncoupled respiration by about 35% and 9%, respectively. The authors concluded that the mechanism involved was independent of the classical genomic effect of steroids or the antioxidant properties but implicated a direct action on the mitochondrial membranes (Morin et al. 2002). DHEA inhibits complex I of the mitochondrial respiratory chain and is neurotoxic *in vitro* and *in vivo* at high concentrations (Safiulina et al. 2006). Using different mitochondrial substrates DHEA suppressed the mitochondrial respiration by inhibiting complex I of the mitochondrial electron transport chain in permeabilized neurons; the half maximal inhibitory concentration was 13 μM. In mice fed with a pellet diet containing 0.6% DHEA for 3 months, there was a significant neuronal loss in the cerebral cortex and hippocampus, a slight decrease in the dopamine/dihydroxyphenylacetic acid ratio, and motor impairment. The authors concluded that high concentrations of DHEA inhibited complex I of the mitochondrial respiratory chain and were neurotoxic *in vitro* and *in vivo* (Safiulina et al. 2006). The effects of DHEA, DHEAS, α-estradiol, and β-estradiol on functions of purified rat brain mitochondria subjected to anoxia–reoxygenation, uncoupling, and apoptosis were investigated. DHEA and α-estradiol (1 μM) lowered the respiratory control ratio (RCR). However, in mitochondria subjected to anoxia–reoxygenation, these concentrations offered partial protection by inhibiting state 3 respiration (Morin et al. 2002).

Using the whole-cell patch clamp technique, it was demonstrated that in the rat medial prefrontal cortex slices, DHEAS inhibited the amplitude of persistent sodium currents at 0.1 μM, reaching the maximum at 1 μM; the effect declined thereafter. The effect was canceled by the Gi protein inhibitor and the protein kinase C inhibitor but not by the protein kinase A inhibitor. It was suggested that DHEAS may presumably protect neurons under ischemia (Cheng et al. 2008). In saponin-permeabilized neurons, DHEA with several other neurosteroids protected the mitochondria against intracellular Ca^{2+} overload by inhibiting Ca^{2+} influx into the mitochondrial matrix (Kaasik et al. 2003).

In Vivo Effects

Treatment of experimental animals with DHEA resulted in hypertrophy of the hepatocytes due to increased proliferation of mitochondria and peroxisomes (Bellei et al. 1992). DHEA administered per os increased the levels of 26 proteins and decreased the levels of 7 proteins in mouse liver. A protein of *M*r approximately 160 K, which was identified as carbamoyl phosphate synthetase-I (CPS-I), decreased markedly by DHEA action. This enzyme, which comprises approximately 15%–20% of mitochondrial matrix protein, is involved in the entry and is a rate-limiting step of the urea cycle (Marrero et al. 1991). Administration of DHEA to rats resulted in lowered body weight, higher liver weights and DNA, RNA, and/or protein content, but lowered lipid and glycogen levels. Activities of a number of liver enzymes involved in lipid and carbohydrate metabolism were altered by this treatment. In addition, net mitochondrial respiration was elevated. Administration of DHEA resulted in increase in the net mitochondrial respiration. Some of these findings may explain DHEA's anti-obesity effect (Cleary 1990). Mitochondrial state 3 and 4 rates expressed per gram liver or per liver with either glutamate–malate or succinate as substrate were higher in DHEA-treated rats than in clofibric acid–treated or control rats (Mohan and Cleary 1989a). Tunez, Munoz, and Montilla (2005) reported that DHEA reduced oxidative stress in synaptosomes isolated from the brain of rats treated with 3-nitropropionic acid (3-NPA) and thus protected mitochondria and maintained synaptic integrity against damage induced by 3-NPA. In lean and obese female Zucker rats, DHEA increased total hepatic mitochondrial protein twofold. The activity of carnitine acyltransferases (mitochondrial carnitine palmitoyltransferase [CPT] and peroxisomal carnitine octanoyl transferase [COT]) also increased. Based on these observations, it was suggested that the hepatic CPT and COT in female Zucker rats are regulated primarily at the transcriptional level by DHEA and clofibrate (Brady, Ramsay, and Brady 1991). It is likely that DHEA suppresses cell growth by altering mitochondrial gene expression, morphology, and functions (Ho et al. 2008).

Feeding 0.4% DHEA decreased hepatic cytosolic selenium–dependent glutathione peroxidase and increased hepatic mitochondrial manganese superoxide dismutase (Schauer et al. 1990). When administered to male F-344 rats for 8 weeks, DHEA produced a significant increase in hepatic ubiquinone-9. Administration of DHEA increased the contents of ubiquinone in liver and, to some extent, in kidney and muscle. No change in corresponding content for heart or brain was observed (Aberg et al. 1996). These findings indicate the potential of high-dose DHEA to modulate ubiquinone in rat hepatic tissue (Khan 2005).

DHEA treatment also stimulated the mitochondrial electron transport functions (Song et al. 1989; Swierczynski et al. 2001). In rats fed a diet containing 0.2% DHEA for short (7 days) and long (100 days) periods, mitochondrial structure changed from expanded to condensed configuration. Consistent with this, in the isolated mitochondria, the rate of coupled respiration increased compared with that in the controls. However, after long-term treatment, the structure of the hepatocytes reverted to normal, but degenerative changes were observed in the mitochondria (Bellei et al. 1992). In rats given 0.6% DHEA in the diet for 4, 20, 32, 70, and 84 weeks, hepatocellular neoplasia was evident. Proliferation of mitochondria was the most prominent feature at all stages (Metzger et al. 1995). Higher doses of DHEA resulted in progressive reduction in state 3 respiration with malate + pyruvate as the substrate system (Chance and McIntosh 1995). In Zucker rats, DHEA treatment for 7 days and 24, 12, and 3 hours had no effect on mitochondrial β oxidation, but state 3 respiration per gram of liver with glutamate–malate was increased except after 7 days. Succinate-supported state 3 respiration per gram of liver was also elevated under all experimental conditions. The authors concluded that mitochondrial respiration is the earliest factor affected by DHEA and may be associated with protein synthesis (Mohan and Cleary 1991). Mitochondria from rats treated for 7 days had lower levels of cardiolipin and phosphatidylethanolamine and an increase in phosphatidylcholine. Changes in fatty acid composition of these phospholipids occurred after 7 days and 24 hours of DHEA treatment (Mohan and Cleary 1991). Dose–response studies (Tables 7.1 and 7.2) from our laboratory showed that while the total phospholipids (TPL) and CHL contents of liver mitochondria were generally unchanged by DHEA treatment, those of the acidic phospholipid phosphatidylinositol (PI) and phosphatidylserine (PS) increased. The contents of other phospholipid components were not changed. By contrast, in the brain, contents of mitochondrial TPL and CHL increased. With respect to the phospholipid contents, the major effect was an increase in the contents of lysophospholipids (Lyso), sphingomyelin (SPM), phosphatidylcholine (PC), phosphatidylinositol (PI), and phosphatidylserine (PS). The contents of phosphatidylethanolamine and diphosphatidylglycerol (DPG) were not affected. At a higher dose (2.0 mg), the observed effects declined (Katyare, Modi,

TABLE 7.1
Effects of Treatment with Dehydroepiandrosterone on Total Phospholipids (TPL) and Cholesterol (CHL) Content of Rat Liver and Brain Mitochondria

	Mitochondria	
Parameter	**Liver**	**Brain**
TPL	—	↑
CHL	—	↑
TPL/CHL (mole:mole)	—	—

↑, increase; —, no effect.

TABLE 7.2
Effects of Treatment with Dehydroepiandrosterone on Contents of Individual Phospholipids of Rat Liver and Brain Mitochondria

	Mitochondria	
Phospholipid Class	**Liver**	**Brain**
Lysophospholipid	—	↑
Sphingomyelin	~	↑
Phosphatidylcholine	—	↑
Phosphatidylinositol	↑	↑
Phosphatidylserine	↑	↑
Phosphatidylethanolamine	↓	—
Diphosphatidylglycerol	~	—

↑, increase; ↓, decrease; ~, marginal change.

and Patel 2006). The observed differences may possibly relate to differences in the dose regimen. In diet-induced obese (DIO) rats fed 0.6% DHEA for 2 weeks, liver mitochondrial state 3 respiration rates per gram and per liver and peroxisomal β-oxidation increased compared with the control rats. In DIO rats, DHEA treatment was shown to have hypolipidemic and hypoinsulinemic effects (Mohan et al. 1990). DHEA (6 g/kg) for 6 weeks increased β-oxidation of fatty acids in peroxisomes (Leighton, Tagliaferro, and Newsholme 1987). Studies from our laboratory have shown that treatment of the old rats with DHEA resulted in significant stimulation of respiratory activity in the liver and the brain mitochondria (Patel and Katyare 2006b; Patel, Modi, and Katyare 2007). Also, treatment with DHEA enhanced the maturation and development of mitochondrial functions in the brain and liver of the growing animals (Patel and Katyare 2006a,c).

Effects on Enzymes Systems

Feeding DHEA to rats induced enhanced formation of liver enzymes, for example, mitochondrial sn-glycerol-3-phosphate dehydrogenase (GPDH) and cytosolic malic enzyme (Lardy et al. 1998). DHEA administered per os resulted in a decrease in CPS-I, (Marrero et al. 1991). In control and diabetic rats, DHEA feeding enhanced the activity of the mitochondrial flavin adenosine dinucleotide–linked GPDH and cytosolic $NADP^+$-linked malate dehydrogenase (MDH) in liver, but not in the parotid gland or pancreatic islets (Giroix et al. 1997). In hepatocytes from thyroidectomized rats (triiodothyronine or DHEA treated), DHEA increased glycealdehyde-3-phosphate dehydrogenase activity. These data are consistent with Ca^{2+} being mobilized by gluconeogenic hormones and glycealdehyde-3-phosphate dehydrogenase being activated by Ca^{2+} so as to permit the transfer of reducing equivalents from the cytosol to the mitochondria (Kneer and Lardy 2000). 3β-hydroxyandrost-5-ene-7,17-dione (7-oxo-DHEA) was found to be more effective than DHEA as an inducer of liver mitochondrial GPDH and cytosolic malic enzyme in rats (Kneer and Lardy 2000; Bobyleva et al. 1993). A similar effect is also seen in Goto-Kakizaki rats (Ladriere et al. 1997). Interestingly, DHEA is a known noncompetitive inhibitor of glucose-6-phosphate dehydrogenase (Shepherd and Cleary 1984).

Effects on FoF_1 ATPase

While FoF_1 ATPase is a target for estradiol, which inhibits the enzyme, DHEA has practically no effect (Mohan and Cleary 1988; Zheng and Ramirez 1999).

Effects on Overall Energy Metabolism

In rats, treatment with DHEA inhibited tumor growth. Also, the metabolism of lipids and glucose changed significantly, switching from an anabolic to a catabolic state following DHEA treatment. DHEA administration to rats also induced an overall increase in energy expenditure. The authors concluded that this energy waste may be related to the inhibitory action of DHEA on tumor growth (Mayer and Forstner 2004). Hampl, Starka, and Jansky (2006) demonstrated that 7-oxo-DHEA is even more effective than DHEA and may also participate in modulating thermogenic effects.

DHEA AND MITOCHONDRIAL ENERGY TRANSDUCTION FUNCTIONS IN YOUNG ADULT, DEVELOPING, AND AGING ORGANISMS

It is clear that DHEA does affect the energy-dependent function and has beneficial effects in human health and diseases. However, the survey of literature indicates that the results are diverse and anomalous. The observed diversity and anomaly may be attributed mainly to the differences in the dose and time regimens used by different researchers for treatment with DHEA. In light of this, it becomes of primary interest to carry out systematic dose–response studies to evaluate the effect(s) of DHEA treatment on mitochondrial energy transduction function. Once the dose–response baseline is established, the studies can be extended further to ascertain if DHEA, the purported "fountain of youth," does play a role in modulating mitochondrial energy metabolism during development and aging.

For the primary purpose of determining the dose–response, we injected young adult rats (8–10 week old) with 0.1, 0.2, 1.0, or 2.0 mg DHEA/kg body weight subcutaneously for 7 consecutive days. This dose regimen was selected based on earlier studies where rats were fed diet containing 0.6% DHEA (Bellei et al. 1992; Su and Lardy 1991). At the end of the treatment period, the animals were killed by decapitation and liver and brain mitochondria were isolated by standard published procedures (Patel and Katyare 2006a,b,c; Patel and Katyare 2007; Patel, Modi, and Katyare 2007). Energy transduction functions were determined in terms of oxidative phosphorylation, state 3 and state 4 respiration rates, dehydrogenases and ATPase activities, and cytochrome contents. The experiments were then extended to evaluate the effects in developing and old rats. The results of these studies have conclusively shown that DHEA in moderate doses does have beneficial effects, whereas higher doses can have somewhat adverse effects. These results are summarized here.

Dose–Response Studies in Young Adult Rats

Liver and Brain Mitochondria

In the liver mitochondria (Patel and Katyare 2007), the state 3 respiration rates were stimulated in a dose-dependent and substrate-specific manner, with doses of up to 1.0 mg/kg body weight showing beneficial effects; a higher dose of 2.0 mg/kg body weight had either inhibitory or no effect depending on the substrate used. The state 4 respiration rates were generally unaffected except at the higher doses of DHEA. However, under no condition, there was uncoupling and adenosine diphosphate/oxygen (ADP/O) ratios were unchanged. Consequently, the energy potential as determined in terms of ADP phosphorylation rates increased depending on the substrate used. In keeping with this trend, the FoF_1 ATPase activity also increased significantly. These results are schematically depicted in Tables 7.3 and 7.4. DHEA treatment also brought about significant increase in the contents of cytochromes aa_3, b, and c+c_1 (Table 7.5), which is consistent with the observed increase in the respiration rates. Increased contents of cytochromes aa_3, b, and c+c_1 would, thus, facilitate the transport of electrons to enhance the rates of respiration. Increased generation of electrons is ensured by increased glutame dehydrogenase (GDH) and succinate 2,6 dichlorophenol indophenol (DCIP) reductase (SDR) activities (Table 7.6). A similar trend was also noted for the brain mitochondria except that the content of cytochrome c+c_1 was unaltered (Tables 7.3 through 7.6).

TABLE 7.3
Effects of Treatment with Dehydroepiandrosterone on State 3 and State 4 Respiration Rates in Rat Liver and Brain Mitochondria

	Liver Mitochondria		Brain Mitochondria	
Animals	**State 3**	**State 4**	**State 3**	**State 4**
Young adult	↑	↑	↑	↑
Developing	↑	↑	↑	↑
Old	↑	↑	↑	↑

↑, increase; —, no effect.

TABLE 7.4
Effects of Treatment with Dehydroepiandrosterone on Aenosine Diphosphate/Oxygen (ADP/O Ratio), Respiratory Control Ratio (RCR), ADP Phosphorylation Rates, and FoF_1 ATPase Activity in Rat Liver and Brain Mitochondria

Animals	Mitochondria	ADP/O Ratio	RCR	ADP Phosphorylation Rate	FoF_1 ATPase
Young adult	Liver	—	—	↑	↑
	Brain	—	—	↑	↑
Developing	Liver	—	—	↑	↑
	Brain	—	—	↑	↑
Old	Liver	—	—	↑	↑
	Brain	—	—	↑	↑

↑, increase; ↓, decrease; —, no effect.

TABLE 7.5
Effect of Treatment with Dehydroepiandrosterone on Cytochrome Content in Rat Liver and Brain Mitochondria

Animals	Mitochondria	Cytochrome aa_3	Cytochrome b	Cytochrome $c{+}c_1$
Young Adult	Liver	↑	↑	↑
	Brain	↑	↑	—
Developing	Liver	↑	↑	↑
	Brain	↑	↑	—
Old	Liver	↑	↑	↑
	Brain	↑	↑	↑↓

↑, increase; ↓, decrease; ↑↓, marginal change.

TABLE 7.6
Effect of Treatment with Dehydroepiandrosterone on Dehydrogenase Activities in Rat Liver and Brain Mitochondria

Animals	Mitochondria	Glutame Dehydrogenase	Succinate DCIP Reductase	Mitochondrial Malate Dehydrogenase	Cytosolic Malate Dehydrogenase
Young Adult	Liver	↑	↑	—	—
	Brain	↑	↑	—	↑
Developing	Liver	↑	↑	—	↑
	Brain	↑	↑	—	↑
Old	Liver	↑	↑	↑	↑
	Brain	↑↑	↑	↑	↑

↑, increase; ↓, decrease; ↑↑, huge increase.

Dose-Dependent Effects in Developing Rats

Liver Mitochondria

Treatment with DHEA (0.2 or 1.0 mg/kg body weight for 7 days) resulted in a progressive dose-dependent increase in the liver weights of the developing animals, with no change in the body weight. In the young adult rats, treatment with 1.0 mg DHEA showed increase only in the body weight. Treatment with DHEA stimulated state 3 and state 4 respiration rates in developing and young adult animals in dose-dependent manner with all the substrates, with the extent of stimulation being age dependent. However, higher dose (1.0 mg) of DHEA resulted in decline in state 3 respiration rate with glutamate and succinate as substrates in young adult rats. Stimulation of state 3 respiration rates was accompanied by an increase in contents of cytochromes aa_3, b, and $c{+}c_1$ and stimulation of FoF_1 ATPase and dehydrogenase activities in a dose- and age-dependent manner (Patel and Katyare 2006c). These results are schematically shown in Tables 7.3 through 7.6.

Brain Mitochondria

Under similar conditions, treatment with DHEA (0.2 or 1.0 mg/kg body weight for 7 days) resulted in increase in the brain weight in the 5-week group. State 3 and state 4 respiration rates with all the substrates increased following DHEA treatment, the effect being more pronounced in the developing rats.

State 4 respiration rates were stimulated to variable extents. Contents of cytochromes aa_3 and b increased following DHEA treatment, and the effect was more pronounced in the developing rats; content of cytochrome c+c_1 changed only marginally. In the developing rats, DHEA treatment also resulted in significant increase in the FoF_1 ATPase activity and the levels of dehydrogenases. The results point out that the treatment with exogenous DHEA accelerated the process of maturation of the brain mitochondria (Patel and Katyare 2006a). The results also emphasize the crucial role of DHEA in brain development during postnatal life. These results are schematically shown in Tables 7.3 through 7.6.

Dose-Dependent Effects in Old Rats

Liver Mitochondria

In the old animals (18–24 months old), there was a progressive increase in the liver weight with increasing dose of DHEA; the body weight was not affected. In the old rats, the state 3 respiration rates were, in general, lower than those in the young rats. The state 3 and state 4 respiration rates increased following DHEA treatment in a dose-dependent manner, bringing them close to values for young adult animals or beyond; the effect was more pronounced at a 1.0 mg dose. Contents of cytochromes aa_3, b, and c+c_1 increased significantly in a dose-dependent manner. DHEA treatment also stimulated the mitochondrial FoF_1 ATPase activity in the old animals. The dehydrogenase activities were considerably low in the old rats compared with young adult animals. Treatment with DHEA stimulated dehydrogenase activities in old rats in a dose-dependent manner, bringing them close to values for the young animals or beyond (Patel and Katyare 2007). These results are schematically depicted in Tables 7.3 through 7.6.

Brain Mitochondria

The state 3 respiration rates in the brain mitochondria in old animals were considerably lower than those in young adults. Treatment with DHEA stimulated state 3 and state 4 respiration rates in both groups of the animals in a dose-dependent manner. Following DHEA treatment, the state 3 respiration rates in the old animals became comparable to or increased more than those in untreated young adults. Compared with the old animals, the extent of stimulation was of greater magnitude in the young adults. However, at higher dose (1.0 mg), the effect declined. Content of cytochrome aa_3 in the brain mitochondria in old rats was significantly low, content of cytochrome b was unchanged, and content of cytochrome c+c_1 increased. Treatment with DHEA increased the content of cytochromes aa_3 and b in old and young adult animals. DHEA in higher dose (1.0 mg) adversely affected the content of cytochrome c+c_1. DHEA treatment only marginally stimulated the FoF_1 ATPase activity in the old rats. The dehydrogenase activities in the old rats were somewhat lower compared with the young adult rats. DHEA treatment stimulated mitochondrial dehydrogenase activities in both the groups. These results suggest that treatment with DHEA can and does improve oxidative energy metabolism parameters even in the brain mitochondria in old rats and thus can have beneficial effects (Patel and Katyare 2006b). These results are schematically shown in Tables 7.3 through 7.6. These results are important in view of the fact that 20% of the total oxygen consumption by the body is utilized in the brain (Erecinska and Silver 2001), which is consistent with the fact that the electrophysiological functions of the brain are known to be energy dependent (Astrup, Sorensen, and Sorensen 1981; Attwell and Laughlin 2001).

SIMILARITIES AND DIFFERENCES IN DHEA EFFECTS ON LIVER AND BRAIN MITOCHONDRIA

In both liver and brain mitochondria, treatment with DHEA resulted in the stimulation of state 3 and state 4 respiration rates in a dose-, age-, and substrate-dependent manner without affecting the RCR or the ADP/O ratio. DHEA treatment also stimulated the FoF_1 ATPase activity and energy potential as viewed in terms of enhanced ADP phosphorylation rates. The contents of cytochromes aa_3 and b increased in both liver and brain mitochondria in animals belonging to all age groups. The same was true for content of cytochrome c+c_1 in the liver mitochondria. However, in case of the brain mitochondria, DHEA treatment had no effect on the content of cytochrome c+c_1 except for in the old animals where lower dose increased and higher dose decreased the content. The GDH and SDR activities increased in mitochondria in both the tissues, with the effect being more pronounced for GDH in the brain mitochondria from old rats. DHEA treatment was effective in stimulating the mitochondrial MDH activity only in the old animals but stimulated cytosolic MDH activity in the liver and the brain mitochondria in all age groups except in the young adult animals. The results, thus, imply that the *in vivo* treatment, in general, had beneficial stimulatory effects rather than the inhibitory or uncoupling effects noted in several earlier studies (cited in the section "*In Vitro* Effects"). This is not surprising because the effects observed under *in vitro* conditions may simply be the nongenomic effects ensuing due to intercalation of steroids into membranes. Nongenomic effects of steroids have been highlighted earlier (Daum 1985; Duval, Desfosses, and Emiliozzi 1980). In fact, it is documented that even CHL intercalates with mitochondrial membranes and inhibits FoF_1 ATPase activity under *in vitro* conditions. Hence, the effects we observe under *in vivo* conditions must be considered as genomic effects (Daum, 1985).

COMMENTS

The observed *in vivo* effects of DHEA treatment deserve some comments. It is widely recognized that DHEA and DHEAS have no known receptors (Nawata et al. 2002). Therefore, one wonders how the genomic effects of DHEA may be mediated. Interesting to note in this context is the fact that the first step of DHEA biosynthesis is localized in the mitochondria (Miller 2002). Second, several tissues, including the brain, synthesize DHEA and the liver converts DHEA to 7-keto and 7-hydroxy derivatives (Robinzon et al. 2003; Robinzon and Prough 2005; Steckelbroeck et al. 2002; Weill-Engerer et al. 2002; Weill-Engerer et al. 2003). It has also been suggested that the 7-keto and 7-hydroxy derivatives are more effective than DHEA in inducing MDH activity (Song et al. 1989). One would, therefore, tend to think that DHEA and/or its 7-keto and 7-hydroxy derivatives could play a role in regulating metabolism, especially mitochondrial energy metabolism.

It is now well established that GDH, MDH—mitochondrial and cytosolic—and succinate dehydrogenase are nuclear gene products. Likewise, it is also recognized that crucial peptides of cytochrome oxidase (cytochrome aa_3), cytochrome b, and FoF_1 ATPase are mitochondrial gene products (Poyton and Mc Ewen 1996). Viewed in this context, our results suggest that DHEA and/or its aforementioned derivatives do activate specific nuclear and mitochondrial genes in a dose-dependent and tissue-specific manner. Such an assumption is also supported by the fact that treatment with DHEA almost uniformly stimulated the GDH, SDR, and cytosolic MDH activities, except in the liver mitochondria in young adult rats. By contrast, DHEA stimulated mitochondrial MDH activity in the liver and brain only in old animals. It is of interest to note that the mitochondrial MDH activity in these tissues of old animals was lower than that in tissues of young adults (Patel and Katyare 2006a; Patel and Katyare 2006b; Patel and Katyare 2006c; Patel and Katyare 2007). Since the levels of DHEA decrease with age, it is possible that in the old animals, exogenously supplied DHEA was effective in restoring the mitochondrial MDH activity. The uniform increase in the contents of cytochromes aa_3 and b and FoF_1 ATPase activity in mitochondria in both the tissues in all age groups would suggest that DHEA and/or its derivatives also activate specific mitochondrial genes. Cytochrome c+c_1 is a specific case where the content increased in liver mitochondria in all age groups. However, in the case of the brain mitochondria, DHEA treatment was ineffective in the young adult and developing animals. In the old animals, the content of cytochrome c+c_1 was relatively high compared with young adults, which may be compensatory mechanism against lowered respiration rates; a high dose of DHEA lowered the content to normalize it to adult value, simultaneously restoring the respiration rates. Thus, in this specific case, one notes a negative nuclear gene regulation. Interesting to note in this connection is the reported presence of dexamethasone binding site in the COX II region of the mitochondrial genome (Demonacos et al. 1995; Tsiriyotis, Spandidos, and Sekeris 1997; Simon et al. 1998). It would be interesting to know if a similar DHEA binding site(s) exists on the mitochondrial genome. Increase in the GDH and SDR activities following DHA treatment is also suggestive of possibility that DHEA binding site(s) may exist on the nuclear genome. These possibilities, however, need to be verified by more direct studies.

CONCLUSIONS

The data presented have addressed three specific questions relating to the role of DHEA in energy metabolism using a rat model. First, even in the young adult group, DHEA treatment had beneficial stimulatory effects of energy metabolism of liver and brain mitochondria. Second, a similar pattern was noted even in the developing rats where DHEA treatment accelerated the process of maturation of mitochondrial functions. Third, and most interestingly, the DHEA treatment of the old animals restored mitochondrial energy transduction functions, bringing them close to or beyond the young adult values. These observations, therefore, give credence to the idea that DHEA is indeed the proverbial "fountain of youth" (Hinson and Raven 1999; Buvat 2003; Celec and Starka 2003; Kim and Morley 2005). The results also point out that the effects are dose-dependent and higher doses can have somewhat adverse effects. The most important point is that DHEA does act by activating/silencing the nuclear and mitochondrial genes.

DHEA is freely available in the United States without any prescription and is also used as a food additive (Hinson and Raven 1999; Baulieu et al. 2000). Equivocal beneficial effects in human health, disease, and memory improvement have been indicated (Sorwell and Urbanski 2010). Viewed in the context of the data presented, it may be suggested that careful and judicious human trials can help in delineating the appropriate dose regimens that can show definitive beneficial effects in humans. Also, a careful evaluation of binding sites for DHEA and/or its derivatives by competitive binding assays can shed more light on the mechanism of its (their) action. Additionally, this can pave way for synthesizing more potent derivatives for management of human health, diseases, improvement of memory functions, and aging.

REFERENCES

Aberg, F., Y. Zhang, H. Teclebrhan, E. L. Appelkvist, and G. Dallner. 1996. Increases in tissue levels of ubiquinone in association with peroxisome proliferation. *Chem Biol Interact* 99:205–18.

Astrup, J., P. M. Sorensen, and H. R. Sorensen. 1981. Oxygen and glucose consumption related to Na^+/K^+ transport in canine brain. *Stroke* 12:726–30.

Attwell, D., and S. B. Laughlin. 2001. An energy budget for signaling in the grey matter of the brain. *J Cereb Blood Flow Metab* 21:1133–45.

Auchus, R. J. 2004. Overview of dehydroepiandrosterone biosynthesis. *Semin Reprod Med* 22:281–8.

Baulieu, E. E., G. Thomas, S. Legrain, N. Lahlou, M. Roger, B. Debuire, V. Faucounau et al. 2000. Dehydroepiandrosterone (DHEA), DHEA sulfate, and aging: Contribution of the DHEAge Study to a sociobiomedical issue. *Proc Natl Acad Sci U S A* 97:4279–84.

Bellei, M., D. Battelli, C. Fornieri, G. Mori, U. Muscatello, H. Lardy, and V. Bobyleva. 1992. Changes in liver structure and function after short-term and long-term treatment of rats with dehydroepiandrosterone. *J Nutr* 122:967–76.

Bobyleva, V., N. Kneer, M. Bellei, D. Battelli, and H. A. Lardy. 1993. Concerning the mechanism of increased thermogenesis in rats treated with dehydroepiandrosterone. *J Bioenerg Biomembr* 25:313–21.

Brady, L. J., R. R. Ramsay, and P. S. Brady. 1991. Regulation of carnitine acyltransferase synthesis in lean and obese Zucker rats by dehydroepiandrosterone and clofibrate. *J Nutr* 121:525–31.

Buvat, J. 2003. Androgen therapy with dehydroepiandrosterone. *World J Urol* 21:346–55.

Celec, P., and L. Starka. 2003. Dehydroepiandrosterone - is the fountain of youth drying out? *Physiol Res* 52:397–407.

Chance, D. S., and M. K. McIntosh. 1995. Hypolipidemic agents alter hepatic mitochondrial respiration in vitro. *Comp Biochem Physiol C Pharmacol Toxicol Endocrinol* 111:317–23.

Cheng, Z. X., D. M. Lan, P. Y. Wu, Y. H. Zhu, Y. Dong, L. Ma, and P. Zheng. 2008. Neurosteroid dehydroepiandrosterone sulphate inhibits persistent sodium currents in rat medial prefrontal cortex via activation of sigma-1 receptors. *Exp Neurol* 210:128–36.

Chiu, K. M., M. J. Schmidt, A. L. Shug, N. Binkley, and S. Gravenstein. 1997. Effect of dehydroepiandrosterone sulfate on carnitine acetyl transferase activity and L-carnitine levels in oophorectomized rats. *Biochim Biophys Acta* 1344:201–9.

Cleary, M. P. 1990. Effect of dehydroepiandrosterone treatment on liver metabolism in rats. *Int J Biochem* 22:205–10.

Corpechot, C., P. Robel, M. Axelson, J. Sjovall, and E. E. Baulieu. 1981. Characterization and measurement of dehydroepiandrosterone sulfate in rat brain. *Proc Natl Acad Sci U S A* 78:4704–7.

Daum, G. 1985. Lipids of mitochondria. *Biochim Biophys Acta* 822:1–42.

Demonacos, C., R. Djordjevic-Mareovic, N. Taswdaroglon, and C. E. Sekeris. 1995. The mitochondrion is a primary site of action of glucocorticoids: The interaction of glucocorticoids receptor with mitochondrial DNA sequence showing partial similarity to the nuclear glucocorticoid responsive elements. *Steroids* 55:43–55.

Duval, D., B. Desfosses, and R. Emiliozzi. 1980. Preparation of dehydroepiandrosterone, testosterone and progesterone antigens through 7-carboxymethyl derivatives: Characteristics of the antisera to testosterone and progesterone. *Steroids* 35:235–49.

Erecinska, M., and I. A. Silver. 2001. Tissue oxygen tension and brain sensitivity to hypoxia. *Respir Physiol* 128:263–76.

George, O., M. Vallee, M. Le Moal, and W. Mayo. 2006. Neurosteroids and cholinergic systems: Implications for sleep and cognitive processes and potential role of age-related changes. *Psychopharmacology (Berl)* 186:402–13.

Giroix, M. H., F. Malaisse-Lagae, B. Portha, A. Sener, and W. J. Malaisse. 1997. Effects of dehydroepiandrosterone in rats injected with streptozotocin during the neonatal period. *Biochem Mol Med* 61:72–81.

Hampl, R., L. Starka, and L. Jansky. 2006. Steroids and thermogenesis. *Physiol Res* 55:123–31.

Hinson, J. P., and P. W. Raven. 1999. DHEA deficiency syndrome: A new term for old age? *J Endocrinol* 163:1–5.

Ho, H. Y., M. L. Cheng, H. Y. Chiu, S. F. Weng, and D. T. Chiu. 2008. Dehydroepiandrosterone induces growth arrest of hepatoma cells via alteration of mitochondrial gene expression and function. *Int J Oncol* 33:969–77.

Kaasik, A., D. Safiulina, A. Kalda, and A. Zharkovsky. 2003. Dehydroepiandrosterone with other neurosteroids preserve neuronal mitochondria from calcium overload. *J Steroid Biochem Mol Biol* 87:97–103.

Katyare, S. S., H. R. Modi, and M. A. Patel. 2006. Dehydroepiandrosterone treatment alters lipid/phospholipid profiles of rat vrain and liver mitochondria. *Curr Neurovasc Res* 3:273–9.

Kazihnitkova, H., H. Tejkalova, O. Benesova, M. Bicikova, M. Hill, and R. Hampl. 2004. Simultaneous determination of dehydroepiandrosterone, its 7-hydroxylated metabolites, and their sulfates in rat brain tissues. *Steroids* 69:667–74.

Khan, S. A. 2005. Effects of dehydroepiandrosterone (DHEA) on ubiquinone and catalase in the livers of male F-344 rats. *Biol Pharm Bull* 28:1301–3.

Kim, M. J., and J. E. Morley. 2005. The hormonal fountains of youth: Myth or reality? *J Endocrinol Invest* 28:5–14.

Kneer, N., and H. Lardy. 2000. Thyroid hormone and dehydroepiandrosterone permit gluconeogenic hormone responses in hepatocytes. *Arch Biochem Biophys* 375:145–53.

Ladriere, L., A. Laghmich, F. Malaisse-Lagae, and W. J. Malaisse. 1997. Effect of dehydroepiandrosterone in hereditarily diabetic rats. *Cell Biochem Funct* 15:287–92.

Lardy, H., N. Kneer, Y. Wei, B. Partridge, and P. Marwah. 1998. Ergosteroids. II: Biologically active metabolites and synthetic derivatives of dehydroepiandrosterone. *Steroids* 63:158–65.

Leighton, B., A. R. Tagliaferro, and E. A. Newsholme. 1987. The effect of dehydroepiandrosterone acetate on liver peroxisomal enzyme activities of male and female rats. *J Nutr* 117:1287–90.

Marrero, M., R. A. Prough, R. S. Putnam, M. Bennett, and L. Milewich. 1991. Inhibition of carbamoyl phosphate synthetase-I by dietary dehydroepiandrosterone. *J Steroid Biochem Mol Biol* 38:599–609.

Mayer, D., and K. Forstner. 2004. Impact of dehydroepiandrosterone on hepatocarcinogenesis in the rat (Review). *Int J Oncol* 25:1021–30.

McIntosh, M. K., J. S. Pan, and C. D. Berdanier. 1993. In vitro studies on the effects of dehydroepiandrosterone and corticosterone on hepatic steroid receptor binding and mitochondrial respiration. *Comp Biochem Physiol Comp Physiol* 104:147–53.

Metzger, C., D. Mayer, H. Hoffmann, T. Bocker, G. Hobe, A. Benner, and P. Bannasch. 1995. Sequential appearance and ultrastructure of amphophilic cell foci, adenomas, and carcinomas in the liver of male and female rats treated with dehydroepiandrosterone. *Toxicol Pathol* 23:591–605.

Milgrom, E. 1990. Steroid hormone. In *Hormones From Molecues to Disease*, ed. E. E. Baulieu and P. A. Kelly, 387–438. New York and London: Hermann Publishers and Hall.

Miller, W. L. 2002. Androgen biosynthesis from cholesterol to DHEA. *Mol Cell Endocrinol* 198:7–14.

Miller, W. L. 2009. Androgen synthesis in adrenarche. *Rev Endocr Metab Disord* 10:3–17.

Mohan, P. F., and M. P. Cleary. 1988. Effect of short-term DHEA administration on liver metabolism of lean and obese rats. *Am J Physiol* 255:E1–8.

Mohan, P. F., and M. P. Cleary. 1989a. Comparison of dehydroepiandrosterone and clofibric acid treatments in obese Zucker rats. *J Nutr* 119:496–501.

Mohan, P. F., and M. P. Cleary. 1989b. Dehydroepiandrosterone and related steroids inhibit mitochondrial respiration in vitro. *Int J Biochem* 21:1103–7.

Mohan, P. F., and M. P. Cleary. 1991. Short-term effects of dehydroepiandrosterone treatment in rats on mitochondrial respiration. *J Nutr* 121:240–50.

Mohan, P. F., J. S. Ihnen, B. E. Levin, and M. P. Cleary. 1990. Effects of dehydroepiandrosterone treatment in rats with diet-induced obesity. *J Nutr* 120:1103–14.

Morin, C., R. Zini, N. Simon, and J. P. Tillement. 2002. Dehydroepiandrosterone and alpha-estradiol limit the functional alterations of rat brain mitochondria submitted to different experimental stresses. *Neuroscience* 115:415–24.

Nawata, H., Y. Toshihoko, G. Kimimobu, and O. K. A. Taijiro. 2002. Mechanism of action of anti-aging DHEA-S and the replacement of DHEA-S. *Mech Age Dev* 123:1101–6.

Oliver, T. W., and K. Clemens. 1999. Action of dehydroepiandrosterone and its sulfate in the central nervous system, effects on cognition and emotion in animals and humans. *Brain Res Rev* 30:264–88.

Parker Jr., C. R. 1999. Dehydroepiandrosterone and dehydroepiandrosterone sulfate production in the human adrenal during development and aging. *Steroids* 64:640–7.

Patel, M. A., and S. S. Katyare. 2006a. Dehydroepiandrosterone (DHEA) treatment stimulates oxidative energy metabolism in the cerebral mitochondria from developing rats. *Int J Dev Neurosci* 24:327–34.

Patel, M. A., and S. S. Katyare. 2006b. Treatment with dehydroepiandrosterone (DHEA) stimulates oxidative energy metabolism in the cerebral mitochondria. A comparative study of effects in old and young adult rats. *Neurosci Lett* 402:131–6.

Patel, M. A., and S. S. Katyare. 2006c. Treatment with dehydroepiandrosterone (DHEA) stimulates oxidative energy metabolism in the liver mitochondria from developing rats. *Mol Cell Biochem* 293:193–201.

Patel, M. A., and S. S. Katyare. 2007. Effect of dehydroepiandrosterone (DHEA) treatment on oxidative energy metabolism in rat liver and brain mitochondria dose–response study. *Clin Biochem* 40:57–65.

Patel, M. A., H. R. Modi, and S. S. Katyare. 2007. Stimulation of oxidative energy metabolism in liver mitochondria from old and young rats by treatment with dehydroepiandrosterone (DHEA). A comparative study. *Age (Dordr)* 29:41–9.

Poyton, R. O., and J. E. Mc Ewen. 1996. Crosstalk between nuclear and mitochondrial genomes. *Annu Rev Biochem* 65:563–607.

Racchi, M., C. Balduzzi, and E. Corsini. 2003. Dehydroepiandrosterone (DHEA) and the aging brain: Flipping a coin in the "fountain of youth." *CNS Drug Rev* 9:21–40.

Ren, J. M., and Y. N. Hou. 2005. Determination of unconjugated neurosteroids in rat brain regions by liquid chromatography-negative atmospheric pressure ionization mass spectrometry. *Yao Xue Xue Bao* 40:262–6.

Robinzon, B., K. K. Michael, S. L. Ripp, S. J. Winters, and R. A. Prough. 2003. Glucocorticoids inhibit interconversion of 7-hydroxy and 7-oxo metabolites of dehydroepiandrosterone: A role for 11beta-hydroxysteroid dehydrogenases? *Arch Biochem Biophys* 412:251–8.

Robinzon, B., and R. A. Prough. 2005. Interactions between dehydroepiandrosterone and glucocorticoid metabolism in pig kidney: Nuclear and microsomal 11beta- hydroxysteroid dehydrogenases. *Arch Biochem Biophys* 442:33–40.

Saad, F., C. E. Hoesl, M. Oettel, J. D. Fauteck, and A. Rommler. 2005. Dehydroepiandrosterone treatment in the aging male—what should the urologist know? *Eur Urol* 48:724–33; discussion 733.

Safiulina, D., N. Peet, E. Seppet, A. Zharkovsky, and A. Kaasik. 2006. Dehydroepiandrosterone inhibits complex I of the mitochondrial respiratory chain and is neurotoxic in vitro and in vivo at high concentrations. *Toxicol Sci* 93:348–56.

Schauer, J. E., A. Schelin, P. Hanson, and F. W. Stratman. 1990. Dehydroepiandrosterone and a beta-agonist, energy transducers, alter antioxidant enzyme systems: Influence of chronic training and acute exercise in rats. *Arch Biochem Biophys* 283:503–11.

Shepherd, A., and M. P. Cleary. 1984. Metabolic alterations after dehydroepiandrosterone treatment in Zucker rats. *Am J Physiol* 246:E123–8.

Simon, N., P. Jolliet, C. Moric, R. Zini, S. Urien, and J. P. Tillerment. 1998. Glucocorticoids decrease cytochrome c oxidase activity of isolated rat kidney mitochondria. *FEBS Lett* 435:25–8.

Song, M. K., D. Grieco, J. E. Rall, and V. M. Nikodem. 1989. Thyroid hormone-mediated transcriptional activation of the rat liver malic enzyme gene by dehydroepiandrosterone. *J Biol Chem* 264:18981–5.

Sorwell, K. G., and H. F. Urbanski. 2010. Dehydroepiandrosterone and age-related cognitive decline. *Age (Dordr)* 32:61–7.

Steckelbroeck, S., M. Watzka, D. Lutjohann, P. Makiola, A. Nassen, V. H. Hans, H. Clusmann et al. 2002. Characterization of the dehydroepiandrosterone (DHEA) metabolism via oxysterol 7alpha-hydroxylase and 17-ketosteroid reductase activity in the human brain. *J Neurochem* 83:713–26.

Su, C. Y., and H. Lardy. 1991. Induction of hepatic mitochondrial glycerophosphate dehydrogenase in rats by dehydroepiandrosterone. *J Biochem (Tokyo)* 110:207–13.

Swierczynski, J., E. Slominska, R. T. Smolenski, and D. Mayer. 2001. Increase in NAD but not ATP and GTP concentrations in rat liver by dehydroepiandrosterone feeding. *Pol J Pharmacol* 53:125–30.

Tsiriyotis, C., D. A. Spandidos, and C. E. Sekeris. 1997. The mitochondrion is a primary site of action of glucocorticoids: Mitochondrial nucleotide sequences, showing similarity to hormone response elements, confer dexamethasone inducibility to chimaeric genes transfected in LATK-cells. *Biochem Biophys Res Commun* 235:349–54.

Tunez, I., M. C. Munoz, and P. Montilla. 2005. Treatment with dehydroepiandrosterone prevents oxidative stress induced by 3-nitropropionic acid in synaptosomes. *Pharmacology* 74:113–8.

Vallee, M., J. D. Rivera, G. F. Koob, R. H. Purdy, and R. L. Fitzgerald. 2000. Quantification of neurosteroids in rat plasma and brain following swim stress and allopregnanolone administration using negative chemical ionization gas chromatography/mass spectrometry. *Anal Biochem* 287:153–66.

Weill-Engerer, S., J. P. David, V. Sazdovitch, P. Liere, B. Eychenne, A. Pianos, M. Schumacher, A. Delacourte, E. E. Baulieu, and Y. Akwa. 2002. Neurosteroid quantification in human brain regions: Comparison between Alzheimer's and nondemented patients. *J Clin Endocrinol Metab* 87:5138–43.

Weill-Engerer, S., J. P. David, V. Sazdovitch, P. Liere, M. Schumacher, A. Delacourte, E. E. Baulieu, and Y. Akwa. 2003. In vitro metabolism of dehydroepiandrosterone (DHEA) to 7 alpha-hyroxy-DHEA and Delta 5 androstene 3 beta, 17 beta-diol in specific regions of the aging brain from Alzheimer's and nondemented patients. *Brain Res* 969:117–25.

Zheng, J., and V. D. Ramirez. 1999. Rapid inhibition of rat brain mitochondrial proton FoF_1-ATPase activity by estrogens: Comparison with Na^+, K^+ -ATPase of porcine cortex. *Eur J Pharmacol* 368:95–102.

8 Dehydroepiandrosterone in Human Immunodeficiency Virus Infection Monitoring and Therapy

Rajesh Kannangai and Gopalan Sridharan

CONTENTS

INTRODUCTION

Human immunodeficiency virus (HIV) infection is characterized by progressive immunosuppression leading to acquired immunodeficiency syndrome (AIDS). The effects of HIV on the immune system are multifold. The cascading effects are principally due to the direct lytic effect on CD4 T-cells and follicular dendritic cells (nursemaid cells). This affects the inductive phase of the immune system by lowering the efficiency of antigen presenting cells and CD4 afferent functions. Sustained loss of this activity in untreated individuals is followed eventually by effector function loss of the cellular limb viz. CD8 effector function. This process results over a period of time; hence, there is an adverse effect on several levels of both innate and acquired immunity. The selective depletion of CD4 lymphocytes is the first observable abnormality related to HIV infection. The HIV infection in its natural course was well studied before the introduction of antiretroviral therapy (ART). These early studies focused on the natural history of infection, which clearly showed the association with several pathological changes affecting multiple organs including the endocrine system. The circulating cytokine levels and more importantly the observations at the lymph node sites based on messenger ribonucleic acid (mRNA) expression revealed interesting changes (Graziosi et al. 1994). It was documented that during HIV disease progression, there was an observable shift from a Th1 to a Th2 cytokine profile (Maggi et al. 1994). The interaction between the endocrine and immune system is well documented. One of the possible reasons for this shift in cytokine profile may be the increase in the production of cortisol and the reduction of dehydroepiandrosterone (DHEA) (Clerici et al. 1997; Clerici et al. 2000).

Morbidity and mortality due to HIV continue to be the foremost problem faced in developing countries where HIV is prevalent (UNAIDS 2005). Estimation of CD4-positive T-cell and the quantitation of plasma HIV-1 RNA continue to be the major prognostic markers of disease progression in HIV-infected individuals. However, in resource-poor countries, facilities for using the above markers are not widely available. The infrastructure cost for starting CD4 testing is very high and hence only very few centers in these countries have this facility. Setting up molecular testing is even more difficult in resource-poor settings and hence HIV-1 RNA quantitation availability is limited. In resource-poor countries, the majority of the infected individuals are from poor socioeconomic background and hence they may not be able to afford these tests, especially the viral load monitoring every 6 months. So, it will be ideal to look for an alternative marker that can be detected by a commonly available testing system like enzyme-linked immunosorbent assay. These markers can be tested along with the CD4 estimation, avoiding the costly viral load testing. Since there is a significant relationship between the CD4 cell count and DHEA levels, the DHEAS level could be considered as a useful monitoring marker in HIV infection (Christeff et al. 1996; Chittiprol et al. 2009).

DHEAS AND CORTISOL LEVEL IN NORMAL HEALTHY INDIVIDUALS

The highest concentration of steroids produced by the adrenal glands is dehydroepiandrosterone (DHEA), and its conjugated form is present in the plasma as DHEA sulfate (DHEAS). The serum concentration of DHEAS is 250–500 times higher than that of DHEA (Kroboth et al. 1999). DHEAS serves primarily as a precursor molecule that is circulated to peripheral tissues throughout the body. In the peripheral tissues and in the adrenal cortex, it is desulfated enzymatically and produces DHEA. Some amount of DHEA in these tissues will also convert back to DHEAS. DHEA will subsequently convert into various estrogenic and androgenic compounds. Since DHEA and DHEAS are interconverted easily, it is commonly referred to as DHEAS. The circulating level of DHEAS can be influenced by various factors. The DHEAS level declines after birth until about the age of five. Subsequently, the levels rise a few years before sexual maturation begins. The maximum concentration is found at about age 30, and thereafter DHEA levels decline to 20%–30% of the peak levels by the age of 70–80. The concentration of DHEAS was observed to be significantly higher in men than in women (Chittiprol et al. 2009). Decreased levels of DHEAS have been associated with a variety of medical conditions, such as rheumatic disease, cardiovascular disease, immune system disorders, and osteoporosis. Increased levels have been observed in individuals with obesity, type II diabetes, and female hirsutism, and individuals subjected to prolonged physical stress (Imrich 2002; Thijs et al. 2003; Chen and Parker 2004; Dharia and Parker 2004; Rosmond 2006).

HIV INFECTION AND DHEA

DHEA and Its Role in the Immune Homeostasis in Relation to the Replication of HIV

There is evidence that DHEA/DHEAS can enhance immunocompetence. A significant protective effect of DHEA was demonstrated in studies of lethal viral infections such as systemic coxsackievirus B4 and herpes simplex type 2 encephalitis in mice (Loria et al. 1988). The molecular basis for the effect of DHEA on the immune system is not clear; however, it is suggested that it may counteract the stress-related immunosuppressive effects of glucocorticoids stimulated by viral infection (Loria et al. 1988). It has been shown that DHEA appears to mediate its protective effect by a mechanism that blocks the toxin-induced production of pathophysiological levels of tumor necrosis factor-alpha and interleukin (IL)-1 (Ben-Nathan, Padgett, and Loria 1999). DHEA can suppress the proliferation of cultures activated by concanavalin A (Con A) or lipopolysaccharides in a dose-dependent manner. The secretion of IL-2 and IL-3 from Con A–activated lymphocytes reflects this phenomenon. These functions were suppressed by DHEA in a dose-dependent manner. The classic immunosuppressive effects of hydrocortisone on Con A–induced lymphocyte proliferation and IL-2 and IL-3 production were unaffected when it was cocultured with DHEA (Loria and Padgett 1998). There was a significant positive correlation between log DHEAS and log interferon (IFN)-γ level and negative association with IL-4 (Loria et al. 1988). When the level of DHEA is decreased, it may allow cortisol to act more effectively, leading to more immunosuppression (Wisniewski et al. 1993). Cortisol suppresses the production of IFN-γ and IL-2, while it enhances the production of IL-4. In contrary to the activity of cortisol, DHEA appears to enhance the immune function. This may be by antagonizing some effects of cortisol on lymphocytes. Hence, this change in the cortisol/DHEA ratio could be predictive of progression to AIDS in HIV-infected individuals (Chittiprol et al. 2009). The drop in DHEA and concomitant elevated cortisol level that accompanies the drop in CD4 T-cell may be due to the inhibition of IL-2 synthesis. Suzuki et al. (1991) has demonstrated the effect of DHEA on the human immune system. In the experiment, T-lymphocytes from healthy adults were stimulated with mitogens or antigens after exposure to DHEA. T-cells pretreated with DHEA after activation with a variety of stimuli produced significantly higher amounts of IL-2 and had a more potent cytotoxicity effect compared with the T-cells activated in the absence of DHEA. The IL-2–enhancing effect of DHEA was limited to fresh $CD4^+$ T-cells and $CD4^+$ clones and not to $CD8^+$ T-cells and $CD8^+$ clones. DHEA may act as a transcriptional enhancer of the IL-2 gene as the enhancing effect of DHEA was also detected at the level of IL-2 mRNA (Suzuki et al. 1991). The effect of DHEA level on the relative dominance of Th1 cytokines versus Th2 cytokines may not be an absolute inverse correlation. At very high levels of DHEA, the balance seems to shift toward Th2 (Caigan et al. 2001).

It is believed that an altered autonomic nervous system and low plasma DHEAS levels might be some of the reasons for the immune deregulation seen in HIV-infected individuals, which tilt the cytokine balance from Th1 to Th2 profile (Schifitto et al. 2000). There exists an association between plasma DHEAS and IFN-γ and IL-4. Data are available for 31 patients on DHEAS, with 24 of the 31 patients having IL-4 and IFN-γ. Log DHEAS showed significant positive association with log IFN-γ ($r = 0.42$, $p = .04$). Though there was negative correlation between log DHEAS and log IL-4, it did not reach statistical significance ($r = -0.23$, $p = .28$) (Schifitto et al. 2000). The cortisol/DHEA ratio, which reflects serum adrenal steroid profile, was found to be progressively increased throughout HIV infection. Changes in cortisol/DHEA ratio appear to be a crucial link between adrenal hormones and immunological and metabolic alterations. There is a good correlation between the increase in this ratio and reduced number of CD4 lymphocytes (Chittiprol et al. 2009). Another possibility is that the adrenal gland may be a target for HIV infection itself or for any opportunistic organisms.

DHEA Level in HIV-Infected Individuals

There are several studies that have shown a significantly lower level of plasma DHEAS in HIV-infected individuals who are not on ART (Ferrando, Rabkin, and Poretsky 1999; Kandathil et al. 2005; Chittiprol et al. 2009). In particular, Kandathil et al. (2005) have shown significantly low levels of DHEA in HIV-infected individuals (Figure 8.1). The DHEA levels estimated in the normal healthy individuals and HIV-infected individuals who are ART naive from different population groups are shown in Table 8.1. These studies have shown similar levels among healthy individuals from different population groups.

In HIV-infected individuals, the decrease in the DHEA level may be due to the deregulation of neuroendocrine functions attributed to the immune suppression. An evaluation of the prevalence and degree of adrenal dysfunction in a longitudinal cohort of HIV-infected individuals found that there is an elevation of plasma adrenocorticotropic hormone (ACTH) level (Findling et al. 1994). Although clinically significant adrenal insufficiency is uncommon, the elevations in plasma ACTH in several patients at the end of a 2-year study suggested that adrenocortical capacity may become compromised (Findling et al. 1994). In the study, the plasma cortisol, aldosterone, and DHEA responses to intravenous cosyntropin were evaluated at 6-month intervals for 2 years and the values were compared to that of normal subjects. The basal and peak cortisol responses to cosyntropin (synthetic derivative of ACTH that is used in the ACTH stimulation test)

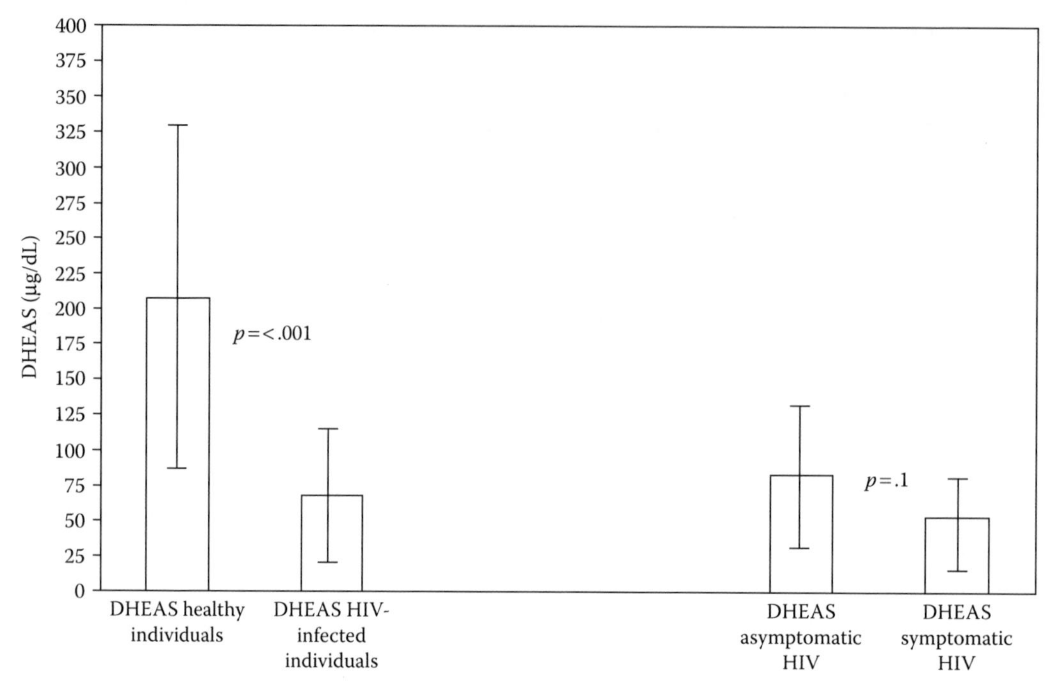

FIGURE 8.1 Mean dehydroepiandrosterone sulfate levels and standard deviations for healthy and human immunodeficiency virus (HIV)–infected individuals from an Indian population, showing a significant difference between HIV-infected individuals and normal healthy controls.

TABLE 8.1
The DHEAS Levels among Normal Healthy Adults and Human Immunodeficiency Virus (HIV)–Infected Antiretroviral Therapy (ART)–Naive Individuals Reported from Different Populations

Population	Healthy Individuals (μg/dL)	HIV-Infected ART Naive	Authors and Year
India	185.1 ± 12.03	81.02 ± 4.9 μg/dL	Chittiprol et al. (2009)
India	144 ± 219	76 ± 49 μg/dL	Kannangai et al. (2008)
India	207 ± 123	83.5 ± 52 μg/dL	Kandathil et al. (2005)
USA	250 ± 164	192 ± 119 μg/dL	Ferrando, Rabkin, and Poretsky (1999)
France	Not available	5.4 (2.3–8.8) μmol/L	Mauboussin et al. (2004)
Switzerland	Not available	6.2 (4.3–9.4) ng/mL	Wunder et al. (2008)

were found to be normal in all HIV patients. It was also observed that there was no difference in the mean basal or stimulated cortisol measurements between different Center for Disease Control classes of HIV-infected individuals. A significantly less aldosterone secretory capacity was seen when the individuals developed AIDS. The mean peak DHEA response to cosyntropin in HIV-infected individuals was significantly less compared with the normal. Basal plasma levels of aldosterone, cortisol, and DHEA were unchanged in 25 HIV patients throughout the study period. However, there was significant increase in the plasma ACTH level in the 25 patients at 24 months (9.7 ± 0.9 pmol/L) compared with that of the baseline values (7.0 ± 0.7 pmol/L). At the end of the study period, 8 of the 25 patients had plasma ACTH concentrations that were higher than the normal range. These observed changes, that is, altered aldosterone and DHEA secretion with normal cortisol production, are similar to the changes seen in seriously ill patients who are HIV negative.

Studies have shown a negative prognostic value for the DHEAS levels during the course of HIV infection (Chittiprol et al. 2009; Ferrando, Rabkin, and Poretsky 1999). One of the longitudinal studies reported that plasma cortisol levels increased significantly in HIV-infected individuals compared with the healthy HIV-negative controls, while there was a significant decrease in DHEAS. This shows up as a significant increase in cortisol/DHEAS ratio (Figure 8.2) (Chittiprol et al. 2009). HIV-infected individuals on follow-up showed a strong positive correlation between DHEAS and CD4 cells ($r = 0.2$, $p < .05$), and a strong negative correlation between cortisol, as well as cortisol/DHEAS ratio ($r = -0.25$, $p < .01$, and $r = -0.31$, $p < .001$, respectively; Figure 8.3).

DHEA protects against acute viral infections and can inhibit HIV-1 latency reactivation as it has a direct inhibitory effect on HIV viral replication (Abrams et al. 2007). DHEA treatment resulted in a modest downregulation of HIV-1 replication in phytohemagglutinin-stimulated peripheral blood lymphocytes as measured by syncytia formation, release of p24 antigen, and accumulation of reverse transcriptase activity. DHEA was also found to reduce syncytia formation in HIV-1–infected SupT1 lymphoblasts (Henderson, Yang, and Schwartz 1992). Irrespective of the HIV-1 subtype, there is a drop in the DHEA level during progression as the changes in DHEAS levels are more an outcome of immune/endocrine function decline.

An association between circulating concentrations of adrenal steroids and the progression of immunosuppression in HIV-infected individuals has been observed (Christeff et al. 1997). The serum concentrations of cortisol and DHEA in HIV-infected patients were different from those of HIV-negative controls. The serum cortisol was found to be significantly elevated at all stages of infection and was especially high in patients with AIDS. There was a correlation between serum

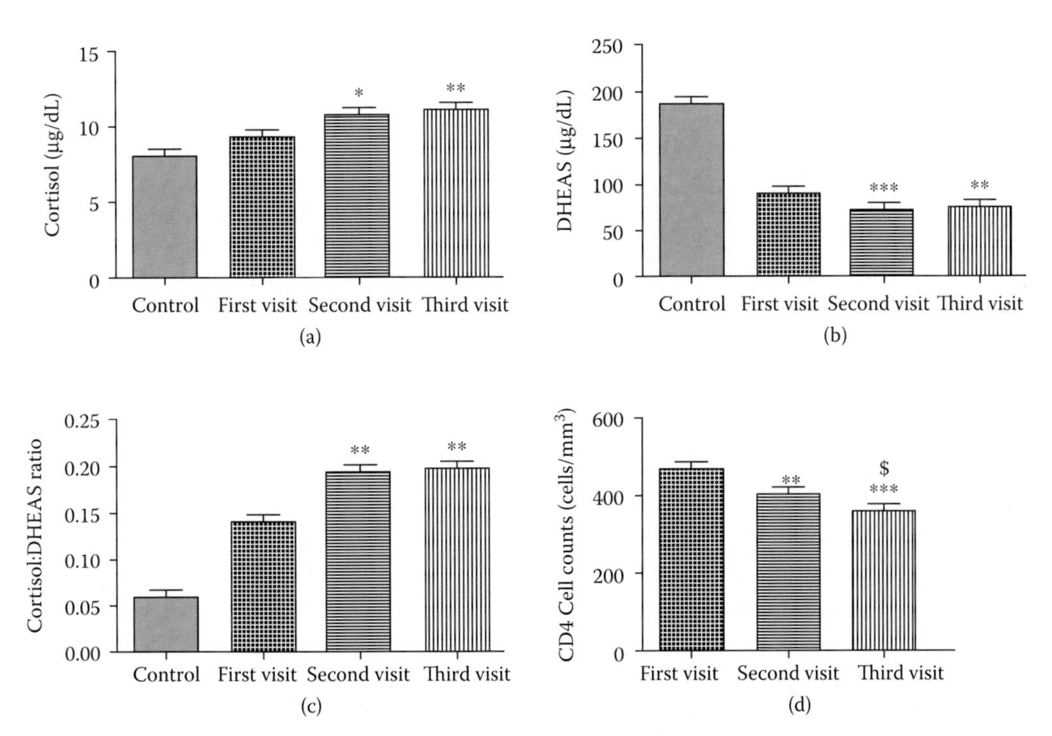

FIGURE 8.2 Plasma levels of cortisol (a), dehydroepiandrosterone sulfate (DHEAS) (b), cortisol/DHEAS ratio (c), and CD_4 cell counts (d) among asymptomatic human immunodeficiency virus seropositives ($n = 84$) during the 2-year follow-up. Statistical significance *: p < .05 **: p < .01 ***: p < .001 is shown in comparison to the 1st visit; $: p < .05 is in comparison to the 2nd visit.

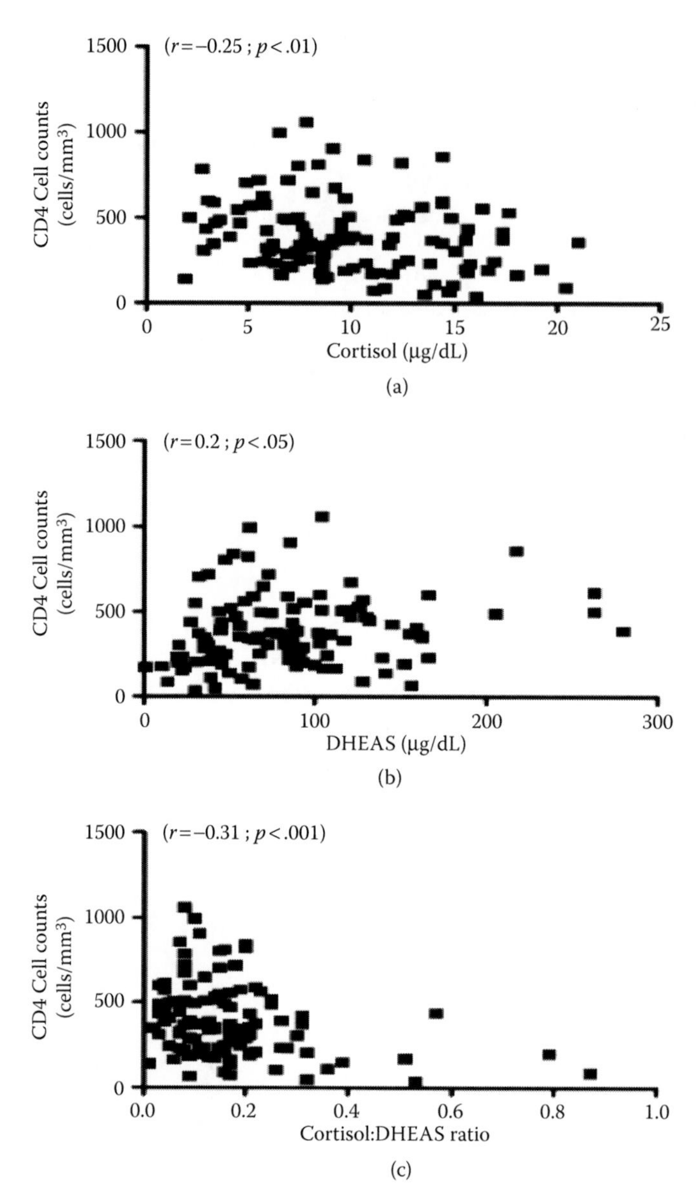

FIGURE 8.3 Correlation between plasma cortisol (a), dehydroepiandrosterone sulfate (DHEAS) (b), cortisol/DHEAS ratio (c), and CD4 counts among asymptomatic human immunodeficiency virus seropositives ($n = 120$) at the time of recruitment in the study (first visit).

DHEA concentration and the stage of HIV infection: it was highest during the early stages of the disease or when the CD4 count was more than 500. There was also a negative linear correlation between the CD4 cell counts and cortisol and a positive linear correlation with DHEA (Christeff et al. 1997).

Significant positive or negative relationships between several markers of malnutrition, adrenal androgens, and cortisol/DHEA ratio have also been reported. There is a negative correlation between body weight loss and DHEA ($r = -0.69$, $p < .0001$) and DHEAS, but a positive correlation with the cortisol/DHEA ratio. A reverse trend was seen with body cell mass (BCM). BCM was positively correlated with DHEA ($r = 0.34$, $p < .04$) and DHEAS ($r = 0.36$, $p < 0.03$), and negatively with the cortisol/DHEA ratio. These data strongly indicated an association between endocrine functions like cortisol/DHEA ratio and malnutrition associated with HIV infection. The decreased DHEA observed in the advanced stages of HIV-infected patients could be associated with increased protein catabolism (Christeff et al. 1999). There is a significant association between markers of immune activation such as cytokines, lipid membrane derivatives, and HIV disease progression. This is also associated with composite measure of autonomic function in HIV-1–infected individuals. The combined measure of autonomic performance was significantly lower in HIV-infected individuals than in controls.

DHEA Measurement in Monitoring HIV

The low DHEAS levels can have a negative prognostic effect during the progression of HIV infection, and the initiation of ART can induce an increase in circulating DHEAS (Ferrando, Rabkin, and Poretsky 1999). A study from southern India investigated the relationship between the lowered levels of DHEAS and HIV infection progression and its effect on the HIV-1 load. There exists a significant negative correlation between the viral load and the DHEAS level ($r = -0.6$, $p < .05$) among HIV-infected individuals as a group (Kandathil et al. 2005). Another study by the same authors looked at the changes in the DHEAS level following ART and antituberculous therapy (ATT) in HIV-1–infected symptomatic individuals and its difference from those with asymptomatic HIV-1 infection and healthy individuals. The study reconfirmed the significant difference in the DHEAS levels between HIV-1–infected individuals and healthy individuals. The DHEAS level also had a significant positive correlation with the $CD4^+$ T-cell count ($r = 0.27$, $p < .01$) and a negative correlation with the HIV-1 plasma RNA level ($r = -0.24$, $p < .01$). Compared with other serum/plasma markers, such as albumin, the researchers could not show any significant change in the group mean levels of DHEAS following ART or ATT. However, when the DHEAS levels in individuals on follow-up were analyzed, 13 (44.8%) in the ART group and 9 (69%) in the ATT group showed an increase in the DHEA level prior to the treatment (Kannangai et al. 2008).

The relationship between serum DHEA and DHEAS levels and subsequent progression to AIDS were investigated in a cohort of HIV-infected men prospectively since 1984 for 5 years (Jacobson et al. 1991). This longitudinal analysis of serum DHEA level in HIV-infected individuals who progress to AIDS and in nonprogressors showed the utility of DHEA as a monitoring marker. DHEA levels were estimated in each of the 41 asymptomatic HIV-1–seropositive subjects who progressed to AIDS within 5 years, HIV-1–seropositive individuals who remained asymptomatic, and HIV-negative controls. The seronegative group had higher DHEA levels at the time of entry in the study (median, 13.3 nmol/L) compared with both HIV-seropositive nonprogressors (median, 9.2 nmol/L, $p = .01$) and the progressors (median, 7.2 nmol/L, $p < .001$). DHEA levels in the progressors were lower than the levels in the nonprogressors (median, 5.6 vs. 8.8 nmol/L, $p = .007$). DHEA level of less than 7 nmol/L was shown to be an independent predictor for disease progression in HIV-1–infected individuals (Mulder et al. 1992).

Coinfection with viruses such as the hepatitis C virus (HCV) has been shown to further reduce the level of DHEA in HIV-infected individuals. A cross-sectional survey was carried out in a cohort of HIV-infected individuals with HCV coinfection. Here, coinfection was defined as the presence of HCV antibody with positive RNA. The DHEAS plasma level was measured in 137 patients, of whom 37 were infected with HCV as well. The median DHEAS level for the 137 patients was 5.5 μmol/L, while that for HCV-infected individuals was lower (2.1 μmol/L; 0.6–6.7 μmol/L) than for those not coinfected (6.6 μmol/L; 3.0–9.1 μmol/L) ($p < .01$). In the variance–covariance analysis of the prognostic factors, HCV coinfection was found to be associated with significant decrease in DHEAS. The subgroup analysis of the DHEAS showed that the age-adjusted mean of the DHEAS level was lower in HCV-coinfected patients for both women (1.3 ± 1.1 μmol/L) and men (4.0 ± 0.7 μmol/L) compared with patients not coinfected with HCV (women, 5.3 ± 0.7 μmol/L; men, 7.2 ± 0.4 μmol/L) ($p < .01$) (Mauboussin et al. 2004).

There is a strong association between neuroendocrine network function and immune function. The level of stress hormone cortisol is inversely correlated with DHEA plasma levels. The DHEA level seems to wane with age and also in viral infections like HIV, which has damaging effects on the immune system. In HIV infections, the progression of disease is strongly related to the decrease in the levels of DHEA. Studies have shown inverse correlation between HIV-1 RNA level and DHEA level. DHEA level parallels the $CD4^+$ T-cell count. This shows that plasma DHEA level is a reliable surrogate marker for immune status assessment of HIV-infected individuals.

SUMMARY

- DHEAS levels are indicative of immune dysfunction in HIV infection.
- Plasma level of DHEA has a strong association with low immune function.
- Low DHEAS levels have an association with the Th2 cytokine profile, a prognosticator of poor immune function in HIV infection.
- Immune status assessed by CD4 levels and clinical status are the two benchmarks for the observed immunosuppression, with strong correlation with DHEAS level even when adjusted for age. A positive correlation exists between the two parameters.
- DHEAS could be a useful inexpensive monitoring tool for assessing immune function in HIV-infected individuals.

REFERENCES

Abrams, D. I., S. B. Shade, P. Couey, J. M. McCune, J. Lo, P. Bacchetti et al. 2007. Dehydroepiandrosterone (DHEA) effects on HIV replication and host immunity: A randomized placebo-controlled study. *AIDS Res Hum Retroviruses* 23:77–85.

Ben-Nathan, D., D. A. Padgett, and R. M. Loria. 1999. Androstenediol and dehydroepiandrosterone protect mice against lethal bacterial infections and lipopolysaccharide toxicity. *J Med Microbiol* 48:425–31.

Caigan, D. U., Q. Guan, M. W. Khalil, and S. Sriram. 2001. Stimulation of Th2 response by high doses of dehydroepiandrosterone in KLH-primed splenocytes. *Exp Biol Med* 226:1051–60.

Chen, C. C., and C. R. Parker Jr. 2004. Adrenal androgens and the immune system. *Semin Reprod Med* 22:369–77.

Chittiprol, S., A. M. Kumar, K. T. Shetty, H. R. Kumar, P. Satishchandra, R. S. Rao et al. 2009. HIV-1 clade C infection and progressive disruption in the relationship between cortisol, DHEAS and CD4 cell numbers: A two-year follow-up study. *Clin Chim Acta* 409:4–10.

Christeff, N., N. Gherbi, O. Mammes, M. T. Dalle, S. Gharakhanian, O. Lortholary et al. 1997. Serum cortisol and DHEA concentrations during HIV infection. *Psychoneuroendocrinology* 22(Suppl 1):S11–8.

Christeff, N., O. Lortholary, P. Casassus, N. Thobie, P. Veyssier, O. Torri et al. 1996. Relationship between sex steroid hormone levels and CD4 lymphocytes in HIV infected men. *Exp Clin Endocrinol Diabetes* 104:130–6.

Christeff, N., J. C. Melchior, O. Mammes, N. Gherbi, M. T. Dalle, and E. A. Nunez. 1999. Correlation between increased cortisol: DHEA ratio and malnutrition in HIV-positive men. *Nutrition* 15:534–9.

Clerici, M., M. Galli, S. Bosis, C. Gervasoni, M. Moroni, and G. Norbiato. 2000. Immunoendocrinologic abnormalities in human immunodeficiency virus infection. *Ann N Y Acad Sci* 917:956–61.

Clerici, M., D. Trabattoni, S. Piconi, M. L. Fusi, S. Ruzzante, C. Clerici et al. 1997. A possible role for the cortisol/anticortisols imbalance in the progression of human immunodeficiency virus. *Psychoneuroendocrinology* 22(Suppl 1):S27–31.

Dharia, S., and C. R. Parker Jr. 2004. Adrenal androgens and aging. *Semin Reprod Med* 22(4):361–8.

Ferrando, S. J., J. G. Rabkin, and L. Poretsky. 1999. Dehydroepiandrosterone sulfate (DHEAS) and testosterone: Relation to HIV illness stage and progression over one year. *J Acquir Immune Defic Syndr* 22:146–54.

Findling, J. W., B. P. Buggy, I. H. Gilson, C. F. Brummitt, B. M. Bernstein, and H. Raff. 1994. Longitudinal evaluation of adrenocortical function in patients infected with the human immunodeficiency virus. *J Clin Endocrinol Metab* 79:1091–6.

Graziosi, C., G. Pantaleo, K. R. Gantt, J. P. Fortin, J. F. Demarest, O. J. Cohen et al. 1994. Lack of evidence for the dichotomy of Th1 and Th2 predominance in HIV-infected individuals. *Science* 265:248–52.

Henderson, E., J. Y. Yang, and A. Schwartz. 1992. Dehydroepiandrosterone (DHEA) and synthetic DHEA analogs are modest inhibitors of HIV-1 IIIB replication. *AIDS Res Hum Retroviruses* 8:625–31.

Imrich, R. 2002. The role of neuroendocrine system in the pathogenesis of rheumatic diseases. *Endocr Regul* 37(1):45–6.

Jacobson, M. A., R. E. Fusaro, M. Galmarini, and W. Lang. 1991. Decreased serum dehydroepiandrosterone is associated with an increased progression of human immunodeficiency virus infection in men with CD4 cell counts of 200–499. *J Infect Dis* 164:864–8.

Kandathil, A. J., R. Kannangai, S. David, R. Selvakumar, V. Job, O. C. Abraham et al. 2005. Human immunodeficiency virus infection and levels of dehydroepiandrosterone sulfate in plasma among Indians. *Clin Diagn Lab Immunol* 12:1117–8.

Kannangai, R., A. J. Kandathil, D. L. Ebenezer, E. Mathai, A. J. Prakash, O. C. Abraham et al. 2008. Usefulness of alternate prognostic serum and plasma markers for antiretroviral therapy for human immunodeficiency virus type 1 infection. *Clin Vaccine Immunol* 15:154–8.

Kroboth, P. D., F. S. Salek, A. L. Pittenger, T. J. Fabian, and R. F. Frye. 1999. DHEA and DHEA-S: A review. *J Clin Pharmacol* 39:327–48.

Loria, R. M., T. H. Inge, S. S. Cook, A. K. Szakal, and W. Regelson. 1988. Protection against acute lethal viral infections with the native steroid dehydroepiandrosterone (DHEA). *J Med Virol* 26:301–14.

Loria, R. M., and D. A. Padgett. 1998. Control of the immune response by DHEA and its metabolites. *Rinsho Byori* 46:505–17.

Maggi, E., M. Mazzetti, A. Ravina, F. Annunziato, M. de Carli, M. P. Piccinni et al. 1994. Ability of HIV to promote a Th1 to TH0 shift and to replicate preferentially in Th2 and TH0 cells. *Science* 265:244–8.

Mauboussin, J. M., A. Mahamat, H. Peyrière, I. Rouanet, P. Fabbro-Peray, J. P. Daures et al. 2004. Low plasma levels of dehydroepiandrosterone sulphate in HIV-positive patients coinfected with hepatitis C virus. *HIV Med* 5:151–7.

Mulder, J. W., P. H. Frissen, P. Krijnen, E. Endert, F. de Wolf, J. Goudsmit et al. 1992. Dehydroepiandrosterone as predictor for progression to AIDS in asymptomatic human immunodeficiency virus-infected men. *J Infect Dis* 165:413–8.

Rosmond, R. 2006. Androgen excess in women—a health hazard? *Med Hypotheses* 67(2):229–34.

Schifitto, G., M. P. McDermott, T. Evans, T. Fitzgerald, J. Schwimmer, L. Demeter et al. 2000. Autonomic performance and dehydroepiandrosterone sulfate levels in HIV-1-infected individuals: Relationship to Th1 and Th2 cytokine profile. *Arch Neurol* 57:1027–32.

Suzuki, T., N. Suzuki, R. A. Daynes, and E. G. Engleman. 1991. Dehydroepiandrosterone enhances IL2 production and cytotoxic effector function of human T cells. *Clin Immunol Immunopathol* 61:202–11.

Thijs, L., R. Fagard, F. Forette, T. Nawrot, and J. A. Staessen. 2003. Are low dehydroepiandrosterone sulphate levels predictive for cardiovascular diseases? A review of prospective and retrospective studies. *Acta Cardiol* 58(5):403–10.

UNAIDS. 2005. Joint United Nations Program on HIV/AIDS. AIDS epidemic update: December 2005. http://unaids.org/epi/2005/ (accessed January 14, 2011).

Wisniewski, T. L., C. W. Hilton, E. V. Morse, and F. Svec. 1993. The relationship of serum DHEA-s and cortisole levels to measures of immune function in human immunodeficiency virus related illness. *Am J Med Sci* 305:79–83.

Wunder, D. M., C. A. Fux, N. A. Bersinger, N. J. Mueller, B. Hirschel, M. Cavassini, L. Elzi, P. Schmid, E. Bernasconi, B. Mueller, and H. Furrer. 2008. Swiss HIV Cohort Study. Androgen and gonadotropin patterns differ in HIV-1–infected men who develop lipoatrophy during antiretroviral therapy: A case-control study. *HIV Med* 9:427–32.

9 Dehydroepiandrosterone Replacement and Bone Mineral Density

Catherine M. Jankowski and Wendy M. Kohrt

CONTENTS

The decline in endogenous dehydroepiandrosterone (DHEA) production associated with aging may contribute to an array of physiological changes that are sex hormone dependent, including the loss of bone mass. Accordingly, increasing circulating DHEA sulfate (DHEAS) and other sex hormone levels to those of younger adults using DHEA replacement therapy may be an effective strategy for increasing bone mineral density (BMD) in older adults. The available evidence to support this contention will be reviewed.

PRELIMINARY STUDIES OF DHEA REPLACEMENT AND BMD

Early studies of DHEA therapy in older women and men used percutaneous cream (Labrie et al. 1997) or oral doses from 25 to 100 mg/day (Casson et al. 1998; Morales et al. 1998; Baulieu et al. 2000; Villareal, Holloszy, and Kohrt 2000) for 6 to 12 months. These studies tended to have small sample sizes and did not necessarily restrict study participants to those with low serum DHEAS levels or those not using sex hormone therapy. Given these limitations, it is not surprising that the effects of DHEA therapy on BMD (as measured by dual-energy x-ray absorptiometry) were equivocal.

In two placebo-controlled, double-blinded studies of older adults (Casson et al. 1998; Morales et al. 1998), DHEA therapy did not generate significant increases in either hip or lumbar spine BMD. Casson et al. (1998) studied healthy postmenopausal women (n = 13) with initially low serum DHEAS (<125 μg/dL) who were randomly assigned to 25 mg/day oral DHEA or placebo for 6 months. Morales et al. (1998) used a randomized crossover design to evaluate the effects of 100 mg/day oral DHEA for 6 months in men (n = 8) and postmenopausal women (n = 8) aged 50–65 years. Notably, because seven of the eight women were using estrogen or estrogen plus progesterone therapy during the trial, the effects of DHEA on BMD may have been masked. In a larger randomized placebo-controlled trial (RCT) of DHEA replacement over a longer duration, Baulieu et al. (2000) investigated changes in hip and forearm BMD in 280 older women and men. The participants were divided into four equally sized (n = 70) sex and age groups of 60–69 and 70–79 years, and randomly assigned to 50 mg/day DHEA or placebo for 12 months. In women younger than 70 years of age, BMD of the femoral neck increased significantly in the DHEA group compared with the placebo. In contrast, only forearm BMD increased in women aged 70 and older in the DHEA group. DHEA treatment had no effects on BMD of men in either age group.

Two early open-label trials found favorable effects of DHEA replacement on BMD. In a single-group study of percutaneous 10% DHEA cream administered to postmenopausal women ($n = 14$) for 12 months (Labrie et al. 1997), there was a significant increase (approximately 2%) in total hip BMD and a trend for lumbar spine BMD to increase (1.2%). The DHEA dose was titrated to reach serum DHEAS concentrations of 20–30 nmol/L, representing a tenfold increase from baseline levels. Villareal, Holloszy, and Kohrt (2000) found a significant increase in total body (1.6 ± 0.6%) and lumbar spine (2.5 ± 0.8%) BMD in older (64–82 years) women and men in response to 50 mg/day oral DHEA for 6 months; hip BMD was unchanged. The baseline DHEAS levels of participants were about 15% of the young adult levels (1.3 ± 0.9 nmol/L and 10.3 ± 2.4 nmol/L, respectively). By 3 months of DHEA administration, DHEAS increased to young adult levels and was sustained at this higher level for 6 months of treatment.

These early studies of the effects of DHEA therapy on BMD confirmed that 50 mg/day DHEA was an appropriate dose to raise serum DHEAS levels in older adults to the normal range of young adults (Arlt et al. 1999, 2001; Morales et al. 1994). Further, the studies provided preliminary evidence that DHEA therapy may increase BMD with few adverse side effects (Baulieu et al. 2000; Villareal, Holloszy, and Kohrt 2000) in older adults.

DHEA REPLACEMENT IN RANDOMIZED PLACEBO-CONTROLLED TRIALS

Four recent double-blinded RCTs (Nair et al. 2006; Jankowski et al. 2006; von Muhlen et al. 2007; Weiss et al. 2009) evaluated the effects of DHEA replacement on BMD in older women and men. BMD was a primary outcome in all these studies, which featured use of oral 50 mg/day DHEA versus placebo as the intervention and treatment duration of 1–2 years. The demographics of the study populations were similar (Table 9.1) in terms of size and sex distribution, age, and baseline serum DHEAS. Presumably, all women were postmenopausal. Stable doses of some medications (e.g., antihypertensives, antidyslipidemic, thyroid) were permitted. In all the studies, average DHEAS levels at study entry were well below young adult levels. Participants were screened for normal

TABLE 9.1
Demographics of Study Populations

	Randomized Placebo-Controlled Trial			
	Nair et al. (2006)	**Jankowski et al. (2006)**	**Von Muhlen et al. (2007)**	**Weiss et al. (2009)**
Sample size, N (female)	117 (57)	140 (70)	225 (115)	113 (58)
Average age (years)[a]	68	68	68	70
	Dehydroepiandrosterone Sulfate (DHEAS; μmol/L)[a]			
Women	0.9	1.3	1.2	1.4
Men	1.8	1.7	2.1	1.9
Screened for low DHEAS	Yes	Yes	No	No
	Excluded if Using			
Sex hormone therapy	Yes	Yes	Yes	Unknown
Antiresorptive therapy	Yes	No	Yes	No[b]

[a] Estimated from the published values.
[b] Data from users of antiresorptive therapy were excluded from the analyses.

blood chemistry, liver and renal function, hematology, prostate specific antigen (PSA) and prostate examination in men, and normal gynecological and breast examinations in women. Additional inclusion and exclusion criteria varied according to the secondary outcomes of each study (e.g., glucose tolerance). Weiss et al. (2009) provided calcium and vitamin D supplements to all participants for the duration of the intervention, whereas Jankowski et al. (2006) provided calcium and vitamin D supplementation for participants with low dietary intake at baseline.

In all four RCTs (Nair et al. 2006; Jankowski et al. 2006; von Muhlen et al. 2007; Weiss et al. 2009), serum DHEAS in the active treatment groups increased to young adult levels and was maintained for the duration of the intervention (Nair et al. 2006; Jankowski et al. 2006; von Muhlen et al. 2007). In Jankowski et al. (2006), the intent-to-treat analysis yielded strong trends for increases in total hip, trochanter, and femoral shaft BMD (1.0%–1.2%, all $p \leq .06$) in response to DHEA replacement in women and men (combined) compared with placebo. The study was not powered to detect sex differences in response to DHEA, but exploratory sex-specific analyses found that lumbar spine BMD increased in women only ($p = .04$). The other three RCTs were apparently powered to evaluate sex-specific responses to DHEA. Two RCTs (von Muhlen et al. 2007; Weiss et al. 2009) reported a significant increase in lumbar spine BMD in DHEA-treated women (Figure 9.1). Although lumbar

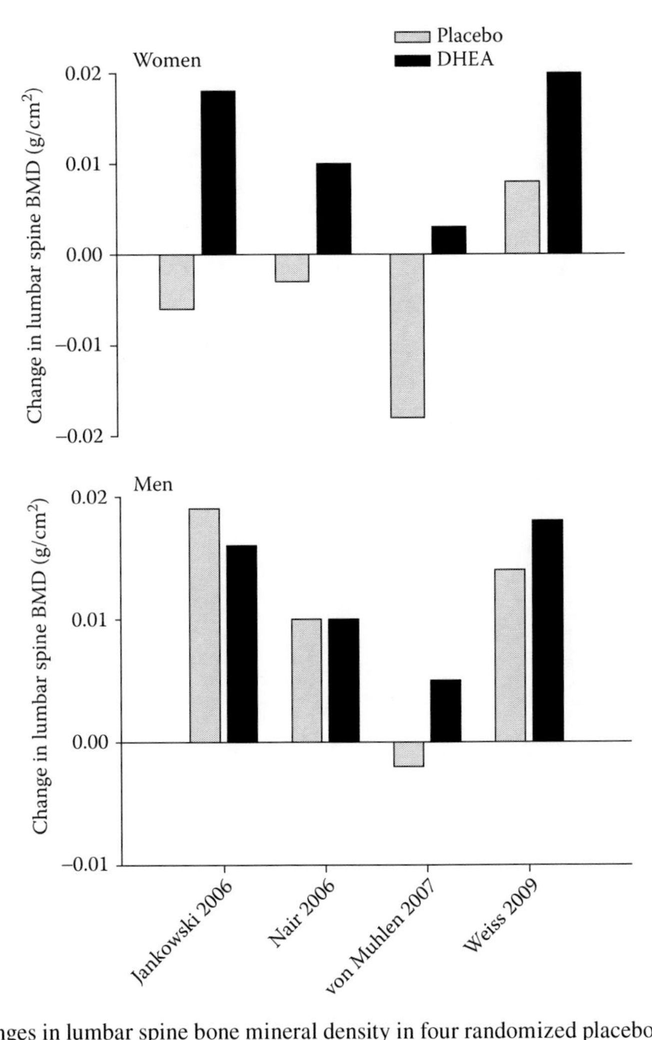

FIGURE 9.1 Changes in lumbar spine bone mineral density in four randomized placebo-controlled trials of dehydroepiandrosterone therapy.

spine BMD tended to increase in men treated with DHEA, the interpretation of this response was confounded by the unexpected trends for BMD to increase in men on placebo (Nair et al. 2006; Jankowski et al. 2006; Weiss et al. 2009). There were no significant effects of DHEA therapy on total hip BMD in women or men (Figure 9.2). Nair et al. (2006) found that the loss of femoral neck BMD in men was attenuated with DHEA replacement and that women on DHEA had a significant increase in ultradistal radius BMD.

Despite the similarities in these four RCTs with respect to the study populations and the intervention approach, there was a lack of concordance regarding the overall and sex-specific effects of DHEA therapy on BMD. This suggests that the variability of the BMD response to DHEA may have been underestimated and that the trials may not have been adequately powered. The most consistent effect of DHEA was an increase in lumbar spine BMD in women (Jankowski et al. 2006; von Muhlen et al. 2007; Weiss et al. 2009).

Jankowski et al. (2006) also completed secondary on-treatment compliance analyses within their RCT. They used the average serum DHEAS during months 3–12 of the intervention to define compliance, which resulted in a sample of 119 compared with 130 in the intent-to-treat analyses. In women and men combined, there were increases in total hip (1.2%), trochanter (1.6%), and femoral shaft (1.6%) BMD in the DHEA group compared with placebo, thus corroborating the trends found in the intent-to-treat analysis described above. The compliance analyses were not powered to detect sex-specific responses, but exploratory analyses revealed increases of approximately 2% in lumbar spine, total hip, trochanter, and femoral shaft BMD in women (all $p < .05$), but not in men.

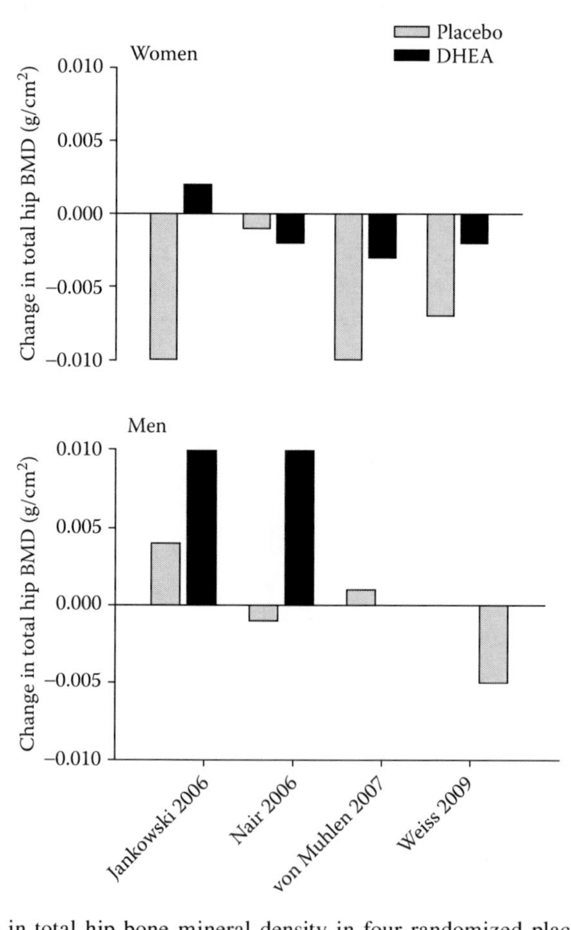

FIGURE 9.2 Changes in total hip bone mineral density in four randomized placebo-controlled trials of dehydroepiandrosterone therapy.

None of the other trials (Nair et al. 2006; von Muhlen et al. 2007; Weiss et al. 2009) reported results based on compliance to treatment. Thus, in contrast to the other RCTs, Jankowski et al. (2006) found that DHEA replacement had positive effects on hip BMD. Consistent with two of the other trials (von Muhlen et al. 2007; Weiss et al. 2009), lumbar spine BMD increased significantly in response to DHEA therapy in women.

Weiss et al. (2009) presented provocative evidence that larger increases in BMD may result from longer duration of DHEA therapy. They included an additional year of open-label observation following the 1-year RCT (Figure 9.3). In the second year, subjects originally assigned to the placebo group were given the option to crossover to active 50 mg/day DHEA, and the DHEA group was invited to continue active treatment. Women who crossed over to DHEA had a significant 2% increase in lumbar spine BMD compared with their 12-month measures (end of placebo treatment). Women who stayed on active DHEA for a second year had a further 2% increase in lumbar spine BMD and a 2-year change of nearly 4%, rivaling the effects of the bisphosphonates risedronate and alendronate (Harris et al. 1999; Cummings et al. 1998). The men who crossed over to DHEA in the second year and the men who were taking DHEA for 2 years both had significant increases (~1.6%) in lumbar spine BMD compared with their respective pretreatment baselines. No significant increases in total body or hip BMD were found in women or men in either the RCT or crossover phase of the study. Overall, lumbar spine BMD appeared to be more responsive to DHEA replacement than the hip BMD, perhaps because of the preponderance of trabecular bone in the vertebrae compared with the hip. Women appeared to be more sensitive to DHEA replacement than men

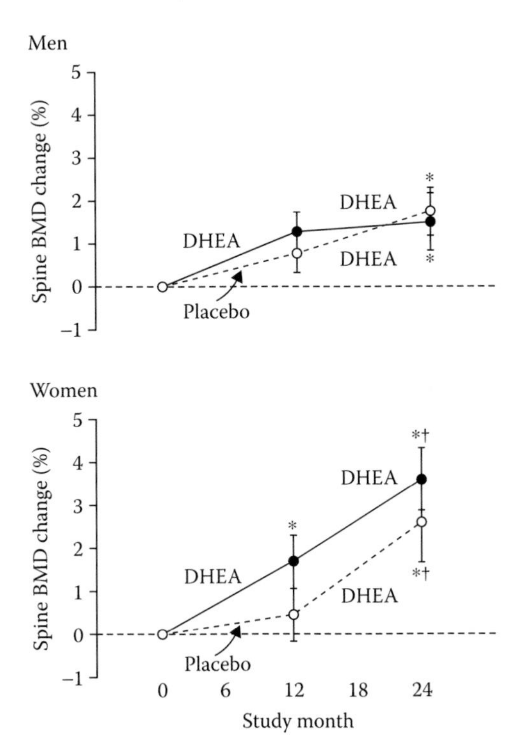

FIGURE 9.3 Mean changes in serum dehydroepiandrosterone sulfate in the 1-year randomized placebo-controlled trial and open-label crossover study in second year. *$p < .0001$ compared with baseline (within group) and †$p \leq .0001$ for the comparison between the placebo or crossover group at the same time point by two-factor repeated measures analysis of variance and Tukey-adjusted post hoc paired comparisons ($p < .0001$ for group-by-time interaction in both women and men). (From Weiss, E. P., K. Shah, L. Fontana et al. 2009. *Am J Clin Nutr* 89:1459–67. With permission.)

TABLE 9.2
Adverse Events in Randomized Controlled Trials of Dehydroepiandrosterone (DHEA) Replacement

			Placebo			DHEA	
	N	Adverse Events	Women	Men	All	Women	Men
Nair et al. (2006)	152	All[a]	30	32		29	30
von Muhlen et al. (2007)	225	Serious[b]	—	—	10	17	16
Jankowski et al. (2006)	140	Serious	0	4		0	3
Weiss et al. (2009)							
Year 1	113	Serious	2	3		1	0
Year 2	90	Serious	1	1		0	4

[a] Serious (reportable to human research oversight committee) and nonserious events.
[b] Events leading to treatment discontinuation.

as evidenced by greater increases in lumbar spine BMD over 1 year of treatment (2% vs. 1%) and further increases (2% per year) during prolonged treatment.

The risk to benefit ratio of DHEA therapy remains undetermined because no study has been sufficiently powered to evaluate safety. Adverse events occurred in all four RCTs (Table 9.2), although the reporting methods were not uniform. Nair et al. (2006), who reported serious and nonserious events, found no difference in the number of events in DHEA- and placebo-treated women and men. Three deaths occurred in the studies: a man taking placebo (Jankowski et al. 2006), a woman taking DHEA (Nair et al. 2006), and a man who crossed over from placebo to DHEA in the second year (Weiss et al. 2009) died. Serum PSA did not change significantly in men taking DHEA for up to 2 years (Weiss et al. 2009), although individual increases in PSA (e.g., >1.4 ng/mL) were found in placebo- and DHEA-treated men (Nair et al. 2006; von Muhlen et al. 2007). With careful screening and safety monitoring, 50 mg/day DHEA resulted in very few serious adverse events. The less serious side effects need to be considered on an individual basis. For example, the modest increase in BMD in response to DHEA may be a reasonable option for women who experience intolerable side effects of bisphosphonates.

These four recent RCTs (Nair et al. 2006; Jankowski et al. 2006; von Muhlen et al. 2007; Weiss et al. 2009) and the study by Baulieu et al. (2000) suggest that DHEA replacement of 50 mg/day for 1–2 years is associated with increased lumbar spine BMD, with less evidence for benefits at the hip. Furthermore, the effects of DHEA replacement on BMD may be sex specific, with more consistent increases found in older women, particularly of the lumbar spine, although attenuation of bone loss may occur in men (Nair et al. 2006). Whether DHEA treatment longer than 2 years confers additional bone mineral accrual and whether starting DHEA therapy at a younger age can prevent or attenuate the age-related decline in BMD are unclear.

POTENTIAL MECHANISMS UNDERLYING THE EFFECTS OF DHEA ON BONE

The mechanisms by which DHEA replacement promotes accrual of BMD in humans are not well defined. DHEA and DHEAS may impart direct effects on bone metabolism (Labrie et al. 2001) or may act indirectly in peripheral tissues (i.e., intracrine effects) via conversion by steroidogenic enzymes to testosterone, and/or subsequent aromatization to estrogens (Dhatariya and Nair 2003). Furthermore, increases in circulating insulin-like growth factor (IGF)-1 in response to DHEA replacement may provide an anabolic stimulus to bone (Villareal, Holloszy, and Kohrt 2000;

Morales et al. 1994). Human studies have relied on the circulating levels of sex hormones and growth factors to explain potential mechanisms of DHEA action. However, if DHEA has intracrine actions in peripheral tissues (Labrie et al. 2001), then circulating levels of markers may or may not be surrogates of tissue-specific activity.

DHEA replacement in older women has been consistently associated with increases of 30%–100% in serum total testosterone (Baulieu et al. 2000; Morales et al. 1994; Nair et al. 2006; Weiss et al. 2009; Genazzani et al. 2001; Jankowski et al. 2008) and 50%–80% increases in estradiol (E_2) (Baulieu et al. 2000; Nair et al. 2006; von Muhlen et al. 2007; Weiss et al. 2009; Jankowski et al. 2008; Flynn et al. 1999). In older men, changes in sex hormones during DHEA therapy were more modest and more variable than in women. Serum testosterone does not appear to increase significantly with DHEA replacement in men (Baulieu et al. 2000; Morales et al. 1998; Arlt et al. 1999; Morales et al. 1994; Nair et al. 2006; von Muhlen et al. 2007; Weiss et al. 2009; Jankowski et al. 2008), whereas serum E_2 increased (10%–50%) in some studies (Baulieu et al. 2000; Nair et al. 2006; Weiss et al. 2009; Jankowski et al. 2008) but not in others (Morales et al. 1998; Baulieu et al. 2000; Arlt et al. 1999; Morales et al. 1994; von Muhlen et al. 2007). In two studies (Nair et al. 2006; Jankowski et al. 2008), trends for serum testosterone to decrease and E_2 to increase in older men in response to DHEA therapy raise the possibility of suppressed endogenous testosterone. Importantly, serum sex hormone binding globulin (SHBG) does not increase in response to oral DHEA therapy (Arlt et al. 2001; Morales et al. 1994); in fact, decreases have been reported (Labrie et al. 1997; Morales et al. 1998; Arlt et al. 1999; Weiss et al. 2009; Jankowski et al. 2008). Thus, the conversion of DHEA to testosterone and/or E_2 without an increase in SHBG increased the bioavailability of hormones that influence bone metabolism.

The evidence for an IGF-1 mechanism of DHEA action on bone is less consistent than that of sex hormones. Serum IGF-1 responses to DHEA replacement generally support an increase in IGF-1 (Morales et al., 1994, 1998; von Muhlen et al. 2007; Weiss et al. 2009; Genazzani et al. 2001; Jankowski et al. 2008). In some studies (von Muhlen et al. 2007; Weiss et al. 2009; Jankowski et al. 2008), serum IGF-1 increased (12%–23%) significantly in older women but not in men undergoing DHEA therapy. IGF binding protein-3 did not increase, suggesting greater bioavailability of IGF-1 in older women following DHEA replacement. It is unclear whether DHEA exerts direct actions on the growth hormone–IGF-1 axis or modulates IGF-1 release from the liver.

The changes in the hormonal milieu during DHEA therapy raise the question of whether it is DHEA, per se, or changes in hormones and growth factors that mediate changes in BMD. Jankowski et al. (2008) found significant associations between serum E_2 and the free E_2 index, but not testosterone (total or free testosterone index) or IGF-1, with increases in total hip, femoral shaft, and trochanter BMD in older women and men. The relative effects of sex hormones on BMD were further assessed using regression models that included DHEA replacement group (i.e., randomization to 50 mg/day DHEA or placebo), serum testosterone, E_2, and the combinations of testosterone + E_2, DHEA + testosterone, DHEA + E_2, and DHEA + testosterone + E_2 as mediators of the changes in BMD (adjusted for baseline BMD). DHEA replacement alone was significantly associated with increased total hip, trochanter, and shaft BMD. For simplicity, only the mediation models for total hip BMD are presented (Table 9.3), although similar results were found using the regional hip BMD regression models. When E_2 was added to the model, DHEA replacement was no longer a significant mediator of changes in total hip BMD, indicating that the effect of DHEA was mediated through the increase in E_2. Further, E_2 remained a significant mediator of hip BMD when combined with DHEA group and/or testosterone. Similar results were obtained using sex-specific models. This analysis revealed the dominant role of serum E_2 in mediating increases in hip (total and regional) BMD in response to DHEA therapy for 12 months in older adults.

The increases in BMD in response to DHEA therapy suggest a change in bone turnover favoring mineral accretion. Measures of serum and urinary bone markers in response to DHEA replacement have yielded mixed results (Baulieu et al. 2000; Kahn and Halloran 2002) due, in part, to technical

TABLE 9.3
Associations of Dehydroepiandrosterone (DHEA) Treatment and 12-Month Serum Sex Hormone Concentrations with Changes in Total Hip Bone Mineral Density[a] (BMD)

Model	Coefficient (*p* value)		
	DHEA	Testosterone[b]	Estradiol[b]
Total Hip BMD			
DHEA	1.14 (.019)	—	—
Testosterone	—	0.00 (.991)	—
E_2	—	—	2.03 (<.001)
Testosterone + E_2	—	–0.31 (.056)	2.44 (<.001)
DHEA + Testosterone	1.18 (.018)	–0.07 (.666)	—
DHEA + E_2	0.22 (.686)	—	1.91 (<.001)
DHEA + Testosterone + E_2	0.17 (.742)	–0.31 (.059)	2.34 (<.001)

Source: Adapted from Jankowski, C. M., W. S. Gozansky, J. M. Kittelson et al. 2008. *J Clin Endocrinol Metab* 93:4767–4773.

[a] Adjusted for baseline BMD.

[b] Log_2.

variability, differences in assay methods, and potential diurnal variations in the serum levels of bone markers. However, a decrease in bone turnover in response to DHEA therapy is supported by significant decreases in serum bone-specific alkaline phosphatase, a marker of bone formation, and/or C-terminal telopeptide of type 1 collagen, a marker of bone resorption (Labrie et al. 1997; von Muhlen et al. 2007; Weiss et al. 2009; Jankowski et al. 2008). Reduced bone turnover is consistent with E_2-mediated effects of DHEA on bone given that E_2 reduces bone resorption, as opposed to the anabolic, bone-forming actions of testosterone or IGF-1.

SUMMARY

There have been several relatively large single-site RCTs of the effectiveness of DHEA therapy to increase BMD in older women and men. Despite this, the effects remain uncertain because of the variability in the findings across studies. The available evidence suggests that the effects of DHEA on bone may be more pronounced in women than in men, more robust at the lumbar spine than at the hip, and less potent than other treatments (e.g., bisphosphonates). It is not known whether BMD would continue to increase with DHEA treatment longer than 2 years, whether DHEA therapy reduces the risk of fragility fractures in older adults, or whether DHEA would be effective in slowing age-related bone loss if started at a younger age. Although DHEA replacement therapy has not been associated with serious adverse events, the RCTs have been too small to effectively evaluate safety. Longer and larger trials of DHEA therapy in older women and men would be needed to evaluate antifracture efficacy and long-term safety.

REFERENCES

Arlt, W., F. Callies, I. Koehler et al. 2001. Dehydroepiandrosterone supplementation in healthy men with an age-related decline of dehydroepiandrosterone secretion. *J Clin Endocrinol Metab* 86:4686–92.

Arlt, W., J. Haas, F. Callies et al. 1999. Biotransformation of oral dehydroepiandrosterone in elderly men: Significant increase in circulating estrogens. *J Clin Endocrinol Metab* 84:2170–6.

Baulieu, E. -E., G. Thomas, S. Legrain et al. 2000. Dehydroepiandrosterone (DHEA), DHEA sulfate, and aging: Contribution of the DHEAge Study to a sociobiomedical issue. *Proc Natl Acad Sci USA* 97:4279–84.

Casson, P. R., N. Santoro, K. E. Elkind-Hirsch et al. 1998. Postmenopausal dehydroepiandrosterone administration increases free insulin-like growth factor-I and decreases high-density lipoprotein: A six-month trial. *Fertil Steril* 70:107–10.

Cummings, S. R., D. M. Black, D. E. Thompson et al. 1998. Effect of alendronate on risk of fracture in women with low bone density but without vertebral fractures. Results from the Fracture Intervention Trial. *JAMA* 280:2077–82.

Dhatariya, K. K., and K. S. Nair. 2003. Dehydroepiandrosterone: Is there a role for replacement? *Mayo Clin Proc* 78:1257–73.

Flynn, M. A., D. Weaver-Osterholtz, K. L. Sharpe-Timms et al. 1999. Dehyrdoepiandrosterone replacement in aging humans. *J Endocrinol Metab* 84:1527–33.

Genazzani, A. D., M. Stomati, C. Strucchi et al. 2001. Oral dehydroepiandrosterone supplementation modulates spontaneous and growth hormone-releasing hormone-induced growth hormone and insulin-like growth factor-1 secretion in early and late postmenopausal women. *Fertil Steril* 76:241–8.

Harris, S. T., N. B. Watts, H. K. Genant et al. 1999. Effects of risedronate treatment on vertebral and non-vertebral fractures in women with postmenopausal osteoporosis: A randomized controlled trial. *JAMA* 282:1344–52.

Jankowski, C. M., W. S. Gozansky, J. M. Kittelson et al. 2008. Increases in bone mineral density in response to oral dehydroepiandrosterone replacement in older adults appear to be mediated by serum estrogens. *J Clin Endocrinol Metab* 93:4767–73.

Jankowski, C. M., W. S. Gozansky, R. S. Schwartz et al. 2006. Effects of dehydroepiandrosterone replacement therapy on bone mineral density in older adults: A randomized, controlled trial. *J Clin Endocrinol Metab* 91:2986–93.

Kahn, A. J., and B. Halloran. 2002. Dehydroepiandrosterone supplementation and bone turnover in middle-aged to elderly men. *J Clin Endocrinol Metab* 87:1544–9.

Labrie, F., P. Diamond, L. Cusan et al. 1997. Effect of 12-month dehydroepiandrosterone replacement therapy on bone, vagina, and endometrium in postmenopausal women. *J Clin Endocrinol Metab* 82:3498–505.

Labrie, F., V. Luu-The, C. Labrie et al. 2001. DHEA and its transformation into androgens and estrogens in peripheral target tissues: Intracrinology. *Front Neuroendocrinol* 22:185–212.

Morales, A. J., R. H. Haubrich, J. Y. Hwang et al. 1998. The effect of six months treatment with a 100 mg daily dose of dehydroepiandrosterone (DHEA) on circulating sex steroids, body composition and muscle strength in age-advanced men and women. *Clin Endocrinol (Oxf)* 49:421–32.

Morales, A. J., J. J. Nolan, J. C. Nelson et al. 1994. Effects of replacement doses of dehydroepiandrosterone in men and women of advancing age. *J Clin Endocrinol Metab* 78:1360–7.

Nair, K. S., R. A. Rizza, P. O'Brien et al. 2006. DHEA in elderly women and DHEA or testosterone on elderly men. *N Engl J Med* 355:1647–59.

Villareal, D. T., J. O. Holloszy, and W. M. Kohrt. 2000. Effects of DHEA replacement on bone mineral density and body composition in elderly women and men. *Clin Endocrinol* 53:561–8.

von Muhlen, D., A. Laughlin, D. Kritz-Silverstein, J. Bergstrom, and R. Bettencourt. 2007. Effect of dehydroepiandrosterone supplementation on bone mineral density, bone markers, and body composition in older adults: The DAWN trial. *Osteoporosis Int* [serial online] 2007;doi: 10.1007.

Weiss, E. P., K. Shah, L. Fontana et al. 2009. Dehydroepiandrosterone replacement therapy in older adults: 1- and 2-y effects on bone. *Am J Clin Nutr* 89:1459–67.

10 DHEA and Vascular Health and Functions

Roxane Paulin, Eric Dumas De La Roque, and Sébastien Bonnet

CONTENTS

INTRODUCTION

Dehydroepiandrosterone (DHEA) is an adrenal steroid abundantly produced in humans (Parker 1999), which circulates as a sulfate-conjugated form, DHEAS. Circulating DHEAS serves as a reservoir for DHEA, and DHEAS plasma concentrations are approximately 20-fold higher than any other steroids (Migeon et al. 1957; Baulieu et al. 1965). DHEAS reaches a maximal plasma level between 15 and 25 years of age and declines continuously after the third decade of life, and therefore differs from other major adrenocorticol steroids like cortisol and aldosterone. The age-related decline in DHEAS (Baulieu 2002; Belanger et al. 1994; Labrie et al. 1997) has suggested that a relative deficiency of this steroid may be causally related to the development of aging-associated diseases. This has led to several hypotheses on the possible role of DHEAS in the pathophysiology of cardiovascular diseases (Barrett-Connor, Khaw, and Yen 1986), insulin resistance (Schriock et al. 1988), cancer (Schwartz, Pashko, and Whitcomb 1986), reduction of the immune defense (Casson et al. 1993), and depression or deterioration in the sensation of well-being (Morales et al. 1994).

DHEA Synthesis

DHEA is primarily secreted by the zona reticularis of the adrenal cortex in humans and other primates. This secretion is controlled by adrenocorticotropin and other pituitary factors (Nieschlag et al. 1973). In all, 75%–90% of the body's DHEA is secreted daily by the adrenal cortex, and the remainder is produced by testes and ovaries (de Peretti and Forest 1978; Nieschlag et al. 1973). DHEAS can be interconverted with DHEA by DHEA sulfotransferase and hydroxysteroid sulfatase (Baulieu 2002), and the interconversion seems to be continuous as suggested by pharmacokinetic analysis following oral DHEA administration (Young et al. 1997; Arlt et al. 1998). Serum DHEAS concentration does not vary throughout the day, whereas DHEA secretion follows a diurnal rhythm similar to that of cortisol. The differences in tissue-specific expression of DHEA sulfotransferase and steroid sulfatase determine the balance between DHEA stored and DHEA metabolized (Allolio and Arlt 2002).

DHEA Metabolism

In humans and higher primates, active sex steroids are, in large part, synthesized locally in peripheral tissues, providing adjustment to local requirement. Adrenal precursor steroids, DHEAS and DHEA, are transformed into androgens and/or estrogens in peripheral target tissues by various steroidogenic and metabolizing enzymes, depending upon the level of each enzyme expression in each tissue (Labrie et al. 2005). The importance of DHEA and DHEAS is illustrated by the finding that approximately 50% of total androgens in the prostate of an adult man derives from these adrenal precursor steroids (Labrie et al. 1993). In women, 75% of estrogen formation depends on these adrenal precursor steroids before menopause and close to 100% after menopause (Labrie 1991).

DHEA Properties

DHEA is a potent uncompetitive inhibitor of mammalian glucose-6-phosphate dehydrogenase (G6PDH), the first enzyme in the pentose phosphate pathway (PPP), which is a major source of ribose-5-phosphate and extramitochondrial nicotinamide adenine dinucleotide phosphate hydrogen (NADPH) (Raineri and Levy 1970; Gordon, Mackow, and Levy 1995). G6PDH is indispensable for the cytosolic pool of NADPH maintenance and is critical for cellular redox balance. Indeed, NADPH is a critical modulator of the cellular redox state and serves as a reductant for several enzymes generating reactive oxygen species (ROS)/reactive nitrogen species (RNS) like nitric oxide (NO) synthase and enzyme of the Fenton reaction catalyzing hydroxyl radical formation from H_2O_2 (Imlay and Linn 1988). Thus, a reduction in the supply of NADPH could have a profound effect on ROS production (Figure 10.1).

In vitro and *in vivo* studies in Sardinian men bearing the Mediterranean variant of G6PDH deficiency support the hypothesis that reduced G6PDH activity has beneficial effects on age-related disease development. Epidemiological studies have found these Sardinian men have a reduced mortality from cerebrovascular and cardiovascular diseases and are more likely to achieve centenarian status than their normal counterparts (Cocco et al. 1998).

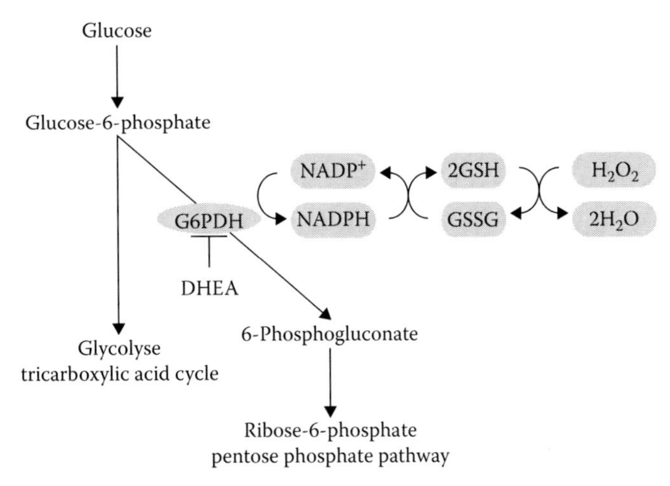

FIGURE 10.1 Glucose-6-phosphate dehydrogenase is the rate-limiting enzyme in the pentose phosphate pathway.

CLINICAL DATA ON DHEA IN HUMAN HEALTH

DHEA and Mortality

In men, an inverse correlation between DHEA or DHEAS plasma concentrations and mortality has been shown in epidemiological studies. The prospective observational study of Barrett-Connor et al. (Rancho Bernardo Study, 1986) involving 1029 men followed for 19 years has shown for the first time an inverse relationship between serum DHEAS levels and mortality risk. This correlation was weak (relative risk of 0.85) but statistically significant. The prospective Personnes Agées QUID (PAQUID) study, including 622 subjects, has also shown higher mortality in men with lower DHEAS plasma concentrations (Berr et al. 1996; Mazat et al. 2001). More recently, in a prospective population-based study including 2644 men (aged 69–81 years), it has been shown that low serum levels of DHEAS can predict death from cardiovascular and ischemic heart diseases in older men (Ohlsson et al. 2010).

In contrast, there is no relationship between serum DHEAS levels and cardiovascular disease mortality in women (Barrett-Connor and Goodman-Gruen 1995a, b, Barrett-Connor and Khaw 1987; Berr et al. 1996; Legrain et al. 1995; Mazat et al. 2001; Tilvis, Kahonen, and Harkonen 1999; Trivedi and Khaw 2001).

DHEA and Cardiovascular Diseases

Prospective studies comparing serum DHEA or DHEAS levels with the prevalence of cardiovascular diseases are inconclusive in both men and women.

Two 5-year follow-up studies found no correlation between low DHEAS levels and carotid atherosclerosis progression or intima/media thickness (Kiechl et al. 2000; Golden et al. 2002). In a recent study, DHEAS was found to be an independent negative factor for carotid atherosclerosis in men but not in women (Yoshida et al. 2010). In ischemic heart disease, high DHEA or DHEAS plasma levels were associated with lower risks of coronary atherosclerosis (Slowinska-Srzednicka et al. 1989; Herrington 1995). A longitudinal 10-year follow-up study found that low plasma concentration of DHEAS is a predictor of cardiovascular mortality in survivors of myocardial infarction (Jansson, Nilsson, and Johnson 1998). Two other studies did not find the same significant association (Newcomer et al. 1994; Contoreggi et al. 1990), and another study involving smaller sample size found increased DHEAS in some myocardial infarction cases (Hautanen et al. 1994).

These discrepancies among studies are still unresolved. Smoking could have a possible confounding effect on DHEAS and cardiovascular diseases. DHEAS plasma levels are higher in smokers and lower in those who had never smoked, and former smokers have DHEAS levels situated between these two groups (Khaw, Tazuke, and Barrett-Connor 1988; Salvini et al. 1992). It has also been hypothesized that differences may be attributable to population variability. In a study by Mazat et al. (2001), the relative risk of an 8-year mortality associated with low DHEAS was 3.4 times higher in men under 70 than in older men (odds ratio of 6.5 vs. 1.9). Finally, most studies have relied on a single measure of DHEAS performed several years before the disease events. Diurnal variations in the levels of DHEAS or possible alterations in the sample over time may have had an impact on the association observed (Wu and von Eckardstein 2003).

DHEA AND VASCULAR FUNCTIONS

Vascular Smooth Muscle Cell Contractile Mechanisms

Calcium-Dependent Contraction

The vascular smooth muscle cell (VSMC) is a highly specialized cell whose primary function is contraction and relaxation. This contractile phenotype is necessary for proper function of many organs including systemic and airway vasculature (Berridge 2008; Owens 2007). Contraction is initiated by changes in the interactions between actin and myosin (Somlyo et al. 1993). Myosin light chain (MLC) kinase is activated by a Ca^{2+}/calmodulin (CaM)-dependent mechanism, which leads to Ser19 phosphorylation of the 20-kDa regulatory MCL, enabling the interaction between myosin and actin.

Numerous Ca^{2+} signaling pathways have been described to regulate VSMC contractility (Figure 10.2). Usually, a neuronal or humoral agonist acts through membrane receptors coupled to phospholipase C (PLC) to produce second messengers inositol-1,4,5-triphosphate (IP_3) and diacylglycerol (DAG). The binding of IP_3 to receptors on the sarcoplasmic reticulum (SR) causes a Ca^{2+} release from SR to cytosol. The primary target protein of this initial release of Ca^{2+} is CaM. Since internal stores are empty, sustained increase in VSMC intracellular Ca^{2+} depends on Ca^{2+} entry

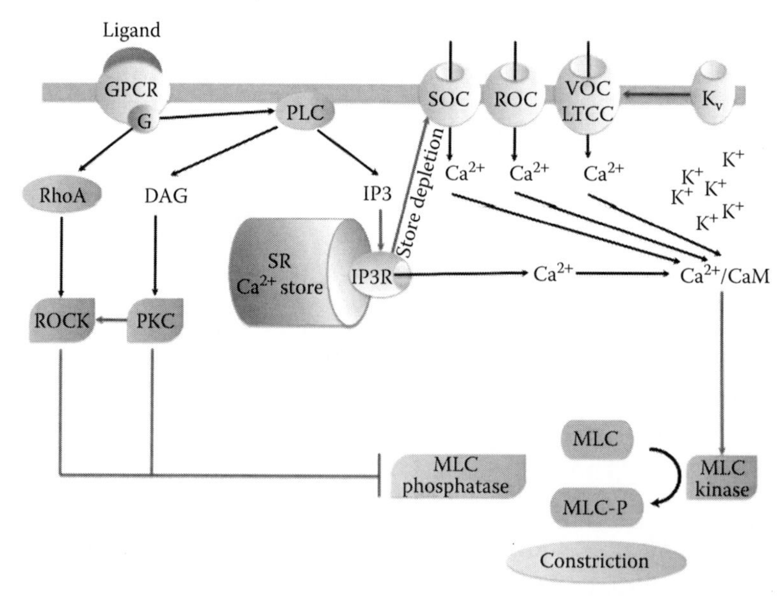

FIGURE 10.2 Simplified view of vascular smooth muscle cell contraction.

through plasma membrane channels that are required for the replenishment of intracellular stores. Extracellular Ca^{2+} enters through either voltage-operated Ca^{2+} channels (VOC L- and T-type calcium channels) or non–voltage-operated Ca^{2+} channels (Guibert, Ducret, and Savineau 2008).

VOC opening is stimulated by membrane depolarization (Tsien, Ellinor, and Horne 1991; van Breemen and Saida 1989). Since the resting membrane potential (E_m) is principally governed by potassium permeability (Yuan 1995), VOC activation is mediated by potassium channels. In vascular smooth muscle, at least five types of K^+ channel have been described, and among these channels, voltage-activated K^+ channel (K_v) are predominantly responsible for resting membrane potential (Nelson and Quayle 1995; Sobey 2001; Korovkina and England 2002). Conductance, open probability, and expression are the three parameters of a K_v channel, which mediates current amplitude. Therefore, inhibition or downregulation of a K_v channel is sufficient to cause a membrane depolarization and to enhance contraction.

Another type of K^+ channel is the Ca^{2+}-activated K^+ channel, subdivided according to its conductance. The large-conductance, Ca^{2+}-activated K^+ channel (BK_{Ca}) is activated by changes in intracellular Ca^{2+} concentrations and contributes to membrane potential maintenance (Ledoux et al. 2006). The efflux of K^+ resulting from BK_{Ca} channel activation can be used to counteract membrane depolarization and vasoconstriction.

In pulmonary arterial hypertension (PAH), sustained pulmonary vasoconstriction is due to both gene expression and functional changes in K_v channels (Michelakis et al. 2002) and in decreased BK_{Ca} current due to a diminished BK_{Ca} protein expression (Bonnet et al. 2003b).

Non-VOC channels consist of receptor-operative Ca^{2+} channels (ROC) and store-operated Ca^{2+} channels (SOC). ROC channels are regulated by agonist–receptor interactions (Guibert, Marthan, and Savineau 2004; Large 2002; Nelson et al. 1990), whereas SOC channels are activated by Ca^{2+} depletion from the SR (Parekh and Penner 1997; Albert and Large 2003).

The reversal of VSMC contractility, called "vasorelaxation or recovery phase," is mediated by mechanisms limiting Ca^{2+} entry (membrane repolarization) and by Ca^{2+} clearance. Clearance depends on plasma membrane Ca^{2+} ATPases (Carafoli et al. 1994), which pump Ca^{2+} out of the cell, and sarcoendoplasmic reticulum Ca^{2+} ATPases (SERCA) (Lytton et al. 1992), which pump cytosolic Ca^{2+} back into SR (Figure 10.3).

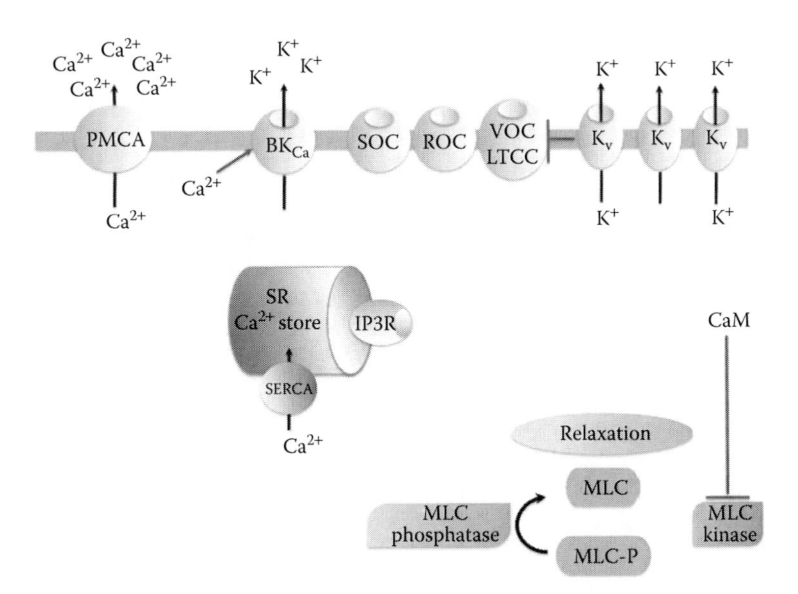

FIGURE 10.3 Simplified view of vascular smooth muscle cell relaxation.

Crosstalk between Ca^{2+} and ROS/RNS Signaling

ROS and RNS represent a large class of molecules regulating many aspects of VSMC biology, including contraction, proliferation, and migration. ROS and RNS are synthesized in mitochondria by NADPH oxidases and NO oxidases (Bryan et al. 2004). Many enzymes involved in ROS and RNS syntheses are Ca^{2+} sensitive, and changes in intracellular Ca^{2+} signals lead to rapid modulation of ROS/RNS production. Reciprocally, ROS/RNS regulates VSMC Ca^{2+} signaling through site-specific modifications of amino acid residues such as oxidation of methionin residues to sulfoxides, nitrosation of cystein residues, carbonylation and deamination of amino acid side chain, or nitration. Functionally, these result in alteration of protein functions or in increased protein susceptibility to degradation and fragmentation.

Although little is known about the role of ROS/RNS in VSMC Ca^{2+} signaling, an emerging paradigm for the action of ROS/RNS on Ca^{2+} signaling from a variety of cell types seems to involve the inhibition of Ca^{2+} pumps and activation of Ca^{2+} release/entry channels, increasing overall cytoplasmic Ca^{2+} concentration (Hidalgo and Donoso 2008; Touyz 2005).

Ca^{2+}-Independent Contraction

Contractility of vascular smooth muscle is regulated not only by intracellular Ca^{2+} but also by Ca^{2+}-independent mechanisms. The state of MLC phosphorylation is further regulated by MLC phosphatase. Some reports have shown that the small G protein RhoA and its downstream target Rho kinase (ROCK) play an important role in the regulation of MLC phosphatase activity (Uehata et al. 1997). ROCK phosphorylates MCL phosphatase and inhibits its activity, thus promoting a phosphorylated state of MCL and leading to contraction (Figure 10.2).

The second messenger, DAG, which is released after PLC activation, enhances protein kinase C (PKC) implicated in specific target protein phosphorylation. In many cases, PKC has constriction-promoting effects such as phosphorylation of many kinases including MLC kinase, extracellular signal–regulated kinases (ERK)1/2, ROCK, and CaM-dependent protein kinase II, as well as various ion channels and ion transporters.

DHEA-Dependent VSMC Vasodilation

DHEA as a K^+ Channel Opener

Several studies have demonstrated that NO mediates VSMC relaxation. As mentioned, NO synthesis is dependent on NADPH levels. It could be postulated that reduction of NADPH by DHEA-dependent PPP inhibition would impair NO-related relaxation. On the other hand, free radicals generated in smooth muscle cells (SMC) and changes in the ratio of cellular reducing factor after inhibition of PPP, including $NADP^+$/NADPH and glutathione/oxidized glutathione (GSH/GSSG), are thought to open K_v and BK_{Ca} channels and hyperpolarize plasma membrane by binding to the channel protein or by regulating the redox state of the cysteine residues (Lee and Tan 1975; Weir and Archer 1995; Michelakis et al. 1997). Following this principle, it was demonstrated that DHEA inhibits hypoxic pulmonary vasoconstriction and opens BK_{Ca} channels in pulmonary VSMC at 100μM (Farrukh et al. 1998; Peng, Hoidal, and Farrukh 1999). Gupte et al. (2002) investigated this paradox of PPP inhibitors using DHEA, among others, and found that inhibition of the major source of NADPH in VSMC induces vasodilation in pulmonary arteries and aorta. By using specific K^+ channel inhibitors, they identified that only K_v channels are positively implicated in DHEA-dependent relaxation of VSMC. DHEA also prevents and reverses chronic hypoxia-induced pulmonary hypertension in rats by BK_{Ca} opening (Bonnet et al. 2003a). Western blot analysis of arterial pulmonary extract showed that the BK_{Ca} subunit expression is upregulated after DHEA treatment in chronic hypoxia rats models compared to control rats without DHEA treatment.

Upregulation of Guanylate Cyclase

Cyclic guanosine-3′,5′-monophosphate (cGMP) is a factor controlling vascular tone and generated in vasculature via two main guanylate cyclases: cytosolic soluble guanylate cyclase (sGC) and membrane-bound particulate guanylate cyclase (Munzel et al. 2003). sGC serves as a receptor for biologically active gas NO (Friebe and Koesling 2003; Moncada and Higgs 2006), and cGMP is generated by sGC following this interaction. The effects of cGMP in vascular tissues are mediated via a number of effectors including cGMP-dependent protein kinase (cGK) (Hofmann et al. 2006), cGMP-regulated phosphodiesterase (PDE) (Conti and Beavo 2007), and cGMP-modulated cation channels (Biel and Michalakis 2007). cGK phosphorylates a number of target proteins including BK_{Ca}, SERCA, and G protein, lowering intracellular calcium and enhancing vasorelaxation (Hofmann et al. 2006).

DHEA has been reported to block hypoxic pulmonary hypertension in rats by increasing sGC protein expression and activity and by improving pulmonary artery vasodilator responsiveness to NO (Oka et al. 2007). Mechanisms implicated in DHEA-induced sGC upregulation have not been fully studied. Nevertheless, H_2O_2 has been reported to stimulate sGC and increase cGMP in vasculature (Sato, Sakuma, and Gutterman 2003; Fujimoto et al. 2001), suggesting that the effects of DHEA on sGC may not be direct but secondary to PPP inhibition and increased H_2O_2 production.

Inhibition of Rho GTPase

As mentioned, activation of the RhoA/ROCK signaling pathway contributes to vasoconstriction in VSMC. Accumulating evidence indicates that this pathway plays an important role in the pathogenesis of PAH and that RhoA inhibitors, like 3-hydroxy-3-methylglutaryl coenzyme A (HMG-CoA) reductase inhibitors, ameliorate PAH in several different rat models (Murata et al. 2005; Nishimura et al. 2003). Also, it has been reported that the protective effect of the PDE-5 inhibitor sildenafil against PAH involves inhibition of the RhoA/ROCK signaling pathway (Guilluy et al. 2005). Considering the fact that DHEA has an inhibitory effect on HMG-CoA reductase (Pascale et al. 1995) and an enhancing effect on sGC (i.e., modulation of PDE-5 activity), it is suggested that DHEA-enhanced relaxation could implicate RhoA/ROCK signaling inhibition. This hypothesis was studied by Homma et al. (2008) in a PAH rat model. They demonstrated that chronic DHEA treatment decreases the pathological RhoA/ROCK signaling pathway activity by multiple mechanisms, including inhibition of HMG-CoA reductase overexpression, preservation of sGC expression, and inhibition of ROCK cleavage.

DHEA AND VASCULAR HEALTH

There is considerable evidence that cardiovascular risks are related to age-associated decline in sex hormone activity like testosterone, DHEA, and DHEAS. The most common explanation is that these hormones exert a direct effect on the arterial tree. Some studies have examined the relationship between androgenic steroids, DHEA included, and arterial wall properties.

DHEA and Arterial Compliance/Stiffness

Increased arterial stiffness (or decreased compliance) is a consequence of age and pathologies like arteriosclerosis. Age-related stiffness results in a reduction of elasticity within the arterial wall and a global stiffening of the arterial tree. This mechanically leads to systolic blood pressure elevation (O'Rourke and Avolio 1980) and left ventricular hypertrophy (Nichols et al. 1985). Arterial stiffness is evaluated by measurement of aortic pulse wave velocity (PWV), which represents the speed of travel of each pulse wave that runs through circulation after each heart contraction. Ishihara et al. (1992) showed an inverse relationship between DHEAS and aortic PWV in men. A similar association was found in 2001 between testosterone (molecule derived from DHEA) and PWV, but this association was lost after age adjustment (Gyllenborg et al. 2001). It also has been demonstrated that testosterone suppression in men with prostate cancer results in increased vascular stiffness, suggesting that loss of androgens might be harmful in men (Smith et al. 2001; Dockery et al. 2003). In women, it was shown that age-associated increases in arterial stiffness are attenuated by postmenopausal estrogen replacement therapy (Nagai et al. 1999; Scuteri et al. 2001). Hougaku et al. (2006) failed to show a significant correlation between androgenic steroids, traditional risk factors, and arterial wall properties simultaneously.

DHEA and Endothelial Functions

It is well recognized that cardiovascular diseases are associated with vascular endothelium dysfunctions (Clarkson et al. 1997; Celermajer et al. 1992, 1996), and numerous actions of estrogen and androgen on vascular endothelium have been described (Nuedling et al. 1999; White et al. 1997; Gilligan et al. 1994; Morales et al. 1995; Chou et al. 1996). Thereafter, more specialized effects of DHEA on endothelial functions were examined (Figure 10.4).

DHEA Enhances NO Production in Vascular Endothelium

The first evidence of physiological action of DHEA on vascular endothelium is the characterization of a high-affinity receptor for DHEA in isolated plasma membranes of vascular endothelial cells (Liu and Dillon 2002). DHEA is not converted to estrogen or androgen by endothelial cells, and its effects are not blocked by antagonists of estrogen, progesterone, glucocorticoid, or androgen receptors, confirming that DHEA acts through a specific receptor (Simoncini et al. 2003). The authors showed that DHEA is able to directly regulate human vascular wall by controlling NO synthesis in endothelial cells. DHEA binding on plasma membrane was linked to endothelial nitric oxide synthase (eNOS) activation and NO production through a G-protein–coupled mechanism in both bovine and human vascular endothelial cells (Liu and Dillon 2004). This mechanism was shown to depend on intracellular kinase activation but not on intracellular calcium rise. At this level, an understanding of the precise mechanisms implicated in signal transduction remains controversial.

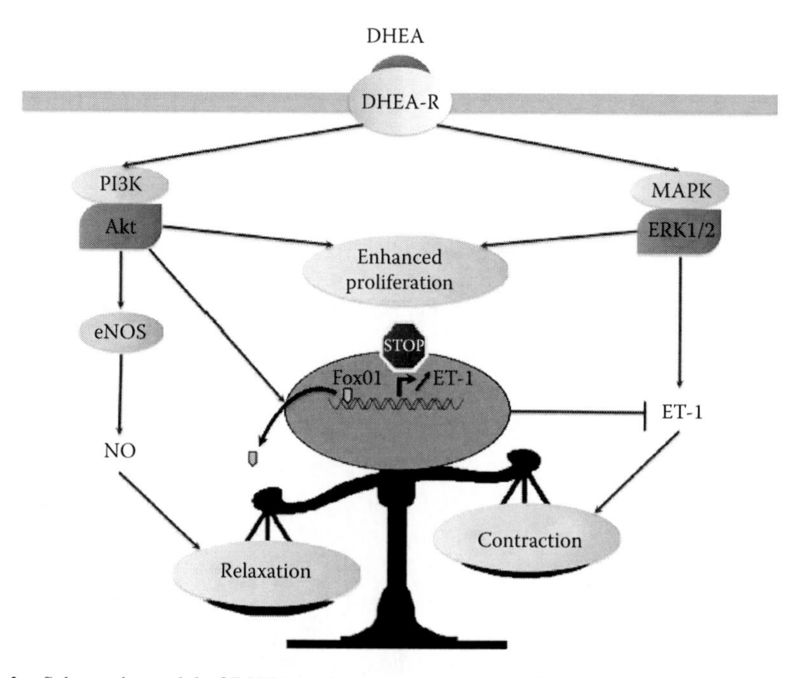

FIGURE 10.4 Schematic model of DHEA actions on vascular endothelial signaling pathways.

Simoncini et al. (2003) established that DHEA administration to human endothelial cells triggers NO synthesis through a nontranscriptional mechanism that depends on ERK1/2-mitogen-activated protein kinase (MAPK) but not on phosphatidylinositol-3-kinase (PI3K)/Akt. Formoso et al. (2006) demonstrated that acute DHEA action causes a significant increase in NO production through activation of a PI3-kinase–dependent pathway and a secondary acute Akt and eNOS phosphorylation. Thus, the authors compared DHEA action to insulin, estrogen, and corticosteroid action on vascular endothelium. They demonstrated that DHEA, like insulin, also mediates endothelin (ET)-1 secretion via MAPK-dependent activation, confirming their first hypothesis. Recently, it has been shown that DHEA administration protects hypertension-induced kidney injury via upregulation of the sigma-1R receptor and stimulation of Akt-eNOS signaling in ovariectomized rats (Bhuiyan et al. 2010), further confirming the results of Formoso et al. (2006). In this case, DHEA stimulates the production of both vasodilators and vasoconstrictors in endothelium via two distinct intracellular signaling pathways. The net vasoactive effect of DHEA may depend on the balance between these pathways. Chen et al. (2008) demonstrated that DHEA stimulates phosphorylation of the transcription factor FoxO1 via the PI3K pathway in endothelial cells, negatively regulating ET-1 promoter activity and secretion. Balance between PI3K-dependent inhibition and MAPK-dependent stimulation of ET-1 secretion in response to DHEA may determine whether DHEA supplementation improves or worsens cardiovascular and metabolic functions. It is also possible that discrepancies between different studies are due to differences in cell type, duration of stimulation with DHEA, and differences in NO production measurement.

DHEA Increases Proliferation and Protects Vascular Endothelium against Apoptosis

Endothelial injury or loss of endothelial cells due to aging-induced apoptosis contributes to development of vascular remodeling diseases such as arteriosclerosis (Hoffmann et al. 2001) and pulmonary hypertension (Voelkel and Tuder 1995). Considering the fact that PI3K/Akt and ERK1/2-MAPK are proliferating pathways enhanced by DHEA binding on specific cell membrane receptors, hypotheses based on possible DHEA abilities to protect vascular endothelial cells against apoptosis were examined. First, DHEA is able to protect cultured endothelial cells from serum starvation–induced apoptosis. This cytoprotective effect of DHEA is at least in part dependent on PI3K/Akt signaling pathway activation, increasing the antiapoptotic Bcl-2 protein expression (Liu et al. 2007). Moreover, evidence of ERK1/2 implication in DHEA-dependent enhanced proliferation has been shown. DHEA significantly increases fetal bovine serum–induced cell proliferation in cultured bovine aortic endothelial cells, associated with an increased ERK1/2-MAPK expression. Antagonists of either androgen or estrogen receptors do not abolish this effect. The eNOS inhibitor Nω-nitro-L-arginine methyl ester (L-NAME) does not affect this DHEA-dependent proliferation, showing that it is the result of a direct DHEA binding on its receptor, and it is not a consequence of a NO-dependent enhanced proliferation through other pathways (Liu et al. 2008). Furthermore, healthy menopausal women (a group with low circulating estrogen and androgen levels) treated with DHEA for 12 weeks exhibited a significant increase in endothelium-mediated vascular reactivity in both large and small blood vessels, without blood pressure or plasma lipid profile changes. DHEA influenced endothelial function in a potentially beneficial manner both *in vitro* and *in vivo* (Williams et al. 2004).

DHEA and Intima/Media Thickness

Under normal conditions, vascular wall thickness is maintained by a fine regulation of the balance between proliferation and apoptosis. When this balance is altered in favor of proliferation, the vascular wall thickness is increased, obstructing the lumen, increasing vascular resistance, and disrupting blood flow. Media and intima are layers particularly affected by thickening. VSMC undergo phenotypic modulation from contractile to proproliferative and antiapoptotic phenotype.

Mechanisms accounting for this phenotype are not fully understood but have been described several times to be, at least in part, dependent on Ca^{2+} signaling molecules and endothelium safety.

Ca^{2+}-Dependent VSMC Proliferation

In quiescent VSMC, the Ca^{2+} signal consists mainly of localized elementary calcium events. Calcium pumps, keeping the cytoplasmic Ca^{2+} concentration low, rapidly reduce global increase in Ca^{2+} concentration. Chronic increase in cytosolic Ca^{2+} concentration and its ubiquitous intracellular receptor CaM are required for cell proliferation. Indeed, some essential enzymes for the transition step of the cell cycle like multifunctional Ca^{2+}/CaM-dependent protein kinase, calcineurin, and spindle pole body protein are Ca^{2+}/CaM-dependent and inactive in the absence of Ca^{2+}/CaM (Means 1994). Intracellular Ca^{2+} pools may also play important roles in cell proliferation independently of CaM. Depletion of IP_3-sensitive intracellular Ca^{2+} with thapsigargin results, for example, in cell cycle arrest (Short et al. 1993). Ca^{2+} is also able to act directly on transcription factors (such as the downstream regulatory element antagonistic modulator [DREAM]) or indirectly through protein phosphatase (calcineurin/nuclear factor of activated T-cell [NFAT]) or kinases (CaM kinases/cyclic adenosine monophosphate responsive element binding protein [CREB], PKC/nuclear factor-κB) to induce the activation of numerous target genes.

Mitogen-Induced VSMC Proliferation

Age- or injury-dependent endothelium loss leads to contact between other layers and circulating molecules such as cytokines and growth factors, increasing the activation of proliferating pathways within the vascular wall.

Platelet-derived growth factors (PDGFs) can induce replication, migration, and contraction on VSMC. These actions are crucial in pathophysiological processes in vasculature. PDGFs, like sustained intracellular Ca^{2+} concentration, seem to be another example of the existing relation between contraction and proliferation and could be an important signal mediator of VSMC phenotype shift. PDGFs bind to specific cell-surface receptor tyrosine kinases and generate a mitogenic signal transduction through the phosphorylation of specific tyrosine. MAPK/ERK1/2 and PI3K/Akt have been reported to be the signaling pathways through which the growth factor receptors elicit their physiological responses (Andrae, Gallini, and Betsholtz 2008; Graf et al. 1997; Kim and Yun 2007).

PDGF-induced MAPK activation is involved in vascular remodeling (Zhan et al. 2003; Iijima et al. 2002) and plays a crucial role in regulating signaling pathways implicated in cellular events leading to restenosis (Pyles et al. 1997; Ohashi et al. 2000).

PI3K/Akt signaling blocks apoptotic cell death (Jung et al. 2000; Datta, Brunet, and Greenberg 1999). PI3K was found to upregulate the expression of several cell cycle proteins in coronary smooth muscle cells after both growth factor stimulation and balloon injury, consistent with a role in restenosis (Braun-Dullaeus et al. 2001). Another study showed that Akt is induced by balloon injury and increases cyclin D1 expression, and that this effect is blocked by the PI3K inhibitor wortmannin (Shigematsu et al. 2000). Once activated, Akt phosphorylates and inactivates several downstream targets that regulate survival pathways in VSMC and induces apoptosis, including glycogen synthase kinase (GSK)-3β (Allard et al. 2008). By inhibiting GSK-3β, Akt promotes NFAT translocation into nucleus, thus increasing the expression of a proproliferative gene (Bcl-2) and decreasing the expression of proapoptotic genes like K_v channels (Bonnet et al. 2007a, b, 2009). NFAT appears to be a key mediator of VSMC proliferation/apoptosis because its activation depends on both calcium signaling and mitogen stimulation.

Mitochondria-Dependent Apoptosis

The mitochondrial-dependent apoptosis pathway is activated by cellular deprivation or stress, and it involves sequential release of cytochrome *c*, recruitment of apoptotic protease activating factor-1, and activation of caspases. Defects in this pathway have been identified in the development of atherosclerosis (Dzau, Braun-Dullaeus, and Sedding 2002), restenosis, and myocardial infraction (Rossi et al. 2004). Proapoptotic factor release is dependent on mitochondrial transition pore (MTP) opening. MTP is both voltage and redox sensitive (Zamzami et al. 2001). For example, mitochondrial membrane potential $\Delta\Psi m$ depolarization and increased ROS production lead to MTP opening and enhanced apoptosis. GSK-3β has also been shown to regulate MTP opening by blocking hexokinase translocation to the mitochondria, where it interacts with the voltage-dependent anion channel (an MTP subunit) and regulates $\Delta\Psi m$ (Pastorino, Hoek, and Shulga 2005). Therefore, GSK-3β inhibition regulates both decreased apoptosis by $\Delta\Psi m$ hyperpolarization and MTP closure, and enhanced proliferation by NFAT activation.

K^+ Channel–Dependent Apoptosis

Activation of K^+ channels, leading to K^+ efflux, is associated with apoptosis through two main mechanisms. First, cell shrinkage (or volume loss), also called "apoptotic volume decrease" (AVD), is an early step of apoptosis, which starts before cell fragmentation. To drive the net efflux of water, which leads to cell shrinkage, release of anions should take place in parallel with K^+ release. Thus, K^+ channels play an important role in apoptosis by modulating K^+ flux across the membrane. K^+ channel activation leads to K^+ efflux, which induces or accelerates AVD and apoptosis (Krick et al. 2001). Second, intracellular K^+ concentration tonically inhibits caspases in a variety of cell types (Remillard and Yuan 2004).

DHEA and Proliferation/Apoptosis Balance

The first report providing evidence that DHEA attenuates VSMC proliferation was published in 1997. It was described that DHEA attenuates serum-induced proliferation of human aortic SMC (Yoneyama et al. 1997). Later, it was demonstrated that DHEA treatment of human aortic SMC inhibits PDGF-induced MAPK activation (Yoshimata et al. 1999). In human internal mammary artery, DHEA significantly decreases PDGF-induced ERK1 kinase activity in a dose-dependent manner and independently of both androgen and estrogen receptors (Williams et al. 2002). Recently, in human carotid VSMC, we demonstrated that DHEA could inhibit PDGF-induced Akt activation. Moreover, DHEA-dependent Akt inhibition was found to be associated with downregulation of NFAT activity, secondary to activation of GSK-3β (Bonnet et al. 2009). These results were confirmed *in vivo* by a decrease of vascular remodeling in a rat model of carotid stenosis treated with DHEA.

Several mechanisms could be implicated in DHEA inhibition of PDGF-induced VSMC proliferation. Urata et al. (2010) have proposed that the level of GSH was increased by DHEA, resulting in a high GSH/GSSG ratio. This result is contrary to expectations since DHEA, as a G6PDH inhibitor, reduces the conversion of GSSG into GSH. Authors showed that an increase in glutaredoxin 1 (GRX1, oxidized GRX1 is reduced by GSH) and γ-glutamylcysteine synthase (γ-GCS, for GSH precursor transformation) is associated with DHEA treatment as a compensatory mechanism, leading to elevated GSH/GSSG ratio. GRX1 plays an important role in PDGF signal regulation by downregulating tyrosine phosphorylation of the PDGF receptor (Kanda et al. 2006). It was demonstrated that PDGF-induced proliferation is inhibited by DHEA through a GSH/GRX1 mechanism (Urata et al. 2010). A promoter analysis of GRX1 and γ-GCS showed that DHEA upregulates transcriptional activity at the peroxisome proliferator–activated receptor (PPAR) response element, suggesting that PPAR-α plays a role in DHEA-dependent GRX1/γ-GCS induction. A previous study has implicated PPAR-α in vascular response to DHEA. The authors showed that DHEAS-dependent inhibition of vascular remodeling and reduced neointima formation after vascular injury are associated with activation of PPAR-α and upregulation of the cyclin-dependent kinase inhibitor p16(INK4a), resulting in G1 cell cycle arrest (Ii et al. 2009) (Figure 10.5).

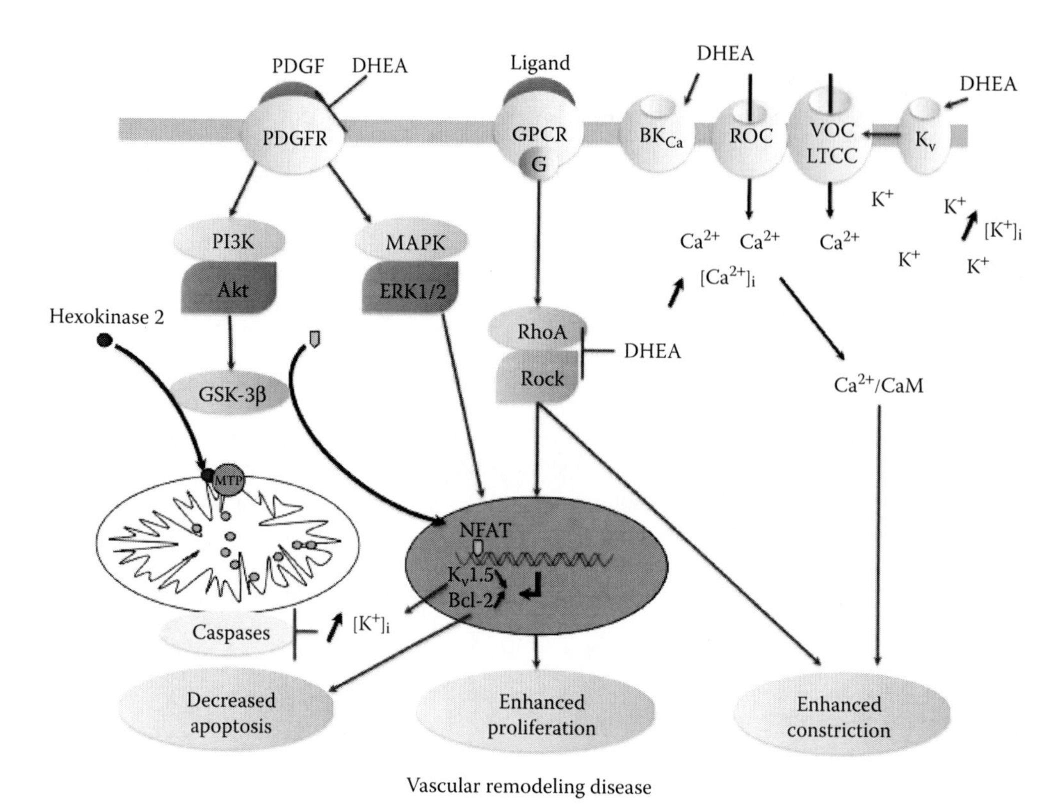

FIGURE 10.5 DHEA prevents vascular remodeling diseases by restoring vascular smooth muscle cell phenotype.

CONCLUSION AND PERSPECTIVES

The mechanisms of DHEA action in vasculature are better understood, but it is still a matter of discussion whether DHEA exerts its effects mainly by direct action or by peripheral bioconversion to androgen and estrogen. DHEA is accepted as a vasodilating agent and an antiproliferative drug. Moreover, DHEA promotes endothelial function, restoring endothelium safety and global vascular protection. Since it is a safe drug, and already clinically available, clinical trials should come in the near future.

A new area in the treatment of cardiovascular diseases has been described with drug-eluting stent implantation. Agents inhibiting cell cycle progression coated onto stents have been tested as inhibitors of VSMC proliferation. Favorable results were obtained with sirolimus, an inhibitor of the mammalian target of rapamycin, and paclitaxel, a chemotherapeutic agent inducing G2/M cell cycle arrest (Wessely, Schomig, and Kastrati 2006), which inhibited neointima formation in animal models. The rationale of this drug delivery system is to inhibit VSMC proliferation but also to promote reendothelization directly on lesion. Since DHEA is a molecule able to do both, it will be interesting to consider DHEA as a new candidate for drug-eluting stents.

REFERENCES

Albert, A. P., and W. A. Large. 2003. Store-operated Ca2+-permeable non-selective cation channels in smooth muscle cells. *Cell Calcium* 33:345–56.

Allard, D., N. Figg, M. R. Bennett, and T. D. Littlewood. 2008. Akt regulates the survival of vascular smooth muscle cells via inhibition of FoxO3a and GSK3. *J Biol Chem* 283:19739–47.

Allolio, B., and W. Arlt. 2002. DHEA treatment: Myth or reality? *Trends Endocrinol Metab* 13:288–94.

Andrae, J., R. Gallini, and C. Betsholtz. 2008. Role of platelet-derived growth factors in physiology and medicine. *Genes Dev* 22:1276–312.

Arlt, W., H. G. Justl, F. Callies, et al. 1998. Oral dehydroepiandrosterone for adrenal androgen replacement: Pharmacokinetics and peripheral conversion to androgens and estrogens in young healthy females after dexamethasone suppression. *J Clin Endocrinol Metab* 83:1928–34.

Barrett-Connor, E., and D. Goodman-Gruen. 1995a. Dehydroepiandrosterone sulfate does not predict cardiovascular death in postmenopausal women. The Rancho Bernardo Study. *Circulation* 91:1757–60.

Barrett-Connor, E., and D. Goodman-Gruen. 1995b. The epidemiology of DHEAS and cardiovascular disease. *Ann N Y Acad Sci* 774:259–70.

Barrett-Connor, E., and K. T. Khaw. 1987. Absence of an inverse relation of dehydroepiandrosterone sulfate with cardiovascular mortality in postmenopausal women. *N Engl J Med* 317:711.

Barrett-Connor, E., K. T. Khaw, and S. S. Yen. 1986. A prospective study of dehydroepiandrosterone sulfate, mortality, and cardiovascular disease. *N Engl J Med* 315:1519–24.

Baulieu, E. E. 2002. Androgens and aging men. *Mol Cell Endocrinol* 198:41–9.

Baulieu, E. E., C. Corpechot, F. Dray, et al. 1965. An adrenal-secreted "androgen": Dehydroisoandrosterone sulfate. Its metabolism and a tentative generalization on the metabolism of other steroid conjugates in man. *Recent Prog Horm Res* 21:411–500.

Belanger, A., B. Candas, A. Dupont, et al. 1994. Changes in serum concentrations of conjugated and unconjugated steroids in 40- to 80-year-old men. *J Clin Endocrinol Metab* 79:1086–90.

Berr, C., S. Lafont, B. Debuire, J. F. Dartigues, and E. E. Baulieu. 1996. Relationships of dehydroepiandrosterone sulfate in the elderly with functional, psychological, and mental status, and short-term mortality: A French community-based study. *Proc Natl Acad Sci U S A* 93:13410–5.

Berridge, M. J. 2008. Smooth muscle cell calcium activation mechanisms. *J Physiol* 586:5047–61.

Bhuiyan, S., and K. Fukunaga. 2010. Stimulation of Sigma-1 receptor by dehydroepiandrosterone ameliorates hypertension-induced kidney hypertrophy in ovariectomized rats. *Exp Biol Med (Maywood)* 235:356–64.

Biel, M., and S. Michalakis. 2007. Function and dysfunction of CNG channels: Insights from channelopathies and mouse models. *Mol Neurobiol* 35:266–77.

Bonnet, S., S. L. Archer, J. Allalunis-Turner et al. 2007a. A mitochondria-K+ channel axis is suppressed in cancer and its normalization promotes apoptosis and inhibits cancer growth. *Cancer Cell* 11:37–51.

Bonnet, S., E. Dumas-de-La-Roque, H. Begueret et al. 2003a. Dehydroepiandrosterone (DHEA) prevents and reverses chronic hypoxic pulmonary hypertension. *Proc Natl Acad Sci U S A* 100:9488–93.

Bonnet, S., R. Paulin, G. Sutendra et al. 2009. Dehydroepiandrosterone reverses systemic vascular remodeling through the inhibition of the Akt/GSK3-{beta}/NFAT axis. *Circulation* 120:1231–40.

Bonnet, S., G. Rochefort, G. Sutendra et al. 2007b. The nuclear factor of activated T cells in pulmonary arterial hypertension can be therapeutically targeted. *Proc Natl Acad Sci U S A* 104:11418–23.

Bonnet, S., J. P. Savineau, W. Barillot et al. 2003b. Role of Ca(2+)-sensitive K(+) channels in the remission phase of pulmonary hypertension in chronic obstructive pulmonary diseases. *Cardiovasc Res* 60:326–36.

Braun-Dullaeus, R. C., M. J. Mann, U. Seay et al. 2001. Cell cycle protein expression in vascular smooth muscle cells in vitro and in vivo is regulated through phosphatidylinositol 3-kinase and mammalian target of rapamycin. *Arterioscler Thromb Vasc Biol* 21:1152–8.

Bryan, N. S., T. Rassaf, R. E. Maloney et al. 2004. Cellular targets and mechanisms of nitros(yl)ation: An insight into their nature and kinetics in vivo. *Proc Natl Acad Sci U S A* 101:4308–13.

Carafoli, E., and T. Stauffer. 1994. The plasma membrane calcium pump: functional domains, regulation of the activity, and tissue specificity of isoform expression. *J Neurobiol* 25:312–24.

Casson, P. R., R. N. Andersen, H. G. Herrod et al. 1993. Oral dehydroepiandrosterone in physiologic doses modulates immune function in postmenopausal women. *Am J Obstet Gynecol* 169:1536–9.

Celermajer, D. S., M. R. Adams, P. Clarkson et al. 1996. Passive smoking and impaired endothelium-dependent arterial dilatation in healthy young adults. *N Engl J Med* 334:150–4.

Celermajer, D. S., K. E. Sorensen, V. M. Gooch et al. 1992. Non-invasive detection of endothelial dysfunction in children and adults at risk of atherosclerosis. *Lancet* 340:1111–5.

Chen, H., A. S. Lin, Y. Li et al. 2008. Dehydroepiandrosterone stimulates phosphorylation of FoxO1 in vascular endothelial cells via phosphatidylinositol 3-kinase- and protein kinase A-dependent signaling pathways to regulate ET-1 synthesis and secretion. *J Biol Chem* 283:29228–38.

Chou, T. M., K. Sudhir, S. J. Hutchison et al. 1996. Testosterone induces dilation of canine coronary conductance and resistance arteries in vivo. *Circulation* 94:2614–9.

Clarkson, P., D. S. Celermajer, A. J. Powe et al. 1997. Endothelium-dependent dilatation is impaired in young healthy subjects with a family history of premature coronary disease. *Circulation* 96:3378–83.

Cocco, P., P. Todde, S. Fornera et al. 1998. Mortality in a cohort of men expressing the glucose-6-phosphate dehydrogenase deficiency. *Blood* 91:706–9.

Conti, M., and J. Beavo. 2007. Biochemistry and physiology of cyclic nucleotide phosphodiesterases: Essential components in cyclic nucleotide signaling. *Annu Rev Biochem* 76:481–511.

Contoreggi, C. S., M. R. Blackman, R. Andres et al. 1990. Plasma levels of estradiol, testosterone, and DHEAS do not predict risk of coronary artery disease in men. *J Androl* 11:460–70.

Datta, S. R., A. Brunet, and M. E. Greenberg. 1999. Cellular survival: A play in three Akts. *Genes Dev* 13:2905–27.

de Peretti, E., and M. G. Forest. 1978. Pattern of plasma dehydroepiandrosterone sulfate levels in humans from birth to adulthood: Evidence for testicular production. *J Clin Endocrinol Metab* 47:572–7.

Dockery, F., C. J. Bulpitt, S. Agarwal, M. Donaldson, and C. Rajkumar. 2003. Testosterone suppression in men with prostate cancer leads to an increase in arterial stiffness and hyperinsulinaemia. *Clin Sci (Lond)* 104:195–201.

Dzau, V. J., R. C. Braun-Dullaeus, and D. G. Sedding. 2002. Vascular proliferation and atherosclerosis: New perspectives and therapeutic strategies. *Nat Med* 8:1249–56.

Farrukh, I. S., W. Peng, U. Orlinska, and J. R. Hoidal. 1998. Effect of dehydroepiandrosterone on hypoxic pulmonary vasoconstriction: A Ca(2+)-activated K(+)-channel opener. *Am J Physiol* 274:L186–95.

Formoso, G., H. Chen, J. A. Kim et al. 2006. Dehydroepiandrosterone mimics acute actions of insulin to stimulate production of both nitric oxide and endothelin 1 via distinct phosphatidylinositol 3-kinase- and mitogen-activated protein kinase-dependent pathways in vascular endothelium. *Mol Endocrinol* 20:1153–63.

Friebe, A., and D. Koesling. 2003. Regulation of nitric oxide-sensitive guanylyl cyclase. *Circ Res* 93:96–105.

Fujimoto, S., T. Asano, M. Sakai et al. 2001. Mechanisms of hydrogen peroxide-induced relaxation in rabbit mesenteric small artery. *Eur J Pharmacol* 412:291–300.

Gilligan, D. M., D. M. Badar, J. A. Panza, A. A. Quyyumi, and R. O. Cannon 3rd. 1994. Acute vascular effects of estrogen in postmenopausal women. *Circulation* 90:786–91.

Golden, S. H., A. Maguire, J. Ding et al. 2002. Endogenous postmenopausal hormones and carotid atherosclerosis: A case-control study of the atherosclerosis risk in communities cohort. *Am J Epidemiol* 155:437–45.

Gordon, G., M. C. Mackow, and H. R. Levy. 1995. On the mechanism of interaction of steroids with human glucose 6-phosphate dehydrogenase. *Arch Biochem Biophys* 318:25–9.

Graf, K., X. P. Xi, D. Yang et al. 1997. Mitogen-activated protein kinase activation is involved in platelet-derived growth factor-directed migration by vascular smooth muscle cells. *Hypertension* 29:334–9.

Guibert, C., T. Ducret, and J. P. Savineau. 2008. Voltage-independent calcium influx in smooth muscle. *Prog Biophys Mol Biol* 98:10–23.

Guibert, C., R. Marthan, and J. P. Savineau. 2004. 5-HT induces an arachidonic acid-sensitive calcium influx in rat small intrapulmonary artery. *Am J Physiol Lung Cell Mol Physiol* 286:L1228–36.

Guilluy, C., V. Sauzeau, M. Rolli-Derkinderen et al. 2005. Inhibition of RhoA/Rho kinase pathway is involved in the beneficial effect of sildenafil on pulmonary hypertension. *Br J Pharmacol* 146:1010–8.

Gupte, S. A., K. X. Li, T. Okada, K. Sato, and M. Oka. 2002. Inhibitors of pentose phosphate pathway cause vasodilation: Involvement of voltage-gated potassium channels. *J Pharmacol Exp Ther* 301:299–305.

Gyllenborg, J., S. L. Rasmussen, K. Borch-Johnsen et al. 2001. Cardiovascular risk factors in men: The role of gonadal steroids and sex hormone-binding globulin. *Metabolism* 50:882–8.

Hautanen, A., M. Manttari, V. Manninen et al. 1994. Adrenal androgens and testosterone as coronary risk factors in the Helsinki Heart Study. *Atherosclerosis* 105:191–200.

Herrington, D. M. 1995. Dehydroepiandrosterone and coronary atherosclerosis. *Ann N Y Acad Sci* 774:271–80.

Hidalgo, C., and P. Donoso. 2008. Crosstalk between calcium and redox signaling: From molecular mechanisms to health implications. *Antioxid Redox Signal* 10:1275–312.

Hoffmann, J., J. Haendeler, A. Aicher et al. 2001. Aging enhances the sensitivity of endothelial cells toward apoptotic stimuli: Important role of nitric oxide. *Circ Res* 89:709–15.

Hofmann, F., R. Feil, T. Kleppisch, and J. Schlossmann. 2006. Function of cGMP-dependent protein kinases as revealed by gene deletion. *Physiol Rev* 86:1–23.

Homma, N., T. Nagaoka, V. Karoor et al. 2008. Involvement of RhoA/Rho kinase signaling in protection against monocrotaline-induced pulmonary hypertension in pneumonectomized rats by dehydroepiandrosterone. *Am J Physiol Lung Cell Mol Physiol* 295:L71–8.

Hougaku, H., J. L. Fleg, S. S. Najjar et al. 2006. Relationship between androgenic hormones and arterial stiffness, based on longitudinal hormone measurements. *Am J Physiol Endocrinol Metab* 290:E234–42.

Ii, M., M. Hoshiga, N. Negoro et al. 2009. Adrenal androgen dehydroepiandrosterone sulfate inhibits vascular remodeling following arterial injury. *Atherosclerosis* 206:77–85.

Iijima, K., M. Yoshizumi, M. Hashimoto et al. 2002. Red wine polyphenols inhibit vascular smooth muscle cell migration through two distinct signaling pathways. *Circulation* 105:2404–10.

Imlay, J. A., and S. Linn. 1988. DNA damage and oxygen radical toxicity. *Science* 240:1302–9.

Ishihara, F., K. Hiramatsu, S. Shigematsu et al. 1992. Role of adrenal androgens in the development of arteriosclerosis as judged by pulse wave velocity and calcification of the aorta. *Cardiology* 80:332–8.

Jansson, J. H., T. K. Nilsson, and O. Johnson. 1998. von Willebrand factor, tissue plasminogen activator, and dehydroepiandrosterone sulphate predict cardiovascular death in a 10 year follow up of survivors of acute myocardial infarction. *Heart* 80:334–7.

Jung, F., J. Haendeler, C. Goebel, A. M. Zeiher, and S. Dimmeler. 2000. Growth factor-induced phosphoinositide 3-OH kinase/Akt phosphorylation in smooth muscle cells: Induction of cell proliferation and inhibition of cell death. *Cardiovasc Res* 48:148–57.

Kanda, M., Y. Ihara, H. Murata et al. 2006. Glutaredoxin modulates platelet-derived growth factor-dependent cell signaling by regulating the redox status of low molecular weight protein-tyrosine phosphatase. *J Biol Chem* 281:28518–28.

Khaw, K. T., S. Tazuke, and E. Barrett-Connor. 1988. Cigarette smoking and levels of adrenal androgens in postmenopausal women. *N Engl J Med* 318:1705–9.

Kiechl, S., J. Willeit, E. Bonora, S. Schwarz, and Q. Xu. 2000. No association between dehydroepiandrosterone sulfate and development of atherosclerosis in a prospective population study (Bruneck Study). *Arterioscler Thromb Vasc Biol* 20:1094–100.

Kim, T. J., and Y. P. Yun. 2007. Antiproliferative activity of NQ304, a synthetic 1,4-naphthoquinone, is mediated via the suppressions of the PI3K/Akt and ERK1/2 signaling pathways in PDGF-BB-stimulated vascular smooth muscle cells. *Vascul Pharmacol* 46:43–51.

Korovkina, V. P., and S. K. England. 2002. Molecular diversity of vascular potassium channel isoforms. *Clin Exp Pharmacol Physiol* 29:317–23.

Krick, S., O. Platoshyn, M. Sweeney, H. Kim, and J. X. Yuan. 2001. Activation of K+ channels induces apoptosis in vascular smooth muscle cells. *Am J Physiol Cell Physiol* 280:C970–9.

Labrie, F. 1991. Intracrinology. *Mol Cell Endocrinol* 78:C113–8.

Labrie, F., A. Belanger, L. Cusan, J. L. Gomez, and B. Candas. 1997. Marked decline in serum concentrations of adrenal C19 sex steroid precursors and conjugated androgen metabolites during aging. *J Clin Endocrinol Metab* 82:2396–402.

Labrie, F., A. Belanger, A. Dupont et al. 1993. Science behind total androgen blockade: From gene to combination therapy. *Clin Invest Med* 16:475–92.

Labrie, F., V. Luu-The, A. Belanger et al. 2005. Is dehydroepiandrosterone a hormone? *J Endocrinol* 187:169–96.

Large, W. A. 2002. Receptor-operated Ca2(+)-permeable nonselective cation channels in vascular smooth muscle: A physiologic perspective. *J Cardiovasc Electrophysiol* 13:493–501.

Ledoux, J., M. E. Werner, J. E. Brayden, and M. T. Nelson. 2006. Calcium-activated potassium channels and the regulation of vascular tone. *Physiology (Bethesda)* 21:69–78.

Lee, K. T., and I. K. Tan. 1975. A general colorimetric procedure using oxidized chlorpromazine hydrochloride for the estimation of enzymes dependent on NADH/NAD+ and NADPH/NADP+ systems. *Mikrochim Acta* 139–50.

Legrain, S., C. Berr, N. Frenoy et al. 1995. Dehydroepiandrosterone sulfate in a long-term care aged population. *Gerontology* 41:343–51.

Liu, D., and J. S. Dillon. 2002. Dehydroepiandrosterone activates endothelial cell nitric-oxide synthase by a specific plasma membrane receptor coupled to Galpha(i2,3). *J Biol Chem* 277:21379–88.

Liu, D., and J. S. Dillon. 2004. Dehydroepiandrosterone stimulates nitric oxide release in vascular endothelial cells: Evidence for a cell surface receptor. *Steroids* 69:279–89.

Liu, D., M. Iruthayanathan, L. L. Homan et al. 2008. Dehydroepiandrosterone stimulates endothelial proliferation and angiogenesis through extracellular signal-regulated kinase 1/2-mediated mechanisms. *Endocrinology* 149:889–98.

Liu, D., H. Si, K. A. Reynolds et al. 2007. Dehydroepiandrosterone protects vascular endothelial cells against apoptosis through a Galphai protein-dependent activation of phosphatidylinositol 3-kinase/Akt and regulation of antiapoptotic Bcl-2 expression. *Endocrinology* 148:3068–76.

Lytton, J., M. Westlin, S. E. Burk, G. E. Shull, and D. H. MacLennan. 1992. Functional comparisons between isoforms of the sarcoplasmic or endoplasmic reticulum family of calcium pumps. *J Biol Chem* 267:14483–9.

Mazat, L., S. Lafont, C. Berr et al. 2001. Prospective measurements of dehydroepiandrosterone sulfate in a cohort of elderly subjects: Relationship to gender, subjective health, smoking habits, and 10-year mortality. *Proc Natl Acad Sci U S A* 98:8145–50.

Means, A. R. 1994. Calcium, calmodulin and cell cycle regulation. *FEBS Lett* 347:1–4.

Michelakis, E. D., M. S. McMurtry, X. C. Wu et al. 2002. Dichloroacetate, a metabolic modulator, prevents and reverses chronic hypoxic pulmonary hypertension in rats: Role of increased expression and activity of voltage-gated potassium channels. *Circulation* 105:244–50.

Michelakis, E. D., H. L. Reeve, J. M. Huang et al. 1997. Potassium channel diversity in vascular smooth muscle cells. *Can J Physiol Pharmacol* 75:889–97.

Migeon, C. J., A. R. Keller, B. Lawrence, and T. H. Shepard 2nd. 1957. Dehydroepiandrosterone and androsterone levels in human plasma: Effect of age and sex; day-to-day and diurnal variations. *J Clin Endocrinol Metab* 17:1051–62.

Moncada, S., and E. A. Higgs. 2006. The discovery of nitric oxide and its role in vascular biology. *Br J Pharmacol* 147 Suppl 1:S193–201.

Morales, D. E., K. A. McGowan, D. S. Grant et al. 1995. Estrogen promotes angiogenic activity in human umbilical vein endothelial cells in vitro and in a murine model. *Circulation* 91:755–63.

Morales, A. J., J. J. Nolan, J. C. Nelson, and S. S. Yen. 1994. Effects of replacement dose of dehydroepiandrosterone in men and women of advancing age. *J Clin Endocrinol Metab* 78:1360–7.

Munzel, T., R. Feil, A. Mulsch et al. 2003. Physiology and pathophysiology of vascular signaling controlled by guanosine 3',5'-cyclic monophosphate-dependent protein kinase [corrected]. *Circulation* 108:2172–83.

Murata, T., K. Kinoshita, M. Hori et al. 2005. Statin protects endothelial nitric oxide synthase activity in hypoxia-induced pulmonary hypertension. *Arterioscler Thromb Vasc Biol* 25:2335–42.

Nagai, Y., C. J. Earley, M. K. Kemper, C. S. Bacal, and E. J. Metter. 1999. Influence of age and postmenopausal estrogen replacement therapy on carotid arterial stiffness in women. *Cardiovasc Res* 41:307–11.

Nelson, M. T., J. B. Patlak, J. F. Worley, and N. B. Standen. 1990. Calcium channels, potassium channels, and voltage dependence of arterial smooth muscle tone. *Am J Physiol* 259:C3–18.

Nelson, M. T., and J. M. Quayle. 1995. Physiological roles and properties of potassium channels in arterial smooth muscle. *Am J Physiol* 268:C799–822.

Newcomer, L. M., J. E. Manson, R. L. Barbieri, C. H. Hennekens, and M. J. Stampfer. 1994. Dehydroepiandrosterone sulfate and the risk of myocardial infarction in US male physicians: A prospective study. *Am J Epidemiol* 140:870–5.

Nichols, W. W., M. F. O'Rourke, A. P. Avolio et al. 1985. Effects of age on ventricular-vascular coupling. *Am J Cardiol* 55:1179–84.

Nieschlag, E., D. L. Loriaux, H. J. Ruder et al. 1973. The secretion of dehydroepiandrosterone and dehydroepiandrosterone sulphate in man. *J Endocrinol* 57:123–34.

Nishimura, T., L. T. Vaszar, J. L. Faul et al. 2003. Simvastatin rescues rats from fatal pulmonary hypertension by inducing apoptosis of neointimal smooth muscle cells. *Circulation* 108:1640–5.

Nuedling, S., S. Kahlert, K. Loebbert et al. 1999. 17 Beta-estradiol stimulates expression of endothelial and inducible NO synthase in rat myocardium in-vitro and in-vivo. *Cardiovasc Res* 43:666–74.

Ohashi, N., A. Matsumori, Y. Furukawa et al. 2000. Role of p38 mitogen-activated protein kinase in neointimal hyperplasia after vascular injury. *Arterioscler Thromb Vasc Biol* 20:2521–6.

Ohlsson, C., F. Labrie, E. Barrett-Connor et al. 2010. Low serum levels of dehydroepiandrosterone sulfate predict all-cause and cardiovascular mortality in elderly swedish men. *J Clin Endocrinol Metab* 95(9):4406–14.

Oka, M., V. Karoor, N. Homma et al. 2007. Dehydroepiandrosterone upregulates soluble guanylate cyclase and inhibits hypoxic pulmonary hypertension. *Cardiovasc Res* 74:377–87.

O'Rourke, M. F., and A. P. Avolio. 1980. Pulsatile flow and pressure in human systemic arteries. Studies in man and in a multibranched model of the human systemic arterial tree. *Circ Res* 46:363–72.

Owens, G. K. 2007. Molecular control of vascular smooth muscle cell differentiation and phenotypic plasticity. *Novartis Found Symp* 283:174–91; discussion 191–3, 238–41.

Parekh, A. B., and R. Penner. 1997. Store depletion and calcium influx. *Physiol Rev* 77:901–30.

Parker, C. R. Jr. 1999. Dehydroepiandrosterone and dehydroepiandrosterone sulfate production in the human adrenal during development and aging. *Steroids* 64:640–7.

Pascale, R. M., M. M. Simile, M. R. De Miglio et al. 1995. Inhibition of 3-hydroxy-3-methylglutaryl-CoA reductase activity and gene expression by dehydroepiandrosterone in preneoplastic liver nodules. *Carcinogenesis* 16:1537–42.

Pastorino, J. G., J. B. Hoek, and N. Shulga. 2005. Activation of glycogen synthase kinase 3beta disrupts the binding of hexokinase II to mitochondria by phosphorylating voltage-dependent anion channel and potentiates chemotherapy-induced cytotoxicity. *Cancer Res* 65:10545–54.

Peng, W., J. R. Hoidal, and I. S. Farrukh. 1999. Role of a novel KCa opener in regulating K+ channels of hypoxic human pulmonary vascular cells. *Am J Respir Cell Mol Biol* 20:737–45.

Pyles, J. M., K. L. March, M. Franklin et al. 1997. Activation of MAP kinase in vivo follows balloon overstretch injury of porcine coronary and carotid arteries. *Circ Res* 81:904–10.

Raineri, R., and H. R. Levy. 1970. On the specificity of steroid interaction with mammary glucose 6-phosphate dehydrogenase. *Biochemistry* 9:2233–43.

Remillard, C. V., and J. X. Yuan. 2004. Activation of K+ channels: An essential pathway in programmed cell death. *Am J Physiol Lung Cell Mol Physiol* 286:L49–67.

Rossi, M. L., N. Marziliano, P. A. Merlini et al. 2004. Different quantitative apoptotic traits in coronary atherosclerotic plaques from patients with stable angina pectoris and acute coronary syndromes. *Circulation* 110:1767–73.

Salvini, S., M. J. Stampfer, R. L. Barbieri, and C. H. Hennekens. 1992. Effects of age, smoking and vitamins on plasma DHEAS levels: A cross-sectional study in men. *J Clin Endocrinol Metab* 74:139–43.

Sato, A., I. Sakuma, and D. D. Gutterman. 2003. Mechanism of dilation to reactive oxygen species in human coronary arterioles. *Am J Physiol Heart Circ Physiol* 285:H2345–54.

Schriock, E. D., C. K. Buffington, G. D. Hubert et al. 1988. Divergent correlations of circulating dehydroepiandrosterone sulfate and testosterone with insulin levels and insulin receptor binding. *J Clin Endocrinol Metab* 66:1329–31.

Schwartz, A. G., L. Pashko, and J. M. Whitcomb. 1986. Inhibition of tumor development by dehydroepiandrosterone and related steroids. *Toxicol Pathol* 14:357–62.

Scuteri, A., E. G. Lakatta, A. J. Bos, and J. L. Fleg. 2001. Effect of estrogen and progestin replacement on arterial stiffness indices in postmenopausal women. *Aging (Milano)* 13:122–30.

Shigematsu, K., H. Koyama, N. E. Olson, A. Cho, and M. A. Reidy. 2000. Phosphatidylinositol 3-kinase signaling is important for smooth muscle cell replication after arterial injury. *Arterioscler Thromb Vasc Biol* 20:2373–8.

Short, A. D., J. Bian, T. K. Ghosh et al. 1993. Intracellular Ca2+ pool content is linked to control of cell growth. *Proc Natl Acad Sci U S A* 90:4986–90.

Simoncini, T., P. Mannella, L. Fornari et al. 2003. Dehydroepiandrosterone modulates endothelial nitric oxide synthesis via direct genomic and nongenomic mechanisms. *Endocrinology* 144:3449–55.

Slowinska-Srzednicka, J., S. Zgliczynski, M. Ciswicka-Sznajderman et al. 1989. Decreased plasma dehydroepiandrosterone sulfate and dihydrotestosterone concentrations in young men after myocardial infarction. *Atherosclerosis* 79:197–203.

Smith, J. C., S. Bennett, L. M. Evans et al. 2001. The effects of induced hypogonadism on arterial stiffness, body composition, and metabolic parameters in males with prostate cancer. *J Clin Endocrinol Metab* 86:4261–7.

Sobey, C. G. 2001. Potassium channel function in vascular disease. *Arterioscler Thromb Vasc Biol* 21:28–38.

Somlyo, A. V., and A. P. Somlyo. 1993. Intracellular signaling in vascular smooth muscle. *Adv Exp Med Biol* 346:31–8.

Tilvis, R. S., M. Kahonen, and M. Harkonen. 1999. Dehydroepiandrosterone sulfate, diseases and mortality in a general aged population. *Aging (Milano)* 11:30–4.

Touyz, R. M. 2005. Reactive oxygen species as mediators of calcium signaling by angiotensin II: Implications in vascular physiology and pathophysiology. *Antioxid Redox Signal* 7:1302–14.

Trivedi, D. P., and K. T. Khaw. 2001. Dehydroepiandrosterone sulfate and mortality in elderly men and women. *J Clin Endocrinol Metab* 86:4171–7.

Tsien, R. W., P. T. Ellinor, and W. A. Horne. 1991. Molecular diversity of voltage-dependent Ca2+ channels. *Trends Pharmacol Sci* 12:349–54.

Uehata, M., T. Ishizaki, H. Satoh et al. 1997. Calcium sensitization of smooth muscle mediated by a Rho-associated protein kinase in hypertension. *Nature* 389:990–4.

Urata, Y., S. Goto, M. Kawakatsu et al. 2010. DHEA attenuates PDGF-induced phenotypic proliferation of vascular smooth muscle A7r5 cells through redox regulation. *Biochem Biophys Res Commun* 396:489–94.

van Breemen, C., and K. Saida. 1989. Cellular mechanisms regulating [Ca2+]i smooth muscle. *Annu Rev Physiol* 51:315–29.

Voelkel, N. F., and R. M. Tuder. 1995. Cellular and molecular mechanisms in the pathogenesis of severe pulmonary hypertension. *Eur Respir J* 8:2129–38.

Weir, E. K., and S. L. Archer. 1995. The mechanism of acute hypoxic pulmonary vasoconstriction: The tale of two channels. *FASEB J* 9:183–9.

Wessely, R., A. Schomig, and A. Kastrati. 2006. Sirolimus and Paclitaxel on polymer-based drug-eluting stents: Similar but different. *J Am Coll Cardiol* 47:708–14.

White, C. R., J. Shelton, S. J. Chen et al. 1997. Estrogen restores endothelial cell function in an experimental model of vascular injury. *Circulation* 96:1624–30.

Williams, M. R., T. Dawood, S. Ling et al. 2004. Dehydroepiandrosterone increases endothelial cell proliferation in vitro and improves endothelial function in vivo by mechanisms independent of androgen and estrogen receptors. *J Clin Endocrinol Metab* 89:4708–15.

Williams, M. R., S. Ling, T. Dawood et al. 2002. Dehydroepiandrosterone inhibits human vascular smooth muscle cell proliferation independent of ARs and ERs. *J Clin Endocrinol Metab* 87:176–81.

Wu, F. C., and A. von Eckardstein. 2003. Androgens and coronary artery disease. *Endocr Rev* 24:183–217.

Yoneyama, A., Y. Kamiya, M. Kawaguchi, and T. Fujinami. 1997. Effects of dehydroepiandrosterone on proliferation of human aortic smooth muscle cells. *Life Sci* 60:833–8.

Yoshida, S., K. I. Aihara, H. Azuma et al. 2010. Dehydroepiandrosterone sulfate is inversely associated with sex-dependent diverse carotid atherosclerosis regardless of endothelial function. *Atherosclerosis* 212(1):310–5.

Yoshimata, T., A. Yoneyama, Y. Jin-no, N. Tamai, and Y. Kamiya. 1999. Effects of dehydroepiandrosterone on mitogen-activated protein kinase in human aortic smooth muscle cells. *Life Sci* 65:431–40.

Young, J., B. Couzinet, K. Nahoul et al. 1997. Panhypopituitarism as a model to study the metabolism of dehydroepiandrosterone (DHEA) in humans. *J Clin Endocrinol Metab* 82:2578–85.

Yuan, X. J. 1995. Voltage-gated K+ currents regulate resting membrane potential and [Ca2+]i in pulmonary arterial myocytes. *Circ Res* 77:370–8.

Zamzami, N., and G. Kroemer. 2001. The mitochondrion in apoptosis: How Pandora's box opens. *Nat Rev Mol Cell Biol* 2:67–71.

Zhan, Y., S. Kim, Y. Izumi et al. 2003. Role of JNK, p38, and ERK in platelet-derived growth factor-induced vascular proliferation, migration, and gene expression. *Arterioscler Thromb Vasc Biol* 23:795–801.

11 DHEA and Ischemia/ Reperfusion Injury Prevention

Hulya Aksoy

CONTENTS

INTRODUCTION

Ischemia is a term formed from the Greek "isch," meaning restriction, and "hema," meaning blood, and is translated as a relative or absolute reduction in the blood supply to an organ. Its impact on the parenchyma depends on its intensity, its duration, the type of cell, and its metabolic needs. In the cell, oxygen is necessary for ATP synthesis by oxidative phosphorylation. ATP is used for many processes such as active transport, muscle contraction, thermoregulation, and synthesis. Within a few seconds after the interruption of blood flow, the cell consumes the oxygen contained in the oxyhemoglobin or myoglobin and adapts its metabolism (Kloner and Jennings 2001).

During ischemia, a decrease in intracellular pH and an increase in the intracellular Na^+ concentration occur. Anaerobic glycolysis provides almost all ATP produced in regions of severe or total ischemia (Jennings et al. 1981), and as a result of anaerobic glycolysis, lactate accumulates. Depending on the increase in lactate and its associated H^+ levels, after minutes of ischemia, the intracellular pH decreases to the range of 5.8–6.0 (Fleet et al. 1985), and the load of intracellular osmotically active particles, such as lactate and inorganic phosphate, increases markedly (Jennings et al. 1986).

In general, there are several mechanisms related to increases in cytosolic Na^+ levels. A great portion of Na^+ enters the cell through the Na^+ channels. Na^+ channels are integral membrane proteins that form ion channels, conducting Na^+ through a cell's plasma membrane. They are classified according to the trigger that opens the channel for such ions, that is, either a voltage-change (voltage-gated Na^+ channels) or binding of a substance (a ligand) to the channel (ligand-gated Na^+ channels). In excitable cells such as neurons, myocytes, and certain types of glia, Na^+ channels are responsible for the rising phase of action potentials. Voltage-gated Na^+ channels have three types of states: (1) deactivated (closed), (2) activated (open), and (3) inactivated (closed). These Na^+ channels may be altered during pathophysiological conditions such as hypoxia and exposure to ischemic metabolites and reactive oxygen species (ROS; Ju, Saint, and Gage 1996). Under such conditions, there is a pronounced late opening (or reopening) of the Na^+ channel up to a few hundreds of milliseconds after depolarization that is referred to as persistent Na^+ channel. Persistent Na^+ channel may therefore represent a major source for increased intracellular Na^+ levels during ischemia (Wang and Wang 2003). Additionally in a cell, ATP is mostly utilized by the Na^+/K^+ ATPase system. Depending on a decrease in the ATP level, the activity of Na^+/K^+ ATPase will decrease, and the intracellular Na^+ concentration will increase. Increased intracellular Na^+ concentration is responsible for initiating a cascade of events leading to cell damage and even death. Already involvement of persistent Na^+ current in the genesis of ischemia/reperfusion (I/R) damage has led to increased interest in the development of selective blockers of persistent Na^+ current as therapeutic tools (Saint 2008).

The Na^+/H^+ antiport is an ion-transport site that regulates intracellular pH in cells. Stimulation of the antiport Na^+/H^+ through cellular acidosis worsens the Na^+ overload and affects the operation of other membrane transporters such as the Na^+/Ca^{2+} antiport (Tani and Neely 1989). Therefore, intracellular Ca^{2+} concentration increases. When Ca^{2+} overloads, calpains are activated. It has been recently shown that calpains can cleave one of the major isoforms of the Na^+/Ca^{2+} antiport that operate Ca^{2+} efflux from neurons. This ultimately leads to an irreversible accumulation of the intracellular Ca^{2+} concentration, and at the end, intracellular Ca^{2+} concentration increases. Although Ca^{2+} signals are necessary for cell communication and survival, abnormal cellular Ca^{2+} load can trigger different cell death programs. Increased intracellular Ca^{2+} concentration is implicated as one of the pivotal events to induce ischemic cell (neuronal) death (Choi 1995). The Ca^{2+} overload may cascade excitotoxic and proapoptotic signaling to produce cysteine proteases including caspase-12 (Berliocchi, Bano, and Nicotera 2005). As a conclusion, depending on the duration of ischemia, apoptosis may occur.

While ischemic injury is precipitated by a lack of oxygen, reperfusion injury is associated with the return of oxygen. The reperfusion of an organ in ischemia is essential for its viability and functional recovery. But there is a harmful side to this coin, because the arrival of blood oxygen will cause a series of lesions; this is known as the phenomenon of I/R. Thus, a balance exists between the benefits of reperfusion to reduce infarct size and reperfusion injury, which depends on the duration of ischemia.

Although ischemia may lead to necrosis of tissue due to the absence of oxygen, during the reperfusion period, a considerable amount of molecular oxygen is supplied to the tissues and abundant amounts of ROS are produced. ROS are molecules having an unpaired electron on their last electron shell such as superoxide anion ($O_2^{\bullet -}$) and hydroxyl radical ($OH^{\bullet}$), as well as the reactive nitrogen oxide species. This makes them aggressive and highly reactive with respect to the various cellular components. They are normal by-products of cellular metabolic processes, but an excess of ROS is implicated in the tissue damage in reperfusion (Bender 2009).

The human body has a complex antioxidant defense system that includes the antioxidant enzymes (superoxide dismutase [SOD], catalase [CAT] and glutathione peroxidase [GPx], etc.) and nonenzymatic antioxidant components such as glutathione, α-tocopherol, ascorbic acid, and β-carotene. These prevent the initiation or propagation of free-radical chain reactions. Oxidative stress occurs when the production of ROS exceeds the capacity of the body's antioxidants to neutralize them. Oxidative stress damages DNA, proteins, lipids, and other macromolecules in many tissues, with the brain being particularly sensitive (Bender 2009).

There are several potential sources of ROS in most cells, including xanthine oxidase, the disruption of the respiratory chain, NADPH oxidase, nitric oxide synthase, cytochrome P450, lipoxygenase/cyclooxygenase pathways, and the auto-oxidation of various substances, particularly catecholamines. In reperfused tissue, NADPH oxidase (Walder et al. 1997) and xanthine oxidase seem to be particularly implicated (Harrison 2004). With the onset of ischemia, ATP is degraded and hypoxanthine is formed, and xanthine dehydrogenase is converted to xanthine oxidase. During reperfusion, xanthine oxidase catalyzes the conversion of hypoxanthine to uric acid with release of the $O_2^{\bullet -}$, which is quickly transformed into H_2O_2 by the SOD, which gives an $OH^{\bullet}$ in the presence of iron or copper (Granger 1988).

In addition to DNA, protein, and lipid damage, ROS may also trigger apoptosis by activating or deactivating some major signaling pathways. ROS also slows the rate of inactivation of the Na^+ channel (persistent Na^+ current), thus contributing to the increase in intracellular Na^+ concentration (Song et al. 2006).

In addition to cellular acidosis, alteration of membrane potential, increase of intracellular Ca^{2+} and Na^+ concentrations, and reduction of ATP, a cell secretes and expresses a multitude of substances and proteins on its surface during ischemia. For example, ischemia stimulates expression on the surface of the endothelial cells of the leukocyte molecules of adhesion such as P-selectin, the secretion of cytokines such as interleukin 6 (IL-6), IL-1β, tumor necrosis factor α (TNF-α), and vasoactive agents such as endothelin (Briaud et al. 2001; Eltzschig and Collard 2004; Yadav et al. 1998). The cytokines initiate the inflammatory response. Also, reperfusion will be accompanied by an important inflammatory response, characterized by the activation of the complement and the polymorphonuclear leukocyte neutrophils. Increased oxidative stress and the production of cytokines with reperfusion will induce apoptosis as well as ischemia. TNF-α plays a key role in reperfusion injury and can initiate a receptor-dependent death pathway by activating downstream caspases (Bajaj and Sharma 2006).

Because it is difficult to compensate for harmful events such as I/R injury, the above mechanisms can damage tissues and organs necessary for life. Therefore, whether several molecules are useful in various tissue ischemia has been the subject of research. One of these molecules is dehydroepiandrosterone (DHEA) or dehydroepiandrosterone sulfate (DHEAS).

DHEA AND DHEAS

DHEA and its sulfate ester DHEAS are the most abundant steroids in humans and mainly produced in zona reticularis of the human adrenal cortex. Already DHEA and DHEAS are produced in adrenal cortex as well as synthesized de novo in various regions of the central and peripheral nervous system (Compagnone and Mellon 2000). Thus, it may be suggested that the brain is a steroidogenic organ, and these hormones are termed as "neurosteroid."

In ischemia and subsequent reperfusion injury, DHEA and/or DHEAS may be useful in various ways. In some studies, preventing the effects of DHEA(S) on ischemia-induced tissue damage of the various organs or tissues such as renal, brain, and muscle are indicated (Aksoy et al. 2004; Aragno et al. 2000; Ayhan et al. 2003). Since the brain utilizes a high percentage of oxygen and is one of the vital organs, it is more affected by I/R injury.

Earlier studies have demonstrated that DHEA(S) appears to be involved in protection from cell I/R-induced damage particularly in the central nervous system (CNS). This protective effect involves multiple pathways and is discussed in the following section.

DHEA(S) Modulate *N*-Methyl-D-Aspartate Receptors

Transient forebrain ischemia causes irreversible neuronal degeneration in highly sensitive regions of the brain. When *N*-methyl-D-aspartate (NMDA) receptors on the cell membrane are activated during ischemia, massive Ca^{2+} influx through NMDA receptors and stimulation of nitric oxide synthetase occur. Excess nitric oxide produced by nitric oxide synthase may play a role in the production of superoxide by inhibiting electron transport chain function. $O_2^{-\bullet}$ increased in ischemia following reperfusion and reacted with nitric oxide to yield peroxynitrite with the potential to damage lipid membranes, proteins, and DNA. All these cause ischemia-induced apoptosis (Zalewska, Ziemka-Nałecz, and Domańska-Janik 2005). Thus, NMDA receptor antagonists prevent apoptosis, and pretreatment with DHEA and DHEAS is found to protect primary hippocampal cultures against NMDA-induced toxicity (Kimonides et al. 1998; Kurata et al. 2004).

Even though it is cited that DHEA blocks NMDA receptors and thereby has a neuroprotective effect, there are also some studies revealing increases in NMDA response or receptor activation

(Debonnel, Bergeron, and de Montigny 1996; Li et al. 2009). Low doses of DHEA potentiate selectively and dose dependently the NMDA response in the CA3 region of the rat dorsal hippocampus pyramidal neurons. According to these results, it can be said that DHEA has a neurotoxic effect (Debonnel, Bergeron, and de Montigny 1996). In a study using a four-vessel occlusion method for transient global cerebral ischemia, it was claimed that the neuroprotective and neurotoxic effects of DHEA might be attributed to the timing of DHEA administration. Specifically, when administered 3–48 hours after ischemia, DHEA exerts neuroprotection, although it shows neurotoxicity when administered 1 hour before or during early ischemia, which displays the importance of the timing of DHEA administration in treating ischemic brain damage (Li et al. 2009). It was demonstrated that the neuroprotective effect of DHEA was due to the σ1 receptor but not the NMDA receptor. On the other hand, the neurotoxic effect of DHEA is attributed to the σ1 receptor and the NMDA receptor. Because during ischemia and early reperfusion NMDA is one of the receptors responsible for the increase in intracellular Ca^+ concentration, administration of an NMDA receptor antagonist prevents DHEA neurotoxicity nearly perfectly, suggesting that the DHEA neurotoxicity here is attributable mainly to NMDA receptor activation (Li et al. 2009).

NMDA receptor complex has heterotetramer protein structure. Among subunits of NMDA receptors, NR2A and NR2B have been most extensively studied because they are broadly expressed in the brain and are believed to play important roles in synaptic plasticity. The tyrosine phosphorylation of NR2 subunits is a crucial component in regulating the activity of the NMDA receptor. A short-term (5-minute) ischemia elevated the tyrosine phosphorylation of NR2A and NR2B. In contrast, ischemia of longer duration (up to 30 minutes) caused a decrease in the levels of protein and tyrosine phosphorylation of both NR2A and NR2B subunits (Zalewska, Ziemka-Nałecz, and Domańska-Janik 2005). Tyrosine phosphorylation of NR2B prevents the degradation of these subunits (Bi et al. 2000). DHEAS improves reduction of NR2B tyrosine phosphorylation, which is σ1 receptor-dependent (Yoon et al. 2010). Therefore, activation of the σ1 receptor by chronically administered DHEAS can stabilize NMDA receptors on the cell surface and thereby increase the NMDA receptor responses.

But memory-enhancing effects of DHEA(S) are required for the protection of memory. The threshold for long-term potentiation in the hippocampal CA1 region is related to NMDA receptor activation; therefore, the NMDA receptor is one of the channels essential for the protection of memory in the ischemic brain. It was demonstrated that while transient (20-minute) incomplete forebrain ischemia led to impairment in the threshold for long-term potentiation in the hippocampal CA1 region, repetitive administrations of DHEAS (20 mg/kg for 3 days) from the first 3 hours of reperfusion prevented the impairment of long-term potentiation produced by ischemia via NR2B tyrosine phosphorylation. Based on these results, it can be said that increased DHEAS levels may have a substantial benefit for the treatment of ischemic stroke (Li et al. 2006).

DHEA(S) Activate the σ1 Receptor

Contrary to the data of Kimonides et al. (1998), it is suggested in another study that the neuroprotective effect of DHEA is related to the σ1 receptor but not to the NMDA receptor (Li et al. 2009). σ1 receptor activation leads to scavenging of free radicals. In a retinal ischemia-reperfusion model investigating the effects of DHEAS against I/R injury in rat retinas, animals with retinal ischemia showed a sustained increase in lactate content and decrease in levels of glucose and ATP. Treatments with DHEAS prevented lactate accumulation in retinal tissue of rats subjected to ischemia and induced an increase in glucose and ATP levels. Also, the effect of DHEAS in retinal tissue was antagonized by a σ1 receptor antagonist, suggesting that most of the neuroprotective action of DHEAS involved in this model is its interaction with the σ1 receptor (Bucolo and Drago 2004).

DHEA(S) Modulate γ-Aminobutyric Acid$_A$ Receptors

γ-Aminobutyric acid$_A$ (GABA$_A$) receptors are the most important inhibitory receptors in the CNS and play a pivotal role in CNS excitability. In an animal model of reversible spinal cord ischemia, DHEAS proved to have neuroprotective effects (Lapchak et al. 2000). These effects were abolished by bicuculline, the GABA$_A$ antagonist, which suggests that GABA$_A$ receptors may mediate the effects of DHEAS.

DHEAS has a modulatory effect on GABA$_A$ receptors. DHEAS induces the expression of α1 and β2 subunits of GABA$_A$ receptors (Xilouri and Papazafiri 2006). In fact, upregulation of GABA$_A$ receptor expression may explain the beneficial role of DHEAS against the deleterious effects of NMDA on cell survival as in cerebral ischemia. Also, naturally occurring neurosteroids, such as DHEA and DHEAS, are potent allosteric modulators of GABA$_A$ receptor function (Rupprecht and Holsboer 1999; Majewska et al. 1986).

DHEA and DHEAS Induce the Expression of the Antiapoptotic Bcl-2 and Bcl-xL Proteins

DHEA and DHEAS have protective effects in rat PC12 sympathoadrenal cells and chromaffin cells against serum deprivation-induced apoptosis (due to absence of growth factors). DHEA and DHEAS induce the expression of the antiapoptotic Bcl-2 and Bcl-xL proteins by activating transcription factors cAMP response element-binding protein and NF-κB (Charalampopoulos et al. 2004). They also activate Bcl-2 protein by activating protein kinase C. DHEA and DHEAS may exert their antiapoptotic effects by activating G-protein-associated membrane binding sites (Charalampopoulos et al. 2006). Kaasik et al. (2001) investigated the effect of DHEAS on the neurodegeneration induced by oxygen-glucose deprivation in cultured cerebellar granule cells of rats and confirmed the inhibitory action of DHEAS on neuronal apoptosis.

DHEA Modulates Extracellular Glutamate Concentration

Glutamate is released to the extracellular region during ischemia, and the extracellular glutamate concentration remains high for hours after reperfusion. Ischemia-induced extracellular glutamate accumulation is due to derivation from transmitter pools in glutamatergic neurons by calcium-dependent exocytotic release or calcium-independent nonexocytotic sources, volume-regulated anion channels, and cellular lysis (Shen et al. 2008). Glutamate transporters are essential for maintaining low concentration by removing the glutamate from the extracellular space, and especially GLT-1 play the dominant role in maintaining extracellular low glutamate level (Zhang et al. 2010).

In the extracellular region, glutamate accumulation excites neurons to death in a process termed "excitotoxicity." By binding NMDA receptors, glutamate stimulates excessive Ca^+ influx into neuronal cells. Then, increased intracellular Ca^+ concentration leads to necrosis and apoptosis. Thus, a GLT-1 activating agent may prevent ischemia-induced apoptosis by the above-mentioned mechanism. DHEA has been shown to upregulate the function of GLT-1 via the activation of the σ1 receptor (Sokabe, Chen, and Furuya 2007).

Sometimes, idiopathic sudden sensorineural hearing loss can be caused by I/R injury (Kim et al. 1999), and glutamate excitotoxicity and free radicals play important roles in I/R injury of the cochlea (Tabuchi et al. 2001; Puel et al. 1994). One study demonstrated that preischemic administration of DHEAS significantly decreased the postischemic cochlear injury induced by transient injury, possibly due to the preventive role of glutamate excitotoxicity and increase in free radicals (Tabuchi et al. 2003).

Reduction in Oxidative Stress by DHEA

Renal ischemia followed by reperfusion increases oxidative stress, as shown by elevated levels of pro-oxidant reactive species and loss of antioxidant systems. It was shown that DHEA administration completely prevented changes in the antioxidant defense mechanism in the kidneys of I/R rats (Aragno et al. 2003). DHEA is known to be an antioxidant molecule. One reason for this effect is to strengthen the antioxidant defense system against free-radical damage (Aragno et al. 2003; Aksoy et al. 2004; Aksoy et al. 2007). Also, DHEA has an inhibitory effect on glucose-6-phosphate dehydrogenase, which is an important enzyme in pentose phosphate way, a major source of ribose 5-phosphate and NADPH, a critical modulator of the cellular redox state and a reductant for several enzymes generating oxygen-free radicals. These include leukocyte NADPH oxidase, nitric oxide synthase, the cytochrome P450 monooxygenases, and Fenton reaction (Schwartz and Pashko 2004). For example, NADPH oxidase catalyzes the production of $O_2^{\bullet-}$ by the one-electron reduction of oxygen, using NADPH as the electron donor, so the concentration of ROS is increased.

DHEA has been reported to change the fatty acid composition of membranes in rats by decreasing the content of some unsaturated fatty acids (Mohan and Clearly 1991; Aragno et al. 2001; Aragno et al. 2000), thus making it more resistant to oxidative stress. DHEA treatment reduces both ROS and nitric oxide production by reducing nitric oxide synthase expression. In the presence of $O_2^{\bullet-}$, nitric oxide interacts with $O_2^{\bullet-}$ to form peroxynitrate radicals. Also, DHEA counteracts the increase of hydroperoxyeicosatetraenoic acids, inhibitors of Na^+/K^+ ATPase activity (Foley 1997), in the kidneys of diabetic rats (Aragno et al. 2001). By the restoration of Na^+/K^+ ATPase activity, DHEA protects lipid membranes. On the basis of this data, it can be said that DHEA decreases the ROS generation and peroxynitrite production in cells rather than directly scavenging superoxide anion.

Testicular damage caused by ischemia followed by reperfusion is assumed to be partly a result of lipid peroxidation, which in turn leads to membrane damage. In the testicular I/R model, administration of DHEA prevented increases in malondialdehyde (a marker of lipid peroxidation) levels and decreases in CAT and SOD activities (Aksoy et al. 2007). This effect of DHEA alternatively may be related to its preventive effect on decreases in some antioxidant enzyme activities such as CAT and SOD, which may explain the preventive function against the development of histological injury. Also, DHEA is reported to increase glutathione levels in rat hepatic tissue (White et al. 1998), which may explain, in part, its protective properties against oxidative stress.

Overproduction of free radicals induced by I/R causes lipid peroxidation in the rabbit kidney, and it was shown that prophylactic administration of DHEA in ischemia condition prevented reperfusion injuries by eliminating oxygen radicals and inhibiting lipid peroxidation by preventing decreases in SOD and CAT activities in the kidney (Aksoy et al. 2004).

Increased ROS activates the transcription factor NF-κB, which is involved in the activation of genes relevant to inflammation, cytokines, cell proliferation, and cell survival as seen in diabetes. Contradictory results have been reported on the effect of NF-κB on apoptosis in various cells (Aragno et al. 2002; Charalampopoulos et al. 2004). Protective or detrimental effects of NF-κB may depend on the onset and duration of its activation (Erl et al. 1999; Charalampopoulos et al. 2004; Schneider et al. 1999). It was speculated that persistent activation of NF-κB mediated by ROS caused cell damage and death, and NF-κB was reduced in diabetic rats treated with DHEA (Aragno et al. 2002).

Although the major effect of I/R is on the relevant organ, it also affects other organs. For example, in orthopedic surgery, the use of the tourniquet on the lower limbs affects both the lung and liver (Laipanov, Petrosyan, and Sergienko 2006). A small intestine I/R induces apoptotic lesions in pulmonary, renal, and cardiac endothelium (Mura et al. 2007). In a study on this subject, DHEA pretreatment appeared to have beneficial effects on antioxidant defenses against hepatic injury after renal I/R in rabbits, possibly by augmenting glutathione levels and lowering malondialdehyde production (Yildirim et al. 2003).

The antioxidant effect of DHEA depends on the dose. While in low doses DHEA is antioxidant, its supraphysiological concentration may be pro-oxidant. It was shown that DHEA caused peroxisomal proliferation. DHEA treatment elevates peroxisome proliferator-activated receptor α (PPAR-α) levels in the cell and increases the expression of other proteins involved in regulation of PPAR-α activation (Poczatková et al. 2007). PPAR-α is the principal mediator of peroxisome proliferation and plays a pivotal role in controlling peroxisomal fatty acid oxidation. In peroxisomes, fatty acid oxidation occurs, and as a result, H_2O_2 is released. On the contrary, PPAR-α also shows an antioxidant feature by inhibiting NF-κB, which is an activatory molecule on inflammation and in turn increases ROS (Poynter and Daynes 1998).

DHEA and DHEAS Act as Anti-Inflammatory Agents by Altering Production of Cytokines

I/R induces significant upregulation of the *TNF*-α gene, which has a stimulatory effect on the expression of other genes important in the inflammatory response (Donnahoo et al. 1999). Increased nitric oxide synthase activity induced by ischemia also plays an important role in I/R injury. DHEA reduces the systemic inflammation, which is measurable via the proinflammatory cytokines such as TNF-α, IL-1β, and IL-6 (Hildebrand et al. 2003). TNF-α also activates NF-κB, and DHEA and DHEAS inhibits both basal and TNF-α-stimulated NF-κB activation possibly through their antioxidant effects in a time- and dose-dependent manner (Iwasaki et al. 2004).

Acute renal failure (ARF) is a frequently seen condition during renal transplantation or suprarenal aortic surgery due to ischemia followed by reperfusion injury occurring during surgery. While ARF develops over a period of hours or days and is potentially reversible, chronic renal failure develops over months or many years and involves the irreversible destruction of kidney tissue leading to end stage renal failure (Meguid El Nahas and Bello 2005), which is associated with high morbidity and mortality rates. In the clinical setting, ischemic ARF is often caused by hypotension followed by resuscitation, the etiology of which is reflected in animal models of renal I/R. Reduced vascular flow in the postischemic kidney leads to the swelling of endothelial cells and results in the "no-reflow" phenomenon (Summers and Jamison 1971), which may initiate and extend renal dysfunction. Vascular endothelial growth factor (VEGF), a protective cytokine, is not only an angiogenesis factor but also a powerful agent increasing vascular permeability; and renal and testicular synthesis of VEGF rapidly increases during renal I/R injury (Vannay et al. 2004; Hashimoto et al. 2009). Another finding that supports this fact is that postischemic VEGF administration enhanced angiogenesis and promoted blood-brain barrier leakage in the rat brain ischemia model (Zhang et al. 2000).

DHEA prevents I/R injury by altering levels of cytokines as it improves postischemic vascular hemodynamics (Lohman et al. 1997) and endothelial cell functions (Simoncini et al. 2003). During renal I/R injury, exogenously administered DHEA decreases renal IL-1β and IL-6 mRNA expression (Vannay et al. 2009) and IL-6 protein synthesis (Kipper-Galperin et al. 1999). Thus, inflammation-induced apoptosis could be prevented in I/R models. On the other hand, IL-1β induces VEGF synthesis mediated by hypoxia-inducible factor 1 (Qian et al. 2004), and IL-6 phosphorylates the eukaryotic initiation factor 4E (Yamagiwa et al. 2004), which increases the translation of VEGF. Thus it can be said that, DHEA reduces I/R injury because of anti-inflammatory properties, but the simultaneous decline in VEGF may be harmful.

Inhibitory Effects of DHEAS on Persistent Na^+ Currents

Ionic imbalance that exists across the neuronal membrane is an important cause of anoxia-induced nerve injury. Ischemia increases intracellular Na^+ concentration, largely via persistent Na^+ currents, which is one of the responsible factors in irreversible cell damage (Hammarström and Gage 2002).

DHEAS has an inhibitory effect on persistent Na^+ currents by either causing a conformational change in the Na^+ channel or inducing a change in intracellular signaling molecules, leading to an inhibition of persistent Na^+ currents. The activation of the σ1 receptor plays a key role in the inhibitory effect of DHEAS on persistent Na^+ currents (Cheng et al. 2008). This mechanism may partly exhibit the importance of DHEAS in neuronal ischemia.

DHEA Prevents Thrombocyte Aggregation and Leukocyte Activation

Adhesion molecules contribute to I/R injury by increasing the endothelial adhesion and extravasation of leukocytes. The initial adhesive event is margination at the site of injury, which is characterized by the rolling of neutrophils along the luminal surface of the vessel wall. A study investigating the effects of presurgical DHEA treatment on microcirculatory hemodynamic parameters and the expression of adhesion molecules in a rat cremaster muscle flap model revealed that intravenous DHEA administration reduced the activation of leukocytes and improved red blood cell velocity and capillary perfusion in the muscle flap microcirculation during I/R injury. Expression of Mac-1 integrin, L-selectin, and CD44 molecules was downregulated by intravenous administration of DHEA, and the protective effect of DHEA could be attributed to this downregulation and consequently delayed expression of these adhesion molecules on leukocytes (Ayhan et al. 2003). In addition, leukocyte activation is reduced, ROS production is reduced, and at the end oxidative stress-induced endothelial damage may not occur. Moreover, it was found that DHEA inhibits arachidonic acid-induced platelet aggregation both *in vitro* and *in vivo* (Jesse et al. 1995).

Lohman et al. investigated the effect of DHEA on microcirculatory hemodynamic acting after I/R injury. In their study, a rat cremaster muscle flap model was used because hemodynamic changes with functional capillary perfusion could easily be evaluated with this preparation. In this study, it was found that subcutaneous injection of DHEA before surgical procedures protects the microcirculation, capillary functional integrity, and muscle structure from ischemia-reperfusion injury. Speculations are that DHEA may also interfere with either thromboxane production or its effects or may cause a significant decrease in the number of adherent and transmigrating leukocytes after DHEA pretreatment (Lohman et al. 1997).

In summary, DHEA and DHEAS are effective steroids in preventing the I/R injury when administered at appropriate doses and time intervals. The mechanisms are very complex and may depend on cell type, drug dose, duration of I/R, and duration of drug administration.

REFERENCES

Aksoy, H., T. Yapanoglu, Y. Aksoy, I. Ozbey, H. Turhan, and N. Gursan. 2007. Dehydroepiandrosterone treatment attenuates reperfusion injury after testicular torsion and detorsion in rats. *J Pediatr Surg* 42:1740–4.

Aksoy, Y., T. Yapanoglu, H. Aksoy, and A. K. Yildirim. 2004. The effect of dehydroepiandrosterone on renal ischemia-reperfusion-induced oxidative stress in rabbits. *Urol Res* 32:93–6.

Aragno, M., J. C. Cutrin, R. Mastrocola et al. 2003. Oxidative stress and kidney dysfunction due to ischemia/reperfusion in rat: Attenuation by dehydroepiandrosterone. *Kidney Int* 64:836–43.

Aragno, M., R. Mastrocola, E. Brignardello et al. 2002. Dehydroepiandrosterone modulates nuclear factor-kappa B activation in hippocampus of diabetic rats. *Endocrinology* 143:3250–8.

Aragno, M., S. Parola, E. Brignardello et al. 2000. Dehydroepiandrosterone prevents oxidative injury induced by transient ischemia/reperfusion in the brain of diabetic rats. *Diabetes* 49:1924–31.

Aragno, M., S. Parola, E. Brignardello et al. 2001. Oxidative stress and eicosanoids in the kidneys of hyperglycemic rats treated with dehydroepiandrosterone. *Free Radic Biol Med* 31:935–42.

Ayhan, S., C. Tugay, S. Norton, B. Araneo, and M. Siemionow. 2003. Dehydroepiandrosterone protects the microcirculation of muscle flaps from ischemia-reperfusion injury by reducing the expression of adhesion molecules. *Plast Reconstr Surg* 111:2286–94.

Bajaj, G., and R. K. Sharma. 2006. TNF-alpha-mediated cardiomyocyte apoptosis involves caspase-12 and calpain. *Biochem Biophys Res Commun* 345:1558–64.

Bender, D. A. 2009. Free radicals and antioxidant nutrients. In *Harper's Illustrated Biochemistry*, ed. R. K. Murray, D. A. Bender, K. M. Botham, P. J. Kennelly, V. W. Rodwell, and P. A. Weil, 482–6. China: The McGraw-Hill Companies.

Berliocchi, L., D. Bano, and P. Nicotera. 2005. Ca^{2+} signals and death programmers in neurons. *Philos Trans R Soc Lond B Biol Sci* 29:2255–8.

Bi, R., Y. Rong, A. Bernard, M. Khrestchatisky, and M. Baudry. 2000. Src-mediated tyrosine phosphorylation of NR2 subunits of N-methyl-D-aspartate receptors protects from calpain-mediated truncation of their C-terminal domains. *J Biol Chem* 275:26477–83.

Briaud, S. A., Z. M. Ding, L. H. Michael, M. L. Entman, S. Daniel, and C. M. Ballantyne. 2001. Leukocyte trafficking and myocardial reperfusion injury in ICAM-1/P-selectin-knockout mice. *Am J Physiol* 280:H60–7.

Bucolo, C., and F. Drago. 2004. Effects of neurosteroids on ischemia-reperfusion injury in the rat retina: Role of sigma1 recognition sites. *Eur J Pharmacol* 498:111–4.

Charalampopoulos, I., V. I. Alexaki, I. Lazaridis et al. 2006. G protein-associated, specific membrane binding sites mediate the neuroprotective effect of dehydroepiandrosterone. *FASEB J* 20:577–9.

Charalampopoulos, I., C. Tsatsanis, E. Dermitzaki et al. 2004. Dehydroepiandrosterone and allopregnanolone protect sympathoadrenal medulla cells against apoptosis via antiapoptotic Bcl-2 proteins. *Proc Natl Acad Sci U S A* 101:8209–14.

Cheng, Z. X., D. M. Lan, P. Y. Wu et al. 2008. Neurosteroid dehydroepiandrosterone sulphate inhibits persistent sodium currents in rat medial prefrontal cortex via activation of sigma-1 receptors. *Exp Neurol* 210:128–36.

Choi, D. W. 1995. Calcium: Still center-stage in hypoxic–ischemic neuronal death. *Trends Neurosci* 18:58–60.

Compagnone, N. A., and S. H. Mellon. 2000. Neurosteroids: Biosynthesis and function of these novel neuromodulators. *Front Neuroendocrinol* 21:1–56.

Debonnel, G., R. Bergeron, and C. de Montigny. 1996. Potentiation by dehydroepiandrosterone of the neuronal response to N-methyl-D-aspartate in the CA3 region of the rat dorsal hippocampus: An effect mediated via sigma receptors. *J Endocrinol* 150(Suppl):S33–42.

Donnahoo, K. K., B. D. Shames, A. H. Harken, and D. R. Meldrum. 1999. The role of tumor necrosis factor in renal ischemia-reperfusion injury. *J Urol* 162:196–203.

Eltzschig, H. K., and C. D. Collard. 2004. Vascular ischaemia and reperfusion injury. *Br Med Bull* 70:71–86.

Erl, W., G. K. Hansson, R. de Martin, G. Draude, K. S. Weber, and C. Weber. 1999. Nuclear factor-kappa B regulates induction of apoptosis and inhibitor of apoptosis protein-1 expression in vascular smooth muscle cells. *Circ Res* 84:668–77.

Fleet, W. F., T. A. Johnson, C. A. Graebner, and L. S. Gettes. 1985. Effect of serial brief ischemic episodes on extracellular K^+, pH, and activation in the pig. *Circulation* 72:922–32.

Foley, T. D. 1997. 5-HPETE is a potential inhibitor of neuronal Na, K-ATPase activity. *Biochem Biophys Res Commun* 235:374–6.

Granger, D. N. 1988. Role of xanthine oxidase and granulocytes in ischemia-reperfusion injury. *Am J Physiol* 255:H1269–75.

Hammarström, A. K., and P. W. Gage. 2002. Hypoxia and persistent sodium current. *Eur Biophys J* 31:323–30.

Harrison, R. 2004. Physiological roles of xanthine oxidoreductase. *Drug Metab Rev* 36:363–75.

Hashimoto, H., T. Ishikawa, K. Yamaguchi, M. Shiotani, and M. Fujisawa. 2009. Experimental ischaemia-reperfusion injury induces vascular endothelial growth factor expression in the rat testis. *Andrologia* 41:216–21.

Hildebrand, F., H. C. Pape, P. Hoevel, C. Krettek, and M. van Griensven. 2003. The importance of systemic cytokines in the pathogenesis of polymicrobial sepsis and dehydroepiandrosterone treatment in a rodent model. *Shock* 20:338–46.

Iwasaki, Y., M. Asai, M. Yoshida, T. Nigawara, M. Kambayashi, and N. Nakashima. 2004. Dehydroepiandrosterone-sulfate inhibits nuclear factor-kappaB-dependent transcription in hepatocytes, possibly through antioxidant effect. *J Clin Endocrinol Metab* 89:3449–54.

Jennings, R. B., K. A. Reimer, M. L. Hill, and S. E. Mayer. 1981. Total ischemia, in dog hearts, in vitro, I: Comparison of high energy phosphate production, utilization and depletion and of adenine nucleotide catabolism in total ischemia in vitro vs severe ischemia in vivo. *Circ Res* 49:892–900.

Jennings, R. B., K. A. Reimer, and C. Steenbergen Jr., 1986. Myocardial ischemia revisited: The osmolar load, membrane damage, and reperfusion. *J Mol Cell Cardiol* 18:769–80.

Jesse, R. L., K. Loesser, D. M. Eich, Y. Z. Qian, M. L. Hess, and J. E. Nestler. 1995. Dehydroepiandrosterone inhibits platelet aggregation in vivo and in vitro. *Ann N Y Acad Sci* 774:281–90.

Ju, Y. K., D. A. Saint, and P. W. Gage. 1996. Hypoxia increases persistent sodium current in rat ventricular myocytes. *J Physiol* 497:337–47.

Kaasik, A., A. Kalda, K. Jaako, and A. Zharkovsky. 2001. Dehydroepiandrosterone sulphate prevents oxygen-glucose deprivation-induced injury in cerebellar granule cell culture. *Neuroscience* 102:427–32.

Kim, J. S., I. Lopez, P. L. DiPatre, F. Liu, A. Ishiyama, and R. W. Baloh. 1999. Internal auditory artery infarction: Clinicopathological correlation. *Neurology* 52:40–4.

Kimonides, V. G., N. H. Khatibi, C. N. Svendsen, M. V. Sofroniew, and J. Herbert. 1998. Dehydroepiandrosterone (DHEA) and DHEA-sulfate (DHEAS) protect hippocampal neurons against excitatory amino acid-induced neurotoxicity. *Proc Natl Acad Sci U S A* 95:1852–7.

Kipper-Galperin, M., R. Galilly, H. D. Danenberg, and T. Brenner. 1999. Dehydroepiandrosterone selectively inhibits production of tumor necrosis factor alpha and interleukin-6 in astrocytes. *Int J Dev Neurosci* 17:765–75.

Kloner, R. A., and R. B. Jennings. 2001. Consequences of brief ischemia: Stunning, preconditioning, and their clinical implications: Part 1. *Circulation* 104:2981–9.

Kurata, K., M. Takebayashi, S. Morinobu, and S. Yamawaki. 2004. Beta-estradiol, dehydroepiandrosterone, and dehydroepiandrosterone sulfate protect against N-methyl-D-aspartate-induced neurotoxicity in rat hippocampal neurons by different mechanisms. *J Pharmacol Exp Ther* 311:237–45.

Laipanov, K. H. I., E. A. Petrosyan, and V. I. Sergienko. 2006. Morphological changes in the liver during experimental modeling of acute ischemia and reperfusion of the limb. *Bull Exp Biol Med* 142:105–7.

Lapchak, P. A., D. F. Chapman, S. Y. Nunez, and J. A. Zivin. 2000. Dehydroepiandrosterone sulfate is neuroprotective in a reversible spinal cord ischemia model: Possible involvement of GABA(A) receptors. *Stroke* 31:1953–6.

Li, Z., S. Cui, Z. Zhang et al. 2009. DHEA-neuroprotection and -neurotoxicity after transient cerebral ischemia in rats. *J Cereb Blood Flow Metab* 29:287–96.

Li, Z., R. Zhou, S. Cui et al. 2006. Dehydroepiandrosterone sulfate prevents ischemia-induced impairment of long-term potentiation in rat hippocampal CA1 by up-regulating tyrosine phosphorylation of NMDA receptor. *Neuropharmacology* 51:958–66.

Lohman, R., R. Yowell, S. Barton, B. Araneo, and M. Siemionow. 1997. Dehydroepiandrosterone protects muscle flap microcirculatory hemodynamics from ischemia/reperfusion injury: An experimental in vivo study. *J Trauma* 42:74–80.

Majewska, M. D., N. L. Harrison, R. D. Schwartz, J. L. Barker, and S. M. Paul. 1986. Steroid hormone metabolites are barbiturate-like modulators of the GABA receptor. *Science* 232:1004–7.

Meguid El Nahas, A., and A. K. Bello. 2005. Chronic kidney disease: The global challenge. *Lancet* 365:331–40.

Mohan, P., and M. Clearly. 1991. Short-term effects of dehydroepiandrosterone treatment in rats on mitochondrial respiration. *J Nutr* 121:240–50.

Mura, M., C. F. Andrade, B. Han et al. 2007. Intestinal ischemia-reperfusion-induced acute lung injury and oncotic cell death in multiple organs. *Shock* 28:227–38.

Poczatková, H., K. Bogdanová, L. Uherková et al. 2007. Dehydroepiandrosterone effects on the mRNA levels of peroxisome proliferator-activated receptors and their coactivators in human hepatoma HepG2 cells. *Gen Physiol Biophys* 26:268–74.

Poynter, M. E., and R. A. Daynes. 1998. Peroxisome proliferator-activated receptor alpha activation modulates cellular redox status, represses nuclear factor-kappaB signaling, and reduces inflammatory cytokine production in aging. *J Biol Chem* 273:32833–41.

Puel, J. L., R. Pujol, F. Tribillac, S. Ladrech, and M. Eybalin. 1994. Excitatory amino acid antagonists protect cochlear auditory neurons from excitotoxicity. *J Comp Neurol* 341:241–56.

Qian, D., H. Y. Lin, H. M. Wang et al. 2004. Normoxic induction of the hypoxic-inducible factor-1 alpha by interleukin-1 beta involves the extracellular signal-regulated kinase 1/2 pathway in normal human cytotrophoblast cells. *Biol Reprod* 70:1822–27.

Rupprecht, R., and F. Holsboer. 1999. Neuroactive steroids: Mechanisms of action and neuropsychopharmacological perspectives. *Trends Neurosci* 22:410–6.

Saint, D. A. 2008. The cardiac persistent sodium current: An appealing therapeutic target? *Br J Pharmacol* 153:1133–42.

Schneider, A., A. Martin-Villalba, F. Weih, J. Vogel, T. Wirth, and M. Schwaninger. 1999. NF-kappaB is activated and promotes cell death in focal cerebral ischemia. *Nat Med* 5:554–9.

Schwartz, A. G., and L. L. Pashko. 2004. Dehydroepiandrosterone, glucose-6-phosphate dehydrogenase, and longevity. *Ageing Res Rev* 3:171–87.

Shen, Y. C., Y. H. Wang, Y. C. Chou et al. 2008. Dimemorfan protects rats against ischemic stroke through activation of sigma-1 receptor-mediated mechanisms by decreasing glutamate accumulation. *J Neurochem* 104:558–72.

Simoncini, T., P. Mannella, L. Fornari, G. Varone, A. Caruso, and A. R. Genazzani. 2003. Dehydroepiandrosterone modulates endothelial nitric oxide synthesis via direct genomic and nongenomic mechanisms. *Endocrinology* 144:3449–55.

Sokabe, M., L. Chen, and K. Furuya. 2007. Potentiation of glutamate transporter GLT-1 by the neurosteroid DHEA in the rat hippocampal dentate gyrus. *J Physiol Sci* 57 (Suppl):S84.

Song, Y., J. C. Shryock, S. Wagner, L. S. Maier, and L. Belardinelli. 2006. Blocking late sodium current reduces hydrogen peroxide-induced arrhythmogenic activity and contractile dysfunction. *J Pharmacol Exp Ther* 318:214–22.

Summers, W. K., and R. L. Jamison. 1971. The no reflow phenomenon in renal ischemia. *Lab Invest* 25:635–43.

Tabuchi, K., K. Oikawa, I. Uemaetomari, S. Tsuji, T. Wada, and A. Hara. 2003. Glucocorticoids and dehydroepiandrosterone sulfate ameliorate ischemia-induced injury of the cochlea. *Hear Res* 180:51–6.

Tabuchi, K., S. Tsuji, I. Zenya, A. Hara, and J. Kusakari. 2001. Does xanthine oxidase contribute to the hydroxyl radical generation in ischemia and reperfusion of the cochlea? *Hear Res* 153:1–6.

Tani, M., and J. R. Neely. 1989. Role of intracellular Na^+ in Ca^{2+} overload and depressed recovery of ventricular function of reperfused ischemic rat hearts. Possible involvement of H^+-Na^+ and Na^+-Ca^{2+} exchange. *Circ Res* 65:1045–56.

Vannay, A., A. Fekete, C. Adori et al. 2004. Divergence of renal vascular endothelial growth factor mRNA expression and protein level in post-ischaemic rat kidneys. *Exp Physiol* 89:435–44.

Vannay, A., A. Fekete, R. Langer et al. 2009. Dehydroepiandrosterone pretreatment alters the ischaemia/reperfusion-induced VEGF, IL-1 and IL-6 gene expression in acute renal failure. *Kidney Blood Press Res* 32:175–84.

Walder, C. E., S. P. Green, W. C. Darbonne et al. 1997. Ischemic stroke injury is reduced in mice lacking a functional NADPH oxidase. *Stroke* 28:2252–58.

Wang, S. Y., and G. K. Wang. 2003. Voltage-gated sodium channels as primary targets of diverse lipid-soluble neurotoxins. *Cell Signal* 15:151–9.

White, E. L., L. J. Ross, S. M. Schmid, G. J. Kelloff, V. E. Steele, and D. L. Hill. 1998. Screening of potential cancer preventing chemicals for induction of glutathione in rat liver cells. *Oncol Rep* 5:507–12.

Xilouri, M., and P. Papazafiri. 2006. Anti-apoptotic effects of allopregnanolone on P19 neurons. *Eur J Neurosci* 23:43–54.

Yadav, S. S., D. N. Howell, W. Gao, D. A. Steeber, R. C. Harland, and P. A. Clavien. 1998. L-selectin and ICAM-1 mediate reperfusion injury and neutrophil adhesion in the warm ischemic mouse liver. *Am J Physiol* 275:G1341–52.

Yamagiwa, Y., C. Marienfeld, F. Meng, M. Holcik, and T. Patel. 2004. Translational regulation of Xlinked inhibitor of apoptosis protein by interleukin-6: A novel mechanism of tumor cell survival. *Cancer Res* 64:1293–8.

Yildirim, A., M. Gumus, S. Dalga, Y. N. Sahin, and F. Akcay. 2003. Dehydroepiandrosterone improves hepatic antioxidant systems after renal ischemia-reperfusion injury in rabbits. *Ann Clin Lab Sci* 33:459–64.

Yoon, S. Y., D. H. Roh, H. S. Seo et al. 2010. An increase in spinal dehydroepiandrosterone sulfate (DHEAS) enhances NMDA-induced pain via phosphorylation of the NR1 subunit in mice: Involvement of the sigma-1 receptor. *Neuropharmacology* 59:460–7. Epub, June 22, 2010.

Zalewska, T., M. Ziemka-Nałecz, and K. Domańska-Janik. 2005. Transient forebrain ischemia effects interaction of Src, FAK, and PYK2 with the NR2B subunit of N-methyl-D-aspartate receptor in gerbil hippocampus. *Brain Res* 1042:214–23.

Zhang, Z. G., L. Zhang, Q. Jiang et al. 2000. VEGF enhances angiogenesis and promotes blood–brain barrier leakage in the ischemic brain. *J Clin Invest* 106:829–38.

Zhang, W., Y. Miao, S. Zhou, B. Wang, Q. Luo, and Y. Qiu. 2010. Involvement of glutamate transporter-1 in neuroprotection against global brain ischemia-reperfusion injury induced by postconditioning in rats. *Int J Mol* Sci 11:4407–16.

12 DHEA and Cancer Risk or Prevention?

Fernand Labrie

CONTENTS

INTRODUCTION

In addition to the sex steroids of gonadal origin that are synthesized in the testicles and ovaries before being released into the blood in order to reach all tissues to exert their action in the classical endocrine manner (Figure 12.1a), it is now well established that an additional source of sex steroids plays a major role in the humans of both sexes (Figure 12.1b). For example, in men who have their testicles surgically removed or who are treated with gonadotropin-releasing hormone (GnRH) agonists to completely block testicular androgen secretion (Labrie et al. 1980), it can be observed that while the blood levels of testosterone are reduced by 95%–97%, the concentrations of intraprostatic dihydrotestosterone (DHT) are only decreased by an average of 40%, as shown by the various data available in the literature (Labrie 2007; Labrie, Dupont, and Bélanger 1985). Such findings clearly indicate the existence of an important local biosynthesis of androgens in the prostate. Most peripheral tissues, including the skin, liver, adipose tissue, and brain also show local biosynthesis of sex steroids that exert their action locally in an intracrine manner (Labrie 1991; Labrie, Luu-The et al. 2005). In other words, the action of the sex steroids made from dehydroepiandrosterone (DHEA) is exerted in the same cells where the steroids are synthesized and inactivated (Labrie 1991).

The role of intracrinology or local biosynthesis of sex steroids from adrenal DHEA is even more dramatic in women where, after menopause, all estrogens and practically all androgens are made from DHEA in a tissue-specific manner by the process of intracrinology (Labrie 2010b; Labrie, Luu-The et al. 2005). The importance of the peripheral formation of estrogens and androgens from

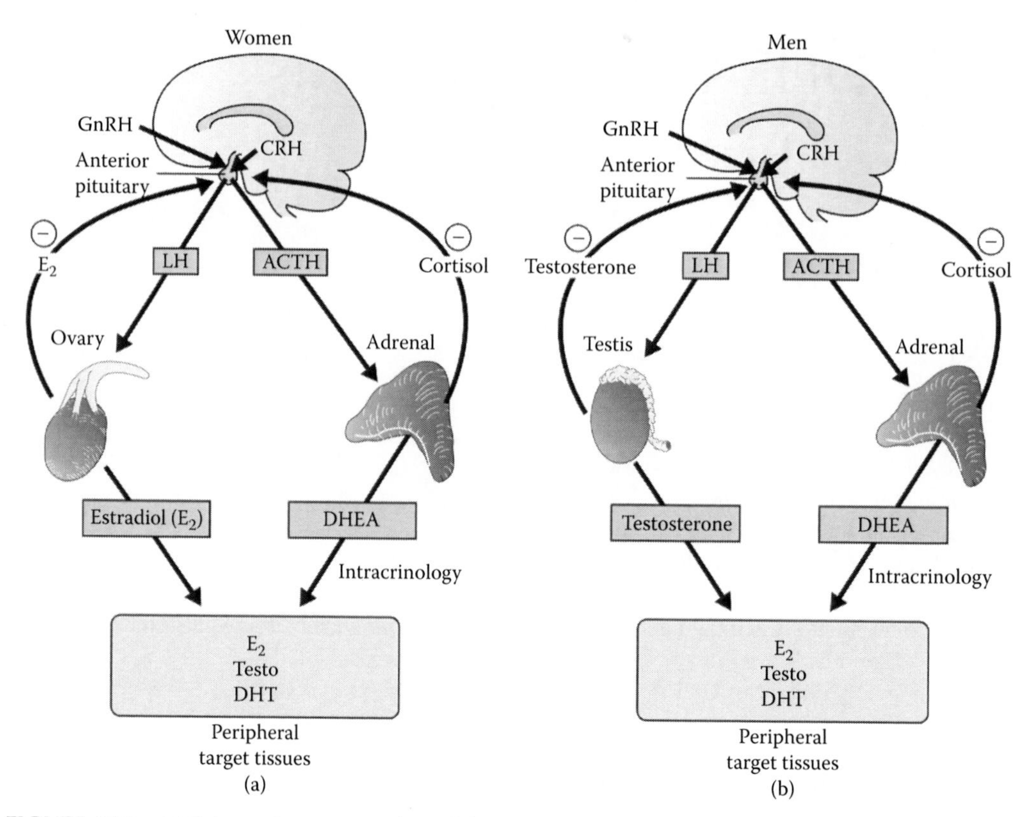

FIGURE 12.1 (a) Schematic representation of the role of the ovarian and adrenal sources of sex steroids in premenopausal women. (b) Schematic representation of the role of testicular and adrenal sources of sex steroids in men. ACTH = adrenocorticotropin; DHEA = dehydroepiandrosterone; Testo = testosterone; DHT = dihydrotestosterone, E_2 = 17β-estradiol; LH = luteinizing hormone; GnRH = gonadotropin-releasing hormone; CRH = corticotropin-releasing hormone.

DHEA has, recently, been clearly demonstrated in postmenopausal women where a rapid correction of all their symptoms and signs of vaginal atrophy accompanied by improved sexual function was achieved rapidly by intravaginal DHEA administration with no change in the blood levels of estrogens and androgens (Labrie et al. 2009a; Labrie et al. 2009b; Labrie et al. 2009c; Labrie et al. 2008b; Labrie et al. 2008a). There are, moreover, a series of data that indicate low DHEA as responsible for the other problems of menopause related to hormonal deficiency (Labrie 2007).

INTRACRINOLOGY: CRUCIAL ROLE OF HIGH-CIRCULATING DHEA EXCLUSIVELY IN HUMANS

Humans, along with the other primates, are unique among animal species in having adrenals that secrete large amounts of the inactive precursor steroids DHEA and especially dehydroepiandrosterone sulfate (DHEAS), which are converted into potent androgens and/or estrogens in peripheral tissues (Labrie 1991; Labrie et al. 1995; Labrie et al. 2001; Labrie et al. 2000; Labrie, Luu-The et al. 1997; Labrie et al. 1996; Luu-The 2001; Figure 12.1).

Adrenal secretion of DHEA and DHEAS increases during adrenarche in children at the ages of 6–8 years. Maximal values of circulating DHEAS are reached between the ages of 20 and 30 years. Thereafter, serum DHEA and DHEAS levels decrease markedly (Bélanger et al. 1994; Labrie, Bélanger et al. 1997; Migeon et al. 1957; Orentreich et al. 1984; Vermeulen et al. 1982; Figure 12.2).

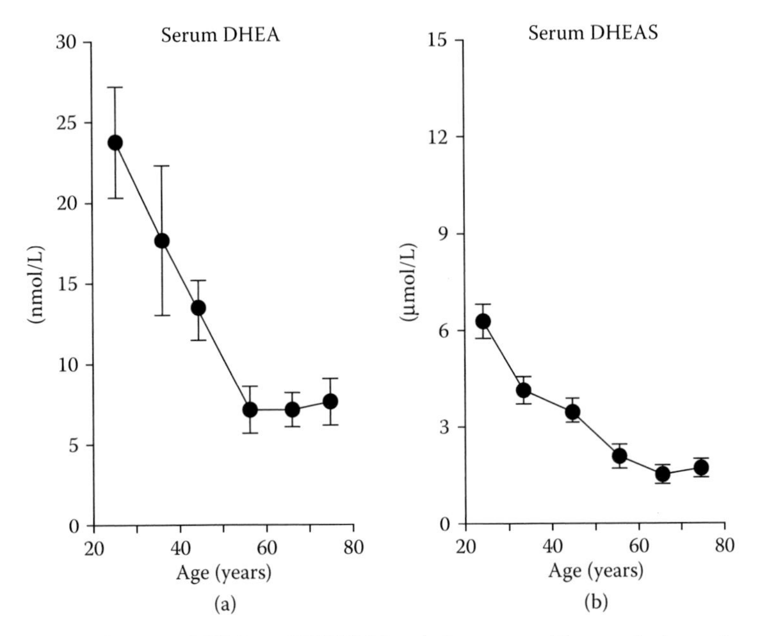

FIGURE 12.2 Effect of age on DHEA and DHEAS levels in women. The graph shows the serum concentrations of (a) DHEA and (b) DHEAS in women aged 20–80 years. Data are expressed as means ± SEM. DHEA = dehydroepiandrosterone; DHEAS = dehydroepiandrosterone sulfate. (From Labrie, F., A. Bélanger, P. Bélanger et al. 2007. *J Steroid Biochem Mol Biol* 103(2):178–88. With permission.)

In fact, at 70 years of age, serum DHEAS levels are decreased to approximately 20% of their peak values, while they can decrease by 95% by the age of 85–90 years (Migeon et al. 1957).

The marked reduction in the formation of DHEAS by the adrenals during aging (Bélanger et al. 1994; Labrie, Bélanger et al. 1997; Migeon et al. 1957; Orentreich et al. 1984; Vermeulen et al. 1982; Vermeulen and Verdonck 1976) results in a dramatic reduction in the formation of androgens and estrogens in peripheral target tissues, a situation that has been proposed to be associated with age-related diseases such as insulin resistance (Coleman, Leiter, and Schwizer 1982; Schriock et al. 1988) and obesity (MacEwen and Kurzman 1991; Nestler et al. 1988; Tchernof et al. 1995). However, much attention has been given to the benefits of DHEA administered to postmenopausal women, especially on the bone, skin, vaginum, and well-being after oral (Baulieu et al. 2000; Morales et al. 1994), percutaneous (Diamond et al. 1996; Labrie, Diamond et al. 1997), or intravaginal (Labrie et al. 2009b; Labrie et al. 2009a) administration.

It is thus remarkable that man, in addition to possessing very sophisticated endocrine and paracrine systems, has largely vested in sex steroid formation in peripheral tissues (Labrie, Bélanger, and Labrie 1988; Labrie 1991; Labrie, Bélanger et al. 1997; Labrie, Dupont, and Bélanger 1985). In fact, while the ovaries and testes are the exclusive sources of androgens and estrogens in lower mammals, the situation is very different in man and higher primates, where active sex steroids are in large part or wholly (postmenopause) synthethized locally in peripheral tissues, thus providing target tissues with the appropriate controls, which adjust the formation and metabolism of sex steroids to local requirements.

Transformation of the adrenal precursor steroids DHEAS and DHEA into androgens and/or estrogens in peripheral target tissues depends on the level of expression of the various steroidogenic and metabolizing enzymes in each cell of these tissues. This sector of endocrinology that focuses on the intracellular hormone formation and action has been called intracrinology (Labrie, Bélanger, and Labrie 1988; Labrie 1991; Figure 12.3). This situation of a high-secretion rate of adrenal precursor sex steroids in men and women is, thus, completely different from all animal models used in the

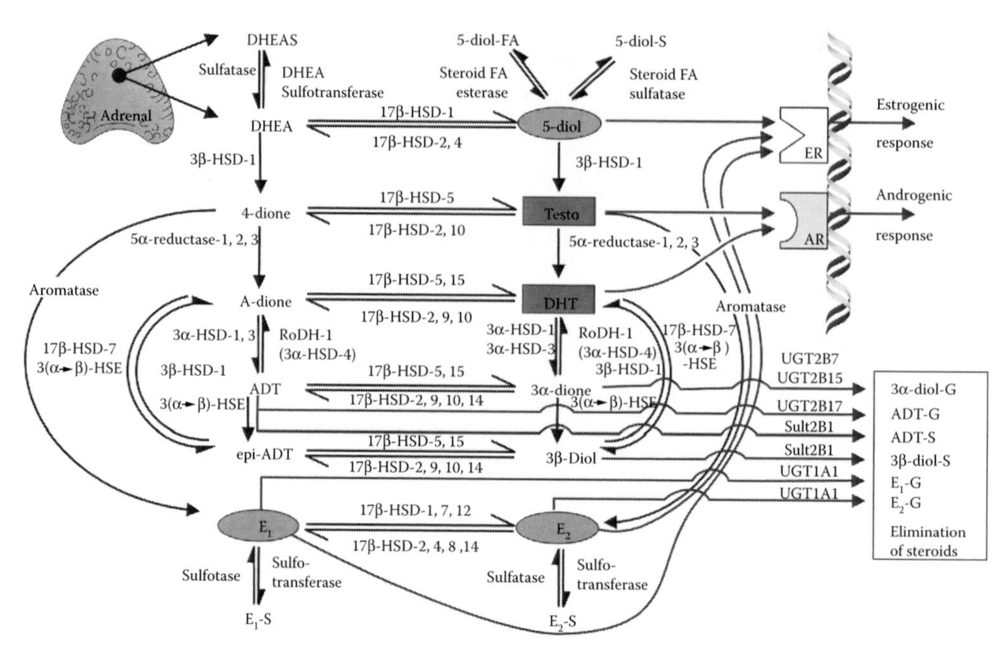

FIGURE 12.3 Human steroidogenic and steroid-inactivating enzymes in peripheral intracrine tissues. 4-dione = androstenedione; A-dione = 5-α-androstane-3,17-dione; ADT = androsterone; epi-ADT = epi-androsterone; E_1 = estrone; E_1-S = estrone sulfate; 5-diol-FA = androst-5-ene-3α, 17β-diol fatty acid; 5-diol-S = androst-5-ene-3α, 17β-diol sulfate; HSD = hydroxysteroid dehydrogenase; Testo = testosterone; RoDH-1 = Ro dehydrogenase 1; ER = estrogen receptor; AR = androgen receptor; UGT2B28 = uridine glucuronosyl transferase 2B28; Sult2B1 = sulfotransferase 2B1; UGT1A1 = uridine glucuronosyl transferase 1A1.

laboratory, namely rats, mice, guinea pigs, and all others (except monkeys), where the secretion of sex steroids takes place exclusively in the gonads (Bélanger et al. 1989; Labrie, Bélanger, and Labrie 1988; Labrie, Bélanger et al. 1997; Labrie, Dupont, and Bélanger 1985). One explanation for the late recognition of the importance of the formation of sex steroids in peripheral target tissues or intracrinology is the fact that the adrenals of the animal models in common use in preclinical research do not secrete significant amounts of adrenal precursor sex steroids, thus focusing all attention on the testes and ovaries as the exclusive sources of androgens and estrogens. The term "intracrinology" was thus coined (Labrie, Bélanger, and Labrie 1988) to describe the synthesis of active steroids in peripheral target tissues in which the action is exerted in the same cells where synthesis takes place without the release of the active steroids in the extracellular space and general circulation (Labrie 1991).

It should be noted that the importance of the intracrine formation of androgens and estrogens extends to nonmalignant diseases such as acne, seborrhea, hirsutism, and androgenic alopecia, as well as to osteoporosis and vaginal atrophy (Cusan et al. 1994; Labrie, Diamond et al. 1997). Another example of the relevance of intracrinology in nonmalignant diseases is endometriosis (Bulun et al. 2000). In this regard, it has recently been demonstrated that aromatase is expressed aberrantly in endometriosis, although this activity is not detectable in the normal endometrium. Furthermore, another abnormality in this disease is the deficient expression of type 2 17β-hydroxysteroid dehydrogenase (17β-HSD), thus impairing the inactivation of estradiol (E_2) into estrone (E_1). Consequently, the increased formation of E_2 by aromatase coupled with the decreased inactivation of E_2 by type 2 17β-HSD leads to increased stimulation of the endometrium and endometriosis.

PHYSIOLOGY OF DHEA AND ITS ROLE IN MEN AND WOMEN

Original Observation of Intracrinology

The unique physiological importance of DHEA in man was first recognized in the early 1980s, when it was discovered that after complete medical castration achieved with GnRH agonists in men suffering from prostate cancer (Labrie et al. 1980), 40%–50% of active androgens were left in the prostate, thus, demonstrating an important extratesticular source of androgens (Labrie, Dupont, and Bélanger 1985). When pooling data obtained by various laboratories around the world, values of residual intraprostatic DHT range from 25% to 50% for an average of 40% of DHT left in the prostate after castration (Labrie et al. 2009). This originally surprising observation is now well explained by the transformation of DHEA into androgens and/or estrogens by specific steroidogenic enzymes in each cell type in each peripheral target tissue according to the process of intracrinology, an expression coined in 1988 (Labrie, Bélanger, and Labrie 1988; Labrie 1991; Labrie, Luu-The et al. 2005; Luu-The and Labrie 2010).

A major problem with DHEA, the only source of sex steroids after menopause, is that there is no feedback control of its secretion. In other words, there is no endogenous mechanism in either women or men to increase DHEA secretion when serum DHEA concentrations become low. Consequently, the only possibility to correct a clinically significant lack of DHEA availability in postmenopausal women is to administer exogenous DHEA in order to replace the amount of DHEA missing in these women. Such a replacement with DHEA can thus improve or even make the symptoms of hormone deficiency disappear as observed recently in women treated for vaginal atrophy, a classical consequence of hormone deficiency during postmenopause (Buster 2009; Labrie et al. 2009c; Labrie et al. 2009b; Labrie et al. 2009a).

Comparison between Men and Women

Although the supply of sex steroids from DHEA decreases in both men and women in a comparable fashion from the age of 30 years, men receive a practically continuous supply of testosterone, E_1, and E_2 from the testicles during their whole life (Figure 12.1), while, in women, E_2 secretion by the ovaries stops at menopause (Figure 12.4). Consequently, after menopause, all estrogens and

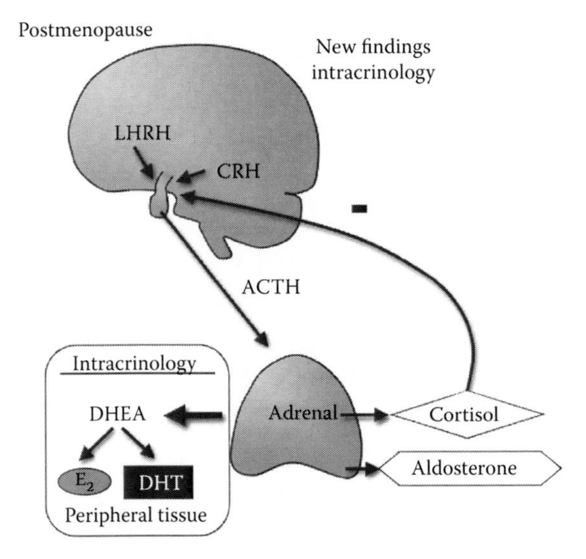

FIGURE 12.4 Schematic representation of the unique source of sex steroids, namely, adrenal DHEA, in postmenopausal women. At menopause, the secretion of E_2 by the ovaries ceases. Consequently, after menopause, all estrogens and practically all androgens are made locally from DHEA in peripheral target intracrine tissues. The amount of sex steroids made in peripheral target tissues depends on the level of the steroid-forming enzymes specifically expressed in each tissue. DHT = dihydrotestosterone; E_2 = 17β-estradiol.

androgens are derived from DHEA, which has already decreased by an average of 60% at the time of menopause (Labrie, Bélanger et al. 2006) and continues to decrease thereafter (Figure 12.2), with some women having barely detectable serum levels of DHEA (Labrie 2010b). Because DHEA is the only source of sex steroids after menopause, it is reasonable to believe that such a decrease in DHEA-derived sex steroid availability, coupled with aging, is at least partially responsible for the numerous symptoms of hormone deficiency observed after menopause. These pertain to vaginal atrophy, bone loss, fat accumulation, type 2 diabetes, skin atrophy, cognition problems, memory loss, and, possibly, Alzheimer's disease (Labrie 2007).

In men, the finding that 25%–50% of androgens are left in the prostate after castration (Bélanger et al. 1989; Labrie, Dupont, and Bélanger 1985; Mostaghel et al. 2007; Nishiyama, Hashimoto, and Takahashi 2004) explains why the addition of a pure (nonsteroidal) antiandrogen to castration achieves a more complete blockade of androgens and has been the first treatment shown to prolong life in prostate cancer (Caubet et al. 1997; Labrie, Bélanger et al. 2005; Labrie, Dupont, and Bélanger 1985; Labrie et al. 1982; Prostate Cancer Triallists' Collaborative Group 2000). The androgens remaining at relatively high levels after castration also explain why combined androgen blockades or the blockade of the androgens of both testicular and adrenal origins at start of treatment can provide a cure for most patients when the treatment is started at the localized stage of the cancer (Akaza 2006; Labrie et al. 2002; Ueno et al. 2006), thus clearly demonstrating the major role of extratesticular androgens or intracrinology in men.

With the knowledge of the major importance of androgens of adrenal origin in men, it is of interest to compare the aforementioned for men with the serum levels of the same steroids measured in intact postmenopausal women. As can be seen in Figure 12.5a and b, the serum levels of testosterone and of the total androgen metabolites are almost superimposable in castrated men and postmenopausal women of comparable age. Most interestingly, it can also be seen that the serum

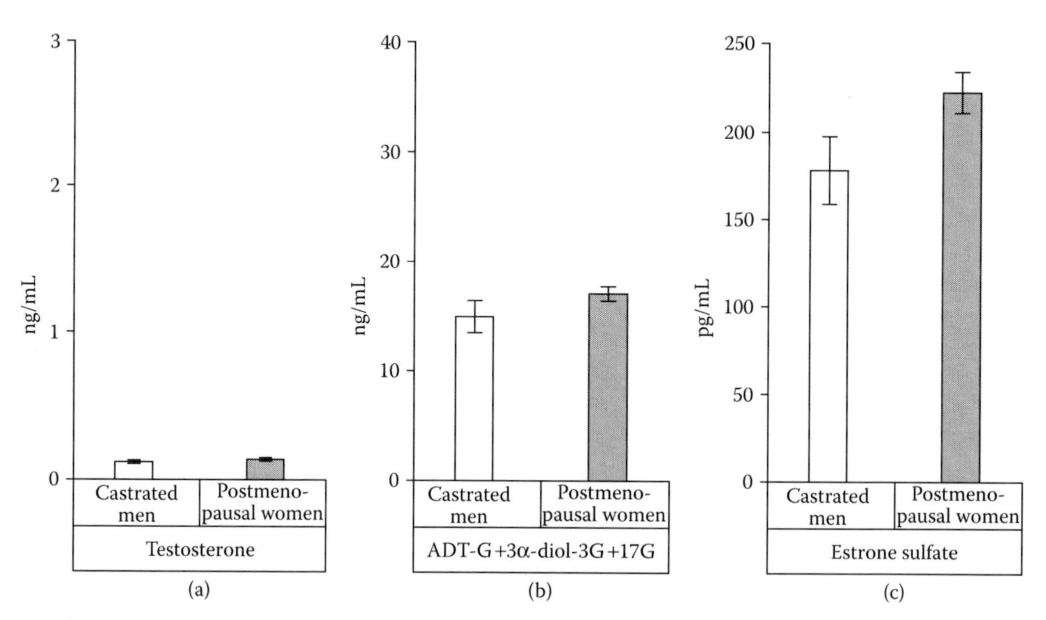

FIGURE 12.5 Comparison of the serum concentrations of (a) testosterone, (b) total androgenic pool (sum of ADT-G, 3α-diol-3G, and 17G), and (c) E_1-S in castrated 69- to 80-year-old men ($n = 34$) and intact 55- to 65-year-old postmenopausal women ($n = 377$). ADT-G = androsterone glucuronide; 3α-diol-3G = androstane-3α, 17β-diol-3 glucuronide; E_1-S = estrone sulfate. (From Labrie, F., R. Poulin, J. Simard, V. Luu-The, C. Labrie, and A. Bélanger. 2006. Androgens, DHEA and breast cancer. In *Androgens and Reproductive Aging*, ed. T. Gelfand, 113–35. Oxsfordshire, UK: Taylor and Francis; Labrie, F., L. Cusan, J. L. Gomez et al. 2009. *J Steroid Biochem Mol Biol* 113:52–6. With permission.)

levels of estrone sulfate (E_1-S) are also comparable (Figure 12.5c). The serum levels of E_1 and E_2 are also comparable, thus indicating that similar amounts of estrogens of adrenal origin are found in both men and women (Labrie et al. 2009).

The above-summarized data show that ~40% of androgens are made in peripheral tissues in the absence of testicles in 69- to 80-year-old men. Because serum DHEA decreases markedly with age starting in the thirties (Labrie, Luu-The et al. 2005), and testicular androgen secretion decreases only slightly, it is most likely that androgens of adrenal origin have an even greater relative and absolute importance at younger ages. The same conclusion applies to women with respect to the androgens synthesized from DHEA.

Most importantly, the summarized aforementioned data show that postmenopausal women synthesize androgens and estrogens in quantities comparable with castrated men of similar age (Figure 12.5). In fact, women secrete approximately 40% as much androgens as men of similar age.

Systemic Estrogens Are Not Physiological after Menopause

Because all women are not exposed to systemic estrogens after menopause, it is reasonable to believe that the nonphysiological situation created by the administration of estrogens could be responsible, up to an unknown extent, for the side effects reported in women receiving traditional estrogen replacement therapy (ERT) and estrogen + progestin replacement therapy (hormone replacement therapy; Beral 2003; Beral, Bull, and Reeves 2005; Grodstein, Manson, and Stampfer 2006; Hsia et al. 2006; Pines et al. 2007; Riman et al. 2002; Rossouw et al. 2002; Ruttimann 2008). The recent observations indicating the risks associated with estrogens should normally help to focus our attention on DHEA, the only physiological source of sex steroids after menopause (Labrie 2010b), and to better understand the mechanism of action of DHEA and its preventive and therapeutic roles.

High Variability of Serum DHEA

An important observation is that despite the arrest of estrogen secretion in women at the time of menopause, not all postmenopausal women suffer from menopausal symptoms. Consequently, the hormonal difference between postmenopausal women suffering from vaginal atrophy symptoms (~75% of women) and those without symptoms (~25%) is not related to estrogens. In fact, the only remaining hormonal difference between these two groups of women is the difference in the availability of DHEA, the exclusive source of sex steroids (Figure 12.4). In addition to markedly decreasing with age (Figure 12.2), the serum levels of DHEA are highly variable with some women having barely detectable levels while others have serum levels of DHEA up to 9–10 ng/mL (Labrie, Bélanger et al. 2006; Figure 12.6).

Rapid Saturation of the Enzymatic Mechanisms Transforming Dehydroepiandrosterone into Sex Steroids: Prevention of Overexposure to Sex Steroids

It should be mentioned that saturation of the enzymatic systems that transform DHEA into active androgens and/or estrogens (Figure 12.3) is observed at serum DHEA levels of about 7 ng/mL, thus protecting women against potential excess levels of sex steroids (Labrie 2010a; Labrie et al. 2007; Figure 12.7). We believe that the presence of this natural saturation of the enzymatic mechanism makes it practically impossible to administer a dose of DHEA that could lead to excess tissue exposure to estrogens and/or androgens. The fact that no serious adverse event has ever been reported with exogenous DHEA (see Tables 1 and 2 in Labrie 2010a) is likely to be explained, up to an unknown extent, by this self-protecting mechanism. In fact, when replacement with DHEA is limited to physiological amounts of this tissue-specific sex steroid prehormone free of intrinsic sex steroid activity, no adverse effect can be logically expected. All our pharmacokinetic studies performed in postmenopausal women indicate that doses of prasterone much larger that those used in our efficacy studies could only have a limited effect on global sex steroid formation (Labrie et al. 2007; Labrie et al. 2008c).

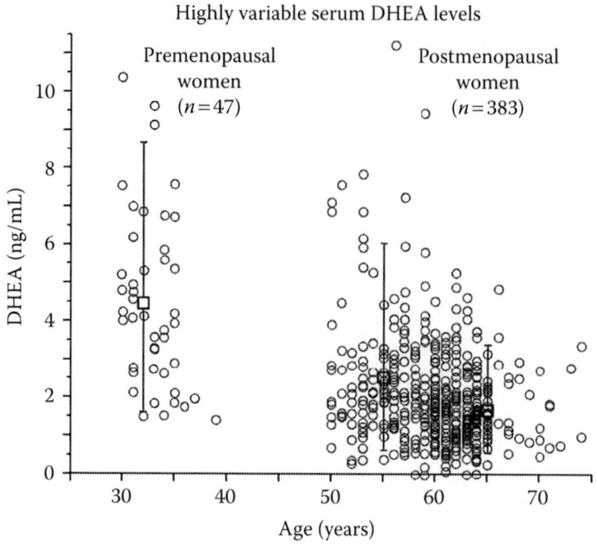

Highly variable serum DHEA levels

Premenopausal women ($n = 47$)

Postmenopausal women ($n = 383$)

DHEA (ng/mL)

Age (years)

FIGURE 12.6 Illustration of the wide variability of serum dehydroepiandrosterone levels in normal women aged 30–40 years and 50–75 years. Data are presented individually as well as means and 5%–95% centiles. (From Labrie, F. 2010b. *Menopause Manag* 19:14–24. With permission.)

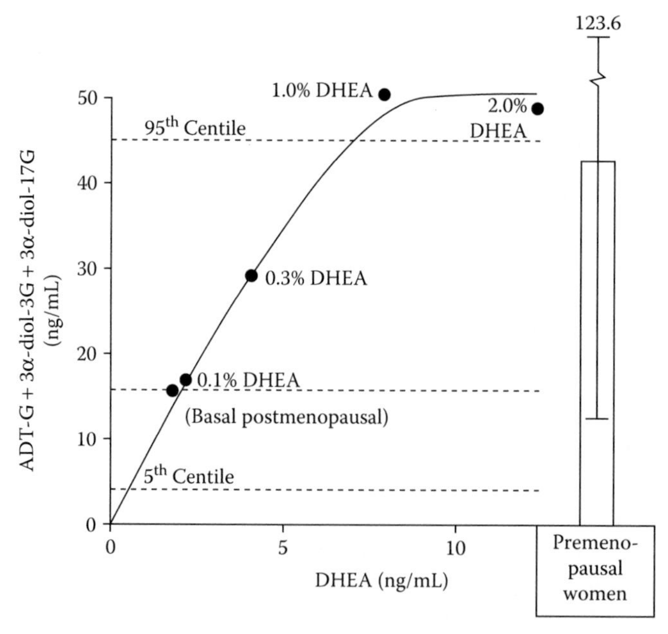

FIGURE 12.7 Effect of increasing serum concentrations of DHEA induced by twice daily percutaneous administration of 3 g of 0% (placebo), 0.1%, 0.3%, 1.0%, or 2.0% DHEA cream on the sum of the serum levels of the androgen metabolites ADT-G, 3α-diol-3G, and 3α-diol-17G expressed in nanograms per milliliter. DHEA = dehydroepiandrosterone; ADT-G = androsterone glucuronide; 3α-diol-3G = androstane-3α, 17β-diol-3 glucuronide; 3α-diol-17G = androstane-3α, 17β-diol-17 glucuronide. (From Labrie, F., A. Bélanger, P. Bélanger et al. 2007. *J Steroid Biochem Mol Biol* 103(2):178–88. With permission.)

ANTICARCINOGENIC ACTIVITY OF DHEA

The absence in rodent peripheral tissues of the uridine diphosphate glucuronosyltransferases responsible, in large part, for the inactivation of estrogens and androgens suggests that in the presence of comparable concentrations of DHEA following its exogenous administration, the intracellular levels of sex steroids should be much higher in rodents than humans. Accordingly, the finding of an anticarcinogenic action of DHEA in the rodent supports the absence of carcinogenic potential of DHEA in the human where lower intratissular concentrations of DHEA-derived steroids should be found due to the high rate of their inactivation by the steroid-inactivating enzymes present in the human target tissues.

DHEA is well known to exert antiproliferative effects (Ho et al. 2008). Much evidence has been obtained on the preclinical chemopreventive efficacy of DHEA (reviewed in [Elmore et al. 2001; Levi, Borne, and Williamson 2001; Schwartz, Lewbart, and Pashko 1992]). In fact, a long series of studies have shown that DHEA, instead of being carcinogenic as shown for estrogens, exerts an opposite effect, namely an anticarcinogenic effect (Boone, Steele, and Kelloff 1992; Kelloff et al. 1996; Kelloff et al. 1994; Ratko et al. 1991; White et al. 1998). Among these, it has been shown that oral DHEA administration in rats and mice inhibits the development of spontaneous breast cancer and various chemically induced tumors of the colon (Nyce et al. 1984; Steele et al. 1994), lung (Schwartz et al. 1981; Schwartz and Tannen 1981), breast (Ratko et al. 1991), liver (Garcea et al. 1987; Moore et al. 1986; Simile et al. 1995; Weber, Moore, and Bannasch 1988), and prostate (Christov et al. 2004; Steele et al. 1994). DHEA has also been found to inhibit chemically induced tumors of the thyroid, skin, and lymphatic tissue, as well as the formation of preneoplastic liver foci and hemangiosarcomas of the liver (Green et al. 2001; Levi, Borne, and Williamson 2001; Schwartz, Lewbart, and Pashko 1992; Schwartz, Pashko, and Whitcomb 1986; Schwartz and Pashko 1993). It was also shown that DHEA confers significant protection against prostate carcinogenesis (Rao et al. 1999; Weber, Moore, and Bannasch 1988). DHEA has also been shown to prevent tumorigenesis in p53 knockout mice (Hursting et al. 1995).

DHEA and its analog 8354 have shown inhibitory effects similar to DHEA on the development of mammary gland carcinogenesis (McCormick et al. 1996; Ratko et al. 1991; Schwartz et al. 1989; Steele et al. 1994), including spontaneous breast cancer in the mouse (Schwartz 1979; Schwartz et al. 1981). We have also observed the inhibitory effects of androgens and DHEA on dimethylbenzanthracene (DMBA)-induced mammary carcinoma in the rat (Dauvois et al. 1989; Li et al. 1993; Luo, Labrie et al. 1997; Luo, Sourla et al. 1997), as well as on human breast cancer ZR-75-1 xenografts in nude mice (Couillard et al. 1998; Dauvois et al. 1991). Topical application of DHEA inhibits the development of DMBA-induced and 12-O-tetradecanoylphorbol-13-acetate (TPA)-promoted skin papillomas and carcinomas in the mouse (Pashko et al. 1985; Pashko et al. 1984).

The anticarcinogenic effect of DHEA in breast cancer can be logically explained by the predominant formation of androgens over estrogens from DHEA in breast tissue, thus providing a higher influence of androgens, which are well-known inhibitors of mammary gland proliferation *in vitro*, as well as *in vivo* in animal models and in women (Labrie et al. 2003). In fact, DHEA has been identified as one of the promising chemoprotective agents by the U.S. National Cancer Institute (NCI), Division of Cancer Prevention and Control. DHEA is part of the ~30 promising compounds across a group of agents chosen by the NCI for clinical chemoprevention trials (Kelloff et al. 1996). It can be mentioned that low DHEA levels have been reported to predispose to certain cancers (Gordon et al. 1993; Gordon, Helzlsouer, and Comstock 1991; Regelson and Kalimi 1994), including an increased risk of breast (Zumoff et al. 1981) and bladder (Gordon, Helzlsouer, and Comstock 1991) cancers.

In addition to its conversion into androgens, other proposed mechanisms are possibly involved. These pertain to inhibition of glucose-6-phosphate dehydrogenase, 3-hydroxy-3-methylglutaryl CoA reductase, glucose metabolism, or mitochondrial gene expression (Ho et al. 2008; Pascale et al. 1995; Tian et al. 1998).

DHEA AND THE BREAST

ANDROGENS INHIBIT BREAST CANCER

Although it is well recognized that estrogens play the predominant stimulatory role in the development and growth of human breast cancer, a series of observations have shown that androgens such as testosterone (Cooperative Breast Cancer Group 1964; Fels 1944; Segaloff et al. 1951; Ulrich 1939), fluoxymesterone (Ingle et al. 1991; Kennedy 1958; Tormey et al. 1983), calusterone (Gordan et al. 1973), and anabolic steroids (Gordan 1976; Segaloff 1977) have an efficacy to inhibit breast cancer comparable with other types of endocrine manipulations.

Evidence from preclinical studies proves the inhibitory effect of androgens on breast cancer (Labrie, Poulin et al. 2006). The above-mentioned clinical data are, in fact, well supported by the observation of a synergistic effect of DHEA and of the pure anti-estrogen EM-800 (a precursor of acolbifene) in preventing the development of DMBA-induced mammary tumors in the rat (Luo, Sourla et al. 1997). Moreover, DHEA shows an almost exclusive androgenic effect on the histomorphology and structure of the rat mammary gland (Sourla et al. 1998), thus suggesting that the inhibitory effect of DHEA results from its transformation into androgens. Moreover, the effect of androgens as direct inhibitors of breast cancer growth is well supported by the presence of androgen receptors (ARs) in a large proportion of human breast cancers (Allegra et al. 1979; Bryan et al. 1984; Miller et al. 1985; Trams and Maass 1977). There is also genetic evidence indicating a protective role of androgens against breast cancer (Lobaccaro et al. 1993; Wooster et al. 1992).

A potent and direct inhibitory effect, at physiological concentrations, of androgens has been observed on the proliferation of human breast cancer cells *in vitro* (Dumont et al. 1989; Poulin, Baker, and Labrie 1988; Poulin, Baker et al. 1989; Poulin, Simard et al. 1989; Simard et al. 1990; Simard et al. 1989). In fact, the first demonstration of a potent and direct inhibitory effect of androgens on human breast cancer growth was obtained in the estrogen-sensitive human breast cancer cell line ZR-75-1 (Poulin, Baker, and Labrie 1988). In that study, DHT not only completely blocked the stimulatory effect of E_2 on cell proliferation but it also further reduced cell growth in the absence of estrogens. At low cell density, DHT completely prevented breast cancer cell growth.

It should be added that treatment of ovariectomized monkeys with testosterone decreased the stimulation of mammary epithelial proliferation induced by E_2 by about 40% (Zhou et al. 2000). Moreover, it is possible that part of the increased risk of breast cancer in *BRCA-1* mutant patients is associated with the decreased efficiency of the mutated *BRCA-1* gene to interact with the AR (Park et al. 2000). It is also pertinent to mention that female athletes and transsexuals taking androgens show atrophy of mammary gland epithelial tissue (Burgess and Shousha 1993; Korkia and Stimson 1997).

In addition to the direct inhibitory effect of androgens on mammary epithelial cell proliferation, it is increasingly apparent that mammary cells possess complex regulatory mechanisms that allow for the strict control of the intracellular levels of both stimulatory and inhibitory sex steroids. For instance, our data show that DHT favors the degradation of E_2 into E_1, thus suggesting that the potent antiproliferative activity of DHT in E_2-stimulated ZR-75-1 human breast cancer cells is, at least partially, exerted on 17β-HSD activity (Adams 1985; Couture et al. 1993; Poulin, Baker, and Labrie 1988; Poulin, Simard et al. 1989). Conversely, we have found that estrogens cause a marked increase in the production of the glucuronidated androgen metabolites, androstane 3α, 17β-diol glucuronide and androsterone glucuronide, in MCF-7 cells, thus decreasing the inhibitory androgenic activity (Roy et al. 1992). In fact, because glucuronidation is the predominant route of androgen inactivation, androgen-inactivating enzymes may constitute an important locus of the regulation of breast cancer growth.

The long series of preclinical and clinical data reviewed in Labrie, Poulin et al. (2006) and Labrie et al. (2003) indicate that proliferation of both the normal mammary gland and breast cancer results from the balance between the stimulatory effect of estrogens and the inhibitory effect of androgens (Figure 12.8). Moreover, the data showing the additive inhibitory action of anti-estrogens

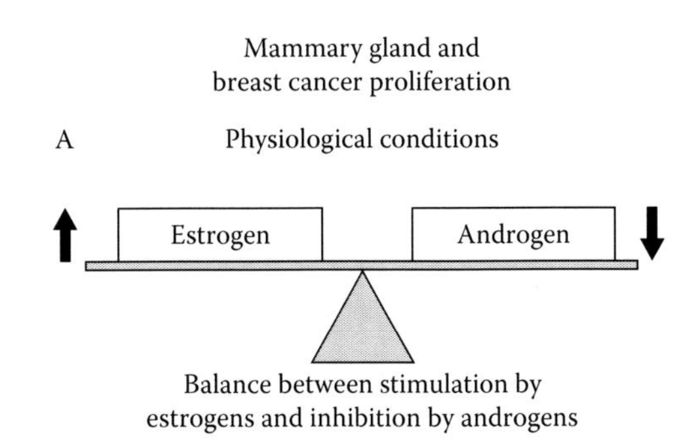

FIGURE 12.8 Schematic representation of the balance between the stimulatory action of estrogens and the inhibitory effect of androgens on mammary gland and breast cancer proliferation.

and androgens suggest that taking advantage of the inhibitory action of androgens on breast cancer proliferation could well improve the efficacy of the currently used and well-tolerated estrogen-deprivation therapies for the treatment and prevention of breast cancer; the most physiological androgen precursor is DHEA, which can lead to formation in a tissue-specific manner.

DHEA Inhibits Breast Cancer

Low-circulating levels of DHEA and DHEAS have also been found in patients with breast cancer (Zumoff et al. 1981) and DHEA has been found to exert anti-oncogenic activity in a series of animal models (Gordon, Shantz, and Talalay 1987; Li et al. 1993; Schwartz, Pashko, and Whitcomb 1986). In fact, DHEA has been shown, in animal model systems, to be an inhibitor of breast, prostate, lung, liver, skin, and thyroid carcinogenesis (Kohama et al. 1997; Li et al. 1994; Lubet et al. 1998; McCormick et al. 1996; Ratko et al. 1991; Shilkaitis et al. 2005). Among a series of data showing its inhibitory effect on breast cancer, DHEA blocked the stimulatory effect of N-nitroso-N-methylurea on mammary tumor development in the rat by a p16-dependent mechanism (Shilkaitis et al. 2005). DHEA has also been shown to have immunomodulatory effects *in vitro* (Suzuki et al. 1991) and *in vivo* in fungal and viral diseases (Rasmussen, Arrowood, and Healey 1992). However, a stimulatory effect of DHEA on the immune system has been described in postmenopausal women (Casson et al. 1993).

In order to investigate the possibility that DHEA and its metabolites could have a preventive effect on the development of mammary carcinoma, we have studied the effect of increasing circulating levels of DHEA, constantly released from Silastic implants, on the development of mammary carcinoma induced by DMBA in the rat. Treatment with increasing doses of DHEA delivered constantly by Silastic implants of increasing length and number caused a progressive inhibition of tumor development (Li et al. 1993). It is of interest that tumor size in the group of animals treated with the highest dose (6 × 3.0-cm-long implants) of DHEA was similar to that found in ovariectomized animals, thus showing a complete blockade of estrogen action by DHEA. Such data clearly demonstrate that circulating levels of the precursor adrenal steroid DHEA comparable with those observed in normal adult premenopausal women exerts a potent inhibitory effect on the development of mammary carcinoma induced by DMBA in the rat.

Prevention of Breast Cancer

DHEA administration in mice and rats inhibits the development of experimental breast, colon, lung, skin, and lymphatic tissue tumors (Levi, Borne, and Williamson 2001; Schwartz, Pashko, and Whitcomb 1986). As an example, in the skin tumorogenesis model in mice, DHEA inhibits tumor initiation as well as tumor promoter-induced epidermal hyperplasia and promotion of papillomas (Schwartz, Pashko, and Whitcomb 1986). In order to investigate the possibility that

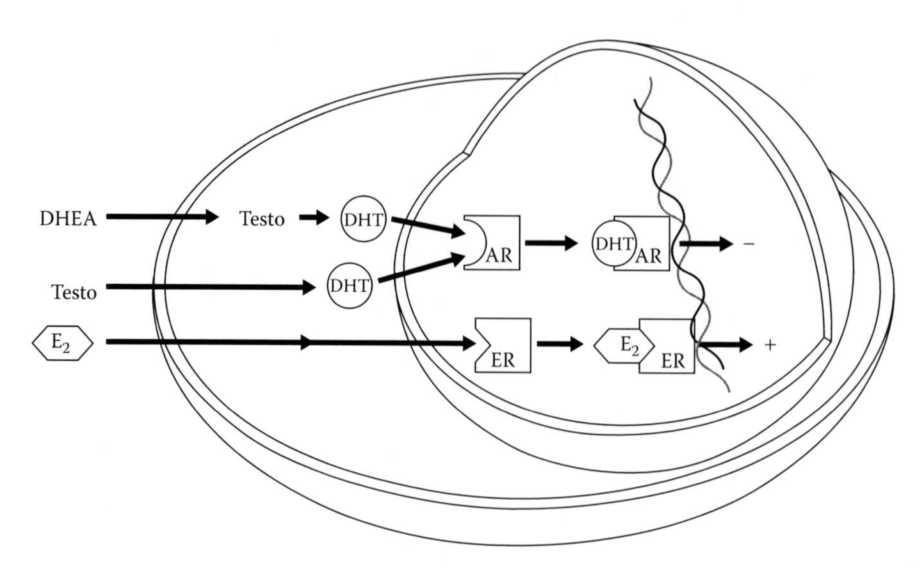

FIGURE 12.9 Antagonism between the inhibitory effects of androgens and dehydroepiandrosterone and the stimulatory effects of estrogens on breast cancer proliferation.

DHEA and its metabolites could have a preventive effect on the development of mammary carcinoma, we have studied the effect of increasing circulating levels of DHEA constantly released from Silastic implants on the development of mammary carcinoma induced by DMBA in the rat. The DMBA-induced mammary carcinoma in the rat has been widely used as a model of hormone-sensitive breast cancer in women (Asselin et al. 1977; Asselin and Labrie 1978; Dauvois et al. 1989; Figure 12.9).

Treatment with increasing doses of DHEA delivered constantly by Silastic implants of increasing length and number caused a progressive inhibition of tumor development (Li et al. 1993). It is of interest to see that tumor size in the group of animals treated with the highest dose (6×3.0-cm-long implants) of DHEA was similar to that found in overiectomized animals, thus showing a complete blockade of estrogen action by DHEA. Such data clearly demonstrate that circulating levels of the precursor adrenal steroid DHEA comparable with those observed in normal adult premenopausal women (Liu et al. 1990) exert a potent inhibitory effect on the development of mammary carcinoma induced by DMBA in the rat. It is of special interest to see that serum levels of DHEA of 7.09 ± 0.64 nM and 17.5 ± 1.1 nM led to a dramatic inhibition of tumor development to 22% and 11% of animals bearing mammary carcinoma compared with 68% in control intact animals. At the highest dose of DHEA used, which corresponds to serum DHEA values of 27.2 ± 2.2 nM, the incidence of tumors was reduced to only 3.8%. It should be mentioned that the serum DHEA levels in normal 20- to 30-year old women ranges between 8.3 and 17.3 nM (Liu et al. 1990).

It might be relevant to mention that treatment with DHEA markedly delayed the appearance of breast tumors in C3H mice that were genetically bred to develop breast cancer (Schwartz 1979).

SAFETY

No serious adverse event related to DHEA has ever been reported in any published data describing DHEA administration in women or men (Allolio, Arlt, and Hahner 2007). The only side effects reported, despite the high doses frequently used, in a small proportion of women are mild facial acne, increased sebum production, and mild changes in hair growth, which, usually, do not require withdrawing from study (Allolio, Arlt, and Hahner 2007; Hunt et al. 2000; Labrie, Diamond et al. 1997). The incidence of these mild side effects is usually reported at about 50% of the same rate observed in placebo-treated subjects (Genelabs, Briefing document, April 19, 2001 FDA Arthritis

Advisory Committee). The higher sebum production is welcome in postmenopausal women who frequently complain of dry skin, while some regrowth of pubic and axillary hair is generally considered positive.

In two Genelabs' placebo-controlled trials and the open-label extension study, which followed completion of the double-blind studies, 387 women have received DHEA, usually at a daily dose of 200 mg, for at least 6 months, 242 for ≥12 months, 138 for ≥18 months, and 36 for ≥24 months. Six hundred and forty one women have been exposed to DHEA (Briefing document FDA, Arthritis Advisory Committee, April 19, 2001).

The patients in the Genelabs' studies were suffering from systemic lupus erythematosus and, in a large proportion, were already treated with relatively high doses of glucocorticoids. In patients treated with 200 mg DHEA, a significant increase versus placebo was observed for acne (36.0% vs. 15.2%) and hirsutism (14.2% vs. 2.3%). The relatively high incidence of acne in both the placebo- and DHEA-treated groups could be related to the fact that many of these patients were taking glucocorticoids. It is of interest that 64% and 53% of the incidence of acne and hirsutism, respectively, were reported during the first 182 days of treatment compared to during the later interval between 183 and 547 days.

It is also pertinent to mention the Memorandum from Claudia B. Karkowski dated March 12, 2001. This document from the FDA provides an overview of postmarketing adverse events reported in association with the use of DHEA, where much higher doses or oral DHEA have been used for long time periods. These cases were retrieved from the Adverse Event Reporting System (AERS), Center for Food Safety and Applied Nutrition (CFSAN)'s postmarketing database, and the medical literature. The conclusion was that no clear safety signals were identified with this product.

This served as a companion document to the review by Parivash Nourjah, PhD, entitled *Epidemiologic Evidence of DHEA in the Etiology of Neoplasia*, which focused on a review of the literature for any published epidemiologic studies that examined cancer risk associated with exogenous DHEA administration. The primary objective in this review was to determine if there were any case reports in our postmarketing databases of the medical literature of neoplasia in association with the use of DHEA. The same single case of worsening of metastatic prostate cancer was described (Jones et al. 1997). In the memorandum (February 20, 2001) of Parivash Nourjah, Division of Post-marketing Drug Risk Assessment 1, HFD-430, it was concluded that "no meaningful conclusion about the association of exogenously administered DHEA and cancer risk can be made based on these epidemiological studies of endogenous levels of DHEA."

"Of the 65 cases identified in the AERS database, SCSAN ARMS database, and the medical literature, there was only one report of neoplasia," namely, as mentioned above, a case of worsening of prostate cancer in a man with metastatic prostate cancer who received a daily dose of 200–700 mg DHEA for the treatment of anemia unresponsive to erythropoietin. "This was already summarized in Dr. Nourjah's review." "His blood cells increased during DHEA eliminating his need for transfusions. However, the patient began to develop facial numbness, increase in prostate size, and difficulty voiding. His prostatic-specific antigen (PSA) levels increased to greater than 10 000 ng/mL (2726 ng/mL prior to DHEA). DHEA was discontinued and DES was initiated with improvement in symptoms and decrease in PSA. Although he exhibited a positive dechallenge after discontinuation of DHEA, his improvement may have been due to treatment with DES" (Jones et al. 1997). There was no report of neoplasia in women.

The recent elucidation of the physiological role of DHEA, especially in postmenopausal women, can explain the very good safety profile of the compound. In fact, DHEA is a physiological and essential precursor of sex steroids, which are required, at various degrees, for the normal functioning of most if not all organs of the human body. Since there is no feedback mechanism to increase DHEA secretion in subjects with low DHEA suffering from the resulting hormone deficiency symptoms, the only physiological means of providing the missing amount of DHEA is local or systemic administration of the sex steroid precursor. Due to the aforementioned normal saturation mechanisms and the relatively small bioavailable DHEA from local or systemic administration compared with the relatively high amounts already present in normal women, it is unlikely that any significant side effect will ever be seen with reasonable DHEA replacement therapy using high-quality pharmaceutical grade product. Most importantly, the intracellular action of DHEA with no significant release of estrogens eliminates the risks associated with estrogen-based replacement therapies, including the increased risk of breast and uterine cancer. Treatment with DHEA simply mimics the physiological situation after menopause where systemic exposure to estrogens has ceased following 500 million years of evolution, and DHEA has become the tissue-specific provider of sex steroids with no systemic exposure to the active compounds, which are inactivated locally before being eliminated.

REFERENCES

Adams, J. B. 1985. Control of secretion and the function of C19-Δ^5 steroids of the human adrenal gland. *Mol Cell Endocrinol* 41:1–17.

Akaza, H. 2006. Trends in primary androgen depletion therapy for patients with localized and locally advanced prostate cancer: Japanese perspective. *Cancer Sci* 97(4):243–7.

Allegra, J. C., M. E. Lippman, E. B. Thompson et al. 1979. Distribution, frequency and quantitative analysis of estrogen, progesterone, androgen and glucocorticoid receptors in human breast cancer. *Cancer Res* 39:1447–54.

Allolio, B., W. Arlt, and S. Hahner. 2007. DHEA: Why, when, and how much—DHEA replacement in adrenal insufficiency. *Ann Endocrinol (Paris)* 68(4):268–73.

Asselin, J., P. A. Kelly, M. G. Caron, and F. Labrie. 1977. Control of hormone receptor levels and growth of 7,12-dimethylbenz(a)anthracene-induced mammary tumors by estrogens, progesterone and prolactin. *Endocrinology* 101:666–71.

Asselin, J., and F. Labrie. 1978. Effects of estradiol and prolactin on steroid receptor levels in 7,12-dimethylbenz(a)anthracene-induced mammary tumors and uterus in the rat. *J Steroid Biochem* 9:1079–82.

Baulieu, E. E., G. Thomas, S. Legrain et al. 2000. Dehydroepiandrosterone (DHEA), DHEA sulfate, and aging: Contribution of the DHEAge Study to a sociobiomedical issue. *Proc Natl Acad Sci USA* 97(8):4279–84.

Bélanger, B., A. Bélanger, F. Labrie, A. Dupont, L. Cusan, and G. Monfette. 1989. Comparison of residual C-19 steroids in plasma and prostatic tissue of human, rat and guinea pig after castration: Unique importance of extra testicular androgens in men. *J Steroid Biochem* 32:695–8.

Bélanger, A., B. Candas, A. Dupont et al. 1994. Changes in serum concentrations of conjugated and unconjugated steroids in 40- to 80-year-old men. *J Clin Endocrinol Metab* 79:1086–90.

Beral, V. 2003. Breast cancer and hormone-replacement therapy in the Million Women Study. *Lancet* 362(9382):419–27.

Beral, V., D. Bull, and G. Reeves. 2005. Endometrial cancer and hormone-replacement therapy in the Million Women Study. *Lancet* 365(9470):1543–51.

Boone, C. W., V. E. Steele, and G. J. Kelloff. 1992. Screening for chemopreventive (anticarcinogenic) compounds in rodents. *Mutat Res* 267(2):251–5.

Bryan, R. M., R. J. Mercer, R. C. Bennett, G. C. Rennie, T. H. Lie, and F. J. Morgan. 1984. Androgen receptors in breast cancer. *Cancer* 54:2436–40.

Bulun, S. E., K. M. Zeitoun, K. Takayama, and H. Sasano. 2000. Estrogen biosynthesis in endometriosis: Molecular basis and clinical relevance. *J Mol Endocrinol* 25(1):35–42.

Burgess, H. E., and S. Shousha. 1993. An immunohistochemical study of the long-term effects of androgen administration on female-to-male transsexual breast: A comparison with normal female breast and male breast showing gynaecomastia. *J Pathol* 170(1):37–43.

Buster, J. E. 2009. Transvaginal dehydroepiandrosterone: An unconventional proposal to deliver a mysterious androgen that has no receptor or target tissue using a strategy with a new name: Hormone precursor replacement therapy (HPRT). *Menopause* 16(5):858–9.

Casson, P. R., R. N. Andersen, H. G. Herrod et al. 1993. Oral dehydroepiandrosterone in physiologic doses modulates immune function in postmenopausal women. *Am J Obstet Gynecol* 169:1536–9.

Caubet, J. F., T. D. Tosteson, E. W. Dong et al. 1997. Maximum androgen blockade in advanced prostate cancer: A meta-analysis of published randomized controlled trials using nonsteroidal antiandrogens. *Urology* 49:71–8.

Christov, K. T., R. C. Moon, D. D. Lantvit et al. 2004. Prostate intraepithelial neoplasia in noble rats, a potential intermediate endpoint for chemoprevention studies. *Eur J Cancer* 40(9):1404–11.

Coleman, D. L., E. H. Leiter, and R. W. Schwizer. 1982. Therapeutic effects of dehydroepiandrosterone (DHEA) in diabetic mice. *Diabetes* 31:830–3.

Cooperative Breast Cancer Group. 1964. Testosterone propionate therapy of breast cancer. *J Am Med Ass* 188:1069–72.

Couillard, S., M. Gutman, C. Labrie, and F. Labrie. 1998. Effect of combined treatment using radiotherapy and the antiestrogen EM-800 on ZR-75-1 human mammary carcinoma growth in nude mice. In *Proceedings 89th American Association Cancer Research*, March 28-April 1. New Orleans, LA.

Couture, P., C. Thériault, J. Simard, and F. Labrie. 1993. Androgen receptor-mediated stimulation of 17β-hydroxysteroid dehydrogenase activity by dihydrotestosterone and medroxyprogesterone acetate in ZR-75-1 human breast cancer cells. *Endocrinology* 132:179–85.

Cusan, L., A. Dupont, J. L. Gomez, R. R. Tremblay, and F. Labrie. 1994. Comparison of flutamide and spironolactone in the treatment of hirsutism: A randomized controlled trial. *Fertil Steril* 61:281–7.

Dauvois, S., C. S. Geng, C. Lévesque, Y. Mérand, and F. Labrie. 1991. Additive inhibitory effects of an androgen and the antiestrogen EM-170 on estradiol-stimulated growth of human ZR-75-1 breast tumors in athymic mice. *Cancer Res* 51:3131–5.

Dauvois, S., S. Li, C. Martel, and F. Labrie. 1989. Inhibitory effect of androgens on DMBA-induced mammary carcinoma in the rat. *Breast Cancer Res Treat* 14:299–306.

Diamond, P., L. Cusan, J. L. Gomez, A. Bélanger, and F. Labrie. 1996. Metabolic effects of 12-month percutaneous DHEA replacement therapy in postmenopausal women. *J Endocrinol* 150:S43–50.

Dumont, M., S. Dauvois, J. Simard, T. Garcia, B. Schachter, and F. Labrie. 1989. Antagonism between estrogens and androgens on GCDFP-15 gene expression in ZR-75-1 cells and correlation between GCDFP-15 and estrogen as well as progesterone receptor expression in human breast cancer. *J Steroid Biochem* 34:397–402.

Elmore, E., T. T. Luc, V. E. Steele, and J. L. Redpath. 2001. Comparative tissue-specific toxicities of 20 cancer preventive agents using cultured cells from 8 different normal human epithelia. *In Vitr Mol Toxicol* 14(3):191–207.

Fels, E. 1944. Treatment of breast cancer with testosterone propionate. A preliminary report. *J Clin Endocrinol* 4:121–5.

Garcea, R., L. Daino, R. Pascale et al. 1987. Inhibition by dehydroepiandrosterone of liver preneoplastic foci formation in rats after initiation-selection in experimental carcinogenesis. *Toxicol Pathol* 15:164–9.

Gordan, G. S. 1976. Anabolic-androgenic steroids. In *Handbook of Experimental Pharmacology*, ed. C. D. Kochakian, vol. 43, 499–513. Berlin, NY: Springer-Verlag.

Gordan, G. S., A. Halden, Y. Horn, J. J. Fuery, R. J. Parsons, and R. M. Walter. 1973. Calusterone (7β, 17α-dimethyltestosterone) as primary and secondary therapy of advanced breast cancer. *Oncology* 28(2):138–46.

Gordon, G. B., K. J. Helzlsouer, A. J. Alberg, and G. W. Comstock. 1993. Serum levels of dehydroepiandrosterone and dehydroepiandrosterone sulfate and the risk of developing gastric cancer. *Cancer Epidemiol Biomarkers Prev* 2(1):33–5.

Gordon, G. B., K. J. Helzlsouer, and G. W. Comstock. 1991. Serum levels of dehydroepiandrosterone and its sulfate and the risk of developing bladder cancer. *Cancer Res* 51:1366–9.

Gordon, G. B., L. M. Shantz, and P. Talalay. 1987. Modulation of growth, differentiation and carcinogenesis by dehydroepiandrosterone. *Adv Enzyme Regul* 26:355–82.

Green, J. E., M. A. Shibata, E. Shibata et al. 2001. 2-difluoromethylornithine and dehydroepiandrosterone inhibit mammary tumor progression but not mammary or prostate tumor initiation in C3(1)/SV40 T/t-antigen transgenic mice. *Cancer Res* 61(20):7449–55.

Grodstein, F., J. E. Manson, and M. J. Stampfer. 2006. Hormone therapy and coronary heart disease: The role of time since menopause and age at hormone initiation. *J Womens Health (Larchmt)* 15(1):35–44.

Ho, H. Y., M. L. Cheng, H. Y. Chiu, S. F. Weng, and D. T. Chiu. 2008. Dehydroepiandrosterone induces growth arrest of hepatoma cells via alteration of mitochondrial gene expression and function. *Int J Oncol* 33(5):969–77.

Hsia, J., R. D. Langer, J. E. Manson et al. 2006. Conjugated equine estrogens and coronary heart disease: The Women's Health Initiative. *Arch Intern Med* 166(3):357–65.

Hunt, P. J., E. M. Gurnell, F. A. Huppert et al. 2000. Improvement in mood and fatigue after dehydroepiandrosterone replacement in Addison's disease in a randomized, double blind trial. *J Clin Endocrinol Metab* 85(12):4650–6.

Hursting, S. D., S. N. Perkins, D. C. Haines, J. M. Ward, and J. M. Phang. 1995. Chemoprevention of spontaneous tumorigenesis in p53-knockout mice. *Cancer Res* 55(18):3949–53.

Ingle, J. N., D. I. Twito, D. J. Schaid et al. 1991. Combination hormonal therapy with tamoxifen plus fluoxymesterone versus tamoxifen alone in postmenopausal women with metastatic breast cancer. A phase II study. *Cancer* 67:886–91.

Jones, J. A., A. Nguyen, M. Straub, R. B. Leidich, R. L. Veech, and S. Wolf. 1997. Use of DHEA in a patient with advanced prostate cancer: A case report and review. *Urology* 50(5):784–8.

Kelloff, G. J., C. W. Boone, J. A. Crowell et al. 1996. New agents for cancer chemoprevention. *J Cell Biochem Suppl* 26:1–28.

Kelloff, G. J., C. W. Boone, V. E. Steele et al. 1994. Mechanistic considerations in chemopreventive drug development. *J Cell Biochem Suppl* 20:1–24.

Kennedy, B. J. 1958. Fluxymesterone therapy in treatment of advanced breast cancer. *N Engl J Med* 259:673–5.

Kohama, T., S. Terada, N. Suzuki, and M. Inoue. 1997. Effects of dehydroepiandrosterone and other sex steroid hormones on mammary carcinogenesis by direct injection of 7,12-dimethylbenz(a) anthracene (DMBA) in hyperprolactinemic female rats. *Breast Cancer Res Treat* 43(2):105–15.

Korkia, P., and G. V. Stimson. 1997. Indications of prevalence, practice and effects of anabolic steroid use in Great Britain. *Int J Sports Med* 18(7):557–62.

Labrie, F. 1991. Intracrinology. *Mol Cell Endocrinol* 78:C113–8.

Labrie, F. 2007. Drug insight: Breast cancer prevention and tissue-targeted hormone replacement therapy. *Nat Clin Pract Endocrinol Metab* 3(8):584–93.

Labrie, F. 2010a. DHEA, important source of sex steroids in men and even more in women. In *Neuroendocrinology, the Normal Neuroendocrine System, Progress in Brain Research*, ed. L. Martini, G. P. Chrousos, F. Labrie, K. Pacak, and D. Pfaff, vol. 182, chap. 6, 97–148. Burlington, MA: Elsevier.

Labrie, F. 2010b. DHEA after menopause - sole source of sex steroids and potential sex steroid deficiency treatment. *Menopause Manag* 19:14–24.

Labrie, F., D. Archer, C. Bouchard et al. 2009a. Effect on intravaginal dehydroepiandrosterone (Prasterone) on libido and sexual dysfunction in postmenopausal women. *Menopause* 16:923–31.

Labrie, F., D. Archer, C. Bouchard et al. 2009b. Intravaginal dehydroepiandrosterone (Prasterone), a physiological and highly efficient treatment of vaginal atrophy. *Menopause* 16:907–22.

Labrie, F., D. Archer, C. Bouchard et al. 2009c. Serum steroid levels during 12-week intravaginal dehydroepiandrosterone administration. *Menopause* 16:897–906.

Labrie, F., A. Bélanger, P. Bélanger et al. 2006. Androgen glucuronides, instead of testosterone, as the new markers of androgenic activity in women. *J Steroid Biochem Mol Biol* 99:182–8.

Labrie, F., A. Bélanger, P. Bélanger et al. 2007. Metabolism of DHEA in postmenopausal women following percutaneous administration. *J Steroid Biochem Mol Biol* 103(2):178–88.

Labrie, F., A. Bélanger, L. Cusan, and B. Candas. 1997. Physiological changes in dehydroepiandrosterone are not reflected by serum levels of active androgens and estrogens but of their metabolites: Intracrinology. *J Clin Endocrinol Metab* 82(8):2403–9.

Labrie, F., A. Bélanger, L. Cusan et al. 1980. Antifertility effects of LHRH agonists in the male. *J Androl* 1:209–28.

Labrie, C., A. Bélanger, and F. Labrie. 1988. Androgenic activity of dehydroepiandrosterone and androstenedione in the rat ventral prostate. *Endocrinology* 123:1412–7.

Labrie, F., A. Bélanger, V. Luu-The et al. 2005. Gonadotropin-releasing hormone agonists in the treatment of prostate cancer. *Endocr Rev* 26(3):361–79.

Labrie, F., A. Bélanger, J. Simard, V. Luu-The, and C. Labrie. 1995. DHEA and peripheral androgen and estrogen formation: Intracrinology. *Ann N Y Acad Sci* 774:16–28.

Labrie, F., B. Candas, J. L. Gomez, and L. Cusan. 2002. Can combined androgen blockade provide long-term control or possible cure of localized prostate cancer? *Urology* 60(1):115–9.

Labrie, F., L. Cusan, J. L. Gomez et al. 2008a. Effect of intravaginal DHEA on serum DHEA and eleven of its metabolites in postmenopausal women. *J Steroid Biochem Mol Biol* 111:178–94.

Labrie, F., L. Cusan, J. L. Gomez et al. 2008b. Corrigendum to: Effect of intravaginal DHEA on serum DHEA and eleven of its metabolites in postmenopausal women. *J Steroid Biochem Mol Biol* 112:169.

Labrie, F., L. Cusan, J. L. Gomez et al. 2008c. Changes in serum DHEA and eleven of its metabolites during 12-month percutaneous administration of DHEA. *J Steroid Biochem Mol Biol* 110(1–2):1–9.

Labrie, F., L. Cusan, J. L. Gomez et al. 2009. Comparable amounts of sex steroids are made outside the gonads in men and women: Strong lesson for hormone therapy of prostate and breast cancer. *J Steroid Biochem Mol Biol* 113:52–6.

Labrie, F., P. Diamond, L. Cusan, J. L. Gomez, A. Belanger, and B. Candas. 1997. Effect of 12-month dehydroepiandrosterone replacement therapy on bone, vagina, and endometrium in postmenopausal women. *J Clin Endocrinol Metab* 82(10):3498–505.

Labrie, F., A. Dupont, A. Bélanger et al. 1982. New hormonal therapy in prostatic carcinoma: Combined treatment with an LHRH agonist and an antiandrogen. *Clin Invest Med* 5:267–75.

Labrie, F., A. Dupont, and A. Bélanger. 1985. Complete androgen blockade for the treatment of prostate cancer. In *Important Advances in Oncology*, ed. V. T. de Vita, S. Hellman, and S. A. Rosenberg, 193–217. Philadelphia, PA: J.B. Lippincott.

Labrie, F., V. Luu-The, A. Bélanger, S.-X. Lin, J. Simard, and C. Labrie. 2005. Is DHEA a hormone? Starling review. *J Endocrinol* 187:169–96.

Labrie, F., V. Luu-The, C. Labrie, and J. Simard. 2001. DHEA and its transformation into androgens and estrogens in peripheral target tissues: Intracrinology. *Front Neuroendocrinol* 22(3):185–212.

Labrie, F., V. Luu-The, C. Labrie et al. 2003. Endocrine and intracrine sources of androgens in women: Inhibition of breast cancer and other roles of androgens and their precursor dehydroepiandrosterone. *Endocr Rev* 24(2):152–82.

Labrie, F., V. Luu-The, S. X. Lin et al. 1997. The key role of 17β-HSDs in sex steroid biology. *Steroids* 62:148–58.

Labrie, F., V. Luu-The, S.-X. Lin et al. 2000. Intracrinology: Role of the family of 17β-hydroxysteroid dehydrogenases in human physiology and disease. *J Mol Endocrinol* 25(1):1–16.

Labrie, F., R. Poulin, J. Simard, V. Luu-The, C. Labrie, and A. Bélanger. 2006. Androgens, DHEA and breast cancer. In *Androgens and Reproductive Aging*, ed. T. Gelfand, 113–35. Oxsfordshire, UK: Taylor and Francis.

Labrie, F., J. Simard, V. Luu-The et al. 1996. The 3β-hydroxysteroid dehydrogenase/isomerase gene family: Lessons from type II 3β-HSD congenital deficiency. In *Signal Transduction in Testicular Cells, Ernst Schering Research Foundation Workshop*, ed. V. Hansson, F. O. Levy, and K. Taskén (Suppl. 2), 185–218. Berlin Heidelberg: Springer-Verlag.

Levi, M. S., R. F. Borne, and J. S. Williamson. 2001. A review of cancer chemopreventive agents. *Curr Med Chem* 8(11):1349–62.

Li, S., X. Yan, A. Bélanger, and F. Labrie. 1993. Prevention by dehydroepiandrosterone of the development of mammary carcinoma induced by 7,12-dimethylbenz(a)anthracene (DMBA) in the rat. *Breast Cancer Res Treat* 29:203–17.

Li, S., X. Yan, A. Belanger, and F. Labrie. 1994. Prevention by dehydroepiandrosterone of the development of mammary carcinoma induced by 7,12-dimethylbenz(a)anthracene (DMBA) in the rat. *Breast Cancer Res Treat* 29(2):203–17.

Liu, C. H., G. A. Laughlin, U. G. Fischer, and S. S. Yen. 1990. Marked attenuation of ultradian and circadian rhythms of dehydroepiandrosterone in postmenopausal women: Evidence for a reduced 17,20-desmolase enzymatic activity. *J Clin Endocrinol Metab* 71(4):900–6.

Lobaccaro, J. M., S. Lumbroso, C. Belon et al. 1993. Androgen receptor gene mutation in male breast cancer. *Hum Mol Genet* 2:1799–1802.

Lubet, R. A., G. B. Gordon, R. A. Prough et al. 1998. Modulation of methylnitrosourea-induced breast cancer in Sprague Dawley rats by dehydroepiandrosterone: Dose-dependent inhibition, effects of limited exposure, effects on peroxisomal enzymes, and lack of effects on levels of Ha-ras mutations. *Cancer Res* 58(5):921–6.

Luo, S., C. Labrie, A. Bélanger, and F. Labrie. 1997. Effect of dehydroepiandrosterone on bone mass, serum lipids, and dimethylbenz(a)anthracene-induced mammary carcinoma in the rat. *Endocrinology* 138:3387–94.

Luo, S., A. Sourla, C. Labrie, A. Bélanger, and F. Labrie. 1997. Combined effects of dehydroepiandrosterone and EM-800 on bone mass, serum lipids, and the development of dimethylbenz(a)anthracene (DMBA)-induced mammary carcinoma in the rat. *Endocrinology* 138:4435–44.

Luu-The, V. 2001. Analysis and characteristics of multiple types of human 17beta-hydroxysteroid dehydrogenase. *J Steroid Biochem Mol Biol* 76(1–5):143–51.

Luu-The, V., and F. Labrie. 2010. The intracrine sex steroid biosynthesis pathways. In *Neuroendocrinology, Pathological Situations and Diseases, Progress in Brain Research*, ed. L. Martini, G. P. Chrousos, F. Labrie, K. Pacak, and D. e. Pfaff, vol. 181, chap. 10, 177–92. Burlington, MA: Elsevier.

MacEwen, E. G., and I. D. Kurzman. 1991. Obesity in the dog: Role of the adrenal steroid dehydroepiandrosterone (DHEA). *J Nutr* 121:S51–5.

McCormick, D. L., K. V. Rao, W. D. Johnson et al. 1996. Exceptional chemopreventive activity of low-dose dehydroepiandrosterone in the rat mammary gland. *Cancer Res* 56(8):1724–6.

Migeon, C. J., A. R. Keller, B. Lawrence, and T. H. Shepart II. 1957. Dehydroepiandrosterone and androsterone levels in human plasma. Effect of age and sex: Day-to-day and diurnal variations. *J Clin Endocrinol Metab* 17:1051–62.

Miller, W. R., J. Telford, J. M. Dixon, and R. A. Hawkins. 1985. Androgen receptor activity in human breast cancer and its relationship with estrogen and progesterone receptor activity. *Eur J Cancer Clin Oncol* 21:539–42.

Moore, M. A., W. Thamavit, A. Ichihara, K. Sato, and N. Ito. 1986. Influence of dehydroepiandrosterone, diaminopropane and butylated hydroxyanisole treatment during the induction phase of rat liver nodular lesions in a short-term system. *Carcinogenesis* 7:1059–63.

Morales, A. J., J. J. Nolan, J. C. Nelson, and S. S. Yen. 1994. Effects of replacement dose of dehydroepiandrosterone in men and women of advancing age. *J Clin Endocrinol Metab* 78:1360–7.

Mostaghel, E. A., S. T. Page, D. W. Lin et al. 2007. Intraprostatic androgens and androgen-regulated gene expression persist after testosterone suppression: Therapeutic implications for castration-resistant prostate cancer. *Cancer Res* 67(10):5033–41.

Nestler, J. E., C. O. Barlascini, J. N. Clore, and W. G. Blackard. 1988. Dehydroepiandrosterone reduces serum low density lipoprotein levels and body fat but does not alter insulin sensitivity in normal men. *J Clin Endocrinol Metab* 66:57–61.

Nishiyama, T., Y. Hashimoto, and K. Takahashi. 2004. The influence of androgen deprivation therapy on dihydrotestosterone levels in the prostatic tissue of patients with prostate cancer. *Clin Cancer Res* 10(21):7121–6.

Nyce, J. W., P. N. Magee, G. C. Hard, and A. G. Schwartz. 1984. Inhibition of 1,2-dimethylhydrazine-induced colon tumorigenesis in Balb/C mice by dehydroepiandrosterone. *Carcinogenesis* 5:57–62.

Orentreich, N., J. L. Brind, R. L. Rizer, and J. H. Vogelman. 1984. Age changes and sex differences in serum dehydroepiandrosterone sulfate concentrations throughout adulthood. *J Clin Endocrinol Metab* 59:551–5.

Park, J. J., R. A. Irvine, G. Buchanan et al. 2000. Breast cancer susceptibility gene 1 (BRCAI) is a coactivator of the androgen receptor. *Cancer Res* 60(21):5946–9.

Pascale, R. M., M. M. Simile, M. R. De Miglio et al. 1995. Inhibition of 3-hydroxy-3-methylglutaryl-CoA reductase activity and gene expression by dehydroepiandrosterone in preneoplastic liver nodules. *Carcinogenesis* 16(7):1537–42.

Pashko, L. L., G. C. Hard, R. J. Rovito, J. R. Williams, E. L. Sobel, and A. G. Schwartz. 1985. Inhibition of 7,12-dimethylbnz(a)anthracene-induced skin papillomas and carcinomas by dehydroepiandrosterone and 3-beta-methylandrost-e-en-17-one in mice. *Cancer Res* 45(1):164–6.

Pashko, L. L., R. J. Rovito, J. R. Williams, E. L. Sobel, and A. G. Schwartz. 1984. Dehydroepiandrosterone (DHEA) and 3 beta-methylandrost-5-en-17-one: Inhibitors of 7,12-dimethylbenz[a]anthracene (DMBA)-initiated and 12-O-tetradecanoylphorbol-13-acetate (TPA)-promoted skin papilloma formation in mice. *Carcinogenesis* 5(4):463–6.

Pines, A., D. W. Sturdee, A. H. MacLennan, H. P. Schneider, H. Burger, and A. Fenton. 2007. The heart of the WHI study: Time for hormone therapy policies to be revised. *Climacteric* 10(4):267–9.

Poulin, R., D. Baker, and F. Labrie. 1988. Androgens inhibit basal and estrogen-induced cell proliferation in the ZR-75-1 human breast cancer cell line. *Breast Cancer Res Treat* 12:213–25.

Poulin, R., D. Baker, D. Poirier, and F. Labrie. 1989. Androgen and glucocorticoid receptor-mediated inhibition of cell proliferation by medroxyprogesterone acetate in ZR-75-1 human breast cancer cells. *Breast Cancer Res Treat* 13:161–72.

Poulin, R., J. Simard, C. Labrie et al. 1989. Down-regulation of estrogen receptors by androgens in the ZR-75-1 human breast cancer cell line. *Endocrinology* 125:392–9.

Prostate Cancer Triallists' Collaborative Group. 2000. Maximum androgen blockade in advanced prostate cancer: An overview of the randomised trials. *Lancet* 355:1491–8.

Rao, K. V., W. D. Johnson, M. C. Bosland et al. 1999. Chemoprevention of rat prostate carcinogenesis by early and delayed administration of dehydroepiandrosterone. *Cancer Res* 59(13):3084–9.

Rasmussen, K. R., M. J. Arrowood, and M. C. Healey. 1992. Effectiveness of dehydroepiandrosterone in reduction of cryptosporidial activity in immunosuppressed rats. *Antimicrob Agents Chemother* 36:220–2.

Ratko, T. A., C. J. Detrisac, R. G. Mehta, G. J. Kelloff, and R. C. Moon. 1991. Inhibition of rat mammary gland chemical carcinogenesis by dietary dehydroepiandorsterone or a fluorinated analogue of dehydroepiandrosterone. *Cancer Res* 51(2):481–6.

Regelson, W., and M. Kalimi. 1994. Dehydroepiandrosterone (DHEA)—the multifunctional steroid. II. Effects on the CNS, cell proliferation, metabolic and vascular, clinical and other effects. Mechanism of action? *Ann N Y Acad Sci* 719:564–75.

Riman, T., P. W. Dickman, S. Nilsson et al. 2002. Hormone replacement therapy and the risk of invasive epithelial ovarian cancer in Swedish women. *J Natl Cancer Inst* 94:497–504.

Rossouw, J. E., G. L. Anderson, R. L. Prentice et al. 2002. Risks and benefits of estrogen plus progestin in healthy postmenopausal women: Principal results from the Women's Health Initiative randomized controlled trial. *JAMA* 288(3):321–33.

Roy, R., S. Dauvois, F. Labrie, and A. Bélanger. 1992. Estrogen-stimulated glucuronidation of dihydrotestosterone in MCF-7 human breast cancer cells. *J Steroid Biochem Mol Biol* 41:579–82.

Ruttimann, J. 2008. The menopause brain effect: Can hormone therapy help? *Endocr News* 15–6.

Schriock, E. D., C. K. Buffington, G. D. Hubert et al. 1988. Divergent correlations of circulating dehydroepiandrosterone sulfate and testosterone with insulin levels and insulin receptor binding. *J Clin Endocrinol Metab* 66:1329–31.

Schwartz, A. G. 1979. Inhibition of spontaneous breast cancer formation in female C3H (Avy/a) mice by long-term treatment with dehydroepiandrosterone. *Cancer Res* 39:1129–32.

Schwartz, A. G., D. K. Fairman, M. Polansky, M. L. Lewbart, and L. L. Pashko. 1989. Inhibition of 7,12-dimethylbenz[a]anthracene-initiated and 12-O-tetradecanoylphorbol-13-acetate-promoted skin papilloma formation in mice by dehydroepiandrosterone and two synthetic analogs. *Carcinogenesis* 10(10):1809–13.

Schwartz, A. G., G. C. Hard, L. L. Pashko, M. Abou Gharbia, and D. Swern. 1981. Dehydroepiandrosterone: An anti-obesity and anti-carcinogenic agent. *Nutr Cancer* 3(1):46–53.

Schwartz, A. G., M. L. Lewbart, and L. L. Pashko. 1992. Inhibition of tumorigenesis by dehydroepiandrosterone and structural analogs. *Canc prev* 443–55.

Schwartz, A. G., and L. L. Pashko. 1993. Cancer chemoprevention with the adrenocortical steroid dehydroepiandrosterone and structural analogs. *J Cell Biochem Suppl* 17G:73–9.

Schwartz, A. G., L. Pashko, and J. M. Whitcomb. 1986. Inhibition of tumor development by dehydroepiandrosterone and related steroids. *Toxicol Pathol* 14:357–62.

Schwartz, A. G., and R. H. Tannen. 1981. Inhibition of 7,12-dimethylbenz(a)anthracene- and urethan-induced lung tumor formation in A/J mice by long-term treatment with dehydroepiandrosterone. *Carcinogenesis* 2:1335–7.

Segaloff, A. 1977. The use of androgens in the treatment of neoplastic disease. *Pharm Ther* C2:33–7.

Segaloff, A., D. Gordon, B. N. Horwitt, J. V. Schlosser, and P. J. Murison. 1951. Hormonal therapy in cancer of the breast. 1. The effect of testosterone propionate therapy on clinical course and hormonal excretion. *Cancer* 4:319–23.

Shilkaitis, A., A. Green, V. Punj, V. E. Steele, R. A. Lubet, and K. Christov. 2005. Dehydroepiandrosterone inhibits the progression phase of mammary carcinogenesis by inducing cellular senescence via a p16-dependent but p53-independent mechanism. *Breast Cancer Res Treatm* 7(6):R1132–8.

Simard, J., S. Dauvois, D. E. Haagensen, C. Lévesque, Y. Mérand, and F. Labrie. 1990. Regulation of progesterone-binding breast cyst protein GCDFP-24 secretion by estrogens and androgens in human breast cancer cells: A new marker of steroid action in breast cancer. *Endocrinology* 126:3223–31.

Simard, J., A. C. Hatton, C. Labrie et al. 1989. Inhibitory effects of estrogens on GCDFP-15 mRNA levels and secretion in ZR-75-1 human breast cancer cells. *Mol Endocrinol* 3:694–702.

Simile, M., R. M. Pascale, M. R. De Miglio et al. 1995. Inhibition by dehydroepiandrosterone of growth and progression of persistent liver nodules in experimental rat liver carcinogenesis. *Int J Canc* 62(2):210–5.

Sourla, A., C. Martel, C. Labrie, and F. Labrie. 1998. Almost exclusive androgenic action of dehydroepiandrosterone in the rat mammary gland. *Endocrinology* 139:753–64.

Steele, V. E., R. C. Moon, R. A. Lubet et al. 1994. Preclinical efficacy evaluation of potential chemopreventive agents in animal carcinogenesis models: Methods and results from the NCI Chemoprevention Drug Development Program. *J Cell Biochem Suppl* 56(Suppl. 20):32–54.

Suzuki, T., N. Suzuki, R. A. Daynes, and E. G. Engleman. 1991. Dehydroepiandrosterone enhances IL2 production and cytotoxic effector function of human T cells. *Clin Immunol Immunopathol* 61:202–11.

Tchernof, A., J. P. Després, A. Bélanger et al. 1995. Reduced testosterone and adrenal C19 steroid levels in obese men. *Metabolism* 44:513–9.

Tian, W. N., L. D. Braunstein, J. Pang et al. 1998. Importance of glucose-6-phosphate dehydrogenase activity for cell growth. *J Biol Chem* 273(17):10609–17.

Tormey, D. C., M. E. Lippman, B. K. Edwards, and J. G. Cassidy. 1983. Evaluation of tamoxifen doses with and without fluoxymesterone in advanced breast cancer. *Ann Intern Med* 98:139–44.

Trams, G., and H. Maass. 1977. Specific binding of estradiol and dihydrotestosterone in human mammary cancers. *Cancer Res* 37:258–61.

Ueno, S., M. Namiki, T. Fukagai, H. Ehara, M. Usami, and H. Akaza. 2006. Efficacy of primary hormonal therapy for patients with localized and locally advanced prostate cancer: A retrospective multicenter study. *Int J Urol* 13(12):1494–500.

Ulrich, P. 1939. Testosterone (hormone mâle) et son role possible dans le traitement de certains cancers du sein. *Acta - Unio Int Contra Cancrum* 4:377–9.

Vermeulen, A., J. P. Deslypene, W. Schelfhout, L. Verdonck, and R. Rubens. 1982. Adrenocortical function in old age: Response to acute adrenocorticotropin stimulation. *J Clin Endocrinol Metab* 54:187–91.

Vermeulen, A., and L. Verdonck. 1976. Radioimmunoassays of 17β-hydroxy-5α-androstan-3-one, 4-androstene-3,17-dione, dehydroepiandrosterone, 17β-hydroxyprogesterone and progesterone and its application to human male plasma. *J Steroid Biochem* 7:1–10.

Weber, E., M. A. Moore, and P. Bannasch. 1988. Phenotypic modulation of hepatocarcinogenesis and reduction in N-nitrosomorpholine-induced hemangiosarcoma and adrenal lesion development in Sprague-Dawley rats by dehydroepiandrosterone. *Carcinogenesis* 9(7):1191–5.

White, E. L., L. J. Ross, S. M. Schmid, G. J. Kelloff, V. E. Steele, and D. L. Hill. 1998. Screening of potential cancer preventing chemicals for induction of glutathione in rat liver cells. *Oncol Rep* 5(2):507–12.

Wooster, R., J. Mangion, R. Eeles et al. 1992. A germline mutation in the androgen receptor in two brothers with breast cancer and Reifenstein syndrome. *Nat Genet* 2:132–4.

Zhou, J., S. Ng, O. Adesanya-Famuiya, K. Anderson, and C. A. Bondy. 2000. Testosterone inhibits estrogen-induced mammary epithelial proliferation and suppresses estrogen receptor expression. *Faseb J* 14(12):1725–30.

Zumoff, B., J. Levin, R. S. Rosenfeld, M. Markham, G. W. Strain, and D. K. Fukushima. 1981. Abnormal 24-hr mean plasma concentrations of dehydroepiandrosterone and dehydroisoandrosterone sulfate in women with primary operable breast cancer. *Cancer Res* 41:3360–3.

13 Dehydroepiandrosterone in Intracellular Infectious Diseases and Related Pathogens

Norma Galindo-Sevilla and Javier Mancilla-Ramírez

CONTENTS

INTRODUCTION

Infections by intracellular microorganisms challenge the defense mechanisms of an organism. Antibodies are unable to bind to foreign intracellular microorganisms, which allows intracellular parasites to avoid the responses initiated by antigen–antibody interactions. To effectively control an intracellular infection, the infected phagocyte, typically a monocyte, macrophage, or dendritic cell (DC), must be activated to a proinflammatory state.

Macrophages have two different activation patterns, M1 and M2. M1 results in a proinflammatory state, enabling the macrophage to deal with intracellular microorganisms. M2 is anti-inflammatory, meaning that the macrophage tolerates the intracellular microorganism; thus, this response is insufficient for controlling an infection (Benoit, Desnues, and Mege 2008). M1 and M2 are defined by the cytokine pattern produced in response to activation: M1 is associated with a Th1 profile and is characterized in part by the secretion of interleukin (IL)-2, interferon (IFN)-γ, and lymphotoxin, whereas M2 is associated with a Th2 profile and the production of IL-4, IL-5, IL-10, and IL-13.

The microenvironment around the macrophage at the moment of activation appears to influence its activation toward M1 or M2, but this effect is not fully understood. The activation pattern might depend on the influence of hormone type and concentration. The hormones that are best known to affect the activation of the immune system are the following: cytokines, such as IL-1, IL-2, IL-4, IL-5, IL-6, IL-10, IL-12, IL-18, tumor necrosis factor (TNF)-α, IFN-γ, T cell growth factor (TCGF-β); adrenal hormones, such as dehydroepiandrosterone (DHEA) and cortisol (Allolio and Arlt 2002); and sexual hormones, such as testosterone, estrogen, and progesterone (Tait, Butts, and Sternberg 2008). Depending on how these hormones interact with each other and influence gene expression, immune regulation can be activated or depressed, resulting in either microorganism clearance or the development of a chronic infection. Understanding the necessary equilibrium for the management of each particular challenge is difficult because even when a scenario seems to be optimal, unexpected problems may arise. Later, it becomes clear that the entire internal milieu was not fully accounted for, and additional important players are discovered.

Nowadays, with the rapid expansion of the knowledge of immunology and infection, it is important to recapitulate what is known and review what is still missing. The possibility that DHEA is a previously overlooked component that is indispensable for shifting the balance toward an efficient phagocytic state to control an intracellular infection will be presented here.

CLINICAL EVIDENCE OF THE IMPORTANCE OF DHEA FOR THE CLEARANCE OF INTRACELLULAR MICROORGANISMS

Intracellular microorganisms have several characteristics that favor the development of chronic infection, such as low growth rate and mechanisms for intracellular survival. Monocyte/macrophage activation helps control intracellular infections by inducing the expression of several genes involved in microorganism destruction and antigen processing, including those encoding inducible nitric oxide synthase (iNOS), an inducible enzyme for nitric oxide production; antigen peptide transporter 2, an antigen-processing protein; and proinflammatory cytokines such as IL-1, IL-12, IFN-γ, and TNF-α. Importantly, the actions of DHEA favor the expression of these genes.

However, DHEA levels are not stable throughout life. DHEA production is high during fetal development, up to eightfold that of a nonstressed adult (Casey and MacDonald 1998), when it is thought to exert an important function in cell differentiation–related processes, such as neocortical organization, axonal outgrowth (Compagnone and Mellon 1998), and immune system development (Brewer et al. 2002). At birth, DHEA production diminishes, and it remains low for the first few years of life (prepubescence). Little is known about the influence of DHEA on infection during this time, except that the high parasite burden in childhood diminishes as soon as DHEA increases during pubertal development (Kurtis et al. 2001). DHEA levels peak between the ages of 20 and 30, when a person is less prone to infections, and its levels gradually decline to reach a nadir in the elderly (Morley et al. 1997), when low DHEA levels are associated with increased susceptibility to infection, inversely correlating with serum values of TNF-α, IL-6, and antibody production. DHEA levels are also associated with Th1/Th2 activation patterns, with Th1 associated with high DHEA and Th2 with low DHEA (Padgett, Sheridan, and Loria 1995).

The severity and recrudescence of chronic intracellular infections are directly correlated with low serum DHEA levels. This association has been demonstrated for HIV-1, for which the development of acquired immunodeficiency syndrome (AIDS), the severest state of the infection, is associated with low serum DHEA and low IFN-γ production and is inversely correlated with $IL\text{-}4^+$ cell number (Mulder et al. 1992; Jacobson et al. 1991; Schifitto et al. 2000; Kandathil et al. 2005). Similarly, both high IL-10 production and reduced IFN-γ production in tuberculosis patients are associated with low plasma DHEA values (Bozza et al. 2007). Furthermore, patients infected with *Leishmania mexicana* develop extended skin lesions, known as diffuse cutaneous leishmaniasis, an extreme presentation of the disease, which is associated with low levels of DHEA, cortisol, and IFN-γ, and high levels of IL-6. Most patients with diffuse leishmaniasis were over the age of 50; however, in young people with diffuse leishmaniasis, some of whom were infected in childhood, researchers also found low serum DHEA levels (Galindo-Sevilla et al. 2007).

Additional factors that influence disease recrudescence and severity include strain pathogenicity, microorganism load, an individual's susceptibility to the infecting species (due to genetic and nutritional background), the kind of immune response developed, age, and sex of the infected individual. The last three factors have been shown to be associated with DHEA serum level.

Based on these observations, some questions arise: Does the natural DHEA decrease with age affect the recrudescence and/or severity of an infection? Does the infection itself alter DHEA production? Can DHEA supplementation effectively control a chronic infectious disease?

An affirmative answer to the first question is supported by evidence from observational studies. For instance, studies of parasitized populations have reported a puberty-associated rise in serum dehydroepiandrosterone sulfate (DHEAS), an albumin-bound metabolite of DHEA. DHEAS constitutes a circulating reservoir of DHEA that is less active at the tissue level (Baulieu et al. 2000); in addition, it downregulates proinflammatory molecules of the immune response and at the same time reduces undernutrition and anemia in a population experiencing a high burden of chronic helminth infections (Coutinho et al. 2007). Furthermore, an increase in host resistance to *Plasmodium falciparum* infection has been described at the beginning of puberty, a stage that is characterized by increased levels of DHEAS. However, because populations in endemic areas are highly exposed to parasites, it is not possible to discard the contribution that immunologic memory might have to increased host resistance (Kurtis et al. 2001).

A positive answer to the second question is supported by studies reporting damage and even destruction of the adrenal glands as a result of organ senescence and/or everyday insults. Adrenal damage is more common than previously thought: recent evidence has shown the presence of nodules in the reticularis zone (the main DHEA producer) known as incidentalomas (Hornsby 2004), which may account for DHEA diminution but at this time are of unknown consequence. Life expectancy has increased as a result of medicinal, physical, chemical, and electronic advances, which have extended organ function beyond that required by evolution (Straub and Besedovsky 2003). Two dangers that are of particular concern for optimal organ function are toxins and infectious agents, both of which may cause organ destruction by themselves or through inflammatory processes.

However, if the suprarenal gland, the main producer of DHEA, is damaged during life, a consequent influence on DHEA production might not be detected; clinical manifestation only occurs when 90% of the gland architecture is lost. Adrenal gland involvement has been documented in as many as two-thirds of patients with AIDS at postmortem examination (Mayo et al. 2002). Furthermore, tuberculosis progression in an animal model of infection with *Mycobacterium tuberculosis* is associated with adrenal hypertrophy (Hernandez-Pando et al. 1995; Baker, Zumla, and Rook 1996).

A few agents toxic to the suprarenal cortex have been described. One is α-[1,4-dioxido-3-methylquinoxalin-2-yl]-*N*-methylnitrone, an antibiotic that causes impaired steroidogenesis, likely by blocking the conversion of cholesterol to pregnenolone. After acute exposure, this chemical causes cytoplasmic vacuolation of the zona fasciculata and zona reticularis (Rosol et al. 2001). Some infectious agents that can cause chronic infections, including *Histoplasma capsulatum* (Macleod et al. 1972), HIV (Mayo et al. 2002), and *L. mexicana* (Figure 13.1), lead to the destruction of the suprarenal gland. Parallel to this process, DHEA and IFN-γ decrease, IL-6 and antibody production increase,

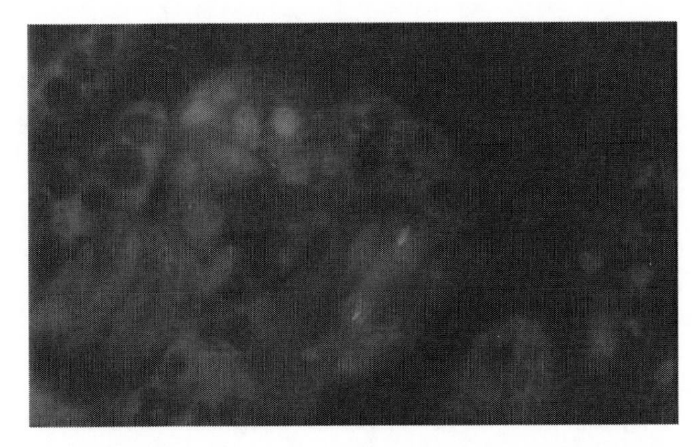

FIGURE 13.1 Immunohistochemistry of suprarenal gland of a mouse infected with *Leishmania mexicana* in a chronic infectious phase, stained with serum of patient with diffuse cutaneous leishmaniasis as first antibody and anti-human fluorescein isothiocyanate conjugated antibody as second antibody. Counterstained with propidium iodine. 1000×.

the Th1 immune response is shifted toward Th2, and, most importantly, the disease recrudesces. Additional research is necessary to confirm whether the disease itself modifies DHEA production.

To answer the third question of whether DHEA supplementation would help resolve chronic infections, more information is available and primarily comes from animal infection models. Experimental administration of DHEA to humans or animal models helps control infections by modifying the expression of cytokines and the activities of immune cells. For instance, DHEA administration to rodents increases IL-2 secretion (Daynes, Dudley, and Araneo 1990) and natural killer (NK) cell cytotoxicity (Corsini et al. 2002; McLachlan, Serkin, and Bakouche 1996; Risdon et al. 1991) and decreases circulating IL-6 (Yang et al. 1998). Studies on the effect of DHEA on the immune system were extended to *in vitro* tests using human mononuclear cells, which showed a DHEA-induced increase in IL-2 secretion (Suzuki et al. 1991) and NK cell cytotoxicity (Solerte et al. 1999) and, conversely, an inhibition of IL-6 release (Gordon, LeBoff, and Glowacki 2001). This impact of DHEA on the immune system should be beneficial against chronic infections, which is one reason why DHEA is being tested in several models of chronic infection. For example, in a rat model of *Trypanosoma cruzi* infection, a reduction in parasite number was observed after DHEA administration, accompanied by increases in serum IL-12 concentration, nitric oxide production by peritoneal macrophages, and splenocyte proliferation. These effects could be explained by DHEA's effect on the trypanocidal properties of macrophages, which involve, in part, the IFN-γ-stimulated expression of iNOS (Caetano et al. 2009). Additionally, DHEA treatment resulted in reduced *T. cruzi* infection of the heart and adrenal glands, as indicated by diminished parasite burden, less inflammatory infiltrate, and less tissue disorganization (Santos et al. 2007). Exogenous DHEA replacement exerts a synergistic effect with orchiectomy, strengthening the immune response by increasing the number of macrophages and concentrations of NO and IFN-γ in the blood; as a result, there is a reduced *T. cruzi* amastigote burden in the heart and a significant decrease in blood parasite levels (Filipin Mdel et al. 2008).

In human tuberculosis, the concentration of DHEA in urine is reportedly low and only returns to its normal value 30–40 days after specific tuberculosis treatment (Rook et al. 1996). Furthermore, in a murine model of tuberculosis, the administration of DHEA to infected rodents restores the Th1 immune response, with increases in IL-1α and TNF-α production and prolonged survival (Hernandez-Pando et al. 1998).

In rodents with *Cryptosporidium parvum* infection, DHEA decreased intestinal parasite colonization and fecal oocyst shedding, resulting in less severe infection (Rasmussen and Healey 1992). In addition, significant protective effects of DHEA against coxsackievirus B4 and herpes simplex type 2 encephalitis have been demonstrated in studies of two lethal viral infection models of mice (Loria et al. 1988).

The protective effect of DHEA was also extended to an acute inflammation model in which mice were subjected to sepsis by cecal ligation and puncture. In these mice, DHEA delayed mortality, restored TNF-α production, and increased TLR2 and TLR4 expression levels in splenic macrophages, even after these genes had been downregulated by the initial infection (Matsuda et al. 2005). Moreover, mouse mortality induced by TNF under lipopolysaccharide stimulation was diminished by DHEA treatment, which was accompanied by a reduction in serum TNF levels (Danenberg et al. 1992).

Although most reports indicate that DHEA stimulates the immune response to adopt a Th1 pattern, there are a few reports of stimuli that shift the response toward Th2, which causes a decrease in IL-12, TNF-α, and iNOS mRNA expression levels (Du et al. 2001). A common feature of these Th2 pattern stimuli is the use of a high dose of DHEA: as high as 2 mg/mouse, whereas the usual dose is 50 μg/mouse (Hernandez-Pando et al. 1998). When a supraphysiologic range of 5–20 μM DHEA is used *in vitro*, a Th2 immune response is favored; in addition, IL-12 production is inhibited in antigen-presenting cells, which is undesirable for the prevention of chronic infection (Du et al. 2001). These effects of high DHEA concentration should be kept in mind when designing experiments to test the effects of DHEA in a physiological context.

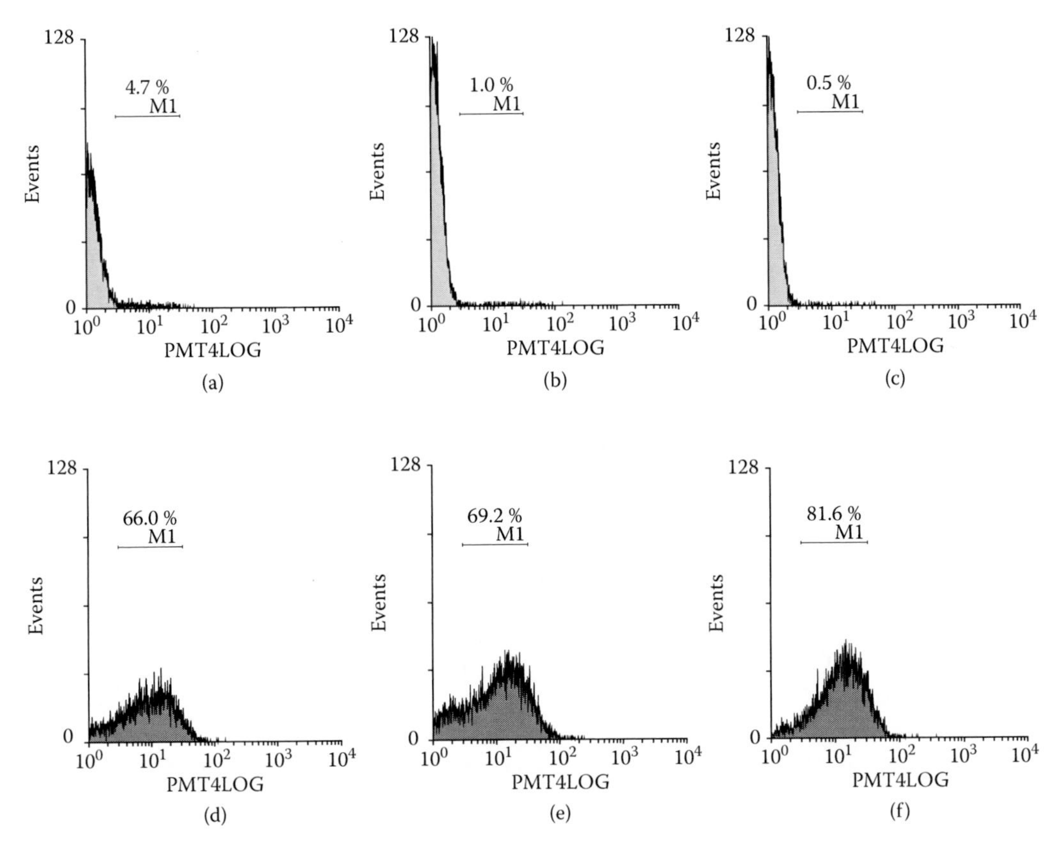

FIGURE 13.2 Analysis of viability determined by flow cytometry for propidium iodine–stained *Leishmania* parasites. In this experiment, the parasites were cultured axenically (a and d for control) or in the presence of DHEA (b and e) or dexamethasone (c and f). (a), (b), and (c) were cultivated at 32°C, and (d), (e), and (f) were exposed 24 hours to 37°C.

Recent results from our laboratory have demonstrated the presence of abundant mast cells in the skin of mice with diffuse cutaneous leishmaniasis (Figure 13.2). A Th2 cytokine pattern favors the multiplication of mast cell precursors and their colonization in tissues. No effective activity in the destruction of *Leishmania* parasites seems to be associated with these cells (Katakura et al. 1993), but their activity, including degranulation, could be contributing substantially to inflammatory events associated with the infection.

BASIC EVIDENCE FOR DHEA'S IMPORTANCE IN MONOCYTE/MACROPHAGE ACTIVATION

The actions of DHEA in an organism can directly impact monocytes and the cells derived from them. The number of monocytes in peripheral blood does not change substantially with age, although there are decreased numbers of macrophage precursors and bone marrow macrophages, and their function diminishes (Larbi et al. 2008). The administration of DHEA can trigger both a rise in the number of macrophages and an increase in NO production (Filipin Mdel et al. 2008). The reduction in class II major histocompatibility expression with age is thought to cause a decreased antigen presentation by macrophages (Larbi et al. 2008). DHEA modulates the expression of important proteins for antigen processing: antigen peptide transporter 2 is upregulated, and β_2-microglobulin is downregulated (Gu et al. 2003). There is no significant effect of DHEA on monocyte–endothelial adhesion or endothelial expression of cell adhesion molecules (Ng et al. 2003). Aged macrophages have impaired respiratory bursts and a decrease in reactive nitrogen intermediates as a result of altered intracellular signaling, rendering them less able to destroy microorganisms. It is possible that DHEA could help improve the function of macrophages, but more experiments should be conducted in this area.

DHEA inhibits the production of IL-6 by bone marrow mononuclear cells from patients with myeloma (Liu et al. 2005), reduces TNF-α production (Danenberg et al. 1992; Daynes, Dudley, and Araneo 1990; Straub et al. 1998), and blocks nuclear factor kappa B activity (Spencer et al. 1997; Straub et al. 1998). Serum TNF-α was not detected in DHEA-treated obese rats, suggesting a reduction in its production (Kimura et al. 1998). Serum IL-4 and immunoglobulin E were diminished after DHEA treatment of mice (Tabata, Tagami, and Terui 1997). These effects have been reported when mononuclear cells are stimulated with 1 μM or less of DHEA.

Effects of DHEA on DCs, including an increase in the production of IL-10 and IL-12, and an influence on the differentiation of DCs to a state of high CD80 (a costimulatory molecule) expression and low CD43 (an antiadhesion molecule and marker of immature DCs) expression (Canning et al. 2000) have been described. However, higher doses of DHEA (e.g., 10–20 μM) have the opposite effect, causing the secretion of the proinflammatory cytokines IFN-γ, IL-12, and TNF-α and the production of NO by splenocytes (Du et al. 2001). Aged DCs are less able to stimulate T and B cells. The age-associated decline in rat immune function reflects an impaired protein kinase C (PKC) signal transduction pathway and, in particular, is related to a defective PKC anchoring system, which is involved in the signal transduction of hormones, neurotransmitters, and cytokines (Nishizuka 1995). There is a direct correlation between circulating DHEA levels and receptor for activated C kinase-1 expression in leukocytes (Corsini et al. 2005).

Human macrophages convert DHEA into the steroid hormones androstenediol, androsterone, testosterone, and estrogen via the expression of a number of steroidogenic enzymes (Schmidt et al. 2000). This steroidogenic ability differs as a function of age; for instance, groups of healthy young men (18–30 years) have more steroidogenic activity than older men (50–70 years) (Hammer et al. 2005). When infusions of 200 mg DHEAS dissolved in 20 mL of 5% dextrose are given to pregnant women at term, a significant increase in the concentrations of plasma circulating estradiol, nitrate, and nitrite was seen (Manabe et al. 1999).

The strong interaction between DHEA and the immune system is further demonstrated by studies of macrophage metabolism of DHEAS to DHEA. This process is also susceptible to regulation by cytokines and is modified to potently inhibit sulfatase activity in murine macrophages by direct exposure to IFN-α, IFN-β, or TNF-α, but not to IL-1, IL-6, granulocyte macrophage colony–stimulating factor, TGF-β, platelet-derived growth factor, or IFN-γ (Hennebold and Daynes 1994). The conversion of DHEAS to DHEA is important because this process is associated with macrophage functions. For example, conversion was found to be positively correlated with more efficient phagocytosis in tests with zymosan, *Escherichia coli* and *Staphylococcus aureus* (Hennebold and Daynes 1994).

EFFECT OF DHEA ON MICROORGANISMS

In addition to its effects on the immune system, DHEA directly influences the activities of microorganisms. For instance, 16α-bromoepiandrosterone (EPI), an analog of DHEA, has antimalarial activity against several strains of *Plasmodium* (Freilich et al. 2000). EPI also induces lipid translocation in the ring stage of *P. falciparum*, enhancing nonopsonic ring phagocytosis and favoring parasite clearing down to undetectable limits (Ayi et al. 2002), which is advantageous to the human host because phagocytosed rings are rapidly digested. In contrast, more mature forms of the parasite, although actively phagocytosed, hinder completion of the phagocytic process because the malarial pigment hemozoin, which is abundantly present in mature parasite forms, is indigestible and severely affects important monocyte functions (Schwarzer et al. 1992).

EPI is also active in a model of progressive pulmonary tuberculosis. In this context, it restores the Th1 immune response, with increased expression of TNF-α, IFN-γ, and iNOS and decreased expression of IL-4, which helps accelerate chemotherapy-induced bacterial clearance (Hernández-Pando et al. 2005).

DHEA favors the growth of *L. mexicana*, but at the same time makes it more susceptible to death by apoptosis induction at 37°C (Figure 13.2). If DHEA favors macrophage elimination of parasites by activating macrophages, the resulting reduction in parasite burden will prevent organ dysfunction due to inflammatory damage.

SHOULD DHEA BE GIVEN PARENTERALLY OR ORALLY?

The immune cell activation of mucosa is limited, with perhaps only an M2/Th2 immune response. This tissue is tolerant to pathogens and normal flora because an adequate coexistence with the external environment is required to permit normal colonization. Therefore, if DHEA is externally supplied orally at high concentrations, the gastric environment could be modified toward the M1/Th1 inflammation process, which might cause severe gastric inflammation, an issue that has not yet been evaluated. Parenteral discharge of DHEA could be a better route of administration; however, because DHEA is not water soluble, several problems with the metabolism of the lipid phase and the absorption or degradation of vehicles make it unsuitable for human use. If DHEA could be reformulated in an aqueous vehicle, it would increase its attractiveness as a commercial product.

One potential vehicle is cyclodextrins, which are cyclic oligosaccharides made from α-1,4-linked glucose units that are able to form soluble noncovalent inclusion complexes with lipophilic substances, such as sterols (Fromming and Szejtly 1993). These complexes also increase cell wall permeability for both steroids and soluble nutrients, disorganize the lipid bilayer, and favor the release of steroid-transforming enzymes in *M. tuberculosis* (Donova et al. 2007). Furthermore, transcutaneous DHEA preparations have been formulated with microemulsions of a DHEA–cyclodextrin complex (Ceschel et al. 2005).

Finally, the decision to administer steroids, including DHEA, must take into account the age of the person and any current treatment regimen. In cases of infection, DHEA may modify the effect of antibiotics; in particular, the cell wall permeability of *M. tuberculosis* is increased by steroids in the presence of antibiotics such as vancomycin, lecithin, glycine, and, polycations (protamine, polymyxin B nonapeptide, and polyethyleneimine) (Donova et al. 2007).

CONCLUSION

DHEA is an important modulator of the innate immune system. It influences the activity of monocytes and derived cells, enhancing phagocytosis, NO production, and cytokine synthesis, which improve the ability of macrophages to manage intracellular infectious disease. DHEA also has deleterious effects, both direct and indirect, on some microorganisms. The effects of DHEA promote the favorable management of intracellular disease, particularly in the chronic phase of infection, with a decrease in microorganism burden and thus reduced organ inflammation. However, adequate dose management is important to obtain the desired effects. In addition, more research is necessary regarding DHEA formulation; our knowledge of its benefits in most animal models comes from direct injections of DHEA dissolved in oil, dimethyl sulfoxide, or ethanol, but information is limited on oral, insoluble doses, which must be used in humans. Nevertheless, DHEA is a promising candidate drug for the treatment of intracellular infectious disease and the reduction mortality from chronic infections.

REFERENCES

Allolio, B., and W. Arlt. 2002. DHEA treatment: Myth or reality? *Trends Endocrinol Metab* 13:288–94.

Ayi, K., G. Giribaldi, A. Skorokhod, E. Schwarzer, P. T. Prendergast, and P. Arese. 2002. 16α-bromoepiandrosterone, an antimalarial analogue of the hormone dehydroepiandrosterone, enhances phagocytosis of ring stage parasitized erythrocytes: A novel mechanism for antimalarial activity. *Antimicrob Agents Chemother* 46:3180–4.

Baker, R. W., A. Zumla, and G. A. W. Rook. 1996. Tuberculosis, steroid metabolism and immunity. *Q J Med* 89:387–94.

Baulieu, E. E., G. Thomas, S. Legrain, N. Lahlou, M. Roger, B. Debuire, V. Faucounau et al. 2000. Dehydroepiandrosterone (DHEA), DHEA sulfate, and aging: Contribution of the DHEAge Study to a sociobiomedical issue. *PNAS* 97:4279–84.

Benoit, M., B. Desnues, and J. L. Mege. 2008. Macrophage polarization in bacterial infections. *J Immunol* 181:3733–9.

Bozza, V. V., L. D'Attilio, C. V. Mahuad, A. A. Giri, A. del Rey, H. Besedovsky, O. Bottasso, and M. L. Bay. 2007. Altered cortisol/DHEA ratio in tuberculosis patients and its relationship with abnormalities in the mycobacterial-driven cytokine production by peripheral blood mononuclear cells. *Scand J Immunol* 66(1):97–103.

Brewer, J. A., O. Kanagawa, B. P. Sleckman, and L. J. Muglia. 2002. Thymocyte apoptosis induced by T cell activation is mediated by glucocorticoids in vivo. *J Immunol* 169:1837–43.

Caetano, L. C., F. H. Santello, M. Del Vecchio Filipin, V. Brazao, L. N. Caetano, M. P. Toldo, J. C. Caldeira, and J. C. do Prado Junior. 2009. *Trypanosoma cruzi*: Dehydroepiandrosterone (DHEA) and immune response during the chronic phase of the experimental Chagas' disease. *Vet Parasitol* 163(1–2):27–32.

Canning, M. O., K. Grotenhuis, H. J. de Wit, and H. A. Drexhage. 2000. Opposing effects of dehydroepiandrosterone and dexamethasone on the generation of monocyte-derived dendritic cells. *Eur J Endocrinol* 143:687–95.

Casey, M. L., and P. C. MacDonald. 1998. Endocrine changes of pregnancy, p 1259/1271. In *Williams Textbook of Endocrinology*, ed. J. D. Wilson, D. W. Foster, H. M. Kronenberg, and P. R. Larsen, 9th ed. Philadelphia, PA: W.B. Saunders Company.

Ceschel, G., V. Bergamante, P. Maffei, S. Lombardi Borgia, V. Calabrese, S. Biserni, and C. Ronchi. 2005. Solubility and transdermal permeation properties of a dehydroepiandrosterone cyclodextrin complex from hydrophilic and lipophilic vehicles. *Drug Deliv* 12:275–80.

Compagnone, N. A., and S. H. Mellon. 1998. Dehydroepiandrosterone: A potential signalling molecule for neocortical organization during development. *Proc Natl Acad Sci* 95:4678–83.

Corsini, E., L. Lucchi, M. Meroni, M. Racchi, B. Solerte, M. Fioravanti, B. Viviani, M. Marinovich, S. Govoni, and C. L. Galli. 2002. In vivo dehydroepiandrosterone restores age-associated defects in the protein kinase C signal transduction pathway and related functional responses. *J Immunol* 168:1753–8.

Corsini, E., M. Racchi, E. Sinforiani, L. Lucchi, B. Viviani, G. E. Rovati, S. Govoni, C. L. Galli, and M. Marinovich. 2005. Age-related decline in RACK-1 expression in human leukocytes is correlated to plasma levels of dehydroepiandrosterone. *J Leukoc Biol* 77:247–56.

Coutinho, H. M., T. Leenstra, L. P. Acosta, R. M. Olveda, S. T. McGarvey, J. F. Friedman, and J. D. Kurtis. 2007. Higher serum concentrations of DHEAS predict improved nutritional status in helminth-infected children, adolescents, and young adults in Leyte, the Philippines. *J Nutr* 137:433–9.

Danenberg, H. D., G. Alpert, S. Lustig, and D. Ben-Nathan. 1992. Dehydroepiandrosterone protects mice from endotoxin toxicity and reduces tumor necrosis factor production. *Antimicrob Agents Chemother* 36:2275–9.

Daynes, R. A., D. J. Dudley, and B. A. Araneo. 1990. Regulation of murine lymphokine production in vivo. II. Dehydroepiandrosterone is a natural enhancer of interleukin-2 synthesis by helper T cells. *Eur J Immunol* 20:793–802.

Donova, M. V., V. M. Nikolayeva, D. V. Dovbnya, S. A. Gulevskaya, and N. E. Suzina. 2007. Methyl-β-cyclodextrin alters growth, activity and cell envelope features of sterol-transforming mycobacteria. *Microbiology* 153:1981–92.

Du, C., Q. Guan, M. W. Khalil, and S. Sriram. 2001. Stimulation of Th2 response by high doses of dehydroepiandrosterone in KLH-primed splenocytes. *Exp Biol Med* 226:1051–60.

Filipin Mdel V., V. Brazao, L. C. Caetano, F. H. Santello, M. P. Toldo, L. N. Caetano, and J. C. do Prado Jr. 2008. Trypanosoma cruzi: Orchiectomy and dehydroepiandrosterone therapy in infected rats. *Exp Parasitol* 120(3):249–54.

Freilich, D., S. Ferris, M. Wallace, L. Leach, A. Kallen, J. Frincke, C. Ahlem, M. Hacker, D. Nelson, and J. Hebert. 2000. 16alpha bromoepiandrosterone, a dehydroepiandrosterone (DHEA) analogue, inhibits *Plasmodium falciparum* and *Plasmodium berghei* growth. *Am J Trop Med Hyg* 63:280–3.

Fromming, K. H., and J. Szejtly. 1993. Cyclodextrins in pharmacy. In *Topics in Inclusion Science*, ed. J. E. D. Davies, vol. 5, 1–81. Dordrecht: Kluwer Academic Publishers.

Galindo-Sevilla, N., N. Soto, J. Mancilla, A. Cerbulo, E. Zambrano, R. Chavira, and J. Huerto. 2007. Low serum levels of dehydroepiandrosterone and cortisol in human diffuse cutaneous leishmaniasis by *Leishmania mexicana*. *Am J Trop Med Hyg* 76:566–72.

Gordon, C. M., M. S. LeBoff, and J. Glowacki. 2001. Adrenal and gonadal steroids inhibit IL-6 secretion by human marrow cells. *Cytokine* 16:178–86.

Gu, S., S. L. Ripp, R. A. Prough, and T. E. Geoghegan. 2003. Dehydroepiandrosterone affects the expression of multiple genes in rat liver including 11β-hydroxysteroid dehydrogenase type 1: A cDNA array analysis. *Mol Pharmacol* 63:722–31.

Hammer, F., D. G. Drescher, S. B. Schneider, M. Quinkler, P. M. Stewart, B. Allolio, and W. Arlt. 2005. Sex steroid metabolism in human peripheral blood mononuclear cells changes with aging. *J Clin Endocrinol Metab* 90:6283–9.

Hennebold, J. D., and R. A. Daynes. 1994. Regulation of macrophage dehydroepiandrosterone sulfate metabolism by inflammatory cytokines. *Endocrinology* 135:67–75.

Hernández-Pando, R., D. Aguilar Leon, H. Orozco, A. Serrano, C. Ahlem, R. Trauger, B. Schramm, C. Reading, J. Frincke, and G. A. W. Rook. 2005. 16α-Bromoepiandrosterone restores T helper cell type 1 activity and accelerates chemotherapy-induced bacterial clearance in model of progressive pulmonary tuberculosis. *J Infect Dis* 191:299–306.

Hernandez-Pando, R., H. Orozco, J. P. Honour, J. Silva, R. Leyva, and G. A. W. Rook. 1995. Adrenal changes in murine pulmonary tuberculosis; a clue to pathogenesis? *FEMS Immunol Med Microbiol* 12:63–72.

Hernandez-Pando, R., M. L. Streber, H. Orozco, K. Arriaga, L. Pavon, O. Marti, S. L. Lightman, and G. A. Rook. 1998. Emergent immunoregulatory properties of combined glucocorticoid and anti-glucocorticoid steroids in a model of tuberculosis. *QJM* 91:755–66.

Hornsby, P. J. 2004. Aging of the human adrenal cortex. *Sci Aging Knowledge Environ* 35:1–8.

Jacobson, M. A., R. E. Fissaro, M. Galmarini, and W. Lang. 1991. Decreased serum dehydroepiandrosterone is associated with an increased progression of human immunodeficiency virus infection in men with CD4 cell counts of 200–499. *J Infect Dis* 164:864–8.

Kandathil, A. J., R. Kannangai, S. David, R. Selvakumar, V. Job, O. C. Abraham, and G. Sridharan. 2005. Human immunodeficiency virus infection and levels of dehydroepiandrosterone sulfate in plasma among Indians. *Clin Diagn Lab Immunol* 12:1117–8.

Katakura, K., S. Saito, A. Hamada, H. Matsuda, and N. Watanabe. 1993. Cutaneous leishmaniasis in mast cell-deficient W/Wv mice. *Infect Immun* 61:2242–4.

Kimura, M., S. Tanaka, Y. Yamada, Y. Kiuchi, T. Yamakawa, and H. Sekihara. 1998. Dehydroepiandrosterone decreases serum tumor necrosis factor-α and restores insulin sensitivity: Independent effect from secondary weight reduction in genetically obese Zucker fatty rats. *Endocrinology* 139:3249–53.

Kurtis, J. D., R. Mtalib, F. K. Onyango, and P. E. Duffy. 2001. Human resistance to *Plasmodium falciparum* increases during puberty and is predicted by dehydroepiandrosterone sulfate levels. *Infect Immun* 69:123–8.

Larbi, A., C. Franceschi, D. Mazzatti, R. Solana, A. Wikby, and G. Pawelec. 2008. Aging of the immune system as a prognostic factor for human longevity. *Physiology* 23:64–74.

Liu, S., H. Ishikawa, F. Li, Z. Ma, K. Otsuyama, H. Asaoku, S. Abroun et al. 2005. Dehydroepiandrosterone can inhibit the proliferation of myeloma cells and the interleukin-6 production of bone marrow mononuclear cells from patients with myeloma. *Cancer Res* 65:2269–76.

Loria, R. M., T. H. Inge, S. S. Cook, A. K. Szakal, and W. Regelson. 1988. Protection against acute lethal viral infections with the native steroid dehydroepiandrosterone (DHEA). *J Med Virol* 26:301–14.

Macleod, W. M., I. G. Murray, J. Davidson, and D. D. Gibbs. 1972. Histoplasmosis: A review, and account of three patients diagnosed in Great Britain. *Thorax* 27:6–17.

Manabe, A., T. Hata, T. Yanagihara, M. Hashimoto, Y. Yamada, S. Irikoma, S. Aoki, S. Masumura, and K. Miyazaki. 1999. Nitric oxide synthesis is increased after dehydroepiandrosterone sulphate administration in term human pregnancy. *Hum Reprod* 14:2116–9.

Matsuda, A., K. Furukawa, H. Suzuki, T. Matsutani, T. Tajiri, and I. H. Chaudry. 2005. Dehydroepiandrosterone modulates toll-like receptor expression on splenic macrophages of mice after severe polymicrobial sepsis. *Shock* 24:364–9.

Mayo, J., J. Collazos, E. Martínez, and S. Ibarra. 2002. Adrenal function in the human immunodeficiency virus-infected patient. *Arch Intern Med* 162:1095–8.

McLachlan, J. A., C. D. Serkin, and O. Bakouche. 1996. Dehydroepiandrosterone modulation of lipopolysaccharide-stimulated monocyte cytotoxicity. *J Immunol* 156:328–35.

Morley, J. E., F. Kaiser, W. J. Raum, H. M. Perry III, J. F. Flood, J. Jensen, A. J. Silver, and E. Roberts. 1997. Potentially predictive and manipulable blood serum correlates of aging in the healthy human male: Progressive decreases in bioavailable testosterone, dehydroepiandrosterone sulfate, and the ratio of insulin-like growth factor 1 to growth hormone. *PNAS* 94:7537–42.

Mulder, J. W., P. H. JosFrissen, P. Krijnen, E. Endert, F. de Wolf, J. Goudsmit, J. G. Masterson, and J. M. Lange. 1992. Dehydroepiandrosterone as predictor for progression to AIDS in asymptomatic human immunodeficiency virus–infected men. *J Infect Dis* 165:413–8.

Ng, M. K. C., S. Nakhla, A. Baoutina, W. Jessup, D. J. Handelsman, and D. S. Celermajer. 2003. Dehydroepiandrosterone, an adrenalandrogen, increases human foam cell formation: A potentially proatherogenic effect. *J Am Coll Cardiol* 42:1967–74.

Nishizuka, Y. 1995. Protein kinase C and lipid signaling for sustained cellular responses. *FASEB J* 9:484–96.

Padgett, D. A., J. Sheridan, and R. M. Loria. 1995. Steroid hormone regulation of a polyclonal Th2 immune response. *Ann N Y Acad Sci* 774:323–5.

Rasmussen, K. R., and M. C. Healey. 1992. Dehydroepiandrosterone induced reduction of *Cryptosporidium parvum* infections in aged Syrian golden hamsters. *J Parasitol* 78:554–7.

Risdon, G., T. A. Moore, V. Kumar, and M. Bennett. 1991. Inhibition of murine natural killer cell differentiation by dehydroepiandrosterone. *Blood* 78:2387–91.

Rook, G. A. W., J. Honour, O. M. Kon, R. J. Wilkinson, R. Davidson, and R. J. Shawn. 1996. Urinary adrenal steroid metabolites in tuberculosis—a new clue to pathogenesis? *QJM* 89:333–41.

Rosol, T. J., J. T. Yarrington, J. Latendresse, and C. C. Capen. 2001. Adrenal gland: Structure, function, and mechanisms of toxicity. *Toxicol Pathol* 29:41–8.

Santos, C. D., M. P. Toldo, A. M. Levy, L. M. Kawasse, S. Zucoloto, and J. C. do Prado Jr. 2007. Dehydroepiandrosterone affects Trypanosoma cruzi tissue parasite burdens in rats. *Acta Trop* 102:143–50.

Schifitto, G., M. P. McDermott, T. Evans, T. Fitzgerald, J. Schwimmer, L. Demeter, and K. Kieburtz. 2000. Autonomic performance and dehydroepiandrosterone sulfate levels in HIV-1–infected individuals: Relationship to T_H1 and T_H2 cytokine profile. *Arch Neurol* 57:1027–32.

Schmidt, M., M. Kreutz, G. Loffler, J. Scholmerich, and R. H. Straub. 2000. Conversion of dehydroepiandrosterone to downstream steroid hormones in macrophages. *J Endocrinol* 164:161–9.

Schwarzer, E., F. Turrini, D. Ulliers, G. Giribaldi, H. Ginsburg, and P. Arese. 1992. Impairment of macrophage functions after ingestion of *Plasmodium falciparum*-infected erythrocytes or isolated malarial pigment. *J Exp Med* 176:1033–41.

Solerte, S. B., M. Fioravanti, G. Vignati, A. Giustina, L. Cravello, and E. Ferrari. 1999. Dehydroepiandrosterone sulfate enhances natural killer cell cytotoxicity in humans via locally generated immunoreactive insulin-like growth factor I. *J Clin Endocrinol Metab* 84:3260–7.

Spencer, N. F., M. E. Poynter, S. Y. Im, and R. A. Daynes. 1997. Constitutive activation of NF-kappa B in an animal model of aging. *Int Immunol* 9:1581–8.

Straub, R. H., and H. O. Besedovsky. 2003. Integrated evolutionary, immunological, and neuroendocrine framework for the pathogenesis of chronic disabling inflammatory diseases. *FASEB J* 17:2176–82.

Straub, R. H., L. Konecna, S. Hrach, G. Rothe, M. Kreutz, J. Scholmerich, W. Falk, and B. Lang. 1998. Serum dehydroepiandrosterone (DHEA) and DHEA sulfate are negatively correlated with serum interleukin-6 (IL-6), and DHEA inhibits IL-6 secretion from mononuclear cells in man in vitro: Possible link between endocrinosenescence and immunosenescence. *J Clin Endocrinol Metab* 83:2012–7.

Suzuki, T., N. Suzuki, R. A. Daynes, and E. G. Engleman. 1991. Dehydroepiandrosterone enhances IL2 production and cytotoxic effector function of human T cells. *Clin Immunol Immunopathol* 61:202–11.

Tabata, N., H. Tagami, and T. Terui. 1997. Dehydroepiandrosterone may be one of the regulators of cytokine production in atopic dermatitis. *Arch Dermatol Res* 289:410–4.

Tait, A. S., C. L. Butts, and E. M. Sternberg. 2008. The role of glucocorticoids and progestins in inflammatory, autoimmune, and infectious disease. *J Leukoc Biol* 84:924–31.

Yang, B. C., C. W. Liu, Y. C. Chen, and C. K. Yu. 1998. Exogenous dehydroepiandrosterone modified the expression of T helper-related cytokines in NZB/NZW F1 mice. *Immunol Invest* 27:291–302.

Section III

Treatment of Diseases and Physiological Disorders by Dehydroepiandrosterone

14 Dehydroepiandrosterone Sulfate and Vascular Remodeling

Masaaki Ii

CONTENTS

INTRODUCTION

Adrenal steroid hormones, dehydroepiandrosterone (DHEA) and its sulfate ester (DHEAS), are the most abundant circulating precursors to androgens and estrogens in humans. Plasma observational studies have indicated that plasma DHEA and DHEAS are inversely correlated with cardiovascular diseases or angiographically defined coronary artery diseases in men (Barrett-Connor and Goodman-Gruen 1995; Mitchell et al. 1994; Slowinska-Srzednicka et al. 1989) but not in women (Porsova-Dutoit, Sulcova, and Starka 2000). These studies suggest that both DHEA and DHEAS may have an antiatherosclerotic or antiarteriosclerotic effect in men.

Going back over a decay in time, hypercholesterolemic animal studies have indicated that DHEA administration significantly reduces atherosclerotic lesions (Arad et al. 1989; Eich et al. 1993; Gordon, Bush, and Weisman 1988); however, without showing the distinct mechanism of the antiatherosclerotic effect by DHEA. Hayashi et al. (2000) reported that estrogen, one of the major metabolites of DHEA, partially (approximately 50%) mediated the indirect favorable effect of DHEA on atherosclerosis. In contrast, Cheng et al. (2009) have demonstrated another mechanism for the anti-atherosclerotic effect by which DHEA reduced inflammation-related molecules, monocyte chemoattractant protein-1 (MCP-1), and vascular cell adhesion molecule-1 (VCAM-1) in aortic lesions, suggesting that the effect of DHEA on anti-atherosclerosis is not due to the indirect effect of DEHA via its conversion to estrogen (Cheng et al. 2009). Other *in vitro* studies also suggested the anti-inflammatory effect of DHEA via reductions of tumor necrosis factor-alpha (TNF-α) (Gutierrez et al. 2007) and CD40/CD40L (Li, Xia, and Wang 2009) productions in human umbilical vein endothelial cells (HUVECs) or inhibition of oxidized low-density lipoprotein–induced inflammatory molecules such as intercellular adhesion molecule-1 (ICAM-1), VCAM-1, and reactive oxygen species expression in HUVECs (Lopez-Marure et al. 2007).

On the other hand, clinical studies of kinetics of DHEAS, a sulfate ester of DHEA, which was originally considered as a mere metabolite of DHEA, have shown that DHEAS levels decline linearly with age (Orentreich et al. 1984; Vermeulen 1995), suggesting that DHEAS can be implicated in the process of aging (e.g., atherosclerosis). Indeed, recent clinical trials have shown negative correlation between serum DHEAS levels and atherosclerosis including its risk factors (e.g., hypercholesterolemia and diabetes) in 75-year-old men (Ponholzer et al. 2010), type 2 diabetic postmenopausal women independent of age (Kanazawa et al. 2008), and type 2 diabetic men (Fukui et al. 2005). There is only one nonsupportive report regarding association of DHEA(S) and development of atherosclerosis (Kiechl et al. 2000). In addition, there are several lines of evidence indicating that DHEAS is not just a reservoir of DHEA but can activate the peroxisome proliferator–activated receptor, (PPAR)-α (Peters et al. 1996), has a more potent inhibitory effect on vascular smooth muscle cell (VSMC) migration than DHEA in rabbit VSMC cell lines (SM-3) (Furutama et al. 1998), reduces chronic hypoxic pulmonary hypertension in rats (Hampl et al. 2003), and inhibits vascular inflammation (Altman et al. 2008). Until now, however, the effects of DHEAS on arterial neointima formation following mechanical injury, which can mimic interventional angioplasty for stenotic coronary/peripheral artery diseases in patients, have never been studied except for our recent animal study (Ii et al. 2009).

In this chapter, I introduce a series of our findings of pathophysiological effects of DHEAS on arterial remodeling in a rabbit carotid balloon injury model, and also those of *in vitro* biological effect of DHEAS on VSMC functions, which play a critical role in the development of not only atherosclerosis development (Ross 1993; Schwartz, deBlois, and O'Brien 1995) but also in-stent restenosis, which is still a major problem after arterial stent implantation (Hoffmann et al. 1996).

EXPERIMENTAL PROTOCOL FOR ASSESSMENT OF THE ROLES OF DHEAS IN VASCULAR REMODELING FOLLOWING ARTERIAL INJURY

Adult male Japanese white rabbits (2.5–3.0 kg) fed with normal chow were assigned to two groups: DHEAS-treated group and saline-treated (control) group. DHEAS (Mylis, Schering-Plough, Japan) was administered intravenously (20 mg/kg/day) via an ear vein 3 days before surgery until sacrifice, and saline was administered as a control vehicle. Carotid balloon injury was performed as described previously (Negoro et al. 1999). The carotid arteries were examined histologically, and blood samples were collected for serological analysis at each time point. The detailed protocols for *in vitro* experiments could be referred to in the original article (Ii et al. 2009).

BIOLOGICAL AND PATHOPHYSIOLOGICAL ROLES OF DHEAS IN VASCULAR REMODELING FOLLOWING ARTERIAL INJURY

Serum Steroid Hormone Levels after DHEAS Administration

Daily administration of DHEAS (20 mg/kg) to rabbits resulted in significant increases in the serum DHEAS levels at 7 days to 1408 ± 453 ng/mL and at 14 days to 1123 ± 448 ng/mL (physiological doses in humans), which then decreased at 28 days to 81 ± 11 ng/mL. All the other hormones (e.g., DHEA, estradiol, and testosterone) and lipid levels that might be affected by conversion of DHEAS did not change significantly despite the dramatic increase in serum DHEAS, suggesting that the DHEAS administered was not converted to other steroid hormones that may also have certain biological effects on vascular remodeling. Since there are no reports on DHEAS-sulfatase, which can convert DHEAS to DHEA, activation in rabbits and evidence of differential DHEAS-sulfatase activity among species (Ruoff and Daniel 1991), we speculate that the significant decrease in serum

DHEAS level on day 28 was a result of DHEAS metabolism by an unknown enzymatic system to other metabolites but not to DHEA or estradiol.

Effect of DHEAS on Neointima Formation following Carotid Injury

DHEAS inhibited neointima formation 28 days after balloon injury in carotid arteries (Figure 14.1a). The intimal area and intima/media ratio in the DHEAS group (n = 10) were significantly reduced compared with the control group (n = 8) (0.48 ± 0.03 mm^2 and 0.29 ± 0.01 vs. 0.73 ± 0.05 mm^2 and 0.61 ± 0.03, $p = .0006$, $n = 8$ and $p < .0001$, $n = 7$, respectively). There was also a difference in the medial area between the control group and DHEAS group, with a thicker medial layer in the DHEAS group (1.20 ± 0.07 vs. 1.66 ± 0.09 mm^2, $p = .000922$) (Figure 14.1b). In addition, striking medial $p16^{INK4a}$ expressions were observed only in the DHEAS-treated injured arteries 3 days after surgery. The merged image exhibits colocalization of $p16^{INK4a}$ and smooth muscle α-actin–positive cells in the DHEAS-treated group (Figure 14.2). Finally, DHEAS significantly suppressed medial cell proliferation (Figure 14.3a) as indicated by a reduced proliferating cell nuclear antigen labeling index in the media 7 days after carotid injury (45.8 ± 2.9% in control, $n = 5$, and 17.8 ± 1.7% in the DHEAS group, $n = 5$, $p = .01116$) (Figure 14.3b).

Another interesting finding is that the medial area at 28 days after injury was significantly increased in the DHEAS-treated injured arteries (Figure 14.1b), inhibiting neointima formation by DHEAS. One possible reason for this phenomenon is an increase of extracellular matrices, because it has been reported that DHEAS promotes collagenase (Sakyo, Ito, and Mori 1987) and hyaluronate (Tanaka et al. 1997) syntheses in rabbit fibroblasts. Indeed, picro-sirius red staining demonstrated the increased medial matrix including collagen fibers in the DEHAS-treated injured arteries (Ii et al. 2009). We speculate that the interstitial space after medial cell apoptosis was replaced with the extracellular matrices produced from resident medial fibroblasts by DHEAS treatment.

Effects of DHEAS on VSMC Functions

Although little effect of DHEAS on human or rat aortic VSMC mitosis was shown in a previous report (Yoneyama et al. 1997), we have demonstrated opposing results with primary cultured rabbit VSMCs stimulated by 10% fetal calf serum/minimum essential medium. When we assessed

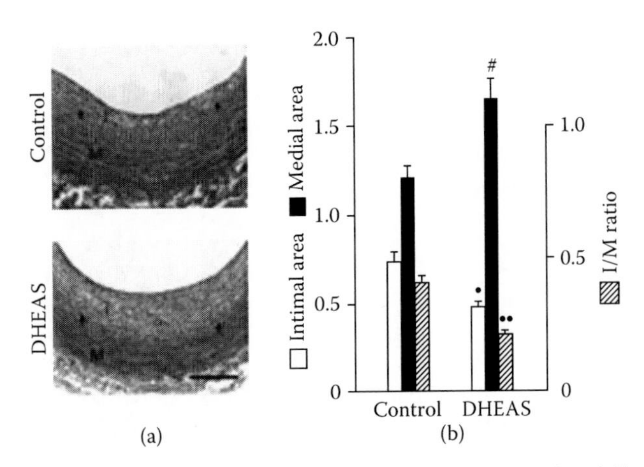

FIGURE 14.1 Dehydroepiandrosterone sulfate reduces neointima formation following carotid balloon injury. (a) Cross sections of Masson's trichrome staining from balloon-injured carotid arteries. I, intima; M, media; arrows, internal elastic lamina (IEL). (b) Morphometric analysis. Comparisons of intimal area, medial area, and intima/media ratio (I/M) between the control group and the DHEAS group. * and **, $p < .01$ and #, $p < .0001$ versus control.

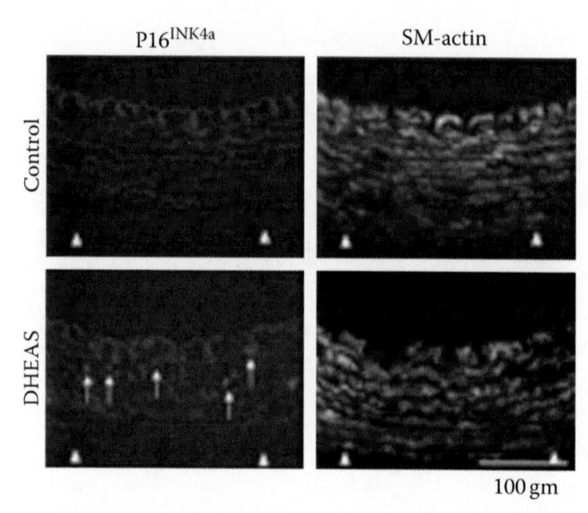

FIGURE 14.2 Dehydroepiandrosterone sulfate upregulates a cyclin-dependent kinase inhibitor $p16^{INK4a}$ expression in medial cells. Double immunofluorescent staining in injured arteries 3 days after surgery. Left panels, p16; right panels, smooth muscle α-actin; nuclear-staining nuclei (4′,6-diamidino-2-phenylindole). Arrowheads indicate external elastic lamina, and arrows indicate p16-positive cells.

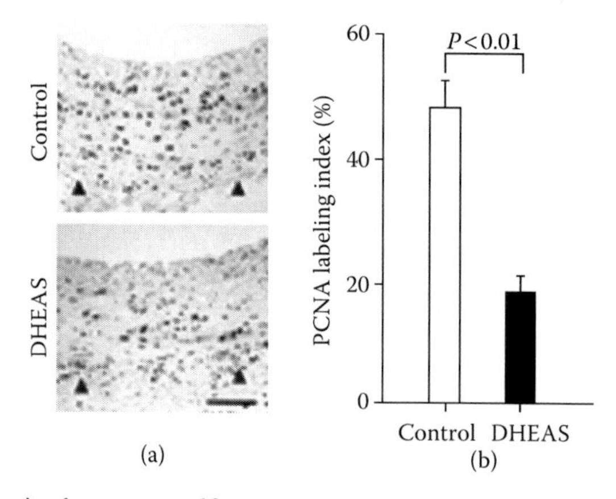

FIGURE 14.3 Dehydroepiandrosterone sulfate reduces medial cell proliferation. Immunostaining for proliferating cell nuclear antigen (PCNA) was performed to evaluate proliferation activity in injured arteries. (a) Control group and DHEAS group. PCNA-positive nuclei are stained in dark grey, and arrowheads indicate the external elastic lamina (EEL). Nuclei are stained with methyl green (in grey). (b) PCNA labeling index as a percentage of PCNA-positive nuclei per total nuclei in the media.

proliferation activity of VSMCs with DHEAS by BrdU incorporation, BrdU labeling index was reduced dose dependently by a maximum of 50% (DHEAS) at a concentration of 10^{-5} mol/L compared with control. Cell cycle analysis by flow cytometry also demonstrated that DHEAS induced G1 cell cycle arrest in proliferating VSMCs (Ii et al. 2009). The discrepancy between the reports might be due to the use of cells derived from different species, the use of a different assay system, and the use of a different form of DHEAS (we used the sodium form of DHEAS).

In migration assay, coincubation of VSMCs with DHEAS for 3 hours decreased platelet-derived growth factor–induced VSMC migration by a maximum of 52% (DHEAS) at a concentration of 10^{-5} mol/L compared with the control. The findings are supported by a previous report in which KCl-induced increases in cytosolic free calcium in rat aortic VSMCs were reversibly inhibited by DHEAS treatment (Barbagallo et al. 1995).

DHEAS Induces Medial Cell Apoptosis in Injured Artery

The effect of DHEAS on cell apoptosis varies depending on the type of cells; for instance, DHEAS has an antiapoptotic effect on mouse thymocytes independent of caspase-3/-6 mechanism (Yan et al. 1999) and on human peripheral blood lymphocytes with association of neither androgen receptor nor estrogen receptor (Takahashi, Nakajima, and Sekihara 2004). On the other hand, DHEAS decreases activated Akt levels and increases apoptosis in rat embryonic neural precursor cells, suggesting that DHEAS plays a role in developmental neurogenesis in association with other prosurvival/antiapoptotic molecules (Zhang et al. 2002).

Since there are no reports on the effect of DHEAS on VSMC apoptosis, we evaluated the effects of DHEAS on medial VSMC apoptosis/necrosis by terminal deoxynucleotidyl transferase-mediated dUTP nick-end labeling (TUNEL) staining 7 and 14 days after carotid injury (Figure 14.4a). TUNEL labeling index was significantly greater in the DHEAS group than in the control group (4.1 ± 0.3% in control, $n = 5$, and 40.8 ± 6.2% in the DHEAS group, $n = 5$, $p = .01116$) (Figure 14.4b). Although there was no remarkable difference in TUNEL positivity in the media, a number of TUNEL-positive cells were observed in the neointima only in the DHEAS group 14 days after carotid injury (Figure 14.4a).

PPAR-α Mediates DHEAS-Induced Apoptosis in VSMCs

It has been reported that PPAR-α activators (fibrate drugs) and their ligands (fish oil, docosahexaenoic acid [DHA]) could inhibit VSMC activation (Staels et al. 1998) and induce apoptosis via p38 mitogen-activated protein (MAP) kinase activation (Diep, Touyz, and Schiffrin 2000), respectively. Based on this evidence, we first tested the hypothesis that DHEAS could enhance PPAR-α upregulation in the presence of DHA that could induce cell apoptosis via p38 MAP kinase signaling in VSMCs. As expected, immunocytochemical analysis showed that only TUNEL-positive cells coexpressed slight PPAR-α by DHA treatment alone (data not shown), and cotreatment of DHA with DHEAS resulted in high expression of PPAR-α only in TUNEL-positive cells (Figure 14.5a). Striking PPAR-α expression was also observed only in the DHEAS-treated injured carotid arteries, and most of the expression was colocalized with SM-2, indicating that PPAR-α was upregulated in injured medial VSMCs (Figure 14.5c). In contrast, when the PPAR-α messenger RNA expression

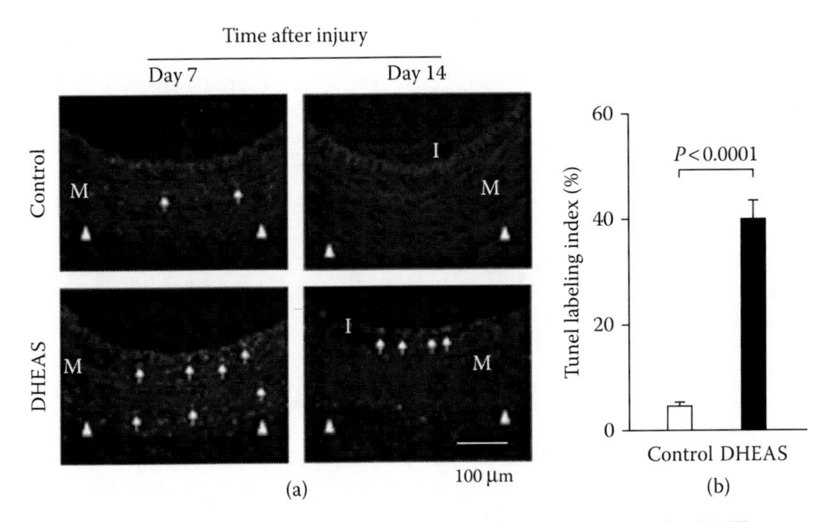

FIGURE 14.4 Dehydroepiandrosterone sulfate induces medial cell apoptosis. (a) Fluorescent terminal deoxynucleotidyl transferase dUTP nick end labeling (TUNEL) staining in injured arteries. Arrows and arrowheads indicate TUNEL-positive cells and external elastic lamina, respectively. Nuclear staining nuclei with 4′,6-diamidino-2-phenylindole; I, intima; M, media. (b) TUNEL labeling index as a percentage of TUNEL-positive nuclei per total nuclei in the media.

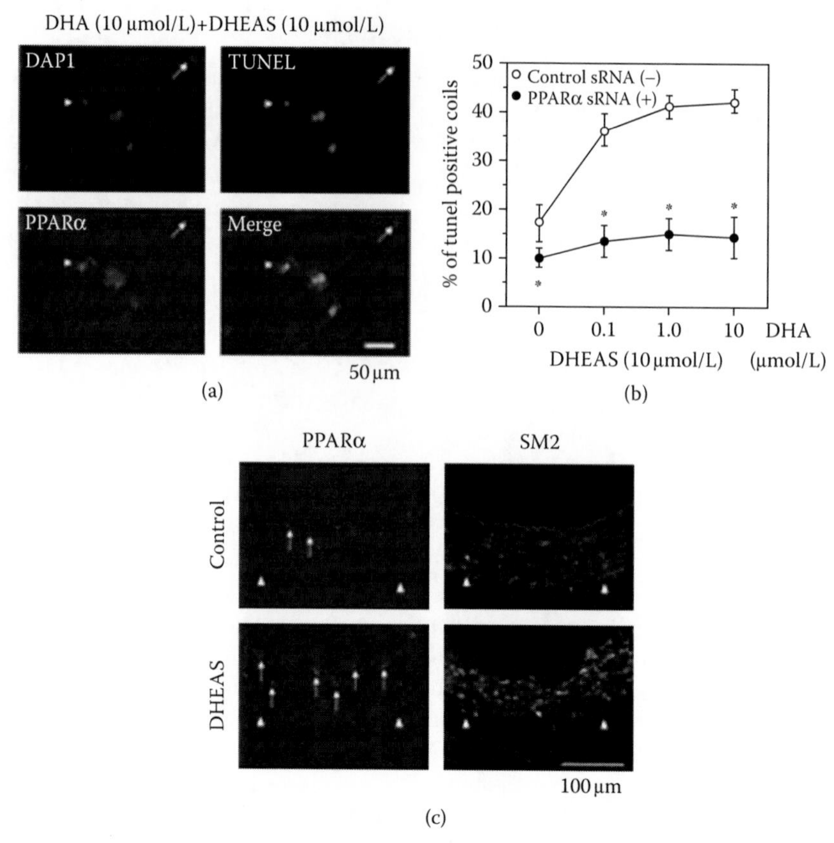

FIGURE 14.5 Dehydroepiandrosterone sulfate (DHEAS) induces peroxisome proliferator–activated receptor (PPAR-α) expression in vascular smooth muscle cells (VSMCs) and injured artery. (a) Cells were cultured in 0.5% fetal calf serum (FCS) with docosahexaenoic acid (DHA; 10 μmol/L) in the presence of DHEAS (10 μmol/L) for 24 hours. Double-immunofluorescent staining: left lower panel, PPAR-α; right upper panel, terminal deoxynucleotidyl transferase-mediated dUTP nick-end labeling (TUNEL); nuclear staining nuclei (4′,6-diamidino-2-phenylindole); arrows, TUNEL-negative cells; arrowheads, TUNEL-positive cells. (b) Apoptosis in control small interference RNA (siRNA)- or PPAR-α siRNA–transfected VSMCs evaluated as a percentage of TUNEL-positive cells to all attached cells. The VSMCs were incubated with DHA (0, control; 0.1, 1.0, and 10 μmol/L) in the presence of DHEAS (10 μmol/L) for 24 hours in 0.5% FCS. *, $p < .05$ versus control siRNA at each concentration of DHA. (c) Injured carotid arteries in the control and the DHEAS group. left panels, PPAR-α; right panels, SM-2; arrows, PPAR-α–positive cells, and arrowheads, external elastic lamina.

was silenced by small interfering RNA technique in VSMCs, DHEAS-induced apoptosis was partially cancelled, which was consistent with the decreased expression of PPAR-α (Figure 14.5b), suggesting that DHEAS might enhance DHA-induced VSMC apoptosis via upregulation of PPAR-α.

Anti-Inflammatory Effect of DHEAS on Injured Artery

Apart from the suppressive effect of DHEAS on VSMC functions *in vitro* and *in vivo*, DHEAS has also been demonstrated to have a potent inhibitory effect on TNF-α–induced inflammatory response including interleukin-8, ICAM-1, and VCAM-1 expressions, as well as nuclear factor-κB activation in aortic endothelial cells (Altman et al. 2008). Indeed, our data of immunohistochemical analysis exhibited the markedly less neutrophil/T-cells in the DEHAS-treated artery compared with that in the control artery 3 days following balloon injury (Figure 14.6), suggesting that DHEAS not only inhibits the endothelial expressions of inflammation-related molecules *in vitro* but also reduces

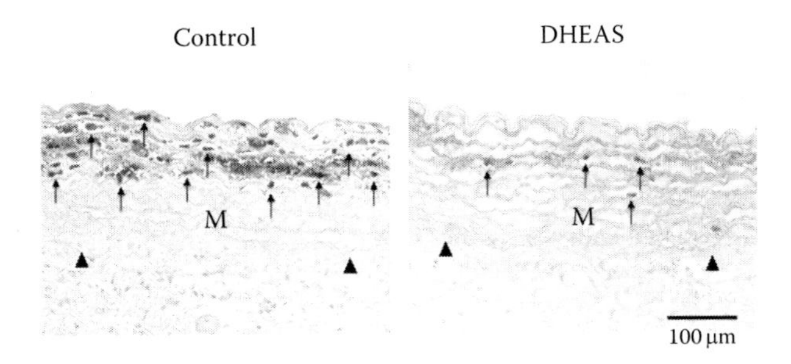

FIGURE 14.6 Dehydroepiandrosterone sulfate (DHEAS) reduces inflammatory cell infiltration in injured artery. Immunostaining of neutrophils/T-cells to evaluate inflammation activity in injured arteries 3 days after surgery. Positive cells are stained dark grey. Arrows, positively stained cells; arrowheads, external elastic lamina (EEL); M, media. Nuclei are stained with methyl green (in dark grey). Only a small number of dark grey cells (neutrophils/T-cells) were observed in the media in the DHEAS-treated injured artery compared with the control.

inflammatory cell infiltration in injured vessel wall. The anti-inflammatory effect of DHEAS could also be another important mechanism for the favorable effect of DHEAS on vascular remodeling after mechanical injury.

SUMMARY

The biological and pathophysiological effects of DHEAS on vascular remodeling after arterial mechanical injury have been introduced in this chapter, demonstrating the following findings: (1) DHEAS exhibits direct effect but not indirect effect by its conversion to other metabolites (i.e., estrogen and DHEA on VSMCs), (2) DHEAS inhibits neointima formation in response to injury by its suppressive effect on VSMC migration/proliferation, (3) cyclin-dependent kinase inhibitor p16^{INK4a} mediates DHEAS-dependent reduction of VSMC functional activity, (4) DHEAS upregulates expression of PPAR-α resulting in increased DHA-induced VSMC apoptosis, and (5) DHEAS reduces vascular inflammation/damage caused by inflammatory cell infiltration after arterial mechanical injury.

Since vascular remodeling following arterial mechanical injury is caused by not only VSMC activation but also endothelial functional alteration, more studies on the effect of DHEAS on endothelial cells in addition to a few previous studies (Altman et al. 2008; Hinson and Khan 2004) would also be crucial to understand the role of DHEAS in the process of vascular remodeling. Moreover, resident or bone marrow–derived stem progenitor cells have recently been shown to contribute to vascular remodeling including neointima/plaque formation, which leads to the development of atherosclerosis (Hoshino et al. 2008; Hristov et al. 2008; Xu 2006). The investigation of the DHEA/DHEAS effect on stem/progenitor cells might also be of interest in the vascular biology research field.

REFERENCES

Altman, R., D. D. Motton, R. S. Kota, and J. C. Rutledge. 2008. Inhibition of vascular inflammation by dehydroepiandrosterone sulfate in human aortic endothelial cells: Roles of PPARalpha and NF-kappaB. *Vascul Pharmacol* 48:76–84.

Arad, Y., J. J. Badimon, L. Badimon, W. C. Hembree, and H. N. Ginsberg. 1989. Dehydroepiandrosterone feeding prevents aortic fatty streak formation and cholesterol accumulation in cholesterol-fed rabbit. *Arteriosclerosis* 9:159–66.

Barbagallo, M., J. Shan, P. K. Pang, and L. M. Resnick. 1995. Effects of dehydroepiandrosterone sulfate on cellular calcium responsiveness and vascular contractility. *Hypertension* 26:1065–9.

Barrett-Connor, E., and D. Goodman-Gruen. 1995. The epidemiology of DHEAS and cardiovascular disease. *Ann N Y Acad Sci* 774:259–70.

Cheng H. H., X. J. Hu, and Q. R. Ruan. 2009. Dehydroepiandrosterone anti-atherogenesis effect is not via its conversion to estrogen. *Acta Pharmacol Sin* 30:42–53.

Diep, Q. N., R. M. Touyz, and E. L. Schiffrin. 2000. Docosahexaenoic acid, a peroxisome proliferator-activated receptor-alpha ligand, induces apoptosis in vascular smooth muscle cells by stimulation of p38 mitogen-activated protein kinase. *Hypertension* 36:851–5.

Eich, D. M., J. E. Nestler, D. E. Johnson et al. 1993. Inhibition of accelerated coronary atherosclerosis with dehydroepiandrosterone in the heterotopic rabbit model of cardiac transplantation. *Circulation* 87:261–9.

Fukui, M., Y. Kitagawa, N. Nakamura et al. 2005. Serum dehydroepiandrosterone sulfate concentration and carotid atherosclerosis in men with type 2 diabetes. *Atherosclerosis* 181:339–44.

Furutama, D., R. Fukui, M. Amakawa, and N. Ohsawa. 1998. Inhibition of migration and proliferation of vascular smooth muscle cells by dehydroepiandrosterone sulfate. *Biochim Biophys Acta* 1406:107–14.

Gordon, G. B., D. E. Bush, and H. F. Weisman. 1988. Reduction of atherosclerosis by administration of dehydroepiandrosterone. A study in the hypercholesterolemic New Zealand white rabbit with aortic intimal injury. *J Clin Invest* 82:712–20.

Gutierrez, G., C. Mendoza, E. Zapata et al. 2007. Dehydroepiandrosterone inhibits the TNF-alpha-induced inflammatory response in human umbilical vein endothelial cells. *Atherosclerosis* 190:90–9.

Hampl, V., J. Bibova, V. Povysilova, and J. Herget. 2003. Dehydroepiandrosterone sulphate reduces chronic hypoxic pulmonary hypertension in rats. *Eur Respir J* 21:862–5.

Hayashi, T., T. Esaki, E. Muto et al. 2000. Dehydroepiandrosterone retards atherosclerosis formation through its conversion to estrogen: The possible role of nitric oxide. *Arterioscler Thromb Vasc Biol* 20:782–92.

Hinson, J. P., and M. Khan. 2004. Dehydroepiandrosterone sulphate (DHEAS) inhibits growth of human vascular endothelial cells. *Endocr Res* 30:667–71.

Hoffmann, R., G. S. Mintz, G. R. Dussaillant et al. 1996. Patterns and mechanisms of in-stent restenosis. A serial intravascular ultrasound study. *Circulation* 94:1247–54.

Hoshino, A., H. Chiba, K. Nagai, G. Ishii, and A. Ochiai. 2008. Human vascular adventitial fibroblasts contain mesenchymal stem/progenitor cells. *Biochem Biophys Res Commun* 368:305–10.

Hristov, M., A. Zernecke, A. Schober, and C. Weber. 2008. Adult progenitor cells in vascular remodeling during atherosclerosis. *Biol Chem* 389:837–44.

Ii, M., M. Hoshiga, N. Negoro et al. 2009. Adrenal androgen dehydroepiandrosterone sulfate inhibits vascular remodeling following arterial injury. *Atherosclerosis* 206:77–85.

Kanazawa, I., T. Yamaguchi, M. Yamamoto et al. 2008. Serum DHEA-S level is associated with the presence of atherosclerosis in postmenopausal women with type 2 diabetes mellitus. *Endocr J* 55:667–75.

Kiechl, S., J. Willeit, E. Bonora, S. Schwarz, and Q. Xu. 2000. No association between dehydroepiandrosterone sulfate and development of atherosclerosis in a prospective population study (Bruneck Study). *Arterioscler Thromb Vasc Biol* 20:1094–100.

Li, Y., Z. Xia, and M. Wang. 2009. Dehydroepiandrosterone inhibits CD40/CD40L expression on human umbilical vein endothelial cells induced by interferon gamma. *Int Immunopharmacol* 9:168–72.

Lopez-Marure, R., C. Huesca-Gomez, J. Ibarra-Sanchez Mde, A. Zentella, and O. Perez-Mendez. 2007. Dehydroepiandrosterone delays LDL oxidation in vitro and attenuates several oxLDL-induced inflammatory responses in endothelial cells. *Inflamm Allergy Drug Targets* 6:174–82.

Mitchell, L. E., D. L. Sprecher, I. B. Borecki et al. 1994. Evidence for an association between dehydroepiandrosterone sulfate and nonfatal, premature myocardial infarction in males. *Circulation* 89:89–93.

Negoro, N., M. Hoshiga, M. Seto et al. 1999. The kinase inhibitor fasudil (HA-1077) reduces intimal hyperplasia through inhibiting migration and enhancing cell loss of vascular smooth muscle cells. *Biochem Biophys Res Commun* 262:211–5.

Orentreich, N., J. L. Brind, R. L. Rizer, and J. H. Vogelman. 1984. Age changes and sex differences in serum dehydroepiandrosterone sulfate concentrations throughout adulthood. *J Clin Endocrinol Metab* 59:551–5.

Peters, J. M., Y. C. Zhou, P. A. Ram et al. 1996. Peroxisome proliferator-activated receptor alpha required for gene induction by dehydroepiandrosterone-3 beta-sulfate. *Mol Pharmacol* 50:67–74.

Ponholzer, A., S. Madersbacher, M. Rauchenwald et al. 2010. Vascular risk factors and their association to serum androgen levels in a population-based cohort of 75-year-old men over 5 years: Results of the VITA study. *World J Urol* 28:209–14.

Porsova-Dutoit, I., J. Sulcova, and L. Starka. 2000. Do DHEA/DHEAS play a protective role in coronary heart disease? *Physiol Res* 49(Suppl 1):S43–56.

Ross, R. 1993. The pathogenesis of atherosclerosis: A perspective for the 1990s. *Nature* 362:801–9.

Ruoff, B. M., and W. L. Daniel. 1991. Comparative biochemistry of mammalian arylsulfatase C and steroid sulfatase. *Comp Biochem Physiol B* 98:313–22.

Sakyo, K., A. Ito, and Y. Mori. 1987. Dehydroepiandrosterone sulfate stimulates collagenase synthesis without affecting the rates of collagen and noncollagen protein syntheses by rabbit uterine cervical fibroblasts. *Biol Reprod* 36:277–81.

Schwartz, S. M., D. deBlois, and E. R. O'Brien. 1995. The intima. Soil for atherosclerosis and restenosis. *Circ Res* 77:445–65.

Slowinska-Srzednicka, J., S. Zgliczynski, M. Ciswicka-Sznajderman et al. 1989. Decreased plasma dehydroepiandrosterone sulfate and dihydrotestosterone concentrations in young men after myocardial infarction. *Atherosclerosis* 79:197–203.

Staels, B., W. Koenig, A. Habib et al. 1998. Activation of human aortic smooth-muscle cells is inhibited by PPARalpha but not by PPARgamma activators. *Nature* 393:790–3.

Takahashi, H., A. Nakajima, and H. Sekihara. 2004. Dehydroepiandrosterone (DHEA) and its sulfate (DHEAS) inhibit the apoptosis in human peripheral blood lymphocytes. *J Steroid Biochem Mol Biol* 88:261–4.

Tanaka, K., T. Nakamura, K. Takagaki et al. 1997. Regulation of hyaluronate metabolism by progesterone in cultured fibroblasts from the human uterine cervix. *FEBS Lett* 402:223–6.

Vermeulen, A. 1995. Dehydroepiandrosterone sulfate and aging. *Ann N Y Acad Sci* 774:121–7.

Xu, Q. 2006. The impact of progenitor cells in atherosclerosis. *Nat Clin Pract Cardiovasc Med* 3:94–101.

Yan, C. H., X. F. Jiang, X. Pei, and Y. R. Dai. 1999. The in vitro antiapoptotic effect of dehydroepiandrosterone sulfate in mouse thymocytes and its relation to caspase-3/caspase-6. *Cell Mol Life Sci* 56:543–7.

Yoneyama, A., Y. Kamiya, M. Kawaguchi, and T. Fujinami. 1997. Effects of dehydroepiandrosterone on proliferation of human aortic smooth muscle cells. *Life Sci* 60:833–8.

Zhang, L., B. Li, W. Ma et al. 2002. Dehydroepiandrosterone (DHEA) and its sulfated derivative (DHEAS) regulate apoptosis during neurogenesis by triggering the Akt signaling pathway in opposing ways. *Brain Res Mol Brain Res* 98:58–66.

15 DHEA in Human Immunodeficiency Virus Infection
Prognostic and Therapeutic Aspects

Carla M. Romero, Emilia P. Liao, Barnett Zumoff, and Leonid Poretsky

CONTENTS

BRIEF HISTORY OF THE HIV EPIDEMIC AND PATHOGENESIS OF HIV INFECTION

In 1981, several homosexual men presented with opportunistic infections that were associated with severe immune deficiency. The most common infection was pneumonia due to *Pneumocystis jirovecii*, formerly *P. carinii*, but there were also cases of cytomegalovirus (CMV) infection and extensive candidiasis, among others. Additionally, aggressive Kaposi's sarcoma (KS) developed in some of these individuals. Later that year, cases of Pneumocystis pneumonia (PCP) were reported in intravenous drug users (Masur et al. 1981). In 1982, reports of similar cases in patients with hemophilia and of transmission of the disease from mother to newborn appeared, and the Centers for Disease Control and Prevention (CDC) named the condition "acquired immune deficiency syndrome" (AIDS).

It was not until 1983 that a novel retrovirus associated with this syndrome, ultimately called "human immunodeficiency virus" (HIV), was discovered (Barre-Sinoussi et al. 1983; Gallo et al. 1983). In 1985, an enzyme-linked immunosorbent assay to detect antibodies against HIV was developed, and all blood donations in the United States have been screened for the virus since then. In 1986, a randomized controlled trial showed that zidovudine was effective against the virus, making this drug the first approved therapy for HIV infection. According to the World Health Organization (WHO), as of 2008, there were about 33.4 million people worldwide living with HIV (WHO 2009). The CDC estimates that over 1 million people in the United States have the disease, but about one in five do not know they have it (CDC 2010).

HIV infection is a blood-borne, sexually transmissible disease, so coinfection with other viruses that share the same routes of transmission is common. Two species of HIV have been identified: HIV-1 and HIV-2. Each has multiple subtypes called "clades." HIV-1 was the type initially discovered and is more prevalent worldwide. HIV-2 was initially isolated from patients with AIDS in West Africa in 1986. Immunodeficiency seems to develop more slowly in patients with HIV-2 infection, although the virus ultimately tends to cause a disease similar to that caused by HIV-1. Global distribution of HIV-1 and HIV-2 differs, which may have implications for the development of future vaccines. In the United States, for example, there have been few reported cases of HIV-2 infection, but this HIV type is more prevalent in West African nations such as Nigeria, Sierra Leone, and Ivory Coast (CDC 1998).

HIV-1 and HIV-2 are retroviruses belonging to the Retroviridae family, Lentivirus genus. They are enveloped, diploid, single-stranded, positive-sense RNA viruses with a DNA intermediate. The virus tends to integrate in areas of active transcription, which makes its eradication more difficult (Pruss, Bushman, and Wolffe 1994; Schröder et al. 2002). Latent proviral genomes can persist without being detected by the immune system and are not targeted by antiviral medications.

HIV infects primarily CD4+ T cells, macrophages, and dendritic cells. The pathology of HIV disease is largely attributable to the decrease in the number of CD4+ cells; their number is inversely correlated with the severity of the immunodeficiency and with the likelihood of opportunistic infections, wasting, and death. There are three mechanisms by which HIV leads to low CD4+ levels: direct viral killing of infected cells, increased rates of apoptosis in infected cells, and killing of infected CD4+ cells by CD8+ cytotoxic lymphocytes that recognize infected cells. When the number of CD4+ cells falls below a critical level, cell-mediated immunity is lost and the infected individual becomes progressively more susceptible to opportunistic infections.

Primary HIV infection can be asymptomatic or associated with an acute febrile illness that develops 2–4 weeks after exposure to the virus. This acute retroviral syndrome may consist of lymphadenopathy, pharyngitis, maculopapular rash, orogenital ulcers, and meningoencephalitis. Profound transient lymphopenia can develop, and opportunistic infections may occur. Primary HIV infection can be identified by newly detectable HIV antibodies or the presence of viral products, such as HIV-RNA, HIV-DNA, and/or ultrasensitive HIV p24 antigen, with a negative or weakly positive HIV antibody level (WHO 2007).

The WHO recognizes four clinical stages of HIV infection in adults and adolescents. The first stage is asymptomatic or can present with persistent generalized lymphadenopathy; the CD4+ cell concentration is in excess of 500 cells/mm^3. In stage 2, with mild immunodeficiency and CD4+ concentration between 350 and 499 cells/mm^3, there may be moderate unexplained weight loss; recurrent respiratory tract infections such as sinusitis, otitis media, and pharyngitis; herpes zoster; recurrent oral ulcers; fungal nail infections; or seborrheic dermatitis. Patients with CD4+ cell concentration between 200 and 349 cells/mm^3 are considered to have stage 3 or advanced HIV disease and can present with unexplained severe weight loss, unexplained chronic diarrhea, persistent fever, persistent oral candidiasis, pulmonary tuberculosis, or severe bacterial infections. Patients with less than 200 CD4+ cells/mm^3 are considered to have AIDS, which is the fourth clinical stage of HIV infection. AIDS is characterized by severe immune deficiency, HIV wasting syndrome, HIV encephalopathy, opportunistic infections such as PCP or CMV infection, among others, and KS (WHO 2007).

The CD4+ cell count is one of the key factors that needs to be taken into account when deciding whether to initiate antiretroviral therapy (ART) and chemoprophylaxis for opportunistic infections. Plasma HIV-RNA (viral load) should be measured at baseline and on a regular basis, especially in the setting of treatment, because it is the most important indicator of response to ART. According to the guidelines issued by the Department of Health and Human Services in December 2009, ART should be initiated in all treatment-naive patients with a history of an AIDS-defining illness or if they have a CD4+ cell count below 350 cells/mm^3. Treatment should also be started in all patients who are pregnant, have HIV-associated nephropathy, or have hepatitis B virus (HBV) coinfection

TABLE 15.1
Classes of Antiretroviral Agents

Class	Examples
Nucleoside/nucleotide reverse transcriptase inhibitors	Zidovudine, didanosine, tenofovir
Nonnucleoside reverse transcriptase inhibitors	Efavirenz, nevirapine
Protease inhibitors	Atazanavir, ritonavir, darunavir, fosamprenavir, lopinavir
Fusion inhibitors	Enfuvirtide
Chemokine (C-C motif) receptor type 5 (CCR5) antagonists	Maraviroc
Integrase strand transfer inhibitors	Raltegravir

when treatment for HBV is indicated. The recommendation to start therapy in patients with CD4+ cell counts between 350 and 500 cells/mm^3 is considered to be strong to moderate. Patients initiating ART should understand the benefits and risks of lifelong treatment and the importance of adherence; they should discuss with the clinician whether to initiate or postpone therapy (Department of Health and Human Services 2009).

There are more than 20 approved antiretroviral drugs divided into six classes according to their mechanism of action (Table 15.1). ART in a treatment-naive patient is initiated with one of the following regimens: a nonnucleoside reverse transcriptase inhibitor plus two nucleoside/nucleotide reverse transcriptase inhibitors (NRTIs); a protease inhibitor (PI) plus two NRTIs; or an integrase strand transfer inhibitor (INSTI) plus two NRTIs (Department of Health and Human Services 2009). The guidelines provide recommendations for preferred, alternative, and acceptable regimens. It is also recommended that the care of patients with HIV, including ART, be carried out by a clinician with HIV expertise, given the complexity of the infection and its treatment.

ENDOCRINOLOGICAL MANIFESTATIONS OF HIV INFECTION

Early in the HIV epidemic, it became apparent that patients with HIV infection may exhibit a number of endocrinological manifestations. Endocrinological dysfunction can occur as a result of direct infection by HIV, infection by opportunistic organisms, or as a side effect of drug therapy.

One of the earliest recognized endocrinological abnormalities in HIV patients was adrenal insufficiency, either primary or secondary (Poretsky, Maran, and Zumoff 1990). Symptoms of adrenal insufficiency do not appear until over 80% of the adrenal gland cortex has been destroyed. The pathophysiology of adrenal insufficiency in HIV infection remains poorly understood, but it may have to do with direct infection of either the pituitary or adrenal gland by HIV. Adrenal insufficiency can also be due to opportunistic infections, such as CMV and tuberculosis.

A diagnosis of adrenal insufficiency in individuals with HIV is not always clear-cut. A study by Stolarczyk et al. (1998) concluded that patients with AIDS may have a suboptimal adrenal reserve as demonstrated by a blunted cortisol response during the adrenocorticotropic hormone (ACTH) stimulation test compared with HIV-negative comparably sick patients or healthy subjects, even in the absence of symptoms of adrenal insufficiency. Another study found that 45% of patients with AIDS and symptoms of adrenal insufficiency exhibited subnormal responses to a low-dose (1 μg) ACTH stimulation test, but normal responses to 250 μg ACTH (Smolyar et al. 2003).

Some studies have demonstrated evidence of cortisol resistance. For example, Norbiato et al. (1994) have reported that some patients with HIV who manifest signs and symptoms of adrenal insufficiency, such as weakness, weight loss, hypotension, chronic fatigue, and intense mucocutaneous melanosis, nevertheless have elevated plasma and urinary cortisol values and increases in ACTH values. The authors reported that the mononuclear leukocytes from these patients showed increased number of glucocorticoid receptors but decreased receptor affinity for glucocorticoids. This acquired corticosteroid resistance in AIDS resembles the genetic primary corticosteroid resistance that has been reported (Kino and Chrousos 2001). The authors also reported subnormal glucocorticoid-induced inhibition of [^{3}H]thymidine incorporation in the mononuclear leukocytes from these patients and an increase in interferon-alpha production *in vivo* (Norbiato et al. 1994). Hyporeninemic hypoaldosteronism, without cortisol deficiency, has also been described in patients with AIDS (Poretsky, Maran, and Zumoff 1990; Etzel, Brocavich, and Torre 1992).

Another reported endocrinological manifestation of HIV infection is hypogonadism, usually of the hypogonadotrophic type, although primary hypogonadism has also been described (Poretsky, Can, and Zumoff 1995). Gonadotrophic dysfunction accounts for 75% of HIV-associated hypogonadism, with prolonged amenorrhea being three times more likely in women with HIV infection (Unachukwu, Uchenna, and Young 2009). The pathogenesis of gonadotrophic axis dysfunction in patients with HIV is thought to be multifactorial and is related to the duration of infection, direct cytopathic effects of the virus, use of drugs, opportunistic infections, malignancies, and malnutrition. In men, reduced levels of testosterone are associated with loss of muscle mass and strength, decreased bone mineral density, lipodystrophy, depression, asthenia, fatigue, and sexual dysfunction (Ponte, Gurgel, and Montenegro 2009).

Abnormalities of calcium metabolism have been reported in HIV infection, particularly in patients with AIDS-related lymphoma or tuberculosis (Poretsky, Maran, and Zumoff 1990). Osteopenia and osteoporosis have also been reported and may be a result of antiviral therapy. Other factors contributing to osteoporosis include long-standing HIV infection, high viral load, disorders of calcium and parathyroid metabolism, and low body weight (Thomas and Doherty 2003).

There have been reports of lymphoma of the thyroid gland and *Pneumocystis jirovecii* thyroiditis in HIV infection (Battan et al. 1991; Gochu et al. 1994). "Sick euthyroid" syndrome and subclinical hypothyroidism have also been observed, particularly in patients with full-blown AIDS (Sellmeyer and Grunfeld 1996; Guttler et al. 1993).

More recently, studies of endocrinological manifestations of HIV infection have focused on the development of insulin resistance and diabetes, which have become common with the use of highly active ART (HAART), particularly PIs (Rao, Mulligan, and Schambelan 2010). Different mechanisms have been implicated, including direct effects of HIV, medications, or opportunistic organisms on glucose transporter type 4 activity; induction of oxidative stress; proteosome inhibition; altered adipokine levels; and increased suppressor of cytokine signaling-1 expression (Rao, Mulligan, and Schambelan 2010; Hruz 2008).

HIV-associated lipodystrophy is a syndrome, characterized by insulin resistance, lipid derangements, and abnormal fat distribution (central fat accumulation and peripheral fat loss), that develops in patients with AIDS on ART, especially those on PIs. Dyslipidemia associated with ART is characterized by increased levels of very-low-density lipoprotein, low-density lipoprotein, and lipoprotein (a), and reduced levels of high-density lipoprotein. In addition, hypertriglyceridemia has been reported in patients with HIV infection before and after the advent of HAART. Lipodystrophy has been associated with steroid hormone alterations, mainly increase in cortisol and decrease in dehydroepiandrosterone (DHEA) levels (Christeff et al. 1999). Furthermore, the metabolic changes associated with the lipodystrophy syndrome may increase the risk of cardiovascular disease (Kramer et al. 2009; Barbaro and Silva 2009).

CIRCULATING DHEA LEVELS IN HIV INFECTION: CLINICAL AND PROGNOSTIC SIGNIFICANCE

There is evidence supporting the use of circulating DHEA levels as a prognostic factor in HIV. Some studies have found elevated levels of DHEA in patients in the asymptomatic stage, but low levels in patients with full-blown AIDS. Low levels of DHEA are associated with high viral loads and low CD4+ cells.

A study by Jacobson et al. (1991) found that among 108 men with HIV infection and with CD4+ cell count of between 200 and 499 cells/mm^3 24 months after entry into the study, levels of DHEA below 180 ng/dL were predictive of subsequent progression to AIDS. Ferrando, Rabkin, and Poretsky (1999) reported an association between dehydroepiandrosterone sulfate (DHEAS), free and total testosterone levels, and HIV illness markers, including viral load, fatigue, and depressed mood. In this study, 169 men with HIV infection were evaluated at baseline, and 6 and 12 months later for levels of DHEAS, total and free testosterone, HIV-RNA, CD4+ levels, HIV symptoms, opportunistic illnesses, fatigue, and depression. Men with AIDS had lower mean levels of DHEAS than those with less advanced HIV disease. Baseline DHEAS was positively correlated with CD4+ cell count and HIV symptom severity, and was inversely correlated with HIV-RNA. Subnormal baseline DHEAS levels were associated with a history of opportunistic infections and/or malignancies, as well as incidence of death over 1 year.

Kannangai et al. (2008) found lower levels of DHEA in 53 patients with HIV infection compared with healthy controls. DHEA levels increased somewhat after ART, but the results were not statistically significant, probably due to a small sample size.

Christeff, Nunez, and Gougeon (2000) found that, in spite of commonly observed elevated cortisol levels in patients with HIV infection in all stages of disease and independent of ART treatment, serum DHEA concentrations, which are elevated in the asymptomatic stage, are subnormal in patients with advanced AIDS. The authors found subnormal levels of DHEA in HAART-treated patients with lipodystrophy, but increased levels of DHEA in HAART-treated patients without lipodystrophy. The cortisol/DHEA ratio was increased in patients with AIDS and in patients who were on HAART and had lipodystrophy. Changes in the cortisol/DHEA ratio were negatively correlated with CD4+ cell counts, with malnutrition markers such as body cell mass and fat mass, and with an increase in circulating lipids (cholesterol, triglycerides, and apolipoprotein B) associated with the lipodystrophy syndrome (Christeff, Nunez, and Gougeon 2000).

Similar to Christeff, Nunez, and Gougeon (2000), other studies have reported evidence of chronic activation of the hypothalamic–pituitary–adrenal (HPA) axis manifested by elevated levels of serum cortisol in patients with HIV infection. The elevation of serum cortisol is usually associated with normal levels of 17-deoxysteroids (corticosterone, deoxycorticosterone, and 18-hydroxy-deoxycorticosterone) and often with subnormal levels of DHEA (Zapanti, Terzidis, and Chrousos 2008). The mechanism for the elevated cortisol and subnormal DHEA levels in individuals infected with HIV remains unclear. One possibility is shunting of adrenal steroid synthesis towards increased cortisol production because of a decrease in the 17,20-lyase activity of the CYP-17 enzyme, while the 17-hydroxylase activity of this enzyme remains normal (Figure 15.1). This has been found to be the case in several studies, including one by Grinspoon et al. (2001) in women with HIV and wasting syndrome.

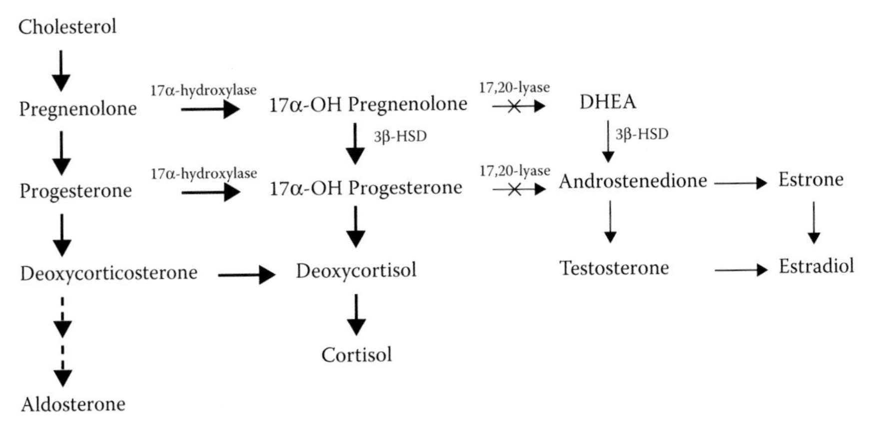

FIGURE 15.1 Shunting of the steroid synthesis pathway toward cortisol.

There have been several proposed mechanisms for the activation of the HPA axis in HIV infection. These include stimulation of corticotropin-releasing hormone activity by proinflammatory cytokines (such as tumor necrosis factor-α, interleukin [IL]-1β, and IL-6). The cytokines can also stimulate ACTH and cortisol release by acting directly on the pituitary and adrenal glands (Zapanti, Terzidis, and Chrousos 2008). It has been proposed that binding of HIV-1 to the limbic area of the brain, where the presence of CD4+ receptors has been documented, may alter HPA axis activity (Kumar et al. 2002).

In individuals without HIV infection, subnormal levels of DHEA have been associated with an increase in pancreatic insulin secretion and hyperinsulinemic insulin resistance (Nestler, Clore, and Blackard 1992). Insulin may reduce serum DHEA and DHEAS levels both by inhibiting production and by increasing clearance of DHEA. In some studies, elevation of insulin levels selectively suppressed adrenal 17,20-lyase activity, thus inhibiting production of adrenal androgens, but not of cortisol (Nestler, Clore, and Blackard 1992). Elevated cortisol levels and subnormal DHEA levels in patients with AIDS may be secondary to the effect of HAART (mainly PIs) on P450 cytochromes involved in steroid synthesis. Some studies suggest that both the lipid and the steroid hormone alterations precede the presence of lipodystrophy (Christeff et al. 1999).

It has been suggested that ART can restore normal DHEA levels in individuals with HIV infection. For example, in the study by Ferrando et al., the initiation of ART (which included PIs) was associated with an increase in DHEAS levels. This increase may be due to suppression of HIV or HIV-related opportunistic infections of the adrenal gland and/or to improvements in general health (Ferrando, Rabkin, and Poretsky 1999).

CLINICAL TRIALS OF DHEA IN INDIVIDUALS WITH HIV INFECTION

There have been several clinical trials of DHEA treatment in patients with HIV infection (Table 15.2). Some of these studies have explored the effects of DHEA treatment on HIV replication and host immunity; others have explored the relationship between treatment with DHEA and quality of life and mental function scores.

Dyner et al. (1993) studied the pharmacokinetics and safety of DHEA in subjects with symptomatic HIV disease and an absolute CD4+ lymphocyte count between 250 and 600 cells/μL, in a phase I, uncontrolled, open-label trial. The authors reported that the drug was well tolerated and that no serious side effects occurred. The most common minor side effects were nasal congestion, headache, fatigue, and nausea. No consistent changes in weight, Karnofsky score (a scale that assesses patients' level of functional impairment), absolute CD8+ cell count, percentage of CD4+ lymphocytes, p24 antigen, or β2-microglubulin were found (Dyner et al. 1993).

Piketty et al. (2001) examined clinical data, virological and immunological surrogate markers of HIV infection, plasma levels of DHEAS, and the Medical Outcomes Study HIV Health Survey quality-of-life scale in patients randomized to receive either 50 mg/day of DHEA for 4 months or placebo. The study found an improvement in mental function scores in patients treated with DHEA and an increase in serum levels of DHEAS, but no change in CD4+ cell counts.

Poretsky et al. (2006) performed a double-blind, placebo-controlled 8-week-long trial of oral DHEA treatment in HIV-positive men. The initial dose of DHEA was 100 mg/day, which was increased weekly by 100 mg/day to a maximum of 400 mg/day in the absence of dose-limiting side effects or clinical improvement, defined as "much" or "very much" improved mood. In the DHEA-treated patients, the authors found significant increases in circulating levels of DHEA and DHEAS, free testosterone, dihydrotestosterone (DHT), androstenedione, and estrone, and decreased levels of sex hormone–binding globulin. There was no difference in body mass index, body cell mass, CD4+ cell count, or HIV viral load. There was also no significant difference in other endocrinological parameters examined, including morning serum cortisol, estradiol, total testosterone, pregnenolone, 17-hydroxypregnenolone, fasting insulin, insulin-like growth factor (IGF)-1, IGF binding factor-1, IGF binding factor-3, adiponectin, growth hormone, or lipid levels. The clinical significance of the

TABLE 15.2
Summary of Selected DHEA Treatment Studies in Patients with Human Immunodeficiency Virus (HIV) Infection

	Type of Study	Subjects	Treatment	Results
Dyner et al. (1993)	Open label, nonrandomized, dose escalation	Adult men with mild symptomatic HIV disease	16 weeks of 250, 500, or 750 mg of DHEA three times a day	No increase in CD4+ cell counts, no reduction in p24 antigenemia or β_2-microglobulin levels; transient decrease in serum neopterin levels and a transient dose-related improvement in lymphocyte response to cytomegalovirus recall antigen
Piketty et al. (2001)	Double blind placebo controlled	Adult patients with advanced HIV	4 months of 50 mg/day of DHEA	Increase in DHEAS levels; significant improvement in the Mental Health and Health Distress dimension of the Medical Outcomes Study HIV Health Survey; no change in CD4+ cell counts
Poretsky et al. (2006)/ Rabkin et al. (2006)	Double blind placebo controlled	Adult men with HIV	8 weeks of DHEA in escalating doses 100–400 mg/day	Increase in DHEA, DHEAS, free testosterone, dihydrotestosterone, androstenedione, and estrone; decreased levels of sex hormone–binding globulin; no change in body mass index, CD4+ cell counts, HIV viral load; improvement in Clinical Global Impression score and Hamilton Depression Rating Scale score
Abrams et al. (2007)	Randomized placebo controlled	Adult patients with HIV, undetectable viral loads, and on a stable antiretroviral regimen	12 weeks of DHEA (100 mg orally twice daily for men, 50 mg orally twice daily for women) or placebo, followed by 12 weeks of open-label DHEA	No change in viral load, immunological markers, lean muscle mass, bone density; improvement in overall quality of life; increase in DHEA, DHEAS, and androstenedione levels
Poretsky et al. (2009)	Randomized placebo controlled	Premenopausal women with HIV	8 weeks of DHEA in escalating doses 100–400 mg/day	Increase in DHEA, DHEAS, testosterone, DHT, androstenedione, and estrone

endocrinological findings remains unclear, pending long-term studies (Poretsky et al. 2006). A finding of significant mood improvement, however, was reported in a separate study by Rabkin et al. (2006) and is discussed later in the chapter.

Abrams et al. (2007) performed a randomized, placebo-controlled trial in men and women with HIV infection who had undetectable viral loads and were on a stable antiretroviral regimen. The patients received 12 weeks of DHEA or placebo, followed by an additional 12 weeks of open-label DHEA. The authors did not find a difference in viral load, immunological markers, lean muscle mass, and bone density; however, there was a significant increase in DHEA, DHEAS, and androstenedione levels and an improvement in overall quality of life.

In a small, randomized, placebo-controlled study, Poretsky et al. (2009) examined the effects of oral DHEA on the HPA and hypothalamic–pituitary–gonadal axes and on insulin sensitivity in 15 premenopausal women with HIV infection. As in the previous study in men by this group, the initial dose of DHEA was 100 mg/day. The daily dose was increased by 100 mg every week in the absence of clinical improvement or dose-limiting side effects, up to a maximum of 400 mg/day, and the patients were followed for a total of 8 weeks. Significant increases were again seen in serum DHEA, DHEAS, testosterone, DHT, androstenedione, and estrone levels. No effect was noted on the pituitary–adrenal axis or on insulin/IGF indices. As in the previously discussed DHEA trial in men with HIV infection, the clinical significance of these findings remains unclear (Poretsky et al. 2009).

Rabkin et al. (2006) reported the effects of DHEA treatment on nonmajor depression in patients with HIV/AIDS. Treatment was initiated at a DHEA dose of 100 mg/day, and the dose was increased weekly by 100 mg/day up to 400 mg/day in the absence of clinical improvement or dose-limiting side effects. The authors found significant improvements in the Clinical Global Impression score and the Hamilton Depression Rating Scale score.

The antidepressant effects of DHEA might be due to several possible mechanisms. First, since activation of the pituitary–adrenal axis and increased cortisol levels have been reported in depression (Kumar et al. 2002; Young, Gallagher, and Porter 2002) and DHEA has been reported to have antiglucocorticoid properties (Karminska et al. 2000), the depression-ameliorating effect of DHEA may be mediated by its effect on cortisol metabolism. Second, *in vitro* and *in vivo* studies have shown that DHEA has an inhibitory effect on monoamine oxidase activity (Perez-Neri, Montes, and Rios 2009), similar to a class of antidepressant drugs. Third, DHEA is readily convertible *in vivo* to testosterone, which has been reported to have antidepressant properties (Zarrouf et al. 2009).

In spite of the high prevalence of depressive symptoms in patients with HIV, some patients and their physicians are reluctant to initiate treatment with antidepressant medications because of concerns about side effects and drug interactions with antiretroviral medications. In such cases, DHEA may be helpful. Patients treated with DHEA for depression should be informed that this is an off-label use of DHEA and that there have been no studies evaluating its safety.

Several long-term studies in women without HIV infection have shown that treatment with DHEA does not result in severe adverse effects, but some women did experience androgenic effects, such as acne and increased hair growth (Allolio, Arlt, and Hahner 2007; Panjari et al. 2009). No significant side effects related to increased estrogenic production have been noted in studies in men without HIV infection (Labrie 2010; Morales et al. 1998; Nair et al. 2006). Similarly, no significant side effects of DHEA were reported in the studies on individuals with HIV infection discussed above. In general, however, oral DHEA therapy is well tolerated, with only mild side effects, including headache, nasal congestion, low energy, irritability, and insomnia.

CONCLUSIONS

Among the many endocrinological derangements found in patients with HIV, subnormal levels of DHEA seem to correlate with low levels of CD4+ cell counts and high viral loads. In this sense, serum DHEA level may serve as a prognostic marker, but there is no clear-cut evidence of a cause and effect relationship.

Although it is still unclear whether DHEA therapy alters the disease course, some studies have shown improvement in depression and mental health scores. Treating depression and dysthymia with DHEA may be worthwhile since the treatment appears to be safe, without major side effects.

Patients treated with DHEA should be informed that DHEA is considered a food supplement and that available DHEA formulations are not regulated by the Food and Drug Administration. As a result, there is wide variation in actual content of the hormone in advertised preparations (Thompson and Carlson 2000). Therefore, serum DHEA and DHEAS levels should be monitored during treatment.

Except for the improvement in symptoms of depression, there is no evidence at this time that treatment with DHEA has beneficial antiviral, immunomodulatory, hormonal, or body composition effects, nor has it been shown to alter HIV disease progression. Larger, long-term randomized controlled studies are needed to assess the clinical relevance of DHEA treatment and its role, if any, in altering the progression of HIV infection.

REFERENCES

Abrams, D. I., S. B. Shade, P. Couey et al. 2007. Dehydroepiandrosterone (DHEA) effects on HIV replication and host immunity: A randomized placebo-controlled study. *AIDS Res Hum Retroviruses* 23(1):77–85.

Allolio, B., W. Arlt, and S. Hahner. 2007. DHEA: Why, when, and how much—DHEA replacement in adrenal insufficiency. *Ann Endocrinol (Paris)* 68(4):268–73.

Barbaro, G., and E. F. Silva. 2009. Cardiovascular complications in the acquired immunodeficiency syndrome. *Rev Assoc Med Bras* 55(5):621–30.

Barre-Sinoussi, F., J. C. Chermann, F. Rey et al. 1983. Isolation of a T-lymphotropic retrovirus from a patient at risk for acquired immune deficiency syndrome (AIDS). *Science* 220:868–70.

Battan, R., P. Mariuz, M. C. Raviglione, M. T. Sabatini, M. P. Mullen, and L. Poretsky. 1991. Pneumocystitis carinii infection of the thyroid in a hypothyroid patient with AIDS: Diagnosis by fine needle aspiration biopsy. *J Clin Endocrionol Metab* 72(3):724–6.

Centers for Disease Control and Prevention. 1998. Human Immunodeficiency Virus Type 2 | Factsheets | CDC HIV/AIDS. http://cdc.gov/hiv/resources/factsheets/hiv2.htm (accessed August 20, 2010).

Centers for Disease Control and Prevention. 2010. HIV in the United States | Factsheets | Resources by Format | CDC HIV/AIDS. http://cdc.gov/hiv/resources/factsheets/us.htm (accessed September 13, 2010).

Christeff, N., J. C. Melchior, P. de Truchis, C. Perronne, E. A. Nunez, and M. L. Gougeon. 1999. Lipodystrophy defined by a clinical score in HIV-infected men on highly active antiretroviral therapy: Correlation between dyslipidaemia and steroid hormone alterations. *AIDS* 13(16):2251–60.

Christeff, N., E. A. Nunez, and M. L. Gougeon. 2000. Changes in cortisol/DHEA ratio in HIV-infected men are related to immunological and metabolic perturbations leading to malnutrition and lipodystrophy. *Ann N Y Acad Sci* 917:962–70.

Department of Health and Human Services. 2009. Panel on Antiretroviral Guidelines for Adults and Adolescents. Guidelines for the use of antiretroviral agents in HIV-1-infected adults and adolescents. http://aidsinfo.nih.gov/ContentFiles/AdultandAdolescentGL.pdf (accessed August 16, 2010).

Dyner, T. S., W. Lang, J. Geaga et al. 1993. An open-label dose-escalation trial of oral dehydroepiandrosterone tolerance and pharmacokinetics in patients with HIV disease. *J Acquir Immune Defic Syndr* 6(5):459–65.

Etzel, J. V., J. M. Brocavich, and M. Torre. 1992. Endocrine complications associated with human immunodeficiency virus infection. *Clin Pharm* 11(8):705–13.

Ferrando, S. J., J. Rabkin, and L. Poretsky. 1999. Dehydroepiandrosterone sulfate (DHEAS) and testosterone: Relation to HIV illness stage and progression over one year. *J Acquir Immune Defic Syndr* 22(2):146–54.

Gallo, R. C., P. S. Sarin, E. P. Gelmann et al. 1983. Isolation of human T-cell leukemia virus in acquired immune deficiency syndrome (AIDS). *Science* 220(4599):865–7.

Gochu, J., B. Piper, J. Montana, H. S. Park, and L. Poretsky. 1994. Lymphoma of the thyroid mimicking thyroiditis in a patient with the acquired immune deficiency syndrome. *J Endocrinol Invest* 17(4):279–82.

Grinspoon, S., C. Corcoran, T. Stanley, J. Rabe, and S. Wilkie. 2001. Mechanisms of androgen deficiency in human immunodeficiency virus-infected women with the wasting syndrome. *J Clin Endocrinol Metab* 86(9):4120–6.

Guttler, R., P. A. Singer, S. G. Axline, T. S. Greaves, and J. J. McGill. 1993. Pneumocystis carinii thyroiditis. Report of three cases and review of the literature. *Arch Intern Med* 153(3):393–6.

Hruz, P. W. 2008. HIV protease inhibitors and insulin resistance: Lessons from in-vitro, rodent and healthy human volunteer models. *Curr Opin HIV AIDS* 3(6):660–5.

Jacobson, M. A., R. E. Fusaro, M. Galmarini, and W. Lang. 1991. Decreased serum dehydroepiandrosterone is associated with an increased progression of human immunodeficiency virus infection in men with CD4 cell counts of 200–499. *J Infect Dis* 164(5):864–8.

Kannangai, R., A. J. Kandathil, D. L. Ebenezer et al. 2008. Usefulness of alternate prognostic serum and plasma markers for antiretroviral therapy for human immunodeficiency virus type 1 infection. *Clin Vaccine Immunol* 15(1):154–8.

Karminska, M., J. Harris, K. Gijsbers, and B. Dubrovsky. 2000. Dehydroepiandrosterone sulfate (DHEAS) counteracts decremental effects of corticosterone on dentate gyrus LTP. Implications for depression. *Brain Res Bull* 52(3):229–34.

Kino, T., and G. P. Chrousos. 2001. Glucocorticoid and mineralocorticoid resistance/hypersensitivity syndromes. *J Endocrinol* 169:437–45.

Kramer, A. S., A. R. Lazzarotto, E. Sprinz, and W. C. Manfroi. 2009. Metabolic abnormalities, antiretroviral therapy and cardiovascular disease in elderly patients with HIV. *Arq Bras Cardiol* 93(5):561–8.

Kumar, M., A. M. Kumar, D. Waldrop, M. H. Antoni, N. Schneiderman, and C. Eisdorfer. 2002. The HPA axis in HIV-1 infection. *J Acquir Immune Defic Syndr* 31(Suppl 2):S89–93.

Labrie, F. 2010. DHEA, important source of sex steroids in men and even more in women. *Prog Brain Res* 182:97–148.

Masur, H., M. A. Michelis, J. B. Greene et al. 1981. An outbreak of community-acquired Pneumocystis carinii pneumonia: Initial manifestation of cellular immune dysfunction. *N Engl J Med* 305(24):1431–8.

Morales, A. J., R. H. Haubrich, J. Y. Hwang, H. Asakura, and S. S. Yen. 1998. The effect of six months treatment with a 100 mg daily dose of dehydroepiandrosterone (DHEA) on circulating sex steroids, body composition and muscle strength in age-advanced men and women. *Clin Endocrinol (Oxf)* 49(4):421–32.

Nair, K. S., R. A. Rizza, P. O'Brien et al. 2006. DHEA in elderly women and DHEA or testosterone in elderly men. *N Engl J Med* 355:1647–59.

Nestler, J. E., J. N. Clore, and W. G. Blackard. 1992. Dehydroepiandrosterone: The "missing link" between hyperinsulinemia and atherosclerosis? *FASEB J* 6(12):3073–5. Review.

Norbiato, G., M. Galli, V. Righini, and M. Moroni. 1994. The syndrome of acquired glucocorticoid resistance in HIV infection. *Baillieres Clin Endocrinol Metab* 8(4):777–87.

Panjari, M., R. J. Bell, F. Jane et al. 2009. A randomized trial of oral DHEA treatment for sexual function, well-being, and menopausal symptoms in postmenopausal women with low libido. *J Sex Med* 6(9):2579–90.

Perez-Neri, I., S. Montes, and C. Rios. 2009. Inhibitory effect of dehydroepiandrosterone on brain monoamine oxidase activity: In vivo and in vitro studies. *Life Sci* 85(17–8):652–6.

Piketty, C., D. Jayle, A. Leplege et al. 2001. Double-blind placebo-controlled trial of oral dehydroepiandrosterone in patients with advanced HIV disease. *Clin Endocrinol (Oxf)* 55(3):325–30.

Ponte, C. M., M. H. Gurgel, and R. M. Montenegro Jr. 2009. Gonadotrophic axis dysfunction in men with HIV-infection/aids. *Arq Bras Endocrinol Metabol* 53(8):983–8. Article in Portuguese.

Poretsky, L., D. J. Brillon, S. Ferrando et al. 2006. Endocrine effects of oral dehydroepiandrosterone in men with HIV infection: A prospective, randomized, double-blind, placebo-controlled trial. *Metabolism* 55(7):858–70.

Poretsky, L., S. Can, and B. Zumoff. 1995. Testicular dysfunction in human immunodeficiency virus-infected men. *Metabolism* 44(7):946–53.

Poretsky, L., A. Maran, and B. Zumoff. 1990. Endocrinologic and metabolic manifestations of the acquired immunodeficiency syndrome. *Mt Sinai J Med* 57(4):236–41.

Poretsky, L., L. Song, D. J. Brillon et al. 2009. Metabolic and hormonal effects of oral DHEA in premenopausal women with HIV infection: A randomized, prospective, placebo-controlled pilot study. *Horm Metab Res* 41(3):244–9.

Pruss, D., F. D. Bushman, and A. P. Wolffe. 1994. Human immunodeficiency virus integrase directs integration to sites of severe DNA distortion within the nucleosome core. *Proc Natl Acad Sci U S A* 91(13):5913–7.

Rabkin, J. G., M. C. McElhiney, R. Rabkin, P. J. McGrath, and S. J. Ferrando. 2006. Placebo-controlled trial of dehydroepiandrosterone (DHEA) for treatment of nonmajor depression in patients with HIV/AIDS. *Am J Psychiatry* 163(1):59–66.

Rao, M. N., K. Mulligan, and M. Schambelan. 2010. HIV infection and diabetes. In *Principles of Diabetes Mellitus*, ed. L. Poretsky, 2nd ed., 617–44. New York: Springer.

Schröder, A. R., P. Shinn, H. Chen, C. Berry, J. R. Ecker, and F. Bushman. 2002. HIV-1 integration in the human genome favors active genes and local hotspots. *Cell* 110(4):521–9.

Sellmeyer, D. E., and C. Grunfeld. 1996. Endocrine and metabolic disturbances in human immunodeficiency virus infection and the acquired immune deficiency syndrome. *Endocr Rev* 17(5):518–32.

Smolyar, D., R. Tirado-Bernardini, R. Landman, M. Lesser, I. Young, and L. Poretsky. 2003. Comparison of 1-micro g and 250-micro g corticotropin stimulation tests for the evaluation of adrenal function in patients with acquired immunodeficiency syndrome. *Metabolism* 52(5):647–51.

Stolarczyk, R., S. I. Rubio, D. Smolyar, I. S. Young, and L. Poretsky. 1998. Twenty-four-hour urinary free cortisol in patients with acquired immunodeficiency syndrome. *Metabolism* 47(6):690–4.

Thomas, J., and S. M. Doherty. 2003. HIV infection—a risk factor for osteoporosis. *J Acquir Immune Defic Syndr* 33(3):281–91.

Thompson, R. D., and M. Carlson. 2000. Liquid chromatographic determination of dehydroepiandrosterone (DHEA) in dietary supplement products. *J Assoc Off Anal Chem Int* 83(4):847–57.

Unachukwu, C. N., D. I. Uchenna, and E. E. Young. 2009. Endocrine and metabolic disorders associated with human immune deficiency virus infection. *West Afr J Med* 28(1):3–9.

World Health Organization. 2007. WHO case definitions of HIV for surveillance and revised clinical staging and immunological classification of HIV-related disease in adults and children. Geneva, Switzerland: WHO Press. http://who.int/hiv/pub/guidelines/hivstaging/en/index.html (accessed August 16, 2010).

World Health Organization. 2009. WHO | AIDS Epidemic Update 2009. http://who.int/hiv/pub/epidemiology/epidemic/en/index.html (accessed September 13, 2010).

Young, A. H., P. Gallagher, and R. J. Porter. 2002. Elevation of the cortisol-dehydroepiandrosterone ratio in drug-free depressed patients. *Am J Psychiatry* 159(7):1237–9.

Zapanti, E., K. Terzidis, and G. Chrousos. 2008. Dysfunction of the hypothalamic-pituitary-adrenal axis in HIV infection and disease. *Hormones (Athens)* 7(3):205–16.

Zarrouf, F. A., S. Artz, J. Griffith, C. Sirbu, and M. Kommor. 2009. Testosterone and depression: Systematic review and meta-analysis. *J Psychiatr Pract* 15(4):289–305.

16 DHEA and Its Metabolites and Analogs

A Role in Immune Modulation and Arthritis Treatment?

Dominick L. Auci, Clarence N. Ahlem, Christopher L. Reading, and James M. Frincke

CONTENTS

INTRODUCTION

Rheumatoid arthritis (RA) is a systemic immune-mediated inflammatory disease (IMID) of the joints and has been associated with multifaceted pathogenic pathways. These may include aberrant T cells, B cells, macrophages, and dendritic cells, as well as an unbalanced network of chemokines and cytokines, with tumor necrosis factor-α (TNF-α) playing a key role (Kuek, Hazleman, and Ostor 2007). Although the pathophysiology of RA is complex, several anti-inflammatory agents provide transient symptomatic relief. These include the use of immunotoxic agents (e.g., methotrexate), nonsteroidal anti-inflammatory drugs (e.g., selective cyclo-oxygenase-2 inhibitors), synthetic glucocorticoids, and biological agents that suppress the activity of TNF-α (Turini and DuBois 2002; Wislowska and Jakubicz 2007; Cronstein 2005; Czock et al. 2005). All existing therapies have significant limitations, which include anergy to prolonged treatment, systemic immune suppression, bone loss, gastrointestinal bleeding, and high cost (Callen 2007; Chen et al. 2009; Kong, Teuber, and Gershwin 2006; Song et al. 2005). Dozens of candidate compounds targeting enzymes, receptors, and signaling molecules critically involved at various points in the inflammatory cascade are either currently in clinical development or have recently been approved (Kukar, Petryna, and Efthimiou 2009; Storage, Agrawal, and Furst 2010; Yurchenko et al. 2010; Serhan, Chiang, and Van Dyke 2008). Some may prove effective, but mechanistic similarities to existing therapies suggest that many of those limitations will emerge, leaving the field of RA therapy open for new, safe, and effective treatment options.

There is considerable data supporting the notion that endogenous dehydroepiandrosterone (DHEA) plays an important role in immune modulation in RA. Plasma concentrations of DHEA are decreased in RA patients (Cutolo 2000; Wilder 1996; Kanik et al. 2000; Hedman, Nilsson, and de la Torre 1992; de la Torre et al. 1993; Feher and Feher 1984; Mirone et al. 1996), and the enzymes that mediate the metabolic conversion of DHEA to potentially more potent anti-inflammatory agents may be altered by inflammatory cytokines (Dulos et al. 2005b), suggesting that intervention with an appropriate DHEA derivative may counter the inflammatory condition. There is also evidence suggesting that alterations in the intrasynovial androgen–estrogen balance are important to pathogenesis in RA, but the role of DHEA in this situation remains unclear (Cutolo et al. 2004).

DHEA has a long history of remarkable effectiveness as an anti-inflammatory and antiarthritic agent both *in vitro* and in rodent models (Dillon 2005; Williams, Jones, and Rademacher 1997; Kobayashi et al. 2003; Rontzsch et al. 2004; Wu et al. 2006). However, clinical trials of DHEA as treatment for rheumatic diseases have been unsuccessful (Giltay et al. 1999; van Vollenhoven 2000; Cameron and Braunstein 2005). This is apparently due to a combination of low oral bioavailability, undesired sex steroid-associated side effects, and possibly inefficient metabolism to active metabolites (e.g., polyhydroxylated molecular species). The failure to translate the biological activity observed in rodents may be termed "The DHEA conundrum" (Herrington 1998). Profound divergences between rodents and humans may form the basis of the problem and hold the key to its solution. Endogenous DHEA is virtually absent in rodent plasma (Wolf and Kirschbaum 1999), and these animals extensively metabolize exogenous DHEA into numerous polyhydroxylated forms (5-androstene diols, triols, and tetrols). It is now suspected that at least some of the anti-inflammatory activity of DHEA can be attributed to these metabolites (Loria and Padgett 1998; Marwah, Marwah, and Lardy 2002; Padgett and Loria 1994). In humans, DHEA is the major product of the adrenal gland zona reticularis and, in contrast to rodents, it is found abundantly in human circulation (Adams 1985; Roberge et al. 2007). Although a small amount of DHEA is metabolized into more highly oxygenated species, the majority is converted into sex steroids in peripheral tissues or conjugated to form dehydroepiandrosterone sulfate (DHEAS), some of which may be converted back in a tissue-specific manner (Labrie et al. 2001). An additional understanding is that DHEA is a ligand-independent peroxisome proliferator (stimulates peroxisome proliferator-activated receptor-α [PPARα]) in rodents, but not in humans (Webb et al. 2006). By contrast, the 7-hydroxy metabolites are not peroxisome proliferators in rodents or humans (Webb et al. 2006). DHEA can activate the pregnane X-receptor (PXR) in rodents, but not in humans, and treatment in rodents results in the induction of various P450 enzymes, which can potentially metabolize it into ligands for nuclear hormone receptors (Webb et al. 2006). Finally, while DHEA itself has several potential avenues to elicit biological responses, including interaction with cell surface (Liu et al. 2006) and nuclear hormone receptors (Peters et al. 1996; Waxman 1996; Kajita et al. 2003), its direct effects have not been differentiated from those attributable to its metabolites in disease models. Because DHEA's activity may largely reside in its metabolites, the DHEA metabolome holds great promise as a new class of anti-inflammatory agents with improved safety and novel mechanisms of action that can be used to treat RA.

Two well-studied, anti-inflammatory DHEA metabolites are 5-androstene-3β,17β-diol (AED) and 5-androstene-3β,7β,17β-triol (β AET; Auci, Reading, and Frincke 2009; Auci et al. 2003; Xiao et al. 2007; Butenandt, Hausmann, and Paland 1938; Reynolds 1966; Hill et al. 2005; Offner et al. 2002; Marcu et al. 2006). These molecules are produced by tissue-specific reactions involving enzymes that are regulated locally, at least in part, in response to inflammatory signals, creating an intracrine system within tissues to control pro- and anti-inflammatory activity (Hennebold and Daynes 1994; Herrmann, Scholmerich, and Straub 2002). AED is formed by the action of 17β-hydroxysteroid dehydrogenase (17β-HSD) on DHEA. β-AET is formed by a series of reactions involving 17β-HSD, CYP 7B1, and 11β-HSD, through intermediates that are described by combinations of 7 and 17-keto, 7α-, 7β-, and 17β-hydroxy functionalized 5-androstenes (Dulos and Boots 2006; Muller et al. 2006b). Definitive studies of the pro- and anti-inflammatory activity associated with all possible di- and trioxygenated isomers have not been performed, but several publications suggest that in the family of 3β-hydroxy-5-androstenes, the 17-keto form of the 7-α-hydroxy derivative may be proinflammatory (Dulos and Boots 2006; Blauer et al. 1991; Dulos et al. 2004), which is not necessarily an undesired property in the proper context. By contrast, the 7β, 17β-hydroxy form is anti-inflammatory (Loria and Padgett 1998; Marcu et al. 2006; Offner et al. 2002; Ahlem et al. 2009), which is potentially useful in clinical indications where excess inflammation is unwanted. Thus, 17β-HSD cooperates with 11β-HSD (an enzyme already closely tied to the balance of anti-inflammatory activity through the reduction of cortisone to cortisol) to locally regulate pro- and anti-inflammatory signals (Robinzon et al. 2003). Expression and function of the steroid dehydrogenases are, in turn, subject to regulation by inflammatory cytokines (Dulos et al. 2005b, Miesel, Hartung, and Kroeger 1996), and they define a regulatory circuit that may be difficult to disrupt and/or once disrupted may make it difficult to reeestablish homeostasis. The implication is that the anti-inflammatory species may be eliminated in the context of the proinflammatory program. Accordingly, in the context of potential RA therapies, protection of 17β-hydroxyl functionality may confer resistance to metabolism and thereby assert an anti-inflammatory action. The anti-inflammatory activity of the DHEA metabolite β-AET and the protected ethynylated derivative, 17α-ethynyl-androst-5-ene-3β,7β,17β-triol (HE3286) is mediated, at least in part, by decreasing the activity of the inflammatory cytokine TNF-α, perhaps via the TNF-α-mitogen-activated protein kinase (MAPK)-NF-κβ signal transduction scaffold (Offner et al. 2009; Wang et al. 2010; Lu et al. 2010; Offner et al. 2002; Auci et al. 2007). Given the central role of TNF-α in RA pathophysiology, compounds like HE3286 may be particularly well suited to provide treatment options to patients with RA and perhaps other IMID driven by this cytokine.

This review will examine the evidence for adrenal insufficiency and altered DHEA metabolism in RA and the role of tissue-specific steroid metabolism in disease. We will also discuss early attempts at therapeutic intervention and review current strategies to use pharmaceutical derivatives of the DHEA metabolome as anti-inflammatory agents. Finally, we will focus attention on potential targets and mechanisms of action. At last, we may realize the promise of DHEA through pharmaceutically engineered compounds that may become part of the pharmacopeia to treat RA and other IMID.

ADRENAL INSUFFICIENCY IN RHEUMATOID ARTHRITIS AND RELATION TO DHEA METABOLISM

A role for adrenocortical insufficiency in RA has long been suspected (Wilder 1996), and its basis has recently been reviewed (Imrich et al. 2010). Low-circulating levels of DHEA for a given age and gender are a simple and reliable indicator of decreased adrenal function. The endogenous production of DHEA is subject to modulation from a variety of signals. In addition to the natural decline in DHEA levels with age (Nafziger et al. 1998), for which the physiological basis is not well understood, adrenal DHEA output can be influenced by inflammation and inflammatory cytokines (Schuld et al. 2000). DHEA production is also decreased by the influence of exogenous glucocorticoids on the hypothalamic-pituitary-adrenal axis: elevated glucocorticoids result in decreased pituitary secretion of adrenocorticotropic hormone, which in turn decreases the stimulus for both DHEA and cortisol synthesis. Considering that DHEA and/or metabolites are reported to counter the deleterious side effects of glucocorticoids, suppression of adrenal DHEA production has the potential to actually exacerbate glucocorticoid side effects, some of which overlap with the features of chronic inflammatory diseases, such as metabolic syndrome, bone loss, and immunosuppression (Loria 1997; Browne et al. 1992; Blauer et al. 1991; Kalimi et al. 1994; Muller, Hennebert, and Morfin 2006a, Cooper and Stewart 2009).

As previously indicated, low levels of DHEA in the serum of RA patients were reported by several groups (Cutolo 2000; Wilder 1996; Kanik et al. 2000; Hedman, Nilsson, and de la Torre 1992; de la Torre et al. 1993; Feher and Feher 1984; Mirone et al. 1996), along with suggestions that replacement therapy could provide benefit. Consistent with this line of thinking, Straub and colleagues hypothesized improved adrenal function correlated with attenuation of joint disease by showing that effective therapeutic strategies such as neutralization of proinflammatory cytokines increased secretion of biologically active adrenal androgens (Straub et al. 2005; Straub et al. 2006). This same group also reported that TNF-α inhibited the conversion of DHEAS to DHEA in RA synovial cells, implying (with *in vitro* observations) that anti-TNF-α strategies in RA might promote DHEA-mediated benefits. Along this same line, in the periphery, Ernestam and colleagues (Ernestam et al. 2007) reported increased DHEAS levels in RA patients after treatment with TNF-α antagonists. They interpreted this observation as evidence for improving adrenal function. In contrast, treatment with the glucocorticoid prednisolone, significantly, decreased DHEAS levels in RA patients (Yukioka et al. 2006).

Altered DHEA Metabolism in Rheumatoid Arthritis Joints

In joints, steroidogenic enzymes are involved in local steroid metabolism as reviewed by Schmidt and colleagues (Schmidt et al. 2006). There is evidence that inflammation in the arthritic joint impacts DHEA metabolism in a way that may be relevant to joint disease. Dulos and colleagues reported that TNF-α enhanced the formation of 7α-hydroxy-DHEA by human fibroblast-like synoviocytes (Dulos et al. 2005a). They showed that proinflammatory cytokines could enhance CYP7B expression and activity in fibroblast-like synoviocytes from patients with RA (Dulos et al. 2005b). They also reported that in knee-joint synovial biopsy samples from arthritic mice, 7α-hydroxy-DHEA levels were fivefold higher than in nonarthritic animals (Dulos et al. 2004). 7α-hydroxy-DHEA is thought to have antiglucocorticoid activity preventing the anti-inflammatory action of endogenous glucocorticoids (Dulos and Boots 2006). It may be important to note that such studies have often focused on 17-ketone derivatives, which may be inactive precursors to 17α- or 17β-triols (Auci, Reading, and Frincke 2009). The deleterious effects of TNF-α in RA may also include decreased NADPH:NADP ratios (Miesel, Hartung, and Kroeger 1996), which can favor deactivating oxidase over activating reductase activity of key steroidogenic enzymes (McCormick, Wang, and Mick 2006; Labrie et al. 1997). Within the RA joint, this may decrease conversion of cortisone to cortisol and 17-keto-DHEA derivatives to anti-inflammatory 17-hydroxy derivatives, thus contributing to the induction and maintenance of an inflammatory state.

Sex Steroids in Rheumatoid Arthritis

The sex steroids have long been recognized as potent regulators of inflammation and immune function. Imbalances of sex hormones (specifically, decreased androgen and increased estrogen) in RA tissues have proved more consistent than observations in serum (Cutolo 2000; Cutolo et al. 1986; Sambrook et al. 1988; Cutolo et al. 1988; Masi 1995). Hedman and colleagues (Hedman, Nilsson, and de la Torre 1992) reported low levels of sulfo-conjugated steroids, including testosterone, in synovial fluid from RA patients. Khalkhali-Ellis and colleagues reported reduced levels of testosterone in synovial fluid of juvenile RA patients (Khalkhali-Ellis, Moore, and Hendrix 1998). Cutolo and colleagues found decreased androgen and increased estrogen formation in adult RA synovial fluid (Cutolo et al. 2004). In particular, they noted that mitogenic 16α-hydroxylated estrogens were increased at the apparent expense of antimitogenic 2-hydroxylated estrogens. In that same year, Rovensky and colleagues (Rovensky et al. 2004) demonstrated increased estrogen:testosterone ratios in synovial fluid from male RA patients. Macho and colleagues (Macho et al. 2007) reported similar findings (a predominance of estrogens compared with androgens as well as increases in proinflammatory cytokines) in knee exudates from RA patients. The role of sex steroids in disease pathogenesis and the relative contribution of steroids of gonadal origin (regulated by the hypothalamus-pituitary-gonadal axis) versus intrasynovial metabolism of DHEA remain controversial. However, intrasynovial TNF-α can increase CYP19 (aromatase) and thus increase concentrations of estradiol, which is proinflammatory (Cutolo et al. 2006). In that regard, DHEA's influence on TNF-α may have additional anti-inflammatory consequences. However, the situation is complex, considering the potential for DHEA to be converted to both anti-inflammatory androgens and proinflammatory estradiol.

ATTEMPTS AT THERAPEUTIC INTERVENTION WITH SEX STEROIDS AND RELATED AGENTS

Early studies showed positive effects of androgen replacement therapy, at least in male RA patients, and particularly as adjuvant treatment (Cutolo 2000). However, sex steroid therapy is not always effective and may involve side effects or serious health risks, such as an increased risk of hormone-sensitive cancers (Banks and Canfell 2009). In the case of estrogen, untoward effects are mediated almost exclusively through estrogen receptor-α (ER-α) and treatment with ER-β selective ligands has been suggested as a potentially safer strategy for RA and other autoimmune diseases (Harris 2007). Development efforts continue to find safer and more effective ER-β agonists (Zhao, Dahlman-Wright, and Gustafsson 2008). Alternatively, the risks associated with ER-α could be minimized by the use of low-dose estrogen that produces synergistic effects when used in combination with another immune regulatory agent (Offner and Polanczyk 2006) or by the development of safer derivatives to be used without estrogen (Offner et al. 2002).

SOLVING THE DHEA CONUNDRUM

In addition to those already discussed, alternative hypotheses for the failure of DHEA in the clinic were advanced including the assertion that metabolic DHEA derivatives are inactive end products or are inactive in humans (Chalbot and Morfin 2006; Lardy 2003; Labrie et al. 2005). However, our initial forays into clinical medicine indicate that androstene derivatives have anti-inflammatory activity in humans (Frincke et al. 2007; Stickney et al. 2007). Translational studies from rodents to primates have shown that primates metabolize steroid precursors into active molecules locally (Labrie et al. 2005). In contrast, rodents metabolize steroids systemically and distribute them into tissues via circulation (Wolf and Kirschbaum 1999). That is to say, rodents are an endocrine species, whereas humans have evolved to utilize a more highly controlled paracrine–autocrine system that has been termed the "intracrine system" (Chalbot and Morfin 2006; Labrie et al. 2005). Perturbed steroidogenesis in the inflamed joint suggests that simple replacement therapy or use of prodrugs, dependent on metabolism for activity, might not provide the desired therapeutic effect and could even exacerbate RA. We hypothesized a solution to the DHEA conundrum that used appropriate pharmacologically active DHEA derivatives that were resistant to and not dependent on steroidogenic enzyme systems. However, because of the historical focus on androgens and estrogens, little was known about the extended DHEA metabolome and its synthetic derivatives.

EARLY EFFORTS

With only limited information on the DHEA metabolome, early efforts to address the DHEA conundrum using synthetic analogs met with mixed results (Schwartz et al. 1988b; Offner et al. 2002; Schwartz and Pashko 2002; Schwartz, Lewbart, and Pashko 1988a; Schwartz, Pashko, and Whitcomb 1986; Auci et al. 2004). Fluasterone (5-androstene-16α-fluoro-17-one) is a synthetic DHEA analog that we previously reported to be effective in the DBA mouse model of collagen induced arthritis (CIA) (Offner et al. 2004). Observations paralleled the activity reported for DHEA in this same model (Rontzsch et al. 2004). Fluasterone retains certain immune-regulating properties and represented a potential improvement over DHEA because it was nontoxic and practically devoid of androgenic or estrogenic side effects. However, far higher doses were required for efficacy, both in our own CIA studies and in the rat model of adjuvant-induced arthritis (AIA) reported by others (Schwartz and Pashko 2002). Further, it was only effective in CIA when given by injection. The combination of low potency and insufficient oral bioavailability ultimately rendered fluasterone unsuitable for continued clinical development.

Novel Agents

Recently, a more complete set of DHEA metabolic products and pathways were identified (Marwah, Marwah, and Lardy 2002). Our premise is that orally active structures that are not readily metabolized into undesired hormones might translate at least some of the preclinical promise of DHEA into humans.

HE3286

HE3286 is a novel, orally bioavailable 17α-ethynylated derivative of β-AET. In mouse CIA, collagen antibody-induced arthritis (CAIA), and rat AIA models (Offner et al. 2009; Auci et al. 2010; Auci et al. 2007), oral treatments at disease onset, or in animals with well-established disease, significantly decreased disease scores in a dose-dependent fashion. Joint inflammation, erosion and synovial proliferation, and proinflammatory cytokine (TNF-α, interleukin [IL]-6, IL-1β, IL-17, and IL-23) mRNA and/or protein expression were all significantly reduced. In CAIA, IL-6 and MMP-3 mRNA levels in diseased joints were significantly reduced. Interestingly, methotrexate, a current mainstay in the treatment of RA, is not effective in the CAIA model while glucocorticoids are effective (Lange et al. 2005). In contrast to glucocorticoids, HE3286 was not found to be immune-suppressive in several classical animal models of immune function (Ahlem et al. 2009), even though it also reduces nuclear factor kappaB (NFκB) activation. Importantly, and perhaps with specific relevance to RA, HE3286 caused the inhibition of the lipopolysaccharide-induced macrophage-activation program *in vitro* primarily by inhibiting TNF-α action (Lu et al. 2010). This activity was associated with significantly decreased phosphorylation of the ikappaB kinase inhibitor (IKK), NFκB, P38, and c-Jun N-terminal kinase. HE3286 treatment was also associated with increased regulatory T cells. An agent that can safely limit TNF-α action would have broad potential beyond RA, extending into many inflammatory diseases.

Tetrols

Our exploration of the C-19 steroid component of the human metabolome led to the discovery of several novel tetrahydroxylated metabolites with pharmaceutical potential. We reported the discovery of androst-5-ene-3β,7β,16α,17β-tetrol (HE3177) and androst-5-ene-3α,7β,16α,17β-tetrol (HE3413) in human plasma and demonstrated their oral bioavailability (Ahlem et al. 2011). These compounds appear to be metabolic end products because they are highly resistant to primary and secondary metabolism both *in vitro* and *in vivo*. In preliminary high-dose evaluations, neither compound displayed acute toxicity, sex-hormone effects, or gave any indication of immune toxicity. In preliminary studies, these compounds demonstrated potent anti-inflammatory activities, suggesting they would be active in RA. In addition to their potential as pharmaceuticals, since both are natural products, either of the compounds may be suitable as a dietary supplement.

TARGETS AND MECHANISM

Many of the known mechanisms through which DHEA itself may elicit its effects have been reviewed (Webb et al. 2006). These include a cell-surface receptor that mediates rapid nongenomic effects (Liu and Dillon 2004). In contrast, no dedicated receptors or signal transduction pathways have been identified for any of the 7-hydroxy C-19 steroids. Potential mechanisms of action have been grouped into a number of broad categories, including gating (ligand inactivation), modulation of ion channels, interaction with atypical receptors, allosteric modification of scaffold proteins, transport proteins and enzymes, and modulation of steroidogenic enzymes (Lathe 2002). Although potential targets within each of these categories are currently under investigation, the relevant protein interactions remain unknown.

Preliminary studies in our laboratory (Harbor Biosciences, unpublished observations) support the hypothesis that DHEA and β-AET modulate 11β-HSD1 gene expression through different pathways (Srivastava 2009). 11β-HSD1 is a key enzyme in regulation of GC signaling and a key pharmaceutical target (Schnackenberg 2008). The effects on 11β-HSD1 may be mediated by modulation of other genes, interactions with proteins integral to tissue-specific transcription complexes such as C/EBP (Balazs et al. 2008) and/or direct effects on enzymatic activity. While our own studies using human microsomes could not demonstrate an effect of HE3286 or β-AET on the enzymatic activity of either 11β-HSD1 or CYP7B (Harbor Biosciences, unpublished observations), it has been reported that 7-hydroxy/keto, 17-keto 5-androstene steroids interact with 11β-HSD1 in cellular assays (Hennebert et al. 2007; Balazs et al. 2009). Three-dimensional-modeling analyses indicated that the 7-keto and 7-hydroxy metabolites of DHEA can occupy the same binding site in 11β-HSD1 as cortisone and cortisol, respectively (Nashev et al. 2007). Thus, within the DHEA metabolome, there is evidence for effects on 11β-HSD enzymatic activity. Kinases and phosphatases (Zabolotny et al. 2008), as well as other enzymes (Cleary 1991), are implicated by the observed biochemical activities. Furthermore, interactions with cell-surface receptors that exert nongenomic activities or signaling via an orphan or other as-yet-uncharacterized nuclear receptor remain possibilities. These are not mutually exclusive, and multiple, simultaneous interactions may be involved, as has been demonstrated for DHEA (Webb et al. 2006). Our own efforts aimed at the identification of protein-binding partner(s) for HE3286 are ongoing collaborations using stable-isotope labeled amino acids in cell culture (SILAC) technology, which has emerged as powerful tool for parallel identification and quantification of proteins binding to ligands (Ong et al. 2009).

CONCLUSIONS

Nonclinical and clinical observations support and expand the potential of the DHEA metabolome as fertile ground in which to identify active core structures around which to build synthetic therapeutic agents. There is evidence that these agents can be modified to at last capture the desirable activities of DHEA and translate them into clinical benefit for patients with RA.

REFERENCES

Adams, J. B. 1985. Control of secretion and the function of C19-delta 5-steroids of the human adrenal gland. *Mol Cell Endocrinol* 41:1–7.

Ahlem, C., D. Auci, K. Mangano, C. Reading, J. Frincke, D. Stickney, and F. Nicoletti. 2009. HE3286: A novel synthetic steroid as an oral treatment for autoimmune disease. *Ann N Y Acad Sci* 1173:781–90.

Ahlem, C. N., T. M. Page, D. L. Auci, M. R. Kennedy, K. Mangano, F. Nicoletti, Y. Ge, Y. Huang, S. K. White, S. Villegas, D. Conrad, A. Wang, C. L. Reading, and J. M. Frincke. 2011. Novel components of the human metabolome: the identification, characterization and anti-inflammatory activity of two 5-androstene tetrols. *Steroids* 76(1–2):145–55.

Auci, D. L., C. Ahlem, M. Li, R. Trauger, C. Dowding, F. Paillard, J. Frincke, and C. L. Reading. 2003. The immunobiology and therapeutic potential of androstene hormones and their synthetic derivatives: Novel anti-inflammatory and immune regulating steroid hormones. *Mod Asp Immunobiol* 3:64–70.

Auci, D., L. Kaler, S. Subramanian, Y. Huang, J. Frincke, C. Reading, and H. Offner. 2007. A new orally bioavailable synthetic androstene inhibits collagen-induced arthritis in the mouse. *Ann N Y Acad Sci* 1110:630–40.

Auci, D., K. Mangano, D. Destiche, S. K. White, Y. Haung, D. Boyle, J. Frincke, C. Reading, and F. Nicoletti. 2010. Oral treatment with HE3286 ameliorates disease in rodent models of rheumatoid arthritis. *Int J Mol Med* 25(4):625–33.

Auci, D., F. Nicoletti, K. Mangano, R. Pieters, S. Nierkens, L. Morgan, I. Schraufstatter, J. Frincke, and C. Reading. 2004. Anti-inflammatory and immune regulatory properties of 5-androsten-3b,17b-diol (HE2100) and synthetic analogue HE3204; implications for treatment of autoimmune diseases. *Autoimmun Rev* 3:87–8.

Auci, D. L., C. L. Reading, and J. M. Frincke. 2009. 7-Hydroxy androstene steroids and a novel synthetic analogue with reduced side effects as a potential agent to treat autoimmune diseases. *Autoimmun Rev* 8:369–72.

Balazs, Z., L. G. Nashev, C. Chandsawangbhuwana, M. E. Baker, and A. Odermatt. 2009. Hexose-6-phosphate dehydrogenase modulates the effect of inhibitors and alternative substrates of 11beta-hydroxysteroid dehydrogenase 1. *Mol Cell Endocrinol* 301:117–22.

Balazs, Z., R. A. Schweizer, F. J. Frey, F. Rohner-Jeanrenaud, and A. Odermatt. 2008. DHEA induces 11 -HSD2 by acting on CCAAT/enhancer-binding proteins. *J Am Soc Nephrol* 19(1): 92–101

Banks, E., and K. Canfell. 2009. Invited Commentary: Hormone therapy risks and benefits—The Women's Health Initiative findings and the postmenopausal estrogen timing hypothesis. *Am J Epidemiol* 170:24–8.

Blauer, K. L., M. Poth, W. M. Rogers, and E. W. Bernton. 1991. Dehydroepiandrosterone antagonizes the suppressive effects of dexamethasone on lymphocyte proliferation. *Endocrinology* 129:3174–9.

Browne, E. S., B. E. Wright, J. R. Porter, and F. Svec. 1992. Dehydroepiandrosterone: Antiglucocorticoid action in mice. *Am J Med Sci* 303:366–71.

Butenandt, A., E. Hausmann, and J. Paland. 1938. Uber delta-androstadien-diol-(3.17). *Berichte d. D. Chem Gessellschaft* 71:1316–22.

Callen, J. P. 2007. Complications and adverse reactions in the use of newer biologic agents. *Semin Cutan Med Surg* 26:6–14.

Cameron, D. R., and G. D. Braunstein. 2005. The use of dehydroepiandrosterone therapy in clinical practice. *Treat Endocrinol* 4:95–114.

Chalbot, S., and R. Morfin. 2006. Dehydroepiandrosterone metabolites and their interactions in humans. *Drug Metabol Drug Interact* 22:1–23.

Chen, Y., E. Bord, T. Tompkins, J. Miller, C. S. Tan, R. P. Kinkel, M. C. Stein, R. P. Viscidi, L. H. Ngo, and I. J. Koralnik. 2009. Asymptomatic reactivation of JC virus in patients treated with natalizumab. *N Engl J Med* 361:1067–74.

Cleary, M. P. 1991. The antiobesity effect of dehydroepiandrosterone in rats. *Proc Soc Exp Biol Med* 196:8–16.

Cooper, M. S., and P. M. Stewart. 2009. 11Beta-hydroxysteroid dehydrogenase type 1 and its role in the hypothalamus-pituitary-adrenal axis, metabolic syndrome, and inflammation. *J Clin Endocrinol Metab* 94:4645–54.

Cronstein, B. N. 2005. Low-dose methotrexate: A mainstay in the treatment of rheumatoid arthritis. *Pharmacol Rev* 57:163–72.

Cutolo, M. 2000. Sex hormone adjuvant therapy in rheumatoid arthritis. *Rheum Dis Clin North Am* 26:881–95.

Cutolo, M., E. Balleari, M. Giusti, M. Monachesi, and S. Accardo. 1986. Sex hormone status in women suffering from rheumatoid arthritis. *J Rheumatol* 13:1019–23.

Cutolo, M., E. Balleari, M. Giusti, M. Monachesi, and S. Accardo. 1988. Sex hormone status of male patients with rheumatoid arthritis: Evidence of low serum concentrations of testosterone at baseline and after human chorionic gonadotropin stimulation. *Arthritis Rheum* 31:1314–7.

Cutolo, M., S. Capellino, A. Sulli, B. Serioli, M. E. Secchi, B. Villaggio, and R. H. Straub. 2006. Estrogens and autoimmune diseases. *Ann N Y Acad Sci* 1089:538–47.

Cutolo, M., B. Villaggio, B. Seriolo, P. Montagna, S. Capellino, R. H. Straub, and A. Sulli. 2004. Synovial fluid estrogens in rheumatoid arthritis. *Autoimmun Rev* 3:193–8.

Czock, D., F. Keller, F. M. Rasche, and U. Haussler. 2005. Pharmacokinetics and pharmacodynamics of systemically administered glucocorticoids. *Clin Pharmacokinet* 44:61–98.

de la Torre, B., M. Hedman, E. Nilsson, O. Olesen, and A. Thorner. 1993. Relationship between blood and joint tissue DHEAS levels in rheumatoid arthritis and osteoarthritis. *Clin Exp Rheumatol* 597–601.

Dillon, J. S. 2005. Dehydroepiandrosterone, dehydroepiandrosterone sulfate and related steroids: Their role in inflammatory, allergic and immunological disorders. *Curr Drug Targets Inflamm Allergy* 4:377–85.

Dulos, J., and A. H. Boots. 2006. DHEA metabolism in arthritis: A role for the p450 enzyme cyp7b at the immune-endocrine crossroad. *Ann N Y Acad Sci* 1069:401–13.

Dulos, J., A. Kaptein, A. Kavelaars, C. Heijnen, and A. Boots. 2005a. Tumour necrosis factor-alpha stimulates dehydroepiandrosterone metabolism in human fibroblast-like synoviocytes: A role for nuclear factor-kappaB and activator protein-1 in the regulation of expression of cytochrome p450 enzyme 7b. *Arthritis Res Ther* 7:R1271–80.

Dulos, J., M. A. Van der vleuten, A. Kavelaars, C. J. Heijnen, and A. M. Boots. 2005b. CYP7B expression and activity in fibroblast-like synoviocytes from patients with rheumatoid arthritis: Regulation by proinflammatory cytokines. *Arthritis Rheum* 52:770–8.

Dulos, J., E. Verbraak, W. M. Bagchus, A. M. Boots, and A. Kaptein. 2004. Severity of murine collagen-induced arthritis correlates with increased CYP7B activity: Enhancement of dehydroepiandrosterone metabolism by interleukin-1beta. *Arthritis Rheum* 50:3346–53.

Ernestam, S., I. Hafstrom, S. Werner, K. Carlstrom, and B. Tengstrand. 2007. Increased DHEAS levels in patients with rheumatoid arthritis after treatment with tumor necrosis factor antagonists: Evidence for improved adrenal function. *J Rheumatol* 34:1451–8.

Feher, K. G., and T. Feher. 1984. Plasma dehydroepiandrosterone, dehydroepiandrosterone sulphate and androsterone sulphate levels and their interaction with plasma proteins in rheumatoid arthritis. *Exp Clin Endocrinol* 84:197–202.

Frincke, J., D. Stickney, N. Onizuka-handa, A. Garsd, C. Reading, S. Krudsood, P. Wilairatana, and S. Looareesuwan. 2007. Reduction of parasite levels in patients with uncomplicated malaria by treatment with HE2000. *Am J Trop Med Hyg* 76:232–6.

Giltay, E. J., D. Van schaardenburg, L. J. Gooren, and B. A. Dijkmans. 1999. Dehydroepiandrosterone sulfate in patients with rheumatoid arthritis. *Ann N Y Acad Sci* 876:152–4.

Harris, H. A. 2007. Estrogen receptor-beta: Recent lessons from in vivo studies. *Mol Endocrinol* 21:1–13.

Hedman, M., E. Nilsson, and B. de la Torre. 1992. Low blood and synovial fluid levels of sulpho-conjugated steroids in rheumatoid arthritis. *Clin Exp Rheumatol* 10:25–30.

Hennebert, O., S. Le mee, C. Pernelle, and R. Morfin. 2007. 5Alpha-androstane-3beta,7alpha,17beta-triol and 5alpha-androstane-3beta,7beta,17beta-triol as substrates for the human 11beta-hydroxysteroid dehydrogenase type 1. *Steroids* 72:855–64.

Hennebold, J. D., and R. A. Daynes. 1994. Regulation of macrophage dehydroepiandrosterone sulfate metabolism by inflammatory cytokines. *Endocrinology* 135:67–75.

Herrington, D. M. 1998. DHEA: A biologic conundrum. *J Lab Clin Med* 131:292–4.

Herrmann, M., J. Scholmerich, and R. H. Straub. 2002. Influence of cytokines and growth factors on distinct steroidogenic enzymes in vitro: A short tabular data collection. *Ann N Y Acad Sci* 966:166–86.

Hill, M., H. Havlikova, J. Vrbikova, R. Kancheva, L. Kancheva, V. Pouzar, I. Cerny, and L. Starka. 2005. The identification and simultaneous quantification of 7-hydroxylated metabolites of pregnenolone, dehydroepiandrosterone, 3beta,17beta-androstenediol, and testosterone in human serum using gas chromatography-mass spectrometry. *J Steroid Biochem Mol Biol* 96:187–200.

Imrich, R., M. Vlcek, J. C. Aldag, J. Kerlik, Z. Radikova, J. Rovensky, M. Vigas, and A. T. Masi. 2010. An endocrinologist's view on relative adrenocortical insufficiency in rheumatoid arthritis. *Ann N Y Acad Sci* 1193:134–8.

Kajita, K., T. Ishizuka, T. Mune, A. Miura, M. Ishizawa, Y. Kanoh, Y. Kawai, Y. Natsume, and K. Yasuda. 2003. Dehydroepiandrosterone down-regulates the expression of peroxisome proliferator-activated receptor gamma in adipocytes. *Endocrinology* 144:253–9.

Kalimi, M., Y. Shafagoj, R. Loria, D. Padgett, and W. Regelson. 1994. Anti-glucocorticoid effects of dehydroepiandrosterone (DHEA). *Mol Cell Biochem* 131:99–104.

Kanik, K. S., G. P. Chrousos, H. R. Schumacher, M. L. Crane, C. H. Yarboro, and R. L. Wilder. 2000. Adrenocorticotropin, glucocorticoid, and androgen secretion in patients with new onset synovitis/rheumatoid arthritis: Relations with indices of inflammation. *J Clin Endocrinol Metab* 85:1461–6.

Khalkhali-Ellis, Z., T. L. Moore, and M. J. Hendrix. 1998. Reduced levels of testosterone and dehydroepiandrosterone sulphate in the serum and synovial fluid of juvenile rheumatoid arthritis patients correlates with disease severity. *Clin Exp Rheumatol* 16:753–6.

Kobayashi, Y., N. Tagawa, K. Muraoka, Y. Okamoto, and M. Nishida. 2003. Participation of endogenous dehydroepiandrosterone and its sulfate in the pathology of collagen-induced arthritis in mice. *Biol Pharm Bull* 26:1596–9.

Kong, J. S., S. S. Teuber, and M. E. Gershwin. 2006. Potential adverse events with biologic response modifiers. *Autoimmun Rev* 5:471–85.

Kuek, A., B. L. Hazleman, and A. J. Ostor. 2007. Immune-mediated inflammatory diseases (IMIDs) and biologic therapy: A medical revolution. *Postgrad Med J* 83:251–60.

Kukar, M., O. Petryna, and P. Efthimiou. 2009. Biological targets in the treatment of rheumatoid arthritis: A comprehensive review of current and in-development biological disease modifying anti-rheumatic drugs. *Biologics* 3:443–57.

Labrie, F., V. Luu-The, A. Belanger, S. X. Lin, J. Simard, G. Pelletier, and C. Labrie. 2005. Is dehydroepiandrosterone a hormone? *J Endocrinol* 187:169–96.

Labrie, F., V. Luu-The, C. Labrie, and J. Simard. 2001. DHEA and its transformation into androgens and estrogens in peripheral target tissues: Intracrinology. *Front Neuroendocrinol* 22:185–212.
Labrie, F., V. Luu-The, S. X. Lin, C. Labrie, J. Simard, R. Breton, and A. Belanger. 1997. The key role of 17 beta-hydroxysteroid dehydrogenases in sex steroid biology. *Steroids* 62:148–58.
Lange, F., E. Bajtner, C. Rintisch, K. S. Nandakumar, U. Sack, and R. Holmdahl. 2005. Methotrexate ameliorates T cell dependent autoimmune arthritis and encephalomyelitis but not antibody induced or fibroblast induced arthritis. *Ann Rheum Dis* 64:599–605.
Lardy, H. 2003. Happily at work. *J Biol Chem* 278:3499–509.
Lathe, R. 2002. Steroid and sterol 7-hydroxylation: Ancient pathways. *Steroids* 67:967–77.
Liu, D., and J. S. Dillon. 2004. Dehydroepiandrosterone stimulates nitric oxide release in vascular endothelial cells: Evidence for a cell surface receptor. *Steroids* 69:279–89.
Liu, D., M. Ren, X. Bing, C. Stotts, S. Deorah, L. Love-Homan, and J. S. Dillon. 2006. Dehydroepiandrosterone inhibits intracellular calcium release in beta-cells by a plasma membrane-dependent mechanism. *Steroids* 71:691–9.
Loria, R. M. 1997. Antiglucocorticoid function of androstenetriol. *Psychoneuroendocrinology* 22:S103–8.
Loria, R. M., and D. A. Padgett. 1998. Control of the immune response by DHEA and its metabolites. *Rinsho Byori* 46:505–17.
Lu, M., D. Patsouris, P. Li, J. Flores-Riveros, J. M. Frincke, S. Watkins, S. Schenk, and J. M. Olefsky. 2010. A new antidiabetic compound attenuates inflammation and insulin resistance in Zucker diabetic fatty rats. *Am J Physiol Endocrinol Metab* 298:E1036–48.
Macho, L., J. Rovensky, Z. Radikova, R. Imrich, O. Greguska, and M. Vigas. 2007. Levels of hormones in plasma and in synovial fluid of knee joint of patients with rheumatoid arthritis. *Cas Lek Cesk* 146:292–6.
Marcu, A. C., N. D. Kielar, K. E. Paccione, R. W. Barbee, H. Carter, R. R. Ivatury, R. F. Diegelmann, K.R. Ward, and R. M. Loria. 2006. Androstenetriol improves survival in a rodent model of traumatic shock. *Resuscitation* 71:379–86.
Marwah, A., P. Marwah, and H. Lardy. 2002. Ergosteroids. VI. Metabolism of dehydroepiandrosterone by rat liver in vitro: A liquid chromatographic-mass spectrometric study. *J Chromatogr B Biomed Sci Appl* 767:285–99.
Masi, A. T. 1995. Sex hormones and rheumatoid arthritis: Cause or effect relationships in a complex pathophysiology? *Clin Exp Rheumatol* 13:227–40.
Mccormick, K. L., X. Wang, and G. J. Mick. 2006. Evidence that the 11 beta-hydroxysteroid dehydrogenase (11 beta-HSD1) is regulated by pentose pathway flux. Studies in rat adipocytes and microsomes. *J Biol Chem* 281:341–7.
Miesel, R., R. Hartung, and H. Kroeger. 1996. Priming of NADPH oxidase by tumor necrosis factor alpha in patients with inflammatory and autoimmune rheumatic diseases. *Inflammation* 20:427–38.
Mirone, L., L. Altomonte, P. D'agostino, A. Zoli, A. Barini, and M. Magaro. 1996. A study of serum androgen and cortisol levels in female patients with rheumatoid arthritis. Correlation with disease activity. *Clin Rheumatol* 15:15–9.
Muller, C., O. Hennebert, and R. Morfin. 2006a. The native anti-glucocorticoid paradigm. *J Steroid Biochem Mol Biol* 100:95–105.
Muller, C., D. Pompon, P. Urban, and R. Morfin. 2006b. Inter-conversion of 7alpha- and 7beta-hydroxy-dehydroepiandrosterone by the human 11beta-hydroxysteroid dehydrogenase type 1. *J Steroid Biochem Mol Biol* 99:215–22.
Nafziger, A. N., S. J. Bowlin, P. L. Jenkins, and T. A. Pearson. 1998. Longitudinal changes in dehydroepiandrosterone concentrations in men and women. *J Lab Clin Med* 131:316–23.
Nashev, L. G., C. Chandsawangbhuwana, Z. Balazs, A. G. Atanasov, B. Dick, F. J. Frey, M. E. Baker, and A. Odermatt. 2007. Hexose-6-phosphate dehydrogenase modulates 11beta-hydroxysteroid dehydrogenase type 1-dependent metabolism of 7-keto- and 7beta-hydroxy-neurosteroids. *PLoS One* 2:e561.
Offner, H., G. S. Firestein, D. L. Boyle, R. Pieters, J. M. Frincke, A. Garsd, S. K. White, C. L. Reading, and D. L. Auci. 2009. An orally bioavailable synthetic analog of an active dehydroepiandrosterone metabolite reduces established disease in rodent models of rheumatoid arthritis. *J Pharmacol Exp Ther* 329:1100–9.
Offner, H., and M. Polanczyk. 2006. A potential role for estrogen in experimental autoimmune encephalomyelitis and multiple sclerosis. *Ann N Y Acad Sci* 1089:343–72.
Offner, H., A. Zamora, H. Drought, A. Matejuk, D. Auci, E. Morgan, A. Vandenbark, and C. Reading. 2002. A synthetic androstene derivative and a natural androstene metabolite inhibit relapsing-remitting EAE. *J Neuroimmunol* 130:128.

Offner, H., A. Zamora, S. Subramanian, M. Polanczyk, A. Krogstad, D. L. Auci, E. E. Morgan, and C. L. Reading. 2004. A synthetic androstene analogue inhibits collagen-induced arthritis in the mouse. *Clin Immunol* 110:181–90.

Ong, S. E., M. Schenone, A. A. Margolin, X. Li, K. Do, M. K. Doud, D. R. Mani et al. 2009. Identifying the proteins to which small-molecule probes and drugs bind in cells. *Proc Natl Acad Sci USA* 106:4617–22.

Padgett, D. A., and R. M. Loria. 1994. In vitro potentiation of lymphocyte activation by dehydroepiandrosterone, androstenediol, and androstenetriol. *J Immunol* 153:1544–52.

Peters, J. M., Y. C. Zhou, P. A. Ram, S. S. Lee, F. J. Gonzalez, and D. J. Waxman. 1996. Peroxisome proliferator-activated receptor alpha required for gene induction by dehydroepiandrosterone-3 beta-sulfate. *Mol Pharmacol* 50:67–74.

Reynolds, J. W. 1966. The identification and quantification of delta-5-androstenetriol (3-beta, 16-alpha, 17-beta-trihydroxyandrost-5-ene) isolated from the urine of premature infants. *Steroids* 8:719–27.

Roberge, C., A. C. Carpentier, M. F. Langlois, J. P. Baillargeon, J. L. Ardilouze, P. Maheux, and N. Gallo-payet. 2007. Adrenocortical dysregulation as a major player in insulin resistance and onset of obesity. *Am J Physiol Endocrinol Metab* 293:E1465–78.

Robinzon, B., K. K. Michael, S. L. Ripp, S. J. Winters, and R. A. Prough. 2003. Glucocorticoids inhibit interconversion of 7-hydroxy and 7-oxo metabolites of dehydroepiandrosterone: A role for 11beta-hydroxysteroid dehydrogenases? *Arch Biochem Biophys* 412:251–8.

Rontzsch, A., K. Thoss, P. K. Petrow, S. Henzgen, and R. Brauer. 2004. Amelioration of murine antigen-induced arthritis by dehydroepiandrosterone (DHEA). *Inflamm Res* 53:189–98.

Rovensky, J., Z. Radikova, R. Imrich, O. Greguska, M. Vigas, and L. Macho. 2004. Gonadal and adrenal steroid hormones in plasma and synovial fluid of patients with rheumatoid arthritis. *Endocr Regul* 38:143–9.

Sambrook, P. N., J. A. Eisman, G. D. Champion, and N. A. Pocock. 1988. Sex hormone status and osteoporosis in postmenopausal women with rheumatoid arthritis. *Arthritis Rheum* 31:973–8.

Schmidt, M., H. Naumann, C. Weidler, M. Schellenberg, S. Anders, and R. H. Straub. 2006. Inflammation and sex hormone metabolism. *Ann N Y Acad Sci* 1069:236–46.

Schnackenberg, C. G. 2008. 11Beta-hydroxysteroid dehydrogenase type 1 inhibitors for metabolic syndrome. *Curr Opin Investig Drugs* 9:295–300.

Schuld, A., J. Mullington, E. Friess, D. M. Hermann, C. Galanos, F. Holsboer, and T. Pollmacher. 2000. Changes in dehydroepiandrosterone (DHEA) and DHEA-sulfate plasma levels during experimental endotoxinemia in healthy volunteers. *J Clin Endocrinol Metab* 85:4624–9.

Schwartz, A. G., M. L. Lewbart, and L. L. Pashko. 1988a. Novel dehydroepiandrosterone analogues with enhanced biological activity and reduced side effects in mice and rats. *Cancer Res* 48:4817–22.

Schwartz, A. G., and L. L. Pashko. 2002. Inhibition of adjuvant-induced arthritis by 16 alpha-fluoro-5-androsten-17-one. *Mil Med* 167:60–3.

Schwartz, A. G., L. Pashko, and J. M. Whitcomb. 1986. Inhibition of tumor development by dehydroepiandrosterone and related steroids. *Toxicol Pathol* 14:357–62.

Schwartz, A. G., J. M. Whitcomb, J. W. Nyce, M. L. Lewbart, and L. L. Pashko. 1988b. Dehydroepiandrosterone and structural analogs: A new class of cancer chemopreventive agents. *Adv Cancer Res* 51:391–424.

Serhan, C. N., N. Chiang, and T. E. Van Dyke. 2008. Resolving inflammation: Dual anti-inflammatory and pro-resolution lipid mediators. *Nat Rev Immunol* 8:349–61.

Song, I. H., R. Gold, R. H. Straub, G. R. Burmester, and F. Buttgereit. 2005. New glucocorticoids on the horizon: Repress, don't activate! *J Rheumatol* 32:1199–1207.

Srivastava, R. A. 2009. Fenofibrate ameliorates diabetic and dyslipidemic profiles in KKAy mice partly via down-regulation of 11beta-HSD1, PEPCK and DGAT2. Comparison of PPARalpha, PPARgamma, and liver x receptor agonists. *Eur J Pharmacol* 607:258–63.

Storage, S. S., H. Agrawal, and D. E. Furst. 2010. Description of the efficacy and safety of three new biologics in the treatment of rheumatoid arthritis. *Korean J Intern Med* 25:1–17.

Stickney, D. R., Z. Noveljic, A. Garsd, D. A. Destiche, and J. M. Frincke. 2007. Safety and activity of the immune modulator HE2000 on the incidence of tuberculosis and other opportunistic infections in AIDS patients. *Antimicrob Agents Chemother* 51:2639–41.

Straub, R. H., P. Harle, F. Atzeni, C. Weidler, M. Cutolo, and P. Sarzi-puttini. 2005. Sex hormone concentrations in patients with rheumatoid arthritis are not normalized during 12 weeks of anti-tumor necrosis factor therapy. *J Rheumatol* 32:1253–8.

Straub, R. H., P. Harle, S. Yamana, T. Matsuda, K. Takasugi, T. Kishimoto, and N. Nishimoto. 2006. Anti-interleukin-6 receptor antibody therapy favors adrenal androgen secretion in patients with rheumatoid arthritis: A randomized, double-blind, placebo-controlled study. *Arthritis Rheum* 54:1778–85.

Turini, M. E., and R. N. Dubois. 2002. CYCLOOXYGENASE-2: A therapeutic target. *Annu Rev Med* 53:35–57.

van Vollenhoven, R. F. 2000. Dehydroepiandrosterone in systemic lupus erythematosus. *Rheum Dis Clin North Am* 26:349–62.

Wang, T., S. Villegas, Y. Huang, S. K. White, C. Ahlem, M. Lu, J. M. Olefsky et al. 2010. Amelioration of glucose intolerance by the synthetic androstene HE3286: Link to inflammatory pathways. *J Pharmacol Exp Ther* 333:70–80.

Waxman, D. J. 1996. Role of metabolism in the activation of dehydroepiandrosterone as a peroxisome proliferator. *J Endocrinol* 150(Suppl):S129–47.

Webb, S. J., T. E. Geoghegan, R. A. Prough, and K. K. Michael miller. 2006. The biological actions of dehydroepiandrosterone involves multiple receptors. *Drug Metab Rev* 38:89–116.

Wilder, R. L. 1996. Adrenal and gonadal steroid hormone deficiency in the pathogenesis of rheumatoid arthritis. *J Rheumatol Suppl* 23:10–2.

Williams, P. J., R. H. Jones, and T. W. Rademacher. 1997. Reduction in the incidence and severity of collagen-induced arthritis in DBA/1 mice, using exogenous dehydroepiandrosterone. *Arthritis Rheum* 40:907–11.

Wislowska, M., and D. Jakubicz. 2007. Preliminary evaluation in rheumatoid arthritis activity in patients treated with TNF-alpha blocker plus methotrexate versus methotrexate or leflunomide alone. *Rheumatol Int* 27:641–7.

Wolf, O. T., and C. Kirschbaum. 1999. Actions of dehydroepiandrosterone and its sulfate in the central nervous system: Effects on cognition and emotion in animals and humans. *Brain Res Brain Res Rev* 30:264–88.

Wu, L. D., H. C. Yu, Y. Xiong, and J. Feng. 2006. Effect of dehydroepiandrosterone on cartilage and synovium of knee joints with osteoarthritis in rabbits. *Rheumatol Int* 27:79–85.

Xiao, M., C. E. Inal, V. I. Parekh, C. M. Chang, and M. H. Whitnall. 2007. 5-Androstenediol promotes survival of gamma-irradiated human hematopoietic progenitors through induction of nuclear factor-kappaB activation and granulocyte colony-stimulating factor expression. *Mol Pharmacol* 72:370–9.

Yukioka, M., Y. Komatsubara, K. Yukioka, T. Toyosaki-maeda, K. Yonenobu, and T. Ochi. 2006. Adrenocorticotropic hormone and dehydroepiandrosterone sulfate levels of rheumatoid arthritis patients treated with glucocorticoids. *Mod Rheumatol* 16:30–5.

Yurchenko, V., S. Constant, E. Eisenmesser, and M. Bukrinsky. 2010. Cyclophilin-CD147 interactions: A new target for anti-inflammatory therapeutics. *Clin Exp Immunol* 160:305–17.

Zabolotny, J. M., Y. B. Kim, L. A. Welsh, E. E. Kershaw, B. G. Neel, and B. B. Kahn. 2008. Protein-tyrosine phosphatase 1B expression is induced by inflammation in vivo. *J Biol Chem* 283:14230–41.

Zhao, C., K. Dahlman-Wright, and J. A. Gustafsson. 2008. Estrogen receptor beta: An overview and update. *Nucl Recept Signal* 6:e003.

17 Asthma and DHEA

Inseon S. Choi

CONTENTS

Asthma is a heterogeneous disease. However, the majority of asthmatic patients (approximately 90%) show evidence of atopic (immunoglobulin E [IgE]-mediated) allergy, and even those with so-called intrinsic asthma exhibit increased local production of IgE in the airways (Barnes 2008). Dr. Kasperska-Zajac will discuss dehydroepiandrosterone (DHEA) in allergic diseases including asthma in a following chapter in this book, so parts of this chapter on asthma may overlap his work. Therefore, I will focus this chapter on my experimental studies on asthma.

FEMALE SEX AND ASTHMA

A recent review article by Chen et al. (2008) scrutinized the relationship between sex hormones and IgE-mediated allergic reactions. In a large cross-sectional study in which 18,659 subjects from 16 countries participated, a female predominance of asthma after puberty and a male predominance before puberty were found consistently in all 16 countries studied (de Marco et al. 2000). Although the sex reversal after puberty in the incidence of asthma seemed to be explained partly by women's smaller airway calibers, the authors suggested that sex hormones could play an important role in this shift as well. Such a change to a female predominance after puberty was also demonstrated not only in asthma but also in other allergic diseases such as allergic rhinitis and eczema in a study in which more than 500,000 patients participated (Osman et al. 2007), and also in urticaria, angioedema, or anaphylaxis in patients admitted to a hospital (Poulos et al. 2007). Among the asthma patients yielding positive responses to skin-prick tests with house dust mite extracts, males exceeded females 3:1 in the first decade, reversing to a female predominance of 1:1.5 and 1:1.6 in the fourth and fifth decades (the child-bearing years) (Wormald 1977). These findings suggest that sex hormones may play a role in the expression of atopic allergies. Actually, it has been demonstrated that skin-test reactivity to allergens (Kalogeromitros et al. 1995; Mandhane et al. 2009), airway hyperresponsiveness (AHR) to adenosine monophosphate (Tan, McFarlane, and Lipworth 1997), and exhaled nitric oxide (NO) level (Mandhane et al. 2009) in asthma change according to the levels of sex hormones during the menstrual cycle, and the associations between the markers of asthma and sex hormones are not observed among women using oral contraceptives (Mandhane et al. 2009; Tan, McFarlane, and Lipworth 1997).

FEMALE SEX HORMONES MODULATING ASTHMA

Recently, Mandhane et al. (2009) found that increases in progesterone levels were associated with increased exhaled NO levels and skin wheal size, while 17β-estradiol levels were inversely associated with them. In this study, levels of exhaled NO, 17β-estradiol, and progesterone were measured daily, and allergy skin-prick tests were performed every other day during the menstrual cycle in asthmatic women. However, other investigators (van den Berge et al. 2009) reported conflicting results; thus, the exact role of each hormone on asthma remains to be determined. Previous studies have shown that progesterone increased total IgE, interleukin (IL)-4, eosinophilic airway inflammation (Mitchell and Gershwin 2007), and histamine H1 receptors (Hamano et al. 1998a); decreased β-adrenergic receptor density (Moawad, River, and Kilpatrick 1982); suppressed histamine secretion from mast cells (Vasiadi et al. 2006); and enhanced the effect of glucocorticoid in fibroblasts (Hackney, Holbrook, and Grasso 1981). Estradiol promoted allergic sensitization and IgE production (Holt, Britten, and Sedgwick 1987) and increased mast cell mediator release (Narita et al. 2007), histamine H1 receptor numbers (Hamano et al. 1998a), and IL-4 and IL-13 production from peripheral blood mononuclear cells (PBMCs; Hamano et al. 1998b). In addition, estrogen inhibited cortisol production (Jamieson et al. 1999). However, estrogen induced FoxP3$^+$ regulatory T (Treg) cells (Tai et al. 2008) and suppressed AHR (Carey et al. 2007). Overall, suppression of estradiol and progesterone increase of the normal luteal phase from the use of oral contraceptives was associated with attenuated cyclic changes in AHR (Tan, McFarlane, and Lipworth 1997). In addition, the risk of asthma decreased with the use of oral contraceptives (Jenkins et al. 2006). Moreover, hormone replacement therapy in postmenopausal women increased the risk of developing asthma (Barr et al. 2004).

TESTOSTERONE AS A MODULATOR OF ASTHMA

It has been shown that testosterone inhibits B cell differentiation (Sthoeger, Chiorazzi, and Lahita 1988), secretion of histamine and serotonin from mast cells (Vliagoftis et al. 1992), and eosinophil adhesion to mucosal endothelial cells and eosinophil viability (Hamano et al. 1998c). Male mice have higher numbers than female mice of CD4$^+$CD25$^+$ T cells and respond to ovalbumin challenges with lower numbers of eosinophils, CD4$^+$ T cells, B cells, levels of IL-4, IL-13, interferon (IFN)-γ, and total and ovalbumin-specific IgE (Melgert et al. 2005). Additionally, medical castration in men decreased serum testosterone, CD4$^+$CD25$^+$ T cells, and CD8$^+$ T cell IFN-γ expression; these responses were prevented by testosterone replacement (Page et al. 2006).

DHEA LEVELS IN ASTHMA

The weak adrenal androgen, DHEA, is metabolized from its precursor DHEA sulfate ester (DHEAS) by DHEA sulfatase in various tissues (peripheral activation). Patients with asthma (Dunn et al. 1984) or atopic dermatitis (AD; Tabata, Tagami, and Terui 1997) have significantly lower levels of serum DHEAS, and the number of cells secreting IFN-γ is related to serum levels of DHEAS (Verthelyi and Klinman 2000). Therefore, it may be desirable to restore this hormone level in these patients. Inhaled steroid treatment in asthmatic children decreased serum DHEAS concentrations in a dose-dependent manner (Kannisto et al. 2001), so monitoring of this hormone level may be useful to follow adrenocortical functions.

DHEA AS A MODULATOR OF ASTHMA

Clinical Trials

A report published in 1987 regarding the clinical experience of DHEA therapy in asthma appears in Medline, but the content is not easily obtainable (Koó et al. 1987). Administration of the adrenal androgen DHEA in women with active systemic lupus erythematosus (SLE) significantly reduced steroid requirements, at a cost of mild acne only (Petri et al. 2002). Therefore, it is promising that DHEA can be applied soon for asthma treatment also. Actually, a phase IIa clinical trial of a synthetic DHEAS (EPI-12323) in an inhaler form showed a significant attenuation of allergen-induced late airway reaction in asthma (Shah 2004), and the inhalation formulations of DHEA were submitted for patent last year (Nyce 2009).

Preclinical Studies

Hygiene/Old Friends Hypothesis

Asthma and other allergic diseases are well known to be associated with T helper type 2 (Th2) immune reactions. The current main concept for asthma pathogenesis, the "hygiene/old friends hypothesis," suggests that T helper type 1 (Th1) and Treg cells counterbalance the Th2 cell reaction, and my colleagues and I have reported in this context that bacille Calmette-Guerin (BCG) is effective in the control of asthma (Han et al. 2010). The Th1 cytokine IFN-γ plays an important role in regulating the Th2-associated allergic airway reaction (Yoshida et al. 2002). Adaptive Treg cells secreting IL-10 or transforming growth factor (TGF)-β suppress the pathogenic immune reactions of asthma (Ryanna et al. 2009). DHEA also may share the mechanisms similar to BCG for its role in asthma.

Actions through T Helper Type 1/Regulatory T Cells and Antiproliferative Effects

DHEA increases the production of IL-2 (Daynes, Dudley, and Araneo 1990), IFN-γ (Araghi-Niknam et al. 1997), IL-10 (Cheng and Tseng 2000), and TGF-β (Wu, Chang, and Tseng 1997); decreases the serum concentrations of IgE and IL-6 (Sudo, Yu, and Kubo 2001); and suppresses allergic airway inflammation in mice (Yu et al. 1999; Yu, Liu, and Chen 2002). My colleagues and I also have shown that DHEA prevents the development of AHR and airway eosinophilia (Cui et al. 2008), and that it treats established asthmatic reactions such as AHR and airway eosinophilia in mice (Lin et al. 2009). The preventive effect of DHEA is associated with increased IFN-γ and decreased IL-4 levels (Cui et al. 2008), and the therapeutic effect with decreased IL-5 levels (Lin et al. 2009).

DHEA reduces IL-4 production in PBMCs from patients with AD (Tabata, Tagami, and Terui 1997). However, my colleagues and I have shown that DHEA decreases not only IL-5 but also IFN-γ and IL-10 production from human PBMCs in a dose-dependent manner (Choi et al. 2008). DHEA suppresses the proliferation of concanavalin A (Con A)-stimulated splenocytes in mice (Loria and Padgett 1998), reduces ATP in association with inhibition of cell proliferation and induction of apoptosis (Yang et al. 2002), and inhibits glucose-6-phosphate dehydrogenase (G6PD) and 3-hydroxy-3-methyl-glutaryl (HMG)-CoA reductase activities that are necessary for DNA synthesis (Yoshida et al. 2003). Therefore, these antiproliferative mechanisms also would work for the suppressive effect of DHEA on asthma. Because ongoing asthma in live animals may offset these antiproliferative effects of DHEA because of its proliferative milieu, DHEA-induced increases in Th1 cytokines may occur in *in vivo* studies (Araghi-Niknam et al. 1997; Cui et al. 2008) in contrast to the results of these effects in *in vitro* studies (Choi et al. 2008; Loria and Padgett 1998; Yang et al. 2002; Yoshida et al. 2003). Moreover, dihydrotestosterone, a metabolite of DHEA, decreases the production of both IL-4 and IFN-γ in splenocytes (Araneo et al. 1991).

In a previous study, my colleagues and I have shown that DHEA downregulates both IFN-γ and IL-5 production while maintaining a Th1 bias, suggesting that the balance between Th1 and Th2 is more important than the actual level of Th1 cytokines in determining the influence of DHEA on asthma (Choi et al. 2008). Because DHEA increases the production of IL-10 (Cheng and Tseng 2000), its suppressive effect may also occur through IL-10. Asthma is associated with diminished IL-10 expression in the allergic airway (Borish and Rosenwasser 2003); thus, the restoration of this cytokine using DHEA (Cheng and Tseng 2000) would be desirable.

Actions on Glucocorticoid Effects and Other Actions

DHEA decreases adenosine (Nyce 2009), tumor necrosis factor (TNF)-α (Kipper-Galperin et al. 1999), and nuclear factor-κB-dependent transcription, possibly through an antioxidant effect (Iwasaki et al. 2004), and modulates endothelial and neutrophil-adhesion molecule expression (Barkhausen et al. 2006). Glucocorticoids decrease IL-2 and increase IL-4 production from murine T cells (Daynes and Araneo 1989), increase the synthesis of IgE by IL-4-stimulated human lymphocytes (Wu et al. 1991), and DHEA antagonizes glucocorticoid-induced Th2 deviation, probably through the peroxisome proliferator–activated receptor α (Hernandez-Pando et al. 1998). DHEA inhibits DNA binding of the transcription factor activator protein (AP)-1 and airway smooth proliferation (Dashtaki et al. 1998). In addition to preventing airway remodeling, Dashtaki et al. (1998) and Kasperska-Zajac (2010) suggested that DHEA may overcome AP-1-associated glucocorticoid resistance and have a steroid-sparing effect.

Factors Affecting the Efficacy of DHEA

Gender

Because DHEA is an androgen, gender difference of DHEA efficacy on asthma would be natural. IFN-γ production in PBMCs has been reported to be positively correlated with plasma DHEA levels and inversely correlated with cortisol:DHEA ratios (Bozza et al. 2007). In a previous study, my colleagues and I have shown that men have a significantly higher level of serum DHEAS and a significantly lower cortisol:DHEAS ratio as compared with women (Choi et al. 2008). Even though increased IFN-γ:IL-5 ratios tended to occur in both genders, DHEA treatment increased the IFN-γ:IL-5 ratio in an inverse relationship to serum DHEAS levels. DHEA in male mice was less effective in suppressing asthmatic reactions than in female mice (Cui et al. 2008; Lin et al. 2009). The higher background levels of DHEA and other androgens associated with the Th1 bias may result in an adaption to DHEA; thus, greater Th1-inducing power or other modality working through a different mechanism may be necessary to overcome this problem in men.

Disease Activity and Inflammation Type

DHEA is an immunomodulating agent, and it upregulates DHEA receptor activity in activated human T lymphocytes (Okabe et al. 1995). Therefore, DHEA would likely be more effective in suppressing inflammatory/immune responses during active allergic diseases; in fact, DHEA-induced cytokine suppression was greater in subjects with relatively more severe atopy or asthma (Choi et al. 2008). Because DHEA inhibits TNF-α (Kipper-Galperin et al. 1999), it would be particularly effective in patients with severe refractory asthma, as the clinical trials with TNF-α inhibitors have been effective for such patients (Matera, Calzetta, and Cazzola 2010). Neutrophilic airway inflammation in some patients with severe asthma is intractable to glucocorticoids therapy, and EPI-12323 (Naturasone) is promising due to its ability to decrease neutrophilic inflammation (Epigenesis Pharmaceuticals).

Dose of DHEA

Supraphysiologic doses of DHEA increase IL-4 and decrease IFN-γ production in splenocyte cultures (Du et al. 2001). My colleagues and I have experienced such a phenomenon in our animal experiments (Cui et al. 2008; Lin et al. 2009), and we believe that an appropriate, not excessive, dose of DHEA should be applied for the treatment of asthma.

Combined Treatment

To minimize the virilizing side effects of DHEA even if they are mild, especially in women, combined treatment with other effective agents in reduced doses of each would be preferable. My colleagues and I have shown that treatment with an appropriate combination of DHEA and BCG has an additive effect in suppressing asthmatic reactions in mice (Cui et al. 2008; Lin et al. 2009). Other investigators (Ribeiro et al. 2007) also found a DHEA-induced enhancement in IFN-γ production during immunization of mice to tuberculosis. Considering that the most effective antiasthma drug at present is an inhaled steroid, combined treatment with an inhaled steroid and DHEA would be ideal according to the suggestions by Dashtaki et al. (1998) and Kasperska-Zajac (2010).

Safety

Although treatment with 200 mg/day of oral DHEA for 7–9 months in women with SLE was associated with mild acne—about twice the incidence as in the control group (Petri et al. 2002)—the possibility of unexpected untoward reactions by systemic administration of DHEA remained. Numerous studies on extrapulmonary diseases are still ongoing, which indicates that our knowledge of DHEA's action on other organs is still insufficient. DHEA inhibits G6PD and HMG-CoA reductase activities (Yoshida et al. 2003), and the accompanying depletion of ubiquinone may result in chronic heart failure following long-term usage (Nyce 2009).

SUMMARY

Generally, female sex hormones seem to increase Th2 cytokine production; thus, women of a childbearing age are prone to developing asthma. On the contrary, DHEA and other androgens attenuate Th2-associated asthmatic reactions through a variety of mechanisms. Naturally, there may be gender differences in the efficacy of DHEA on asthma. Overdose and systemic administration may also follow untoward reactions. Therefore, appropriate dosage, administration route (inhaled), and combination with other agents (especially inhaled steroid) in appropriate clinical settings (active, refractory, or neutrophilic asthma and patients with decreased serum DHEAS level) should be considered for each gender for ideal asthma treatment with DHEA.

REFERENCES

Araghi-Niknam, M., Z. Zhang, S. Jiang, O. Call, C. D. Eskelson, and R. R. Watson. 1997. Cytokine dysregulation and increased oxidation is prevented by dehydroepiandrosterone in mice infected with murine leukemia retrovirus. *Pro Soc Exp Biol Med* 216:386–91.

Araneo, B. A., T. Dowell, M. Diegel, and R. A. Daynes. 1991. Dihydrotestosterone exerts a depressive influence on the production of interleukin-4 (IL-4), IL-5, and γ-interferon, but not IL-2 by activated murine T cells. *Blood* 78:688–99.

Barkhausen, T., B. M. Westphal, C. Pütz, C. Krettek, and M. van Griensven. 2006. Dehydroepiandrosterone administration modulates endothelial and neutrophil adhesion molecule expression *in vitro*. *Crit Care* 10:R109.

Barnes, P. J. 2008. Asthma. In *Harrison's Principles of Internal Medicine*, 17th ed., ed. A. S. Fauci, E. Braunwald, D. L. Kasper, S. L. Hauser, D. L. Longo, J. L. Jameson, and J. Loscalzo, 1596–607. New York: McGraw-Hill.

Barr, R. G., C. C. Wentowski, F. Grodstein et al. 2004. Prospective study of postmenopausal hormone use and newly diagnosed asthma and chronic obstructive pulmonary disease. *Arch Intern Med* 164:379–86.

Borish, L., and L. J. Rosenwasser. 2003. Cytokines in allergic inflammation. In *Middleton's Allergy Principles & Practice*, 6th ed., ed. N. F. Adkinson Jr., J. W. Yunginger, W. W. Busse, B. S. Bochner, S. T. Holgate, and F. E. Simons, 135–57. Philadelphia, PA: Mosby.

Bozza, V. V., L. D'Attilio, C. V. Mahuad et al. 2007. Altered cortisol/DHEA ratio in tuberculosis patients and its relationship with abnormalities in the mycobacterial driven cytokine production by peripheral blood mononuclear cells. *Scand J Immunol* 66:97–103.

Carey, M. A., J. W. Card, J. A. Bradbury et al. 2007. Spontaneous airway hyperresponsiveness in estrogen receptor-α-deficient mice. *Am J Respir Crit Care Med* 175:126–35.

Chen, W., M. Mempel, W. Schober, H. Behrendt, and J. Ring. 2008. Gender difference, sex hormones, and immediate type hypersensitivity reactions. *Allergy* 63:1418–27.

Cheng, G. F., and J. Tseng. 2000. Regulation of murine interleukin-10 production by dehydroepiandrosterone. *J Interferon Cytokine Res* 20:471–8.

Choi, I. S., Y. Cui, Y. A. Koh, H. C. Lee, Y. B. Cho, and Y. H. Won. 2008. Effects of dehydroepiandrosterone on Th2 cytokine production in peripheral blood mononuclear cells from asthmatics. *Korean J Intern Med* 23:176–81.

Cui, Y., I. S. Choi, Y. A. Koh, X. H. Lin, Y. B. Cho, and Y. H. Won. 2008. Effects of combined BCG and DHEA treatment in preventing the development of asthma. *Immunol Invest* 37:191–202.

Dashtaki, R., A. R. Whorton, T. M. Murphy, P. Chitano, W. Reed, and T. P. Kennedy. 1998. Dehydroepiandrosterone and analogs inhibit DNA binding of AP-1 and airway smooth muscle proliferation. *J Pharmacol Exp Ther* 285:876–83.

Daynes, R. A., and B. A. Araneo. 1989. Contrasting effects of glucocorticoids on the capacity of T cells to produce the growth factors interleukin 2 and interleukin 4. *Eur J Immunol* 19:2319–25.

Daynes, R. A., D. J. Dudley, and B. A. Araneo. 1990. Regulation of murine lymphokine production *in vivo*. II. Dehydroepiandrosterone is a natural enhancer of interleukin 2 synthesis by helper T cells. *Eur J Immunol* 20:793–802.

de Marco, R., F. Locatelli, J. Sunyer, and P. Burney. 2000. Differences in incidence of reported asthma related to age in men and women. A retrospective analysis of the data of the European Respiratory Health Survey. *Am J Respir Crit Care Med* 162:68–74.

Du, C., Q. Guan, M. W. Khalil, and S. Sriram. 2001. Stimulation of Th2 response by high doses of dehydroepiandrosterone in KLH-primed splenocytes. *Exp Biol Med* 226:1051–60.

Dunn, P. J., C. B. Mahood, J. F. Speed, and D. R. Jury. 1984. Dehydroepiandrosterone sulphate concentrations in asthmatic patients: Pilot study. *N Z Med J* 97:805–8.

Epigenesis Pharmaceuticals. 2010. *Corporate Overview*. http://www.epigene.com/about/overview.html. Accessed June 10, 2010.

Hackney, J. F., N. J. Holbrook, and R. J. Grasso. 1981. Progesterone as a partial glucocorticoid agonist in L929 mouse fibroblasts: Effects on cell growth, glutamine synthetase induction and glucocorticoid receptors. *J Steroid Biochem* 14:971–7.

Hamano, N., N. Terada, K. Maesako et al. 1998a. Effect of female hormones on the production of IL-4 and IL-13 from peripheral blood mononuclear cells. *Acta Otolaryngol Suppl* 537:27–31.

Hamano, N., N. Terada, K. Maesako et al. 1998b. Expression of histamine receptors in nasal epithelial cells and endothelial cells-the effects of sex hormones. *Int Arch Allergy Immunol* 115:220–7.

Hamano, N., N. Terada, K. Maesako, T. Numata, and A. Konno. 1998c. Effect of sex hormones on eosinophilic inflammation in nasal mucosa. *Allergy Asthma Proc* 19:263–9.

Han, E. R., I. S. Choi, S. H. Eeom, and H. J. Kim. 2010. Preventive effects of mycobacteria and their culture supernatants against asthma development in BALB/c mice. *Allergy Asthma Immunol Res* 2:34–40.

Hernandez-Pando, R., M. De La Luz Streber, and H. Orozco et al. 1998. The effects of androstenediol and dehydroepiandrosterone on the course and cytokine profile of tuberculosis in BALB/c mice. *Immunology* 95:234–41.

Holt, P. G., D. Britten, and J. D. Sedgwick. 1987. Suppression of IgE responses by antigen inhalation: Studies on the role of genetic and environmental factors. *Immunology* 60:97–102.

Iwasaki, Y., M. Asai, M. Yoshida, T. Nigawara, M. Kambayashi, and N. Nakashima. 2004. Dehydroepiandrosterone-sulfate inhibits nuclear factor-κB-dependent transcription in hepatocytes, possibly through antioxidant effect. *J Clin Endocrinol Metab* 89:3449–54.

Jamieson, P. M., M. J. Nyirenda, B. R. Walker, K. E. Chapman, and J. R. Seckl. 1999. Interactions between oestradiol and glucocorticoid regulatory effects on liver-specific glucocorticoid-inducible genes: Possible evidence for a role of hepatic 11β-hydroxysteroid dehydrogenase type 1. *J Endocrinol* 160:103–9.

Jenkins, M. A., S. C. Dharmage, L. B. Flander et al. 2006. Parity and decreased use of oral contraceptives as predictors of asthma in young women. *Clin Exp Allergy* 36:609–13.

Kalogeromitros, D., A. Katsarou, M. Armenaka, D. Rigopoulos, M. Zapanti, and I. Stratigos. 1995. Influence of the menstrual cycle on skin-prick test reactions to histamine, morphine and allergen. *Clin Exp Allergy* 25:461–56.

Kannisto, S., M. Korppi, K. Remes, and R. Voutilainen. 2001. Serum dehydroepiandrosterone sulfate concentration as an indicator of adrenocortical suppression in asthmatic children treated with inhaled steroids. *J Clin Endocrinol Metab* 86:4908–12.

Kasperska-Zajac, A. 2010. Asthma and dehydroepiandrosterone (DHEA): Facts and hypotheses. *Inflammation* 33(5):320–4.

Kipper-Galperin, M., R. Galilly, H. D. Danenberg, and T. Brenner. 1999. Dehydroepiandrosterone selectively inhibits production of tumor necrosis factor α and interleukin-6 in astrocytes. *Int J Dev Neurosci* 17:765–75.

Koó, E., G. K. Fehér, T. Fehér, and G. Füst. 1987. Experience with dehydroepiandrosterone therapy in intrinsic steroid-dependent bronchial asthma. [In Hungarian.] *Orv Hetil* 128:1995–7.

Lin, X. H., I. S. Choi, Y. A. Koh, and Y. Cui. 2009. Effects of combined Bacille Calmette-Guérin and dehydroepiandrosterone treatment on established asthma in mice. *Exp Lung Res* 35:250–61.

Loria, R. M., and D. A. Padgett. 1998. Control of the immune response by DHEA and its metabolites. *Rinsho Byori* 46:505–17.

Mandhane, P. J., S. E. Hanna, M. D. Inman et al. 2009. Changes in exhaled nitric oxide related to estrogen and progesterone during the menstrual cycle. *Chest* 136:1301–7.

Matera, M. G., L. Calzetta, and M. Cazzola. 2010. TNF-α inhibitors in asthma and COPD: We must not throw the baby out with the bath water. *Pulm Pharmacol Ther* 23:121–8.

Melgert, B. N., D. S. Postma, I. Kuipers et al. 2005. Female mice are more susceptible to the development of allergic airway inflammation than male mice. *Clin Exp Allergy* 35:1496–503.

Mitchell, V. L., and L. J. Gershwin. 2007. Progesterone and environmental tobacco smoke act synergistically to exacerbate the development of allergic asthma in a mouse model. *Clin Exp Allergy* 37:276–86.

Moawad, A. H., L. P. River, and S. J. Kilpatrick. 1982. The effect of estrogen and progesterone on beta-adrenergic receptor activity in rabbit lung tissue. *Am J Obstet Gynecol* 144:608–13.

Narita, S., R. M. Goldblum, C. S. Watson et al. 2007. Environmental estrogens induce mast cell degranulation and enhance IgE-mediated release of allergic mediators. *Environ Health Perspect* 115:48–52.

Nyce, J. W. 2009. Patent application title: Liquid inhalation formulations of dehydroepiandrosterone derivatives. http://www.faqs.org/patents/app/20090053143. Accessed June 9, 2010.

Okabe, T., M. Haji, R. Takayanagi et al. 1995. Up-regulation of high-affinity dehydroepiandrosterone binding activity by dehydroepiandrosterone in activated human T lymphocytes. *J Clin Endocrinol Metab* 80:2993–6.

Osman, M., A. L. Hansell, C. R. Simpson, J. Hollowell, and P. J. Helms. 2007. Gender-specific presentations for asthma, allergic rhinitis and eczema in primary care. *Prim Care Respir J* 16:28–35.

Page, S. T., S. R. Plymate, W. J. Bremner et al. 2006. Effect of medical castration on $CD4^+$ $CD25^+$ T cells, $CD8^+$ T cell IFN-gamma expression, and NK cells: A physiological role for testosterone and/or its metabolites. *Am J Physiol Endocrinol Metab* 290:E856–63.

Petri, M. A., R. G. Lahita, R. F. Van Vollenhoven et al. 2002. Effects of prasterone on corticosteroid requirements of women with systemic lupus erythematosus: A double-blind, randomized, placebo-controlled trial. *Arthritis Rheum* 46:1820–9.

Poulos, L. M., A. M. Waters, P. K. Correll, R. H. Loblay, and G. B. Marks. 2007. Trends in hospitalizations for anaphylaxis, angioedema, and urticaria in Australia, 1993-1994 to 2004-2005. *J Allergy Clin Immunol* 120:878–84.

Ribeiro, F., R. P. Lopes, C. P. Nunes, F. Maito, C. Bonorino, and M. E. Bauer. 2007. Dehydroepiandrosterone sulphate enhances IgG and interferon-gamma production during immunization to tuberculosis in young but not aged mice. *Biogerontology* 8:209–20.

Ryanna, K., V. Stratigou, N. Safinia, and C. Hawrylowicz. 2009. Regulatory T cells in bronchial asthma. *Allergy* 64:335–47.

Shah, S. 2004. European respiratory society - 14th Annual Congress. Drug highlights. *IDrugs* 7:914–6.

Sthoeger, Z. M., N. Chiorazzi, and R. G. Lahita. 1988. Regulation of the immune response by sex hormones. I. In vitro effects of estradiol and testosterone on pokeweed mitogen-induced human B cell differentiation. *J Immunol* 141:91–8.

Sudo, N., X. N. Yu, and C. Kubo. 2001. Dehydroepiandrosterone attenuates the spontaneous elevation of serum IgE level in NC/Nga mice. *Immunol Lett* 79:177–9.

Tabata, N., H. Tagami, and T. Terui. 1997. Dehydroepiandrosterone may be one of the regulators of cytokine production in atopic dermatitis. *Arch Dermatol Res* 289:410–4.

Tai, P., J. Wang, H. Jin et al. 2008. Induction of regulatory T cells by physiological level estrogen. *J Cell Physiol* 214:456–64.

Tan, K. S., L. C. McFarlane, and B. J. Lipworth. 1997. Modulation of airway reactivity and peak flow variability in asthmatics receiving the oral contraceptive pill. *Am J Respir Crit Care Med* 155:1273–7.

van den Berge, M., H. I. Heijink, A. J. van Oosterhout, and D. S. Postma. 2009. The role of female sex hormones in the development and severity of allergic and non-allergic asthma. *Clin Exp Allergy* 39:1477–81.

Vasiadi, M., D. Kempuraj, W. Boucher, D. Kalogeromitros, and T. C. Theoharides. 2006. Progesterone inhibits mast cell secretion. *Int J Immunopathol Pharmacol* 19:787–94.

Verthelyi, D., and D. M. Klinman. 2000. Sex hormone levels correlate with the activity of cytokine-secreting cells *in vivo*. *Immunology* 100:384–90.

Vliagoftis, H., V. Dimitriadou, W. Boucher et al. 1992. Estradiol augments while tamoxifen inhibits rat mast cell secretion. *Int Arch Allergy Immunol* 98:398–409.

Wormald, P. J. 1977. Age-sex incidence in symptomatic allergies: An excess of females in the child-bearing years. *J Hyg (Lond)* 79:39–42.

Wu, M. F., H. L. Chang, and J. Tseng. 1997. Dehydroepiandrosterone induces the transforming growth factor-beta production by murine macrophages. *Int J Tissue React* 19:141–8.

Wu, C. Y., M. Sarfati, C. Heusser et al. 1991. Glucocorticoids increase the synthesis of immunoglobulin E by interleukin 4-stimulated human lymphocytes. *J Clin Invest* 87:870–977.

Yang, N. C., K. C. Jeng, W. M. Ho, and M. L. Hu. 2002. ATP depletion is an important factor in DHEA-induced growth inhibition and apoptosis in BV-2 cells. *Life Sci* 70:1979–88.

Yoshida, S., A. Honda, Y. Matsuzaki et al. 2003. Anti-proliferative action of endogenous dehydroepiandrosterone metabolites on human cancer cell lines. *Steroids* 68:73–83.

Yoshida, M., R. Leigh, K. Matsumoto et al. 2002. Effect of interferon-γ on allergic airway responses in interferon-γ-deficient mice. *Am J Respir Crit Care Med* 166:451–6.

Yu, C. K., Y. H. Liu, and C. L. Chen. 2002. Dehydroepiandrosterone attenuates allergic airway inflammation in *Dermatophagoides farinae*-sensitized mice. *J Microbiol Immunol Infect* 35:199–202.

Yu, C. K., B. C. Yang, H. Y. Lei et al. 1999. Attenuation of house dust mite *Dermatophagoides farinae*-induced airway allergic responses in mice by dehydroepiandrosterone is correlated with down-regulation of Th2 response. *Clin Exp Allergy* 29:414–22.

18 DHEA in Allergy, Asthma, and Urticaria/Angioedema

Alicja Kasperska-Zajac

CONTENTS

INTRODUCTION

Dehydroepiandrosterone (DHEA) is a weak androgenic hormone largely found in the circulation in its sulfated form of the dehydroepiandrosterone-sulfate (DHEAS). DHEAS is converted to the biologically active form—DHEA—by sulfatase, present in a variety of cells. Apart from the predominant secretion by the adrenal gland, DHEAS is synthesized by the gonadal tissue as well as cells of the central nervous system (CNS). Although DHEA has been extensively studied over the last decades, its physiological significance and possible role in human diseases is still unclear (Kroboth et al. 1999; Mellon 2007). Interestingly, it has been suggested that DHEA may act as a "buffer hormone" to preserve the homeostatic balance (Regelson, Loria, and Kalimi 1988).

In vivo and *in vitro* studies revealed its multifunctional activity, including the immunomodulating properties (Casson et al. 1993; Kroboth et al. 1999). DHEA may modulate both the cellular and humoral immune responses in animals and humans (Casson et al. 1993). Although DHEA exerts different effects, the mechanism of its performance remains unclear. So far, no nuclear steroid receptor has been found for DHEA. DHEA may exert its action either indirectly, following conversion into some more potent sex steroids (androgens and estrogens), or directly, via interaction with the neurotransmitter receptors in the CNS (as a neurosteroid) or as some high-affinity binding sites in the peripheral target tissues (Kroboth et al. 1999; Mellon 2007). Functionally active high-affinity binding sites for DHEA have been described in murine (Meikle et al. 1992) and human (Okabe et al. 1995) T cells.

The ability of DHEA to regulate T cell proliferation and differentiation as well as cytokine production suggests its potential role in the immune-inflammatory diseases (Daynes, Dudley, and Araneo 1990; Suzuki et al. 1991; Araghi-Niknam et al. 1997; Kroboth et al. 1999), including allergy and asthma. Allergy and asthma are common immune-mediated disorders characterized by deficiency of regulatory T cells (Treg) and $CD4^+$ T helper (Th) cells type 2 (Th2)-driven inflammation (Akbari et al. 2003; Umetsu and DeKruyff 2006). In animals, it has been demonstrated that DHEA is able to attenuate allergic reactions in airways (Yu et al. 1999; Yu, Liu, and Chen 2002) and skin (Sudo, Yu, and Kubo 2001), downregulating the Th2 response.

DHEA IN ALLERGIC RESPONSE AND ASTHMA

DHEA and the Switch from T Helper Type 2 to T Helper Type 1 Pattern

Allergic disorders and asthma are characterized by a shift in Th1/Th2 balance toward a Th2-mediated immunity in genetically susceptible individuals (Akbari et al. 2003; Umetsu and DeKruyff 2006).

Following activation by antigens, $CD4^+$ T helper cells differentiate into two functionally distinct subsets, Th1 and Th2, which produce distinct profiles of cytokines and regulate different immune responses. Through secretion of interferon-γ (IFN-γ) and interleukin-2 (IL-2), Th1 cells are involved in the delayed type hypersensitivity reactions and inhibit Th2 activity. On the contrary, due to secretion of IL-4, IL-5, IL-9, and IL-10, Th2 cells are essential for the development of allergic responses, in which (1) IL-4 and IL-13 are potent in activating B cells to secrete immunoglobulin E (IgE), (2) IL-5 is crucial for differentiation of eosinophils from bone marrow precursor cells and also prolongs the eosinophil survival, (3) IL-4, IL-9, and IL-10 enhance mast cell growth, and (4) IL-9 and IL-13 directly enhance the mucus hypersecretion and airway hyperreactivity (AHR; Umetsu and DeKruyff 1997, 2006).

DHEA has been proven to regulate T cell proliferation and differentiation into two functionally distinctive subsets: Th1 and Th2 (Du, Guan, Khalil, and Sriram 2001). The accumulated data based on either *in vivo* administration of DHEA to mice, or direct *in vitro* exposure of murine and human T cells, suggest that by the increased IL-2 and IFN-γ production, DHEA induces a Th1-type immune response (Daynes, Dudley, and Araneo 1990; Suzuki et al. 1991; Araghi-Niknam et al. 1997; Zhang et al. 1999). The enhancing effect of DHEA was detected at the level of IL-2 mRNA, suggesting that DHEA may act as a transcriptional enhancer of the IL-2 gene in $CD4^+$ T cells (Suzuki et al. 1991). However, production of IL-2 and IFN-γ might also be suppressed by DHEA under different experimental conditions *in vitro* (Moynihan et al. 1998). The studies by Daynes's group (Araneo et al. 1991) also suggest that DHEA might suppress Th2-type immune responses by downregulating the production of IL-4, IL-5, and IL-6. In general, favoring the development of Th1 response, DHEA may play a role in regulating the Th1/Th2 immune response. Stimulating Th1 cytokine production and inhibiting Th2 cytokine synthesis, DHEA counteracts a form of the antigen-specific immune suppression. However, scarce data are available on the effect of DHEA on such a response in allergy and asthma (Dillon 2005; Kasperska-Zajac, Brzoza, and Rogala 2008).

DHEA May Prevent and Attenuate Th2-Mediated Immune Response in the Airways

Asthma is a complex and chronic inflammatory disorder of the airways, involving many cells driven by T cell activation. Th2 cells and their cytokines are thought to play a role in the pathophysiology of allergic as well as nonallergic asthma (Humbert et al. 1996).

It has been reported that DHEA may prevent or attenuate allergic inflammatory responses (Yu et al. 1999; Yu, Liu, and Chen 2002) as well as reduce blood eosinophilia and serum concentrations of IgE in the animal model of asthma (Yu et al. 1999; Yu, Liu, and Chen 2002). DHEA was capable of decreasing both Th2 cytokine (IL-5) and Th1 cytokine (IFN-γ) production in cultured peripheral blood mononuclear cells (PBMCs) from asthmatic patients and skewed the Th1/Th2 balance toward Th1, suggesting that DHEA supplementation may bring some therapeutic benefits for the asthmatic patients (Choi et al. 2008).

Interestingly, Yu et al. (1999) and Yu, Liu, and Chen (2002) reported that DHEA administration may be effective in both preventing allergen sensitization and suppressing the progression of allergic airway inflammation in a *Dermatophagoides farinae* (*D. farine*)-induced asthma model. Supplementation of mice diet with DHEA prior to and during sensitization to *D. farine* attenuated the mite-induced airway eosinophilic inflammation, which was associated with the decrease in IL-4, IL-5, IFN-γ, and total IgE antibody concentrations in bronchoalveolar lavage (BAL) fluids or serum (Yu et al. 1999). Similar effects were observed when *D. farine*-sensitized mice were fed with DHEA after the onset of airway inflammation, suggesting a therapeutic effect of DHEA also on the established allergic inflammation (Yu, Liu, and Chen 2002). In the DHEA-fed mice, airway inflammation, blood eosinophilia, and serum concentrations of IL-4, IL-5, and IFN-γ were significantly reduced, which was not the case for total IgE antibody concentration in serum and BAL (Yu, Liu, and Chen 2002). Taken together, DHEA may prevent and attenuate allergic inflammatory response in the airway, but the mechanisms underlying this process are unclear. On one hand, DHEA might modulate allergen sensitization and allergic inflammation through suppression of Th2 response; yet, on the other hand, downregulation of Th1 (IFN-γ) cytokine production was observed, suggesting that its effect might be mediated by a mechanism other than modification of Th1/Th2 balance (Yu et al. 1999; Yu, Liu, and Chen 2002). Alternatively, DHEA may downregulate both Th1 (IFN-γ) and Th2 (IL-5) cytokines production while maintaining a Th1 bias, which would suggest that the Th1/Th2 balance is more important than the actual level of Th1 cytokines when determining the influence of DHEA on asthma (Humbert et al. 1996). However, it has not yet been recognized whether DHEA would regulate the same cytokine balance in humans.

DHEA and Other Cytokines Involved in Allergy and Asthma

As mentioned in the section "DHEA in Allergic Response and Asthma," cytokine imbalance is involved in the pathogenesis of asthma and allergy. Complex interactions exist between proinflammatory cytokines and DHEA. On one hand, DHEA is a potent regulator of cytokines production, while on the other hand, cytokines released during the course of an immune response appear to play a role in regulation of DHEA secretion. For example, transforming growth factor-β (TGF-β) was found to inhibit basal and adrenocorticotropic hormone (ACTH)-stimulated DHEA production by cultured human adrenal cells, suggesting that TGF-β may play an autocrine/paracrine role in the human adrenal (Stankovic, Dion, and Parker 1994).

Tumor necrosis factor-α (TNF-α) is a proinflammatory cytokine, abundant in asthmatic airways and playing an important role in the disease. As DHEA may reduce TNF-α production (Danenberg et al. 1992; Di Santo et al. 1996; Araghi-Niknam et al. 1997), it is likely that this hormone might cause a protective effect on inflammation in asthma. Some other proinflammatory cytokines, including IL-6 and macrophage inhibitory factor (MIF) are also downregulated by DHEA, except for the vascular endothelial growth factor (VEGF) or the fibroblast growth factor-5 (FGF-5; Harding et al. 2006).

Regulatory T cells in Allergy and Asthma—A Role for DHEA

Tregs, including natural Tregs ($CD4^+CD25^+FoxP3^+$) and antigen-specific adaptive Tregs, are key players in suppression of Th2-biased responses, whereas impaired expansion of Tregs is hypothesized to lead to the development of allergy and asthma (Akbari et al. 2003). It is known that allergen-specific immunotherapy (SIT) that enhances the development of allergen-specific Tregs may be effective in limiting allergic inflammation as well as providing specific and long-lasting control of such disorders (Umetsu and DeKruyff 2006; Elkord 2008). Interestingly, it has been demonstrated that DHEA is able to upregulate the expression of IL-10 (Del Prete et al. 1993) and TGF-β (Wu, Chang, and Tseng 1997). Auci, Kaler, and Subramanian (2007) observed that treatment with HE3286, a synthetic analog of an active DHEA metabolite, is associated with the increasing number of Tregs, suggesting that androstene hormones may be natural regulators of Tregs. In addition, oral DHEA replacement restored normal levels of Tregs and led to increased forkhead P3 (FoxP3) expression (Coles et al. 2005). Taken together, it may be speculated that DHEA or its analogs, alone or combined with SIT, can induce immune tolerance to allergens and reverse the established Th2 responses, depending on the increasing number and functions of Tregs. Such combination may improve both the safety of SIT, by reducing IgE-mediated reactions associated with immunotherapy, and the rapidity by which this therapy induces Tregs.

DHEA—An Inhibitor of Airway Smooth Muscle Proliferation

Airway remodeling is a key aspect of asthma, including the structural changes, such as increased airway smooth muscle mass, subepithelial fibrosis, neovascularization, contributing to fixed airflow obstruction and clinical severity (Munakata 2006). The mechanisms of such processes in asthma are poorly understood. Apart from its immunomodulatory properties, DHEA has also been described as reducing proliferation of rat airway smooth muscle cells and preventing remodeling, probably due to its ability to inhibit the DNA binding of activator protein-1 (AP-1; Dashtaki et al. 1998). So far, however, no data has indicated that airway remodeling in asthmatic patients may be regulated by DHEA.

DHEA as an "Antiglucocorticoid" Hormone

DHEA is widely considered a physiologic antagonist of certain glucocorticoid (GC) activities, including the immunoregulatory effect (Blauer et al. 1991; Clerici et al. 1997). GCs may mediate the Th2 shift by suppressing the production of antigen-presenting cells (APCs) and Th1-cytokines as well as directly upregulating the Th2 response (Elenkov 2004), which may appear unfavorable upon GCs therapy of allergic diseases. It has been demonstrated that corticosterone-induced depressions in IL-2 synthesis by T cells could be reversed *in vitro* and *in vivo* by concomitant treatment with DHEA (Daynes, Dudley, and Araneo 1990). In addition, DHEA was able to overcome the suppressive effect of dexamethasone upon T and B lymphocytes functions *in vivo*, but not *in vitro* (Blauer et al. 1991). Importantly, DHEA did not hamper the immunosuppressive influence of GCs on the production of proinflammatory cytokines (Padgett and Loria 1998). Therefore, it may be speculated that a combination of GCs and DHEA or its analogs might have a beneficial effect in the therapy of allergic diseases, at least partially by counteracting GCs-mediated pro-Th2 immune response.

DHEA—A Steroid-Sparing Agent

There are several therapeutic strategies for management of GC-dependent or GC-resistant asthma, based on the use of alternative anti-inflammatory (steroid-sparing) treatments or reversing the molecular mechanisms of GCs resistance. Asthma can be controlled by inhaling GCs, which is the first-line treatment for patients suffering from mild to severe asthma. Approximately 5%–10% of asthmatic patients require the maximum inhaled dose of GCs to control their symptoms. In case of refractory asthma or one poorly responsive to inhaled GCs therapy, patients may need regular treatment with oral GCs, the so-called GC-dependent asthma. Few patients show no clinical improvement after the treatment, even with a high dose of oral GCs, pointing to the termed GC-resistant asthma (Barnes and Adcock 2009).

Because of the anti-inflammatory and immunomodulatory properties, free of any catabolic effect, including the deleterious skeletal effects, observed at GCs use, DHEA might be useful as a steroid-sparing agent in asthma therapy. It addition, it has been hypothesized that DHEA (and its analogs) might be useful in reversing the acquired failure to respond to GCs (Dashtaki et al. 1998).

Several mechanisms have been proposed to account for the GCs resistance, including cytokine-induced increased expression of transcription factors, such as AP-1. These would reduce the number of activated GC receptors (GRs) within the nucleus available for binding to specific sites on DNA—termed GC response elements (GREs; Lane et al. 1998; Adcock et al. 1995). The mechanism of beneficial DHEA effect in this process might result from binding AP-1 by DHEA, thereby enabling free activated GRs to attach to GREs. DHEA also inhibits activation and translocation of the transcription factor nuclear factor kappa B (NF-κB; Du, Khalil, and Sriram 2001). In this manner, DHEA might overcome the relative GC resistance of the inflammatory state in different disorders (Dashtaki et al. 1998), including asthma. DHEA, as a steroid-sparing agent with an added anti-inflammatory effect and reversing GC resistance, might allow for reduction in the high-GCs dose needed to control asthma symptoms. The steroid-sparing effect exerted by DHEA may be due to its ability to affect the function of the immune system and transcription factors (Harding et al. 2006), and it seems not to be associated with any alteration of GCs pharmacokinetics (Meno-Tetang et al. 1999). In addition, the combined DHEA and GCs therapy applied in severe and difficult-to-treat asthma might allow for reduced GCs dosage, therefore avoiding the undesirable side effects.

Lower Serum DHEA Concentration in Asthmatic Patients

Although many chronic inflammatory diseases are associated with lower serum DHEAS concentrations, information regarding their regulation and functional significance in the disorders is not fully understood (Kasperska-Zajac, Brzoza, and Rogala 2008). Few studies have demonstrated some decrease in serum DHEAS concentrations in asthmatic patients (Feher, Koo, and Feher 1983; Dunn et al. 1984). Dunn et al. (1984) have reported decreased serum DHEAS concentrations in 40% of asthmatic patients admitted for treatment of severe bronchospasm. The group included patients who had earlier received the oral treatment (71.4%), those who inhaled GCs (34.8%), and patients without a history of such therapy (20.7%), suggesting that DHEA concentrations may appear decreased also in patients on non-GCs therapy due to asthma (Dunn et al. 1984). Similar observations were made during the stable phase of the disease. Feher, Koo, and Feher (1983) observed low serum DHEAS concentrations and low urinary excretion of metabolites in female asthma patients remaining without GCs therapy for at least 3 months. In addition, significantly lower serum DHEAS concentrations were found in postmenopausal asthmatic women, as compared with the postmenopausal, nonasthmatic women, regardless of any earlier medication (Weinstein et al. 1996). Together, these suggest that part of the patients suffering from bronchial asthma show lower concentration of the hormone due not only to the inhaled or oral GCs but also to other factors. So far, neither the mechanisms nor the long-term consequences of such changes in the DHEA milieu of asthmatic patients have been recognized. Similar to other diseases, stress, infection, inflammation, or a chronic course of a disease might play a certain role here. Low-circulating concentration of DHEA has been postulated as a concomitant of chronic illness, infection, and stress (Semple, Gray, and Beastall 1987; Straub, Schölmerich, and Zietz 2000). Another possible explanation of the decreased DHEAS concentrations in asthma is the anomalous sympathic control of adrenal androgen production, which might be proved by test results of Weinstein et al. (1996), which pointed to a 42% increase in mean serum DHEAS concentration in patients with asthma, following adrenergic stimulation by β-adrenegic agonist (albuterol). Interestingly, it has been suggested that lower concentrations of circulating adrenal androgens in asthmatics may point to inadequate hormone supply of the target organ (Feher, Koo, and Feher 1983). However, DHEAS concentration in serum may not reflect the actual concentrations of DHEA in the target tissue, which could appear much higher than the values observed in the circulation because of conversion to DHEAS by sulfatase (Du, Guan, Khalil, and Sriram 2001).

TABLE 18.1
Possible Significance of DHEAS and Its Analogs in Allergy and Asthma

DHEA Effects
Reduction of immune dysfunction by overcoming cytokine dysregulation through modification of Th1/Th2 balance toward Th1 immunity and by the increase in the number and functions of Tregs
Prevention or attenuation of allergic inflammatory airway response
Steroid-sparing effect
Prevention of airway smooth muscle proliferation and remodeling
Reversal of the acquired glucocorticoid resistance

As for now, the only practical aspect of evaluating serum DHEAS concentration is a possible application as a screening test to detect adrenocortical function suppression in asthmatic children treated with medium to high doses of inhaled GCs (Dorsey et al. 2006).

Considering, however, that DHEA concentration may be influenced by a variety of factors, it should be emphasized that basal serum DHEAS concentrations may not be accurate enough to reflect the adrenal suppression or to predict the cortisol response to provocative testing. Further research is therefore necessary to precisely define the hypothalamic-pituitary-adrenal axis (HPA) function in asthmatic patients (Littley et al. 1990; Dorsey et al. 2006).

General Remarks

Keeping in mind the series of experiments performed, it seems that DHEA tends to increase Th1 (Daynes, Dudley, and Araneo 1990; Suzuki et al. 1991; Araghi-Niknam et al. 1997; Zhang et al. 1999) and to decrease Th2 cytokine production (Araneo et al. 1991). Observations using the experimental model of asthma make us believe that, much as in animals, some of the immunopathological changes associated with allergic response may either be prevented or reversed by oral administration of DHEA (Yu et al. 1999; Yu, Liu, and Chen 2002). It is known, however, that information obtained from animal models must be very cautiously extrapolated to human asthma and allergy. Apart from this, some of the contradictory findings brought by the studies carried out may be due to methodological differences, including *in vivo* versus *in vitro* approach, the species studied, or the dose of DHEAS used. In addition, DHEAS actions may depend on the cytokine milieu, the costimulatory signals, antigen concentration, as well as other factors—for example, stress. At the present stage, it should be concluded that the role of DHEAS in the cascade of processes associated with allergy and asthma is a complex phenomenon, which remains poorly recognized. The possible significance of DHEAS and its analogs in such disorders is presented in Table 18.1.

DHEA IN ATOPIC DERMATITIS

Atopic dermatitis (AD) is a chronic relapsing inflammatory skin disease often associated with the so-called atopic triad disorders, such as asthma and allergic rhinitis. Some immunological abnormalities found in AD patients are considered a result of the predominant production of Th2 cytokines. Information regarding the role of DHEA in AD patients appears scarce and inconsistent. Since in an animal model of human AD, DHEA profoundly suppressed the spontaneous elevation of serum IgE, it has been suggested that this hormone promotes a shift in Thl/Th2 balance toward Thl-dominant immunity (Danenberg et al. 1992). Tabata, Tagami, and Terui (1997) showed significantly lower serum DHEA concentration in male patients with mild-to-moderate AD, as compared with the healthy subjects. However, Ebata et al. (1996) demonstrated that in female and male patients suffering from the disease, serum DHEAS concentrations did not differ from the controls.

Similarly, Kasperska-Zajac, Brzoza, and Rogala (2007) failed to detect any significant changes in DHEAS serum concentrations between female patients suffering from severe AD as compared with the healthy, nonatopic women.

Tabata, Tagami, and Terui (1997) suggested that DHEA may be an endogenous immunomodulator involved in the pathogenesis of AD where controlling of the Th1/Th2 balance and lower DHEA would affect the shift from Th1- to Th2-dominant state. However, no significant correlation between DHEA concentration and the disease severity or serum IgE concentration has been found (Tabata, Tagami, and Terui 1997). Also Kasperska-Zajac, Brzoza, and Rogala (2007) did not observe any decrease in DHEAS serum concentration in AD patients with severe skin lesions. Thus, lower DHEAS concentration may be only a secondary phenomenon, appearing as a response to different stimuli, including stress, and may not contribute in any way to pathogenesis of the disease.

POTENTIAL USE OF DHEA IN THE THERAPY OF ALLERGIC DISEASES AND ASTHMA

A number of strategies aimed at modifying the cytokine imbalance are currently being applied to the treatment of allergy and asthma. Among these, a therapeutic effect of DHEA on the disorders has been suggested. Considering the beneficial immunomodulatory and anti-inflammatory properties of DHEA, this hormone might represent a reasonable therapeutic option to reverse the cytokine imbalance in allergy and asthma. In addition, immune side effects of GCs might be reversed by the coincident *in vivo* treatment with DHEA. This hormone also might be useful as a steroid-sparing agent in therapy of AD and asthma. However, at present, there is no scientific evidence to recommend DHEA replacement in these disorders. Importantly, it has been shown that at appropriate doses, DHEA or its derivative androstenediol promote Th1 cytokine pattern, while too low (Hernandez-Pando et al. 1998) or too high (supraphysiologic) levels (Du, Guan, Khalil, and Sriram 2001) may permit a switch to Th2. The supraphysiologic level of DHEA favored Th2 immune response *in vitro* by inhibition of IL-12 production from APCs and/or stimulation of Th2 proliferation during the interactions of T cells with APCs (Du, Guan, Khalil, and Sriram 2001). These data challenge the hypothesis that DHEA may differentially regulate Th1-like versus Th2-like cytokine production dependent on the dose used. Any future investigations should, therefore, examine the doses of DHEA optimal to asthmatic and allergic patients who might benefit from such therapy.

Since DHEAS has both low potency and androgenic or estrogenic side effects, resulting from metabolism to sex steroids, the inconveniences might be reduced by aerosolizing the drug into the airways or by oral administration of bioavailable synthetic analog, including potent 16α-fluorinated analogs of DHEA devoid of androgenic or estrogenic activity in animals (Dashtaki et al. 1998).

Interestingly, skin treatment with DHEA enhanced IL-2 production by activated T cells obtained from the lymph nodes receiving drainage from the steroid treatment sites indicate that DHEA after topical skin application may exert immunomodulatory activity (Araneo et al. 1991). Thus, DHEA or its analogs might provide either an alternative to GCs or an adjunctive in the therapy of asthma and allergic diseases.

ANY ROLE FOR DHEA IN URTICARIA/ANGIOEDEMA?

Interestingly, it has been demonstrated that DHEA inhibits both spontaneous and immune complex-induced complement activation, interfering probably with the internal activation of C1 (Hidvégi et al. 1984). Therefore, it may be suggested that DHEA plays a role in the diseases associated with complement activation or the effect of anaphylatoxins (C3a, C5a), including serum sickness (a type III hypersensitivity reaction mediated by immune complex deposition), angioedema caused by C1 esterase inhibitor (C1-INH) deficiency, or autoimmune chronic urticaria (CU).

Angioedema due to Inhibitor of the C1 Esterase Inhibitor Deficiency

Recurrent angioedema associated with C1-INH deficiency is seen in hereditary and acquired form, with the latter resulting from the effect of autoantibodies against this protein or concomitant diseases, which are responsible for accelerated catabolism of C1-INH. Such disorder is characterized by recurrent episodes of increased vascular permeability in different body sites, including the gut and the upper airways (Cugno et al. 2009). Interestingly, it has been demonstrated that DHEA upregulates the gene expression and secretion of C1-INH from different cell lines (Falus et al. 1990). Moreover, serum DHEA concentration in hereditary angioedema patients is markedly lower than that in healthy subjects (Thon et al. 2007). Despite some encouraging results pointing to the DHEAS therapy as potentially effective to control the hereditary angioedema attacks (Koo et al. 1983; Hidvégi et al. 1984) and to moderately increase the serum concentration of C1-INH (Koo et al. 1983), no such form of therapy has found extensive use in treatment of the disease. However, it is known that another medicine—danazol, successfully used in treatment of hereditary angioedema—is capable of increasing serum concentration of DHEAS (Murakami et al. 1993).

Lower DHEA Serum Concentration in Chronic Urticaria

CU is an extremely distressing disorder, reducing the quality of life and affecting more than 2% of the population. Similarly to allergic diseases, mast cells also play the major role in the pathogenesis, and histamine is the key mediator responsible for increased vascular permeability. This disorder is characterized by itchy hives, sometimes accompanied by angioedema, etiology of which is difficult to determine. Allergic reactions very rarely cause CU/angioedema as atopy is not frequent in this group of patients. Nevertheless, many patients claim that the disease they suffer from is caused by some sort of allergen. About 40%–50% of patients are considered as autoimmune cases due to histamine-releasing autoantibodies, while the remaining ones usually show chronic idiopathic urticaria (Maurer and Grabbe 2008; Kaplan and Greaves 2009). It seems that, in some cases, endocrinological and neurological alterations may play a role in the pathogenesis of this disease. In addition, compared to men, women are twice as vulnerable to develop in CU while the sex hormones, including androgens, might be related to such disparity (Kasperska-Zajac, Brzoza, and Rogala 2008a). Lower serum DHEA concentration has been observed in symptomatic urticaria patients and recovered the value observed in the healthy subjects during remission of the urticarial symptoms, suggesting that this hormone is involved in the pathomechanism of the skin lesions (Kasperska-Zajac, Brzoza, and Rogala 2006; Kasperska-Zajac, Brzoza, and Rogala 2008b). There are several hypothetical explanations of this phenomenon. On one hand, DHEAS deficiency might facilitate or induce complement activation involved in pathogenesis of the disease (Kasperska-Zajac, Brzoza, and Rogala 2006a); yet, on the other hand, it has been proposed that DHEAS is able to increase the vascular permeability through histamine release mediated by $G_{q/11}$-protein coupled neurosteroid receptor in mast cells (Ueda et al. 2001; Uchida et al. 2003; Mizota et al. 2005). Therefore, the observed DHEA reduction could have been a defense response aimed at reduction of mast cell activation and leading, therefore, to progression of the urticarial reaction. A different hypothesis could also be considered, namely, that reduced DHEA may only reflect a chronic illness or mental distress, rather than effecting directly upon pathophysiology in CU (Kasperska-Zajac, Brzoza, and Rogala 2006a; Brzoza et al. 2008; Kasperska-Zajac, Brzoza, and Rogala 2008b). The mechanism of DHEA action and its significance in CU seems to be an interesting problem worth further exploration.

CONCLUSIONS

So far, there has been no explicit data defining the role of DHEAS in allergy, asthma, or urticaria/angioedema. Despite a remarkably growing number of studies devoted to DHEAS research, numerous questions concerning its mechanisms and the potential effect in such disorders have not been answered yet. Whether or not endogenous circulating concentrations of DHEAS are abnormal, it is possible that supplementation with exogenous DHEAS or the analog could have a beneficial role in these disorders by suppression of proinflammatory mediators, by immunomodulatory and steroid-sparing effects, or by reversion of acquired GC resistance. Nevertheless, further investigations are required in this realm.

REFERENCES

Adcock, I. M., S. J. Lane, C. R. Brown, T. H. Lee, and P. J. Barnes. 1995. Abnormal glucocorticoid receptor-activator protein 1 interaction in steroid-resistant asthma. *J Exp Med* 182:1951–8.

Akbari, O., P. Stock, R. H. DeKruyff, and D. T. Umetsu. 2003. Role of regulatory T cells in allergy and asthma. *Curr Opin Immunol* 15:627–33.

Araghi-Niknam, M., Z. Zhang, S. Jiang, O. Call, C. D. Eskelson, and R. R. Watson. 1997. Cytokine dysregulation and increased oxidation is prevented by dehydroepiandrosterone in mice infected with murine leukemia retrovirus. *Proc Soc Exp Biol Med* 216:386–91.

Araneo, B. A., T. Dowell, M. Diegel, and R. A. Daynes. 1991. Dihydrotestosterone exerts a depressive influence on the production of interleukin-4 (IL-4), IL-5, and gamma-interferon, but not IL-2 by activated murine T cells. *Blood* 78:688–99.

Auci, D., L. Kaler, S. Subramanian et al. 2007. A new orally bioavailable synthetic androstene inhibits collagen-induced arthritis in the mouse: Androstene hormones as regulators of regulatory T cells. *Ann N Y Acad Sci* 1110:630–40.

Barnes, P. J., and I. M. Adcock. 2009. Glucocorticoid resistance in inflammatory diseases. *Lancet* 373:1905–17.

Blauer, K. L., M. Poth, W. M. Rogers, and E. W. Bernton. 1991. Dehydroepiandrosterone antagonizes the suppressive effects of dexamethasone on lymphocyte proliferation. *Endocrinology* 129:3174–9.

Brzoza, Z., A. Kasperska-Zajac, K. Badura-Brzoza, J. Matysiakiewicz, R. T. Hese, and B. Rogala. 2008. Decline in dehydroepiandrosterone sulphate observed in chronic urticaria is associated with psychological distress. *Psychosom Med* 70:723–8.

Casson, P. R., R. N. Andersen, H. G. Herrod et al. 1993. Oral dehydroepiandrosterone in physiologic doses modulates immune function in postmenopausal women. *Am J Obstet Gynecol* 169:1536–9.

Choi, I. S., Y. Cui, Y. A. Koh, H. C. Lee, Y. B. Cho, and Y. H. Won. 2008. Effects of dehydroepiandrosterone on Th2 cytokine production in peripheral blood mononuclear cells from asthmatics. *Korean J Intern Med* 23:176–81.

Clerici, M., D. Trabattoni, S. Piconi et al. 1997. A possible role for the cortisol/anticortisols imbalance in the progression of human immunodeficiency virus. *Psychoneuroendocrinology* 22:S27–31.

Coles, A. J., S. Thompson, A. L. Cox, S. Curran, E. M. Gurnell, and V. K. Chatterjee. 2005. Dehydroepiandrosterone replacement in patients with Addison's disease has a bimodal effect on regulatory ($CD4^+CD25hi$ and $CD4^+FoxP3^+$) T cells. *Eur J Immunol* 35:3694–703.

Cugno, M., A. Zanichelli, F. Foieni, S. Caccia, and M. Cicardi. 2009. C1-inhibitor deficiency and angioedema: Molecular mechanisms and clinical progress. *Trends Mol Med* 15:69–78.

Danenberg, H. D., G. Alpert, S. Lustig, and D. Ben-Nathan. 1992. Dehydroepiandrosterone protects mice from endotoxin toxicity and reduces tumor necrosis factor production. *Antimicrob Agents Chemother* 36:2275–9.

Dashtaki, R., A. R. Whorton, T. M. Murphy, P. Chitano, W. Reed, and T. P. Kennedy. 1998. Dehydroepiandrosterone and analogs inhibit DNA binding of AP-1 and airway smooth muscle proliferation. *J Pharmacol Exp Ther* 285:876–83.

Daynes, R. A., D. J. Dudley, and B. A. Araneo. 1990. Regulation of murine lymphokine production in vivo: II. Dehydroepiandrosterone is a natural enhancer of interleukin 2 synthesis by helper T cells. *Eur J Immunol* 20:793–802.

Del Prete, G., M. De Carli, F. Almerigogna, M. G. Giudizi, R. Biagiotti, and S. Romagnani. 1993. Human IL-10 is produced by both type 1 helper (Th1) and type 2 helper (Th2) T cell clones and inhibits their antigen-specific proliferation and cytokine production. *J Immunol* 150:353–60.

Dillon, J. S. 2005. Dehydroepiandrosterone, dehydroepiandrosterone sulfate and related steroids: Their role in inflammatory, allergic and immunological disorders. *Curr Drug Targets Inflamm Allergy* 4:377–85.

Di Santo, E., M. Sironi, T. Mennini et al. 1996. A glucocorticoid receptor-independent mechanism for neurosteroid inhibition of tumor necrosis factor production. *Eur J Pharmacol* 299:179–86.

Dorsey, M. J., L. E. Cohen, W. Phipatanakul, D. Denufrio, and L. C. Schneider. 2006. Assessment of adrenal suppression in children with asthma treated with inhaled corticosteroids: Use of dehydroepiandrosterone sulfate as a screening test. *Ann Allergy Asthma Immunol* 97:182–6.

Du, C., Q. Guan, M. W. Khalil, and S. Sriram. 2001. Stimulation of Th2 response by high doses of dehydroepiandrosterone in KLH-primed splenocytes. *Exp Biol Med (Maywood)* 226:1051–60.

Du, C., M. W. Khalil, and S. Sriram. 2001. Administration of dehydroepiandrosterone suppresses experimental allergic encephalomyelitis in SJL/J mice. *J Immunol* 167:7094–101.

Dunn, P. J., C. B. Mahood, J. F. Speed, and D. R. Jury. 1984. Dehydroepiandrosterone sulphate concentrations in asthmatic patients: Pilot study. *N Z Med J* 97:805–8.

Ebata, T., R. Itamura, H. Aizawa, and M. Niimura. 1996. Serum sex hormone levels in adult patients with atopic dermatitis. *J Dermatol* 23:603–5.

Elenkov, I. J. 2004. Glucocorticoids and the Th1/Th2 balance. *Ann N Y Acad Sci* 1024:138–46.

Elkord, E. 2008. Novel therapeutic strategies by regulatory T cells in allergy. *Chem Immunol Allergy* 94:150–7.

Falus, A., G. K. Fehér, E. Walcz et al. 1990. Hormonal regulation of complement biosynthesis in human cell lines–I. Androgens and gamma-interferon stimulate the biosynthesis and gene expression of C1 inhibitor in human cell lines U937 and HepG2. *Mol Immunol* 27:191–5.

Feher, K. G., E. Koo, and T. Feher. 1983. Adrenocortical function in bronchial asthma. *Acta Med Hung* 40:125–31.

Harding, G., Y. T. Mak, B. Evans, J. Cheung, D. MacDonald, and G. Hampson. 2006. The effects of dexamethasone and dehydroepiandrosterone (DHEA) on cytokines and receptor expression in a human osteoblastic cell line: Potential steroid-sparing role for DHEA. *Cytokine* 36:57–68.

Hernandez-Pando, R., M. De La Luz Streber, H. Orozco et al. 1998. The effects of androstenediol and dehydroepiandrosterone on the course and cytokine profile of tuberculosis in BALB/c mice. *Immunology* 95:234–41.

Hidvégi, T., G. K. Fehér, T. Fehér, E. Koó, and G. Füst. 1984. Inhibition of the complement activation by an adrenal androgen, dehydroepiandrosterone. *Complement* 1:201–6.

Humbert, M., S. R. Durham, S. Ying et al. 1996. IL-4 and IL-5 mRNA and protein in bronchial biopsies from patients with atopic and nonatopic asthma: Evidence against "intrinsic" asthma being a distinct immunopathologic entity. *Am J Respir Crit Care Med* 154:1497–504.

Kaplan, A. P., and M. Greaves. Pathogenesis of chronic urticaria. 2009. *Clin Exp Allergy* 39:777–87.

Kasperska-Zajac, A., Z. Brzoza, and B. Rogala. 2006. Lower serum concentration of dehydyroepiandrosterone sulphate in patients with chronic idiopathic urticaria. *Allergy* 61:1489–90.

Kasperska-Zajac, A., Z. Brzoza, and B. Rogala. 2006a. Serum concentration of dehydroepiandrosterone sulphate in female patients with chronic idiopathic urticaria. *J Dermatol Sci* 41:80–1.

Kasperska-Zajac, A., Z. Brzoza, and B. Rogala. 2007. Serum concentration of dehydroepiandrosterone sulfate and testosterone in women with severe atopic eczema/dermatitis syndrome. *J Investig Allergol Clin Immunol* 17:160–3.

Kasperska-Zajac, A., Z. Brzoza, and B. Rogala. 2008. Dehydroepiandrosterone and dehydroepiandrosterone sulphate in atopic allergy and chronic urticaria. *Inflammation* 31:141–5.

Kasperska-Zajac, A., Z. Brzoza, and B. Rogala. 2008a. Sex hormones and urticaria. *J Dermatol Sci* 52:79–86.

Kasperska-Zajac, A., Z. Brzoza, and B. Rogala. 2008b. Lower serum dehydroepiandrosterone sulfate (DHEA-S) concentration in chronic idiopathic urticaria: A secondary transient phenomenon? *Br J Dermatol* 159:743–4.

Koo, E., K. G. Feher, T. Feher, and G. Fust. 1983. Effect of dehydroepiandrosterone on hereditary angioedema. *Klin Wochenschr* 61:715–7.

Kroboth, P. D., F. S. Salek, A. L. Pittenger, T. J. Fabian, and R. F. Frye. 1999. DHEA and DHEA-S: A review. *J Clin Pharmacol* 39:327–48.

Lane, S. J., I. M. Adcock, D. Richards, C. Hawrylowicz, P. J. Barnes, and T. H. Lee. 1998. Corticosteroid-resistant bronchial asthma is associated with increased c-fos expression in monocytes and T lymphocytes. *J Clin Invest* 102:2156–64.

Littley, M. D., A. Pollock, J. Kane, and S. M. Shalet. 1990. Basal serum dehydroepiandrosterone sulphate concentration does not predict the cortisol response to provocative testing. *Ann Clin Biochem* 27:557–61.

Maurer, M., and J. Grabbe. 2008. Urticaria: Its history-based diagnosis and etiologically oriented treatment. *Dtsch Arztebl Int* 105:458–66.

Meikle, A. W., R. W. Dorchuck, B. A. Araneo et al. 1992. The presence of a dehydroepiandrosterone-specific receptor binding complex in murine T cells. *J Steroid Biochem Mol Biol* 42:293–304.

Mellon, S. H. 2007. Neurosteroid regulation of central nervous system development. *Pharmacol Ther* 116:107–24.

Meno-Tetang, G. M., Y. Y. Hon, S. Van Wart, and W. J. Jusko. 1999. Pharmacokinetic and pharmacodynamic interactions between dehydroepiandrosterone and prednisolone in the rat. *Drug Metabol Drug Interact* 15:51–70.

Mizota, K., A. Yoshida, H. Uchida, R. Fujita, and H. Ueda. 2005. Novel type of Gq/11 protein-coupled neurosteroid receptor sensitive to endocrine disrupting chemicals in mast cell line (RBL-2H3). *Br J Pharmacol* 145:545–50.

Moynihan, J. A., T. A. Callahan, S. P. Kelley, and L. M. Campbell. 1998. Adrenal hormone modulation of type 1 and type 2 cytokine production by spleen cells: Dexamethasone and dehydroepiandrosterone suppress interleukin-2, interleukin-4, and interferon-gamma production in vitro. *Cell Immunol* 184:58–64.

Munakata, M. 2006. Airway remodeling and airway smooth muscle in asthma. *Allergol Int* 55:253–43.

Murakami, K., T. Nakagawa, G. Yamashiro, K. Araki, and K. Akasofu. 1993. Adrenal steroids in serum during danazol therapy, taking into account cross-reactions between danazol metabolites and serum androgens. *Endocr J* 40:659–64.

Okabe, T., M. Haji, R. Takayanagi et al. 1995. Up-regulation of high-affinity dehydroepiandrosterone binding activity by dehydroepiandrosterone in activated human T lymphocytes. *J Clin Endocrinol Metab* 80:2993–6.

Padgett, D. A., and R. M. Loria. 1998. Endocrine regulation of murine macrophage function: Effects of dehydroepiandrosterone, androstenediol, and androstenetriol. *J Neuroimmunol* 84:61–8.

Regelson, W., R. Loria, and M. Kalimi. 1988. Hormonal intervention: "Buffer hormones" or "state dependency". The role of dehydroepiandrosterone (DHEA), thyroid hormone, estrogen and hypophysectomy in aging. *Ann N Y Acad Sci* 521:260–73.

Semple, C. G., C. E. Gray, and G. H. Beastall. 1987. Adrenal androgens and illness. *Acta Endocrinol (Copenh)* 116:155–60.

Stankovic, A. K., L. D. Dion, and C. R. Parker Jr. 1994. Effects of transforming growth factor-beta on human fetal adrenal steroid production. *Mol Cell Endocrinol* 99:145–51.

Straub, R. H., J. Schölmerich, and B. Zietz. 2000. Replacement therapy with DHEA plus corticosteroids in patients with chronic inflammatory diseases—substitutes of adrenal and sex hormones. *J Rheumatol* 59:108–18.

Sudo, N., X. N. Yu, and C. Kubo. 2001. Dehydroepiandrosterone attenuates the spontaneous elevation of serum IgE level in NC/Nga mice. *Immunol Lett* 79:177–9.

Suzuki, T., N. Suzuki, R. A. Daynes, and E. G. Engleman. 1991. Dehydroepiandrosterone enhances IL2 production and cytotoxic effector function of human T cells. *Clin Immunol Immunopathol* 61:202–11.

Tabata, N., H. Tagami, and T. Terui. 1997. Dehydroepiandrosterone may be one of the regulators of cytokine production in atopic dermatitis. *Arch Dermatol Res* 289:410–4.

Thon, V., P. Härle, J. Schölmerich, P. Kuklinek, J. Lokaj, and R. H. Straub. 2007. Lack of dehydroepiandrosterone in type I and II hereditary angioedema and role of danazol in steroid hormone conversion. *Allergy* 62:1320–5.

Uchida, H., K. Mizuno, A. Yoshida, and H. Ueda. 2003. Neurosteroid-induced hyperalgesia through a histamine release is inhibited by progesterone and p,p'-DDE, an endocrine disrupting chemical. *Neurochem Int* 42:401–7.

Ueda, H., M. Inoue, A. Yoshida et al. 2001. Metabotropic neurosteroid/sigma-receptor involved in stimulation of nociceptor endings of mice. *J Pharmacol Exp Ther* 298:703–10.

Umetsu, D. T., and R. H. DeKruyff. 1997. Th1 and Th2 $CD4^+$ cells in human allergic diseases. *J Aller Clin Immunol* 100:1–6.

Umetsu, D. T., and R. H. DeKruyff. 2006. The regulation of allergy and asthma. *Immunol Rev* 212:238–55.

Weinstein, R. E., C. A. Lobocki, S. Gravett et al. 1996. Decreased adrenal sex steroid levels in the absence of glucocorticoid suppression in postmenopausal asthmatic women. *J Allergy Clin Immunol* 97:1–8.

Wu, M. F., H. L. Chang, and J. Tseng. 1997. Dehydroepiandrosterone induces the transforming growth factor-beta production by murine macrophages. *Int J Tissue React* 19:141–8.

Yu, C. K., Y. H. Liu, and C. L. Chen. 2002. Dehydroepiandrosterone attenuates allergic airway inflammation in dermatophagoides farinae-sensitized mice. *J Microbiol Immunol Infect* 35:199–202.

Yu, C. K., B. C. Yang, H. Y. Lei et al. 1999. Attenuation of house dust mite dermatophagoides farinae-induced airway allergic responses in mice by dehydroepiandrosterone is correlated with down-regulation of TH2 response. *Clin Exp Allergy* 29:414–22.

Zhang, Z., M. Araghi-Niknam, B. Liang et al. 1999. Prevention of immune dysfunction and vitamin E loss by dehydroepiandrosterone and melatonin supplementation during murine retrovirus infection. *Immunology* 96:291–7.

19 The Role of DHEA in Mental Disorders

Iván Pérez-Neri and Camilo Ríos

CONTENTS

INTRODUCTION

Dehydroepiandrosterone (DHEA) and its sulfate ester, dehydroepiandrosterone sulfate (DHEAS), modulate several neurotransmitter systems (Maninger et al. 2009; Pérez-Neri et al. 2008) involved in the pathophysiology of psychiatric disorders such as depression, dementia, schizophrenia, anxiety, and mania. Some studies have found an association between endogenous DHEA levels and the incidence and course of those mental disorders. Also, several controlled clinical trials have reported beneficial effects of DHEA administration.

In spite of an increasing body of evidence in this regard, the actual role of DHEA in mental disease is yet to be completely elucidated. This review summarizes published evidence regarding the possible role of DHEA and DHEAS in psychiatric disorders.

DEPRESSIVE DISORDER

Major depressive disorder is one of the most devastating mental diseases (Alexopoulos and Kelly Jr. 2009). Depressive symptoms include negative affect, sleep disturbance, feelings of guilt, and suicidal ideation, among others (Gotlib and Joormann 2010). Prevalence of depression throughout life has been estimated around 20% in some populations, and the rate of relapse may be as high as 75% (Gotlib and Joormann 2010). The mechanism for antidepressant action is partially understood and a therapeutic response is not achieved in every case (Katz, Bowden, and Frazer 2010).

Several studies have described abnormal DHEA or DHEAS levels in depressive disorders. Plasma DHEA concentration was increased in depressed (Heuser et al. 1998) and psychotic depressed (Maayan et al. 2000) patients, but salivary (Eser et al. 2006b; Goodyer et al. 2001b; Michael et al. 2000) and urinary (Poór et al. 2004) levels were decreased in other studies. Decreased (Jozuka et al. 2003; Maninger et al. 2009; Morgan et al. 2010) and unchanged (Kahl et al. 2006; Maninger et al. 2009; Young, Gallagher, and Porter 2002) blood DHEA levels have also been reported.

Changes in DHEA and DHEAS salivary and blood concentrations are relevant to central nervous system function as those levels are positively correlated to their cerebrospinal fluid (CSF) counterparts (Goodyer et al. 2001b; Guazzo et al. 1996); however, it is possible that the brain content of the steroids is differently altered or even unchanged in spite of a different level in the extracellular environment. In fact, DHEA content in cingulate and parietal cortices from depressed patients was not significantly different from controls (Marx et al. 2006a), although other brain regions were not studied.

DHEA may be associated not only to the incidence of the disease, but also to the severity of depressive symptoms. Morning salivary DHEA levels were inversely correlated to the severity of depression in some studies (Eser et al. 2006b; Michael et al. 2000), although there was no correlation in patients with burning mouth disorder (Fernandes et al. 2009), healthy elderly (Fukai et al. 2009), or psychotic depressed patients (Maayan et al. 2000).

Moreover, it is possible that salivary DHEA concentration is not altered by the chronicity of the disease because it was not different in boys with chronic major depression compared with those who recovered from a depressive episode (Goodyer, Park, and Herbert 2001a). Thus, DHEA may be altered from the first depressive episode and remain altered throughout the course of the disease independently of remission. This hypothesis is supported by the lack of association between steroid levels and the effect of antidepressants. The therapeutic effect of repetitive transcranial magnetic stimulation was not accompanied by changes in plasma DHEA concentration in depressed patients (Padberg et al. 2002). However, low DHEA levels were associated with the antidepressant effect of sleep deprivation (Schüle et al. 2003).

The role of DHEA as the cause or the consequence of depression remains a matter of debate. Changes in steroid levels should be found before the disease onset if it is involved in the development of the disorder. However, changes in DHEA concentration were absent before the onset of major depression. Also, steroid levels were not significantly correlated to mood scores in adolescents at high risk of developing depressive disorders (Goodyer et al. 2000a). Furthermore, there was no significant difference in DHEA concentration between adolescents at high and low risk for depression (Goodyer et al. 2000a). However, those results may be influenced by the fact that not every high-risk case will finally develop depressive illness (Goodyer et al. 2000a). Actually, an increased DHEA concentration at baseline was significantly associated to the onset of major depression in adolescents at follow-up (Goodyer et al. 2000a,b, 2001b), although this result was not replicated in adults (Harris et al. 2000).

Even if an altered DHEA concentration is the cause or the consequence of depressive disorders, an increasing body of evidence supports a therapeutic effect of the steroid. Several studies have found beneficial effects of DHEA administration for depressive symptoms (Binello and Gordon 2003; Bovenberg, van Uum, and Hermus 2005; Brooke et al. 2006; Dubrovsky 2005; Eser et al. 2006a; Maninger et al. 2009; Ravindran et al. 2009; Schmidt et al. 2005) or psychological well-being (Brooke et al. 2006; Dubrovsky 2005; Maninger et al. 2009; Nawata et al. 2002; Schumacher et al. 2003). In placebo-controlled, double-blind clinical trials, DHEA administration to healthy subjects improves mood (Arlt et al. 1999). The steroid reduces symptom severity in depressed patients (Bloch et al. 1999; Eser et al. 2006b; Schmidt et al. 2005; Wolkowitz et al. 1999), and this effect also occurs in other diseases such as adrenal insufficiency (Binder et al. 2009; Hunt et al. 2000; Maninger et al. 2009), schizophrenia (Strous et al. 2003), and human immunodeficiency virus infection (Rabkin et al. 2006).

However, some studies have failed to replicate those results (Arlt et al. 2001; Kritz-Silverstein et al. 2008), but it should be noted that increased blood DHEAS levels were associated to an antidepressant response after DHEA treatment (Bloch et al. 1999; Rabkin et al. 2006); thus, the failure to increase DHEAS (and possibly DHEA) levels in some patients may be responsible for the absence of a clinical response to DHEA supplementation.

Regarding DHEAS, it is possible that reduced levels of this steroid favor the development of a depressive episode. Low DHEAS concentration is associated to an enhanced negative emotional reaction following social rejection (Akinola and Mendes 2008). However, increased salivary (Assies et al. 2004; Maninger et al. 2009) and urinary (Eser et al. 2006b) concentrations were reported in depressed patients. Some authors have reported reduced DHEAS concentration in patients with depression (Eser et al. 2006b; Maninger et al. 2009) or dysthymia (Markianos et al. 2007). No difference in DHEAS plasma levels was found in other studies (Jozuka et al. 2003; Paslakis et al. 2010).

Supporting the role of DHEAS deficiency in depression, the steroid was inversely correlated to the severity of depressive symptoms according to some studies (Brzoza et al. 2008; Haren et al. 2007; Maninger et al. 2009; Nagata et al. 2000), although no significant correlations have been reported (Adali et al. 2008; Hsiao 2006; Maayan et al. 2000; Schüle et al. 2009). Also, DHEAS levels were positively correlated to mood scores, showing a better sense of well-being at increased steroid concentration (Valtysdottir, Wide, and Hallgren 2003). Depressive symptomatology in elderly women was associated to low DHEAS levels (Berr et al. 1996).

Even though an increased DHEAS concentration following DHEA administration is associated with an antidepressant response (Bloch et al. 1999; Rabkin et al. 2006), an increased baseline level may interfere with that effect. Depressed patients with high DHEAS levels do not respond to electroconvulsive therapy (Eser et al. 2006a,b) or pharmacological treatment (Schüle et al. 2009).

Thus, some studies suggest that an increased DHEAS baseline level may be detrimental for an antidepressant response; however, changes in DHEAS concentration from baseline are likely associated to the clinical efficacy of antidepressants. Reduction in symptom severity was positively correlated to the decrease in DHEAS levels according to some studies (Fabian et al. 2001; Schüle et al. 2009). Also, DHEA and DHEAS levels decrease following remission from depression (Fabian et al. 2001).

In summary, it may be suggested that DHEA levels are increased before the onset of depression and that those levels decrease when the disease is established. Both DHEA and DHEAS deficiency correlate to an increased symptom severity, and the restoration of DHEAS levels is associated to an antidepressant response; however, an increased baseline DHEA or DHEAS concentration may reduce the antidepressant effect of drugs and electroconvulsive therapy. In spite of the contrasting results regarding endogenous steroid levels, an increasing body of evidence supports the hypothesis that DHEA is reduced in major depression and steroid supplementation reduces symptom severity in this disorder (Table 19.1).

TABLE 19.1
Summary of Studies Reporting Altered DHEA or DHEAS Levels in Patients with Depressive Disorders

Reference	Patients	Diagnosis	Biological Sample	Results
		DHEA		
Heuser et al. (1998)	15 male, 47.7 ± 14.8 years; 11 female, 48.2 ± 18.1 years	Major depressive disorder	Plasma	Increased DHEA levels
Maayan et al. (2000)	7 men, 10 female; 40.4 ± 3.1 years	Major depression with psychotic features ($n = 2$), schizophrenia with comorbid depression ($n = 10$), schizoaffective disorder with depressive symptoms ($n = 5$)	Plasma	Increased DHEA levels

(Continues)

TABLE 19.1 (*Continued*)
Summary of Studies Reporting Altered DHEA or DHEAS Levels in Patients with Depressive Disorders

Reference	Patients	Diagnosis	Biological Sample	Results
		DHEA		
Morgan et al. (2010)	16 female; 54.5 ± 4.9 years	Major depressive disorder	Serum	Decreased DHEA levels
Jozuka et al. (2003)	8 male, 9 female; 40.3 ± 15.1 years	Major depressive disorder	Blood	Decreased DHEA levels
Michael et al. (2000)	12 male, 32 female; 20–64 years	Major depressive disorder	Saliva	Decreased DHEA levels
Poór et al. (2004)	9 male, 46.6 ± 9.9 years; 11 female, 35.3 ± 12.9 years	Major depressive disorder	Urine	Decreased DHEA levels
Kahl et al. (2006)	12 female; 26.3 ± 5.1 years	Major depressive disorder comorbid with borderline personality disorder	Serum	Unchanged DHEA levels
Young, Gallagher, and Porter (2002)	15 male, 29 female; 33 ± 11 years	Major depressive disorder	Saliva	Unchanged DHEA levels
		DHEAS		
Maayan et al. (2000)	7 men, 10 female; 40.4 ± 3.1 years	Major depression with psychotic features ($n = 2$), schizophrenia with comorbid depression ($n = 10$), schizoaffective disorder with depressive symptoms ($n = 5$)	Plasma	Increased DHEAS levels
Assies et al. (2004)	3 male, 10 female; 39.8 ± 11.3 years	Major depressive disorder	Saliva	Increased DHEAS levels
Markianos et al. (2007)	18 male, 47.1 ± 13.3 years; 43 female, 45.2 ± 13.9 years	Dysthymic disorder	Plasma	Decreased DHEAS levels
Jozuka et al. (2003)	8 male, 9 female; 40.3 ± 15.1 years	Major depressive disorder	Blood	Unchanged DHEAS levels
Paslakis et al. (2010)	22 male, 48 female; 51.0 ± 14.8 years	Major depressive disorder	Blood	Unchanged DHEAS levels

DHEA = dehydroepiandrosterone; DHEAS = dehydroepiandrosterone sulfate.

DEMENTIA

Dementia is a cognitive disorder characterized by amnesia that also includes altered abstract thinking, judgment, and behavior among other disturbances. Dementia is an increasing health problem worldwide (Schumacher et al. 2003) that is most frequently present as Alzheimer's disease (AD; Galimberti and Scarpini 2010; Henderson 2010), but it may also be associated with stroke (Pendlebury 2009) or frontal lobar degeneration (Galimberti and Scarpini 2010). The prevalence of AD has been estimated to be 5% after 65 years of age (Galimberti and Scarpini 2010) and its treatment remains challenging.

It has been reported that DHEAS levels are reduced in the striatum, cerebellum, and hypothalamus from AD patients (Kim et al. 2003; Maninger et al. 2009; Weill-Engerer et al. 2002; Wojtal, Trojnar, and Czuczwar 2006). Those levels were also reduced in cognitively impaired elderly (Ulubaev et al. 2009) and multi-infarct dementia patients (Azuma et al. 1999; Kim et al. 2003; Maninger et al. 2009), suggesting that this alteration may be associated to cognitive dysfunction rather than to a specific disease. That decrease may be related to the degenerative process in AD because serum DHEAS levels were correlated to hippocampal volume (Maninger et al. 2009) and were also associated to the development of AD in women with Down syndrome (trisomy 21; Schupf et al. 2006).

The steroid may accumulate in the brain due, at least in part, to a reduced metabolism because expression of CYP7B, the gene encoding 7α-hydroxylase that converts DHEA to its 7α-hydroxylated metabolite, was reduced in the hippocampus (Hampl and Bicíková 2010; Maninger et al. 2009; Yau et al. 2003); also, plasma 7α-hydroxydehydroepiandrosterone concentration was reduced in AD patients (Maninger et al. 2009).

Additionally, decreased DHEAS content may result from a reduced sulfotransferase activity because DHEA content is increased in the CSF, hypothalamus, hippocampus, and frontal cortex of AD patients (Brown et al. 2003; Kim et al. 2003; Maninger et al. 2009; Marx et al. 2006b; Naylor et al. 2008). Interestingly, CSF DHEA concentration positively correlates with the content of the steroid in the temporal cortex (Naylor et al. 2008).

Some studies have reported that plasma DHEA and DHEAS levels were decreased in AD patients compared with healthy controls (Bernardi et al. 2000; Ferrari et al. 2001a; Hillen et al. 2000; Nawata et al. 2002). Those results may be associated to a reduced adrenocorticotropic hormone release (Näisman et al. 1996). Similar findings have been reported in vascular dementia (Bernardi et al. 2000; Ferrari et al. 2001a; Nawata et al. 2002).

Although some studies have reported that serum DHEA and DHEAS concentrations are positively correlated to cognitive performance in healthy subjects (Maninger et al. 2009; Ulubaev et al. 2009), some other studies have failed to replicate in AD those previous results (Brown et al. 2003; Carlson, Sherwin, and Chertkow 1999; Ferrari et al. 2001a; Fuller, Tan, and Martins 2007; Hoskin et al. 2004; Rasmuson et al. 1998; Schneider, Hinsey, and Lyness 1992). Also, the association of the steroid to cognitive function was not replicated in elderly subjects, as measured by the correlation between steroid levels and cognitive scale scores (Carlson and Sherwin 1999; Ferrari et al. 2001b; Fuller, Tan, and Martins 2007; Maninger et al. 2009; Schumacher et al. 2003; Ulubaev et al. 2009). DHEAS levels were not associated with minimental state examination (MMSE) scores or the incidence of dementia in either the elderly (Berr et al. 1996; de Bruin et al. 2002) or AD patients (Rasmuson et al. 1998). Even inverse correlations between DHEAS levels and cognitive performance in the elderly have been reported (Fuller, Tan, and Martins 2007; Maninger et al. 2009).

However, among AD patients, those with high plasma DHEAS levels performed better in some cognitive tasks compared with those with low steroid levels (Carlson, Sherwin, and Chertkow 1999; Fuller, Tan, and Martins 2007). Plasma 7αOH-DHEA was positively correlated to MMSE scores (Maninger et al. 2009).

Regarding steroid supplementation, cognitive scale scores improve in some studies following DHEAS administration (Azuma et al. 1999; Maninger et al. 2009). Thus, both endogenous and administered DHEA and DHEAS have been associated to cognitive performance in AD and other dementias. Those results suggest that, although DHEA is increased in AD, DHEAS deficiency is related to cognitive dysfunction, and thus, steroid supplementation is beneficial in this disorder (Table 19.2).

TABLE 19.2
Summary of Studies Reporting Altered DHEA or DHEAS Levels in Patients with Dementia

Reference	Patients	Diagnosis	Biological Sample	Results
		DHEA		
Brown et al. (2003)	4 male, 5 female; 74.6 ± 7.2 years	AD	CSF	Increased DHEA levels
Naylor et al. (2008)	25 patients; 81 years	AD	CSF	Increased DHEA levels
Kim et al. (2003)	7 male, 7 female; 75.1 ± 9.8 years	Probable AD	CSF	Increased DHEA levels
Kim et al. (2003)	4 male, 4 female; 78.5 ± 4.8 years	Vascular dementia	CSF	Increased DHEA levels
Brown et al. (2003)	6 male, 6 female; 74.6 ± 7.2 years	AD	Brain tissue	Increased DHEA levels
Marx et al. (2006b)	14 male; 83 years	AD	Brain tissue	Increased DHEA levels
Bernardi et al. (2000)	5 male, 7 female; 64–84 years	AD	Serum	Decreased DHEA levels
Bernardi et al. (2000)	6 male, 6 female; 65–82 years	Vascular dementia	Serum	Decreased DHEA levels
Brown et al. (2003)	5 male; 80.0 ± 6.9 years	AD	Serum	Unchanged DHEA levels
		DHEAS		
Kim et al. (2003)	7 male, 7 female; 75.1 ± 9.8 years	Probable AD	CSF	Decreased DHEAS levels
Azuma et al. (1999)	4 male, 3 female; 69.4 ± 6. years	Multi-infarct dementia	CSF	Decreased DHEAS levels
Kim et al. (2003)	4 male, 4 female; 78.5 ± 4.8 years	Vascular dementia	CSF	Decreased DHEAS levels
Weill-Engerer et al. (2002)	1 male, 4 female; 86.2 ± 3.7 years	AD	Brain tissue	Decreased DHEAS levels
Hillen et al. (2000)	7 male, 7 female; 87.2 ± 1.9 years	AD	Plasma	Decreased DHEAS levels
Azuma et al. (1999)	4 male, 3 female; 69.4 ± 6. years	Multi-infarct dementia	Serum	Unchanged DHEAS levels

DHEA = dehydroepiandrosterone; DHEAS = dehydroepiandrosterone sulfate; AD = Alzheimer's disease; CSF = cerebrospinal fluid.

SCHIZOPHRENIA

Schizophrenia is a mental disorder characterized by psychotic, cognitive, and affective symptoms (Simpson, Kellendonk, and Kandel 2010). Its prevalence has been estimated around 1% worldwide (Stevens 2002). In spite of the scientific efforts to elucidate the disease, its etiology remains unclear and its therapeutics limited (Ritsner 2010; Simpson, Kellendonk, and Kandel 2010). Several factors are involved in the pathophysiology of this disorder: these include genes, environment, and hormones. In this regard, some studies suggest that DHEA has a role in this disorder (Ritsner 2010) although its relevance to the onset, course, and treatment of the disease remains to be completely elucidated.

Some abnormalities in DHEA and DHEAS levels have been reported in schizophrenia. Plasma DHEA concentration was increased in schizophrenic patients compared with healthy patients independently of antipsychotic treatment (di Michele et al. 2005; Maninger et al. 2009; Ritsner 2010; Strous et al. 2004). Also, the content of DHEA was increased in the posterior cingulate cortex from those patients (Maninger et al. 2009; Marx et al. 2006a).

Similar to those of DHEA, DHEAS levels were increased in schizophrenic patients (Oades and Schepker 1994; Strous et al. 2004), and they were associated to symptom severity. DHEAS concentration was positively associated to cognitive performance in schizophrenic patients while DHEA was inversely correlated (Harris, Wolkowitz, and Reus 2001; Ritsner 2010; Ritsner and Strous 2010; Silver et al. 2005). In another study, serum DHEA concentration was positively correlated with working memory performance (Harris, Wolkowitz, and Reus 2001).

In spite of the studies showing an increased DHEA concentration in schizophrenia, steroid supplementation exerted a therapeutic effect. DHEA administration to schizophrenic patients, along with antipsychotic medication, significantly reduced the severity of negative symptoms (Strous et al. 2003). Thus, DHEA may influence the response to antipsychotics; but antipsychotics, in turn, influence DHEAS levels; it has been reported that antipsychotic medication reduces DHEAS concentration in schizophrenic patients (Baptista, Reyes, and Hernández 1999).

Medication-induced side effects are also an important issue during the course of an antipsychotic treatment because those effects may severely compromise patients' health. In this regard, it has been reported that DHEA administration reduced antipsychotic-induced extrapyramidal symptoms in schizophrenic patients (Ritsner 2010), which is the most frequent side effect of first-generation antipsychotics.

In summary, DHEA levels are increased in blood and brain tissue from schizophrenic patients. In spite of those increased levels, high DHEA concentration is associated to a reduced severity of psychiatric symptoms and steroid supplementation leads to a beneficial effect, especially regarding cognitive symptoms and extrapyramidal side effects. It remains to be determined if increased DHEA concentration in schizophrenia is associated to the positive symptoms in this disorder because a further increase is beneficial to the negative symptoms only.

ANXIETY

The term "anxiety" involves a group of mental disorders characterized by feelings of fearfulness that may include panic, psychological complaints, and autonomic symptoms (Tyrer and Baldwin 2006). Its prevalence has been estimated around 30% (Nandi, Beard, and Galea 2009), but it is higher in women than in men (McLean and Anderson 2009). Several anxiolytic drugs are currently in use, but clinical response is achieved in less than half of cases (Tyrer and Baldwin 2006).

Some studies support an association of endogenous or administered DHEA to the incidence or treatment of anxiety disorders. Plasma DHEA concentration was increased in patients with panic (Brambilla et al. 2005; Maninger et al. 2009) and posttraumatic stress disorders (Maninger et al. 2009). Steroid levels were not different between patients and controls in other studies (Brambilla et al. 2003; Eser et al. 2006b; Laufer et al. 2005; Maninger et al. 2009; Semeniuk, Jhangri, and Le Mellédo 2001). Moreover, DHEA levels increase following experimentally induced panic attacks in humans (Eser et al. 2006b).

Interestingly, DHEA concentration was positively correlated to the severity of panic and phobia symptoms and negatively correlated to anxiety symptoms, according to some studies (Brambilla etal. 2003; Luz et al. 2003). DHEAS, in turn, was negatively correlated to the severity of anxiety in patients with chronic urticaria (Brzoza et al. 2008) but was positively correlated to anxiety scores in depressed patients (Hsiao 2006). However, DHEA levels were not correlated to anxiety scores in patients with panic disorder (Brambilla et al. 2005), social phobia (Laufer et al. 2005), or victims of intimate-partner violence (Pico-Alfonso et al. 2004).

Several studies have found beneficial effects of DHEA supplementation for anxiety or psychological distress (Binder et al. 2009). Administration of DHEA, but not estrogens, reduced anxiety in female patients with anorexia nervosa compared with baseline scores (Gordon et al. 2002). Also, DHEA, along with antipsychotic medication, reduced anxiety in schizophrenic patients (Eser et al. 2006b; Strous et al. 2003).

In summary, some studies have found that DHEA concentration is increased in anxiety disorders, that it further increases following panic attacks, and that it is positively correlated to phobia symptoms. In contrast, both DHEA and DHEAS levels were inversely correlated to anxiety symptoms in other studies. It is possible that DHEA is differently involved in phobia and anxiety; the steroid may increase with increasing severity of phobia and panic symptoms, but, by reducing anxiety, the steroid may contribute to control the behavioral response to those symptoms. This issue remains speculative and awaits further investigation; however, some studies support the therapeutic role of DHEA supplementation for anxiety.

AGGRESSIVE BEHAVIOR

Aggression is a complex behavior, displayed by several animal species, that is intended to establish dominance for survival (Soma et al. 2008), but it may also involve a pathological background.

Several studies have associated aggressive behavior to estradiol, testosterone, and other anabolic-androgenic substances, but adrenal steroids also seem to be involved (Soma et al. 2008; Talih, Fattal, and Malone 2007). Some studies have found associations between aggression, but not testosterone, and DHEAS in children (Soma et al. 2008; van Goozen et al. 1998). It is possible that the lower testosterone levels in children compared with adults accounts for that apparent discrepancy. Also, adolescent females with congenital adrenal hyperplasia, leading to increased DHEAS levels, show aggressive behavior (Soma et al. 2008); pharmacologic reduction of DHEAS levels in those patients reduces aggression (Soma et al. 2008). DHEAS levels increase according to the intensity of aggression in 7- to 11-year-old boys (Butovskaya et al. 2005).

However, some studies have failed to replicate the associations between aggression scores and either testosterone or DHEA in 5-year-old boys (Azurmendi et al. 2006; Sánchez-Martín et al. 2009); rather androstenedione is associated in that population (Azurmendi et al. 2006). The relationship between DHEA or DHEAS and aggression in adults is likely to be different. DHEAS levels are lower in highly aggressive patients, compared with controls, following alcohol withdrawal (Ozsoy and Esel 2008).

Several animal models show that DHEA administration reduces aggressive behavior (Soma et al. 2008). Taken together, those results suggest that DHEAS increases, while DHEA decreases, aggressive behavior and, thus, the sulfated and unsulfated steroid lead to opposite effects.

MANIA

Mania is characterized by irritability and euphoria that may be accompanied by high self-esteem, racing thoughts and speech, and increased goal-directed activity; psychotic features are present in some cases. Mania is the main component of bipolar disorder (Mansell and Pedley 2008).

It has been reported that DHEA levels are increased in the posterior cingulate and parietal cortices from patients with bipolar disorder (Marx et al. 2006a). Also, DHEA consumption has been associated to the development of episodes of mania (Dean 2000; Kline and Jaggers 1999; Markowitz, Carson, and Jackson 1999; Vacheron-Trystam et al. 2002), The psychostimulating-like effect of DHEA has been observed after administration of high doses (up to 300 mg/day) during several weeks or months (more than 3 months; Markowitz, Carson, and Jackson 1999), and it remains to be determined if this effect involves DHEA conversion to androgens since anabolic steroid consumption has been associated with mania (Talih, Fattal, and Malone 2007).

Also, it is yet to be elucidated whether DHEA consumption could induce mania in women. In fact, mood-stabilizers (valproic acid) increase the expression of P450scc and P450c17, as well as the synthesis of DHEA and androstenedione, in ovarian theca cells (Nelson-DeGrave et al. 2004); thus, it is possible that DHEA is involved in the therapeutic effect of those drugs.

In summary, case reports of DHEA-induced mania are anecdotic and may involve androgen formation. However, some studies suggest that DHEA may be involved in the mechanism of action of mood-stabilizers.

SUMMARY

Endogenous DHEA levels are altered in psychiatric disorders as shown by several studies. Some studies suggest that DHEA deficiency may be involved in the pathophysiology of mental disease, but increased steroid levels have been reported before the onset of depression and after that of dementia, schizophrenia, and anxiety. Also, although an increase in DHEA concentration is involved in the effect of some neuroleptics, high steroid levels at baseline may interfere with their therapeutic effect.

DHEA levels were inversely correlated to disease severity according to several studies, suggesting that, in spite of a possible baseline increase, a further increase is beneficial. However, DHEA concentration was positively correlated to the severity of phobia and panic symptoms; thus, the role of the steroid in anxiety remains to be elucidated.

Controlled clinical trials consistently show beneficial effects of DHEA supplementation for several psychiatric disorders. Thus, even though the involvement of DHEA in the pathophysiology of psychiatric disorders remains controversial, the therapeutic effect of steroid administration is supported by an increasing body of evidence.

ACKNOWLEDGMENTS

I. Pérez-Neri receives a grant from CONACyT (83521).

REFERENCES

Adali, E., R. Yildizhan, M. Kurdoglu, A. Kolusari, T. Edirne, H. Sahin et al. 2008. The relationship between clinicobiochemical characteristics and psychiatric distress in young women with polycystic ovary syndrome. *J Int Med Res* 36:1188–96.

Akinola, M., and W. B. Mendes. 2008. The dark side of creativity: Biological vulnerability and negative emotions lead to greater artistic creativity. *Pers Soc Psychol Bull* 34:1677–86.

Alexopoulos, G. S., and R. E. Kelly Jr. 2009. Research advances in geriatric depression. *World Psychiatry* 8:140–9.

Arlt, W., F. Callies, I. Koehler, J. C. van Vlijmen, M. Fassnacht, and C. J. Strasburger. 2001. Dehydroepiandrosterone supplementation in healthy men with an age-related decline of dehydroepiandrosterone secretion. *J Clin Endocrinol Metab* 86:4686–92.

Arlt, W., F. Callies, J. C. van Vlijmen, I. Koehler, M. Reincke, M. Bidlingmaier et al. 1999. Dehydroepiandrosterone replacement in women with adrenal insufficiency. *N Engl J Med* 341:1013–20.

Assies, J., I. Visser, N. A. Nicolson, T. A. Eggelte, E. M. Wekking, J. Huyser et al. 2004. Elevated salivary dehydroepiandrosterone-sulfate but normal cortisol levels in medicated depressed patients: Preliminary findings. *Psychiatry Res* 128:117–22.

Azuma, T., Y. Nagai, T. Saito, M. Funauchi, T. Matsubara, and S. Sakoda. 1999. The effect of dehydroepiandrosterone sulfate administration to patients with multi-infarct dementia. *J Neurol Sci* 162:69–73.

Azurmendi, A., F. Braza, A. García, P. Braza, J. M. Muñoz, and J. R. Sánchez-Martín. 2006. Aggression, dominance, and affiliation: Their relationships with androgen levels and intelligence in 5-year-old children. *Horm Behav* 50:132–40.

Baptista, T., D. Reyes, and L. Hernández. 1999. Antipsychotic drugs and reproductive hormones: Relationship to body weight regulation. *Pharmacol Biochem Behav* 62:409–17.

Bernardi, F., A. Lanzone, R. M. Cento, R. S. Spada, I. Pezzani, A. D. Genazzani et al. 2000. Allopregnanolone and dehydroepiandrosterone response to corticotropin-releasing factor in patients suffering from Alzheimer's disease and vascular dementia. *Eur J Endocrinol* 142:466–71.

Berr, C., S. Lafont, B. Debuire, J. F. Dartigues, and E. E. Baulieu. 1996. Relationships of dehydroepiandrosterone sulfate in the elderly with functional, psychological, and mental status, and short-term mortality: A French community-based study. *Proc Natl Acad Sci U S A* 93:13410–5.

Binder, G., S. Weber, M. Ehrismann, N. Zaiser, C. Meisner, M. B. Ranke et al. 2009. Effects of dehydroepiandrosterone therapy on pubic hair growth and psychological well-being in adolescent girls and young women with central adrenal insufficiency: A double-blind, randomized, placebo-controlled phase III trial. *J Clin Endocrinol Metab* 94:1182–90.

Binello, E., and C. M. Gordon. 2003. Clinical uses and misuses of dehydroepiandrosterone. *Curr Opin Pharmacol* 3:635–41.

Bloch, M., P. J. Schmidt, M. A. Danaceau, L. F. Adams, and D. R. Rubinow. 1999. Dehydroepiandrosterone treatment of midlife dysthymia. *Biol Psychiatry* 45:1533–41.

Bovenberg, S. A., S. H. M. van Uum, and A. R. M. M. Hermus. 2005. Dehydroepiandrosterone administration in humans: Evidence based? *Neth J Med* 63:300–4.

Brambilla, F., G. Biggio, M. G. Pisu, L. Bellodi, G. Perna, V. Bogdanovich-Djukic et al. 2003. Neurosteroid secretion in panic disorder. *Psychiatry Res* 118:107–16.

Brambilla, F., C. Mellado, A. Alciati, M. G. Pisu, R. H. Purdy, S. Zanone et al. 2005. Plasma concentrations of anxiolytic neuroactive steroids in men with panic disorder. *Psychiatry Res* 135:185–90.

Brooke, A. M., L. A. Kalingag, F. Miraki-Moud, C. Camacho-Hübner, K. T. Maher, D. M. Walker et al. 2006. Dehydroepiandrosterone improves psychological well-being in male and female hypopituitary patients on maintenance growth hormone replacement. *J Clin Endocrinol Metab* 91:3773–9.

Brown, R. C., Z. Han, C. Cascio, and V. Papadopoulos. 2003. Oxidative stress-mediated DHEA formation in Alzheimer's disease pathology. *Neurobiol Aging* 24:57–65.

Brzoza, Z., A. Kasperska-Zajac, K. Badura-Brzoza, J. Matysiakiewicz, R. T. Hese, and B. Rogala. 2008. Decline in dehydroepiandrosterone sulfate observed in chronic urticaria is associated with psychological distress. *Psychosom Med* 70:723–8.

Butovskaya, M. L., E. Y. Boyko, N. B. Selverova, and I. V. Ermakova. 2005. The hormonal basis of reconciliation in humans. *J Physiol Anthropol Appl Human Sci* 24:333–7.

Carlson, L. E., and B. B. Sherwin. 1999. Relationships among cortisol (CRT), dehydroepiandrosterone-sulfate (DHEAS), and memory in a longitudinal study of healthy elderly men and women. *Neurobiol Aging* 20:315–24.

Carlson, L. E., B. B. Sherwin, and H. M. Chertkow. 1999. Relationships between dehydroepiandrosterone sulfate (DHEAS) and cortisol (CRT) plasma levels and everyday memory in Alzheimer's disease patients compared to healthy controls. *Horm Behav* 35:254–63.

de Bruin, V. M. S., M. C. M. Vieira, M. N. M. Rocha, and G. S. B. Viana. 2002. Cortisol and dehydroepiandosterone sulfate plasma levels and their relationship to aging, cognitive function, and dementia. *Brain Cogn* 50:316–23.

Dean, C. E. 2000. Prasterone (DHEA) and mania. *Ann Pharmacother* 34:1419–22.

di Michele, F., C. Caltagirone, G. Bonaviri, E. Romeo, and G. Spalletta. 2005. Plasma dehydroepiandrosterone levels are strongly increased in schizophrenia. *J Psychiatr Res* 39:267–73.

Dubrovsky, B. O. 2005. Steroids, neuroactive steroids and neurosteroids in psychopathology. *Prog Neuropsychopharmacol Biol Psychiatry* 29:169–92.

Eser, D., C. Schüle, T. C. Baghai, E. Romeo, D. P. Uzunov, and R. Rupprecht. 2006a. Neuroactive steroids and affective disorders. *Pharmacol Biochem Behav* 84:656–66.

Eser, D., C. Schüle, E. Romeo, T. C. Baghai, F. di Michele, A. Pasini et al. 2006b. Neuropsychopharmacological properties of neuroactive steroids in depression and anxiety disorders. *Psychopharmacology* 186:373–87.

Fabian, T. J., M. A. Dew, B. G. Pollock, C. F. Reynolds III, B. H. Mulsant, M. A. Butters et al. 2001. Endogenous concentrations of DHEA and DHEA-S decrease with remission of depression in older adults. *Biol Psychiatry* 50:767–74.

Fernandes, C. S. D., F. G. Salum, D. Bandeira, J. Pawlowski, C. Luz, and K. Cherubini. 2009. Salivary dehydroepiandrosterone (DHEA) levels in patients with the complaint of burning mouth: A case-control study. *Oral Surg Oral Med Oral Pathol Oral Radiol Endod* 108:537–43.

Ferrari, E., D. Casarotti, B. Muzzoni, N. Albertelli, L. Cravello, M. Fioravanti, S. B. Solerte, and F. Magri. 2001a. Age-related changes of the adrenal secretory pattern: Possible role in pathological brain aging. *Brain Res Rev* 37:294–300.

Ferrari, E., L. Cravello, B. Muzzoni, D. Casarotti, M. Paltro, S. B. Solerte et al. 2001b. Age-related changes of the hypothalamic-pituitary-adrenal axis: Pathophysiological correlates. *Eur J Endocrinol* 144:319–29.

Fukai, S., M. Akishita, S. Yamada, T. Hama, S. Ogawa, K. Iijima et al. 2009. Association of plasma sex hormone levels with functional decline in elderly men and women. *Geriatr Gerontol Int* 9:282–9.

Fuller, S. J., R. S. Tan, and R. N. Martins. 2007. Androgens in the etiology of Alzheimer's disease in aging men and possible therapeutic interventions. *J Alzheimers Dis* 12:129–42.

Galimberti, D., and E. Scarpini. 2010. Genetics and biology of Alzheimer's disease and frontotemporal lobar degeneration. *Int J Clin Exp Med* 3:129–43.

Goodyer, I. M., J. Herbert, A. Tamplin, and P. M. E. Altham. 2000a. First-episode major depression in adolescents: Affective, cognitive and endocrine characteristics of risk status and predictors of onset. *Br J Psychiatry* 176:142–9.

Goodyer, I. M., J. Herbert, A. Tamplin, and P. M. E. Altham. 2000b. Recent life events, cortisol, dehydroepiandrosterone and the onset of major depression in high-risk adolescents. *Br J Psychiatry* 177:499–504.

Goodyer, I. M., R. J. Park, and J. Herbert. 2001a. Psychosocial and endocrine features of chronic first-episode major depression in 8–16 year olds. *Biol Psychiatry* 50:351–7.

Goodyer, I. M., R. J. Park, C. M. Netherton, and J. Herbert. 2001b. Possible role of cortisol and dehydroepiandrosterone in human development and psychopathology. *Br J Psychiatry* 179:243–9.

Gordon, C. M., E. Grace, S. J. Emans, H. A. Feldman, E. Goodman, K. A. Becker et al. 2002. Effects of oral dehydroepiandrosterone on bone density in young women with anorexia nervosa: A randomized trial. *J Clin Endocrinol Metab* 87:4935–41.

Gotlib, I. H., and J. Joormann. 2010. Cognition and depression: Current status and future directions. *Annu Rev Clin Psychol* 6:285–312.

Guazzo, E. P., P. J. Kirkpatrick, I. M. Goodyer, H. M. Shiers, and J. Herbert. 1996. Cortisol, dehydroepiandrosterone (DHEA), and DHEA sulfate in the cerebrospinal fluid of man: Relation to blood levels and the effects of age. *J Clin Endocrinol Metab* 81:3951–60.

Hampl, R., and M. Bicíková. 2010. Neuroimmunomodulatory steroids in Alzheimer dementia. *J Steroid Biochem Mol Biol* 119:97–104.

Haren, M. T., T. K. Malmstrom, W. A. Banks, P. Patrick, D. K. Miller, and J. E. Morley. 2007. Lower serum DHEAS levels are associated with a higher degree of physical disability and depressive symptoms in middle-aged to older African American women. *Maturitas* 57:347–60.

Harris, T. O., S. Borsanyi, S. Messari, K. Stanford, S. E. Cleary, H. M. Shiers et al. 2000. Morning cortisol as a risk factor for subsequent depressive disorder in adult women. *Br J Psychiatry* 177:505–10.

Harris, D. S., O. M. Wolkowitz, and V. I. Reus. 2001. Movement disorder, memory, psychiatric symptoms and serum DHEA levels in schizophrenic and schizoaffective patients. *World J Biol Psychiatry* 2:99–102.

Henderson, V. W. 2010. Action of estrogens in the aging brain: Dementia and cognitive aging. *Biochim Biophys Acta* 1800:1077–83.

Heuser, I., M. Deuschle, P. Luppa, U. Schweiger, H. Standhardt, and B. Weber. 1998. Increased diurnal plasma concentrations of dehydroepiandrosterone in depressed patients. *J Clin Endocrinol Metab* 83:3130–3.

Hillen, T., A. Lun, F. M. Reischies, M. Borchelt, E. Steinhagen-Thiessen, and R. T. Schaub. 2000. DHEA-S plasma levels and incidence of Alzheimer's disease. *Biol Psychiatry* 47:161–3.

Hoskin, E. K., M. X. Tang, J. J. Manly, and R. Mayeux. 2004. Elevated sex-hormone binding globulin in elderly women with Alzheimer's disease. *Neurobiol Aging* 25:141–7.

Hsiao, C. 2006. Positive correlation between anxiety severity and plasma levels of dehydroepiandrosterone sulfate in medication-free patients experiencing a major episode of depression. *Psychiatry Clin Neurosci* 60:746–50.

Hunt, P. J., E. M. Gurnell, F. A. Huppert, C. Richards, A. T. Prevost, J. A. Wass et al. 2000. Improvement in mood and fatigue after dehydroepiandrosterone replacement in Addison's disease in a randomized, double blind trial. *J Clin Endocrinol Metab* 85:4650–6.

Jozuka, H., E. Jozuka, S. Takeuchi, and O. Nishikaze. 2003. Comparison of immunological and endocrinological markers associated with major depression. *J Int Med Res* 31:36–41.

Kahl, K. G., S. Bens, K. Ziegler, S. Rudolf, L. Dibbelt, A. Kordon et al. 2006. Cortisol, the cortisol-dehydroepiandrosterone ratio, and pro-inflammatory cytokines in patients with current major depressive disorder comorbid with borderline personality disorder. *Biol Psychiatry* 59:667–71.

Katz, M. M., C. L. Bowden, and A. Frazer. 2010. Rethinking depression and the actions of antidepressants: Uncovering the links between the neural and behavioral elements. *J Affect Disord* 120:16–23.

Kim, S. B., M. Hill, Y. T. Kwak, R. Hampl, D. H. Jo, and R. Morfin. 2003. Neurosteroids: Cerebrospinal fluid levels for Alzheimer's disease and vascular dementia diagnostics. *J Clin Endocrinol Metab* 88:5199–206.

Kline, M. D., and E. D. Jaggers. 1999. Mania onset while using dehydroepiandrosterone. *Am J Psychiatry* 156:970.

Kritz-Silverstein, D., D. von Mühlen, G. A. Laughlin, and R. Bettencourt. 2008. Effects of dehydroepiandrosterone supplementation on cognitive function and quality of life: The DHEA and Well-Ness (DAWN) trial. *J Am Geriatr Soc* 56:1292–8.

Laufer, N., R. Maayan, H. Hermesh, S. Marom, R. Gilad, R. Strous et al. 2005. Involvement of $GABA_A$ receptor modulating neuroactive steroids in patients with social phobia. *Psychiatry Res* 137:131–6.

Luz, C., F. Dornelles, T. Preissler, D. Collaziol, I. M. da Cruz, and M. E. Bauer. 2003. Impact of psychological and endocrine factors on cytokine production of healthy elderly people. *Mech Ageing Dev* 124:887–95.

Maayan, R., Y. Yagorowski, D. Grupper, M. Weiss, B. Shtaif, M. A. Kaoud et al. 2000. Basal plasma dehydroepiandrosterone sulfate level: A possible predictor for response to electroconvulsive therapy in depressed psychotic inpatients. *Biol Psychiatry* 48:693–701.

Maninger, N., O. M. Wolkowitz, V. I. Reus, E. S. Epel, and S. H. Mellon. 2009. Neurobiological and neuropsychiatric effects of dehydroepiandrosterone (DHEA) and DHEA sulfate (DHEAS). *Front Neuroendocrinol* 30:65–91.

Mansell, W., and R. Pedley. 2008. The ascent into mania: A review of psychological processes associated with the development of manic symptoms. *Clin Psychol Rev* 28:494–520.

Markianos, M., J. Tripodianakis, D. Sarantidis, and J. Hatzimanolis. 2007. Plasma testosterone and dehydroepiandrosterone sulfate in male and female patients with dysthymic disorder. *J Affect Disord* 101:255–8.

Markowitz, J. S., W. H. Carson, and C. W. Jackson. 1999. Possible dihydroepiandrosterone-induced mania. *Biol Psychiatry* 45:241–2.

Marx, C. E., R. D. Stevens, L. J. Shampine, V. Uzunova, W. T. Trost, M. I. Butterfield et al. 2006a. Neuroactive steroids are altered in schizophrenia and bipolar disorder: Relevance to pathophysiology and therapeutics. *Neuropsychopharmacology* 31:1249–63.

Marx, C. E., W. T. Trost, L. J. Shampine, R. D. Stevens, C. M. Hulette, D. C. Steffens et al. 2006b. The neurosteroid allopregnanolone is reduced in prefrontal cortex in Alzheimer's disease. *Biol Psychiatry* 60:1287–94.

McLean, C. P., and E. R. Anderson. 2009. Brave men and timid women? A review of the gender differences in fear and anxiety. *Clin Psychol Rev* 29:496–505.

Michael, A., A. Jenaway, E. S. Paykel, and J. Herbert. 2000. Altered salivary dehydroepiandrosterone levels in major depression in adults. *Biol Psychiatry* 48:989–95.

Morgan, M. L., A. J. Rapkin, G. Biggio, M. Serra, M. G. Pisu, and N. Rasgon. 2010. Neuroactive steroids after estrogen exposure in depressed postmenopausal women treated with sertraline and asymptomatic postmenopausal women. *Arch Womens Ment Health* 13:91–8.

Näisman, B., T. Olsson, M. Fagerlund, S. Eriksson, M. Viitanen, and K. Carlström. 1996. Blunted adrenocorticotropin and increased adrenal steroid response to human corticotropin-releasing hormone in Alzheimer's disease. *Biol Psychiatry* 39:311–8.

Nagata, C., H. Shimizu, R. Takami, M. Hayashi, N. Takeda, and K. Yasuda. 2000. Serum concentrations of estradiol and dehydroepiandrosterone sulfate and soy product intake in relation to psychologic well-being in peri- and postmenopausal Japanese women. *Metabolism* 49:1561–4.

Nandi, A., J. R. Beard, and S. Galea. 2009. Epidemiologic heterogeneity of common mood and anxiety disorders over the lifecourse in the general population: A systematic review. *BMC Psychiatry* 9:31.

Nawata, H., T. Yanase, K. Goto, T. Okabe, and K. Ashida. 2002. Mechanism of action of anti-aging DHEA-S and the replacement of DHEA-S. *Mech Ageing Dev* 123:1101–6.

Naylor, J. C., C. M. Hulette, D. C. Steffens, L. J. Shampine, J. F. Ervin, V. M. Payne et al. 2008. Cerebrospinal fluid dehydroepiandrosterone levels are correlated with brain dehydroepiandrosterone levels, elevated in Alzheimer's disease, and related to neuropathological disease stage. *J Clin Endocrinol Metab* 93:3173–8.

Nelson-DeGrave, V. L., J. K. Wickenheisser, J. E. Cockrell, J. R. Wood, R. S. Legro, J. F. Strauss III, and J. M. Mcallister. 2004. Valproate potentiates androgen biosynthesis in human ovarian theca cells. *Endocrinology* 145:799–808.

Oades, R. D., and R. Schepker. 1994. Serum gonadal steroid hormones in young schizophrenic patients. *Psychoneuroendocrinology* 19:373–85.

Ozsoy, S., and E. Esel. 2008. Hypothalamic–pituitary–adrenal axis activity, dehydroepiandrosterone sulphate and their relationships with aggression in early and late alcohol withdrawal. *Prog Neuropsychopharmacol Biol Psychiatry* 32:340–7.

Padberg, F., F. di Michele, P. Zwanzger, E. Romeo, G. Bernardi, C. Schüle et al. 2002. Plasma concentrations of neuroactive steroids before and after repetitive transcranial magnetic stimulation (rTMS) in major depression. *Neuropsychopharmacology* 27:874–8.

Paslakis, G., P. Luppa, M. Gilles, D. Kopf, B. Hamann-Weber, F. Lederbogen, and M. Deuschle. 2010. Venlafaxine and mirtazapine treatment lowers serum concentrations of dehydroepiandrosterone-sulfate in depressed patients remitting during the course of treatment. *J Psychiatr Res* 44:556–60.

Pérez-Neri, I., S. Montes, C. Ojeda-López, J. Ramírez-Bermúdez, and C. Ríos. 2008. *Prog Neuropsychopharmacol Biol Psychiatry* 32:1118–30.

Pendlebury, S. T. 2009. Stroke-related dementia: Rates, risk factors and implications for future research. *Maturitas* 64:165–71.

Pico-Alfonso, M. A., M. I. Garcia-Linares, N. Celda-Navarro, J. Herbert, and M. Martinez. 2004. Changes in cortisol and dehydroepiandrosterone in women victims of physical and psychological intimate partner violence. *Biol Psychiatry* 56:233–40.

Poór, V., S. Juricskay, A. Gáti, P. Osváth, and T. Tényi. 2004. Urinary steroid metabolites and 11β-hydroxysteroid dehyrogenase activity in patients with unipolar recurrent major depression. *J Affect Disord* 81:55–9.

Rabkin, J. G., M. C. McElhiney, R. Rabkin, P. J. McGrath, and S. J. Ferrando. 2006. Placebo-controlled trial of dehydroepiandrosterone (DHEA) for treatment of nonmajor depression in patients with HIV/AIDS. *Am J Psychiatry* 163:59–66.

Rasmuson, S., B. Näsman, S. Eriksson, K. Calström, and T. Olsson. 1998. Arenal responsivity in normal aging and mild to moderate Alzheimer's disease. *Biol Psychiatry* 43:401–7.

Ravindran, A. V., R. W. Lam, M. J. Filteau, F. Lespérance, S. H. Kennedy, S. V. Parikh et al. 2009. Canadian Network for Mood and Anxiety Treatments (CANMAT) clinical guidelines for the management of major depressive disorder in adults: V. Complementary and alternative medicine treatments. *J Affect Disord* 117:S54–64.

Ritsner, M. S. 2010. Pregnenolone, dehydroepiandrosterone, and schizophrenia: Alterations and clinical trials. *CNS Neurosci Ther* 16:32–44.

Ritsner, M. S., and R. D. Strous. 2010. Neurocognitive deficits in schizophrenia are associated with alterations in blood levels of neurosteroids: A multiple regression analysis of findings from a double-blind, randomized, placebo-controlled, crossover trial with DHEA. *J Psychiatric Res* 44:75–80.

Sánchez-Martín, J. R., A. Azurmendi Imaz, E. Fano Ardanaz, F. Braza Lloret, J. M. Muñoz Sánchez, and M. R. Carreras de Alba. 2009. Niveles de andrógenos, estilos parentales y conducta agresiva en niños y niñas de 5-6 años de edad. *Psicothema* 21:57–62.

Schmidt, P. J., R. C. Daly, M. Bloch, M. J. Smith, M. A. Danaceau et al. 2005. Dehydroepiandrosterone monotherapy in midlife-onset major and minor depression. *Arch Gen Psychiatry* 62:154–62.

Schneider, L. S., M. Hinsey, and S. Lyness. 1992. Plasma dehydroepiandrosterone sulfate in Alzheimer's disease. *Biol Psychiatry* 31:205–8.

Schüle, C., T. C. Baghai, D. Eser, M. Schwarz, B. Bondy, and R. Rupprecht. 2009. Effects of mirtazapine on dehydroepiandrosterone-sulfate and cortisol plasma concentrations in depressed patients. *J Psychiatric Res* 43:538–45.

Schüle, C., F. di Michele, T. Baghai, E. Romeo, G. Bernardi, P. Zwanzger et al. 2003. Influence of sleep deprivation on neuroactive steroids in major depression. *Neuropsychopharmacology* 28:577–81.

Schumacher, M., S. Weill-Engerer, P. Liere, F. Robert, R. J. M. Franklin, L. M. Garcia-Segura et al. 2003. Steroid hormones and neurosteroids in normal and pathological aging of the nervous system. *Prog Neurobiol* 71:3–29.

Schupf, N., S. Winsten, B. Patel, D. Pang, M. Ferin, W. B. Zigman et al. 2006. Bioavailable estradiol and age at onset of Alzheimer's disease in postmenopausal women with Down syndrome. *Neurosci Lett* 406:298–302.

Semeniuk, T., G. S. Jhangri, and G. M. Le Mellédo. 2001. Neuroactive steroid levels in patients with generalized anxiety disorder. *J Neuropsychiatry Clin Neurosci* 13:396–8.

Silver, H., G. Knoll, V. Isakov, C. Goodman, and Y. Finkelstein. 2005. Blood DHEAS concentrations correlate with cognitive function in chronic schizophrenia patients. A pilot study. *J Psychiatr Res* 39:569–75.

Simpson, E. H., C. Kellendonk, and E. Kandel. 2010. A possible role for the striatum in the pathogenesis of the cognitive symptoms of schizophrenia. *Neuron* 65:585–96.

Soma, K. K., M. A. L. Scotti, A. E. M. Newman, T. D. Charlier, and G. E. Demas. 2008. Novel mechanisms for neuroendocrine regulation of aggression. *Front Neuroendocrinol* 29:476–89.

Stevens, J. R. 2002. Schizophrenia: Reproductive hormones and the brain. *Am J Psychiatry* 159:713–9.

Strous, R. D., R. Maayan, R. Lapidus, L. Goredetsky, E. Zeldich, M. Kotler, and A. Weizman. 2004. Increased circulatory dehydroepiandrosterone and dehydroepiandrosterone-sulphate in first-episode schizophrenia: Relationship to gender, aggression and symptomatology. *Schizophr Res* 71:427–34.

Strous, R. D., R. Maayan, R. Lapidus, R. Stryjer, M. Lustig, M. Kotler, and A. Weizman. 2003. Dehydroepiandrosterone augmentation in the management of negative, depressive, and anxiety symptoms in schizophrenia. *Arch Gen Psychiatry* 60:133–41.

Talih, F., O. Fattal, and D. Malone Jr. 2007. Anabolic steroid abuse: Psychiatric and physical costs. *Cleve Clin J Med* 74:341–52.

Tyrer, P., and D. Baldwin. 2006. Generalised anxiety disorder. *Lancet* 368:2156–66.

Ulubaev, A., D. M. Lee, N. Purandare, N. Pendleton, and F. C. W. Wu. 2009. Activational effects of sex hormones on cognition in men. *Clin Endocrinol* 71:607–23.

Vacheron-Trystam, M. N., S. Cheref, J. Gauillard, and J. Plas. 2002. À propous d'un cas de manie sous DHEA. *L'Encephale* 28:563–6.

Valtysdottir, S. T., L. Wide, and R. Hallgren. 2003. Mental wellbeing and quality of sexual life in women with primary Sjögren's syndrome are related to circulating dehydroepiandrosterone sulphate. *Ann Rheum Dis* 62:875–9.

van Goozen, S. H. M., W. Matthys, P. T. Cohen-Kettenis, J. H. H. Thijssen, and H. van Engeland. 1998. Adrenal androgens and aggression in conduct disorder prepubertal boys and normal controls. *Biol Psychiatry* 43:156–8.

Weill-Engerer, S., J. P. David, V. Sazdovitch, P. Liere, B. Eychenne, A. Pianos et al. 2002. Neurosteroid quantification in human brain regions: Comparison between Alzheimer's and nondemented patients. *J Clin Endocrinol Metab* 87:5138–43.

Wojtal, K., M. K. Trojnar, and S. J. Czuczwar. 2006. Endogenous neuroprotective factors: Neurosteroids. *Pharmacol Rep* 58:335–40.

Wolkowitz, O. M., V. I. Reus, A. Keebler, N. Nelson, M. Friedland, L. Brizendine, and E. Roberts. 1999. Double-blind treatment of major depression with dehydroepiandrosterone. *Am J Psychiatry* 156:646–9.

Yau, J. L. W., S. Rasmuson, R. Andrew, M. Graham, J. Noble, T. Olsson et al. 2003. Dehydroepiandrosterone 7-hydroxylase CYP7B: Predominant expression in primate hippocampus and reduced expression in alzheimer's disease. *Neuroscience* 121:307–14.

Young, A. H., P. Gallagher, and R. J. Porter. 2002. Elevation of the cortisol-dehydroepiandrosterone ratio in drug-free depressed patients. *Am J Psychiatry* 159:1237–9.

20 DHEA, Androgen Receptors, and Their Potential Role in Breast Cancer

Zeina Nahleh and Nishant Tageja

CONTENTS

INTRODUCTION

Dehydroepiandrosterone (DHEA) is an endogenous steroid that has been implicated in a broad range of biological effects in humans and other mammals (Schulman and Dean 2007). DHEA is produced by the adrenal glands, gonads, and the brain (Mo, Lu, and Simon 2006). Dehydroepiandrosterone sulfate (DHEAS) is the sulfated version of DHEA. In the blood, most DHEA is found as DHEAS with levels that are about 300 times higher than those of free DHEA. Plasma DHEAS levels in adult women are 10,000 times higher than those of testosterone and 3,000–30,000 times higher than those of estradiol (E2), thus providing a large reservoir of substrate for conversion into androgens and/or estrogens in the peripheral tissues, which possess the enzymatic mechanisms necessary to transform DHEA into active sex steroids (NIH National Library of Medicine).

DHEA acts as a precursor to approximately 30%–50% of circulating androgens in men and 100% of circulating estrogens in postmenopausal women (Labrie et al. 1997; Arlt et al. 1999). Notably, DHEA secretion declines with age, a phenomenon referred to as the "adrenopause" (Parker et al. 1997). This DHEA reduction occurs in both sexes and is associated with a reduction in the size of the zona reticularis. In women, estradiol plasma levels decrease by 90% after menopause (Russo and Russo 2006), and the main estrogen is estrone, resulting from the aromatization of androgens in adipose tissue (Gruber et al. 2002). The aromatase activity increases to maintain high concentrations of estrogens in the body (Somboonporn and Davis 2004).

Despite this compensatory mechanism, it has been suggested that the DHEA reduction may have other implications for health in old age, and its effects on immune cell function and cytokine productions have been reported (Hazeldine, Arlt, and Lord 2010). Also, of particular interest would be the effect of DHEA reduction on decreasing androgen levels and the implications of adrenopause on the risk of hormonally driven cancers like breast cancer. Although men are sheltered from the age-related decline in serum DHEA by the continuous testicular secretion of androgens, women depend solely on adrenal DHEA for their production of androgens. The 70%–95% reduction in serum DHEA after menopause leads, therefore, to major androgen deficiency in postmenopausal women. Estrogens are known to directly stimulate the proliferation of breast cells, whereas the effect of androgens on breast tissue is more complex and still unclear. Elucidating the role of DHEA, androgens, and androgen receptors (ARs) in breast cancer may unravel, as of yet, unexplored territories in the management of this disease. Understanding the effects of androgen and the AR in women, as well as the mechanisms of action of DHEA and its interaction with the AR both directly and through its metabolites, may be a reasonable first step.

ANDROGENS IN BREAST CANCER

Historical Use for Breast Cancer Treatment

In vivo studies have shown that androgens may affect the growth of breast carcinoma in animals (Smith and King 1972). Pharmacologic administration of androgens to rats bearing dimethylbenzanthracene-induced breast carcinoma leads to tumor regression. Tumor proliferation in human mammary carcinoma also is significantly altered by androgens (Lippman, Bolan, and Huff 1976). Historically, androgens have been used successfully as hormonal therapy for advanced breast cancer. Approximately 20% of patients with metastatic breast carcinoma may experience tumor regression after treatment with androgens (AMA 1960; Goldenberg et al. 1973). However, androgen therapy (e.g., fluoxymesterone or testosterone) has not gained popularity due to a high incidence of undesirable, virilizing side effects. Also, the advent of estrogen receptor (ER)-targeted therapy and aromatase inhibitors (AIs) for the treatment of ER^+ breast cancer has focused hormonal therapy on those agents. Of particular interest is the role of AIs, which block the conversion of adrenal steroids (mainly androgens) into estrogens in the treatment of breast cancer (Assikis and Buzdar 2002; Brueggemeier 2002; Miller et al. 1973; Nimrod and Ryan 1975; Winer et al. 2002). This would also underscore the important role of androgens (albeit in an indirect way, through estrogens) in the stimulation of human mammary carcinoma growth. Thus, androgens can have either stimulatory or inhibitory effects on tumor growth. These seemingly paradoxical effects may depend on carcinoma cell type and/or may be related to the presence or absence of other steroid receptors, such as ER and progesterone receptor (PR). In addition, the heterogeneity of carcinoma cells in terms of steroid receptor positivity and the proportional distribution of each steroid receptor among carcinoma cells may influence the activity of androgens in either a proliferative or inhibitory direction.

Paradoxical Effect: Stimulatory or Inhibitory?

Androgens have a predominantly inhibitory effect on the growth of breast cancer cells, both *in vitro* and *in vivo* (Birrell et al. 1995; de Launoit et al. 1991; Dauvois et al. 1991; Greeve et al. 2004; Hackenberg et al. 1991; Ortmann et al. 2002), potentially through induction of apoptosis (Hardin et al. 2007; Kandouz et al. 1999; Lapointe et al. 1999). However, preclinical studies have suggested that androgen action in breast cancer cell lines could be cell type-specific and has been reported to result in either stimulation or inhibition of proliferation as noted in the previous section, "Historical Use for Breast Cancer Treatment" (Birrell et al. 1995).

Clinically, it has been suggested that the balance between androgenic and estrogenic stimuli drives the proliferation of breast tumors. The overwhelming clinical evidence for tumor regression observed in 20%–50% of pre- and postmenopausal breast cancer patients treated with various androgens favors the view that naturally occurring androgens might constitute, as mentioned in the previous section, an overlooked, direct inhibitory control of mammary cancer cell growth (Adair et al. 1949; Gordan et al. 1973; Ingle et al. 1991; Segaloff et al. 1951; Tormey et al. 1983). In that regard, it has been found that Western women with breast cancer who have a low excretion of adrenal androgenic metabolites respond more poorly to endocrine therapy and have a shorter survival time (Zumoff et al. 1981). Also, in a prospective study in this field, levels of androgen metabolites in urine were found to be abnormally reduced in premenopausal women who subsequently developed breast cancer (Bulbrook, Hayward, and Spicer 1971), indicating a protective role of androgens on the breast. In contrast, other studies have led to contradictory data (Bulbrook, Hayward, and Spicer 1971; Eliassen et al. 2006; Page et al. 2004). A prospective study of premenopausal women found no association between plasma adrenal androgen levels and risk of breast cancer (Page et al. 2004). In the Nurses' Health Study II, no correlation was found between DHEA and DHEAS levels and breast cancer risk overall, but interestingly, among premenopausal women, there was a positive association, especially for tumors that express both ERs and PRs (Bulbrook, Hayward, and Spicer 1971). Also, among premenopausal women, higher levels of testosterone and androstendione were associated with increased risk of invasive ER^+/PR^+ tumors, although with a nonstatistically significant increase in overall risk of breast cancer (Eliassen et al. 2006). In postmenopausal women, similarly, epidemiological studies showed that elevated serum levels of both estrogens and androgens contribute to a greater risk of breast cancer (Berrino et al. 1996; Dorgan et al. 1996), and a meta-analysis of nine prospective studies revealed that breast cancer risk increases with increasing concentrations of almost all sex hormones (Key et al. 2002). None of these studies manage, however, to disconnect the risk associated with increased estradiol levels from the androgen component. This is a major confounding factor in independently assesing the role of androgen from known cancer-promoting estrogen effect since androgens are the obligate precursors for estradiol synthesis.

Limitation of Androgen Assays

Several epidemiological studies have examined the correlation of circulating androgens, such as testosterone, and the risk for breast cancer. Some of the potential limitations that prevented the clear identification of a role for naturally occurring androgens in association with many diseases, including breast cancer, include the design of these trials. This includes comparison of normal control subjects with patients already having breast cancer and, frequently, too small number of patients in case control studies. But a major limitation of many studies is the lack of reliability of serum steroid levels measured by radioimmunoassay. First, the androgen assays used were developed primarily to measure the higher levels found in men, and they lack reliability in the low ranges found in normal women (Lobo 2001). Second, testosterone and androstenedione levels are the most commonly measured, but they demonstrate substantial daily variability, while most of the epidemiological data are based on a single blood sample collected at nonstandard times. Third, using serum testosterone levels to gauge androgenic effects at the tissue level is problematic because the circulating testosterone is tightly bound to sex-hormone-binding globulin (SHBG), while only the free hormone is bioactive. SHBG and, thus, total testosterone levels, vary widely based on genetic, metabolic, and endocrine influences, and it is now suggested that measurement of free or bioavailable testosterone might predict androgenic effects more accurately than total testosterone levels (Vermeulen, Verdonck, and Kaufman 1999). But more importantly, because the androgens synthesized locally in peripheral tissues from the precursor DHEA do not originate from circulating testosterone, one could reasonably expect that the measurement of the serum testosterone levels is of questionable biological and clinical significance as a marker of androgenic activity. Androgens made locally in large amounts act in the same cells where synthesis takes place and are not released in significant amounts in the circulation, thus limiting the reliability of the measurement of serum testosterone levels as a marker of total androgenic activity.

It has been recently suggested that a more practical and probably more valid measure of androgenic activity in women is measuring the glucuronide derivatives of androgens, the obligatory route of elimination of all androgens (Labrie et al. 2006). Measurement of the total pool of androgens reflected by the serum levels of androsterone glucuronide (ADT-G), and androstenediol glucuronide (3α-diol-G), can be done using a validated liquid chromatography tandem mass spectrometry technique (Labrie et al. 2006). While not permitting the assessment of androgenic activity in specific tissues, measurement of the glucuronide derivatives of ADT and 3α-diol-G by validated mass spectrometry techniques would permit a precise measure of the total pool of androgens in the whole organism.

In conclusion, a clear association between androgens and clinical situations affecting women's health including breast cancer has remained somewhat elusive despite the long series of cohort studies performed during the last 20 years. Identifying a reliable and valid test of androgenic activity and function is a crucial first step to better elucidate the role of androgens in any clinical situation believed to be under androgen control, particularly in women. Measuring serum levels of ADT-G and 3α-diol-G might be a more reliable measure to assess androgenic activity compared with serum testosterone or any other steroid, including DHEA or DHEAS.

THE ANDROGEN RECEPTOR AS A POTENTIAL THERAPEUTIC TARGET IN BREAST CANCER

Androgen Receptor–Dependent Androgenic Action

The AR is a member of the steroid receptor subfamily also containing the glucocorticoid receptor (GR), PR, and mineralocorticoid receptor, and it binds to the same response elements as these receptors (Beato and Klug, 2000). There is emerging evidence that the androgen-signaling pathway plays a critical role in breast carcinogenesis (Birrell, Hall, and Tilley 1998; Brys 2000; Langer et al. 1990; Liao and Dickson 2002). The AR is expressed in more than 70% of breast cancer and has been implicated in the pathogenesis of this disease (Birrell, Hall, and Tilley 1998; Brys 2000; Hackenberg and Schulz 1996; Hall et al. 1996; Hall et al. 1998; Honma et al. 2003; Isola 1993; Kuenen-Boumeester et al. 1996; Lea, Kvinnsland, and Thorsen 1989; Langer et al. 1990; Liao and Dickson 2002; Lundgren, Soreide, and Lea 1994; Moinfar et al. 2003; Riva et al. 2005; Spinder et al. 1989; Soreide and Kvinnsland 1991). This could be through the activation of a number of estrogen responsive genes (Nantermet et al. 2005). However, many pathological studies have demonstrated that direct AR-mediated action of androgens is the major mechanism used by androgens to influence the growth of breast carcinomas, independent of the estrogen and PRs (Doane et al. 2006; Labrie et al. 2003; Liao and Dickson 2002).

Birrell et al. have run a series of experiments using androgenic agents, dihydrotestosterone (DHT) and mibolerone, on six human breast cancer cell lines (Birrell et al. 1995). Their data suggests that androgens inhibit the proliferation of T47-D and ZR-75-1 cells via an interaction with the AR. However, in the case of MDA-MB-453 and MCF-7 breast cancer cells, androgen-induced stimulation of proliferation was observed, and both AR-dependent and AR-independent pathways appear to be involved. Two other cell lines examined, MDA-MB-231 and BT-20, which expressed very low or undetectable levels of AR, were not affected by androgens. All stimulatory and inhibitory proliferative responses were reversed by androgen antagonists (hydroxyflutamide or anandron); however, the androgen antagonists alone had no significant effect on cell proliferation. This observation suggests that the androgens' interaction through AR may primarily cause inhibitory growth on cancer cells; however, AR-independent activity may also occur, and that is influenced by the presence or absence of other receptors such as ER. Other studies have shown that activation of AR-independent pathways could result from the action of active metabolites of DHT that have estrogenic-like actions (Hackenberg et al. 1991). One of the metabolites of DHT, 5α androstane-3B, 17β-diol, was shown to increase proliferation of the MCF-7 cell line via interaction with ER (Hackenberg et al. 1991). One could, therefore, hypothesize that in the absence of ER, as observed by Birell et al., androgenic action may be mediated mostly via interaction of DHT metabolites with AR. However, in breast cancer cells expressing ER, such as the MCF-7, ZR-75-1, and T47-D cell lines, androgenic action is executed via interaction of DHT metabolites with ER. This interaction may explain the differential androgenic effect on different cell types and the paradoxical effect observed in some preclinical studies.

Androgen Receptor Frequency in Breast Cancer

Moinfar et al. have studied the frequency of AR expression in 200 cases of breast carcinoma (Moinfar et al. 2003). Sixty percent of invasive carcinoma and 82% of ductal carcinoma in situ (DCIS) were AR^+. Also, 46% of all ER^- invasive carcinomas were AR^+; and among the poorly differentiated invasive carcinomas, 39% were ER^- and PR^- but AR^+. Among noninvasive carcinomas, 68% were ER^- but AR^+. It is, therefore, possible that breast tumors known as ER^- and/or PR^- may not be truly hormone insensitive and exploration of androgen-based hormone therapy using AR as a target may be warranted in this population. It is clear that the frequent expression of AR in breast carcinoma cells, as observed in multiple studies outlined in the previous section, "Androgen Receptor–Dependent Androgenic Action", raises the important question of the interaction between androgens and human breast carcinoma. While the expression of the AR is necessary for androgens to modulate the growth of breast cancer cells *in vitro*, additional cellular factors, such as the interaction with ER, may determine whether cell proliferation will be stimulated or inhibited in the presence of androgens. Further research should attempt to determine those factors.

DHEA IN BREAST CANCER

DHEA's Action through the Androgen Receptor

Some *in vitro* studies have found DHEA to have both antiproliferative and apoptotic effects on cancer cell lines (Loria 2002; Schulz et al. 1992; Tworoger et al. 2006; Yang et al. 2002). The clinical significance of these findings has largely remained unclear.

In order to investigate the effect of DHEA and its metabolites on mammary carcinoma, Li and his colleagues studied the effect of increasing circulating levels of DHEA constantly released from Silastic implants on the development of mammary carcinoma induced by 7,12-dimethylbenz(a) anthracene (DMBA) in rats. Treatment with increasing doses of DHEA caused a progressive inhibition of tumor development (Li et al. 1993). It is of interest to see that tumor size in the group of animals treated with the highest dose (6 by 3.0-cm-long implants) of DHEA was similar to that found in ovariectomized animals, thus showing a complete blockade of estrogen action by DHEA.

More recently, in a series of experiments conducted by Hardin et al., three human ER^-/PR^- breast cancer cell lines (HCC 1937, 1954, and 38) were treated with DHEAS (Hardin et al. 2007). HCC cell lines 1954 and 1937 had a strong expression of AR, whereas HCC 38 was weakly positive. Methylthiotetrazole proliferation assay analysis showed DHEAS-induced decreases in cell proliferation of 47% in HCC 1937, 27% in HCC 1954, and 0.4% in HCC 38. It appears, therefore, that the cell lines that demonstrated a strong AR expression showed a decrease in cell proliferation after treatment with DHEAS for 7 days compared with untreated cells, whereas cells that have a barely detectable expression of AR were unaffected by DHEAS treatment. Ten days of culturing HCC 1954 cells after the removal of DHEAS resulted in a 3.5-fold increase in growth. Continuous treatment for the same duration induced a 2.8-fold decrease in growth. Parallel experiments showed no significant changes in HCC 38 cultures. Terminal deoxynucleotidyl transferase dUTP nick end labeling (TUNEL) assays showed DHEAS-induced 2.8-fold increases in apoptosis in HCC 1937, 1.9 in HCC 1954, and no significant difference in HCC 38 cultures. It is worth noting that these cell lines were pretreated with anastrazole to prevent any conversion of DHEAS to estrogens. Quantitative RT-PCR of HCC 1954 cells showed a sixfold DHEAS-induced decrease in AR gene expression at 4 hours. Upon cotreatment with the AR antagonist bicalutamide, the downregulatory effect on the AR by DHEAS was not observed, thus localizing the effect of DHEAS to the AR.

DHEA as an Androgenic Treatment for Estrogen Receptor–Negative Breast Cancer

We could hypothesize that a subset of ER^- and PR^- breast carcinomas may respond to hormonal manipulation with an endogenous precursor of androgens and estrogens, like DHEA.

Experiments by Garreau and colleagues have suggested that ER^- and PR^- breast cancer cells respond to hormonal therapy using DHEAS, provided there is AR expression (Garreau et al. 2006). First, $ER^-/PR^-/AR^-$ HCC 1806 breast cancer cells were shown to be unaffected by treatment with DHEAS and an AI. These cells were then transfected with an AR expression vector and treated with DHEAS/AI for 2 days. Growth inhibition of these cells was compared with that of transfected cells treated with only AI or with nontransfected cells treated with DHEAS/AI. Cell death rates of 53.5% ($p = .001$) and 40.1% ($p = .006$) were seen in transfected cells treated with DHEAS/AI compared with controls for days 1 and 2, respectively. Nontransfected cells were unaffected by treatment. The above preclinical data are also well supported by other studies confirming the inhibitory effect of DHEA on mammary tumors almost exclusively through the androgenic component of its action (Sourla et al. 1998) and suggesting additional potential roles of DHEA in mammary tumors through synergistic effects with antiestrogens (Luo et al. 1997).

These studies suggest that DHEA may be potentially explored as an intervention for the treatment of breast cancer. Its inhibitory effects should be further defined in the different subsets. A subset of ER^-/PR^- breast cancers may respond to hormonal manipulation with DHEA acting through AR.

CONCLUSION AND FUTURE DIRECTIONS

Most androgenic activity in women originates from the peripheral conversion of precursors such as DHEA into androgens within the cells of target tissues, and this activity will not be detected by measurement of traditional circulating androgens like testosterone levels. Better assays and measurement of androgenic activity should be refined and adopted in clinical trials. The effect of DHEA as potential direct inhibitors of breast cancer growth needs to be evaluated. The role of DHEA declines with age; androgen insufficiency and a relative imbalance of sex steroid hormones in favor of estrogens may potentially contribute to the increased risk of breast cancer with age, and this association should be further explored. The role of AR in breast cancer is well supported by preclinical evidence and suggested by the presence of AR in a large proportion of human breast cancers (Nahleh 2008).

Multiple questions may come to mind: Is it possible that postmenopausal women, by losing 70%–95% of DHEA, acquire an additional risk factor for breast cancer through, possibly, the loss of androgenic properties of DHEA and possibly the loss of its direct interaction with AR? AR^+ breast tumors have relatively better prognosis than AR^- tumors. Could this be related to the inhibitory effect of androgens maintained through its interaction with AR? If that is the case, could this effect be enhanced by DHEA replacement in postmenopausal women, therefore, leading to decreased risk of breast cancer recurrence and improved outcome especially in AR^+ tumors? Can DHEA have a wider spectrum of preventive properties across some groups of women and, therefore, decrease the incidence of breast cancer? All these are valid questions that anxiously await validated answers. Multiple, currently ongoing clinical trials are attempting to answer some of these questions and determine the role of DHEA and AR in breast cancer (cancer.gov WSU-2008-012, NCT00972023; cancer.gov MSKCC-07022, 07-022, NCT00468715; cancer.gov OHSU-e2109, NCT00516542).

REFERENCES

Adair, F. E., R. C. Mellors, J. H. Farrow et al. 1949. The use of estrogens and androgens in advanced mammary cancer. *J Am Med Assoc* 15:1193–2000.

AMA Committee on Research. 1960. Androgens and estrogens in the treatment of disseminated mammary carcinoma. *J Am Med Assoc* 172:1271–4.

Arlt, W., J. Haas, F. Callies, et al. 1999. Biotransformation of oral dehydroepiandrosterone in elderly men: Significant increase in circulating estrogens. *J Clin Endocrinol Metab* 84(6):2170–6.

Assikis, V. J., and A. Buzdar. 2002. Recent advances in aromatase inhibitor therapy for breast cancer. *Semin Onco* 29(3 Suppl 11):120–8.

Beato, M., and J. Klug. 2000. Steroid hormone receptors: An update. *Hum Reprod Update* 6:225–36.

Berrino, F., P. Muti, A. Micheli et al. 1996. Serum sex hormone levels after menopause and subsequent breast cancer. *J Natl Cancer Inst* 88:291–6.

Birrell, S. N., J. M. Bentel, T. E. Hickey et al. 1995. Androgens induce divergent proliferative responses in human breast cancer cell lines. *J Steroid Biochem Mol Biol* 52:459–67.

Birrell, S. N., R. E. Hall, and W. D. Tilley. 1998. Role of the androgen receptor in human breast cancer. *J Mammary Gland Biol Neoplasia* 3:95–103.

Brueggemeier, R. W. 2002. Overview of the pharmacology of the aromatase inactivator exemestane. *Breast Cancer Res Treat* 74:177–85.

Bry, S. 2000. Androgens and androgen receptor: Do they play a role in breast cancer? *Med Sci Monit* 6:433–8.

Bulbrook, R. D., J. L. Hayward, and C. C. Spicer. 1971. Relation between urinary androgen and corticoid excretion and subsequent breast cancer. *Lancet* 2:395–398.

Cancer.gov Phase I Pilot Study of Neoadjuvant Dehydroepiandrosterone (DHEA) in Women With Estrogen Receptor-Negative, Progesterone Receptor-Negative, HER2/neu-Negative, and Androgen Receptor-Positive Stage I-III Adenocarcinoma of the Breast. WSU-2008-012, NCT00972023.

Cancer.gov Phase II Study of Bicalutamide in Patients With Androgen Receptor-Positive and Estrogen Receptor- and Progesterone Receptor-Negative Metastatic Breast Cancer. MSKCC-07022, 07-022, NCT00468715.

Cancer.gov Phase I Study of Dehydroepiandrosterone and Letrozole in Patients With Androgen Receptor-Positive and Estrogen Receptor- and Progesterone Receptor-Negative Metastatic Breast Cancer. OHSU-e2109, NCT00516542.

Dauvois, S., C. S. Geng, C. Levesque et al. 1991. Additive inhibitory effects of an androgen and the antiestrogen EM-170 on estradiol-stimulated growth of human ZR-75–1 breast tumors in athymic mice. *Cancer Res* 51:3131–5.

De Launoit, Y., S. Dauvois, M. Dufour et al. 1991. Inhibition of cell cycle kinetics and proliferation by the androgen 5 alpha-dihydrotestosterone and antiestrogen N, n-butyl-N-methyl-11-[16'alpha-chloro-3',17 beta-dihydroxy-estra- 1',3',5'-(10')triene-7'alpha-yl] undecanamide in human breast cancer ZR-75–1 cells. *Cancer Res* 51:2797–802.

Doane, A. S., M. Danso, P. Lal et al. 2006. An estrogen receptor-negative breast cancer subset characterized by a hormonally regulated transcriptional program and response to androgen. *Oncogene* 25:3994–4008.

Dorgan, J. F., C. Longcope, H. E. Stephenson Jr et al. 1996. Relation of prediagnostic serum estrogen and androgen levels to breast cancer risk. *Cancer Epidemiol Biomarkers Prev* 5:533–9.

Eliassen, A. H., S. A. Missmer, S. S. Tworoger et al. 2006. Endogenous steroid hormone concentrations and risk of breast cancer among premenopausal women. *J Natl Cancer Inst* 98:1406–15.

Garreau, J. R., P. Muller, R. Pommier et al. 2006. Transgenic introduction of androgen receptor into estrogen-receptor-, progesterone-receptor-, and androgen-receptor-negative breast cancer cells renders them responsive to hormonal manipulation. *Am J Surg* 191(5):576–80.

Goldenberg, I. S., N. Waters, R. S. Ravdin et al. 1973. Androgenic therapy for advanced breast cancer in women. A report of the Cooperative Breast Cancer Group. *JAMA* 3(223):1267–8.

Gordan, G. S., A. Halden, Y. Horn et al. 1973. (7b, 17a-dimethyltestosterone) as primary and secondary therapy of advanced breast cancer. *Oncology* 28:138–46.

Greeve, M. A., R. K. Allan, J. M. Harvey et al. 2004. Inhibition of MCF-7 breast cancer cell proliferation by 5alpha-dihydrotestosterone; A role for p21(Cip1/Waf1). *J Mol Endocrino* 32:793–810.

Gruber, C. J., W. Tschugguel, C. Schneeberger et al. 2002. Production and actions of estrogens. *N Engl J Med* 346:340–52.

Hackenberg, R., S. Luttchens, J. Hofmann et al. 1991. Androgen sensitivity of the new human breast cancer cell line MFM-223. *Cancer Res* 51:5722–7.

Hackenberg, R., and K. D. Schulz. 1996. Androgen receptor mediated growth control of breast cancer and endometrial cancer modulated by antiandrogen- and androgen-like steroids. *J Steroid Biochem Mol Biol* 56:113–7.

Hall, R. E., J. O. Aspinall, D. J. Horsfall et al. 1996. Expression of the androgen receptor and an androgen-responsive protein, apolipoprotein D, in human breast cancer. *Br J Cancer* 74:1175–80.

Hall, R. E., J. A. Clements, S. N. Birrel et al. 1998. Prostate-specific antigen and gross cystic disease fluid protein-15 are co-expressed in androgen receptor-positive breast tumors. *Br J Cancer* 78:360–5.

Hardin, C., R. Pommier, K. Calhoun et al. 2007. New hormonal therapy for estrogen receptor–negative breast cancer. *World J Surg* 31(5):1432–2323.

Hazeldine, J., W. Arlt, and J. M. Lord. 2010. Dehydroepiandrosterone as a regulator of immune cell function. *J Steroid Biochem Mol Biol* 120(2–3):127–36. Epub 2010 Jan 12.

Honma, N., G. Sakamoto, F. Akiyama et al. 2003. Breast carcinoma in women over the age of 85: Distinct histological pattern and androgen, oestrogen, and progesterone receptor status. *Histopathology* 42:120–7.

Ingle, J. N., D. I. Twito, D. J. Schaid et al. 1991. Combination hormonal therapy with tamoxifen plus fluoxymesterone versus tamoxifen alone in postmenopausal women with metastatic breast cancer. A phase II study. *Cancer* 67:886–91.

Isola, J. J. 1993. Immunohistochemical demonstration of androgen receptor in breast cancer and its relationship to other prognostic factors. *J Pathol* 170(1):31–5.

Kandouz, M., A. Lombet, J. Y. Perrot et al. 1999. Proapoptotic effects of antiestrogens, progestins and androgen in breast cancer cells. *J Steroid Biochem Mol Biol* 69:463–71.

Key, T., P. Appleby, I. Barnes et al. 2002. Endogenous Hormones and Breast Cancer Collaborative Group: Endogenous sex hormones and breast cancer in postmenopausal women: Reanalysis of nine prospective studies. *J Natl Cancer Inst* 94:606–16.

Kuenen-Boumeester, T. H., C. C. Van der Kwast, M. P. Claassen et al. 1996. The clinical significance of androgen receptors in breast cancer and their relation to histological and cell biological parameters. *Eur J Cancer* A32:1560–5.

Labrie, F., A. Bélanger, P. Bélanger et al. 2006. Androgen glucuronides, instead of testosterone, as the new markers of androgenic activity in women. *J Steroid Biochem Mol Biol* 99(4–5):182–8. Epub 2006 Apr 18.

Labrie, A., L. Belanger, J. Cusan et al. 1997. Marked decline in serum concentrations of adrenal C19 sex steroid precursors and conjugated androgen metabolites during aging, *J Clin Endocrinol Metab* 82(8):2396–402.

Labrie, F., V. L. The, C. Labrie et al. 2003. Endocrine and intracrine sources of androgens in women: Inhibition of breast cancer and other roles of androgens and their precursor dehydroepiandrosterone. *Endocrine Reviews* 24(2):152–82.

Langer, M., E. Kubista, M. Schemper et al. 1990. Androgen receptors, serum androgen levels and survival of breast cancer patients. *Arch Gynecol Obstet* 247:203–9.

Lapointe, J., A. Fournier, V. Richard et al. 1999. Androgens down-regulate bcl-2 protooncogene expression in ZR-75–1 human breast cancer cells. *Endocrinology* 140:416–21.

Lea, O. A., S. Kvinnsland, and T. Thorsen. 1989. Improved measurement of androgen receptors in human breast cancer. *Cancer Res* 49:7162–7.

Li, S., X. Yan, A. Bélanger et al. 1993. Prevention by dehydroepiandrosterone of the development of mammary carcinoma induced by 7, 12-dimethylbenz(a)anthracene (DMBA) in the rat. *Breast Cancer Res Treat* 29:203–17.

Liao, D. J., and R. B. Dickson. 2002. Roles of androgens in the development, growth, and carcinogenesis of the mammary gland. *J Steroid Biochem Mol Biol* 20:175–89.

Lippman, M., G. Bolan, and K. Huff. 1976. The effects of androgens and antiandrogens on hormone-responsive human breast cancer in long-term tissue culture. *Cancer Res* 36:4610–18.

Lobo, R. A. 2001. Androgens in postmenopausal women: Production, possible role, and replacement options. *Obstet Gynecol Surv* 56:361–76.

Loria, R. M. 2002. Immune up-regulation and tumor apoptosis by androstene steroids. *Steroids* 67(12):953–66.

Lundgren, S., J. A. Soreide, and O. A. Lea. 1994. Influence of tamoxifen on the tumor content of steroid hormone receptors (ER, PgR and AR) in patients with primary breast cancer. *Anticancer Res* 14:1313–6.

Luo, S., A. Sourla, C. Labrie et al. 1997. Combined effects of dehydroepiandrosterone and EM-800 on bone mass, serum lipids, and the development of dimethylbenz(a)anthracene (DMBA)-induced mammary carcinoma in the rat. *Endocrinology* 138:4435–44.

Miller, W. R., D. McDonald, A. P. Forrest et al. 1973. Metabolism of androgens by human breast tissue. *Lancet* 1:912–3.

Mo, Q., S. F. Lu, and N. G. Simon. 2006. Dehydroepiandrosterone and its metabolites: Differential effects on androgen receptor trafficking and transcriptional activity. *J Steroid Biochem Mol Biol* 99(1):50–8. doi:10.1016/j.jsbmb.2005.11.011. PMID 16524719.

Moinfar, F., M. Okcu, O. Tsybrovskyy et al. 2003. Androgen receptors frequently are expressed in breast carcinomas: Potential relevance to new therapeutic strategies. *Cancer* 98:703–11.

Nahleh, Z. 2008. Androgen receptor as a target for the treatment of hormone receptor-negative breast cancer: An unchartered territory. *Future Oncol* 4(1):15–21.

Nantermet, P. V., P. Masarachia, M. A. Gentile et al. 2005. Androgenic induction of growth and differentiation in the rodent uterus involves the modulation of estrogen-regulated genetic pathways. *Endocrinology* 146(2):564–78.

NIH National Library of Medicine—Dehydroepiandrosterone. Last modified November 18, 2010. http://www.nlm.nih.gov/medlineplus/druginfo/natural/patient-dhea.html. Accessed February 28, 2011.

Nimrod, A., and K. J. Ryan. 1975. Aromatization of androgens by human abdominal and breast fat tissue. *J Clin Endocrinol Metab* 40:367–72.

Ortmann, J., S. Prifti, M. K. Bohlmann et al. 2002. Testosterone and 5 alpha-dihydrotestosterone inhibit in vitro growth of human breast cancer cell lines. *Gynecol Endocrinol* 16:113–20.

Page, J. H., G. A. Colditz, N. Rifai et al. 2004. Plasma adrenal androgens and risk of breast cancer in premenopausal women. *Cancer Epidemiol Biomarkers Prev* 13:1032–6.

Parker Jr, C. R., R. L. Mixon, R. M. Brissie et al. 1997. Aging alters zonation in the adrenal cortex of men. *J Clin Endocrinol Metab* 82(11):3898–901.

Riva, C., E. Dainese, G. Caprara et al. 2005. Immunohistochemical study of androgen receptors in breast carcinoma. Evidence of their frequent expression in lobular carcinoma. *Virchows Arch* 447:695–700.

Russo, J., and I. H. Russo. 2006. The role of estrogen in the initiation of breast cancer. *J Steroid Biochem Mol Biol* 102:89–96.

Schulman, R., and C. Dean. 2007. Solve it with supplements. New York: Rodale Books.

Schulz, S., R. C. Klann, S. Schönfeld et al. 1992. Mechanisms of cell growth inhibition and cell cycle arrest in human colonic adenocarcinoma cells by dehydroepiandrosterone: Role of isoprenoid biosynthesis. *Cancer Res* 52(5):1372–6. PMID 1531325.

Segaloff, A., D. Gordon, B. N. Horwitt et al. 1951. Hormonal therapy in cancer of the breast. 1. The effect of testosterone propionate therapy on clinical course and hormonal excretion. *Cancer* 4:319–23.

Smith, J. A., and R. J. King. 1972. Effects of steroids on growth of an androgen-dependent mouse mammary carcinoma in cell culture. *Exp Cell Res* 73:351–9.

Somboonporn, W., and S. R. Davis. 2004. Postmenopausal testosterone therapy and breast cancer risk. *Maturitas* 49:267–75.

Soreide, O. A., and S. Kvinnsland. 1991. Progesterone-binding cyst protein (PBCP = GCDFP-24) and steroid hormone receptors as markers of differentiation in breast cancer. Inverse relation of distribution in normal and malignant tissue of the same breast. *Anticancer Res* 11:1323–6.

Sourla, A., C. Martel, C. Labrie et al. 1998. Almost exclusive androgenic action of dehydroepiandrosterone in the rat mammary gland. *Endocrinology* 139:753–76.

Spinder, J. J., J. G. Spijkstra, C. W. van den Tweel et al. 1989. The effects of long term testosterone administration on pulsatile luteinizing hormone secretion and on ovarian histology in eugonadal female to male transsexual subjects. *J Clin Endocrinol Metab* 69:151–7.

Tormey, D. C., M. E. Lippman, B. K. Edwards et al. 1983. Evaluation of tamoxifen doses with and without fluoxymesterone in advanced breast cancer. *Ann Intern Med* 98:139–44.

Tworoger, S., S. A. Missmer, A. H. Eliassen et al. 2006. The association of plasma DHEA and DHEA sulfate with breast cancer risk in predominantly premenopausal women. *Cancer Epidemiol Biomarkers Prev* 15(5):967–71. doi:10.1158/1055-9965.EPI-05-0976. PMID 16702378.

Vermeulen, A., L. Verdonck, and J. M. Kaufman. 1999. A critical evaluation of simple methods for the estimation of free testosterone in serum. *J Clin Endocrinol Metab* 84(10):3666–72.

Winer, E. P., C. Hudis, H. J. Burstein et al. 2002. American Society of Clinical Oncology technology assessment on the use of aromatase inhibitors as adjuvant therapy for women with hormone receptor-positive breast cancer: Status report. *J Clin Oncol* 20:3317–27.

Yang, N. C., K. C. Jeng, W. M. Ho et al. 2002. ATP depletion is an important factor in DHEA-induced growth inhibition and apoptosis in BV-2 cells. *Life Sci* 70(17):1979–88. doi:10.1016/S0024-3205(01)01542-9. PMID 12148690.

Zumoff, B., J. Levin, R. S. Rosenfeld et al. 1981. Abnormal 24-hr mean plasma concentrations of dehydroepiandrosterone and dehydroisoandrosterone sulfate in women with primary operable breast cancer. *Cancer Res* 41:3360–3.

21 DHEAS and Periodontal Status in Older Japanese

Akihiro Yoshida and Toshihiro Ansai

CONTENTS

ETIOLOGY AND FEATURES OF PERIODONTITIS

Periodontal disease is a general term used to describe diseases that affect the gingiva, the supporting connective tissue, and alveolar bones, which anchor the teeth in the jaws (Figure 21.1). Periodontal diseases are among the most common chronic disorders that have plagued humans for centuries (Williams 1990).

CLINICAL FEATURES OF PERIODONTITIS

Periodontitis is an infectious disease, suspected to be caused primarily by periodontopathic bacteria that bring about destructive changes, which ultimately leads to loss of bone and connective tissue attachment. The schemata of normal and periodontitis-affected periodontium are shown in Figure 21.2. Of the various forms of periodontitis, adult periodontitis is the most common form. The characteristics of adult periodontitis are listed in Table 21.1, and a representative X-ray image is shown in Figure 21.3. Adult periodontitis is characterized by an age of onset of 35 years or more. The presence of microbial deposits is commensurate with the amount of periodontal destruction, along with generalized or localized bone loss. The flora of the periodontal pockets is characterized by a complex of gram-negative microorganisms. Clinical features include little

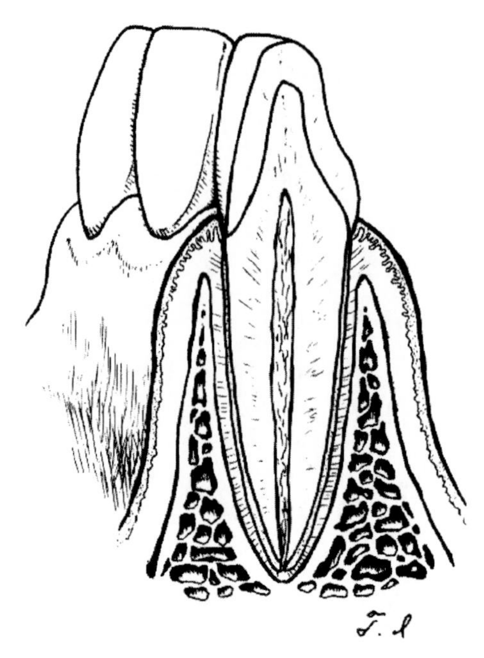

FIGURE 21.1 Schematic illustration of the periodontium. The tooth is held within the alveolar socket by the attachment structures of the periodontium. The gingiva covers the attachment structures of the alveolar bone, periodontal ligament, and cementum.

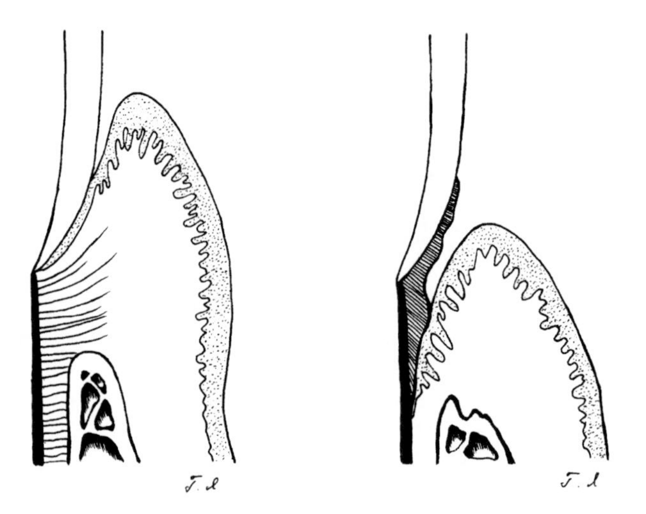

FIGURE 21.2 The schema of healthy gingival sulcus (left) and periodontal pocket (right). Junctional epithelium and gingival collagen fibers are observed in healthy gingiva (left). In contrast, calculus on the root surface and a deepened periodontal pocket are observed in the periodontal pocket (right).

or no proliferation of marginal gingival tissue, although some inflammation may be present. The gingival tissue may be thickened or misshapen; gingival recession sometimes presents. In most cases of untreated adult periodontitis, the amount of plaque and calculus is commensurate with the amount of pocket formation and bone loss. Open interdental contacts and malposed teeth are frequently observed.

TABLE 21.1
Features of Adult Periodontitis

Age of onset is usually 30–35 years or more.
The molars and incisors are more commonly and severely affected than are the canines and premolars.
Conditions enhancing plaque accumulation are present, and amounts of microbial deposits are consistent with the severity of the lesions.
The extent and distribution of bone loss are highly variable; both vertical and horizontal patterns may be seen.
Acute destructive exacerbation can occur at one or more sites.

Source: Modified from Schroeder, H. E., and R. C. Page, *Periodontal Diseases*, Lea & Febiger, Philadelphia, 1990.

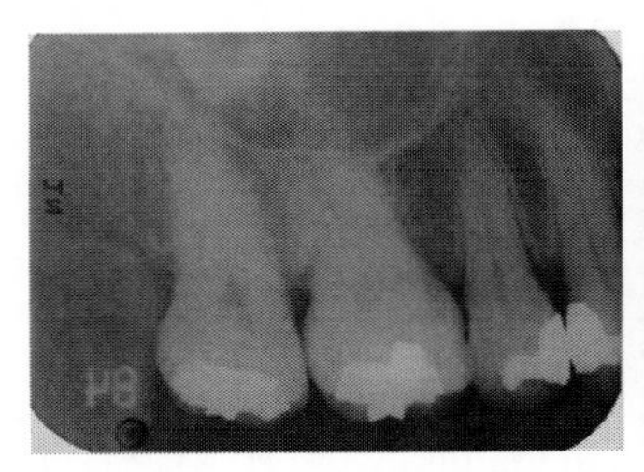

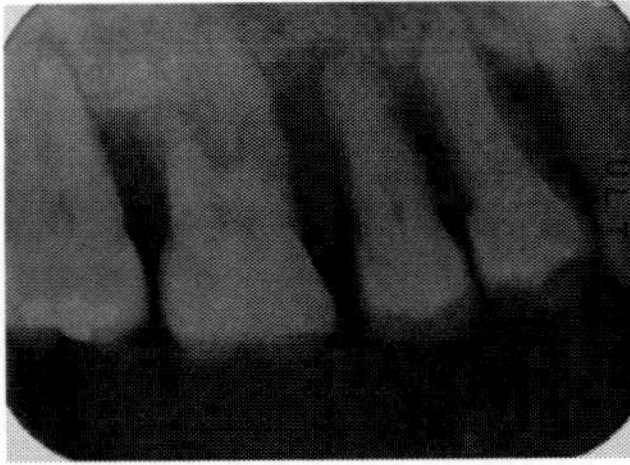

FIGURE 21.3 X-ray images of right upper molar of patients without (left) or with (right) periodontitis. Advanced alveolar bone loss is observed in the right image.

Epidemiology of Periodontitis

A survey of employed adults and elderly people conducted by the National Institute of Dental Research in 1985–1986 indicated that overall prevalence of some loss of attachment was high, with 80% of employed men and 73% of working women having a loss of 2 mm or more involving one or more teeth (National Institutes of Health 1987). The prevalence of periodontitis is thought to be high worldwide, but percentages differ widely among the studies reported (Papapanou 1996; Locker, Slade, and Muray 1998).

Etiology of Periodontitis

The picture generally presented of the etiology of inflammatory periodontal disease is one of complex interactions of local oral factors with systemic, emotional, and environmental conditions (Genco et al. 1998). Although both local tooth deposits and constitutional or systemic factors undoubtedly are related etiologically to the periodontal diseases, efforts to simplify the etiologic problem and to evaluate the relative importance of various local and systemic factors by experimental observation have been fraught with difficulties. Of local oral factors, the association of specific bacteria is generally suspected. Of these bacteria, *Porphyromonas gingivalis*, *Tannerella forsythia* (black-pigmented, gram-negative rods), and *Treponema denticola* (helical oral spirochete) are strongly implicated as major pathogens in the etiology of periodontitis (Socransky, Smith, and Haffajee 2002). Numerous bacterial products are present in the periodontal pocket and may exert direct effect (e.g., hystolysis, enzymes, endotoxins, and exotoxins). In addition to microbiological factors, local adjunctive factors have been reported in terms of anatomical relationships, restorative and prosthetic considerations, trauma from occlusion, and other factors, such as diet, saliva, and preexisting pathogenic conditions.

Periodontitis and Systemic Disease

Secondary etiologic factors exert their effects by causing the periodontal tissue to be less resistant to bacterial challenge. Of these factors, diabetes mellitus is one of the most suspected systemic diseases (Nishimura et al. 2003). There is a widely held belief that periodontitis is more prevalent and manifests a much more rapid progression in individuals with diabetes mellitus than in normal individuals. Pregnancy and puberty are also suspected to be periodontitis-accelerating factors. Several studies have reported the prevalence and severity of gingival inflammation during pregnancy and puberty. These studies showed that, although plaque scores remain unchanged, the prevalence and severity of gingival inflammation increased during pregnancy. Similarly, enhanced gingival inflammation without increased plaque accumulation occurred at puberty. Several investigators have suggested that hormonal changes associated with pregnancy may be responsible. Additionally, several investigators have suggested that the altered levels of steroid hormones may cause gingival tissues to become more sensitive to microbial challenge.

ORAL STATUS OF ELDERLY PEOPLE

Effects of Aging on Periodontium

Age-associated changes in gingival connective tissue are comparable to changes in similar tissues found elsewhere in the body. The number of fibroblasts in the gingival connective tissue and periodontal ligament decreases with age (Ryan, Toto, and Gargiulo 1974), whereas collagen content of the gingival connective tissue appears to increase with age (Hill 1984). The loss of alveolar bone with age is associated with the increase in tooth loss and decrease in alveolar bone height over time as subjects age (Russell and Ship 2008). Bone loss that occurs over time due to chronic periodontitis should not be seen, however, as a natural aging process. Indeed, periodontitis, infectious disease caused by oral bacteria, commonly seen as loss of periodontal attachment in the elderly, occurs at all ages.

Gingivitis is an inflammatory response of the gingival tissues to bacterial plaque. It has been reported that in the absence of oral hygiene, gingivitis will develop more rapidly in older (65–78 years) than in younger (20–24 years) individuals (Holm-Pedersen, Agerbaek, and Theilade 1975). The reason for this difference is most likely related to the concomitant increased plaque accumulation seen in the older group. However, while gingivitis developed more rapidly in the older group, when oral hygiene was resumed, both older and younger groups returned to a clinically healthy state in a matter of days. The greater amount of plaque that accumulated in the older group over time was related to a greater oral surface area available for plaque retention due to a greater degree of gingival recession in older persons.

Albandar and Kingman (1999) have reported that gingival recession is closely linked to aging. Additionally, significant attachment loss is the result of increased gingival recession and not of increased pocket depth (PD) (Holm-Pedersen et al. 2006; Figure 21.4). The increased recession observed occurs as a result of a combination of factors over time: a predisposition to develop recession due to tissue morphology, combined with periodontal disease, periodontal treatment, and oral hygiene practices, may contribute to recession in elderly people.

Periodontitis in Older Patients

Periodontitis is an infectious disease that causes destructive changes, ultimately leading to the loss of bone and connective tissue attachment. As a result, deep periodontal pockets and radiographic bone destruction are observed. Periodontitis is generally irreversible, and because its measurement reflects the cumulative effects of the disease over time, older populations manifest greater levels of the disease. However, the amount of disease measured may be an underestimate of the actual

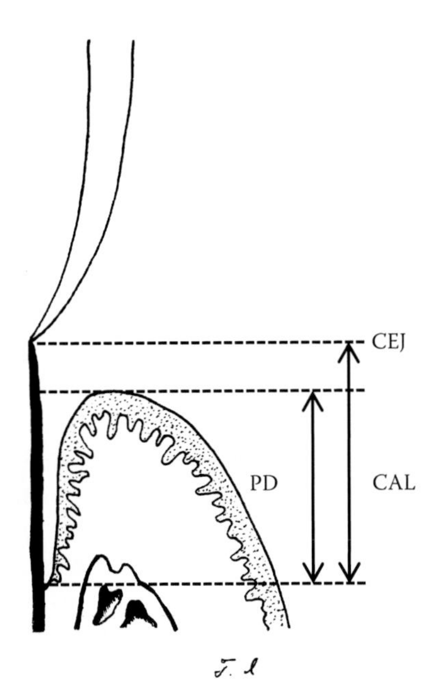

FIGURE 21.4 The definition of pocket depth (PD) and clinical attachment level (CAL). CEJ: cement–enamel junction.

disease experienced. The prevalence of periodontal disease in the United States has been estimated to be 68%–91% in those over the age of 65 who have at least one site with at least 4 mm of attachment loss, and between 30% and 71% in those who have at least one site with at least 6 mm of attachment loss (Katz, Neely, and Morse 1996) (Figure 21.4). As in younger populations, severe PD and severe bone loss are uncommon, although in older populations, the rarity of severe periodontitis reflects survival of teeth with less periodontitis.

Saliva and Salivary Glands in the Elderly

Saliva is produced by three major paired salivary glands: the parotid, submandibular, and sublingual. Accessory glands scattered throughout the oral cavity make smaller contributions. These glands and their secretions are critical in maintaining both oral and systemic health (Atkinson and Wu 1994). A prime function of saliva is to buffer acids and inhibit the demineralization of teeth. A significant complementary function is remineralization, and this is facilitated by the availability of salivary calcium and phosphate ions. Antibacterial activities of saliva are mediated through salivary immunoglobulin A, lactoferrin, lysozyme, peroxidase systems, and histatins (Vissink, Spijkervet, and Van Nieuw Amerongen 1996). Saliva serves as a lubricant, aiding in mastication, softening foods, and swallowing. Food digestion begins in the oral cavity, with the ability of salivary amylase to metabolize carbohydrates. Taste is expedited because saliva dissolves chemical tastants within food and delivers these to taste bud receptors.

With aging, a significant change occurs in the histological features of the salivary glands. Parenchymal depletion develops with a corresponding increase in fibroadipose tissue and the number of dilated ducts (Drummond and Chisholm 1994). Additionally, 30%–40% decreases in parenchymal volumes of individual salivary glands (Baum 1992) and acinar loss have been observed in computed tomography scans of the parotid gland, along with the histological evidence. A significant decrease in parotid gland density, which reflects loss of parenchyma and increased presence of fibroadipose tissue, has been observed (Drummond, Newton, and Abel 1995). Despite the age-related loss of secreting acini, no meaningful decrease in salivary production occurs (Fox 1997; Ship, Pillemer, and Baum 2002; Nagler 2004). The probable explanation for this is that the parotid and submandibular glands have idling nonfunctioning secretory reserve capacity. This reserve is activated when the secreting parenchyma present in young individuals atrophies with age (Baum 1992). Aging brings about modest decreases in submandibular and subgingival salivary volumes, but the magnitude is unlikely to cause symptoms (Longman et al. 1995). When the glandular reserve is reduced in the elderly, their glands become more susceptible to the effects of medications, radiation, and systemic diseases, particularly autoimmune diseases.

PSYCHOLOGICAL FACTORS IN PERIODONTAL DISEASE

The biological plausibility for an association is supported by studies showing that psychosocial conditions, such as depression and exposure to stressors, may affect the host immune response, making the individual more susceptible to the development of unhealthy conditions and affecting periodontal health. Previous studies have reported that psychological stress can downregulate the cellular immune response through at least three mechanisms. First, stress-induced response is transmitted to the hypothalamic–pituitary–adrenal (HPA) axis and promotes the release of corticotropin-releasing hormone from the pituitary gland and glucocorticoids from the adrenal cortex. Glucocorticoids released into the cortex of the suprarenals decrease the production of proinflammatory cytokines (interleukins [IL], prostaglandins, tumor necrosis factor). Second, exposure to stressors can induce the sympathetic nervous system to release adrenaline and noradrenaline from the adrenal medulla and can exert an immunosuppressive effect. This phenomenon indirectly provokes periodontal tissue breakdown. Third, stress can induce the release of neuropeptides from sensory nerve fibers (neurogenic inflammation), and the presence of neuropeptides has been widely documented as a possible influence on chronic inflammatory processes modulating the activity of the immune system and release of cytokines (Monterio de Silva et al. 1998). The impact of stress on the immune system has been widely reported and is a possible influence on chronic inflammatory periodontal disease. Individuals with high stress levels tend to adopt habits that are harmful to periodontal health (negligent oral hygiene, nicotine consumption, changes in eating habits with negative effects on the immune system).

Of the psychological factors, depression and anxiety have been suggested as factors that lead to neglecting oral hygiene (Moulton et al. 1952). Investigations have revealed that depressed patients accumulated calculus more quickly than patients with other types of mental disorders. The investigations offered two possible explanations: (1) depression may reduce patients' willingness to perform physical activities, leading them to pay less attention to their mouths and (2) depression may cause chemical changes in mouth secretions, which, in turn, increase calculus formation (Preston 1941). More recently, it has been appreciated that depression and anxiety are involved in stress as reactions to stressors impinging on individuals. In this context, the theory that patients with depression and anxiety tend to neglect oral hygiene seems reasonable.

STRESS HORMONES AND PERIODONTITIS

Stress Hormones and Saliva

The two primary neuroendocrine systems associated with human stress are the HPA system, with the secretion of cortisol and dehydroepiandrosterone (DHEA), and the sympathetic adrenomedullary (SAM) system, with the secretion of catecholamine. In the HPA system, cortisol is the main product of adrenocortical activity in response to adrenocorticotropic hormone (ACTH). It is well known that salivary cortisol levels increase and reach a peak 30–60 minutes after waking up. Salivary cortisol levels are closely correlated to blood cortisol levels and reliably reflect HPA activity. Various kinds of psychological stress activate the HPA system and consequently induce significant increases in salivary cortisol levels. In the SAM system, direct measurements of salivary catecholamine do not reflect SAM activity. Chromogranin A (CgA), an acidic glycoprotein, is stored in and coreleased by exocytosis with catecholamines from the adrenal medulla and sympathetic nerve endings; thus, it is considered to be a sensitive and important index of SAM. Salivary Cg is produced by human submandibular glands and secreted into the saliva. Salivary cortisol, salivary amylase, and CgA, which can be sampled noninvasively, are evaluated as stress biomarkers (Michael et al. 2000; Hironaka et al. 2008).

Cortisol and Periodontitis

A relationship between periodontitis and psychosocial stress has been proposed by many investigators. However, the psychoimmunological mechanism association for an etiological role of periodontitis is poorly understood. Recently, a significant correlation between alveolar bone loss and salivary cortisol level in a population aged 50 and over was reported (Hilgert et al. 2006, Hugo et al. 2006). However, another study reported no significant correlation among IL-1b and IL-6, cortisol, and stress (Mengel, Bacher, and Flores-de-Jacoby 2002). To our knowledge, there are only three reported studies, including our own, on the relationship between cortisol levels and severity of periodontitis (Mengel, Bacher, and Flores-de-Jacoby 2002; Hilgert et al. 2006; Ishisaka et al. 2008). Johannsen et al. (2006, 2007) reported that the amount of plaque was significantly higher in women (42.0 ± 9.3 [mean age ± SD]) with stress-related depression and exhaustion compared with the controls (42.0 ± 9.3).

Ishisaka et al. (2008) reported the comparison of levels of salivary cortisol between the group with and without extensive periodontitis, which are defined by the number of teeth with ≥5 mm of PD or ≥6 mm of clinical attachment loss (CAL; Table 21.2). The authors divided the subjects into three categories, according to the number of teeth with maximum PD of >5 mm or a maximum

TABLE 21.2
Median Values of Salivary Cortisol in the Presence or Absence of Extensive Periodontitis Separated by Pocket Depth and Clinical Attachment Loss

	Not Extensive Periodontitis	Extensive Periodontitis	Severely Extensive Periodontitis	*p* Value[a]
Separated by PD[b]				
Subjects (*n*)	87	63	21	
Cortisol (ng/mL)				
Median	1.75	2.22	2.66	.004
25th, 75th percentile	1.36, 2.55	1.46, 3.03	2.03, 3.32	
Separated by CAL[c]				
Subjects (*n*)	93	50	28	
Cortisol (ng/mL)				
Median	1.68	2.33	2.66	<.001
25th, 75th percentile	1.31, 2.43	1.56, 3.06	1.83, 3.62	

[a] Kruskal–Wallis test.
[b] Pocket depth (PD), defined by the number of teeth (no teeth, less than three teeth, and three or more teeth) with PD of ≥5 mm.
[c] Clinical attachment loss (CAL), defined by the number of teeth (no teeth, less than three teeth, and three or more teeth) with CAL of ≥6 mm.

CAL of ≥6 mm, using the mean value of each cutoff point. Three categories for PD are as follows: not extensive, no teeth with PD of ≥5 mm; extensive, fewer than three teeth with PD of ≥5 mm; and severely extensive, more than three teeth with PD of ≥5 mm. There were significant differences among the three categories in cortisol levels (Table 21.2). The categories of CAL are as follows: not extensive, no teeth with CAL of ≥6 mm; extensive, fewer than three teeth with CAL of ≥6 mm; and severely extensive, more than three teeth with CAL of ≥6 mm. Significant differences in cortisol levels were found among three categories (Ishisaka et al. 2008). The authors also performed multiple regression analysis with adjustments for various confounding variables (Model 2 adjusted for age and gender, and Model 3 adjusted for age, gender, smoking status, diabetes, oral hygiene habits, salivary flow rate, and bleeding on probing) shown by univariate analysis (Table 21.3). The level of salivary cortisol was associated, independently and significantly, with extensive and severely extensive periodontitis assessed by PD and CAL, even after adjustment for potential confounding factors.

In Japanese elderly subjects who had never smoked, Ishisaka et al. (2008) reported a positive association between serum cortisol level and periodontal severity assessed by CAL. However, no significant association between cortisol and periodontitis severity, assessed by PD, was observed. This finding may be explained as follows. Severe chronic diseases are disorders related to hyperfunction of the HPA axis responses (Chrousos and Gold 1992). Because CAL can be regarded as a result of an inflammatory burden from the past into the present and while the PD level reflects the current pathophysiological status of the periodontitis, these findings can be attributed to dysregulation of the stress system, in which the HPA axis is chronically activated in patients with severe periodontitis.

TABLE 21.3
Multiple Regression Analysis of the Effects of Explanatory Variables Including Pocket Depth (PD) and Clinical Attachment Loss (CAL) on Cortisol

	Explanatory Variables					
	PD			CAL		
Independent Variables	β[a]	*t*	*p*	β	*t*	*p*
Subjects (*n*)		171			171	
Crude						
Extensive periodontitis	0.171	2.178	.031	0.228	2.957	.004
Severely extensive periodontitis	0.177	2.247	.026	0.251	3.256	.001
Model 2[b]						
Extensive periodontitis	0.168	2.084	.039	0.223	2.861	.005
Severely extensive periodontitis	0.168	2.082	.039	0.240	2.889	.004
Model 3[c]						
Extensive periodontitis	0.177	2.144	.034	0.233	2.900	.004
Severely extensive periodontitis	0.181	2.119	.036	0.225	2.624	.010

[a] β is the regression coefficient.
[a] Model 2 adjusted for age and gender.
[b] Model 3 adjusted for age, gender, smoking status, diabetes, oral hygiene habits, salivary flow rate, and bleeding on probing. Reference is not extensive periodontitis (i.e., no teeth with PD of ≥5 mm and CAL of ≥6 mm).

DHEA and Periodontitis

Another ACTH-dependent hormone, DHEA, also known as DHES-sulfate (DHEAS), is also affected by dysregulation of the stress system (Hauser et al. 1998; Mengel et al. 2002). A recent study reported that the cortisol/DHEAS molar ratio was a useful marker of anxiety and depressive illness (Ristener, Gibel, and Maayan 2007). A relationship between DHEA and inflammatory disease has been reported by many groups. However, the relationship between DHEA and severity of periodontitis has been reported by our group alone (Ishisaka et al. 2008), whereas the relationship between cortisol and severity of periodontitis has been reported by four groups (Hilgert, Hugo, and Bandeira 2006; Ishisaka et al. 2007; Ansai et al. 2009; Rosania et al. 2009). Initially, the patients who visited Kyushu Dental College Hospital, Kitakyushu, Japan, were divided into three categories (not extensive, extensive, and severely extensive), as shown in the section "Cortisol and Periodontitis." Next, levels of salivary DHEA between the groups with and without extensive periodontitis were compared. Significant differences among three categories defined by PD and CAL were observed (Table 21.4). Furthermore, to analyze whether elevated levels of DHEA were associated with the severity and extent of periodontitis, multiple regression analyses were performed. The level of salivary DHEA showed a strong association with severely extensive periodontitis in all the regression models, whereas only marginally significant associations were found between DHEA level and extensive periodontitis (Table 21.5; Ishisaka et al. 2007).

TABLE 21.4
Median Values of Salivary DHEA in the Presence or Absence of Extensive Periodontitis Separated by Pocket Depth and Clinical Attachment Loss

	Not Extensive Periodontitis	Extensive Periodontitis	Severely Extensive Periodontitis	*p* Value[a]
Separated by PD[b]				
Subjects (*n*)	87	63	21	
DHEA (pg/mL)				
Median	43.55	60.87	103.01	<.001
25th, 75th percentile	28.50, 72.01	31.96, 93.74	76.68, 126.22	
Salivary Flow Rate (mL/minute)				
Median	1.00	1.00	1.17	.828
25th, 75th percentile	0.67, 1.40	0.67, 1.50	0.67, 1.62	
Separated by CAL[c]				
Subjects (*n*)	93	50	28	
DHEA (pg/mL)				
Median	51.70	63.08	79.67	.022
25th, 75th percentile	30.11, 77.77	33.47, 100.36	40.95, 121.12	
Salivary Flow Rate (mL/minute)				
Median	1.00	1.20	0.83	.042
25th, 75th percentile	0.67, 1.42	0.81, 1.59	0.43, 1.22	

[a] Kruskal–Wallis test.
[b] Pocket depth (PD), defined by the number of teeth (no teeth, less than three teeth, and three or more teeth) with PD of ≥5 mm.
[c] Clinical attachment loss (CAL), defined by the number of teeth (no teeth, less than three teeth, and three or more teeth) with CAL of ≥6 mm.

TABLE 21.5
Multiple Regression Analysis of the Effects of Explanatory Variables Including Pocket Depth (PD) and Clinical Attachment Loss (CAL) on DHEA

	Explanatory Variables					
	PD			CAL		
Independent Variables	β[a]	*t*	*p*	β	*t*	*p*
Subjects (*n*)		171			171	
Crude						
Extensive periodontitis	0.141	1.848	.066	0.151	1.929	.055
Severely extensive periodontitis	0.330	4.330	<.001	0.238	3.050	.003
***Model 2*[b]**						
Extensive periodontitis	0.161	2.062	.041	0.159	2.019	.045
Severely extensive periodontitis	0.348	4.468	<.001	0.271	3.243	.001
***Model 3*[c]**						
Extensive periodontitis	0.132	1.711	.089	0.138	1.773	.078
Severely extensive periodontitis	0.308	3.857	<.001	0.198	2.379	.019

[a] β is the regression coefficient.

[b] Model 2 adjusted for age and gender.

[c] Model 3 adjusted for age, gender, smoking status, diabetes, oral hygiene habits, salivary flow rate, and bleeding on probing. Reference is not extensive periodontitis (i.e., no teeth with PD of ≥5 mm and CAL of ≥6 mm).

No significant difference in serum DHEA levels was found for PD and CAL in any smoking status group (Tables 21.6 and 21.7). Significant differences were found only among the CAL tertiles with regard to cortisol level (Table 21.7) in individuals who have never smoked, as described in the section "Cortisol and Periodontitis." However, no significant association was found with regard to the cortisol/DHEAS molar ratio, regardless of PD and CAL levels (Tables 21.6 and 21.7). With regard to the levels of cortisol and DHEAS, there were significant associations among the three stratifications by smoking, which indicated that smoking caused an increase in cortisol and DHEAS. Thus, to determine whether levels of cortisol, DHEAS, and the cortisol/DHEAS molar ratio were associated with the severity and extent of periodontitis, multiple regression analyses, stratified by smoking status, with adjustment for various potential confounding variables, were performed.

For PD, there was no significant association between the levels of cortisol and DHEAS, including the cortisol/DHEAS molar ratio, in any of the models, regardless of smoking status. In contrast, for CAL, the cortisol level was significantly associated with both the second and third tertiles in both models in people who have never smoked, even after adjusting for confounding factors, and a stronger association was seen with the third tertile. Significant associations were found between the cortisol/DHEAS molar ratio and second tertile in people who have never smoked in the final model (adjusted for age, gender, diabetes, frequency of toothbrushing per day, and bleeding on probing). There was no significant association between periodontal status and serum DHEAS level in any of the models.

TABLE 21.6
Median Values for Serum Stress Hormones in the Presence or Absence of Extensive Periodontitis, Separated by Pocket Depth (PD) across Smoking Status

	Current				Past				Never			
	Sites with PD ≥4 mm[a]											
	None	Low	High	*p* Value[b]	None	Low	High	*p* Value[b]	None	Low	High	*p* Value[b]
Number	16	13	21		34	42	45		107	107	80	
Cortisol (nmol/L)												
Median	393.16	333.84	441.4	.29	441.44	460.75	460.8	.51	364.19	342.12	389.02	.11
25th, 75th percentile	303.2, 525.6	198.6, 491.1	289.7, 586.3		320.7, 517.3	357.3, 568.4	314.5, 589.0		273.1, 485.6	270.4, 458.0	325.6, 524.2	
DHEAS (nmol/L)												
Median	6106.50	3148.2	4858.06	.10	4559.52	4410.25	4125.28	.48	2667.86	2768.28	2632.6	.65
25th, 75th percentile	3446.8, 7694.2	2534.9, 5034.5	2917.6, 7368.5		3304.3, 5862.2	3080.4, 6737.5	2095.2, 5916.5		1769.5, 3473.9	1948.7, 3908.2	1842.1, 3867.5	
***Cortisol/DHEAS Ratio* (×100)**												
Median	7.67	8.11	9.03	.55	9.55	10.28	11.89	.48	14.57	12.99	15.84	.19
25th, 75th percentile	5.1, 8.9	6.7, 15.9	5.3, 16.5		6.4, 16.7	7.1, 17.0	7.1, 19.3		9.7, 20.2	8.6, 18.1	10.7, 24.6	

[a] Defined by the percent with sites (tertiles) with ≥4 mm.
[b] Kruskal–Wallis test.

TABLE 21.7
Median Values for Serum Stress Hormones in the Presence or Absence of Extensive Periodontitis, Separated by Clinical Attachment Level (CAL) across Smoking Status

	Current				Past				Never			
	Sites with CAL ≥5 mm[a]											
	None	Low	High	*p* Value[b]	None	Low	High	*p* Value[b]	None	Low	High	*p* Value[b]
Number	17	11	22		34	36	51		136	83	75	
Cortisol (nmol/L)												
Median	419.37	386.26	413.9	.84	437.30	437.16	460.75	.77	332.46	389.02	391.78	.001
25th, 75th percentile	306.2, 492.5	289.7, 554.6	234.5, 550.4		323.5, 573.2	330.4, 549.7	361.4, 571.1		249.0, 460.1	320.0, 535.2	311.8, 524.2	
DHEAS (nmol/L)												
Median	5020.90	5563.70	4722.36	.90	4559.5	4545.95	4016.7	0.87	2609.5	2714.00	2876.84	.55
25th, 75th percentile	2575.6, 6703.6	2795.4, 6703.6	2890.4, 7232.8		3087.2, 6255.8	3161.8, 5726.5	2518.6, 6296.5		1789.9, 3514.6	1709.8, 3935.3	1959.5, 3826.7	
***Cortisol/DHEAS Ratio* (×100)**												
Median	8.11	8.27	8.20	.74	10.03	7.99	12.38	0.23	12.83	15.25	14.16	.10
25th, 75th percentile	6.1, 19.0	4.7, 13.7	5.4, 15.4		7.3, 16.0	6.1, 14.4	7.1, 19.6		8.3, 18.9	10.2, 22.7	11.0, 20.9	

[a] Defined by the percent with sites (tertiles) with ≥4 mm.
[b] Kruskal–Wallis test.

CONCLUDING REMARKS AND FUTURE DIRECTIONS

Cross-sectional epidemiological studies have shown that the levels of salivary DHEA had a stronger association with severely extensive periodontitis than those of cortisol. Salivary DHEA level may be a more appropriate marker of extensive periodontitis than salivary cortisol level. However, serum cortisol level, and the cortisol/DHEAS ratio, is strongly associated with severity of periodontitis in elderly subjects who had never smoked, whereas serum DHEA had no association with periodontitis severity. Smoking is not only one of the major risk factors for severe periodontal disease (Tomar and Asma 2000), but it also is known to be associated with elevated cortisol and DHEA levels. In a recent report that compared cortisol profiles of smokers and nonsmokers over 1 day, cortisol levels were elevated daily among smokers compared with nonsmokers, and the differences in values were quite substantial, averaging 35% or more (Steptoe and Ussher 2006). Thus, we should take smoking status into consideration in analyzing the relationship between stress-related hormone and periodontitis.

In conclusion, significant associations between salivary DHEA and severity of periodontitis were shown in elderly subjects. However, these investigations have just begun and further studies are required to clarify the etiology of periodontal disease.

ACKNOWLEDGMENTS

We thank Dr. Takako Ichiki, Kyushu Dental College, Kitakyushu, Japan, for the professionally drawn illustrations (Figures 21.1, 21.2, and 21.4).

REFERENCES

Albandar, J. M., and A. Kingman. 1999. Gingival recession, gingival bleeding, and dental calculus in adults 30 years of age and older in the United States, 1988–1994. *J Periodontol* 70:30–43.

Ansai, T., I. Soh, A. Ishisaka et al. 2009. Determination of cortisol and dehydroepiandrosterone levels in saliva for screening of periodontitis in older Japanese adults. *Int J Dent* 2009:280737.

Atkinson, J. C., and A. J. Wu. 1994. Salivary gland dysfunction: Causes, symptoms, treatment. *J Am Dent Assoc* 125:409–16.

Baum, B. J. 1992. Age-related vulnerability. *Otolaryngol Head Neck Surg* 106:730–2.

Chrousos, G. P., and P. W. Gold. 1992. The concepts of stress and stress disorders. Overview of physical and behavioral homeostasis. *JAMA* 267:1244–52.

Drummond, J. R., and D. M. Chisholm. 1994. A qualitative and quantitative study of the ageing human labial salivary glands. *Arch Oral Biol* 29:151–5.

Drummond, J. R., J. P. Newton, and R. W. Abel. 1995. Tomographic measurements of age changes in the human parotid gland. *Gerodontology* 12:26–30.

Epidemiology of periodontal disease among older adults: A review. *Periodontology 2000* 16:16–33.

Fox, P. C. 1997. Management of dry mouth. *Dent Clin North Am* 41:863–75.

Genco, R. J., A. W. Ho, J. Kopman et al. 1998. Models to evaluate the role of stress in periodontal disease. *Ann Periodontol* 3:288–302.

Heuser, I., M. Deuschle, P. Luppa et al. 1998. Increased diurnal plasma concentrations of dehydroepiandrosterone in depressed patients. *J Clin Endocrinol Metabol* 83:3130–3.

Hilgert, J. B., F. N. Hugo, D. R. Bandeira et al. 2006. Stress, cortisol, and periodontitis in a population aged 50 years and over. *J Dent Res* 85:324–8.

Hill, M. W. 1988. Influence of age on the morphology and transit time of murine stratified squamous epithelia. *Arch Oral Biol* 33:221–9.

Hironaka, M., T. Ansai, I. Soh et al. 2008. Association between salivary levels of chromogranin A and periodontitis in older Japanese. *Biomed Res* 29:125–30.

Holm-Pedersen, P., N. Agerbaek, and E. Theilade. 1975. Experimental gingivitis in young and elderly individuals. *J Clin Periodontol* 2:14–24.

Holm-Pedersen, P., S. L. Russell, K. Avlund et al. 2006. Periodontal disease in the oldest-old living in Kungsholmen, Sweden: Findings from the KEOHS project. *J Clin Periodontol* 33:376–84.

Hugo, F. N., J. B. Hilgert, M. C. Bozzetti et al. 2006. Chronic stress, depression, and cortisol levels as risk indicators of elevated plaque and gingivitis levels in individuals aged 50 years and older. *J Periodontol* 77:1008–14.

Ishisaka, A., T. Ansai, I. Soh et al. 2007. Association of salivary levels of cortisol and dehydroepiandrosterone with periodontitis in older Japanese adults. *J Periodontol* 78:1767–73.

Ishisaka, A., T. Ansai, I. Soh et al. 2008. Association of cortisol and dehydroepiandrosterone sulphate levels in serum with periodontal status in older Japanese adults. *J Clin Periodontol* 35:853–61.

Johannsen, A., I. Rydmark, B. Söder et al. 2007. Gingival inflammation, increased periodontal pocket depth and elevated interleukin-6 in gingival crevicular fluid of depressed women on long-term sick leave. *J Periodontal Res* 42:546–52.

Johannsen, A., G. Rylander, B. Söder et al. 2006. Dental plaque, gingival inflammation, and elevated levels of interleukin-6 and cortisol in gingival crevicular fluid from women with stress-related depression and exhaustion. *J Periodontol* 77:1403–9.

Katz, R. V., A. L. Neely, and D. E. Morse. 1996. The epidemiology of oral diseases in older adults. In *Textbook of Geritric Dentistry*, ed. P. Holm-Pederson and H. Loe, 263–301. Copenhagen: Munksgaard.

Longman, L. P., S. M. Higham, K. Rai et al. 1995. Salivary gland hypofunction in elderly patients attending a xerostomia clinic. *Gerodontology* 12:67–72.

Mengel, R., M. Bacher, and L. Flores-de-Jacoby. 2002. Interactions between stress, interleukin-1b, interleukin-6 and cortisol in periodontally diseased patients. *J Clin Periodontol* 29:1012–22.

Michael, A., A. Jenaway, E. S. Paykel et al. 2000. Altered salivary dehydroepiandrosterone levels in major depression in adults. *Biological Psychiatry* 48:989–95.

Monterio de Silva, A. M., H. N. Newman, D. A. Oakley et al. 1998. Psychosocial factors, dental plaque levels and smoking in periodontitis patients. *J Clin Periodontol* 25:517–23.

Moulton, R., S. Ewen, and W. Thieman 1952. Emotional factors in periodontal disease. *Oral Surgery* 5:833–60.

Nagler, R. M. 2004. Salivary glands and the aging process: Mechanistic aspects, health-status and medicinal-efficacy monitoring. *Biogerontology* 5:223–33.

National Institute of Dental Research 1987. Oral health of United States adults. National findings. Bethesda, MD: US Department of Health and Human Services. (NIH Publication No. 87-2868)

Nishimura, F., Y. Iwamoto, J. Mineshiba et al. 2003. Periodontal disease and diabetes mellitus: The role of tumor necrosis factor-a in a 2-way relationship. *J Periodontol* 74:97–102.

Papapanou, P. N. 1996. Periodontal diseases: Epidemiology. *Ann Periodontol* 1:1–36.

Preston, J. L. 1941. Dental treatment of focal infection in mental disease. *Texas Dental J* 59:65–8.

Ristener, M., A. Gibel, R. Maayan et al. 2007. State and trait related predictors of serum cortisol to DHEA(S) molar ratios and hormone concentrations in schizopherenia patients. *European Neuropsychopharmacol* 17:257–64.

Rosania, A. E., K. G. Low, C. M. McCormick et al. 2009. Stress, depression, cortisol, and periodontal disease. *J Periodontol* 80:260–6.

Russell, S. L., and J. A. Ship. 2008. Normal oral mucosal, dental, periodontal, and alveolar bone changes associated with aging. In *Improving Oral Health for the Elderly, An Interdisciplinary Approach*, ed. I. B. Lamster and M. E. Northridge, 233–46. New York: Springer Science and Business Media, LLC.

Ryan, E. J., P. D. Toto, and A. W. Gargiulo. 1974. Aging in human attached gingival epithelium. *J Dent Res* 53:74–6.

Schroeder, H. E., and R. C. Page. 1990. Diseases of the periodontium. In *Periodontal Diseases*, ed. S. Schluger, R. Yuodelis, R. C. Page, and R. H. Johnson, 53–71. Philadelphia: Lea & Febiger.

Ship, J. A., S. R. Pillemer, and B. J. Baum. 2002. Xerostomia and the geriatric patient. *J Am Geriatr Soc 2002* 50:535–43.

Socransky, S. S., C. Smith, and A. D. Haffajee. 2002. Subgingival microbial profiles in refractory periodontal disease. *J Clin Periodontol* 29:260–8.

Steptoe, A., and M. Ussher. 2006. Smoking, cortisol and nicotine. *Int J Psychophysiol* 59:228–35.

Tomar, S. L., and S. Asma. 2000. Smoking-attribute periodontitis in the United States: Findings from NHANES III. National Health and Nutrition Examination Survey. *J Periodontol* 71:743–51.

Vissink, A., F. K. Spijkervet, and A. Van Nieuw Amerongen. 1996. Aging and saliva: A review of the literature. *Spec Care Dentist* 16:95–103.

Williams, R. C. 1990. Periodontal disease. *Lancet* 322:373–82.

22 DHEA Levels and Increased Risk of Atherosclerosis and Cardiovascular Disease

Ippei Kanazawa and Toru Yamaguchi

CONTENTS

INTRODUCTION

Age-related hormones have recently attracted widespread attention due to their beneficial antiaging effects. Dehydroepiandrosterone (DHEA) is produced by the adrenal gland and has a weak androgen action after peripheral conversion to more potent androgens, testosterone and dihydrotestosterone. DHEA is known to decrease with aging; this decline is referred to as "adrenopause." Studies have shown that DHEA has protective effects against a series of diseases generally associated with aging (Barrett-Connor and Edelstein 1994; Haffner et al. 1996). DHEA-sulfate (DHEAS), which is converted to its active form, DHEA, is the circulating hormonal pool of DHEA and is a good marker for DHEA availability. Thus, maintaining DHEA and DHEAS in the circulation at sufficient levels might be important for preventing these aging-related diseases.

Numerous *in vitro* and *in vivo* studies have shown that DHEA has a beneficial effect on the formation and progression of atherosclerosis. DHEA promotes fibrinolysis (Beer et al. 1996), decreases lipid accumulation in cultured mouse foam cells (Taniguchi et al. 1996), limits the proliferation and migration of vascular smooth muscle cells (Furutama et al. 1998), and decreases macrophage infiltration into atherosclerotic plaques (Yamakawa et al. 2009). Moreover, it has been reported that DHEA administration decreases atherosclerotic lesions in experimental animals (Gordon, Bush, and Weisman 1988; Yamakawa et al. 2009). Barrett-Connor, Khaw, and Yen (1986) for the first time conducted a prospective population study on the relationship between DHEAS and mortality in men and showed that decreased DHEAS levels were associated with a higher mortality rate due to cardiovascular diseases (CVD) and other causes. Since then, a lot of longitudinal, cross-sectional, retrospective, and prospective studies have been carried out to examine the relationship between DHEAS and various aspects of CVD.

In this review, we summarize the studies on the association of serum DHEA and DHEAS levels with CVD events and atherosclerosis parameters and the studies on the effect of DHEA supplementation on atherosclerosis parameters.

ASSOCIATION BETWEEN DHEA AND ATHEROSCLEROSIS IN MEN

Dozens of studies have been published on the association of circulating DHEA and DHEAS with the presence of CVD and the parameters of atherosclerosis in men (Table 22.1). Four of seven case–control studies indicated that DHEAS levels in patients with atherosclerosis and coronary artery disease (CAD) were significantly lower than those in healthy controls. Slowinska-Srzednicka et al. (1989) reported a significant decrease in serum DHEAS levels in 32 men aged 26–40 years who had myocardial infarction 3–4 months prior to the study and who had angiographically demonstrated coronary occlusion, compared with 76 healthy men aged 25–40 years. Mitchell et al. (1994) conducted a retrospective study on 49 male survivors (<56 years) of premature myocardial infarction and 49 age-matched male controls. They found that serum DHEAS levels were significantly lower in the patients than in the control subjects and that this association remained statistically significant even after the adjustment for risk factors of myocardial infarction. Cao et al. (2010) performed a relatively larger study in 139 male CAD patients and 400 healthy controls and showed that serum DHEAS levels in the CAD group were significantly reduced (Cao et al. 2010). In contrast, other studies have shown that a predictive effect of low DHEAS levels on subsequent CVD risk was not significant (Kajinami et al. 2004; Hauner et al. 1991; Naessen et al. 2010). On the other hand, Tedeschi-Reiner et al. (2009) have reported the association between serum DHEAS and atherosclerosis of retinal arteries in a case–control study. They recruited 101 male patients with atherosclerotic changes in the retinal vessels, which were identified by direct ophthalmoscopy and were graded on a scale of 1 to 4 according to Scheie's classification, and 47 age-matched subjects with healthy retinal vessels. Retinal vessel atherosclerosis was inversely correlated with serum DHEAS levels, suggesting that the lower serum DHEAS level was associated with the more advanced stage of retinal vessel atherosclerosis.

A number of longitudinal studies have also been conducted to examine the association of serum DHEA and DHEAS levels with CVD events in men (Table 22.2). Five studies found that decreased DHEAS levels were associated with higher mortality rates due to CVD. Barrett-Connor, Khaw, and Yen (1986) conducted a 12-year follow-up study in 242 men and showed that the age-adjusted relative risks when serum DHEAS levels were below 140 μg/dL were 1.5-fold for death due to any causes, 3.3-fold for death due to CVD, and 3.2-fold for death due to CAD. Furthermore, in multivariate analyses, they found that an increase in DHEAS level by 100 μg/dL was associated with a 36% reduction in mortality due to any causes and a 48% reduction in mortality due to CVD, after adjustment for age, systolic blood pressure, serum cholesterol level, obesity, fasting plasma glucose level, smoking status, and personal history of heart disease. They also analyzed the same cohort with a larger scale and a longer follow-up period of 19 years (Barrett-Connor and Goodman-Gruen 1995a). Multiple adjusted models showed that serum DHEAS levels were significantly and inversely associated with the risk of fatal CVD or CAD when CVD or CAD death was compared with 19-year survivors (relative risk 0.85). LaCroix et al. (1992) performed an 18-year follow-up study in 714 men and showed that age-adjusted DHEAS levels were lower in fatal cases of CAD than in controls, and that the odds ratio for fatal CAD comparing a 100 μg/dL difference in DHEAS level was 0.46 after adjustment for conventional coronary risk factors. Moreover, Trivedi and Khaw (2001) examined the relationship between DHEAS levels and subsequent all-cause and cardiovascular mortality in 963 men aged 65–76 years and followed for 7.4 years. They found that all-cause and CVD mortality rates were highest in the lowest DHEAS quartile, independent of age, smoking habit, systolic blood pressure, body mass index, blood cholesterol, or steroid use. Feldman et al. (2001) reported that middle-aged men with serum DHEAS in the lowest quartile at baseline (<160 μg/dL) were significantly more likely to suffer CAD events by follow-up periods (odds ratio 1.60), independently of a comprehensive set of known risk factors including age, obesity, diabetes, hypertension, smoking, serum lipids, alcohol intake, and physical activity. Taken together, although controversy still exists, a number of longitudinal and cross-sectional studies suggest that serum DHEA and DHEAS levels were inversely associated with CVD and CAD events in men and that the hormonal levels could potentially assess the risk of the atherosclerosis-related diseases.

TABLE 22.1
Cross-Sectional Studies on the Association of DHEA and DHEAS with Atherosclerosis in Men

Study	Subjects	Outcome Measurement	Results
Slowinska-Srzednicka et al. (1989)	32 cases and 76 controls	CAD	DHEAS lower in cases
Herrington et al. (1990)	103 middle-aged men undergoing elective coronary angiography	CAD	DHEAS lower in coronary artery stenosis
Hauner et al. (1991)	200 cases and 74 controls	CAD	NS
Ishihara et al. (1992)	69 men without overt CAD	AC	DHEA and DHEAS lower in patients with AC
Mitchell et al. (1994)	49 cases and 49 controls	MI	DHEAS lower in cases
Phillips, Pinkemell, and Jing (1994)	55 men undergoing coronary angiography	CAD	NS
Herrington (1995)	101 middle-aged men undergoing elective coronary angiography	CAD	DHEAS lower in coronary artery stenosis
Kiechl et al. (2000)	371 men	IMT, CVD	NS
Hak et al. (2002)	504 nonsmokers aged 55 years and older	AC	NS
Dockery et al. (2003)	55 men (mean age 71.1 years)	PWV	DHEAS was inversely associated with PWV[a]
van den Beld et al. (2003)	403 men aged 73–94 years	IMT	NS
Kajinami et al. (2004)	236 cases and 143 controls	Stable CAD	NS
Fukui et al. (2005)	206 men with type 2 DM	IMT, PS, CVD	DHEAS was inversely associated with PWV and IMT; DHEAS lower in cases with CVD[b]
Svartberg et al. (2006)	1482 men aged 25–84 years	IMT	NS
Hougaku et al. (2006)	206 men (mean age 68.1 years)	IMT, arterial stiffness index	DHEAS was inversely associated with arterial stiffness index[a]
Fukui et al. (2007)	268 men with type 2 DM	PWV	DHEAS was inversely associated with PWV[b]
Michos et al. (2008)	978 men aged 45–84 years	AC	NS
Kanazawa et al. (2008)	148 men with type 2 DM	IMT, PWV	DHEAS was inversely associated with PWV and IMT[a]
Tedeschi-Reiner et al. (2009)	101 cases and 47 controls	Atherosclerosis of retinal vessels	DHEAS lower in cases
Cao et al. (2010)	139 cases and 400 controls	CAD	DHEAS lower in cases
Naessen et al. (2010)	18 cases and 77 controls aged 70 years	CVD	NS

Note: CAD, coronary artery disease; AC, aortic calcification; PWV, pulse wave velocity; IMT, intima–media thickness; DM, diabetes mellitus; PS, plaque score; CVD, cardiovascular disease; NS, not significant; DHEA, dehydroepiandrosterone; DHEAS, DHEA sulfate; MI, myocardial infarction.

[a] Not significant after adjusting for age.

[b] Not adjusted for age.

TABLE 22.2
Longitudinal Studies on the Association of DHEA and DHEAS with CVD Event and Mortality

Study	Follow-Up Duration (Years)	Subjects	Outcome Measurement	Results
		Men		
Barrett-Connor, Khaw, and Yen (1986)	12	242 men aged 50–79 years	CVD and CAD mortality	DHEAS lower in CVD or CAD death
Contoreggi et al. (1990)	9.5	46 cases and 124 controls	CAD	NS
LaCroix et al. (1992)	18	238 cases and 476 controls	Fatal and nonfatal CAD	DHEAS lower in CAD death
Hautanen et al. (1994)	4	62 cases and 97 controls	Nonfatal CAD	DHEAS higher in CAD
Newcomer et al. (1994)	5	169 cases and 169 controls	Fatal and nonfatal CAD	NS
Barrett-Connor and Goodman-Gruen (1995)	19	1029 men aged 30–88 years	CVD and CAD mortality	DHEAS lower in CVD or CAD death
Haffner et al. (1996)	5	41 cases and 82 controls	CAD mortality	NS
Tilvis, Kahonen, and Harkonen (1999)	5	150 men aged 75–85 years	CVD mortality	NS
Trivedi and Khaw (2001)	7.4	963 men aged 65–76 years	CVD mortality	DHEAS lower in CVD death
Feldman et al. (2001)	2	1167 men aged 40–70 years	CAD	DHEAS lower in CAD
Akishita et al. (2010)	6.4	171 men aged 30–69 years	CVD	NS
		Women		
Barrett-Connor and Goodman-Gruen (1995b)	19	942 women aged 30–88 years	CVD and CAD mortality	NS
Barrett-Connor and Goodman-Gruen (1995b)	19	942 postmenopausal women	CVD and CAD mortality	NS
Haffner et al. (1996)	5	40 cases and 80 controls with DM	CAD mortality	DHEAS lower in cases
Tilvis et al. (1999)	5	421 postmenopausal women aged 75–80 years	CVD mortality	NS
Trivedi	7.4	1171 postmenopausal women aged 65–76 years	CVD mortality	NS

Note: CVD, cardiovascular disease; CAD, coronary artery disease; DM, diabetes mellitus; NS, not significant; DHEA, dehydroepiandrosterone; DHEAS, DHEA sulfate.

Many cross-sectional studies have also been performed using the parameters of atherosclerosis such as pulse wave velocity (PWV), carotid intima–media thickness (IMT), and aortic calcification (AC) (Table 22.1). Although several studies indicated that serum DHEA and DHEAS levels were correlated with those parameters by linear correlation analysis, the statistical significance of the association was not found when adjusting for the participants' age (Dockery et al. 2003; Hougaku et al.

2006; Kanazawa et al. 2008). On the other hand, Fukui et al. (2005) reported that serum DHEAS levels were negatively correlated with IMT and plaque score evaluated by ultrasonography in 206 men with type 2 diabetes mellitus (DM), independently of body mass index, systolic blood pressure, total cholesterol, and hemoglobin A1c (HbA1c). Fukui et al. (2007) also reported a negative relationship between serum DHEAS levels and PWV in 268 men with DM. However, in their studies, these associations were not adjusted for age. Ishihara et al. (1992) reported that after adjusting for age, DHEA and DHEAS levels were lower in subjects with AC than in those without AC, and that PWV was slower in the latter. Therefore, it is still controversial whether serum DHEA and DHEAS levels are negatively associated with atherosclerosis parameters independent of age in men.

ASSOCIATION BETWEEN DHEA AND ATHEROSCLEROSIS IN WOMEN

Cross-sectional studies on DHEA and atherosclerosis parameters in women are summarized in Table 22.3. The data on the association between DHEA and atherosclerosis are fewer in women than in men. Although DHEA does not have intrinsic estrogenic or androgenic activity, DHEA is known to be converted into active androgens and estrogens in peripheral target tissue (Labrie 1991). It is well known that the hormonal status of the women dramatically changes after menopause, when the relative contribution of the adrenal gland to total C19 steroid production is known to increase in order to compensate for the loss of ovarian follicular androgen synthesis. Thus, the clinical effect of impaired adrenal androgen synthesis might be more profound in postmenopausal women than in premenopausal women. Thus, studies performed on women should consider the menstrual state, duration of menopause, and history of hormone replacement therapy (HRT). Among studies summarized in Table 22.3, eight studies were done in postmenopausal women, four in pre- and postmenopausal women, and one in premenopausal women. Most studies excluded the subjects with HRT, whereas in one study, of the 881 postmenopausal participants recruited, 301 women had HRT (Michos et al. 2008). Three studies did not describe HRT history (Herrington 1995; Ishihara et al. 1992; Kiechl et al. 2000).

Most of the large and medium-sized studies in postmenopausal women indicated no significant relationship between serum DHEA and DHEAS levels and atherosclerotic parameters, although a small-scale study by Naessen et al. (2010) showed that DHEAS levels were significantly lower in 8 CVD cases than in 64 controls. Thus, circulating DHEA or DHEAS seems not to be associated with atherosclerosis in postmenopausal women. On the other hand, we conducted a study on postmenopausal women with type 2 DM and found that serum DHEAS levels were inversely associated with parameters of atherosclerosis, PWV, and IMT (Kanazawa et al. 2008). This association was significant even after multiple regression analysis was adjusted for age, duration of diabetes, body mass index, HbA1c, systolic blood pressure, low-density lipoprotein (LDL)-cholesterol, renal function, and smoking. It has been reported that DM patients have significantly lower plasma concentrations of DHEA than non-DM controls because of their low 17,20-lyase activity in the adrenal steroidogenic enzymes (Barrett-Connor 1992; Ueshiba et al. 2002). Thus, the negative correlation between serum DHEAS levels and atherosclerosis found in our study might be specific for DM postmenopausal women, but not for non-DM counterparts.

In pre- and postmenopausal women, two studies showed that lower serum DHEA and DHEAS levels were associated with aortic calcification and coronary artery sclerosis (Herrington 1995; Ishihara et al. 1992). However, the studies did not describe whether or not the participants had histories of HRT therapy. Bernini et al. (1999) also demonstrated that serum DHEAS was inversely correlated with IMT in women, but this relationship was dependent on age. Only one study has been conducted in premenopausal women until now. A small-scale study by Savastano et al. (2003) showed that DHEAS was inversely associated with IMT in 17 premenopausal women with severe obesity (average body mass index was 43.5 kg/m^2). Thus, little is known about whether serum DHEAS is associated with atherosclerosis in premenopausal women, and large-scale studies are necessary to clarify the issue.

TABLE 22.3
Cross-Sectional Studies on the Association of DHEA and DHEAS with Atherosclerosis in Women

Study	Subjects	Outcome Measurement	Results
	Postmenopausal Women		
Bernini et al. (2001)	44 postmenopausal women	IMT	NS
Hak et al. (2002)	528 nonsmoking postmenopausal women	AC	NS
Golden et al. (2002)	182 cases with severe IMT and 182 controls	IMT	NS
Debing et al. (2007)	56 cases and 56 controls	Carotid artery atherosclerosis	NS
Kanazawa et al. (2008)	106 postmenopausal women with type 2 DM	PWV, IMT	DHEAS was inversely associated with PWV and IMT
Michos et al. (2008)	881 postmenopausal women	AC	NS
Ouyang et al. (2009)	1947 postmenopausal women	AC, IMT	NS
Naessen et al. (2010)	8 cases and 64 controls aged 70 years	CVD	DHEAS lower in cases
	Pre- and Postmenopausal Women		
Ishihara et al. (1992)	119 women without overt CAD	AC	DHEA and DHEAS lower in patients with AC
Herrington (1995)	103 women undergoing elective coronary angiography	CAD	DHEAS lower in cases
Bernini et al. (1999)	101 women	IMT	DHEAS was inversely associated with IMT[a]
Kiechl et al. (2000)	379 women	IMT, CVD	NS
	Premenopausal Women		
Savastano et al. (2003)	17 premenopausal women with severe obesity	IMT	DHEAS was inversely associated with IMT

Note: IMT, intima–media thickness; AC, aortic calcification; DM, diabetes mellitus; PWV, pulse wave velocity; CVD, cardiovascular disease; CAD, coronary artery disease; NS, not significant; DHEA, dehydroepiandrosterone; DHEAS, DHEA sulfate.

[a] Not significant after adjusting for age.

Five longitudinal studies on DHEA and CVD mortality were reported in women (Table 22.2). Among them, four studies indicated no significant association between serum DHEAS levels and mortality due to CVD and CAD. In contrast, Haffner et al. (1996) conducted a prospective case–control study on the relationship between DHEAS and CAD mortality in DM patients. In their study, lower serum levels of DHEAS predicted CAD mortality, and the result was essentially unchanged after adjustment for other risk factors. DHEA is believed to be associated with insulin sensitivity, glucose metabolism, fat accumulation, and plasma lipid levels (Buffington, Givens, and Kitabchi 1991; Macewen and Kurzman 1991; Nestler et al. 1988). Thus, the longitudinal study by Haffner et al. (1996) and the cross-sectional study by us (Kanazawa et al. 2008) in DM subjects suggest that the significant and negative correlation between DHEAS levels and atherosclerosis-related events or parameters is more likely to be found in DM women than in non-DM counterparts.

DHEA ADMINISTRATION THERAPY

DHEA replacement therapy seems to be an attractive method for preventing atherosclerosis and CVD since previous *in vitro* and *in vivo* studies suggest that DHEA has a beneficial effect on vascular function (Gordon, Bush, and Weisman 1988; Yamakawa et al. 2009). Although DHEA preparations have been available on the market in several countries since the 1990s, only several small-scale studies have been published on the effects of DHEA on atherosclerosis in humans (Table 22.4). It was demonstrated that intravenous DHEA administration increased blood flow in uterine and ophthalmic arteries in full-term pregnant women (Hata, Senoh et al. 1995; Hata, Hashimoto et al. 1995). Kawano et al. (2003) examined long-term effects of DHEA supplementation on vascular function. In 24 men with hypercholesterolemia, they found that DHEA supplementation for 12 weeks significantly improved vascular endothelial function and insulin sensitivity, and decreased plasma concentration of plasminogen activator inhibitor type 1 (PAI-1), a suppressor of fibrinolysis with a pathogenic role in CAD. Moreover, Williams et al. (2004) showed that DHEA administration increased flow-mediated dilation in healthy postmenopausal women. Martina et al. (2006) showed that DHEA treatment for 2 months increased platelet cyclic guanosine monophosphate concentration, a marker of nitric oxide production, and decreased PAI-1 and LDL-cholesterol in healthy elderly men. Thus, these studies documented the positive effect of DHEA on endothelial function. In contrast, several studies indicated no significant effect of DHEA on the parameters of atherosclerosis (Marder et al. 2010; Rice et al. 2009; Silvestri et al. 2005). Although the reason for this discrepancy is unclear, it may be caused by differences in the criteria for enrolling participants or in the route or dosage of DHEA administration. Indeed, the cutoff serum DHEAS level for diagnosing DHEA deficiency, the effective dose of DHEA, and the route of DHEA administration (oral or intravenous) for treating deficiency are still unclear. In addition, fewer studies have been performed on the effects of DHEA treatment on the incidence or mortality of CVD. Therefore, further studies are needed to clarify these issues.

TABLE 22.4
Effects of DHEA and DHEAS Administration on Parameters of Atherosclerosis

Study	Subjects	Treatment	Outcome Measurement	Results
Hata, Senoh et al. (1995)	10 pregnant women	DHEAS, 200 mg iv	Doppler ultrasonography	Increased blood flow in uterine artery
Hata, Hashimoto et al. (1995)	14 pregnant women	DHEAS, 200 mg iv	Doppler ultrasonography	Increased blood flow in ophthalmic artery
Kawano et al. (2003)	12 men with hypercholesterolemia	DHEA, 25 mg/day for 12 weeks	FMD, PAI-1	Increased FMD and decreased PAI-1
Williams et al. (2004)	36 postmenopausal women	DHEA, 100 mg/day for 3 months	FMD	Increased FMD
Silvestri et al. (2005)	16 postmenopausal women	DHEA, 50 mg/day for 4 weeks	FMD	NS
Martina et al. (2006)	12 men	DHEA, 50 mg/day for 2 months	cGMP, PAI-1	Increased cGMP and decreased PAI-1
Rice et al. (2009)	20 Addison's and 20 hypopituitarism	DHEA, 50 mg/day for 12 weeks	PWV	NS
Marder et al. (2010)	13 postmenopausal women with SLE	DHEA, 200 mg/day for 22 weeks	FMD, NMD	Tendency to increase FMD

Note: FMD, flow-mediated dilation; PAI-1, plasminogen activator inhibitor type 1; cGMP, cyclic guanosine monophosphate; PWV, pulse wave velocity; SLE, systemic lupus erythematosus; NMD, nitroglycerin-mediated dilatation; iv, intravenous; NS, not significant; DHEA, dehydroepiandrosterone; DHEAS, DHEA sulfate.

CONCLUSION

Although controversy still exists, a number of cross-sectional and longitudinal studies show that serum DHEA and DHEAS levels are inversely associated with CVD and CAD events in men. In women, this association has not been found except in DM patients. Some small-scale clinical trials indicate that DHEA administration could have a beneficial effect on vascular endothelial function in both men and women. Thus, some but not all studies suggest that DHEA has a potential protective effect against atherosclerosis and its related disorders.

REFERENCES

Akishita, M., M. Hashimoto, Y. Ohike et al. 2010. Low testosterone level as a predictor of cardiovascular events in Japanese men with coronary risk factors. *Atherosclerosis* 210:232–6.

Barrett-Connor, E. 1992. Lower endogenous androgen levels and dyslipidemia in men with non-insulin-dependent diabetes mellitus. *Ann Intern Med* 117:807–11.

Barrett-Connor, E., and S. L. Edelstein. 1994. A prospective study of dehydroepiandrosterone sulfate and cognitive function in an older population: The Rancho Bernardo Study. *J Am Geriatr Soc* 42:420–3.

Barrett-Connor, E., and D. Goodman-Gruen. 1995a. The epidemiology of DHEAS and cardiovascular disease. *Ann N Y Acad Sci* 774:259–70.

Barrett-Connor, E., and D. Goodman-Gruen. 1995b. Dehydroepiandrosterone sulfate does not predict cardiovascular death in postmenopausal women. The Rancho Bernardo Study. *Circulation* 91:1757–60.

Barrett-Connor, E., K. T. Khaw, and S. S. Yen. 1986. A prospective study of dehydroepiandrosterone sulfate, mortality, and cardiovascular disease. *N Engl J Med* 315:1519–24.

Beer, N. A., D. J. Jakubowicz, D. W. Matt, R. M. Beer, and J. E. Nestler. 1996. Dehydroepiandrosterone reduces plasma plasminogen activator inhibitor type 1 and tissue plasminogen activator antigen in men. *Am J Med Sci* 311:205–10.

Bernini, G. P., A. Moretti, M. Sgro et al. 2001. Influence of endogenous androgens on carotid wall in postmenopausal women. *Menopause* 8:43–50.

Bernini, G. P., M. Sgro, A. Moretti et al. 1999. Endogenous androgens and carotid intimal-medial thickness in women. *J Clin Endocrinol Metab* 84:2008–12.

Buffington, C. K., J. R. Givens, and A. E. Kitabchi. 1991. Opposing actions of dehydroepiandrosterone and testosterone on insulin sensitivity. In vivo and in vitro studies of hyperandrogenic females. *Diabetes* 40:693–700.

Cao, J., H. Zou, B. P. Zhu et al. 2010. Sex hormones and androgen receptor: risk factors of coronary heart disease in elderly men. *Chi Med Sci J* 25:44–9.

Contoreggi, C. S., M. R. Blackman, R. Andres et al. 1990. Plasma levels of estradiol, testosterone, and DHEAS do not predict risk of coronary artery disease in men. *J Androl* 11:460–70.

Debing, E., E. Peeters, W. Duquet, K. Poppe, B. Velkeniers, and P. van den Brande. 2007. Endogenous sex hormone levels in postmenopausal women undergoing carotid artery endarterectomy. *Eur J Endocrinol* 156:687–93.

Dockery, F., C. J. Bulpitt, M. Donaldson, S. Fernandez, and C. Rajkumar. 2003. The relationship between androgens and arterial stiffness in older men. *J Am Geriatr Soc* 51:1627–32.

Feldman, H. A., C. B. Johannes, A. B. Araujo, B. A. Mohr, C. Longcope, and J. B. McKinlay. 2001. Low dehydroepiandrosterone and ischemic heart disease in middle-aged men: Prospective results from the Massachusetts Male Aging Study. *Am J Epidemiol* 153:79–89.

Fukui, M., Y. Kitagawa, N. Nakamura et al. 2005. Serum dehydroepiandrosterone sulfate concentration and carotid atherosclerosis in men with type 2 diabetes. *Atherosclerosis* 181:339–44.

Fukui, M., H. Ose, Y. Kitagawa et al. 2007. Relationship between low serum endogenous androgen concentrations and arterial stiffness in men with type 2 diabetes mellitus. *Metabolism* 56:1167–73.

Furutama, D., R. Fukui, M. Amakawa, and N. Ohsawa. 1998. Inhibition of migration and proliferation of vascular smooth muscle cells by dehydroepiandrosterone sulfate. *Biochim Biophys Acta* 1406:107–14.

Golden, S. H., A. Maguire, J. Ding et al. 2002. Endogenous postmenopausal hormones and carotid atherosclerosis: A case-control study of the atherosclerosis risk in communities cohort. *Am J Epidemiol* 155:437–45.

Gordon, G. B., D. E. Bush, and H. F. Weisman. 1988. Reduction of atherosclerosis by administration of dehydroepiandrosterone. A study in the hypercholesterolemic New Zealand white rabbit with aortic intimal injury. *J Clin Invest* 82:712–20.

Haffner, S.M., S. E. Moss, B. E. Klein, and R. Klein. 1996. Sex hormones and DHEA-SO4 in relation to ischemic heart disease mortality in diabetic subjects. The Wisconsin Epidemiologic Study of Diabetic Retinopathy. *Diabetes Care* 19:1045–50.

Hak, A. E., J. C. Witteman, F. H. de Jong, M. I. Geerlings, A. Hofman, and H. A. Pols. 2002. Low levels of endogenous androgens increase the risk of atherosclerosis in elderly men: the Rotterdam study. *J Clin Endocrinol Metab* 87:3632–9.

Hata, T., M. Hashimoto, D. Senoh, K. Hata, M. Kitano, and S. Masumura. 1995. Effects of dehydroepiandrosterone sulfate on ophthalmic artery flow velocity waveforms in full-term pregnant women. *Am J Perinatol* 12:135–7.

Hata, T., D. Senoh, K. Hata, M. Kitano, and S. Masumura. 1995. Effects of dehydroepiandrosterone sulfate on uterine artery flow velocity waveforms in term pregnancy. *Obstet Gynecol* 85:118–21.

Hauner, H., K. Stangl, K. Burger, U. Busch, H. Blomer, and E. F. Pfeiffer. 1991. Sex hormone concentrations in men with angiographically assessed coronary artery disease-relationship to obesity and body fat distribution. *Klin Wochenschr* 69:664–8.

Hautanen, A., M. Manttari, V. Manninen et al. 1994. Adrenal androgens and testosterone as coronary risk factors in the Helsinki Heart Study. *Atherosclerosis* 105:191–200.

Herrington, D. M. 1995. Dehydroepiandrosterone and coronary atherosclerosis. *Ann N Y Acad Sci* 774:271–80.

Herrington, D. M., G. B. Gordon, S. C. Achuff et al. 1990. Plasma dehydroepiandrosterone and dehydroepiandrosterone sulfate in patients undergoing diagnostic coronary angiography. *J Am Coll Cardiol* 16:862–70.

Hougaku, H., J. L. Fleg, S. S. Najjar et al. 2006. Relationship between androgenic hormones and arterial stiffness, based on longitudinal hormone measurements. *Am J Physiol Endocrinol Metab* 290:E234–42.

Ishihara, F., K. Hiramatsu, S. Shigematsu et al. 1992. Role of adrenal androgens in the development of arteriosclerosis as judged by pulse wave velocity and calcification of the aorta. *Cardiology* 80:332–8.

Kajinami, K., K. Takeda, N. Takekoshi et al. 2004. Imbalance of sex hormone levels in men with coronary artery disease. *Coron Artery Dis* 15:199–203.

Kanazawa, I., T. Yamaguchi, M. Yamamoto et al. 2008. Serum DHEA-S level is associated with the presence of atherosclerosis in postmenopausal women with type 2 diabetes mellitus. *Endocrin J* 55:667–75.

Kawano, H., H. Yasue, A. Kitagawa et al. 2003. Dehydroepiandrosterone supplementation improves endothelial function and insulin sensitivity in men. *J Clin Endocrinol Metab* 88:3190–5.

Kiechl, S., J. Willeit, E. Bonora, S. Schwarz, and Q. Xu. 2000. No association between dehydroepiandrosterone sulfate and development of atherosclerosis in a prospective population study (Bruneck Study). *Arterioscler Thromb Vasc Biol* 20:1094–100.

Labrie, F. 1991. Intracrinology. *Mol Cell Endocrinol* 78:C113–8.

LaCroix, A. Z., K. Yano, and D. M. Reed. 1992. Dehydroepiandrosterone sulfate, incidence of myocardial infarction, and extent of atherosclerosis in men. *Circulation* 86:1529–35.

Macewen, E. G., and I. D. Kurzman. 1991. Obesity in the dog: role of the adrenal steroid dehydroepiandrosterone (DHEA). *J Nutr* 121:S51–5.

Marder, W., E. C. Somers, M. J. Kaplan, M. R. Anderson, E. E. Lewis, and W. J. McCune. 2010. Effects of prasterone (dehydroepiandrosterone) on markers of cardiovascular risk and bone turnover in premenopausal women with SLE: A pilot study. *Lupus* (Epub ahead of print).

Martina, V., A. Benso, V. R. Gigliardi et al. 2006. Short-term endothelial function treatment increases platelet cGMP production in elderly male subjects. *Clin Endocrinol* 64:260–4.

Michos, E. D., D. Vaidya, S. M. Gapstur et al. 2008. Sex hormones, sex hormone binding globulin, and abdominal aortic calcification in women and men in the multi-ethnic study of atherosclerosis (MESA). *Atherosclerosis* 200:432–8.

Mitchell, L. E., D. L. Sprecher, I. B. Borecki, T. Rice, P. M. Laskarzewski, and D. C. Rao. 1994. Evidence for an association between dehydroepiandrosterone sulfate and nonfatal, premature myocardial infarction in males. *Circulation* 89:89–93.

Naessen, T., U. Sjogren, J. Bergquist, M. Larsson, L. Lind, and M. M. Kushnir. 2010. Endogenous steroids measured by high-specificity liquid chromatography-tandem mass spectrometry and prevalent cardiovascular disease in 70-year-old men and women. *J Clin Endocrinol Metab* 95:1889–97.

Nestler, J. E., C. O. Barlascini, J. N. Clore, and W. G. Blackard. 1988. Dehydroepiandrosterone reduces serum low density lipoprotein levels and body fat but does not alter insulin sensitivity in normal men. *J Clin Endocrinol Metab* 66:57–61.

Newcomer, L. M., J. E. Manson, R. L. Barbieri, C. H. Hennekens, and M. J. Stampfer. 1994. Dehydroepiandrosterone sulfate and the risk of myocardial infarction in US male physicians: A prospective study. *Am J Epidemiol* 140:870–5.

Ouyang, P., D. Vaidya, A. Dobs et al. 2009. Sex hormone levels and subclinical atherosclerosis in postmenopausal women: the Multi-Ethnic Study of Atherosclerosis. *Atherosclerosis* 204:255–61.

Phillips, G. B., B. H. Pinkemell, and T. Y. Jing. 1994. The association of hypotestosteronemia with coronary artery disease in men. *Arterioscler Thromb* 14:701–6.

Rice, S. P., N. Agarwal, H. Bolusani et al. 2009. Effects of dehydroepiandrosterone replacement on vascular function in primary and secondary adrenal insufficiency: A randomized crossover trial. *J Clin Endocrinol Metab* 94:1966–72.

Savastano, S., R. Valentino, A. Belfiore et al. 2003. Early carotid atherosclerosis in normotensive severe obese premenopausal women with low DHEA(S). *J Endocrinol Invest* 26:236–43.

Silvestri, A., M. Gambacciani, C. Vitale et al. 2005. Different effect of hormone replacement therapy, DHEAS and tibolone on endothelial function in postmenopausal women with increased cardiovascular risk. *Maturitas* 50:305–11.

Slowinska-Srzednicka, J., S. Zgliczynski, M. Ciswicka-Sznajderman et al. 1989. Decreased plasma dehydroepiandrosterone sulfate and dihydrotestosterone concentrations in young men after myocardial infarction. *Atherosclerosis* 79:197–203.

Svartberg, J., D. von Muhlen, E. Mathiesen, O. Joakimsen, K. H. Bonaa, and E. Stensland-Bugge. 2006. Low testosterone levels are associated with carotid atherosclerosis in men. *J Intern Med* 259:576–82.

Taniguchi, S., T. Yanase, K. Kobayashi, R. Takayanagi, and H. Nawata. 1996. Dehydroepiandrosterone markedly inhibits the accumulation of cholesteryl ester in mouse macrophage J774-1 cells. *Atherosclerosis* 126:143–54.

Tedeschi-Reiner, E., R. Ivekovic, K. Novak-Laus, and Z. Reiner. 2009. Endogenous steroid sex hormones and atherosclerosis of retinal arteries in men. *Med Sci Monit* 15:CR211–6.

Tilvis, R. S., M. Kahonen, and M. Harkonen. 1999. Dehydroepiandrosterone sulfate, disease and mortality in a general aged population. *Aging* 11:30–4.

Trivedi, D. P., and K. T. Khaw. 2001. Dehydroepiandrosterone sulfate and mortality in elderly men and women. *J Clin Endocrinol Metab* 86:4171–7.

Ueshiba, H., Y. Shimizu, N. Hiroi et al. 2002. Decreased steroidogenic enzyme 17, 20-lyase and increased 17-hydroxylase activities in type 2 diabetes mellitus. *Eur J Endocrinol* 146:375–80.

van den Beld, A. W., M. L. Bots, J. A. Jassen, H. A. Pols, S. W. Lamberts, and D. E. Grobbee. 2003. Endogenous hormones and carotid atherosclerosis in elderly men. *Am J Epidemiol* 157:25–31.

Williams, M. R., T. Dawood, S. Ling et al. 2004. Dehydroepiandrosterone increases endothelial cell proliferation in vitro and improves endothelial function in vivo by mechanisms independent of androgen and estrogen receptors. *J Clin Endocrinol Metab* 89:4708–15.

Yamakawa, T., K. Ogihara, M. Nakamura et al. 2009. Effect of dehydroepiandrosterone on atherosclerosis in apolipoprotein E-deficient mce. *J Atheroscler Thromb* 16:501–8.

23 DHEAS and Related Factors in Autism

Jan Croonenberghs, Katelijne Van Praet, Dirk Deboutte, and Michael Maes

CONTENTS

CORTISOL/DHEAS RATIO AND STRESS RESPONSE

DHEAS, CORTISOL, HYPOTHALAMO-PITUITARY-ADRENAL (HPA) AXIS, AND SEROTONINE

Dehydroepiandrosterone (DHEA) and dehydroepiandrosterone sulfate (DHEAS) are the most abundant steroids in the human circulatory system (Baulieu and Robel 1996). DHEAS is the proposed storage form of active DHEA (Strous et al. 2005). DHEAS levels are 300- to 500-fold higher than DHEA levels.

The normal range of DHEAS shows a broad, mainly genetically determined variation (Rotter et al. 1985). The conversion of DHEA to DHEAS by sulfotransferases takes place in the adrenal cortex and a multitude of other tissues (Falany 1997). There is evidence for a production of DHEA as a neurosteroid in the brain (Baulieu 1998) and for its direct testicular production (de Peretti and Forest 1978). Plasma and saliva levels of DHEA correlate highly with levels in the ventricular cerebrospinal fluid (Guazzo et al. 1996). The production in the adrenal gland of pregnelolone, which is converted to DHEA or cortisol, is regulated by HPA-axis hormones and hence in part by serotonine (5 HT; Maes et al. 1995; Meltzer and Maes 1994).

Under normal circumstances, DHEA and cortisol are secreted synchronously by the adrenal cortex in response to adrenocorticotropic hormone (ACTH; Nieschlag et al. 1973). However, independent control mechanisms exist for cortisol and DHEAS (Kroboth et al. 1999). For example, unlike cortisol, concentrations of DHEAS vary with age (Orentreich et al. 1984). The diurnal rhythm in salivary DHEA is much less than that in cortisol. Therefore, the cortisol/DHEA ratio changes during the day from about 5–7 in the morning to about 2 at 20 hours (Goodyer et al. 2001).

Events during either prenatal or early postnatal development may influence levels of both neurosteroids. During the first year of infant life, the responsiveness of the HPA axis and, hence, the production of stress hormones evolves from a highly labile and responsive system to a system characterized by less responsiveness to nonspecific general stressors. This buffering of the adrenocortical response to stress during the first year of life is characterized by an adaptation of the stress response to normal life events (Gunnar et al. 1996; Lewis and Ramsay 1995). The brain is sensitive to excess exposure to cortisol, with potential impairments in mental and behavioral function.

The importance of the serotonergic function on autism and stress regulation (through the HPA axis and cortisol) is one of the most consistent findings in the pathophysiology of autism. Serotonine (5-HT) plays an important role in the developmental disorders because of its role as a morphogenetic agent and as a neurotransmitter in the central nervous system (Azmitia and Withaker-Azmitia 1995). The release of the HPA-axis hormones is mediated in part with 5-HT.

Cortisol/DHEAS Ratio and Stress Response

Cortisol and DHEAS play a pivotal role in the regulation of the stress response caused by external stressors.

Under conditions of acute stress, DHEAS acts like an antiglucocorticoid.

1. Its anabolic action antagonizes the catabolic effects of cortisol (Goodyer et al. 2001; Azmitia and Withaker-Azmitia 1995; Hu et al. 2000; Kalimi et al. 1994; Kimonides et al. 1999). For example, it antagonizes the immunosuppressant and lympholytic actions of cortisol (Blauer et al. 1991).
2. DHEA protects hippocampal function and may influence processes of cognition and memory. The hippocampus may be protected by DHEAS from the neurotoxic effects of both glutamate analogs and corticoids (Mao and Barger 1998).
3. Stress-activated protein kinase/c-Jun N-terminal kinase (SAPK/JNK) has been reported to phosphorylate c-Jun in striatal neuronal cultures following glutamate treatment (Schwarzschild, Cole, and Hymann 1997). The activation of SAPK3 by corticosterone was attenuated by DHEA (Kimonides et al. 1999).

Chronic exposure to high levels of glucocorticoid (as a result of chronic stress) may have a detrimental effect on the brain by at least three mechanisms (Goodyer et al. 2001; Wolkowitz et al. 2001).

1. A genomic effect, by altering the production of neurotransmitters and the expression of receptors in brain regions, such as the hippocampus, septum, amygdala, and prefrontal cortex, where corticosteroid receptors are densely located. Corticoids may alter the stress-dependent expression of some immediate-early genes as well as of corticotropin-releasing factor and vasopressin.
2. Direct interaction with neuronal cell surface receptors ($GABA_A$ and N-methyl-D-aspartate [NMDA]-receptors). The hippocampus has one extraordinary feature: active neurogenesis occurs in the dentate gyrus well into adult life (Altman and Das 1965), not only in rats but also in monkeys and humans (Eriksson et al. 1998). Furthermore, this process seems to be modulated both by NMDA receptors and glucocorticoids. Corticoids potentiate neurodegeneration induced by such agents as anoxia, glutamate analogs such as kainic acid, or cholinergic agents (Hortnagl et al. 1993).
3. A morphological, neurotoxic effect on hippocampal neurons.

In chronic stress, however, DHEAS levels may decline and thus the ability to counteract the catabolic effects of glucocorticoids becomes impaired (Bruce and McEwen 2004).

Consequently, the cortisol/DHEAS ratio is an important index reflecting the integrative role for these two adrenal steroids in neurotoxic processes in the brain. A high cortisol/DHEAS ratio may reflect an increased neurotoxic risk to the brain (Bruce and McEwen 2004) and therefore may play a role in abnormal psychological processes and psychiatric disturbances (Goodyer et al. 2001; Hechter, Grossman, and Chatterton 1997).

DHEA levels are reduced in major depressive disorders in both adolescents and adults, and an increased cortisol/DHEA ratio (together with intercurrent life events) predicts delayed recovery. DHEA may have a role in the treatment of depression. Together, these findings suggest that altered steroidal environment, whether induced by stress or aging, can have appreciable results on the cellular structure of the brain as well as on its function (Herbert 1998).

AUTISM AND STRESS

Autism is a psychiatric disturbance characterized by qualitative impairment in communication and social interactions, and repetitive and stereotyped patterns of behavior or interests (American Psychiatric Association 1994). Children with autism have been described as experiencing difficulties in the perception of novelty and environmental stressors (Kanner 1943). Changes in the environment and the anticipation and experience of stressful events result mostly in an aggravation of the symptomatology. The occurrence of pre- and perinatal and early life stressors, which are known to play a role in the etiology of autism (Hultman, Sparen, and Cnattingius 2002; Kolevzon, Gross, and Reichenberg 2007), may explain a lowered buffering of the adrenocortical response to stress in autism (Croonenberghs et al. 2008).

The limbic system and hippocampus, which are both very sensitive to the neurotoxic effects of glucocorticoids (Bastianetto et al. 1999; Cardounel, Regelson, and Kalimi 1999; Compagnone and Mellon 1998; Diamond et al. 1995; Kimonides et al. 1998), are involved in autism (Kern and Jones 2006). Neuropsychological and neuroanatomical research shows evidence for disturbances of the medial-temporal region and the presence of memory dysfunctions corresponding to this region, in subgroups of autistic children with mental retardation and/or neurological disturbances. Bachevalier et al. (Bachevalier 1994; Bachevalier and Mishkin 1994) studied newborn rhesus monkeys and observed severe disturbances of social contact and communication and the appearance of stereotyped behavior, all key symptoms of the autistic disorder, after the infliction of bilateral lesions in the amygdalo–hippocampal complex.

AUTISM, DHEAS, AND TESTOSTERONE

Sulfation of DHEA to DHEAS

DHEA is a key initial regulatory metabolite in the androgen synthesis pathway. It is at the DHEA location in the androgen synthesis pathway that DHEA can either be converted further down the androgen pathway toward testosterone by being converted to androstenedione or androstenediol, or toward the normally favored storage molecule of DHEAS. The conversion of DHEA to DHEAS by the enzyme hydroxysteroid sulfotransferase is dependent on sulfation and is inhibited by inflammation (Kim et al. 2004; Ryan and Carrol 1976). This enzyme requires glutathione as a cofactor in rats (Ryan and Carrol 1976).

Methionine Cycle

Glutathione is produced from cysteine, which is made out of cystathione, as a result of a conversion from homocysteine. Homocysteine can also be converted in methionine by the enzyme betaine-homocysteine methyltransferase. 5-methyltetrahydrofolate (5-MTHF), the most active form of the B-vitamin folic acid, is the methylated derivate of tetrahydrofolate. It is generated by methylenetetrahydrofolate reductase (MTHFR). 5-MTHF functions, in concert with vitamin B12, as a methyl group donor involved in the conversion of homocysteine to methionine. Methionine on its turn can be converted to homocysteine.

The MTHFR gene is of significant importance in the folate pathway, and single nucleotide polymorphisms (SNPs) in the MTHFR gene have been shown to reduce its functional capacity and hence reduce the ability of methionine synthase to recycle homocysteine to methionine. The same genetic mutation causes elevations in homocysteine (Boris et al. 2004).

Decreases in total glutathione levels can cause a marked shift toward the androgen synthesis pathway out of DHEA (Aarum et al. 2003).

Methionine Cycle Transsulfuration and Androgen Pathways

Testosterone and possibly other androgen metabolites may have an influence on the methionine cycle-transsulfuration pathways and hence may have an influence on the balance between the transsulfuration and androgen pathways (Aarum et al. 2003).

Excess androgen interferes with the conversion of homocysteine to cystathione, a major source of glutathione. Thus, the excess androgen can increase homocysteine levels and decrease the levels of glutathione, a major antioxidant in brain (Giltay et al. 1998; Giltay et al. 2003). Vrbikova et al. (2003) have shown significant positive correlations between homocysteine and androstenedione levels and glutathione and DHEAS levels in humans.

Androgens and the Brain

Levels of testosterone differ with age, pubertal stadium, medication, certain diseases, and stress (Croonenberghs et al. 2010).

Androgen receptors have been demonstrated in the hypothalamus, hippocampus, preoptic area, amygdala, medial hypothalamus, and frontal lobe areas (Cooke 2006; DonCarlos et al. 2003; DonCarlos et al. 2006; Henderson et al. 2006; MacLusky et al. 2006; Yang et al. 2002). It is known that both testosterone and estrogen, at basal levels, are neuroprotective and play a significant role in neuronal development, migration, dendritic outgrowth, and synaptogenesis (Leranth, Petnehazy, and Maclusky 2003; Weiland 1992). Chronic elevation of testosterone activates microglia, triggering the release of a number of neurotoxic elements, including the excitotoxins, glutamate and quinolinic acid, and inflammatory cytokines (Aarum et al. 2003; Estrada, Uhlen, and Ehrlich 2006; Estrada, Varshney, and Ehrlich 2006; Komuro and Rakie 1993; Lieberherr and Grosse 1994; Lin and Constantine-Paton 1998; Spitzer 2000).

This combination of inflammatory cytokines, androgens, and excitatory neurotransmitters would precipitate chronic neurodegeneration and alter progenitor cell differentiation and maturation, dendritic outgrowth and arborization, synaptic development and stabilization, and neuronal migration (Blaylock 2009).

Role of Androgens in the Pathophysiology of Autism

Elements indicating that testosterone plays a role in the pathophysiology of autism are as follows:

- Autism is four times more common in boys than in girls (Croonenberghs 2003).
- Eye contact and language development, which are impaired by autism, are negatively correlated with prenatal testosterone levels (Lutchmaya, Baron-Cohen, and Raggat 2002). Girls make significantly more eye contact than boys and learn to speak (and read) earlier and develop a greater vocabulary (Lutchmaya, Baron-Cohen, and Raggat 2002; Higley et al. 1996; Sanchez-Martin et al. 2000).

Fetal testosterone (Baron-Cohen et al. 2005) is negatively correlated to the quality of social relationships by girls and boys and is positively correlated with restricted interests in boys.

Autistic disorder is characterized by extreme systemizing and very low empathizing (Baron-Cohen 2002). Boys are better at systemizing than at empathizing, whereas girls are better at empathizing than at systemizing.

In people with autism, low 2° digit/4° digit ratios have been found and this is influenced by fetal testosterone (Ingudomnukul et al. 2007; Manning et al. 2001).

Some neuroanatomical studies comparing the brains of individuals with and without an autism spectrum disorder (ASD) reveal structural differences associated with high levels of fetal testosterone, including hemispheric asymmetries (Herbert et al. 2005).

Chronic elevation of testosterone activates the microglia, which directs neuronal precursor cell migration and differentiation. Activated microglia can increase neuronal numbers significantly. This may explain the hypercellularity observed in certain areas of the autistic brain, particularly in the amygdala (Aarum et al. 2003).

In autism, the expression of symptoms often changes after puberty.

In some boys with autism, pubertas praecox is observed, and this goes together with increased testosterone levels (Baron-Cohen 2002; Tordjman et al. 1997).

Girls with abnormally high levels of fetal testosterone as a result of congenital adrenal hyperplasia have a higher number of autistic traits than their unaffected sisters (Knickmeyer, Baron-Cohen, and Fane 2006). Girls with an ASD show a significant delay in the onset of menarche (Knickmeyer, Wheelwright, Hoekstra, and Baron-Cohen 2006) and are more likely to display elevated rates of testosterone-related disorders than controls (Ingudomnukul et al. 2007).

Studies have shown a link between elevated prenatal testosterone (de Bruin et al. 2006), postnatal serum testosterone (Geier et al. 2006a), and autism spectrum disorders. Geier and Geier found that patients with ASD diagnosis, with a mean age of 10.8 years, have significantly increased androgen metabolites: serum testosterone, serum free testosterone, DHEA, and androstenedione. The level of follicle-stimulating hormone is significantly decreased, possibly indicating an attempt to downregulate pituitary-controlled androgen synthesis. Antiandrogen may be of benefit in some autistic clients (Geier et al. 2006b; Geier et al. 2007b). Improvement was seen in sociability, cognitive awareness, and aggressive behavior. In a heterogenic population of pre- and postpubertal subjects, however (Tordjman et al. 1995), testosterone was not significantly different in autism compared with the control group. In a very homogeneous group of postpubertal youngsters with autism, the morning serum testosterone was found to be significantly lower compared with that of the control group (Croonenberghs et al. 2010).

Role of DHEAS and the Cortisol/DHEAS Ratio in the Pathophysiology of Autism

Elements indicating that the regulation of the adrenal stress hormones, cortisol and DHEAS, and the cortisol/DHEAS ratio play a role in the pathophysiology of autism are as follows:

In a study performed in our center (Croonenberghs et al. 2008), we measured the cortisol and DHEAS responses to administration of L-5-hydroxytryptophan (L-5-HTP), the direct precursor of 5-HT, and the cortisol/DHEAS ratio in 18 male, postpubertal, Caucasian autistic patients (age 13–19 years; I.Q. > 55) and 22 matched healthy volunteers. Serum cortisol and DHEAS were determined on two consecutive days, 45 and 30 minutes before administration of L-5-HTP (4 mg/kg in nonenteric-coated tablets) or an identical placebo in a single-blind order and, thereafter, every 30 minutes during a 3-hour period.

The hormonal responses to a challenge with L-5-HTP, the direct precursor of 5-HT, offer an index for the functional state of the central serotonergic system (Maes et al. 1995; Meltzer and Maes 1994).

The L-5-HTP-induced increase in serum cortisol was significantly lower in autistic patients than in controls, whereas the 5-HTP-induced DHEAS response was significantly higher in autistic patients than in controls.

The cortisol/DHEAS ratio was significantly lower in autistic patients than in controls during the L-5-HTP challenge test.

Individuals with ASD have significant decreases in transsulfuration metabolites including cysteine, glutathione (Geier et al. 2006a; Geier et al. 2007b; James et al. 2004; James et al. 2006), and sulfate (Waring and Klovrza 2000; Waring et al. 1997), impaired sulfation (Alberti et al. 1999), as well as significant increases in common polymorphic variants known to modulate the transsulfuration pathway (James et al. 2006). This possibly results in a marked shift toward DHEA and subsequent metabolites in the androgen pathway.

DISCUSSION AND HYPOTHESIS

The aforementioned studies suggest an important role for DHEAS, cortisol, testosterone, and the serotonergic function in the aethiopathophysiology of autism.

A putative link between DHEAS and testosterone possibly consists of the cyclical interaction between the methionine cycle transsulfuration and the pathways producing, on the one hand, DHEAS and, on the other hand, the androgens out of DHEA. This interaction is possibly disturbed in some children with ASD (Geier et al. 2006a). In ASD homocysteine, S-adenosylhomocysteine or adenosine are significantly increased (James et al. 2004; James et al. 2006; Pasca et al. 2006).

Oxidized glutathione levels were found to be twofold higher in autistic subjects, which strongly indicate oxidative stress (James et al. 2004; Chauhan and Chauhan 2006; Kern and Jones 2006; McGinnis 2004; Mutter et al. 2005; Vargas et al. 2005). Several of the enzymes utilized in the methionine cycle, such as methionine synthase, betaine-homocysteine methyltransferase, and methionine adenosyltransferases, are known to be redox-sensitive enzymes (Avilla et al. 1998; Gulati et al. 1997).

Differences in allele frequency and/or significant gene–gene interactions were found for relevant genes encoding the reduced folate carrier, transcobalamin II, catechol-O-methyltransferase, MTHFR, and GST M1. Increased vulnerability to oxidative stress (endogenous or environmental) may contribute to the development and clinical manifestations of autism (James et al. 2006). Children with ASD have been shown to have a significant increase in SNPs in this MTHFR gene. It was reported that only 2% of autistic patients did not have at least one SNP in their MTHFR gene (Boris et al. 2004). Studies have shown abnormal absorption of vitamin B12 from the ileum of autistic children (Wakefield et al. 1998).

Vitamin B12 deficiency is well known to have neuropsychiatric consequences in adults (Zucker et al. 1981) and adversely affects neurodevelopment during infancy (Graham, Arvela, and Wise 1992). In toddlers, severe vitamin B12 deficiency has been associated with developmental regression similar to that observed in a subgroup of autistic children (Grattan-Smith et al. 1997).

OH-steroid sulfotransferase is shown to be necessary for appropriate function of bile salts (Radominska et al. 1990). This may contribute to malabsorption and the high prevalence of gastrointestinal disease found in ASD patients (White 2003). Furthermore, impaired sulfation may also play an important role in other common biochemical abnormalities found in ASD cases, involving neurotransmitters, peptides, glycosaminoglycans, amines, and/or phenols.

A major finding of our study (Croonenberghs et al. 2008) is the significantly higher baseline cortisol/DHEAS ratio in autistic patients compared with normal volunteers. The ratio in the latter group is an index representing the anabolic/catabolic state with regard to the effects of DHEAS versus cortisol. Therefore, the increased cortisol/DHEAS ratio may represent important consequences for neurotoxicity and brain development in autism.

Another finding of our study is that in autistic patients, the L-5-HTP-induced plasma DHEAS values were significantly higher, the increase in serum cortisol was significantly lower, and the cortisol/DHEAS ratio also was significantly lower compared with that of controls during the L-5-HTP challenge test. This seems in contrast to the baseline data. A possible explanation consists in disequilibrium in the serotonergic metabolism in autism.

Studies (Croonenberghs et al. 2007; Cohen, Shaywitz, and Johnson 1974; Cook and Leventhal 1996; Croonenberghs et al. 2005) suggest a central serotonergic hyporesponsivity in autism. Other findings showed that L-5-HTP increased the peripheral 5-HT levels significantly more in autism than in controls (Croonenberghs et al. 2005; Croonenberghs et al. 2007) pointing to an enhanced peripheral metabolism of 5-HT in autism.

This disequilibrium could explain the differences in the effects of L-5-HTP on plasma cortisol versus DHEAS, the former being under the influence of central serotonergic circuits, whereas the latter being more affected by peripheral circuits. In other words, the impact of the peripheral hyperactivity of 5-HT turnover could explain higher L-5-HTP-induced increases in plasma DHEAS. This effect is underscored by findings that DHEAS levels can increase independently from pituitary stimulation by ACTH (Kalimi et al. 1994). These peripheral mechanisms may include the testicular production of DHEAS (de Peretti and Forest 1978) and peripheral serotonergic inputs. Another possible mechanism, which could explain the findings of our study, is an exaggerated response of DHEAS to L-5-HTP challenge by an immature HPA axis in autism (Gunnar et al. 1996). According to this hypothesis, the increased release of DHEAS possibly is a hyperreactive, compensatory mechanism for a (relatively small) increase in cortisol.

This finding is of direct importance to the neuropathology of autism because the limbic system and hippocampus are involved in this illness, and because both structures are very sensitive to the neurotoxic effects of glucocorticoids (Bastianetto et al. 1999; Kern and Jones 2006). Indeed, DHEA and DHEAS show potent neuroprotective qualities and play an important role in neurodevelopment (Bastianetto et al. 1999; Cardounel, Regelson, and Kalimi 1999; Compagnone and Mellon 1998; Lapchak and Araujo 2001). "Early life stressors," for example, pre-, peri-, or postnatal risk factors, which are known to play a role in the etiology of autism (Kolevzon, Gross, and Reichenberg 2007), may explain abnormalities in the turnover of 5-HT, testosterone, and the cortisol/DHEAS ratio and hence in the "buffering" of the adrenocortical response to stress.

REFERENCES

Aarum, J., K. Sandberg, S. L. Budd Haeberlein, and M. A. Persson. 2003. Migration and differentiation of neural precursor cells can be directed by microglia. *Proc Natl Acad Sci USA* 100(26):15983–8.

Alberti, A., P. Pirrone, M. Elia, R. H. Waring, and C. Romano. 1999. Sulphation deficit in "low functioning" autistic children: A pilot study. *Biol Psychiatry* 46:420–4.

Altman, J., and G. D. Das. 1965. Autoradiographic and histological evidence of postnatal hippocampal neurogenesis in rats. *J Compr Neurol* 124:319–35.

American Psychiatric Association. 1994. *Diagnostic and Statistical Manual of Mental Disorders*. 4th ed. Washington, DC: American Psychiatric Association.

Avilla, M. A., M. V. Carretero, E. N. Rodriguez, and J. M. Mato. 1998. Regulation by hypoxia of methionine adensyltransferase activity and gene expression in rat hepatocytes. *Gastroenterology* 114(2):364–71.

Azmitia, E. C., and P. M. Withaker-Azmitia. 1995. Anatomy, cell biology and plasticity of the serotonergic system. In *Psychopharmacology: The Fourth Generation of Progress*, ed. F. E. Bloom and D. J. Kupfer, 443–90. New York: Raven Press, Ltd.

Bachevalier, J. 1994. Medial temporal lobe structures and autism: A review of clinical and experimental findings. *Neuropsychologia* 32(6):627–48.

Bachevalier, J., and M. Mishkin. 1994. Effects of selective neonatal temporal lobe lesions on visual recognition memory in rhesus monkeys. *J Neurosci* 14(4):2128–39.

Baron-Cohen, S. 2002. The extreme male brain theory of autism. *Trends Cogn Sci* 6(6):248–54.

Baron-Cohen, S., R. C. Knickmeyer, and M. K. Belmonte. 2005. Sex differences in the brain: Implications for explaining autism. *Science* 310:819–23.

Bastianetto, S., C. Ramassamy, J. Poirier, and R. Quirion. 1999. Dehydroepiandrosterone (DHEA) protects hippocampal cells from oxidative stress-induced damage. *Brain Res Mol Brain Res* 66:35–4.

Baulieu, E. E. 1998. Neurosteroids, a novel function of the brain. *Psychoneuroendocrinology* 23(8):963–87.

Baulieu, E. E., and P. Robel. 1996. Dehydroepiandrosterone and dehydroepiandrosterone sulfate as neuroactive neurosteroids. *J Endocrinol* 150:S221–39.

Blauer, K. L., M. Poth, W. M. Rogers et al. 1991. Dehydroepiandrosterone antagonizes the suprressive effects of dexamethasone on lymphocyte proliferation. *Endocrinology* 129:3174–9.

Blaylock, R. L. 2009. A possible central mechanism in autism spectrum disorders, part 2: Immunoexcitotoxicity. Review. *Altern Ther Health Med* 15(1):60–7.

Boris, M., A. Glodblatt, J. Galanko, and J. James. 2004. Association of MTHFR gene variants with autism. *J Am Phys Surg* 9(4):106–8.

Bruce, S., and B. McEwen. 2004. Protection and damage from acute and chronic stress. Allostasis and allostatic overload and relevance to the pathophysiology of psychiatric disorders. *Ann N Y Acad Sci* 1032:1–7.

Cardounel, A., W. Regelson, and M. Kalimi. 1999. Dehydroepiandrosterone protects hippocampal neurons against neurotoxin-induced cell death: Mechanism of action. *Proc Soc Exp Biol Med* 222(2):45–9.

Chauhan, A., and V. Chauhan. 2006. Oxidative stress in autism. *Pathophysiology* 13:171–81.

Cohen, D. J., B. A. Shaywitz, and W. T. Johnson. 1974. Biogenic amines in autistic and atypical children. *Arch Gen Psychiatry* 31:845–53.

Compagnone, N. A., and S. H. Mellon. 1998. Dehydroepiandrosterone: A potential signalling molecule for neocortical organization during development. *Proc Natl Acad Sci USA* 95(8):4678–83.

Cook, E., and B. L. Leventhal. 1996. The 5-HT system in Autism. *Curr Opin Pediatr* 8:348–54.

Cooke, B. M. 2006. Steroid-dependent plasticity in the medial amygdala. *Neuroscience* 138(3):997–1005.

Croonenberghs, J. 2003. Biological studies in the involvement of the serotonergic and immunological system in the autistic disorder. Publication University of Antwerp, Belgium.

Croonenberghs, J., K. Spaas, A. Wauters et al. 2008. Faulty serotonin--DHEA interactions in autism: Results of the 5-hydroxytryptophan challenge test. *Neuro Endocrinol Lett* 29(3):385–90.

Croonenberghs, J., S. Van Grieken, A. Wauters et al. 2010. Serum testosterone concentration in male autistic youngsters. *Neuro Endocrinol Lett* 31(4):483–8.

Croonenberghs, J., R. Verkerk, S. Scharpe, D. Deboutte, and M. Maes. 2005. Serotonergic disturbances in autistic disorder: L-5-hydroxytryptophan administration to autistic youngsters increases the blood concentrations of serotonin in patients but not in controls. *Life Sci* 76:2171–83.

Croonenberghs, J., A. Wauters, D. Deboutte, R. Verkerk, S. Scharpe, and M. Maes. 2007. Central serotonergic hypofunction in autism: Results of the 5-hydroxy-tryptophan challenge test. *Neuro Endocrinol Lett* 28(4):449–55.

de Bruin, E. I., F. Verheij, T. Wiegman, and R. F. Ferdinand. 2006. Differences in finger length ratio between males with autism, pervasive developmental disorder not otherwise specified, ADHD, and anxiety disorders. *Dev Med Child Neurol* 48(12):962–5.

de Peretti, E., and M. G. Forest. 1978. Pattern of plasma dehydroepiandrosterone sulfate levels in humans from birth to adulthood: Evidence for testicular production. *J Clin Endocrinol Metab* 47(3):572–7.

Diamond, D. M., B. J. Branch, M. Fleshner, and G. M. Rose. 1995. Effects of dehydroepiandrosterone sulfate and stress on hippocampal electrophysiological plasticity. *Ann N Y Acad Sci* 774:304–7.

DonCarlos, L. I., D. Garcia-Ovejero, S. Sarkey, I. M. Garcia-Segura, and I. Azcoitia. 2003. Androgen receptor immunoreactivity in forebrain axons and dendrites in the rat. *Endocrinology* 144(8):3632–8.

DonCarlos, L. I., S. Sarkey, B. Lorenz et al. 2006. Novel cellular phenotypes and subcellular sites for androgen action in the forebrain. *Neuroscience* 138(3):801–7.

Eriksson, P. S., E. Perfilieva, T. Bjork-Eriksson et al. 1998. Neurogenesis in the adult human hippocampus. *Nat Med* 4:1313–7.

Estrada, M., P. Uhlen, and B. E. Ehrlich. 2006. Ca2+ oscillations induced by testosterone enhance neurite outgrowth. *J Cell Sci* 119(Pt 4):733–43.

Estrada, M., A. Varshney, and B. E. Ehrlich. 2006. Elevated testosterone induces apoptosis in neuronal cells. *J Biol Chem* 281(35):25492–501.

Falany, C. N. 1997. Enzymology of human cytosolic sulfotransferases. *Faseb J* 11:206–26.

Geier, D. A., and M. R. Geier. 2006a. A clinical and laboratory evaluation of methionine cycle-transsulfuration and androgen pathway markers in children with autistic disorders. *Horm Res* 66(4):182–8.

Geier, D. A., and M. R. Geier. 2006b. A clinical trial of combined anti-androgen and anti-heavy metal therapy in autistic disorders. *Neuro Endocrinol Lett* 27(6):833–8.

Geier, D. A., and M. R. Geier. 2007a. A case series of children with apparent mercury toxic encephalopathies manifesting with clinical symptoms of regressive autistic disorders. *J Toxicol Environ Health A* 70:837–51.

Geier, D. A., and M. R. Geier. 2007b. A prospective assessment of androgen levels in patients with autistic spectrum disorders: Biochemical underpinnings and suggested therapies. *Neuro Endocrinol Lett* 28(5):565–73.
Giltay, E. J., E. K. Hoogeveen, J. M. Elbers, L. J. Gooren, H. Asscheman, and C. D. Stehouwer. 1998. Effects of sex steroids on plasma total homocysteine levels: A study in transsexual males and femals. *J Clin Endocrinol Metab* 83:550–3.
Giltay, E. J., P. Verhoef, L. J. Gooren, J. M. Geleijnse, E. G. Schouten, and C. D. Stehouwer. 2003. Oral and transdermal estrogens both lower plasma total homocysteine in male-to-female transsexuals. *Atherosclerosis* 168:139–46.
Goodyer, I. M., R. J. Park, C. M. Netherton, and J. Herbert. 2001. Possible role of cortisol and dehydroepiandrosterone in human development and psychopathology. *Br J Psychiatry* 79:243–9.
Graham, S. M., O. M. Arvela, and G. A. Wise. 1992. Long-term neurologic consequences of nutritional vitamin B12 deficiency in infants. *J Pediatr* 121:710–4.
Grattan-Smith, P. J., B. Wilcken, P. G. Procopis, and G. A. Wise. 1997. The neurological syndrome of infantile cobalamine deficiency: Developmental regression and involuntary movements. *Mov Disord* 12:39–46.
Guazzo, E. P., P. J. Kirkpatrick, J. J. V. I. Goodyer, H. M. Shiers, and J. Herbert. 1996. Cortisol, dehydroepiandrosterone (DHEA), and DHEA sulfate in the cerebrospinal fluid of man: Relation to blood levels and the effects of age. *J Clin Endocrinol Metab* 81(11):3951–60.
Gulati, S., Z. Chen, L. C. Brody, D. S. Rosenblatt, and R. Banerjee. 1997. Defects in auxilliary redox protein leads to funcional methionine synthase deficiency. *J Biol Chem* 272(31):19171–5.
Gunnar, M. R., L. Brodersen, K. Krueger, and J. Rigatuso. 1996. Dampening of adrenocortical responses during infancy: Normative changes and individual differences. *Child Dev* 67:877–89.
Hechter, O., A. Grossman, and R. T. Chatterton Jr. 1997. Relationship of dehydroepiandrosterone and cortisol in disease. *Med Hypotheses* 49:85–91.
Henderson, L. P., C. A. A. Penatti, B. L. Jones, P. Yang, and A. S. Clark. 2006. Anabolic androgen steroids and forebrain GABAnergic transmission. *Neuroscience* 138(3):793–9.
Herbert, J. 1998. Neurosteroids, brain damage, and mental illness. *Exp Gerontol* 33:731–27.
Herbert, M. R., D. A. Ziegler, C. K. Deutsch et al. 2005. Brain asymmetries in autism and developmental language disorder: A nested whole brain analysis. *Brain* 128:213–26.
Higley, J., P. Mehlman, R. Poland et al. 1996. CSF testosterone and 5-HIAA correlate with different types of aggressive behaviour. *Biol Psychiatry* 40:1067–82.
Hortnagl, H., M. L. Berger, L. Havelec et al. 1993. Role of glucocorticoids in the cholinergic degeneration in rat hippocampus induced by ethylcholine aziridinium. *J Neurosci* 13:2939–45.
Hu, Y., A. Cardounel, E. Gursoy, P. Anderson, and M. Kalimi. 2000. Anti-stress effects of dehydroepiandrosterone: Protection of rats against repeated immobilixation stress-induced weight loss, glucocorticoid receptor production, and lipid peroxidation. *Biochem Pharmacol* 59:753–62.
Hultman, C. M., P. Sparen, and S. Cnattingius. 2002. Perinatal risk factors for infantile autism. *Epidemiology* 13(4):417–23.
Ingudomnukul, E., S. Baron-Cohen, S. Wheelwright, and R. Knickmeyer. 2007. Elevated rates of testosterone-related disorders in women with autism spectrum conditions. *Horm Behav* 51:597–604.
James, S. J., P. Cutler, S. Melnyk et al. 2004. Metabolic biomarkers of increased oxidative stress and impaired methylation capacity in children with autism. *Am J Clin Nutr* 80:1611–7.
James, S. J., S. Melnyk, S. Jernigan et al. 2006. Metabolic endophenotype and related genotypes are associated with oxidative stress in children with autism. *Am J Med Genet B Neuropsychiatr Genet* 141:947–56.
Kalimi, M., Y. Shafagoj, R. Loria, D. Padgett, and W. Regelson. 1994. Anti-glucocorticoid effects of dehydroepiandrosterone (DHEA). *Mol Cell Biochem* 131:99–104.
Kanner, L. 1943. Autistic disturbances of affective contact. *Nerv Child* 2:217–50.
Kern, J. K., and A. M. Jones. 2006. Evidence of toxicity, oxidative stress, and neuronal insult in autism. *J Toxicol Environ Health B Crit Rev* 9(6):485–99.
Kim, M. S., J. Shigenaga, A. Moser, C. Grunfeld, and K. R. Feingold. 2004. Suppression of DHEA sulfotransferase (Sult2A1) during the acute-phase response. *Am J Physiol Endocrinol Metab* 287:E731–8.
Kimonides, V. G., N. H. Khatibi, C. N. Svendsen, M. V. Sofroniew, and J. Herbert. 1998. Dehydroepiandrosterone (DHEA) and DHEA-sulfate (DHEA-S) protect hippocampal neurons against excitatory amino acid-induced neurotoxicity. *Proc Natl Acad Sci U S A* 95:852–7.
Kimonides, V. G., M. G. Spillantini, M. V. Sofroniew, J. W. Fawcett, and J. Herbert. 1999. Dehydroepiandrosterone antagonizes the neurotoxic effects of corticosterone and translocation of stress-activated protein kinase 3 in hippocampal primary cultures. *Neuroscience* 89:429–36.
Knickmeyer, R., S. Baron-Cohen, B. A. Fane et al. 2006. Androgens and autistic traits: A study of individuals with congenital adrenal hyperplasia. *Horm Behav* 50:148–53.

Knickmeyer, R. C., S. Wheelwright, R. Hoekstra, and S. Baron-Cohen. 2006. Age of menarche in females with autism spectrum conditions. *Dev Med Child Neurol* 48:1007–8.

Kolevzon, A., R. Gross, and A. Reichenberg. 2007. Prenatal and perinatal risk factors for autism: A review and integration of findings. *Arch Pediatr Adolesc Med* 161(4):326–33.

Komuro, H., and P. Rakie. 1993. Modulation of neuronal migration by NMDA receptors. *Science* 260(5104):95–7.

Kroboth, P. D., F. S. Salek, A. L. Pittenger, T. J. Fabian, and R. F. Frye. 1999. DHEA and DHEA-S: A review. *J Clin Pharmacol* 39(4):327–48.

Lapchak, P. A., and D. M. Araujo. 2001. Preclinical development of neurosteroids as neuroprotective agents in the treatment of neurodegenerative diseases. *Int Rev Neurobiol* 46:379–97.

Leranth, C., O. Petnehazy, and N. J. Maclusky. 2003. Gonadal hormones affect spine synaptic density in the CA1 hippocampal subfield of male rats. *J Neurosci* 23(5):1588–92.

Lewis, M., and D. Ramsay. 1995. Developmental change in infant's response to stress. *Child Dev* 66:657–70.

Lieberherr, M., and B. Grosse. 1994. Androgens increase intracellular calcium concentrations and inositol 1,4,5-trophosphate and diacylglycerol formation via a pertussis toxin-sensitive G-protein. *J Biol Chem* 269(10):7217–23.

Lin, S. Y., and M. Constantine-Paton. 1998. Suppression of sprouting: An early function of NMDA receptors in the absence of AMPA/kainate receptor activity. *J Neurosci* 18(10):3725–37.

Lutchmaya, S., S. Baron-Cohen, and P. Raggat. 2002. Foetal testosterone and vocabulary size in 18 and 24 month old infants. *Infant Behav Dev* 24:418–24.

MacLusky, N. J., T. Hajszan, J. Prange Kiel, and C. Leranth. 2006. Androgen modulation of hippocampal synaptic plasticity. *Neuroscience* 138(3):957–65.

Maes, M., P. D'Hondt, H. Y. Meltzer, P. Cosyns, and P. Blockx. 1995. Effects of serotonin agonists on the negative feedback by glucocorticoids on the hypothalamic-pituitary-adrenal axis in depression. *Psychoneuroendocrinol* 20:49–67.

Manning, J., S. Baron-Cohen, S. Wheelwright, and G. Sanders. 2001. The 2nd to 4th digit ratio and autism. *Dev Med Child Neurol* 43:160–4.

Mao, X., and S. W. Barger. 1998. Neuroprotection by dehydroepiandrosterone-sulfate: Role of an NF kappaB-like factor. *Neuroreport* 9:759–63.

McGinnis, W. R. 2004. Oxidative stress in autism. *Altern Ther Health Med* 10:22–36.

Meltzer, H. Y., and M. Maes. 1994. Effect of pindolol on the L-5-HTP-induced increase in plasma prolactin and cortisol concentrations in man. *Psychopharmacology* 114:635–43.

Mutter, J., J. Naumann, R. Schneider, H. Walach, and B. Haley. 2005. Mercury and autism: Accelerating evidence? *Neuro Endocrinol Lett* 26:439–46.

Nieschlag, E., D. L. Loriaux, H. J. Ruder, I. R. Zucker, M. A. Kirschner, and M. B. Lipsett. 1973. The secretion of dehydroepiandrosterone and dehydroepiandrosterone sulphate in man. *J Clin Endocrinol* 57:23–34.

Orentreich, N., J. L. Brin, R. L. Rizer, and J. H. Volgelman. 1984. Age changes and sex differences in serum dehydroepiandrosterone sulfate concentrations from puberty through adulthood. *J Clin Endocrinol Metab* 59:55–555.

Pasca, S. P., B. Nemes, L. Vlase et al. 2006. High levels of homocysteine and low serum paraxonase 1 arylesterase activity in children with autism. *Life Sci* 78:2244–8.

Radominska, A., K. A. Comer, P. Zimniak, J. Falany, M. Iscan, and C. N. Falany. 1990. Human liver steroid sulphotransferase sulfates bile acids. *Biochem J* 272:597–604.

Rotter, J. I., F. L. Wong, E. T. Lifrak, and L. N. Parker. 1985. A genetic component to the variation of dehydroepiandrosterone sulfate. *Metabolism* 34:731–6.

Ryan, R. A., and J. Carrol. 1976. Studies on a 3 beta-hydroxysteroid sulphotransferse from rat liver. *Biochim Biophys Acta* 429:391–402.

Sanchez-Martin, J., E. Fano, L. Ahedo, J. Cardas, P. Brain, and A. Aspiroz. 2000. Relating testosterone levels and free play social behaviour in male and female preschool children. *Psychoneuroendocrinology* 25:773–83.

Schwarzschild, M. A., R. L. Cole, and S. E. Hymann. 1997. Glutamate, but not dopamine, stimulates stress-activated protein kinase and AP-I-mediated transcription in striatal neurons. *J Neurosci* 17:3455–66.

Spitzer, N. C., N. J. Lautermilch, R. D. Smith, and T. M. Gomez. 2000. Coding of neuronal differentiation by calcium transients. *Bioessays* 22(9):811–7.

Strous, R. D., P. Golubehik, R. Maayan et al. 2005. Lowered DHEA-S plasma levels in adult individuals with autistic disorder. *Eur Neuropsychopharmacol* 15(3):305–9.

Tordjman, S., G. Anderson, P. McBride et al. 1995. Plasma androgens in autism. *J Autism Dev Disord* 25:295–304.

Tordjman, S., P. Ferrari, V. Sulmont, M. Duyme, and P. Roubertoux. 1997. Androgenic activity in autism. *Am J Psychiatry* 154:1626–7.

Vargas, D. L., C. Nascimbene, C. Krishnan, A. W. Zimmerman, and C. A. Pardo. 2005. Neuroglial activation and neuroinflammation in the brain of patients with autism. *Ann Neurol* 57:67–81.

Vrbikova, J., J. Tallova, M. Bleikova, K. Dvorakova, M. Hill, and L. Starka. 2003. Plasma thiols and androgen levels in polycystic ovary syndrome. *Clin Chem Lab Med* 41(2):216–21.

Wakefield, A. J., S. H. Murch, A. Anthony et al. 1998. Ileal-lymphoid-nodular hyperplasia, non-specific colitis, and pervasive developmental disorder in children. *Lancet* 351(9103):637–41.

Waring, R. H., and L. V. Klovrza. 2000. Sulphur metabolism in autism. *J Nutr Environ Med* 10:25–32.

Waring, R. H., J. M. Ngong, L. Klovrza, S. Green, and H. Sharp. 1997. Biochemical parameters in autistic children. *Dev Brain Dysfunct* 10:40–3.

Weiland, N. G. 1992. Estradiol selectivity regulates agonist binding sites in the N-methyl-D-aspartate receptor complex in the CA1 region of the hippocampus. *Endocrinology* 131(2):662–8.

White, J. F. 2003. Intestinal pathology in autism. *Exp Biol Med (Maywood)* 228:639–49.

Wolkowitz, O. M., E. S. Epel, and V. I. Reus. 2001. Stress hormone-related psychopathology: Pathophysiological and treatment implications. *World J Biol Psychiatry* 2:115–43.

Yang, S. H., E. Perez, J. Cutright et al. 2002. Testosterone increases neurotoxicity of glutamate in vitro and ischemia-reperfusion injury in an animal model. *J Appl Physiol* 92(1):195–201.

Zucker, D. K., R. L. Livingston, R. Nakra, and P. J. Clayton. 1981. B12 deficiency and psychiatric disorders: Case report and literatue review. *Biol Psychiatry* 16:197–205.

Section IV

Animal and In Vitro *Model Studies*

Future Uses of DHEA

24 DHEA Antiviral Properties in Cell Cultures and Animal Models

Viviana Castilla and Mónica B. Wachsman

CONTENTS

INTRODUCTION

Compared with the antibiotic therapy available for bacterial infections, there are still many viral diseases for which no effective drugs exist. Viruses are obligate intracellular parasites, so antiviral drugs should specifically inhibit one or more steps of virus replication without causing unacceptable side effects. Since 1959, when the first antiviral drug idoxuridine (5-iodo-2′-deoxyuridine) was licensed for clinical use, about 50 drugs have been approved for the treatment of human virosis, half of which are used in human immunodeficiency virus (HIV) infections (De Clercq 2010). Limitations of current antiviral compounds include a narrow antiviral spectrum, ineffectiveness against latent viruses, selection of drug-resistant mutants, and toxic side effects. To overcome these disadvantages, more effective compounds are being developed, and hundreds of virus inhibitors of different chemical structure are continuously being isolated from natural sources or laboratory synthesized. However, it has proved difficult to find compounds that can selectively block viral replication without interference to the normal cellular processes and, thus, without significant toxicity to the host. In the last decades, the antiviral activity of natural and synthetic steroids has been reported (Castilla, Ramírez, and Coto 2010) and among the mammalian steroids tested, dehydroepiandrosterone (DHEA) exhibits inhibitory action against a broad range of viral infections.

Here, we will discuss several aspects about the ability of DHEA to block virus replication both in cell cultures and animal models.

ANTIVIRAL PROPERTIES OF DHEA IN CELL CULTURES: FIRST STUDIES

Antiviral activity of DHEA was reported for the first time in studies performed in cell cultures (*in vitro* studies) with the Epstein–Barr virus (EBV), which is an oncogenic herpes virus associated with a number of human malignancies, including Burkitt's lymphoma, Hodgkin's lymphoma, and post-transplantation lymphoproliferative diseases. Interestingly, it was shown that DHEA inhibits EBV-induced morphologic transformation and stimulation of DNA synthesis in human lymphocytes (Henderson et al. 1981).

Twenty-five years after the demonstration of the therapeutic activity of azidothymidine (AZT), the first antiretroviral drug used in the clinic for HIV infection, highly active antiretroviral therapy (HAART) provides durable control of viral replication in many patients. Nevertheless, HAART treatment is not devoid of unwanted secondary effects, some of which are now affecting aging populations under long-term treatment. The emergence of multidrug resistance and transmission of drug-resistant HIV strains also limits the clinical efficacy of current therapy. Furthermore, HIV infection cannot be cured, so patients are destined to undergo drug therapy for life. Further simplification of treatment and identification of new drugs and new drug targets will be important steps toward achieving virus control and, ultimately, finding a cure (Esté and Cihlar 2010).

After 1990, several reports supported the hypothesis that DHEA might play a role in HIV infections and progression to acquired immunodeficiency syndrome (AIDS). Sera from AIDS patients have lower than normal levels of DHEA (Jacobson et al. 1991; Mulder et al. 1992), thus it was proposed that protection against disease progression might be ascribed to DHEA modulation of immune response. Correlation between AIDS progression and DHEA levels in human serum encouraged investigators to evaluate the ability of DHEA as inhibitor of HIV-1 replication in cell cultures. Henderson et al. demonstrated that DHEA treatment resulted in a modest downregulation of HIV-1 replication in phytohemagglutinin-stimulated peripheral blood lymphocytes (Henderson, Yang, and Schwartz 1992).

As was mentioned above, AZT is an effective drug in the treatment of HIV-infected patients with advanced disease. However, prolonged treatment with AZT leads not only to the emergence of drug-resistant viruses but also to undesirable side effects such as bone marrow suppression, inflammatory myopathy, myalgia, nausea, tremor, fatigue, and headache. Based on the results obtained earlier, Henderson and collaborators decided to investigate the effect of DHEA on the *in vitro* replication of strains of HIV-1 resistant to AZT. This study showed that DHEA, at concentrations between 50 and 100 μM, has the ability to suppress both AZT-sensitive and AZT-resistant strains of HIV-1 in MT-2 cell cultures (Table 24.1; Yang, Schwartz, and Henderson 1994).

In most individuals, the initial infection with HIV-1 usually results in the establishment of a latent or chronic infection before eventual progression toward AIDS. HIV-1 can establish a latent or persistent infection in some T cell lines that show minimal constitutive virus expression. Antigens, mitogens, cytokines, and various gene products from other viruses can induce the activation that leads to the enhancement of HIV-1 replication. DHEA proved to be effective on impairing HIV-1 reactivation from chronically infected cell lines (Yang, Schwartz, and Henderson 1993), and this observation is relevant taking into account that reactivation of latent HIV-1 harbored in chronically infected T lymphocytes, monocytes, or macrophages plays an important role in the pathogenesis of AIDS.

Bradley et al. examined the effect of DHEA on the replication of another retrovirus, feline immunodeficiency virus (FIV). This virus, which is found worldwide in domestic cats and wild felines, attacks the immune system, and as a result, the animals are unable to fight off various infections. It was demonstrated that DHEA affects FIV replication in chronically infected cells at levels where cellular viability and DNA synthesis were not affected (Bradley et al. 1995).

ANIMAL MODELS AND CONTRASTING RESULTS WITH STUDIES PERFORMED IN CELL CULTURES

In 1988, Loria et al. examined the effect of DHEA in two lethal viral infection models in mice: systemic coxackievirus (CV) and herpes simplex virus (HSV) encephalitis. HSV, an enveloped virus with DNA genome, is a frequent human pathogen, most commonly associated with orolabial,

TABLE 24.1
Antiviral Action of DHEA against Different Viruses in Cell Cultures

Virus	Family	Main Characteristics	EC_{50} (μM)	Reference
HIV	*Retroviridae*	Enveloped particles, RNA positive single-stranded genome	50	Yang, Schwartz, and Henderson (1994)
JEV	*Flaviviridae*	Enveloped particles, RNA positive single-stranded genome	50	Chang et al. (2005)
JUNV	*Arenaviridae*	Enveloped particles, RNA ambisense single-stranded genome	100	Acosta et al. (2008)
TCRV	*Arenaviridae*	Enveloped particles, RNA ambisense single-stranded genome	204	Acosta et al. (2008)
PICV	*Arenaviridae*	Enveloped particles, RNA ambisense single-stranded genome	110	Acosta et al. (2008)
VSV	*Rhabdoviridae*	Enveloped particles, RNA negative single-stranded genome	60	Romanutti et al. (2009)
ADV	*Adenoviridae*	Nonenveloped particles, double-stranded DNA genome	88	Romanutti et al. (2010)

EC_{50} values were calculated as the compound concentration that reduced by 50% the virus-induced cytopathic effect (HIV and JEV) or compound concentration required to inhibit by 50% infectious virus released to the culture supernatant (JUNV, TCRV, PICV, VSV, and ADV).

EC_{50} = Effective concentration 50; HIV = Human immunodeficiency virus; JEV = Japanese encephalitis virus; JUNV = Junin virus; TCRV = Tacaribe virus; PICV = Pichinde virus; VSV = Vesicular stomatitis virus; ADV = Adenovirus.

genital, and ocular infections. HSV primary infection in immunocompetent individuals is usually mild or even asymptomatic and results in lifelong latent infections in sensory ganglia and the central nervous system (CNS). On the contrary, herpes simplex encephalitis (HSE) is a life-threatening consequence of HSV infection of the CNS. Mortality rate reaches 70% in the absence of therapy and only a minority of affected individuals returned to normal. HSE is the result of HSV primary infection or viral reactivation from latent state. On the other hand, CV is a nonenveloped virus member of the *Picornaviridae* family, having a single strand of RNA as its genetic material. CVs are separable into two groups, A and B: type A viruses cause herpangina and conjunctivitis and type B viruses cause epidemic pleurodynia, which is also termed Bornholm disease. Less frequently, both type A and type B viruses can cause meningitis, myocarditis, and pericarditis. DHEA was protective in both CV B4 and HSV-2 models and histopathological analysis, leukocyte counts, and numbers of spleen antibody-forming cells in the CV B4 model suggest that DHEA acts by maintaining or potentiating the immune competence of mice, counteracting the stress-related immunosuppressive effects induced by viral infection. Further support for this hypothesis resulted from experiments carried out in genetically immunodeficient mice because DHEA could not exert protection in these animals. By contrast, DHEA did not demonstrate antiviral activity against CV B4 and HSV-2 replication in cell cultures (Loria et al. 1988).

In accordance with these investigations, a couple of years later, Ben-Nathan and coworkers reported that DHEA had a significant protective effect in mice infected with enveloped viruses, such as West Nile virus (WNV), Sindbis virus (SINV), and Semliki Forest virus (SFV), spread by insect vectors such as mosquitoes. WNV, a member of the *Flaviviridae* family, is an enveloped virus with positive-sense, single-stranded RNA genome. WNV is amplified during periods of adult mosquito blood-feeding by continuous transmission between mosquito vectors and bird reservoir hosts. People, horses, and most other mammals are probably "dead-end" or incidental hosts. Infection with WNV usually has no symptoms or mild symptoms (fever, headache, body aches), but in certain occasions, this virus can cause encephalitis or meningitis. On the other hand, SINV and SFV, members of genus Alphavirus in the family *Togaviridae*, are enveloped viruses with single-stranded positive-polarity genome. Alphaviruses distributed around the world are able to infect various vertebrates and can cause human diseases, with infectious arthritis, encephalitis, rashes, and fever the most commonly observed.

Mice injected subcutaneously with DHEA on the same day or a day pre- or postinfection with WNV resulted in 40%–50% mortality as compared with 100% in control mice. Furthermore, DHEA injection reduced viremia and death rate, significantly delayed the onset of the disease, and the authors found that titers of antivirus antibodies in surviving mice were very high. However, DHEA had no effect on WNV growth in baby hamster kidneys (BHK) or Vero cell lines obtained from baby hamster or monkey kidneys, respectively. Therefore, as in the cases of CV B4 and HSV-2, DHEA-mediated protection against WNV seems to be related with upregulation of humoral immune response (Ben-Nathan et al. 1991).

DHEA ANTIVIRAL ACTION AND MECHANISMS: NEW EVIDENCE

In the studies mentioned in the previous section, DHEA seemed to be unable to inhibit the replication of the flavivirus WNV in cell cultures; however, more recent studies showed that replication of Japanese encephalitis virus (JEV), also a member of the *Flaviviridae* family, was strongly affected by treatment of infected cell cultures with DHEA (Chang et al. 2005). Furthermore, in the last years, our research group proved that DHEA displays *in vitro* inhibitory action against other RNA viruses (Table 24.1): the rhabdovirus vesicular stomatitis virus (VSV; Romanutti et al. 2009) and the arenaviruses Junin virus (JUNV), Tacaribe virus (TCRV), and Pichinde virus (PICV; Acosta et al. 2008).

JEV is an enveloped virus with RNA positive-stranded genome. JEV infection, which commonly affects children, is a major cause of acute encephalopathy in several parts of Southeast Asia, and there is no specific antiviral therapeutic available for the treatment of this viral pathology. Infection by JEV can cause acute encephalitis with a high mortality rate in humans and increasing evidence suggests that both neuronal destruction and dysfunction might partly explain the manifestation of the disease. Although the mechanisms by which JEV directly induces the death of infected neurons remain largely unknown, Chang et al. showed that JEV infection resulted in the alteration of cell signaling networks. Cell signaling is part of a complex system of communication that governs basic cellular activities and coordinates cell actions including cell survival and proliferation. In particular, JEV infection induces modifications on mitogen-activated protein kinase (MAPK)-signaling cascades, leading to apoptotic cell death, and studies performed in DHEA-treated, JEV-infected cells suggested that DHEA antiviral action could be attributed to the ability of this compound to restore MAPK signaling functions (Chang et al. 2005).

In the last 10 years, our research group has been evaluating the ability of steroids from diverse origin to block virus replication in different cell cultures (Wachsman et al. 2004; Castilla et al. 2005,

Castilla, Ramírez, and Coto 2010; Romanutti et al. 2007). These studies have been focused on JUNV and VSV sensitivity to plant or animal steroids. JUNV, an enveloped RNA virus with ambisense coding strategy, causes a severe disease, the Argentine hemorrhagic fever (AHF), in humans. Other arenaviruses, such as Lassa virus, Machupo virus, Guanarito virus, and Sabia virus, also cause severe hemorrhagic diseases in man. Although these viruses have critical importance as human pathogens and new arenaviruses have continued to emerge in North and South America, to date, ribavarin is the only drug that has shown efficacy for Lassa fever infection and only partial success in treatement of AHF patients. In addition, undesirable side effects were recorded for ribavirin treatment, thus the current therapy for AHF is based on the early administration of immune plasma (Acosta et al. 2008).

On the other hand, VSV, an enveloped negative-sense, single-stranded RNA virus, causes a disease in cattle, horses, pigs, and sheep, leading to vesiculation and ulceration of the tongue and oral epithelia and, sometimes, the appearance of lesions on the feet and teats. VSV causes severe economic losses due to quarantine, trade barriers, and lowered productivity.

Once it was demonstrated that DHEA efficiently reduced the production of infective virus in Vero cell cultures infected with JUNV or VSV, we decided to analyze the effect of DHEA on different steps of JUNV and VSV replication cycles that have several features in common (Figure 24.1). Briefly, both viruses recognize a specific receptor at the cell surface and are then internalized by endocytosis. The acidic pH within endosomal vesicles triggers the fusion between viral and

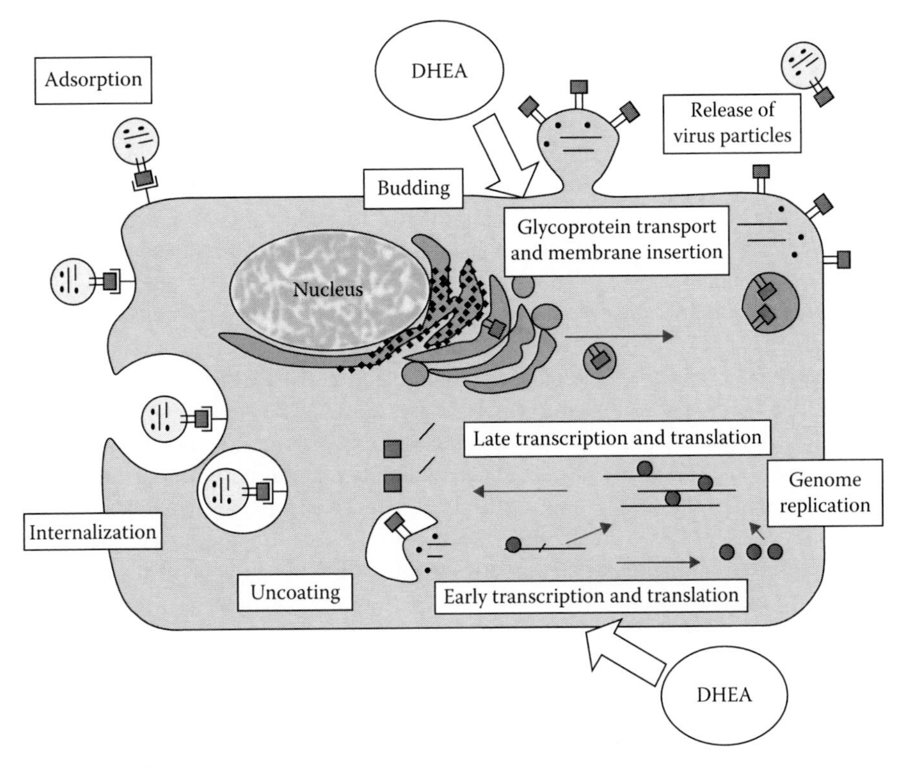

FIGURE 24.1 Steps of Vesicular stomatitis virus (VSV) and Junin virus (JUNV) replicative cycle affected by DHEA. VSV and JUNV enter cells through receptor-mediated endocytosis. Viral particles lose their envelope in late endosomes, and the viral genome is released in the cytoplasm. After early transcription and translation of viral proteins, genome replication takes place. Viral components are transported to the cell membrane, where virus assembly occurs and mature virus particles are finally released by budding from the cell membrane. DHEA affects the early synthesis of viral RNA and proteins, and the insertion of glycoproteins into the cell membrane.

endosomal membranes leading to the release of the viral nucleocapsid (RNA and internal proteins) into the cytoplasm. Early transcription and translation take place allowing the replication of virus genome followed by transcription and translation of structural proteins. Viral glycoproteins are transported to the cell membrane, where the assembly of viral components takes place. Mature virus particles are formed after the acquisition of the envelope by budding from the plasma membrane (Figure 24.1). The studies performed to elucidate the DHEA mode of action revealed that treatment with the steroid causes an important reduction of the synthesis of primary transcripts and a strong inhibition on early viral protein synthesis in JUNV- and VSV-infected DHEA-treated cells (Acosta et al. 2008; Romanutti et al. 2009).

Although little is known about host components required for JUNV and VSV replication, increasing evidence indicates that successful viral replication relies on the evolution of strategies that modulate host cell signaling pathways. Many viruses, including influenzavirus A, can trigger PI3K/Akt cell signaling pathway activation to favor their replication and specific inhibitors of Akt phosphorylation lead to a reduction of viral RNA and protein synthesis (Shin et al. 2007). It has been reported that DHEA impairs Akt phosphorylation, and although the effects of this steroid appear to be determined by the cell type, these experimental evidences would support the speculation that the inhibitory effect of DHEA might be related to cell signaling pathway modulation (Jiang et al. 2005). Interestingly, it has been recently demonstrated that, early during infection, JUNV activates Akt phosphorylation in Vero cells triggering the PI3K/Akt signaling pathway, and it has been proposed that this early activation of cellular genes is required for a productive JUNV infection (Linero and Scolaro 2009). Therefore, inhibition of early JUNV transcription and translation processes in JUNV-infected cells treated with DHEA might be a consequence of the inhibitory effect of the steroid on JUNV-mediated activation of Akt.

It was also shown that DHEA partially reduced the expression of JUNV and VSV glycoproteins at the surface of treated, infected cells (Acosta et al. 2008; Romanutti et al. 2009). This effect would be related to the capacity of steroid hormones to intercalate into cell membrane bilayer and alter membrane fluidity (Whiting, Restall, and Brain 2000). This property of the steroid would affect the insertion of viral glycoproteins further inhibiting the assembly of virus particles (Figure 24.1).

The effect of steroids on membrane-dependent processes was also assessed with a DHEA-derivative named immunor (IM28). In contrast to JUNV or VSV, HIV enters the cell by direct fusion to the cell membrane and some viral strains exhibit the ability to induce the formation of multinucleated cells because virus glycoprotein that is newly synthesized in the infected cells mediates membrane fusion between infected and noninfected cells. IM28 inhibited cell–cell fusion mediated by HIV-1 in a T cell line that stably expressed HIV-1 glycoprotein (Mavoungou et al. 2005), indicating that the steroid impedes the dissemination of viral infection to neighboring cells.

The antiviral activity of DHEA has also been related to its ability to promote nitric oxide (NO) production (Simoncini et al. 2003). NO would exert its antiviral action through S-nitrosylation of cysteine residues contained in viral enzymes (such as proteases, reverse transcriptases, ribonucleotide reductases) or in transcriptional factors involved in viral replication (Benz et al. 2002).

In an effort to gain more insight on DHEA spectrum and mechanism of antiviral action, we recently examined the effect of DHEA on the replication of adenoviruses (ADVs) in Vero cell cultures (Romanutti et al. 2010). ADVs are a group of viruses responsible for a spectrum of respiratory disease as well as gastroenteritis, conjunctivitis, cystitis, or rash. ADVs are nonenveloped, icosohedral viruses containing double-stranded DNA. There are 49 immunologically distinct types of ADVs that can cause human infections. Although several compounds of dissimilar chemical structures were evaluated for their effectiveness *in vitro* and in virus-infected patients, currently, there is no formally approved antiviral therapy for ADV infections. We demonstrated that DHEA inhibited the multiplication of ADV in confluent A549 cells in a dose-dependent manner, and we determined that DHEA prevents ADV-protein synthesis at concentrations that do not affect cellular protein synthesis (Table 24.1). We have mentioned above that in certain cell lines DHEA induces extracellular regulated kinase (ERK) phosphorylation (Chang et al. 2005) and several viruses, including ADVs, not only induce ERK phosphorylation but also seem to depend on it, and it was proved that the levels of ADV protein synthesis are strongly reduced by inhibitors of ERK activation (Schümann and Dobbelstein 2006), so a possible explanation for anti-ADV DHEA activity may be related to the ability of the steroid to modulate MAPK activation induced by viral infection.

INHIBITION OF VIRUS GROWTH BY DHEA ANALOGS

Despite the fact that DHEA is a native steroid that has been used clinically with minimal side effects, administration of the steroid for extended periods increases circulating testosterone and dihydrotestosterone manifold above normal levels, especially in women, and may cause masculinization (Labrie et al. 2003). Therefore, the evaluation of the antiviral activity of structural related compounds is useful in the search of an effective treatment for virus infections.

ANTIVIRAL ACTIVITY OF DHEA DERIVATIVES IN CELL CULTURES

It has been reported that, like DHEA, the synthetic analog 3α-hydroxy-16α-bromo-5α-androstan-17-one (Figure 24.2) prevented EBV-cell transformation of human lymphocytes (Henderson et al. 1981).

However, compounds 3β-hydroxy-16α-fluoro-5α-androstan-17-one and 3β-hydroxy-16α-fluoro-androst-5-en-17-one (Figure 24.2) inhibited HIV-1 replication in human lymphocytes (Henderson, Yang, and Schwartz 1992). Furthermore, a remarkable result was obtained in a rat syncytiotrophoblast cell line treated with dehydroepiandrosterone sulfate (DHEAS) and AZT. AZT is used for the prevention of mother-to-child transmission of HIV-1 and is transplacentally transferred to the fetus. It was shown that the presence of DHEAS promoted the apical uptake of AZT by syncytiotrophoblasts, increasing AZT cellular accumulation (Nishimura et al. 2008). This effect of DHEAS treatment on AZT incorporation into the cell might be functional for delivery of AZT into the placenta in the clinical context.

The 17-ketosteroid epiandrosterone (3β-hydroxy-5α-androstan-17-one; EA) is an inactive isomer of androsterone (Table 24.2); it is formed in peripheral tissues, from which it is released into the circulation and is ultimately excreted in the urine. As DHEA, it is also found in testicular and ovarian

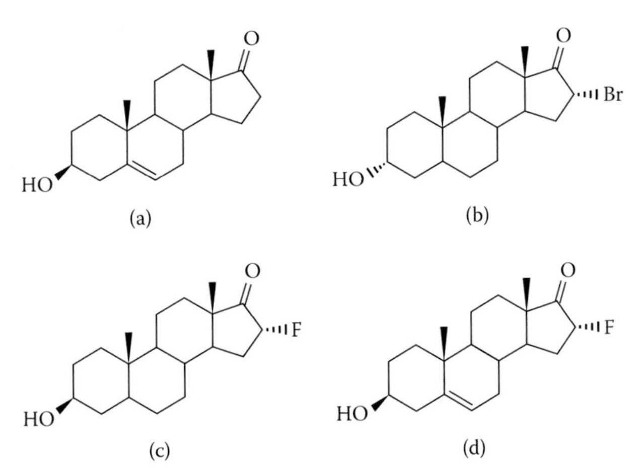

FIGURE 24.2 DHEA analogs with antiviral activity. (a) 3β-hydroxyandrost-5-en-17-one (DHEA), (b) 3α-hydroxy-16α-bromo-5α-androstan-17-one, (c) 3β-hydroxy-16α-fluoro-5α-androstan-17-one, and (d) 3β-hydroxy-16α-fluoro-androst-5-en-17-one.

TABLE 24.2
Selective Inhibitory Action of DHEA and EA Derivatives against JUNV, VSV, and ADV

	SI		
Compound	**JUNV**	**VSV**	**ADV**
3β-hydroxyandrost-5-en-17-one (DHEA)	30	50	63
21-norpregna-5,17(20)-dien-3β,16α-diyl-diacetate	I	37.3	79
3β-hydroxy-5α-androstan-17-one (EA)	83.3	72.6	83
17-oxo-5α-androstan-3β,16α-diyl diacetate	21	22.6	21
17,17-ethylendioxy-5α-androst-15-en-3β-ol	121	43.8	24
3β-hydroxy-5α-androst-15-en-17-one	I	24.7	42
Ribavirin	12	14	ND
Cidofovir	ND	ND	67

SI: ratio CC_{50}/EC_{50}.
CC_{50}: compound concentration required to reduce cell viability by 50%.
EC_{50}: compound concentration required to reduce virus yield by 50% in Vero cells (for VSV and JUNV) or in A549 cells (for ADV) (Acosta et al. 2008; Romanutti et al. 2009, 2010).
SI = selectivity index; ND = not done; I = inactive; VSV = Vesicular stomatitis virus; JUNV = Junin virus; ADV = adenovirus.

tissue, but with the difference that EA is only a weak androgen. We have studied the effect of several synthetic compounds derived from DHEA and EA on the replication of JUNV, VSV, and ADV type 5. To this end, we first evaluate the cytotoxicity of different concentrations of the compounds, and we determined the cytotoxic concentration 50 (CC_{50}) as the concentration of the drug that reduced cell viability by 50%. Then, non-cytotoxic concentrations of each compound were assessed for antiviral activity and the effective concentration 50 (EC_{50}), concentration of the compound that inhibited virus production by 50%, was determined. These parameters allow us to calculate the selectivity of action of each compound expressed as a selectivity index (SI) or the ratio CC_{50}/EC_{50}. A high SI indicates that the compound inhibits viral replication and exhibits low levels of cytotoxicity. As can be seen in Table 24.2, EA was even more active than DHEA against the viruses tested. The DHEA derivative 21-norpregna-5, 17(20)-dien-3β, 16α-diyl-diacetate (Table 24.2) showed antiviral action against VSV and ADV exhibiting a very good SI especially for ADV, but this compound was inactive against JUNV. In a similar way, one of the EA derivatives, 3β-hydroxy-5α-androst-15-en-17-one, was able to inhibit VSV and ADV replication at inhibitory levels similar to those seen for DHEA, but this derivative could not prevent JUNV replication.

There were two EA derivatives that displayed antiviral activity against JUNV, VSV, and ADV: the compound 17,17-ethylendioxy-5α-androst-15-en-3β-ol, which was more effective with JUNV than with VSV or ADV, and the compound 17-oxo-5α-androstan-3β, 16α-diyl diacetate that exhibited SI values similar to that of DHEA (Acosta et al. 2008; Romanutti et al. 2009, 2010).

Antiviral Activity of DHEA Derivatives in Animal Models

Pedersen and coworkers examined the effect of the compound 3β-hydroxy-16α-bromo-5α-androstan-17-one on the acute stage of FIV infection in laboratory cats. FIV infection in cats has three stages, just like HIV infection in humans: the initial acute stage characterized by fever, swollen lymph nodes, and susceptibility to skin or intestinal infections; a second latent or subclinical stage without signs of disease; and the final, or AIDS-like, stage. Results obtained with DHEA-synthetic analog in FIV-infected cats support the protective effects previously reported for this and related androsterones in a number of infection models. Treatment with the analog 3β-hydroxy-16α-bromo-5α-androstan-17-one during the first 4 weeks of experimental infection caused viremia to appear earlier and at higher levels compared with untreated control animals. However, after this initial burst, virus levels in treated cats actually decreased to levels significantly lower than in untreated animals. Furthermore, FIV-infected cats that were treated with the analog exhibited significantly higher levels of antibody responses (specific antibodies reactive to viral proteins) than control cats receiving vehicle alone. Associated with the protective effect, treatment with the steroid significantly decreased the virus-induced suppression of mRNA levels of certain cytokines in peripheral blood mononuclear cells (Pedersen et al. 2003).

The studies described in the previous paragraph were in accordance with earlier observations in mice infected with the retrovirus murine leukemia virus (MLV). Like aging, leukemia is a condition with dysregulated cytokine production, and it has been demonstrated that DHEAS restored normal cytokine production in MLV-infected old mice (Araghi-Niknam et al. 1997). These results were reinforced by further experiments in mice infected with the LP-BM5 leukemia retrovirus. Treatment with DHEA prevented the reduction of B- and T-cell proliferation, as well as Th1 cytokine secretion caused by retrovirus infection. DHEA administration also suppressed the elevated production of Th2 cytokines stimulated by retrovirus infection (Zhang et al. 1999).

CONCLUSIONS

DHEA is a native steroid that has been used clinically with minimal undesired effects, thus the utility of DHEA in the therapeutic modulation of acute and chronic viral infections deserves intensive study. The precise mechanism for the protective effect of DHEA and synthetic analogs remains to be determined, but the results obtained, up to now, in animal model infections indicate that protection involves significant changes in viral-specific antibody responses, increases in certain protective cell populations, and improvements in impaired Th1 cytokine responses (Table 24.3). The broad antiviral spectrum of action of DHEA or DHEA derivatives was also demonstrated in studies performed in cell cultures; however, each virus-cell system analyzed revealed that DHEA is able to, adversely, affect different steps of viral replication (Table 24.3). The elucidation of DHEA mode of action against viruses with different replication strategies requires a deeper knowledge of the involvement of cellular factors and cell signaling pathways on virus multiplication, taking into account the ability of this hormone to modulate the activation of key effectors of cell signalization.

TABLE 24.3
DHEA Antiviral Mode of Action

Proposed Mechanism of Action	Effect on Viral Infection
Modulation of cell signaling pathways (MAPK, PI3K)	Inhibition of virus-induced cell death Inhibition of viral transcription and translation
Alteration of cell membrane fluidity and fusion processes	Blockage of virus entry Altered membrane insertion of virus glycoproteins during assembly
Induction of nitric oxide synthesis	S-nitrosylation and inactivation of viral enzymes
Upregulation of immune system (viral-specific antibodies, protective cell populations, and cytokines)	Protection in virus-infected animals

The ability of DHEA to modulate different cell functions and physiological processes offers a possible explanation for the effects of DHEA treatment on different aspects of virus infections.

DHEA = dehydroepiandrosterone.

SUMMARY POINTS

- Studies performed in cell cultures (*in vitro*) and animal models (*in vivo*) show that DHEA exhibits antiviral action against different viruses that cause human and animal diseases.
- Although the molecular basis for DHEA's effect is still poorly understood, protection of lethal virus infection in animal models is related to the ability of DHEA to promote the upregulation of immune response.
- The analysis of DHEA's inhibitory action against a great variety of viruses in cell cultures reveals that this steroid can affect different steps of viral replication: virus entry, RNA and protein synthesis, and, in the case of enveloped viruses, proper insertion of virus glycoproteins into the cell membrane, thus leading to a failure in virus particle formation and virus dissemination.
- Although viruses display a great variety of replication strategies, more or less, all of them depend on several metabolic and signaling cellular pathways. These cell functions are main targets of DHEA biological effects, offering a possible explanation for the broad spectrum of antiviral action of this hormone.
- Many investigations are currently underway to identify DHEA analogs that may be responsible for its beneficial effects and are unable to be transformed into sexual hormones; some analogs have proven to be promising antiviral compounds tested both *in vitro* and *in vivo.*

REFERENCES

Acosta, E. G., A. C. Bruttomesso, J. A. Bisceglia et al. 2008. Dehydroepiandrosterone, epiandrosterone and synthetic derivatives inhibit Junin virus replication in vitro. *Virus Res* 135:203–12.

Araghi-Niknam, M., B. Liang, Z. Zhang et al. 1997. Modulation of immune dysfunction during murine leukaemia retrovirus infection in old mice by dehyroepiandrosterone sulphate (DHEAS). *Immunology* 90:344–9.

Ben-Nathan, D., B. Lachmi, S. Lustig et al. 1991. Protection by dehydroepiandrosterone in mice infected with viral encephalitis. *Arch Virol* 120:263–71.

Benz, D., P. Cadet, K. Mantione et al. 2002. Tonal nitric oxide and health: Anti-bacterial and -viral actions and implications for HIV. *Med Sci Monit* 8:RA27–31.

Bradley, W., L. Graus, R. Good et al. 1995. Dehydroepiandrosterone inhibits replication of feline immunodeficiency virus in chronically infected cells. *Vet Immunol Immunopathol* 46:159–68.

Castilla, V., M. Larzábal, N. A. Sgalippa et al. 2005. Antiviral mode of action of a synthetic brassinosteroid against Junin virus replication. *Antiviral Res* 68:88–95.

Castilla, V., J. A. Ramírez, and C. E. Coto. 2010. Plant and animal steroids a new hope to search for antiviral agents. *Curr Med Chem* 17:1858–73.

Chang, C., Y. Ou, S. Raung, and C. Chen. 2005. Antiviral effect of dehydroepiandrosterone on Japanese encephalitis virus infection. *J Gen Virol* 86:2513–23.

De Clercq, E. 2010. Highlights in the discovery of antiviral drugs: A personal retrospective. *J Med Chem* 53:1438–50.

Esté, J. A., and T. Cihlar. 2010. Current status and challenges of antiretroviral research and therapy. *Antiviral Res* 85:25–33.

Henderson, E., A. Schwartz, L. Pashko et al. 1981. Dehydroepiandrosterone and 16 alpha-bromo-epiandrosterone: Inhibitors of Epstein-Barr virus-induced transformation of human lymphocytes. *Carcinogenesis* 2:683–6.

Henderson, E., J. Y. Yang, and A. Schwartz. 1992. Dehydroepiandrosterone (DHEA) and synthetic DHEA analogs are modest inhibitors of HIV-1 IIIB replication. *AIDS Res Hum Retroviruses* 8:625–31.

Jacobson, M. A., R. E. Fusaro, M. Galmarini et al. 1991. Decreased serum dehydroepiandrosterone is associated with an increased progression of human immunodeficiency virus infection in men with CD4 cell counts of 200–499. *J Infect Dis* 164:864–8.

Jiang, Y., T. Miyazaki, A. Honda et al. 2005. Apoptosis and inhibition of the phosphatidylinositol 3-kinase/Akt signaling pathway in the anti-proliferative actions of dehydroepiandreosterone. *J Gastoenterol* 40:460–97.

Labrie, F., V. Luu-The, C. Labrie et al. 2003. Endocrine and intracrine sources of androgens in women: Inhibition of breast cancer and other roles of androgens and their precursor dehydroepiandrosterone. *Endocr Rev* 24:152–82.

Linero, F. N., and L. A. Scolaro. 2009. Participation of the phosphatidylinositol 3-kinase/Akt pathway in Junin virus replication in vitro. *Virus Res* 145:166–70.

Loria, R. M., T. H. Inge, S. S. Cook et al. 1988. Protection against acute lethal viral infections with the native steroid dehydroepiandrosterone (DHEA). *J Med Virol* 26:301–14.

Mavoungou, D., V. Poaty-Mavoungoun, M. Akoume et al. 2005. Inhibition of human immunodeficiency virus type-1 (HIV-1) glycoprotein mediated cell–cell fusion by immunor (IM28). *Virol J* 2:2–9.

Mulder, J. W., P. H. Frissen, P. Krijnen et al. 1992. Dehydroepiandrosterone as predictor for progression to AIDS in asymptomatic human immunodeficiency virus-infected men. *J Inf Dis* 165:413–8.

Nishimura, T., Y. Seki, K. Sato et al. 2008. Enhancement of zidovudine uptake by dehydroepiandrosterone sulfate in rat syncytiotrophoblast cell line TR-TBT 18d-1. *Drug Metab Dispos* 36:2080–5.

Pedersen, N. C., T. W. North, R. Rigg et al. 2003. 16alpha-Bromo-epiandrosterone therapy modulates experimental feline immunodeficiency virus viremia: Initial enhancement leading to long-term suppression. *Vet Immunol Immunopathol* 94:133–48.

Romanutti, C., A. C. Bruttomesso, V. Castilla et al. 2009. *In vitro* antiviral activity of dehydroepiandrosterone and its synthetic derivatives against vesicular stomatitis virus. *Vet J* 182:327–35.

Romanutti, C., A. C. Bruttomesso, V. Castilla et al. 2010. Anti-adenovirus activity of epiandrosterone and dehydroepiandrosterone derivatives. *Chemotherapy* 56:158–65.

Romanutti, C., V. Castilla, C. E. Coto et al. 2007. Antiviral effect of a synthetic brassinosteroid on the replication of vesicular stomatitis virus in Vero cells. *Int J Antimicrob Agents* 29:311–6.

Schümann, M., and M. Dobbelstein. 2006. Adenovirus induced extracellular signal-regulated kinase phosphorylation during the late phase of infection enhances viral protein levels and virus progeny. *Cancer Res* 66:1282–8.

Shin, Y., Q. Liu, S. K. Tikoo et al. 2007. Effect of the phosphatidylinositol 3-kinase/Akt pathway on influenza a virus propagation. *J Gen Virol* 88:942–50.

Simoncini, T., P. Mannella, L. Fornari et al. 2003. Dehydroepiandrosterone modulates endothelial nitric oxide synthesis via direct genomic and nongenomic mechanisms. *Endocrinology* 144:3449–55.

Wachsman, M. B., J. A. Ramírez, L. B. Talarico et al. 2004. Antiviral activity of natural and synthetic brassinosteroids. *Curr Med Chem- Anti-Infective Agents* 3:163–79.

Whiting, K. P., C. J. Restall, and P. F. Brain. 2000. Steroid hormone-induced effects on membrane fluidity and their potential roles in non-genomic mechanisms. *Life Sci* 7:743–57.

Yang, J. Y., A. Schwartz, and E. E. Henderson. 1993. Inhibition of HIV-1 latency reactivation by dehydroepiandrosterone (DHEA) and an analog of DHEA. *AIDS Res Hum Retroviruses* 9:747–54.

Yang, J. Y., A. Schwartz, and E. E. Henderson. 1994. Inhibition of 3′azido-3′deoxythymidine-resistant HIV-1 infection by dehydroepiandrosterone in vitro. *Biochem Biophys Res Commun* 201:1424–32.

Zhang, Z., M. Araghi-Niknam, B. Liang et al. 1999. Prevention of immune dysfunction and vitamin E loss by dehydrepiandrosterone and melatonin supplementation during murine retrovirus infection. *Immmunology* 96:291–7.

25 Adipose Tissue as a Target for Dehydroepiandrosterone and Its Sulfate

Fátima Pérez-de Heredia, Juan Jose Hernández-Morante, and Marta Garaulet

CONTENTS

AN INTRODUCTION TO ADIPOSE TISSUE AND OBESITY

Obesity, characterized by adipose tissue hypertrophy, is well recognized to be associated with higher morbidity and mortality through its relationship with hypertension and increased risk of type 2 diabetes and of cardiovascular disease (CVD). Although the association between obesity and cardiovascular morbidity and mortality can seem of low magnitude compared with the impact of other well-known risk factors such as smoking, dyslipidemia, or hypertension, the continuing increase in this condition's prevalence and its epidemic proportions worldwide have reinforced the need for research in this field.

Epidemiological data suggest that adipose tissue distribution per se might be an indicator of CVD risk, with intra-abdominal obesity being more detrimental than peripheral obesity. It is known that body fat distribution is related to insulin resistance, and that adipose tissue, besides playing an endocrine role as a target for sex hormones, produces modulator substances, such as leptin, adiponectin, tumor necrosis factor-α (TNF-α), or interleukins, which exert endocrine, autocrine, and paracrine effects.

Sex steroids, the metabolism of which is influenced by adipose tissue, have a strong impact on the regional development of this tissue, and play a primordial role, together with other factors, in the expansion of abdominal fat (Garaulet et al. 2000). Regional growth of adipose tissue is mainly dependent on the metabolism of mature fat cells, and is determined by the capacity of the adipocyte to accumulate and mobilize triacylglycerides, that is, adipocyte size. Studies on the relationships between hormones and adipose tissue function and morphology have shown the influence of sex steroids on the accumulation of fat in adipocytes of different adipose regions (Garaulet et al. 2000). It has been reported that sexual hormones can be determinant in the rate of two important physiological processes implicated in the accumulation of body fat: on the one hand, the differentiation of preadipocytes into adipocytes, and on the other hand, the proliferation of preadipocytes and adipocytes to originate new fat cells. Both processes are determinant in regional body fat accumulation and in obesity features.

DHEA, A "FOUNTAIN OF LEANNESS"

Evidence from Studies in Humans

Humans and other primates are unique in having elevated circulating levels of the sex steroid dehydroepiandrosterone (DHEA) and its sulfated ester, dehydroepiandrosterone sulfate (DHEAS), both precursors for androgens and estrogens, as has been already explained in this book. In humans, DHEAS concentrations are 100- to 500-fold higher than those of testosterone and 1,000–10,000 times greater than those of estradiol. In young adults, DHEA levels are 10%–20% higher in men than in women.

Although perhaps the most important reason why DHEA has attracted so much attention is its dramatic relationship to aging, along these years many authors have demonstrated an antiobesity effect for the hormone. Our own research group reported that visceral fat correlated inversely with plasma levels of DHEAS in males (Garaulet et al. 2000), suggesting that in obese men, the greater the DHEAS production, the lower the abdominal fat accumulation. However, in spite of the relatively high number of studies, data are still contradictory, as shown in Table 25.1. Authors have reported significant reductions in fat mass after DHEA administration, at either lower (Villareal, Holloszy, and Kohrt 2000; Villareal and Holloszy 2004) or higher doses (Nestler et al. 1988) in both men and women (Villareal, Holloszy, and Kohrt 2000; Villareal and Holloszy 2004), while others observed no effect under similar experimental conditions (Mortola and Yen 1990; Percheron et al. 2003; Welle, Jozefowicz, and Statt 1990).

These discrepancies could be due to the characteristics of the subjects (gender, age, hormonal status, presence and degree of obesity, etc.), or to experimental conditions (treatment lengths, form of administration, etc.). Body fat distribution could also be a confounding factor. Indeed, as we have already reported, in a study performed by our group, we demonstrated an inverse association between the internal fat (visceral fat) and plasma DHEAS concentrations in men (Hernández-Morante et al. 2008). In contrast, in women, DHEAS correlated inversely with subcutaneous abdominal fat (Garaulet et al. 2000), although earlier studies had found no relationship between these two measures.

After many years of work on this field, data collected in our lab from cross-sectional, *in vitro* (human adipose tissue cultures), and *in vivo* (rats) studies, together with the work by other authors, consistently show the antiobesity properties of DHEA and its sulfate, through the stimulation of lipolysis in adipose tissue depots, the reduction of fat accumulation, and the decrease in energy intake. We also observed a protective role of DHEA(S) against features of the metabolic syndrome, as will be explained along this chapter.

DHEA Administration Reduces Body Fat in Rodents

Results from animal experiments, mainly rodents, are sounder in relation to the antiobesity properties of DHEA. Since as early as the 1980s, a number of works have studied the effects of the administration of DHEA to rats, and the authors generally observed a reduction in total body weight and body fat content (Table 25.2).

TABLE 25.1
Summary of the Main Clinical Studies on DHEA and Obesity-Related Variables

Author(s)	Dose (mg/day)	Duration (Weeks)	Population	Main Outcome
Kawano et al. (2003)	25	12	24 M	↑ Insulin sensitivity
Casson et al. (1998)	25	24	13 F (Post-men*)	↓ HDLc and ApoA1*
Løvås et al. (2003)	25	36	39 F (Ad-Ins)	No effect
Jedrzejuk et al. (2003)	50	12	12 M	No effect
Barnhart et al. (1999)	50	12	60 F	No effect
Dhatariya, Bigelow, and Nair (2005)	50	12	28 F (Ad-Ins*)	↑ Insulin sensitivity
Callies et al. (2001)	50	16	24 F (Ad-Ins)	No effect
Arlt et al. (2001)	50	16	22 M	No effect
Arlt et al. (1999)	50	16	24 F (Ad-Ins)	↓ Cholesterol No effect of body weight
Villareal, Holloszy, and Kohrt (2000)	50	24	8 M 10 F	↓ Fat mass and trunk fat
Morales et al. (1994)	50	24	13 M 17 F	↓ HDLc in women
Villareal and Holloszy (2004)	50	24	56 M+F (Elderly)	↓ VAT* and SAT* areas ↑ Insulin sensitivity
Percheron et al. (2003)	50	52	140 M 140 F	No effect
Vogiatzi et al. (1996)	80	8	13 (obese adolescents)	No effect on fat mass
Flynn et al. (1999)	100	12	39 M	↓ Cholesterol and HDLc*
Diamond et al. (1996)	300–500†	52	15 F (Post-men)	↓ Skinfold thickness, total cholesterol, and HDLc
Mortola and Yen (1990)	1600	4	6 F (Post-men)	↓ Cholesterol No effect of body weight
Welle, Jozefowicz, and Statt (1990)	1600	4	8 M	No effect on fat mass
Nestler et al. (1988)	1600	4	10 M	↓ Fat mass, cholesterol, and LDLc*

*Ad-Ins = adrenal insufficiency; DHEA = dehydroepiandrosterone; ApoA1 = apolipoprotein A1; F = females; HDLc = high-density lipoprotein cholesterol; LDLc = low-density lipoprotein cholesterol; M = males; post-men = postmenopausal; SAT = subcutaneous adipose tissue; VAT = visceral adipose tissue.

†Transdermally administered.

Initially, this effect was linked with a reduction of food intake (Abadie et al. 2001; Pham et al. 2000; Ryu et al. 2003). Previous work on other aspects of DHEA action had shown that the hormone could act as a neurotransmitter, interacting with receptors such as NMDA (N-methyl-D-aspartic) and gamma-amino-butyric acid (GABA; Bergeron, de Montigny, and Debonnel 1996; Majewska et al. 1990). These findings led other authors to propose that DHEA could participate in food intake control at the central nervous system (Pham et al. 2000; Navar et al. 2006). However, some studies found no changes in food intake in the rats as a result of DHEA treatment (Lea-Currie, Wen, and McIntosh 1997), whereas others declared that reduction in body mass exceeded what could be expected if only energy intake was considered (Abadie et al. 2001; Pérez de Heredia et al. 2007; Richards, Porter, and Svec 2000; Ryu et al. 2003). Besides, the anorectic effect of DHEA seemed to be only an acute response, not being observed in the longer term (Pérez de Heredia et al. 2007; Porter and Svec 1995).

TABLE 25.2
Major Studies on DHEA Effects on Animal Adipose Tissue *In Vivo*

Author(s)	Dose	Duration	Model	Main Outcome
Mauriège et al. (2003)	30 mg/day (p.c.)	27 weeks	SD rats; ovariectomized	↓ LPL and ↑ HSL activities
Pérez de Heredia et al. (2009)	0.5% (diet)	13 weeks	SD rats; HF diet	Changes in adipose tissue fatty acid composition
Sánchez et al. (2008)	0.5% (diet)	13 weeks	SD rats; HF diet	↑ Adiponectin and resistin expression in visceral fat
Wang, Sun, and Qiu (2004)	s.c.	20 days	SD rats	↑ Resistin expression and insulin resistance
Ryu et al. (2003)	0.6% (diet)	17 days	OLETF rats	Induction of UCP1 and UCP3
Karbowska and Kochan (2005)	0.6% (diet)	2 weeks	Wistar rats	↑ Adiponectin, AdipoR1, carnitine-palmitoyl transferase, and HSL expression
Kochan and Karbowska (2004)	0.6% (diet)	2 weeks	Wistar rats	↑ PPAR-α and resistin expression
Ishizuka et al. (2007)	0.4% (diet)	2 weeks	OLETF rats	↑ GLUT4 translocation and 2DOG uptake
Kajita et al. (2003)	0.4% (diet)	2 weeks	OLETF rats	↓ PPAR-γ content
Lea-Currie, Wu, and McIntosh (1997)	DHEAS 10, 100 μg/mL (water)	2 weeks	SD rats; HF diet	↓ WAT mass ↑ BAT mass and UCP content
Apostolova et al. (2005)	0.2% (diet)	12 days	C57BL/6J mice	Counteraction of glucocorticoid actions
Tagliaferro et al. (1995)	0.4% (diet)	—	Wistar rats	↓ WAT mass, cell size and number ↑ Glycerol release (lipolysis)

2DOG = 2-deoxyglucose; AdipoR1 = adiponectin receptor 1; BAT = brown adipose tissue; GLUT4 = glucose transporter; HF = high-fat; HSL = hormone-sensitive lipase; LPL = lipoprotein lipase; p.c. = percutaneously; PPAR = peroxisome proliferator-activated receptor; s.c. = subcutaneously; SD = Sprague-Dawley; UCP = uncoupling protein; WAT = white adipose tissue.

Another explanation for DHEA's slimming effect on rodents was an increase in energy expenditure through stimulation of mitochondrial and peroxisomal proliferation in liver, leading to enhanced fatty acid oxidation (Bellei et al. 1992; Frenkel et al. 1990; Tagliaferro et al. 1986). In addition to these hypotheses, the evidence for significant reductions of not only body weight but also specifically body fat in animals receiving DHEA suggested a direct effect on adipose tissue, on both fat cell size and number. Research in the last two decades has unequivocally confirmed the adipocyte as a target for the actions of DHEA and its sulfate.

MODULATING FAT CELL FUNCTION

Experimental Models for the Study of DHEAS

The role of DHEA as a regulator of adipocyte function has been studied both *in vivo* in animal models and *in vitro*. For the *in vivo* experiments, rat is the most used model, be it to observe the effect of DHEA on adipose tissue under normal conditions, or to specifically assess the antiobesity properties of the hormone. Different models of obesity have been used for the latter purpose: genetic (Zucker and OLEFT rat strains), surgical (e.g., by ovariectomy), or dietary (high-fat diets, cafeteria diets, etc.). The preferred form of administration of the hormone is DHEA rather than its sulfate (the two forms seem to occasionally differ in their properties, as will be commented on further in this chapter). It is usually administered orally in the diet (0.3%–0.6% weight of the food consumed), but also as subcutaneous injections (see Table 25.2). It is worth noting that rodents, unlike humans and primates in general, do not produce significant amounts of DHEA and DHEAS, despite which their tissues are responsive to the steroid. This makes them suitable models for the study of the effects of DHEA(S), albeit it must be kept in mind that any dose exogenously administered to rodents will not be physiological but pharmacological, an important issue to be considered when researchers aim to repeat the experiments in humans.

In the *in vitro* studies, the most frequently used systems are mouse 3T3-L1 cells—a fibroblast cell line with a high capacity to differentiate into adipocytes—as well as primary cultures and explants from both human and animal fat depots. Tables 25.3 and 25.4 present the most relevant work performed in human and animal adipose tissue *in vitro*. For the animal-derived cultures, DHEA is preferred to DHEAS, and the dose usually used is 100 μM. Studies conducted on human adipose tissue cultures are still scarce, especially for the visceral region, because the introduction of laparoscopy techniques in the surgery process has limited the accessibility to adipose tissue. The usual treatment dose for human samples varies from 100 nM to 10 μM, with 1 μM being the dose with the strongest effect (Hernández-Morante et al. 2006; Piñeiro et al. 1999; Tan et al. 2007). The results from

TABLE 25.3
Major Studies on the Effects of DHEA and DHEAS on Human Adipose Tissue *In Vitro*

Author(s)	Hormone	Cell Types	Main Outcomes
Hernández-Morante et al. (2010)	DHEAS	Adipocytes (SAT and VAT)	Changes in fatty acid composition
Hernández-Morante et al. (2008)	DHEAS	Adipocytes (SAT and VAT)	↑ Lipolysis (gender- and depot-specific)
Hernández-Morante etal. (2006)	DHEAS	Adipocytes (SAT and VAT)	↑ Adiponectin expression
Saraç et al. (2006)	DHEA	SVC (SAT and VAT)	(–) Cell proliferation and differentiation
Perrini et al. (2004)	DHEA	Adipocytes (SAT)	↑ Glucose uptake by stimulating GLUT1 and -4 translocation
Machinal-Quélin et al. (2002)	DHEA	Adipocytes (SAT)	↑ Leptin expression and secretion
Piñeiro et al. (1999)	DHEAS	Adipocytes (VAT)	(–) Leptin production (only in females)
McIntosh et al. (1998)	DHEA	SVC (SAT)	(+) Preadipocyte differentiation

DHEA = dehydroepiandrosterone; DHEAS = dehydroepiandrosterone sulfate; GLUT = glucose transporter; SAT = subcutaneous adipose tissue; SVC = stromal-vascular cells; VAT = visceral adipose tissue.

TABLE 25.4
Major Studies on DHEA Effects on Animal Adipose Tissue *In Vitro*

Author(s)	Treatment	Model	Main Outcomes
Ishizuka et al. (2007)	1 μM (30 minutes)	eWAT (Wistar rats)	↑ GLUT4 translocation, 2DOG uptake
Kajita et al. (2003)	1 μM	eWAT (Wistar rats)	↓ TG, PPAR-γ, aP2, and SREBP
Apostolova et al. (2005)	12.5–100 μM (48 hours)	C57BL/6J mice	(–) Glucocorticoid actions
Perrini et al. (2004)	Various doses, mainly 100 μM	3T3-L1 cells	↑ GLUT1 and -4 translocation, 2DOG uptake
McIntosh et al. (1998)	15, 50, 150 μM 5, 25, 75 μM	SVF (Pigs; SD rats)	↓ Preadipocyte proliferation, differentiation ↓ C/EBP-α expression
Park et al. (2006)	100 μM	3T3-L1 cells	(–) Inflammatory response, oxidative stress.
Marwah et al. (2006)	100 μM (48 h)	3T3-L1 cells	DHEA was metabolized to androstenediol, 7α-OH-DHEA and 7α-androstenetriol
McCormick, Wang, and Mick (2006)	100 μM	Adipocyte (rat gonadal AT)	↓11β-HSD1 oxo-reductase activity and NADPH content.
Gómez et al. (2002)	100 μM	3T3-L1 cells	↓ Lipid accumulation, SCD levels, and activity ↑ C18:1 and C18:0, ↓ C16:1 ↑ Basal and insulin-stimulated glucose uptake
Ishizawa et al. (2001)	1 μM	eWAT (Wistar rats)	↑ 2DOG uptake ↑ PLD activity and DAG production
Kajita et al. (2000)	100 nM	eWAT (Wistar rats)	↑ 2DOG uptake (+) PI3K and PKC-β activities
Lea-Currie et al. (1998)	5, 25, 50, 100 μM, (4 days)	3T3-L1 cells	↓ Preadipocyte proliferation, differentiation

DHEA = dehydroepiandrosterone; 11β-HSD1 = 11β-hydroxysteroid dehydrogenase 1; 2DOG = 2-deoxyglucose; aP2 = adipocyte-specific fatty acid binding protein; C/EBP = CCAAT-enhancer binding protein; DAG = diacylglycerol; eWAT = epididymal white adipose tissue; GLUT = glucose transporter; PDL = phospholipase D; PI3K = phophoinositol-3 kinase; PKC-β = protein kinase C-β; PPAR = peroxisome proliferator-activated receptor; SCD = stearoyl-CoA-desaturase; SD = Sprague-Dawley; SREBP = sterol regulatory element binding protein; SV = stromal-vascular; TG = triglycerides.

the different systems—animals and cell/tissue cultures—are reasonably coherent, offering reliable models to be extrapolated to humans *in vivo* and used for future clinical studies.

Data collected from all those studies are summarized in Tables 25.3 and 25.4, and they will be presented in more detail in the following sections ("DHEA and Adipogenesis," "Regulation of Fat Storage," "Effects on Fatty Acid Composition," "Insulin-Sensitizing Actions," and "Regulation of Adipokine Secretion"), where it will be discussed how DHEA and its sulfate can regulate different aspects of the adipose function, from cell proliferation to metabolism and adipokine expression.

DHEA Reduces Adipogenesis

In the late 1990s, the group of McIntosh and coworkers found that DHEA could impair preadipocyte proliferation and differentiation in different culture systems, like the stromal-vascular fraction from pig adipose tissue, the rat subcutaneous adipose tissue (McIntosh et al. 1998), and the murine 3T3-L1 cells (Lea-Currie, Wen, and McIntosh 1998). The inhibition of (pre)adipocyte proliferation due to DHEA was dose-dependent and did not appear to be due to the cytotoxic effect of the steroid. In contrast, no effect was observed for the sulfate conjugate (Lea-Currie, Wen, and McIntosh 1998), so the authors hypothesized that DHEAS was either not taken up or not metabolized by rodent (pre)adipocytes, and therefore, it needed to be converted into DHEA or another metabolite prior to uptake by these cells.

One proposed explanation for the *antiproliferative action* of DHEA on adipose tissue is that DHEA acts as an uncompetitive inhibitor of glucose-6-phosphate dehydrogenase (Marks and Banks 1960), potentially limiting the availability of NADPH, a coenzyme for redox reactions and a source of H^+ that is required for de novo synthesis of intermediates of the pentose phosphate pathway, in particular of ribose. In consequence, the low availability of NADPH would impair the synthesis of new DNA molecules, and delay the cell division process.

Regarding *inhibition of preadipocyte differentiation*, there are several explanations:

- The effect of DHEA could be mediated by interfering with cholesterol and isoprene biosynthesis, thus inhibiting the production of compounds essential for cell proliferation (as proposed by Lea-Currie, Wen, and McIntosh 1998). It has been shown that blocking this metabolic pathway impairs 3T3-L1 differentiation (Nishio et al. 1996), and that DHEA exerts this inhibitory effect on other cell types (Pascale et al. 1995), which allows the thinking that this could be also the case for preadipocytes.
- In addition, DHEA can reduce the expression of the transcription factor CCAAT-enhancer binding protein-α (C/EBP-α; McIntosh et al. 1998), which is involved in the adipogenic process and is necessary for preadipocyte differentiation (see Figure 25.1).

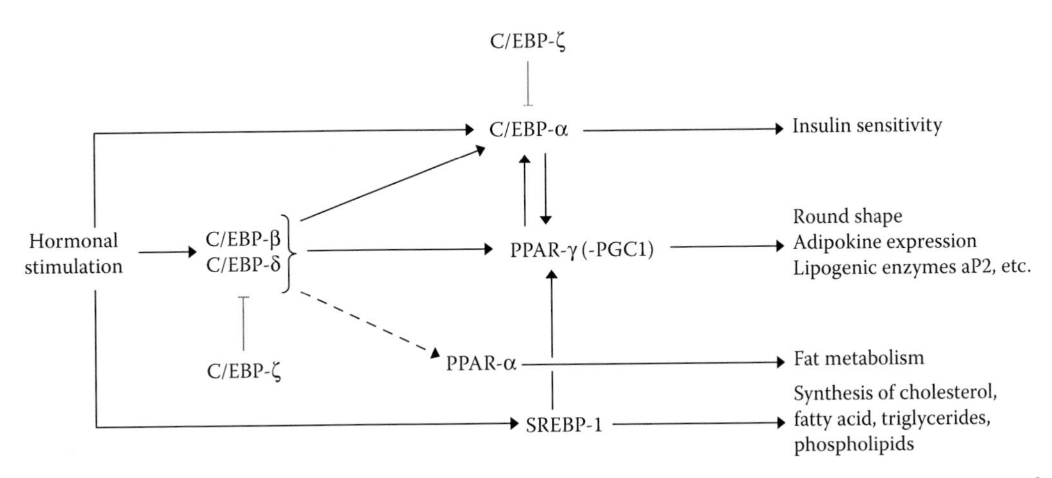

FIGURE 25.1 Scheme of adipogenesis. The C/EBP and PPAR families are the two master regulators of preadipocyte differentiation. In response to adequate hormonal stimulation (e.g., glucocorticoids or insulin), C/EBP-β and -δ are activated, stimulating the expression of the key adipogenic factors, C/EBP-α and PPAR-γ. Other transcription factors, such as PPAR-α and SREBP-1, will also contribute to adipocyte differentiation by stimulating the expression of lipogenic enzymes. C/EBP-ζ cannot induce gene expression, but it will form heterodimers with other C/EBPs, constituting an endogenous negative-feedback mechanism. C/EBP = CCAAT-enhancer binding protein; PGC1 = PPAR-γ coactivator 1; PPAR = peroxisome proliferator-activated receptor; SREBP = Sterol regulatory element binding protein.

- The peroxisome proliferator-activated receptor-γ (PPAR-γ) could also be implicated in the effect of DHEA on adipocyte differentiation. It has been proposed that a decreased action of this transcription factor, which is essential for adipocyte differentiation (Kajita et al. 2003), could be contributing to the antiadipogenic effect of DHEA.
- DHEA seems to reduce the gene expression of other adipogenic factors, such as sterol regulatory element binding protein (SREBP), and of aP2 (an adipocyte-specific fatty acid binding protein), a marker of adipocyte function (Figure 25.1; Kajita et al. 2003). By contrast, DHEA increases the expression of other factors necessary for adipogenesis, namely PPAR-γ coactivator 1, C/EBP-β, and PPAR-α (Apostolova et al. 2005; Kochan and Karbowska 2004; Figure 25.1). Through the enhancement of PPAR-α, DHEA may promote fatty acid utilization within the adipocyte, as this transcription factor stimulates the expression of peroxisomal, mitochondrial, and microsomal enzymes that activate β-oxidation of fatty acids (Brandt, Djouadi, and Kelly 1998; Muoio et al. 2002).
- Another possible way by which DHEA may impair adipogenesis is through inhibition of lipid storage. 3T3-L1 preadipocytes cultured with DHEA showed less lipid accumulation and, significantly, lower stearoyl-coenzyme A-desaturase (SCD) levels and activity (Gómez et al. 2002). SCD is the enzyme responsible for adding the first double bond to saturated fatty acids (SFA) palmitic (C16:0) and stearic (C18:0), in order to produce palmitoleic and oleic acids, respectively, and hence, it is a key enzyme for fatty acid synthesis and storage. In addition, DHEA treatment resulted in increased carnitine-palmitoyl transferase activity, implying higher fatty acid oxidation and a possible thermogenic effect (Gómez et al. 2002).

Regulation of Fat Storage within the Adipocyte by DHEAS

As indicated in the previous section, "DHEA Reduces Adipogenesis," DHEA treatment in animals and cell cultures seems to enhance lipid mobilization and utilization and reduce lipid storage. In 1995, Tagliaferro and coworkers published that glycerol release (a consequence of lipolysis) was higher in adipose tissue from Wistar rats treated with DHEA, indicating increased lipolytic activity (Tagliaferro et al. 1995). A decade later, new studies proved that DHEA supplementation actually increased the gene expression of the hormone-sensitive lipase and the carnitine-palmitoyl transferase (Karbowska and Kochan 2005) in rat adipose tissue, enzymes that are involved in triglyceride and fatty acid catabolism, respectively.

Other works showed that following ovariectomy in rats, the response to lipolytic stimuli was diminished while lipoprotein-lipase activity (and, therefore, fatty acid uptake into the adipocyte) was increased (Lemieux et al. 2003; Mauriège et al. 2003). DHEA administration to ovariectomized rats prevented these changes (Lemieux et al. 2003; Mauriège et al. 2003).

Furthermore, the nature of DHEAS as a potent stimulator of lipolysis has also been demonstrated in primary cultures of human adipocytes. Our group observed that DHEAS significantly increased lipolytic activity in subcutaneous fat in women, and in visceral adipose tissue in men (Hernández-Morante et al. 2008), also reflecting that, similar to other sex hormones (de Pergola et al. 1996), DHEA exerts a gender- and tissue-specific effect. This differential action of DHEA(S) will be discussed further in this chapter.

Effects of DHEA on Adipose Tissue Fatty Acid Composition

Besides its action on lipid storage, DHEA affects lipid metabolism on another level. Abadie and coworkers reported that DHEA-treated rats showed altered fatty acid profiles in their muscles (Abadie et al. 2001). They did not find an effect on adipose tissue, but this was probably due to the length of the experiment (7 days), as fatty acid turnover in adipose tissue is slower than in other tissues. In contrast, 3T3-L1 adipocytes cultured with DHEA experienced a shift toward higher proportions of oleic (C18:1) and stearic acids (C18:0) and lower proportions of palmitic (C16:0; Gómez et al. 2002). In our laboratory, we observed that rats treated chronically with DHEA (13 weeks) reproduced some of the changes observed *in vitro*, as they had higher proportions of stearic and oleic acids, and lower proportions of palmitoleic in their adipose tissue (Pérez de Heredia et al. 2009). Total n-6 polyunsaturated fatty acids (PUFA) were also decreased while n-3 PUFA were increased, leading to significant reductions in n-6/n-3 PUFA ratios in DHEA-treated rats (Pérez de Heredia et al. 2009). This change could be linked to some of the beneficial properties attributed to DHEA because high n-6/n-3 PUFA ratios have been associated with greater body fat accumulation, increased cardiovascular risk, and inflammatory processes (Ailhaud et al. 2006; Ghafoorunissa, Ibrahim, and Natarajan 2005; Riediger et al. 2008).

We later confirmed these results *in vitro* in human subcutaneous and visceral adipose tissue samples treated for 24 hours with 1 μM DHEAS. The effects of DHEAS on fatty acid profile seemed to be similar in the two depots: there was a reduction in palmitic (16:0) and total SFA, total n-6 PUFA, and in n-6/n-3 PUFA ratios after DHEAS treatment, whereas monounsaturated fatty acids (MUFA) increased (Hernández-Morante et al. 2010). Recently, we have analyzed plasma fatty acids from menopausal women treated orally with DHEAS (100 mg) for 3 months, and similar results have been found, particularly, for the increase in MUFA and the decrease in SFA.

Additionally, markers of δ-6-desaturase activity were significantly increased by DHEA(S) in both rats and humans (Hernández-Morante et al. 2010; Pérez de Heredia et al. 2009). δ-6-desaturase is the enzyme that allows conversion of linoleic acid (C18:2 n-6) into other n-6 fatty acids such as γ-linolenic (C18:3) and arachidonic (C20:4), and its activity has been positively associated with insulin-stimulated glucose-uptake (Das 2005); therefore, increased estimated δ-6-desaturase activity might reflect a mechanism by which DHEA and DHEAS improve insulin sensitivity, an action we will discuss immediately.

In addition to these changes, DHEA was shown to augment the proportion of n-3 PUFA in rat interscapular brown adipose tissue. The content of n-3 fatty acids in brown fat is related to the levels of uncoupling protein 1 (UCP1; Takahashi and Ide 2000), a mitochondrial enzyme, which uncouples β-oxidation from ATP production, so that catabolism of fatty acids produces energy dissipation and heat—a process known as thermogenesis. Thus, DHEA could act on thermogenesis at two levels, by stimulating fatty acid oxidation, as previously commented, and by increasing n-3 PUFA proportions in adipose tissue.

Insulin-Sensitizing Actions of DHEA on Adipose Tissue

DHEA and DHEAS *act as modulators of glucose metabolism* as well. It has been reported that circulating DHEA levels are associated with glucose tolerance in humans (Haffner and Valdez 1994), and its administration has been shown to reduce glucose and/or insulin concentrations in obese or hyperinsulinemic rodents (Kimura et al. 1998; Richards, Porter, and Svec 2000; Sánchez et al. 2008). The protective action of DHEA against insulin resistance can result from a combination of different mechanisms, like a secondary effect of reduced fat accumulation, a counteraction of glucocorticoid action (Apostolova et al. 2005), or the direct stimulation of glucose uptake (Ishizuka et al. 2007; Perrini et al. 2004). The changes in fatty acid profiles discussed in the section the previous section, "Effects of DHEA on Adipose Tissue Fatty Acid Composition," can provide a support for the first possible mechanism; for instance, DHEA(S) reduced the proportions of palmitic acid, which is a precursor of ceramides, which in turn are known to induce peripheral insulin resistance (Powell et al. 2004). In addition, SFA alter membrane fluidity and, in consequence, insulin receptor number and/or affinity (Field et al. 1990; Grunfeld, Baird, and Kahn 1981).

DHEA may also, directly, enhance insulin sensitivity in adipose tissue, similar to what was observed in rat liver and skeletal muscle (Campbell et al. 2004). Studies *in vivo* in obese rats, and *in vitro* in human and rodent adipocytes, have found that DHEA stimulated the uptake of the analogous 2-deoxy-glucose (Perrini et al. 2004). DHEA mimics insulin action by inducing the

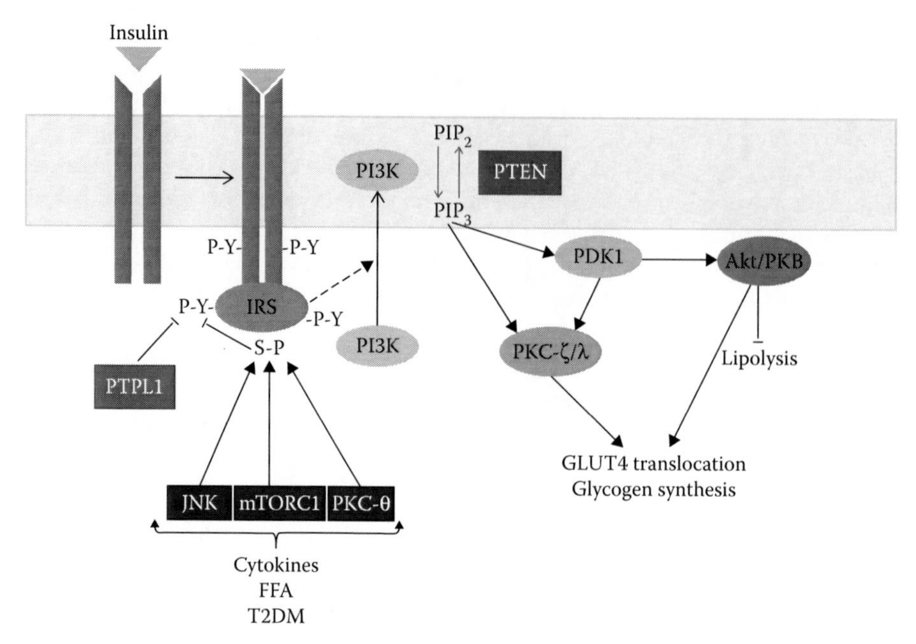

FIGURE 25.2 An overview of the insulin signaling pathway. Insulin binding to its receptor leads to autophosphorylation in tyrosine (Y) residues and to association with the insulin receptor substrate (IRS). Phosphorylated IRS activates PI3K, which translocates to the plasma membrane, where it converts PIP_2 into PIP_3. This molecule will activate the kinases PDKI, Akt/PKB, and PKC-ζ/λ, which in turn stimulate translocation of glucose transporters to the plasma membrane and in consequence glucose uptake. Insulin signal can be terminated or inhibited by different enzymes, such as phosphatases (PTEN, PTPLI) or serine-kinases (JNK, mTORC, PKC-θ). Akt/PKB = protein kinase B; JNK = c-Jun N-terminal kinase; mTORC = mammalian target of rapamycin complex; PCK-ζ/λ = a typical protein kinase C; PI3K = phosphatidylinositol-3-kinase; PDKI = phosphoinositide-dependent kinase; PIP_2 = phosphoinositol byphosphate; PIP_3 = phosphoinositol byphosphate; PKC-θ = protein kinase C zeta; GLUT = glucose transporter; PTEN = phosphatase and tensin homolog; PTPLI = protein tyrosine phosphatase L1; S = serine residues.

phosphorylation of the insulin receptor substrates IRS-1 and -2 (Perrini et al. 2004), activating phosphatidylinositol-3-kinase (PI3K; Field et al. 1990; Ishizawa et al. 2001; Ishizuka et al. 2007; Kajita et al. 2000)—a key enzyme for insulin-stimulated glucose uptake (Figure 25.2)—or by increasing the activity of two members of the protein kinase C (PKC) family: PKC-β and PKC-ζ (Grunfeld, Baird, and Kahn 1981; Perrini et al. 2004).

Interestingly, DHEA and DHEAS do not always exert the same effects. For example, DHEAS has been reported to have no effect on glucose uptake in cultured cells (Nakashima et al. 1995), but it appears to be a more potent activator of fatty acid oxidation than DHEA (Yamada, Sakuma, and Suga 1992). Therefore, the suitability of DHEA or DHEAS as modulators of cell metabolism might depend on the cell type, the target tissue, and/or the species considered.

DHEA Regulates Adipokine and Cytokine Secretion from Adipocytes

The classic paradigm of adipose tissue as a mere storage depot has been long overcome in favor of a new picture as an endocrine organ that secretes a number of bioactive molecules, collectively termed "adipokines" (Trayhurn and Wood 2004). This constitutes a very diverse group regarding structure and function, and includes cytokines (interleukins, TNF-α), growth factors (TGF-β), proteins involved in angiogenesis (VEGF), hemostasis, blood pressure (PAI-1, angiotensin), or inflammation (MCP1, haptoglobin, CRP), to mention a few. DHEA and DHEAS have been shown to regulate the secretion of some of these adipokines, indicating additional mechanisms to favor insulin sensitivity and weight loss.

- Leptin was the first adipokine to be discovered (Zhang et al. 1994). It participates in the regulation of food intake and energy expenditure, and its gene expression and circulating levels are positively correlated with body fat. In rodent adipocytes, DHEA has been shown to downregulate the expression of the leptin messenger (Apostolova et al. 2005; Kochan and Karbowska 2004). There is not a total agreement regarding its effects in humans, although it is worth mentioning that it exhibits a gender dimorphism: in men, neither DHEA nor its derivatives estradiol or testosterone affect leptin expression; in women the results are controversial, as some authors have found that DHEA and other steroids inhibit leptin expression and secretion (Piñeiro et al. 1999), whereas others have reported an elevation in these variables (Machinal-Quélin et al. 2002).
- Another crucial adipose-derived hormone is adiponectin. This adipokine is, after fatty acids, the most abundant secretion from adipose tissue and has recognized insulin-sensitizing, anti-inflammatory, and antiatherogenic properties (Guerre-Millo 2008). DHEA(S) induces adiponectin gene expression in rat and human adipose tissue (Hernández-Morante et al. 2006; Karbowska and Kochan 2005; Sánchez et al. 2008) and that of the adiponectin receptor, AdipoR1, in rat adipose tissue (Karbowska and Kochan 2005). In our laboratory, we observed that aged rats treated with DHEA had higher adiponectin messenger levels in visceral adipose tissue that the controls, partially counteracting the diminution observed with aging (Sánchez et al. 2008). This observation represents a link between the aging-related decline in DHEA levels, adiponectin function, and elevated incidence of CVD and insulin resistance. Actually, we found that among all variables affected by DHEA treatment in aged rats, adiponectin was the most influencing one (Pérez de Heredia et al. 2008). In human adipocytes, we observed that adiponectin expression was differentially regulated in the subcutaneous and visceral depots, the latter being the only responsive tissue to DHEAS stimulation (Hernández-Morante et al. 2006). This suggests that the beneficial effects of DHEAS treatment on features of the metabolic syndrome might be exerted, in part, through upregulation of adiponectin expression in visceral fat.
- Resistin is another adipokine that seems to be upregulated by DHEA (Kochan and Karbowska 2004; Sánchez et al. 2008; Wang, Sun, and Qiu 2004). It has been associated with insulin resistance (Steppan et al. 2001), so these findings were initially surprising, given the insulin-sensitizing actions of DHEA. It was proposed that this elevation of resistin gene expression was a secondary effect of DHEA-induced upregulation of PPAR-α (Kochan and Karbowska 2004), and it has been reported that resistin can inhibit adipocyte differentiation (Villena, Kim, and Sul 2002), adding another potential explanation to DHEA's slimming properties.

DEPOT-SPECIFIC EFFECTS OF DHEA(S) ON ADIPOSE TISSUE

The different adipose depots vary in their response to DHEA(S) stimulation. Data from our and other labs demonstrate a depot- and gender-specificity in DHEA(S)'s actions at different levels. For instance, a work conducted in men by Bélanger and coworkers found that DHEA concentration was higher in the omental adipose tissue than in the subcutaneous (Bélanger et al. 2006); unfortunately, there were no data available in women.

We observed than DHEA in rats and DHEAS in human adipose tissue stimulated adiponectin gene expression specifically in the visceral depots, whereas there was no change in the subcutaneous one (Hernández-Morante et al. 2006; Sánchez et al. 2008). In contrast, the effect of DHEA on rat adipose tissue fatty acid profile was more marked in the subcutaneous than in the visceral depots (Pérez de Heredia et al. 2009), while in human samples, the different depots showed a similar response regarding fatty acid composition (Hernández-Morante et al. 2010). Considering the effect of gender as well, we observed that circulating DHEAS levels were positively associated with subcutaneous but not visceral adipocyte size in men, and no association of this kind was found in women (Garaulet et al. 2000). However, in human adipocytes *in vitro*, DHEAS stimulated lipolysis in the visceral depot in men, whereas in women, it was more evident in the subcutaneous one (Hernández-Morante et al. 2008). Saraç and coworkers, in contrast, reported that female visceral adipose tissue was more sensitive to DHEA than the subcutaneous depot, in relation to proliferation and differentiation of stromal-vascular cells (Saraç et al. 2006).

It needs to be clarified which factors may contribute to these depot-related differences in DHEA's action and metabolism. It could be the result of heterogeneous distribution of DHEA(S) receptor(s), similarly to other steroids; for instance, estrogen receptor density is higher in subcutaneous than in visceral fat (Pedersen et al. 2001). However, as we will discuss later, the existence and nature of specific receptors for DHEA or DHEAS still remains a question.

Independently of the cause, this hormone being more active on one region or another is of great importance for the clinical practice, as it has been widely shown than visceral and subcutaneous adipose tissues are different from the morphological, physiological, and pathological point of view. Visceral fat appears to be more detrimental, with increased release of free fatty acids into the portal circulation, and higher expression of inflammation-related factors, such as interleukin (IL)-6, TNF-α, angiotensinogen, and so on (Wajchenberg 2000) that renders individuals with preferential visceral fat accumulation more prone to develop metabolic and cardiovascular complications.

HOW CAN DHEA(S) WORK ON ADIPOSE TISSUE?

There is various evidence and different hypotheses to help explain how DHEA and DHEAS can exert their actions within adipose tissue. The actual mechanisms are not unraveled yet, and they could be a combination of some or all of the proposed ones.

DHEA is the precursor for most androgens and estrogens, as well as sulfate, hydroxylation, and oxidation derivatives, and thus it could act through conversion into other steroids. Gaining insight into DHEA's derivatives can be crucial to understanding the physiological paths by which DHEA exerts its action because each derivative seems to have its own physiological role (Bazin et al. 2009; Gómez et al. 2002). Marwah and coworkers analyzed the culture medium of 3T3-L1 adipocytes incubated with DHEA for 48 hours and detected significant levels of DHEA itself, androstenediol, 7α-OH-DHEA, 7α-androstenetriol, and smaller amounts of testosterone, showing that adipocytes can metabolize DHEA (Marwah et al. 2006). However, no estrogens could be detected and androstendione levels were very low (Marwah et al. 2006).

In this line, although both DHEA and estradiol were reported to prevent obesity by decreasing fat deposition in rats, only estradiol decreased cholesterolemia, whereas DHEA showed a greater effect on insulin sensitivity (Lemieux et al. 2003). In addition, estradiol is a potent stimulator of leptin *in vitro* (Casabiell et al. 1998), whereas DHEA reduces its expression (Apostolova et al. 2005). These results indicate that the two hormones may act through different pathways.

Similarly, metabolization of DHEA into androgens is not likely to explain DHEA-induced upregulation of adiponectin, as testosterone was found to inhibit adiponectin gene expression (Xu et al. 2005). Furthermore, Perrini and coworkers reported that the stimulation of 2-deoxy-glucose uptake induced by DHEA *in vitro* was not mimicked by estradiol, progesterone, androstendione, testosterone, or dihydrotestosterone (Perrini et al. 2004). Thus, although we cannot rule out that some of DHEA's effects may be because of its transformation into other steroids, it seems also plausible for DHEA to act directly on adipocytes.

Among the possible direct actions of DHEA, we can find an antagonist effect of glucocorticoid action, through downregulation of 11β-hydroxysteroid dehydrogenase 1 (11β-HSD1) expression and activity (Apostolova et al. 2005; McCormick, Wang, and Mick 2006). This enzyme is crucial in the synthesis and activation of glucocorticoids, and appears to influence body weight regulation, as indicated by the fact that knockout mice for the 11β-HSD1 gene are resistant to body weight gain and fat accumulation when fed on a high-fat diet (Morton et al. 2004). DHEA might inhibit

11β-HSD1 by decreasing C/EBP-α and hexose-6-phosphate dehydrogenase expression, with a concomitant reduction of NADPH availability, necessary for glucocorticoid synthesis by 11β-HSD1 (Apostolova et al. 2005; McCormick, Wang, and Mick 2006).

It is not known whether DHEA, or DHEAS, can interact with the receptors for other androgens or estrogens, in a similar way as it has been shown to interact with receptors for neurotransmitters (Bergeron, de Montigny, and Debonnel 1996; Majewska et al. 1990), and no specific receptors for DHEA(S) have been isolated from human adipocytes so far. However, data support a receptor-dependent basis for DHEA(S)'s direct physiological action (Widstrom and Dillon 2004), and there is evidence for a high-affinity binding site for DHEA in other cell types, such as T-lymphocytes, skeletal muscle, or endothelium (Liu and Dillon 2002; Meikle et al. 1992; Tsuji et al. 1999). This putative receptor could be coupled to a membrane G-protein (Liu and Dillon 2002), but its nature still needs to be determined, as does its tissue distribution. However, studies have demonstrated the capacity of DHEA and some of its metabolites to bind or activate to both membrane and intracellular receptors in adipose tissue, such as PPARs, pregnane X receptor (PXR), constitutive androstane receptor (CAR), or estrogen receptor-β (reviewed in Webb et al. 2006).

In conclusion, the data discussed in this chapter confirm the antiobesity properties of DHEA(S), partly because of a significant effect on adipose tissue function. However, they also reveal that we are still far from understanding their precise mechanisms of action. The isolation and identification of a specific receptor for DHEA and/or DHEAS may undoubtedly help to understand the physiological role of these hormones and, in consequence, enable the development of new analogous and specific drugs, hopefully less androgenic, for the treatment of disorders associated with adipose tissue, especially for the longer term.

SUMMARY

- DHEA and its sulfate, DHEAS, are potent antiobesity steroids, capable of acting centrally on energy intake regulation, and peripherally on adipose tissue function. Adipocyte and preadipocytes have been revealed as target cells in the actions of DHEAS.
- DHEA and DHEAS are capable of inhibiting adipogenesis, reducing preadipocyte differentiation and proliferation. In addition, they regulate lipid and glucose metabolism within the adipocyte, stimulating fatty acid oxidation, modifying fatty acid profiles, and enhancing basal and insulin-stimulated glucose uptake.
- DHEA and DHEAS can also modulate adipokine expression and production toward an anti-inflammatory profile. All these actions contribute to the beneficial properties of DHEA, namely antiaging, antiobesity, insulin-sensitizing, or cardiovascular-protective.
- DHEA(S) show specificity of action, which depends greatly on the gender and on the adipose region considered.
- DHEA may exert its effects either through conversion into other steroids, or directly, as there is evidence for a receptor-dependent basis for DHEAS actions.

REFERENCES

Abadie, J. M., G. T. Malcom, J. R. Porter, and F. Svec. 2001. Dehydroepiandrosterone alters Zucker rat soleus and cardiac muscle lipid profiles. *Exp Biol Med* 226:782–9.

Ailhaud, G., F. Massiera, P. Weill, P. Legrand, J. M. Alessandri, and P. Guesnet. 2006. Temporal changes in dietary fats: Role of n-6 polyunsaturated fatty acids in excessive adipose tissue development and relationship to obesity. *Prog Lipid Res* 45:203–36.

Apostolova, G., R. A. S. Schweizer, Z. Balazs, R. M. Kostadinova, and A. Odermatt. 2005. Dehydroepiandrosterone inhibits the amplification of glucocorticoid action in adipose tissue. *Am J Physiol Endocrinol Metab* 288:E957–64.

Arlt, W., F. Callies, I. Koehler et al. 2001. Dehydroepiandrosterone supplementation in healthy men with an age-related decline of dehydroepiandrosterone secretion. *J Clin Endocrinol Metab* 86(10):4686–792.

Arlt, W., F. Callies, J. C. van Vlijmen et al. 1999. Dehydroepiandrosterone replacement in women with adrenal insufficiency. *N Engl J Med* 341(14):1013–20.

Barnhart, K. T., E. Freeman, J. A. Grisso et al. 1999. The effect of dehydroepiandrosterone supplementation to symptomatic perimenopausal women on serum endocrine profiles, lipid parameters, and health-related quality of life. *J Clin Endocrinol Metab* 84(11):3896–902.

Bazin, M. A., L. El Kihel, M. Boulouard, V. Bouët, and S. Rault. 2009. The effects of DHEA, 3beta-hydroxy-5alpha-androstane-6,17-dione, and 7-amino-DHEA analogues on short term and long term memory in the mouse. *Steroids* 74(12):931–7.

Bélanger, C., F. S. Hould, S. Lebel, S. Biron, G. Brochu, and A. Tchernof. 2006. Omental and subcutaneous adipose tissue steroid levels in obese men. *Steroids* 71(8):674–82.

Bellei, M., D. Batelli, C. Fornieri et al. 1992. Changes in liver structure and function after short-term and long-term treatment of rats with dehydroepiandrosterone. *J Nutr* 122:967–76.

Bergeron, R., C. de Montigny, and G. Debonnel. 1996. Potentiation of neuronal NMDA response induced by dehydroepiandrosterone and its suppression by progesterone: Effects mediated via sigma receptors. *J Neurosci* 16:1193–202.

Brandt, J. M., F. Djouadi, and D. P. Kelly. 1998. Fatty acids activate transcription of the muscle carnitine palmitoyltransferase I gene in cardiac myocytes via the peroxisome proliferator-activated receptor alpha. *J Biol Chem* 273:23786–92.

Callies, F., M. Fassnacht, J. C. van Vlijmen et al. 2001. Dehydroepiandrosterone replacement in women with adrenal insufficiency: Effects on body composition, serum leptin, bone turnover, and exercise capacity. *J Clin Endocrinol Metab* 86(5):1968–72.

Campbell, C. S., L. C. Caperuto, A. E. Hirata et al. 2004. The phosphatidylinositol/AKT/atypical PKC pathway is involved in the improved insulin sensitivity by DHEA in muscle and liver of rats in vivo. *Life Sci* 76:57–70.

Casabiell, X., V. Piñeiro, R. Peinó et al. 1998. Gender differences in both spontaneous and stimulated leptin secretion by omental adipose tissue in vitro: Dexamethasone and estradiol stimulate leptin release in women but not in men samples. *J Clin Endocrinol Metab* 83:2149–55.

Casson, P. R., N. Santoro, K. Elkind-Hirsch et al. 1998. Postmenopausal dehydroepiandrosterone administration increases free insulin-like growth factor-I and decreases high-density lipoprotein: A six-month trial. *Fertil Steril* 70(1):107–10.

Das, U. N. 2005. A defect in the activity of Delta6 and Delta5 desaturases may be a factor predisposing to the development of insulin resistance syndrome. *Prostaglandins Leukot Essent Fatty Acids* 72:343–50.

de Pergola, G., M. Zamboni, M. Sciaraffia et al. 1996. Body fat accumulation is possibly responsible for lower dehydroepiandrosterone circulating levels in premenopausal obese women. *Int J Obes Relat Metab Disord* 20:1105–10.

Dhatariya, K., M. L. Bigelow, and K. S. Nair. 2005. Effect of dehydroepiandrosterone replacement on insulin sensitivity and lipids in hypoadrenal women. *Diabetes* 54(3):765–9.

Diamond, P., L. Cusan, J. L. Gomez, A. Bélanger, and F. Labrie. 1996. Metabolic effects of 12-month percutaneous dehydroepiandrosterone replacement therapy in postmenopausal women. *J Endocrinol* 150(Suppl):S43–50.

Field, C. J., E. A. Ryan, A. B. Thomson et al. 1990. Diet fat composition alters membrane phospholipid composition, insulin binding, and glucose metabolism in adipocytes from control and diabetic animals. *J Biol Chem* 265:11143–50.

Flynn, M. A., D. Weaver-Osterholtz, K. L. Sharpe-Timms, S. Allen, and G. Krause. 1999. Dehydroepiandrosterone replacement in aging humans. *J Clin Endocrinol Metab* 84(5):1527–33.

Frenkel, R. A., C. A. Slaughter, K. Orth et al. 1990. Peroxisomal proliferation and induction of peroxisomal enzymes in mouse and rat liver by dehydroepiandrosterone feeding. *J Steroid Biochem* 35:333–42.

Garaulet, M., F. Pérez-Llamas, T. Fuente, S. Zamora, and F. J. Tébar. 2000. Anthropometric, computed tomography and fat cell data in an obese population: Relationship with insulin, leptin, tumor necrosis factor-alpha, sex hormone-binding globulin and sex hormones. *Eur J Endocrinol* 143:657–66.

Ghafoorunissa, A. Ibrahim, and S. Natarajan. 2005. Substituting dietary linoleic acid with alpha-linolenic acid improves insulin sensitivity in sucrose fed rats. *Biochim Biophys Acta* 1733:67–75.

Gómez, E. F., M. Miyazaki, Y. C. Kim et al. 2002. Molecular differences caused by differentiation of 3T3-L1 preadipocytes in the presence of either dehydroepiandrosterone (DHEA) or 7-oxo-DHEA. *Biochemistry* 41:5473–82.

Grunfeld, C., K. L. Baird, and C. R. Kahn. 1981. Maintenance of 3T3-L1 cells in culture media containing saturated fatty acids decreases insulin binding and insulin action. *Biochem Biophys Res Commun* 103:219–26.

Guerre-Millo, M. 2008. Adiponectin: An update. *Diabetes Metab* 34:12–8.

Haffner, S. M., and R. A. Valdez. 1994. Decreased testosterone and dehydroepiandrosterone sulphate concentrations are associated with increased insulin and glucose concentrations in nondiabetic men. *Metabolism* 43:599–603.

Hernández-Morante, J. J., D. Cerezo, R. M. Cruz, E. Larqué, S. Zamora, and M. Garaulet. 2011. Dehydroepiandrosterone-sulphate modifies human fatty acid composition of different adipose tissue depots. *Obes Surg* 21(1):102–11. doi: 10.1007/s11695-009-0064-8.

Hernández-Morante, J. J., F. Milagro, J. A. Gabaldon, J. A. Martinez, S. Zamora, and M. Garaulet. 2006. Effect of DHEA-sulfate on adiponectin gene expression in adipose tissue from different fat depots in morbidly obese humans. *Eur J Endocrinol* 155(4):593–600.

Hernández-Morante, J. J., F. Pérez-de-Heredia, J. A. Luján, S. Zamora, and M. Garaulet. 2008. Role of DHEA-S on body fat distribution: Gender- and depot-specific stimulation of adipose tissue lipolysis. *Steroids* 73(2):209–15.

Ishizawa, M., T. Ishizuka, K. Kajita et al. 2001. Dehydroepiandrosterone (DHEA) stimulates glucose uptake in rat adipocytes: Activation of phospholipase D. *Comp Biochem Physiol B Biochem Mol Biol* 130(3):359–64.

Ishizuka, T., A. Miura, K. Kajita et al. 2007. Effect of dehydroepiandrosterone on insulin sensitivity in Otsuka Long-Evans Tokushima-fatty rats. *Acta Diabetol* 44(4):219–26.

Jedrzejuk, D., M. Medras, A. Milewicz, and M. Demissie. 2003. Dehydroepiandrosterone replacement in healthy men with age-related decline of DHEA-S: Effects on fat distribution, insulin sensitivity and lipid metabolism. *Aging Male* 6(3):151–6.

Kajita, K., T. Ishizuka, A. Miura, M. Ishizawa, Y. Kanoh, and K. Yasuda. 2000. The role of atypical and conventional PKC in dehydroepiandrosterone-induced glucose uptake and dexamethasone-induced insulin resistance. *Biochem Biophys Res Commun* 277(2):361–7.

Kajita, K., T. Ishizuka, T. Muna et al. 2003. Dehydroepiandrosterone down-regulates the expression of peroxisome proliferator-activated receptor-γ in adipocytes. *Endocrinology* 144:253–9.

Karbowska, J., and Z. Kochan. 2005. Effect of DHEA on endocrine functions of adipose tissue. The involvement of PPARγ. *Biochem Pharmacol* 70:249–57.

Kawano, H., H. Yasue, A. Kitagawa et al. 2003. Dehydroepiandrosterone supplementation improves endothelial function and insulin sensitivity in men. *J Clin Endocrinol Metab* 88(7):3190–5.

Kimura, M., S. Tanaka, Y. Yamada, Y. Kiuchi, T. Yamakawa, and H. Sekihara. 1998. Dehydroepiandrosterone decreases serum tumor necrosis factor-alpha and restores insulin sensitivity: Independent effect from secondary weight reduction in genetically obese Zucker fatty rats. *Endocrinology* 139(7):3249–53.

Kochan, Z., and J. Karbowska. 2004. Dehydroepiandrosterone up-regulates resistin gene expression in white adipose tissue. *Mol Cell Endocrinol* 218:57–64.

Lea-Currie, Y. R., P. Wen, and M. K. McIntosh. 1997. Dehydroepiandrosterone-sulphate (DHEAS) reduces adipocyte hyperplasia associated with feeding rats a high-fat diet. *Int J Obes Relat Metab Disord* 21(11):1058–64.

Lea-Currie, Y. R., P. Wen, and M. K. McIntosh. 1998. Dehydroepiandrosterone reduces proliferation and differentiation of 3T3-L1 preadipocytes. *Biochem Biophys Res Commun* 248:497–504.

Lea-Currie, Y. R., S. M. Wu, and M. K. McIntosh. 1997. Effects of acute administration of dehydroepiandrosterone-sulphate on adipose tissue mass and cellularity in male rats. *Int J Obes Relat Metab Disord* 21:147–54.

Lemieux, C., F. Picard, F. Labrie, D. Richard, and Y. Deshaies. 2003. The estrogen antagonist EM-652 and dehydroepiandrosterone prevent diet- and ovariectomy-induced obesity. *Obes Res* 11:477–90.

Liu, D., and J. S. Dillon. 2002. Dehydroepiandrosterone activates endotelial cell nitric-oxide synthase by a specific plasma membrane receptor coupled to Gαi2,3. *J Biol Chem* 277:21379–88.

Løvås, K., G. Gebre-Medhin, T. S. Trovik et al. 2003. Replacement of dehydroepiandrosterone in adrenal failure: No benefit for subjective health status and sexuality in a 9-month, randomized, parallel group clinical trial. *J Clin Endocrinol Metab* 88(3):1112–8.

Machinal-Quélin, F., M. N. Dieudonne, R. Pecquery, M. C. Leneveu, and Y. Giudicelli. 2002. Direct in vitro effects of androgens and estrogens on ob gene expression and leptin secretion in human adipose tissue. *Endocrine* 18(2):179–84.

Majewska, M. D., S. Demirgoren, C. E. Spivak, and E. D. London. 1990. The neurosteroid dehydroepiandrosterone sulfate is an allosteric antagonist of the GABAA receptor. *Brain Res* 526:143–6.

Marks, P. A., and J. Banks. 1960. Inhibition of mammalian glucose-6-phosphate dehydrogenase by steroids. *Proc Natl Acad Sci U S A* 46(4):447–52.

Marwah, A., F. E. Gomez, P. Marwah, J. M. Ntambi, B. G. Fox, and H. Lardy. 2006. Redox reactions of dehydroepiandrosterone and its metabolites in differentiating 3T3-L1 adipocytes: A liquid chromatographic-mass spectrometric study. *Arch Biochem Biophys* 456(1):1–7.

Mauriège, P., C. Martel, D. Langin et al. 2003. Chronic effects of dehydroepiandrosterone on rat adipose tissue metabolism. *Metabolism* 52(3):264–72.

McCormick, K. L., X. Wang, and G. J. Mick. 2006. Evidence that the 11β-hydroxysteroid dehydrogenase (11β-HSD1) is regulated by pentose pathway flux. Studies in rat adipocytes and microsomes. *J Biol Chem* 281:341–7.

McIntosh, M., D. Hausman, R. Martin, and G. Hausman. 1998. Dehydroepiandrosterone attenuates preadipocyte growth in primary cultures of stromal-vascular cells. *Am J Physiol Endocrinol Metab* 275:E285–93.

Meikle, A. W., R. W. Dorchuck, B. A. Araneo et al. 1992. The presence of a dehydroepiandrosterone-specific receptor binding complex in murine T cells. *J Steroid Biochem Mol Biol* 42:293–304.

Morales, A. J., J. J. Nolan, J. C. Nelson, and S. S. Yen. 1994. Effects of replacement dose of dehydroepiandrosterone in men and women of advancing age. *J Clin Endocrinol Metab* 78(6):1360–7.

Mortola, J. F., and S. S. Yen. 1990. The effects of oral dehydroepiandrosterone on endocrine-metabolic parameters in postmenopausal women. *J Clin Endocrinol Metab* 71(3):696–704.

Morton, N. M., J. M. Paterson, H. Masuzaki et al. 2004. Novel adipose tissue-mediated resistance to diet-induced visceral obesity in 11 beta-hydroxysteroid dehydrogenase type 1-deficient mice. *Diabetes* 53:931–8.

Muoio, D. M., J. M. Way, C. J. Tanner et al. 2002. Peroxisome proliferator-activated receptor-alpha regulates fatty acid utilization in primary human skeletal muscle cells. *Diabetes* 51:901–9.

Nakashima, N., M. Haji, Y. Sakai, Y. Ono, F. Umeda, and H. Nawata. 1995. Effect of dehydroepiandrosterone on glucose uptake in cultured human fibroblasts. *Metabolism* 44:543–8.

Navar, D., D. Saulis, C. Corll, F. Svec, and J. R. Porter. 2006. Dehydroepiandrosterone (DHEA) blocks the increase in food intake caused by neuropeptide Y (NPY) in the Zucker rat. *Nutr Neurosci* 9(5–6):225–32.

Nestler, J. E., C. O. Barlascini, J. N. Clore, and W. G. Blackard. 1988. Dehydroepiandrosterone reduces serum low density lipoprotein levels and body fat but does not alter insulin sensitivity in normal men. *J Clin Endocrinol Metab* 66(1):57–61.

Nishio, E., K. Tomiyama, H. Nakata, and Y. Watanabe. 1996. 3-Hydroxy-3-methylglutaryl coenzyme A reductase inhibitor impairs cell differentiation in cultured adipogenic cells (3T3-L1). *Eur J Pharmacol* 301(1–3):203–6.

Park, J., S. S. Choe, A. H. Choi et al. 2006. Increase in glucose-6-phosphate dehydrogenase in adipocytes stimulates oxidative stress and inflammatory signals. *Diabetes* 55(11):2939–49.

Pascale, R. M., M. M. Simile, M. R. De Miglio et al. 1995. Inhibition of 3-hydroxy-3-methylglutaryl-CoA reductase activity and gene expression by dehydroepiandrosterone in preneoplastic liver nodules. *Carcinogenesis* 16(7):1537–42.

Pedersen, S. B., J. M. Bruun, F. Hube, K. Kristensen, H. Hauner, and B. Richelsen. 2001. Demonstration of estrogen receptor subtypes alpha and beta in human adipose tissue: Influences of adipose cell differentiation and fat depot localization. *Mol Cell Endocrinol* 182:27–37.

Percheron, G., J. Y. Hogrel, S. Denot-Ledunois et al. 2003. Effect of 1-year oral administration of dehydroepiandrosterone to 60- to 80-year-old individuals on muscle function and cross-sectional area: A double-blind placebo-controlled trial. *Arch Intern Med* 163(6):720–7.

Pérez de Heredia, F., D. Cerezo, S. Zamora, and M. Garaulet. 2007. Effect of dehydroepiandrosterone (DHEA) on protein and fat digestibility, body protein and muscular composition in high-fat-diet-fed old rats. *Br J Nutr* 97:464–70.

Pérez de Heredia, F., E. Larqué, S. Zamora, and M. Garaulet. 2009. DHEA modifies rat fatty acid composition of serum and different adipose tissue depots towards an insulin-sensitizing profile. *J Endocrinol* 201:67–74.

Pérez de Heredia, F., J. Sánchez, T. Priego et al. 2008. Adiponectin is involved in the protective effect of DHEA against metabolic risk in aged rats. *Steroids* 73:1128–36.

Perrini, S., A. Natalicchio, L. Laviola et al. 2004. Dehydroepiandrosterone stimulates glucose uptake in human and murine adipocytes by inducing GLUT1 and GLUT4 translocation to the plasma membrane. *Diabetes* 53:41–52.

Pham, J., J. Porter, D. Svec, C. Eiswirth, and F. Svec. 2000. The effect of dehydroepiandrosterone on Zucker rats selected for fat food preference. *Physiol Behav* 70:431–41.

Piñeiro, V., X. Casabiell, R. Peinó et al. 1999. Dihydrotestosterone, stanozolol, androstenedione and dehydroepiandrosterone sulphate inhibit leptin secretion in female but not in male samples of omental adipose tissue in vitro: Lack of effect of testosterone. *J Endocrinol* 160(3):425–32.

Porter, J., and F. Svec. 1995. DHEA diminishes fat food intake in lean and obese Zucker rats. *Ann NY Acad Sci* 774:329–31.

Powell, D. J., S. Turban, A. Gray et al. 2004. Intracellular ceramide synthesis and protein kinase Czeta activation play an essential role in palmitate-induced insulin resistance in rat L6 skeletal muscle cells. *Biochem J* 382:619–29.

Richards, R. J., J. R. Porter, and F. Svec. 2000. Serum leptin, lipids, free fatty acids and fat-pads in long-term dehydroepiandrosterone-treated Zucker rats. *Proc Soc Exp Biol Med* 223:258–62.

Riediger, N. D., R. Othman, E. Fitz, G. N. Pierce, M. Suh, and M. H. Moghadasian. 2008. Low n-6:n-3 fatty acid ratio, with fish- or flaxseed oil, in a high fat diet improves plasma lipids and beneficially alters tissue fatty acid composition in mice. *Eur J Nutr* 47:153–60.

Ryu, J. W., M. S. Kim, C. H. Kim et al. 2003. DHEA administration increases brown fat uncoupling protein 1 levels in obese OLETF rats. *Biochem Biophys Res Commun* 303:726–31.

Sánchez, J., F. Pérez-Heredia, T. Priego et al. 2008. Dehydroepiandrosterone prevents age-associated alterations, increasing insulin sensitivity. *J Nutr Biochem* 19:809–18.

Saraç, F., S. Yildiz, F. Saygili et al. 2006. Dehydroepiandrosterone and human adipose tissue. *J Endocrinol Invest* 29(5):393–8.

Steppan, C. M., S. T. Bailey, S. Bhat et al. 2001. The hormone resistin links obesity to diabetes. *Nature* 409:307–12.

Tagliaferro, A. R., J. R. Davis, S. Truchon, and N. Van Hamont. 1986. Effects of dehydroepiandrosterone acetate on metabolism, body weight and composition of male and female rats. *J Nutr* 116:1977–83.

Tagliaferro, A. R., A. M. Ronan, J. Payne, L. D. Meeker, and S. Tse. 1995. Increased lipolysis to β-adrenergic stimulation after dehydroepiandrosterone treatment in rats. *Am J Physiol* 268 (Regulatory Integrative Comp Physiol 37):R1374–80.

Takahashi, Y., and T. Ide. 2000. Dietary n-3 fatty acids affect mRNA level of brown adipose tissue uncoupling protein 1, and white adipose tissue leptin and glucose transporter 4 in the rat. *Brit J Nutr* 84:175–84.

Tan, B. K., J. Chen, H. Lehnert, R. Kennedy, and H. S. Randeva. 2007. Raised serum, adipocyte, and adipose tissue retinol-binding protein 4 in overweight women with polycystic ovary syndrome: Effects of gonadal and adrenal steroids. *J Clin Endocrinol Metab* 92(7):2764–72.

Trayhurn, P., and I. S. Wood. 2004. Adipokines: Inflammation and the pleiotropic role of white adipose tissue. *Br J Nutr* 92:347–55.

Tsuji, K., D. Furutama, M. Tagami, and N. Ohsawa. 1999. Specific binding and effects of dehydroepiandrosterone sulphate (DHEA-S) on skeletal muscle. Possible implications for DHEA-S replacement therapy in patients with miotonic dystrophy. *Life Sci* 65:17–26.

Villareal, D. T., and J. O. Holloszy. 2004. Effect of DHEA on abdominal fat and insulin action in elderly women and men: A randomized controlled trial. *JAMA* 292(18):2243–8.

Villareal, D. T., J. O. Holloszy, and W. M. Kohrt. 2000. Effects of DHEA replacement on bone mineral density and body composition in elderly women and men. *Clin Endocrinol (Oxf)* 53(5):561–8.

Villena, J. A., K. H. Kim, and H. S. Sul. 2002. Pref-1 and ADSF/resistin: Two secreted factors inhibiting adipose tissue development. *Horm Metab Res* 34:664–70.

Vogiatzi, M. G., M. A. Boeck, E. Vlachopapadopoulou, R. el-Rashid, and M. I. New. 1996. Dehydroepiandrosterone in morbidly obese adolescents: Effects on weight, body composition, lipids, and insulin resistance. *Metabolism* 45(8):1011–5.

Wajchenberg, B. L. 2000. Subcutaneous and visceral adipose tissue: Their relation to the metabolic syndrome. *Endocr Rev* 21(6):697–738.

Wang, Y., Y. Sun, and H. Qiu. 2004. Expression of resistin mRNA in adipose tissue of rat model with polycystic ovarian syndrome and its implication. *J Huazhong Univ Sci Technolog Med Sci* 24(6):621–4.

Webb, S. J., T. E. Geoghegan, R. A. Prough, and K. K. Michael Miller. 2006. The biological actions of dehydroepiandrosterone involve multiple receptors. *Drug Metab Rev* 38(1–2):89–116.

Welle, S., R. Jozefowicz, and M. Statt. 1990. Failure of dehydroepiandrosterone to influence energy and protein metabolism in humans. *J Clin Endocrinol Metab* 71(5):1259–64.

Widstrom, R. L., and J. S. Dillon. 2004. Is there a receptor for dehydroepiandrosterone or dehydroepiandrosterone sulfate? *Semin Reprod Med* 22:289–96.

Xu, A., K. W. Chan, R. L. Hoo et al. 2005. Testosterone selectively reduces the high molecular weight form of adiponectin by inhibiting its secretion from adipocytes. *J Biol Chem* 280:18073–80.
Yamada, J., M. Sakuma, and T. Suga. 1992. Induction of peroxisomal B-oxidation enzymes by dehydroepiandrosterone and its sulfate in primary cultures of rat hepatocytes. *Biochim Biophys Acta* 1137:231–6.
Zhang, Y. Y., R. Proenca, M. Maffei, M. Barone, L. Leopold, and J. M. Friedman. 1994. Positional cloning of the mouse obese gene and its human homolog. *Nature* 372:425–32.

26 Dehydroepiandrosterone and Cell Differentiation

Alexander W. Krug

CONTENTS

Dehydroepiandrosterone (DHEA), produced by the adrenal glands and the brain, is the most abundant hormone in the human body and one of the least understood. In contrast to other adrenal steroids or gonad-derived estrogens and androgens, circulating DHEA levels follow a trend during development and aging. DHEA is a major product of the adrenal primordium, and DHEA and dehydroepiandrosterone sulfate (DHEAS) levels are very high in fetal adrenals. Directly after birth, DHEA levels decrease and remain low for the first five years of life, followed by a rise during adrenarche at 5–6 years of age, reaching a peak in early adulthood before declining again (Rainey et al. 2002). These observations suggest a significant role for DHEA in development and aging, affecting cell proliferation and differentiation processes. However, the mechanisms regulating these DHEA secretion patterns, a specific DHEA receptor, and the exact physiological and pathophysiological roles of DHEA have not been defined yet. This chapter provides an overview of the effects of DHEA on cell differentiation.

DHEA AFFECTS NEURONAL AND NEUROENDOCRINE CELL DIFFERENTIATION

Neurons and glia cells express the necessary enzymes to synthesize DHEA, and several studies associate decreased neurosteroid levels with neuronal degeneration and dysfunction in diseases such as major depression, Alzheimer's, and Parkinson's. Besides affecting neuronal cell survival and protection, DHEA is known to affect neurogenesis (Charalampopoulos et al. 2008). A limited number of studies have examined the effects of DHEA on neuronal cell differentiation. Shiri et al. (2009) demonstrated that neuronal-competent bone marrow mesenchymal stem cells acquire features of neurons after treatment with DHEA *in vitro*. Using mouse carcinoma cell and human embryonic stem cell-derived neural progenitor cells, the same group demonstrated that DHEA reduced apoptosis and activated neurogenesis, significantly increasing dopaminergic cells. Compagnone and Mellon (1998) showed that DHEA, in concentrations normally found in the brain, affects parameters of neuronal differentiation in primary cells of mouse embryonic neocortical neurons via activation of *N*-methyl-D-aspartate (NMDA) receptors, suggesting a role for DHEA in brain development. Similarly, another *in vitro* study using human neural stem cells derived from the fetal cortex showed that DHEA, but not DHEA precursor molecules, promotes neurogenesis via NMDA receptors in an epidermal growth factor (EGF)/leukemia inhibitory factor-dependent way, indicating that DHEA is involved in the division and neuronal differentiation of human neural stem cells (Suzuki et al. 2004). In rodents, DHEA administration increased the number of newly formed neurons in the dentate gyrus of the hippocampus of male rats, antagonizing the suppressive effects of glucocorticoids on neurogenesis. In the same study, DHEA also affected neurogenesis in a small number of aged (12 months) animals, suggesting a potential role for DHEA in neuronal cell renewal and neurogenesis in aged animals despite declining DHEA levels (Karishma and Herbert 2002).

DHEA also seems to play an important role in adrenal development and function. Patients with 21-hydroxylase (21-OH) deficiency and hyperandrogenism, a pathological state that is associated with altered DHEA levels, also show severe chromaffin cell dysfunction. Chromaffin cells are catecholamine-producing cells of the adrenal medulla (Merke et al. 2002; Merke et al. 2000). Restoring 21-OH activity by gene transfer in the 21-OH knockout mouse not only reverses androgen excess but also normalizes functional and morphological adrenomedullary alterations, demonstrating that androgens, such as DHEA, are important for normal adrenomedullary cell development and function (Tajima et al. 1999). We demonstrated age-dependent sensitivity of juvenile and adult primary bovine chromaffin cells to DHEA. DHEA reduced the proliferation in juvenile and adult cells, whereas DHEAS exclusively increased the proliferation provoked by EGF in adult cell cultures, indicating differential effects of DHEA in cells of different ages (Sicard et al. 2007).

Taken together, the above observations from *in vitro*, animal, and human studies suggest that DHEA is crucially involved in neuronal and neuroendocrine cell differentiation and function, and may have implications in possible future treatments of neuronal diseases such as Parkinson's and Alzheimer's (Azizi et al. 2009). In different rodent animal models of Parkinson's disease, DHEA administration can prevent dopamine depletion, underscoring the importance of DHEA for protection of differentiated neuronal cells (D'Astous et al. 2003; Tomas-Camardiel et al. 2002). However, it is still far from being proven which factors are exactly involved in the various steps of neuronal progenitor cell differentiation, and what exact role DHEA might play in this orchestrated interplay of various factors.

MOLECULAR MECHANISMS OF DHEA ACTION ON CELL DIFFERENTIATION

The exact cellular mechanisms of DHEA on cell differentiation are only poorly understood. Using rat pheochromocytoma PC12 cells, a model of proliferation and mitotic competent chromaffin cells, we demonstrated that nerve growth factor (NGF)-promoted cell survival in serum-deprived cells, and neuronal differentiation of the cells was affected by DHEA (Ziegler et al. 2006). In a dose-dependent way, DHEA reduced neuronal markers, such as SNAP-25 and VAMP-2, in NGF-stimulated cells. This reduction in neuronal cell parameters was associated with a shift of the cells to a more endocrine phenotype, as shown by increased chromogranin A expression, and elevated catecholamine release of DHEA-treated cells (Ziegler et al. 2008). Another group has shown that DHEA directly stimulates tyrosine hydroxylase expression in PC12 cells, a crucial factor of the neuroendocrine phenotype mediating catecholamine secretion (Charalampopoulos et al. 2005). On a molecular level, DHEA decreased NGF-induced extracellular signal–regulated kinase (ERK1/2) phosphorylation, an effect that was even more pronounced by the DHEA-sulfate ester, DHEAS. Interestingly, in the absence of NGF, the neurosteroids did not affect ERK1/2 activation. This shows that DHEA and DHEAS can modify the activation of a crucial signaling mechanism involved in neuroendocrine differentiation of chromaffin cells (Krug et al. 2009), an effect previously believed to be mediated exclusively by glucocorticoids (Huber et al. 2002). Our observations are in accordance with findings from glucocorticoid receptor–deficient mice, which develop normal numbers of chromaffin cells (Finotto et al. 1999).

Studies investigating the effects of DHEA on cell proliferation are limited. Several studies have shown that DHEA affects neuronal stem and progenitor cell differentiation, and it has been proposed that age- and disease-associated decline in DHEA levels may represent a causative factor in diseases such as Alzheimer's and Parkinson's. DHEA also seems to be involved in adrenal disease and development. In contrast to other steroids, DHEA levels follow a characteristic pattern during prenatal development and throughout life. DHEA is shown to impact adrenomedullary chromaffin cell proliferation and differentiation, affecting catecholamine secretion. However, the exact functions of DHEA on cell differentiation in most tissues and its physiological and pathological significance are far from being understood. Future research is needed to further elucidate the role of DHEA in health and disease.

REFERENCES

Azizi, H., N. Z. Mehrjardi, E. Shahbazi, K. Hemmesi, M. K. Bahmani, and H. Baharvand. 2009. Dehydroepiandrosterone stimulates neurogenesis in mouse embryonal carcinoma cell- and human embryonic stem cell-derived neural progenitors and induces dopaminergic neurons. *Stem Cells Dev* 19(6):809–18.

Charalampopoulos, I., E. Dermitzaki, L. Vardouli et al. 2005. Dehydroepiandrosterone sulfate and allopregnanolone directly stimulate catecholamine production via induction of tyrosine hydroxylase and secretion by affecting actin polymerization. *Endocrinology* 146:3309–18.

Charalampopoulos, I., E. Remboutsika, A. N. Margioris, and A. Gravanis. 2008. Neurosteroids as modulators of neurogenesis and neuronal survival. *Trends Endocrinol Metab* 19:300–7.

Compagnone, N. A., and S. H. Mellon. 1998. Dehydroepiandrosterone: A potential signalling molecule for neocortical organization during development. *Proc Natl Acad Sci U S A* 95:4678–83.

D'Astous, M., M. Morissette, B. Tanguay, S. Callier, and P. T. Di. 2003. Dehydroepiandrosterone (DHEA) such as 17beta-estradiol prevents MPTP-induced dopamine depletion in mice. *Synapse* 47:10–4.

Finotto, S., K. Krieglstein, A. Schober et al. 1999. Analysis of mice carrying targeted mutations of the glucocorticoid receptor gene argues against an essential role of glucocorticoid signalling for generating adrenal chromaffin cells. *Development* 126:2935–44.

Huber, K., S. Combs, U. Ernsberger, C. Kalcheim, and K. Unsicker. 2002. Generation of neuroendocrine chromaffin cells from sympathoadrenal progenitors: Beyond the glucocorticoid hypothesis. *Ann N Y Acad Sci* 971:554–9.

Karishma, K. K., and J. Herbert. 2002. Dehydroepiandrosterone (DHEA) stimulates neurogenesis in the hippocampus of the rat, promotes survival of newly formed neurons and prevents corticosterone-induced suppression. *Eur J Neurosci* 16:445–53.

Krug, A. W., H. Langbein, C. G. Ziegler, S. R. Bornstein, G. Eisenhofer, and M. Ehrhart-Bornstein. 2009. Dehydroepiandrosterone-sulphate (DHEA-S) promotes neuroendocrine differentiation of chromaffin pheochromocytoma PC12 cells. *Mol Cell Endocrinol* 300:126–31.

Merke, D. P., S. R. Bornstein, N. A. Avila, and G. P. Chrousos. 2002. NIH conference. Future directions in the study and management of congenital adrenal hyperplasia due to 21-hydroxylase deficiency. *Ann Intern Med* 136:320–34.

Merke, D. P., G. P. Chrousos, G. Eisenhofer et al. 2000. Adrenomedullary dysplasia and hypofunction in patients with classic 21-hydroxylase deficiency. *N Engl J Med* 343:1362–8.

Rainey, W. E., B. R. Carr, H. Sasano, T. Suzuki, and J. I. Mason. 2002. Dissecting human adrenal androgen production. *Trends Endocrinol Metab* 13:234–9.

Shiri, E. H., N. Z. Mehrjardi, M. Tavallaei, S. K. Ashtiani, and H. Baharvand. 2009. Neurogenic and mitotic effects of dehydroepiandrosterone on neuronal-competent marrow mesenchymal stem cells. *Int J Dev Biol* 53:579–84.

Sicard, F., M. Ehrhart-Bornstein, D. Corbeil et al. 2007. Age-dependent regulation of chromaffin cell proliferation by growth factors, dehydroepiandrosterone (DHEA), and DHEA sulfate. *Proc Natl Acad Sci U S A* 104:2007–12.

Suzuki, M., L. S. Wright, P. Marwah, H. A. Lardy, and C. N. Svendsen. 2004. Mitotic and neurogenic effects of dehydroepiandrosterone (DHEA) on human neural stem cell cultures derived from the fetal cortex. *Proc Natl Acad Sci U S A* 101:3202–7.

Tajima, T., T. Okada, X. M. Ma, W. Ramsey, S. Bornstein, and G. Aguilera. 1999. Restoration of adrenal steroidogenesis by adenovirus-mediated transfer of human cytochromeP450 21-hydroxylase into the adrenal gland of 21-hydroxylase-deficient mice. *Gene Ther* 6:1898–903.

Tomas-Camardiel, M., M. C. Sanchez-Hidalgo, M. J. Sanchez del Pino, A. Navarro, A. Machado, and J. Cano. 2002. Comparative study of the neuroprotective effect of dehydroepiandrosterone and 17beta-estradiol against 1-methyl-4-phenylpyridium toxicity on rat striatum. *Neuroscience* 109:569–84.

Ziegler, C. G., F. Sicard, P. Lattke, S. R. Bornstein, M. Ehrhart-Bornstein, and A. W. Krug. 2008. Dehydroepiandrosterone induces a neuroendocrine phenotype in nerve growth factor-stimulated chromaffin pheochromocytoma PC12 cells. *Endocrinology* 149:320–8.

Ziegler, C. G., F. Sicard, S. Sperber, M. Ehrhart-Bornstein, S. R. Bornstein, and A. W. Krug. 2006. DHEA reduces NGF-mediated cell survival in serum-deprived PC12 cells. *Ann N Y Acad Sci* 1073:306–11.

27 DHEA, Oxidative Stress, and Akt

Maria Helena Vianna Metello Jacob, Alex Sander da Rosa Araújo, Maria Flavia M. Ribeiro, and Adriane Belló-Klein

CONTENTS

INTRODUCTION

Dehydroepiandrosterone (DHEA), the most abundant steroid hormone in circulation, is widely consumed as a drug for a multiple range of therapeutic actions, including hormonal replacement (Legrain and Girard 2003), treatment of weight loss (Kurzman, Macewen, and Haffa 1990), and the improvement of aging-related diseases (Wolkowitz et al. 2003). Another issue to consider is that DHEA is also used at pharmacological or supraphysiological doses by fitness apprentices and athletes for muscle-building purposes (Labrie et al. 2006). Observations in human population trials, animal models, and *in vitro* findings support the potential utility of DHEA as a possible therapeutic intervention. Yet, DHEA was considered a "fountain of youth" hormone worldwide (Baulieu 1996). A significant clinical interest in this steroid is based on many observations, including the decline of DHEA production after early adulthood and evidence in literature showing changes in DHEA levels associated with multiple pathologies. However, a great controversy on this subject still remains, and no sufficient data are available in literature to support its secure recommendation.

DHEA may act both directly or through its metabolites (including androstenediol and androstenedione), which can undergo further conversion to produce testosterone and estradiol (Mo, Lu, and Simon 2006; Maninger et al. 2009). DHEA can act as a modulator of neurotransmitter receptors, such as γ-aminobutyric acid (GABAA), *N*-methyl-D-aspartic acid (NMDA), and sigma-1 receptors (Baulieu and Robel 1998; Schumacher et al. 2000). DHEA acts genomically through the androgen receptor in addition to its well-known effects on cell surface receptors (Mo, Lu, and Simon 2006). DHEA's protective effects against oxidative stress could be through both genomic and nongenomic pathways (Simoncini and Genazzani 2003). However, a better and deeper understanding of its mechanisms of action—either over oxidative stress and/or over cell signaling modulation—is crucial to elucidate the beneficial power of DHEA because it is widely used in humans with controversial effects.

OXIDATIVE STRESS AND DHEA

The free radical theory (Harman 1956) proposes that physiological iron and other metals cause reactive oxygen species (ROS) formation, promoting oxidative damage. These reactive species may initiate a great deal of deleterious peroxidative reactions in genetic cell apparatus, cell membranes, proteins, and lipids. In order to protect cells, defense mechanisms use enzymatic and nonenzymatic antioxidants avoiding and/or retarding cellular damage arising from oxidative stress—an imbalance between the production of free radicals and antioxidant systems (Dröge 2002). Oxidative stress plays an important role in the pathogenesis of many diseases, such as cancer, diabetes mellitus, and cardiovascular and inflammatory bowel diseases (Sun 1990; Grisham 1994; Vendemiale, Grattagliano, and Altomare 1999; Valko et al. 2007).

Several steroids have been considered antioxidants, including DHEA (Yildirim et al. 2003; Yoshimura 2006). There can be dual effects of exogenous DHEA administration, antioxidant and pro-oxidant, probably reflecting differences in dosages and time protocols regarding the experimental models used (Pelissier et al. 2006; Aksoy et al. 2004).

An antioxidant activity has been recognized for DHEA. Reducing the oxidant burden in asbestos-induced alveolitis of the lower respiratory tract, Rom and Harkin (1991) proposed a therapeutic role for DHEA. Also, there are results showing a protective DHEA action against lipid peroxidation (LPO) and necrosis in the liver with *in vivo* and *in vitro* models (Aragno et al. 1993, 1994). Tamagno et al. (1998) found that DHEA or its breakdown products act as metal chelators. These results indicate that the DHEA protection role is not given by the steroid itself, but that DHEA or its metabolites are able to modulate in the Haber-Weiss reaction, generating less ROS by chelating metals. Notwithstanding, these authors also suggest that even in very high concentrations (10, 50, or 100 μmol/L), DHEA addition to microsomes does not exert a protective effect on LPO (Tamagno et al. 1998).

Gallo et al. (1999), in an *in vitro* experiment (Chang liver cells), pointed out that slightly higher concentrations of DHEA (0.1 μmol/L)—above those found in human tissues—protected cells from LPO induced by oxidative stress (cells received cumene hydroperoxide 0.5 mmol/L, a pro-oxidant stimulus). However, at pharmacological doses (10–50 μmol/L), DHEA displayed a pro-oxidant activity.

DHEA can be a pro-oxidant agent per se. Goldfarb, McIntosh, and Boyer (1996) induced oxidative stress by DHEA or exercise to determine whether vitamin E could protect male Sprague-Dawley rat hearts. Catalase (CAT) activity was increased by DHEA at rest, and the results have shown that aerobic exercise or DHEA could be mild oxidative stressors to the heart.

When administered intraperitoneally in short term (6 or 24 hours), distinct doses (1, 10, and 50 mg/kg) of DHEA promoted a pro-oxidant effect in an *in vivo* model with healthy Wistar rat hearts (Jacob et al. 2008).

Investigating DHEA chronic (10 mg/kg, subcutaneously, for 5 weeks) effects over oxidative stress markers in erythrocytes of 3-, 13-, and 18-month old Wistar rats, Jacob et al. (2010) demonstrated that this xenobiotic exerted pro-oxidant effects in all ages studied, especially in 13-month-old rats. DHEA promoted a marked increase in LPO over the 13-month-old group, and this effect might be a consequence of enhanced ROS, which, in turn, would be a stimulus or a source to higher SOD, CAT, and GST activities in this group, demonstrating an adaptative change of the antioxidant system.

DHEA has been shown to exert its pro-oxidant effects via the activation of peroxisome proliferator-activated receptor alpha (PPAR-α; Hayashi et al. 1994; Zhou and Waxman 1998). PPARs play a role in transcription control of many cellular processes such as inflammation, glucose homeostasis, cell differentiation, extracellular matrix remodeling, and lipid metabolism (Duez, Fruchart, and Staels 2001). PPARs are greatly expressed in the heart, liver, and kidneys—tissues that possess high β-oxidation rates. Mastrocola et al. (2003) have shown that at high doses (50 and 100 mg, 7 days) DHEA acts as a PPARs regulator. They found that DHEA enhances β-oxidation, demonstrated by an increase in cytochrome P450 4A content and acyl-CoA-oxidase activity. Indeed, Morgan (2001) found that P450 enzyme activation contributes to elevated ROS levels—superoxide anion and hydrogen peroxide. Nevertheless, when administered at lower doses (4 mg/day) chronically (21 days), DHEA was unable to affect PPARs activation or β-oxidation. Moreover, DHEA had an important protective effect on ADP/Fe^{2+}-induced LPO (Mastrocola et al. 2003), acting, by this way, as an antioxidant. These authors demonstrated that at lower levels (physiological concentration), this steroid may affect the activation of redox-sensitive transcriptional factors and exert a beneficial influence on responsiveness to oxidative damage. It means that a possible mechanism of

DHEA action is the modulation of oxidative imbalance. Products of normal cellular metabolism, ROS concentration, and redox imbalance are well recognized for playing a crucial role as modulators of a number of cellular signaling systems (Valko et al. 2007).

Jones (2006) has suggested a more modern concept of oxidative stress as a disruption of redox signaling and control. In fact, ROS function as important intra- and intercellular second messengers to modulate downstream signaling molecules, such as ion channels, protein tyrosine phosphatases and kinases, transcription factors, and mitogen-activated protein kinases (MAPK). The different chemical properties of individual ROS have distinct implications on their role in cellular signaling. Thus, ROS, depending on their chemical structure, concentration, and site of production may activate different signaling pathways, which may lead to different cell phenotypes (Paravicini and Touyz 2006).

AKT—A REDOX SENSITIVE SIGNALING PROTEIN

One important redox-sensitive signaling pathway is Akt (Valko et al. 2007). Akt—also called PKB—is a central player in the signal transduction pathways activated in response to growth factors or insulin and is believed to contribute to diverse cellular functions including cell growth, nutrient metabolism, and apoptosis. Akt inhibits the apoptotic cycle and stimulates growth pathways (Hanada, Feng, and Hemmings 2004). The involvement of Akt in cell survival is a complex process that demands a long cascade of intracellular events (Mullonkal and Toledo-Pereira 2007).

There is evidence that DHEA can act as a survival factor in endothelial cells by triggering the Galphai-PI3K/Akt-Bcl-2 pathway to protect bovine aortic endothelial cells against apoptosis (Liu et al. 2007). DHEA exerted a vascular protective effect resulting in a rapid and dose-dependent phosphorylation of Akt. To explore the antiproliferative mechanisms of DHEA in human hepatoblastoma cells, Jiang et al. (2005) studied the induction of apoptosis through the inhibition of the Akt signaling pathway after 24 hours of incubation with DHEA (100 and 200 μmol/L). When activated, Akt phosphorylates a range of intracellular substrates that regulate growth, metabolism, and survival (DeBosch et al. 2006). Janner et al. (2010) studied DHEA effects on Akt signaling modulation in the central nervous system of young and aged healthy rats. The results show that acute (50 mg/kg) and chronic (10 mg/kg) DHEA injections modulate total p-Akt levels. Such effects were dose and time dependent and corroborate the idea that Akt is a protein kinase, which is related to be DHEA modulated. After 24 hours, the 50 mg/kg DHEA treatment to healthy Wistar rat hearts resulted in an elevation (47%) in protein levels of the p-Akt/Akt ratio as compared to control (Jacob et al. 2008). Moreover, a positive correlation between p-Akt/Akt ratio and HNE-Michael adducts (LPO products) was found, suggesting that membrane phospholipids oxidation would trigger the downstream signaling net (Jacob et al. 2008). Corroborating this finding, Wang et al. (2000) have shown that ROS, such as hydrogen peroxide, are stimulants for Akt phosphorylation, which generates a signal to protect cells from oxidative stress. Jacob et al. (2009) evaluated DHEA chronic effects over myocardial Akt protein expression associated with oxidative stress markers during aging in Wistar rats (3 and 18 months). The major outcome of this study was to demonstrate that p-Akt expression was enhanced in the groups treated with DHEA compared with controls, independent of age. On the other hand, Akt ratio (p-Akt/Akt) was increased only in the treated 18-month-old group compared to the control group of the same age. The authors suggest that Akt pathway activation by DHEA could be associated with a redox-sensitive modulation.

DHEA can play an important role in glucose utilization in peripheral tissues. Some studies have already shown that androgens influence glucose metabolism by accelerating the induction of Akt phosphorylation via the activation of PI3-kinase (Ishizuka et al. 1999; Kang et al. 2004). Jahn et al. (2010) have studied the chronic DHEA effect (10 mg/kg once a week for 5 weeks) over skeletal muscle of diabetic rats. DHEA proved to decrease blood glucose, and it may be beneficial; however, the decrease in Akt expression caused by this xenobiotic displayed an environment favorable to redox imbalance.

CONCLUSION

The overall data suggest that DHEA is likely to exert beneficial and/or toxic actions on cells not through its potential as an anti- or pro-oxidant, but rather through its modulation of signaling cascades. DHEA interactions with intracellular signaling could have relevant outcomes and must be related to aging, cell type, and stimulus, and must be disease focused (Jacob et al. 2009). Further studies leading to a deeper understanding of the net effect of DHEA on oxidative stress parameters and its relation to Akt would contribute to reaching conclusions about the safety and efficacy of DHEA replacement and use.

REFERENCES

Aksoy, Y., T. Yapanoglu, H. Aksou, and A. K. Yildirim. 2004. The effect of dehydroepiandrosterone on renal ischemia-reperfusion-induced oxidative stress in rabbits. *Urol Res* 32(3):93–6.

Aragno, M., E. Tamagno, G. Boccuzzi, E. Brignardello, E. Chiarpotto, A. Pizzini et al. 1993. Dehydroepiandrosterone pretreatment protects rats against prooxidant and necrogenic effects of carbon tetrachloride. *Biochem Pharmacol* 46:1689–94.

Aragno, M., E. Tamagno, G. Poli, G. Boccuzzi, E. Brignardello, and O. Danni. 1994. Prevention of carbon tetrachloride-induced lipid peroxidation in liver microsomes from dehydroepiandrosterone-pretreated rats. *Free Radic Res* 21:427–35.

Baulieu, E. E. 1996. Dehydroepiandrosterone: A fountain of youth? *J Clin Endocrinol Metab* 81(9):3147–51.

Baulieu, E. E., and P. Robel. 1998. Dehydroepiandrosterone (DHEA) and dehydroepiandrosterone sulfate (DHEAS) as neuroactive neurosteroids. *Proc Natl Acad Sci* 95:1089–91.

DeBosch, B., I. Treskov, T. S. Lupu, C. Weinheimer, A. Kovacs, M. Courtois et al. 2006. Akt1 is required for physiological cardiac growth. *Circulation* 113(17):2097–104.

Dröge, W. 2002. The plasma redox state and ageing. *Ageing Res Rev* 1(2):257–78.

Duez, H., J. C. Fruchart, and B. Staels. 2001. PPARs in inflammation, atherosclerosis and thrombosis. *J Cardiovasc Risk* 8:187–94.

Gallo, M., M. Aragno, V. Gatto, E. Tamagno, E. Brignardello, R. Manti et al. 1999. Protective effect of dehydroepiandrosterone against lipid peroxidation in a human liver cell line. *Eur J Endocrinol* 141:35–9.

Goldfarb, A. H., M. K. McIntosh, and B. T. Boyer. 1996. Vitamin E attenuates myocardial oxidative stress induced by DHEA in rested and exercised rats. *J Appl Physiol* 80(2):486–90.

Grisham, M. B. 1994. Oxidants and free radicals in inflammatory bowel disease. *Lancet* 344:859–61.

Hanada, M., J. Feng, and B. A. Hemmings. 2004. Strucutre, regulation and function of PKB/AKT—a major therapeutic target. *Biochim Biophys Acta* 1697:3–16.

Harman, D. 1956. Aging: A theory based on free radicals and radiation biology. *J Gerontol* 11:298–300.

Hayashi, F., H. Tamura, J. Yamada, H. Kasai, and T. Suga. 1994. Characteristics of the hepatocarcinogenesis caused by DHEA, as a peroxisome proliferator, in male F-344 rats. *Carcinogenesis* 15:2215–9.

Ishizuka, T., K. Kajita, A. Miura, M. Ishizawa, Y. Kanoh, S. Itaya et al. 1999. DHEA improves glucose uptake via activations of protein kinase C and phosphatidylinositol 3-kinase. *Am J Physiol* 276:E196–204.

Jacob, M. H. V. M., D. R. Janner, A. Belló-Klein, S. F. Llesuy, and M. F. M. Ribeiro. 2008. Dehydroepiandrosterone modulates antioxidant enzymes and Akt signaling in healthy Wistar rat hearts. *J Steroid Biochem Mol Biol* 112(1–3):138–44.

Jacob, M. H. V. M., D. R. Janner, M. P. Jahn, L. C. R. Kucharski, A. Belló-Klein, and M. F. M. Ribeiro. 2009. DHEA effects on myocardial Akt signaling modulation and oxidative stress changes in aged rats. *Steroids* 74:1045–50.

Jacob, M. H. V. M., D. R. Janner, M. P. Jahn, L. C. Kucharski, A. Belló-Klein, and M. F. M. Ribeiro. 2010. Age-related effects of DHEA on peripheral markers of oxidative stress. *Cell Biochem Funct* 28:52–7.

Jahn, M. P., M. H. V. M. Jacob, L. F. Gomes, R. Duarte, A. S. R. Araújo, A. Belló-Klein et al. 2010. The effect of long-term DHEA treatment on glucose metabolism, hydrogen peroxide and thioredoxin levels in the skeletal muscle of diabetic rats. *J Steroid Biochem Mol Biol* 120:38–44.

Janner, D. D., M. H. Jacob, M. P. Jahn, L. C. Kucharski, and M. F. Ribeiro. 2010. Dehydroepiandrosterone effects on Akt signaling modulation in central nervous system of young and aged healthy rats. *J Steroid Biochem Mol Biol.* doi:122(4):142–8.

Jiang, Y., T. Miyazaki, A. Honda, T. Hirayama, S. Yoshida, N. Tanaka et al. 2005. Apoptosis and inhibition of the phosphatidylinositol 3-kinase/Akt signaling pathway in the anti-proliferative actions of dehydroepiandrosterone. *J Gastroenterol* 40(5):490–7.

Jones, D. P. 2006. Redefining oxidative stress. *Antioxid Redox Signal* 8(9–10):1865–79.

Kang, H. Y., C. L. Cho, K. L. Huang, J. C. Wang, Y. C. Hu, H. K. Lin et al. 2004. Nongenomic androgen activation of phosphatidylinositol 3-kinase/Akt signaling pathway in MC3T3-E1 osteoblasts. *J Bone Miner Res* 19:1181–90.

Kurzman, I. D., E. G. Macewen, and A. L. M. Haffa. 1990. Reduction in body weight and cholesterol in spontaneously obese dogs by dehydroepiandrosterone and clofibrate. *Int J Obes* 14:95–104.

Labrie, F., V. Luu-The, C. Martel, A. Chernomoretz, E. Calvo, J. Morissette et al. 2006. Dehydroepiandrosterone (DHEA) is an anabolic steroid like dihydrotestosterone (DHT), the most potent natural androgen, and tetrahydrogestrinone (THG). *J Steroid Biochem Mol Biol* 100:52–8.

Legrain, S., and L. Girard. 2003. Pharmacology and therapeutic effects of dehydroepiandrosterone in older subjects. *Drugs Aging* 20(13):949–67.

Liu, D., H. Si, K. A. Reynolds, W. Zhen, Z. Jia, and J. S. Dillon. 2007. Dehydroepiandrosterone protects vascular endothelial cells against apoptosis through a Galphai protein-dependent activation of phosphatidylinositol 3-kinase-Akt and regulation of antiapoptotic Bcl-2 expression. *Endocrinology* 148(7):3068–76.

Maninger, N., O. M. Wolkowitz, V. I. Reus, E. S. Epell, and S. H. Mellon. 2009. Neurobiological and neuropsychiatric effects of dehydroepiandrosterone (DHEA) and DHEA sulfate (DHEAS) *Neuroendocrinology* 30(1):65–9.

Mastrocola, R., M. Aragno, S. Betteto, E. Brignardello, M. G. Catalano, O. Danni et al. 2003. Pro-oxidant effect of dehydroepiandrosterone in rats is mediated by PPAR activation. *Life Sci* 73:289–99.

Mo, Q., S. Lu, and N. G. Simon. 2006. Dehydroepiandrosterone and its metabolites: Differential effects on androgen receptor trafficking and transcriptional activity. *J Steroid Biochem Mol Biol* 99(1):50–8.

Morgan, E. T. 2001. Regulation of cytochrome P450 by inflammatory mediators: Why and how? *Drug Metab Dispos* 29:207–12.

Mullonkal, C. J., and L. H. Toledo-Pereira. 2007. Akt in ischemia and reperfusion. *J Invest Surg* 20(3):195–203.

Paravicini, T. M., and R. M. Touyz. 2006. Redox signaling in hypertension. *Cardiovasc Res* 71(2):247–58.

Pelissier, M. A., C. Muller, M. Hill, and R. Morfin. 2006. Protection against dextran sodium sulfate-induced colitis by dehydroepiandrosterone and 7alpha-hydroxy-dehydroepiandrosterone in the rat. *Steroids* 71(3):240–8.

Rom, W., and T. Harkin. 1991. Dehydroepiandrosterone inhibits the spontaneous release of superoxide radical by alveolar macrophages *in vitro* in asbestosis. *Environ Res* 55:145–56.

Schumacher, M., Y. Akwa, R. Guennoun, F. Robert, F. Labombarda, F. Désarnaud et al. 2000. Steroids synthesis and metabolism in the nervous system: Trophic and protective effects. *J Neurocytol* 29:307–26.

Simoncini, T., and A. R. Genazzani. 2003. Non-genomic actions of sex steroids hormones. *Eur J Endocrinol* 148:281–92.

Sun, Y. 1990. Free radicals, antioxidant enzymes, and carcinogenesis. *Free Radic Biol Med* 8:583–99.

Tamagno, E., M. Aragno, G. Boccuzzi, M. Gallo, S. Parola, B. Fubini et al. 1998. Oxygen free radical scavenger properties of dehydroepiandrosterone. *Cell Biochem Funct* 16(1):57–63.

Valko, M., D. Leibfritz, J. Moncol, M. T. D. Cronin, M. Mazur, and J. Telser. 2007. Free radicals and antioxidants in normal physiological functions and human disease. *Int J Biochem Cell Biol* 39:44–84.

Vendemiale, G., I. Grattagliano, and E. Altomare. 1999. An update on the role of free radicals and antioxidant defense in human disease. *Int J Clin Lab Res* 29:49–55.

Wang, X., K. D. McCullogh, T. F. Franke, and N. J. Holbrook. 2000. Epidermal growth factor receptor-dependent Akt activation by oxidative stress enhances cell survival. *J Biol Chem* 275(19):14624–31.

Wolkowitz, O. M., J. H. Kramer, V. I. Reus, M. M. Costa, K. Yaffe, P. Walton et al. 2003. DHEA treatment of Alzheimer's disease—A randomized, double-blind, placebo-controlled study. *Neurology* 60:1071–6.

Yildirim, A., M. Gumus, S. Dalga, Y. N. Sahin, and F. Akcay. 2003. Dehydroepiandrosterone improves hepatic antioxidant systems after renal ischemia-reperfusion injury in rabbits. *Ann Clin Lab Sci* 33(4):459–64.

Yoshimura, M. 2006. Cardiac aldosterone. *Nippon Rinsho* 64(5):837–42.

Zhou, Y. C., and D. J. Waxman. 1998. Activation of peroxisome proliferator-activated receptors by chlorinated hydrocarbons and endogenous steroids. *Environ Health Perspect* 106:983–8.

28 Evidence for a Cellular DHEA Receptor

Brianne O'Leary and Joseph S. Dillon

CONTENTS

INTRODUCTION

The molecular mechanisms of action of dehydroepiandrosterone (DHEA) and its sulfated form, dehydroepiandrosterone sulfate (DHEAS), remain unclear despite extensive studies over many decades. The predominant mechanistic hypothesis for the actions of DHEA is that the steroid is metabolized to potent androgens and estrogens, which activate the intracellular estrogen receptors (ERs) and androgen receptors (ARs; Labrie et al. 2005). Additional proposed molecular mechanisms include the activation of (1) other ligand-dependent intracellular receptors (e.g., peroxisome proliferator-activated receptor-α, pregnane X receptor, or an uncharacterized DHEA-specific receptor); (2) other classes of receptors (e.g., sigma-1 receptor); (3) ion channels (e.g., γ-amino butyric acid A receptor chloride channel); or (4) enzymes (e.g., protein kinase C or glucose-6-phosphate dehydrogenase; comprehensively reviewed in a study by Webb et al. [2006]). Additionally, recent studies have focused on plasma membrane–initiated rapid intracellular signaling related to DHEA. Notable in these studies is that binding and signaling are shown to occur at DHEA concentrations that are within or close to the physiological concentration range of circulating DHEA in humans, suggesting that these mechanisms may be active *in vivo*.

This review will describe studies that support the concept of plasma membrane–initiated DHEA signaling and the evidence supporting a receptor-based mechanism for this signaling. We will operationally define a "DHEA receptor" as a cellular protein that interacts with DHEA in a specific binding reaction at physiological circulating concentrations of DHEA, resulting in activation of an intracellular signaling cascade. The evidence that we will review related to a possible cell surface receptor for DHEA includes DHEA-binding studies *in vitro* that include affinity and ligand-specificity data, cellular signaling studies at physiological or near physiological circulating concentrations of DHEA (≤100 nM); evidence of involvement of key intracellular mediators of receptor-dependent signaling (e.g., G-proteins); and direct-visualization studies of DHEA analog binding to cell surface binding sites. We will note some key differences between the published data on these possible binding sites. Our focus in this review on rapid plasma membrane–initiated actions of physiological concentrations of DHEA does not negate the possibility that DHEA has other mechanisms of action, depending on the cellular context and local concentration of the steroid.

DHEA-BINDING STUDIES

Evidence supporting specific binding of DHEA to cell surface or other cellular binding sites has been presented in recent years by multiple research groups in a variety of cells and tissues (Table 28.1). Williams et al. (2002) presented data on the specific binding of DHEA to human internal mammary artery vascular smooth muscle cells. Based on their Scatchard analysis of binding at 37°C in intact cells, the dissociation constant (Kd) for this receptor was 14 nM. Liu and Dillon (2002) showed evidence supporting a DHEA-specific receptor in vascular endothelial cells from both human umbilical vein and bovine aorta. The studies by Liu involved the plasma membrane fractions and caveolar fractions, specifically. The receptor Kd, determined from binding to membrane fractions at 4°C, was 0.05 nM. Charalampopoulos et al. (2006) demonstrated similar findings in ligand-binding studies, which were performed at 37°C for 30 minutes on plasma membranes of rat neural crest-derived PC12 cells (Kd = 0.9 nM), primary rat hippocampus cells (Kd = 61.9 nM), and human adrenal chromaffin cells (Kd = 0.1 nM). Alexaki et al. (2009) also demonstrated DHEA binding to the human keratinocyte cell line HaCaT with a Kd of 7.2 nM.

Although all of these examples concur reasonably well on affinity, there are differences in ligand-specificity data. Liu and Dillon (2002) demonstrated that the DHEA binding to isolated plasma membranes of bovine aortic endothelial cells was highly specific for DHEA. There was no evident inhibition of [^{3}H]DHEA binding to plasma membranes by DHEAS, estradiol, testosterone, 17α-hydroxy pregnenolone, or androstenedione, up to a competing ligand concentration of 1 μM (Liu and Dillon 2002). In contrast, studies of Charalampopoulos et al. (2006) and Alexaki et al.

TABLE 28.1
Pharmacological Evidence for High-Affinity DHEA Binding to Whole Cells and Plasma Membranes

Reference	Cell/Tissue Type	Subcellular Fraction	K_d(nM)	B_{max}, fmol/mg (Binding Sites per Cell)	DHEA Selectivity
Williams et al. (2002)	VSMC	WC	14	(37,000)	No effect on ER or AR antagonists
Liu and Dillon (2002)	BAEC	PM	0.05	500	No binding inhibition by DHEAS, E, T, 17OH P, or Adione
	HUVEC	PM	—	616	Not tested
Charalampopoulos et al. (2006)	PC12	PM	0.9	21	DHEA = DHEAS > CORT and DEX > T and DHT with no displacement by Allo, DES, or ORG
	Human chromaffin	PM	0.1	35	Not tested
	Rat hippocampus	PM	61.9	93	Not tested
Alexaki et al. (2009)	HaCaT	PM	7.2	113	DHEA = DHEAS = DHEA-7-BSA; no displacement by DES

DHEA = dehydroepiandrosterone; 17OH P = 17-hydroxyprogesterone; Adione = androstenedione; Allo = allopregnanolone; AR = androgen receptor; BAEC = bovine aortic endothelial cells; CORT = cortisol; DES = diethylstilbestrol; DEX = dexamethasone; DHT = dihydrotestosterone; E = estradiol; ER = estrogen receptor; HUVEC = human umbilical vein endothelial cells; ORG = ORG2058; PM = plasma membranes; T = testosterone; VSMC = vascular smooth muscle cells; WC = whole cells.

(2009) in plasma membranes of PC12 cell or HaCaT keratinocytes have consistently shown that there is equal affinity for DHEA and DHEAS. They find the relative binding affinities of steroids for this receptor to be DHEA = DHEAS > corticosterone = dexamethasone > testosterone = dihydrotestosterone, with no apparent affinity of the pregnane, allopregnanolone, the synthetic estrogen, diethyl stilbestrol, or the progesterone analog, ORG2058, at concentrations of up to 1 μM. The fluorescence microscopy studies of Lemcke et al. (2010) using the human neuroblastoma-derived SH-SY5Y cells also demonstrate competition of DHEAS for the DHEA-binding site (see the next section, "Visualization of DHEA binding").

These ligand-selectivity studies demonstrate an apparent difference in the structure–activity relationships determined by different research groups. Liu et al. (2002) demonstrated similar biological activity and binding affinity for DHEA analogs that had molecular substitutions at the 17-position (Figure 28.1).Thus, DHEA-17-carboxymethyloxime (CMO) bovine serum albumin (BSA) activated endothelial nitric oxide synthase (eNOS) in intact bovine aortic endothelial cells, similar to the parent steroid (Liu and Dillon 2004). Furthermore, the addition of benzophenone and biotin groups at the 17-position resulted in an analog, which retained high-affinity binding to the DHEA-specific plasma membrane binding site (Kd = 3.5 nM), and activated eNOS in bovine aortic endothelial cells at similar potency to DHEA (Liu et al. 2010).

On the contrary, Liu and Dillon (2002) demonstrated that DHEAS or other C-3 substituted steroids failed to activate eNOS or inhibit binding of [^{3}H]DHEA to the endothelial plasma membrane DHEA-binding site.

Unlike the data from Liu et al. (2002), however, data from Charalampopoulos et al. (2006), Alexaki et al. (2009), and Lemcke et al. (2010) in PC12, human adrenal medulla, rat hippocampus, HaCaT, and SH-SY5Y cells demonstrate that DHEAS is approximately equipotent with DHEA in binding to the putative DHEA receptor (IC_{50}= 0.4–1.3 nM). Consistent with those binding data, DHEAS is equipotent to DHEA in antiapoptotic activity in serum-starved PC12 (Charalampopoulos et al. 2004) and HaCaT cells (Alexaki et al. 2009). Additionally, DHEA-7-CMO-BSA conjugate, although not

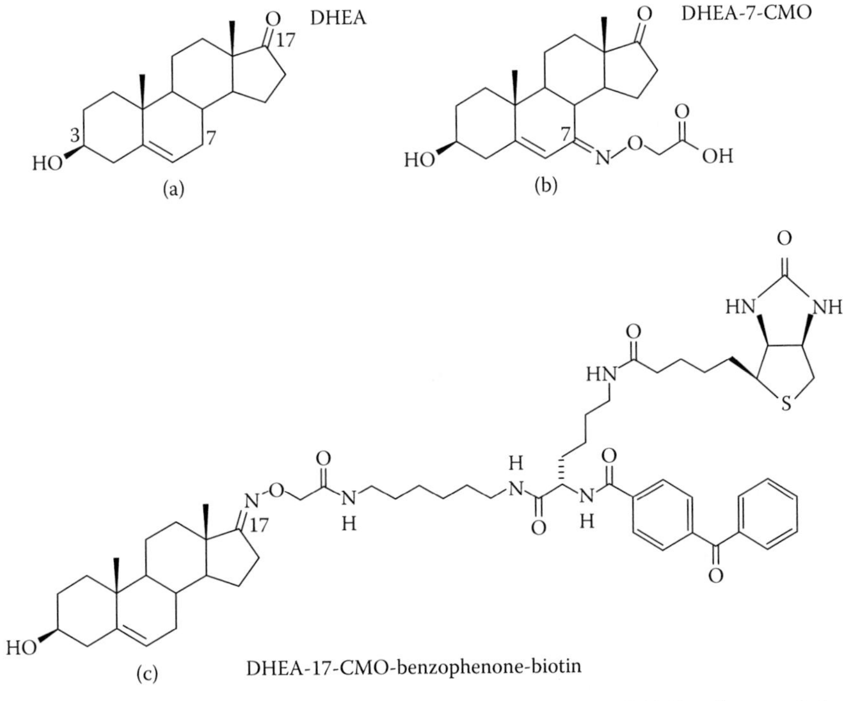

DHEA

(a)

DHEA-7-CMO

(b)

(c) DHEA-17-CMO-benzophenone-biotin

FIGURE 28.1 Structure of (a) dehydroepiandrosterone (DHEA), (b) DHEA-7-carboxymethyloxime, and (c) DHEA-17-carboxymethyloxime-benzophenone-biotin.

7-hydroxy-DHEA, retains the antiapoptotic activity of DHEA in PC12 cells (Charalampopoulos et al. 2004). Furthermore, Lemcke et al. (2010) demonstrated binding of a novel fluorescent analog, DHEA-7-bodipy, equipotently with native DHEA, by inverted live cell microscopy. They also demonstrated biological activity of this 7-substituted DHEA analog in a serum-deprivation apoptosis assay and in a fluorescence microscopic assessment of focal adhesion contacts and stress fiber assembly in SH-SY5Y cells. In their studies, DHEAS inhibited the binding of DHEA-7-bodipy to PC12 cell plasma membranes. Thus, the relative importance of the third position of DHEA, whether hydroxy or sulfate, is quite different between the studies of Liu et al. (2002) and other studies evaluating this issue. Although the molecular basis for this difference is unknown, cell-specific factors could play a role. For example, the studies do use different cells (neuronal cells and keratinocytes vs. vascular endothelial cells) that could have differential expression of members of a receptor family (Hannon et al. 2002) or differential expression of receptor-interacting proteins that alter the ligand affinity (McLatchie et al. 1998).

DHEA-SIGNALING STUDIES

Activation of intracellular ARs or ERs by DHEA requires time-dependent metabolism of DHEA to androgen or estrogen and interaction with specific cytoplasmic ligand-dependent transcription factor receptors, leading to the initiation of specific gene expression and protein translation. Plasma membrane receptors initially and rapidly activate intracellular signaling cascades. The rapidity of these signaling events is in contrast to the slower signaling achieved with activation of the intracellular AR and ER. DHEA has recently been linked with rapid activation of multiple signaling pathways by various research groups (Table 28.2).

The major intracellular signaling cascades that have been shown to be activated by DHEA include the Raf1/MEK1/2/ERK1/2 (Williams et al. 2002; Simoncini et al. 2003; Williams et al. 2004; Formoso et al. 2006; Liu et al. 2008; Charalampopoulos, Margioris, and Gravanis 2008), PI3K/Akt (Formoso et al. 2006; Liu et al. 2007; Chen et al. 2008; Charalampopoulos, Margioris, and Gravanis 2008), and cAMP/PKA/CREB (Chen et al. 2008; Charalampopoulos, Margioris, and Gravanis 2008). Variable effects of DHEA have been documented for other mitogen-activated protein kinases, for example, no effect on JNK or p38 kinase in vascular smooth muscle cells (Williams et al. 2002); no effect on p38 in bovine aortic endothelial cells (Chen et al. 2008); or a decrease in p38 activity in human lymphoblastic cell line PEER (Ashida et al. 2005).

These signaling cascades activated by DHEA have been linked to eNOS activation (Liu et al. 2008; Simoncini et al. 2003; Formoso et al. 2006), cellular proliferation (Liu et al. 2008; Williams et al. 2004), protection from apoptosis (Liu et al. 2007; Charalampopoulos et al. 2004, Lemcke et al. 2010; Alexaki et al. 2009), and endothelin 1 secretion (Formoso et al. 2006). There are, however, some differences in the linkage of signaling cascades to these end points. For example, Formoso et al. (2006) demonstrate that PI3K/Akt activation, but not ERK1/2 activation, is linked to eNOS, whereas Simoncini et al. (2003) and Liu et al. (2008) link eNOS to ERK1/2 activation. Furthermore, most, although not all, groups have documented rapid intracellular signaling in a pertussis toxin-sensitive manner, suggesting a dependence of this signaling on G proteins of the Gi or Go family. Thus, DHEA-induced eNOS activation and NO synthesis are inhibited by pertussis toxin in the studies of Simoncini et al. (2003) and Liu and Dillon (2002) in human umbilical vein and bovine aortic endothelial cells. Additionally, the activation of ERK1/2 and Akt and their downstream effects on proliferation and apoptosis in vascular endothelial cells are pertussis toxin sensitive (Liu and Dillon 2002, 2004; Liu et al. 2008; Liu et al. 2007). The inhibition of carbachol-induced intracellular calcium release in the insulin-secreting cell line INS1 is also inhibited by preincubation with pertussis toxin (Liu, Ren, and Dillon 2006). Pertussis toxin blocks the antiapoptotic effect of DHEA in PC12 cells and in the human keratinocyte cell line HaCaT (Charalampopoulos et al. 2006; Alexaki et al. 2009) and additionally blocks the phosphorylation of Src and activation of PI3K/Akt, PKA, and PKC-α/β/ERK1/2 in PC12 cells (Charalampopoulos, Margioris, and Gravanis 2008).

TABLE 28.2
Rapid Cellular Signaling at Physiological or Near Physiological Circulating Concentrations of DHEA

Reference	G-protein Coupling	Intracellular Signaling	Time of Signal Assessment	Ligand Concentration (M)
Liu and Dillon (2002)	PTX inhibited; ↑ [^{35}S] GTP-γS binding to $G_{\alpha i2,3}$	↑ eNOS phosphorylation; ↑ NO synthesis	5 minutes	10^{-12} to 10^{-6}
Williams et al. (2002)	Not tested	↓ PDGF-induced ERK1/2 phosphorylation; ↓ proliferation	4 hours, 20 hours	10^{-10} to 10^{-7}
Williams et al. (2004)	Not tested	↑ ERK1/2 phosphorylation; ↑ eNOS expression; ↑cell proliferation	4 hours, 16 hours	10^{-9} to 10^{-7}
Simoncini et al. (2003)	PTX inhibited	↑ Raf-1/ERK1/2 phosphorylation; ↑ NO synthesis	30 minutes	10^{-9} to 10^{-6}
Liu and Dillon (2004)	PTX inhibited	↑ NO synthesis; ↑ cGMP	5 minutes	10^{-12} to 10^{-6}
Liu et al. (2006)	PTX inhibited	↓ Carbachol-induced Ca^{2+} release	0.5 minute	10^{-12} to 10^{-6}
Charalampopoulos et al. (2006)	PTX inhibited; ↑ [^{35}S] GTP-γS binding	↑ Src phosphorylation	5 minutes	10^{-7}
Formoso et al. (2006)	Not tested	↑ PI3K/Akt activation; ↑ eNOS phosphorylation; ↑ NO synthesis; ↑ ERK1/2 activation; ↑ ET-1 secretion	5 minutes	10^{-7}
Chen et al. (2008)	Pertussis toxin insensitive	↑ PI3K activation; ↑ PKA activation; ↑ cAMP; ↑ FoxO1 phosphorylation; ↓ ET-1 transactivation and secretion	30 minutes	10^{-7}
Liu et al. (2007)	PTX inhibited	↑ Akt phosphorylation; ↓ apoptosis	15 minutes	10^{-11} to 10^{-7}
Liu et al. (2008)	PTX inhibited	↑ ERK1/2 phosphorylation/nuclear translocation; ↑ P90RSK phosphorylation; ↑ cell proliferation and angiogenesis	15 minutes	10^{-10} to 10^{-8}
Lemcke et al. (2010)	Not tested	↑ Stress fiber assembly; ↑ formation of focal adhesion contacts	30 minutes	10^{-8}

DHEA = dehydroepiandrosterone; ERK = extracellular regulated kinase; eNOS = endothelial nitric oxide synthase; ET-1 = endothelin 1; NO = nitric oxide; p90RSK = p90 ribosomal S6 kinase; PDGF = platelet-derived growth factor; PKA = protein kinase A; PI3K = phosphoinositide-3-kinase; PTX = pertussis toxin.

Liu and Dillon (2002) showed that Gi-α2,3, but not Gi-α1 or Go, were specifically involved in DHEA-induced rapid activation of eNOS in vascular endothelial cells. However, using bovine aortic endothelial cells, Chen et al. (2008) found that pertussis toxin had no effect on the DHEA-induced phosphorylation of FoxO1 forkhead transcription factor, although they did not test the effect of pertussis toxin on the upstream kinases responsible for FoxO1 phosphorylation (PI3K/Akt and PKA).

VISUALIZATION OF DHEA BINDING

Various groups have demonstrated the binding of DHEA to the cell surface of different cells by confocal and fluorescence microscopy using differently labeled DHEA analogs or conjugates. Charalampopoulos et al. (2006) have synthesized a DHEA-7-CMO-BSA-fluoroscein isothiocyanate (DHEA-BSA-FITC), based on similar reagents that have been developed for other steroids, to analyze cell surface binding. They have used this reagent to visualize cell surface binding of DHEA to the rat adrenal medulla cell line PC12 and human keratinocyte cell line HaCaT, by confocal microscopy (Charalampopoulos et al. 2006; Alexaki et al. 2009). Additionally, this reagent has been used in detection of membrane-binding sites by fluorescence-activated cell sorting in PC12 cells and HaCaT (Charalampopoulos et al. 2006; Alexaki et al. 2009). Finally, Alexaki et al. (2009) have also used this agent in the staining of human skin biopsies to show the expression of plasma membrane–binding sites in tissues.

Lemcke et al. have extended these studies to live cells with the use of a novel analog, with the fluorescent molecule bodipy attached at the DHEA-7 position through a 7-carbon linker. Using this analog and fluorescence-inverted microscopy, they demonstrated cell surface binding of the DHEA ligand to PC12 cells (Lemcke et al. 2010). This binding was inhibited by DHEA, DHEAS, and androstenedione. Additionally, they demonstrated subcellular localization of the steroid analog to endoplasmic reticulum, mitochondria, and nucleus of the ERα-expressing human neuroblastoma cell line, SK-ERα. The novel analog was functional in cells, inhibiting serum deprivation–induced apoptosis and increasing stress fiber and focal adhesion contacts in a neuroblastoma cell line (SH-SY5Y) in a manner similar to DHEA. The excellent image definition available with DHEA-bodipy will facilitate careful analysis of the fate of DHEA and its putative cell surface receptor.

TABLE 28.3
Imaging of DHEA Analog or Conjugate Binding to Cells

Reference	Cell Type	Microscopy	Ligand	Concentration (M)	Competition	Subcellular Localization
Charalampopoulos et al. (2006)	PC12	Confocal	DHEA-BSA-FITC	10^{-7}	DHEA-BSA	Cell surface
Alexaki et al. (2009)	HaCaT	Confocal	DHEA-BSA-FITC	10^{-7}	DHEA	Cell surface
Lemcke et al. (2010)	PC12; SK-ER-α; SH-SY5Y	Fluorescence	DHEA-bodipy	10^{-6}	DHEA, DHEAS, Adione	Cell surface, cytoplasm, ER, nucleus, mitochondria
Liu et al. (2010)	BAEC	Confocal	DHEA-BP-Bt, Neutravidin-FITC	10^{-7}	DHEA	Cell surface

DHEA = dehydroepiandrosterone; Adione = androstenedione; DHEA-BSA-FITC = DHEA-7-carboxymethyloxime-bovine serum albumin-fluorescein isothiocyanate conjugate; DHEA-BP-Bt = DHEA-17-CMO-benzophenone-biotin.

Olivo et al. (2010) and Liu et al. (2010) have developed and used a DHEA-17-CMO-benzophenone-biotin (DHEA-BP-Bt) analog, in combination with neutravidin-FITC, for confocal microscopic analysis. They demonstrated DHEA-specific plasma membrane binding of the analog to bovine aortic endothelial cells (Liu et al. 2010; Table 28.3).

NEW REAGENTS FOR DHEA RECEPTOR IDENTIFICATION

As of the time of writing, no fully identified and validated DHEA receptor has been isolated. However, attempts to isolate the putative DHEA plasma membrane receptor and to examine the rapid plasma membrane–initiated intracellular signaling of DHEA have benefitted from the development of novel DHEA analogs such as DHEA-BP-Bt (Olivo et al. 2010; Liu et al. 2010). DHEA-BP-Bt, which retains high-affinity binding for the plasma membrane DHEA-binding site, has a photoreactive benzophenone group for high-efficiency cross-linking to adjacent proteins and a biotin group allowing for affinity isolation using avidin-based methods. Liu et al. have used it to characterize the apparent molecular weight of DHEA-binding proteins from plasma membranes of bovine aortic endothelial cells. They have reported that proteins of apparent molecular weight 55, 80, and 130 kD bind to the DHEA analog, in a DHEA-inhibited manner, on ultraviolet light exposure of the analog incubated with live endothelial cells or isolated plasma membrane fractions (Liu et al. 2010). These proteins have not, yet, been further characterized.

ACKNOWLEDGMENTS

This chapter is based on work supported by the Department of Veterans Affairs, Veterans Health Administration, Office of Research and Development, and Biomedical Laboratory Research and Development Program. The contents do not represent the views of the Department of Veterans Affairs or the U.S. Government. The studies were also supported by grants from the American Heart Association and the NIH (AG18928).

REFERENCES

Alexaki, V. I., I. Charalampopoulos, M. Panayotopoulou, M. Kampa, A. Gravanis, and E. Castanas. 2009. Dehydroepiandrosterone protects human keratinocytes against apoptosis through membrane binding sites. *Exp Cell Res* 315:2275–83.

Ashida, K., K. Goto, Y. Zhao, T. Okabe, T. Yanase, R. Takayanagi, M. Nomura, and H. Nawata. 2005. Dehydroepiandrosterone negatively regulates the p38 mitogen-activated protein kinase pathway by a novel mitogen-activated protein kinase phosphatase. *Biochim Biophys Acta* 1728:84–94.

Charalampopoulos, I., V. I. Alexaki, I. Lazaridis, E. Dermitzaki, N. Avlonitis, C. Tsatsanis, T. Calogeropoulou, A. N. Margioris, E. Castanas, and A. Gravanis. 2006. G protein-associated, specific membrane binding sites mediate the neuroprotective effect of dehydroepiandrosterone. *FASEB J* 20:577–9.

Charalampopoulos, I., A. N. Margioris, and A. Gravanis. 2008. Neurosteroid dehydroepiandrosterone exerts anti-apoptotic effects by membrane-mediated, integrated genomic and non-genomic pro-survival signaling pathways. *J Neurochem* 107:1457–69.

Charalampopoulos, I., C. Tsatsanis, E. Dermitzaki, V. I. Alexaki, E. Castanas, A. N. Margioris, and A. Gravanis. 2004. Dehydroepiandrosterone and allopregnanolone protect sympathoadrenal medulla cells against apoptosis via antiapoptotic Bcl-2 proteins. *Proc Natl Acad Sci U S A* 101:8209–14.

Chen, H., A. S. Lin, Y. Li, C. E. Reiter, M. R. Ver, and M. J. Quon. 2008. Dehydroepiandrosterone stimulates phosphorylation of FoxO1 in vascular endothelial cells via phosphatidylinositol 3-kinase- and protein kinase A-dependent signaling pathways to regulate ET-1 synthesis and secretion. *J Biol Chem* 283:29228–38.

Formoso, G., H. Chen, J. A. Kim, M. Montagnani, A. Consoli, and M. J. Quon. 2006. Dehydroepiandrosterone mimics acute actions of insulin to stimulate production of both nitric oxide and endothelin 1 via distinct phosphatidylinositol 3-kinase- and mitogen-activated protein kinase-dependent pathways in vascular endothelium. *Mol Endocrinol* 20:1153–63.

Hannon, J., C. Nunn, B. Stolz, C. Bruns, G. Weckbecker, I. Lewis, T. Troxler, K. Hurth, and D. Hoyer. 2002. Drug design at peptide receptors: somatostatin receptor ligands. *J Mol Neurosci* 18:15–27.

Labrie, F., V. Luu-The, A. Belanger, S. X. Lin, J. Simard, G. Pelletier, and C. Labrie. 2005. Is dehydroepiandrosterone a hormone? *J Endocrinol* 187:169–96.

Lemcke, S., C. Honnscheidt, G. Waschatko, A. Bopp, D. Lutjohann, N. Bertram, and K. Gehrig-Burger. 2010. DHEA-Bodipy—a functional fluorescent DHEA analog for live cell imaging. *Mol Cell Endocrinol* 314:31–40.

Liu, D., and J. S. Dillon. 2002. Dehydroepiandrosterone activates endothelial cell nitric-oxide synthase by a specific plasma membrane receptor coupled to Galpha (i2,3). *J Biol Chem* 277:21379–88.

Liu, D., and J. S. Dillon. 2004. Dehydroepiandrosterone stimulates nitric oxide release in vascular endothelial cells: Evidence for a cell surface receptor. *Steroids* 69:279–89.

Liu, D., M. Iruthayanathan, L. L. Homan, Y. Wang, L. Yang, Y. Wang, and J. S.Dillon. 2008. Dehydroepiandrosterone stimulates endothelial proliferation and angiogenesis through extracellular signal-regulated kinase 1/2-mediated mechanisms. *Endocrinology* 149:889–98.

Liu, D., B. O'leary, M. Iruthayanathan, L. Love-Homan, N. Perez-Hernandez, H. F. Olivo, and J. S. Dillon. 2010. Evaluation of a novel photoactive and biotinylated dehydroepiandrosterone analog. *Mol Cell Endocrinol* 328:56–62.

Liu, D., M. Ren, X. Bing, C. Stotts, S. Deorah, L. Love-Homan, and J. S. Dillon. 2006. Dehydroepiandrosterone inhibits intracellular calcium release in beta-cells by a plasma membrane-dependent mechanism. *Steroids* 71:691–9.

Liu, D., M. Ren, and J. Dillon. 2006. Dehydroepiandrosterone inhibits agonist-induced intracellular calcium release in INS-1 cells by a non-genomic mechanism. *Steroids* 71:691–9.

Liu, D., H. Si, K. A. Reynolds, W. Zhen, Z. Jia, and J. S. Dillon. 2007. Dehydroepiandrosterone protects vascular endothelial cells against apoptosis through a Galphai protein-dependent activation of phosphatidylinositol 3-kinase/Akt and regulation of antiapoptotic Bcl-2 expression. *Endocrinology* 148:3068–76.

Mclatchie, L., N. Fraser, M. Main, A. Wise, J. Brown, N. Thompson, R. Solari, M. Lee, and S. Foord. 1998. RAMPs regulate the transport and ligand specificity of the calcitonin-receptor-like receptor. *Nature* 393:333–9.

Olivo, H., N. Perez-Hernandez, D. Liu, M. Iruthayanathan, B. O'leary, L. Homan, and J. Dillon. 2010. Synthesis and application of a photoaffinity analog of dehydroepiandrosterone (DHEA). *Bioorg Med Chem Lett* 20(3):1153–5.

Simoncini, T., P. Mannella, L. Fornari, G. Varone, A. Caruso, and A. R. Genazzani. 2003. Dehydroepiandrosterone modulates endothelial nitric oxide synthesis via direct genomic and nongenomic mechanisms. *Endocrinology* 144:3449–55.

Webb, S. J., T. E. Geoghegan, R. A. Prough, and K. K. Michael Miller. 2006. The biological actions of dehydroepiandrosterone involves multiple receptors. *Drug Metab Rev* 38:89–116.

Williams, M. R., T. Dawood, S. Ling, A. Dai, R. Lew, K. Myles, J. W. Funder, K. Sudhir, and P. A. Komesaroff. 2004. Dehydroepiandrosterone increases endothelial cell proliferation in vitro and improves endothelial function in vivo by mechanisms independent of androgen and estrogen receptors. *J Clin Endocrinol Metab* 89:4708–15.

Williams, M. R., S. Ling, T. Dawood, K. Hashimura, A. Dai, H. Li, J. P. Liu, J. W. Funder, K. Sudhir, and P. A. Komesaroff. 2002. Dehydroepiandrosterone inhibits human vascular smooth muscle cell proliferation independent of ARs and ERs. *J Clin Endocrinol Metab* 87:176–81.

Section V

DHEA and Mechanisms of Action in Humans

29 Dehydroepiandrosterone and Testosterone
Effects on Erectile Function

Ahmed I. El-Sakka

CONTENTS

INTRODUCTION

Dehydroepiandrosterone (DHEA) is a precursor sex steroid hormone synthesized from cholesterol in the zona reticularis of the adrenal cortex, the gonads, adipose tissue, brain, and skin. Together with its sulfate version, DHEA sulfate (DHEAS), it is the most abundant steroid in humans. DHEAS, generated in the liver from the parent adrenal steroid DHEA, circulates in the blood of men in relatively huge quantities. In young people in the second and third decades of age, the highest concentrations of these steroids are observed, and gradually decrease by approximately 10% per decade (Orentreich et al. 1984). Serum DHEA steadily declines and by the age of 70 years, serum DHEA levels are approximately 20%–23% of their peak values (Davison et al. 2005; Labrie et al. 1998). The mechanism underlying this physiological decline is unknown. Consistently, in one of our previous studies we have demonstrated age-related decline in testosterone level throughout 4 years of follow-up in patients with erectile dysfunction (ED). Patients with decreasing testosterone levels were older than patients with a steady testosterone level (El-Sakka and Hassoba 2006). Therefore, this steady decrease in circulating DHEAS concentrations with age has prompted speculation that DHEA therapy might have potential benefits in several diseases associated with aging. Clinical studies have also found associations between low levels of serum DHEA or DHEAS and diseases or deterioration in various physiological functions. Particularly, aging; neurological functions including decreased well-being, cognition, and memory; increased depression; aggressiveness; and dementia have involved changes in body composition, decreased bone mineral density, obesity, diabetes, increased cardiovascular morbidity, ED, and decreased libido (Gurnell and Chatterjee 2001; Johnson, Bebb, and Sirrs 2002). Supporting this result, some trials of DHEA supplementation in healthy, middle-aged, and elderly subjects have reported improvements in different aspects of well-being (Morales et al. 1994).

DHEA AND UNDERLYING MECHANISMS OF SEXUAL FUNCTION

DHEA and Endothelial Function

Innovative research shows that DHEA is involved in vascular smooth muscle relaxation. Furthermore, DHEA has its own receptors, primarily on endothelial cells. It is thus important to note that the positive effects of DHEA on sexual function are due to the actions of DHEA alone as well as its function as a precursor of multiple androgens, especially testosterone and estradiol (Baulieu 1996; Labrie et al. 1997). DHEA primarily exists as DHEAS at levels that are approximately 250 times higher than those of free DHEA in serum. In target tissues such as the brain, bone, breast, and adipose, DHEAS is converted to DHEA by the sulfatase enzyme, which may then be further metabolized to androstenediol, androstenedione, estrone, testosterone, dihydrotestosterone (DHT), and 17β-estradiol (Labrie et al. 1997; Labrie et al. 2001).

Endothelial dysfunction contributes to the pathogenesis of atherosclerosis and cardiovascular disease and precedes the development of insulin resistance. DHEA administration improved flow-mediated dilation of the brachial artery, an endothelium-dependent process; reduced plasminogen activator inhibitor type 1 (PAI-1), a suppressor of fibrinolysis with a pathogenic role in coronary artery disease; and improved insulin sensitivity (Kawano et al. 2003). The improvement of endothelial function with DHEA was subsequently confirmed in postmenopausal women (Williams et al. 2004).

Several studies supported the notion that DHEA has important actions on the vasculature. A number of *in vitro* experiments have shown the potential for antiatherosclerotic actions of DHEA (Williams et al. 2004; Simoncini et al. 2003). These data support the action of DHEA on the endothelium, based on the rapidity of action on endothelial nitric oxide synthase (eNOS) and failure to block endothelial cell activation with selective estrogen or testosterone receptor antagonists (Williams et al. 2004; Simoncini et al. 2003). Supporting studies have extended these observations to demonstrate roles of DHEA in vascular endothelial cell survival (Liu et al. 2007), proliferation/angiogenesis (Liu et al. 2008), and activation, including transcriptional regulation of endothelin-1 (Chen et al. 2008). Despite the convincing evidence for vasculoprotective actions of DHEA from *in vitro* studies, human clinical trials of DHEA replacement on metabolic and vascular function have shown conflicting results. The epidemiological data have demonstrated either an inverse (Trivedi and Khaw 2001) or no (Tilvis, Kähö nen, and Härkö nen 1999) relationship between cardiovascular mortality and circulating DHEAS levels in men.

A specific DHEA receptor on the plasma membrane of bovine aortic endothelial cells was identified (Liu and Dillon 2002). This receptor is functionally coupled to the G protein family and primarily to Ga12 and Ga13 subtypes. Activation of these G proteins activates eNOS. Another supporting study demonstrated the existence of a DHEA-specific receptor in human vascular smooth muscle cells (VSMC) involving ERK1 signaling pathways: VSMC proliferation contributes to remodeling of blood vessels and may be implicated in the pathogenesis of atherosclerosis (Williams et al. 2002). DHEA also inhibits *in vitro* VSMC proliferation through a mechanism independent of its transformation into estrogens or androgens because this effect is unaffected by antiestrogens and antiandrogens. DHEA shows minimal affinity for estrogen and androgen receptors found in VSMC cells, but binds specifically to putative receptors in the same cells. Although not opposing the mechanism of action through conversion into testosterone and estradiol, these findings will have an important impact on our interpretations of the biological actions of DHEA and the control of sexual function, which involves many vascular mechanisms.

DHEA and the Nitric Oxide Pathway

It is reported that DHEA and DHEAS have additional biological functions besides being precursors of testosterone. DHEA is shown to dilate arteries, block hypoxia-induced vasoconstriction by activating potassium channels via soluble guanylate cyclase activation (Farrukh et al. 1998), and enhance endothelial function through increased nitric oxide (NO) synthesis (Simoncini et al. 2003). The physiological concentrations of DHEA acutely increase NO release from intact vascular endothelial cells, by a plasma membrane initiated mechanism (Liu and Dillon 2004).

The effects of adrenalectomy on penile NOS in castrated rats coincide with those of castration alone in terms of the decrease in cytosolic NOS enzyme activity (Lugg, Rajfer, and Gonzalez-Cadavid 1995). However, penile eNOS content remains unaffected after 1 week. Adrenalectomy combined with castration significantly reduced penile nNOS content, in contrast to what was found with castration or total ablation of androgen binding in the penis (Penson et al. 1996), where NOS activity appeared to be inhibited in the presence of constant NOS levels. The cause of this reduction in penile nNOS content is unknown. It may be related to a loss of nerve terminals or a true NOS downregulation (Rand and Li 1995). The rat adrenal gland contributes to the maintenance of the erectile mechanism and may affect neuronal NOS content in the rat penis (Penson et al. 1997).

In pulmonary artery tissue from DHEA-treated rats, soluble guanylate cyclase, but not eNOS, levels were increased (Oka et al. 2007). However, in other experiments, DHEA/DHEAS did not affect relaxation induced by acetylcholine or sodium nitroprusside, and DHEA/DHEAS-induced relaxation responses were not changed by treatment with methylene blue, a known inhibitor of guanylyl cyclase and a potential inhibitor of endothelial-mediated cavernous relaxation (Lee et al. 2009).

DHEA and Erectile Dysfunction

It has also been speculated that DHEA plays a role in the process of erection. The Massachusetts Male Aging Study investigated 17 hormones, and DHEAS was the only one strongly (and inversely) correlated to the ED prevalence (Feldman et al. 1994). This data was partly confirmed by another study, which demonstrated that DHEAS levels were significantly lower in men with ED but otherwise healthy in comparison to age-matched normal controls (Reiter et al. 2000). The same authors evaluated the effects of DHEA replacement in 40 men with ED and demonstrated that DHEA treatment was associated with higher mean scores for each of the five domains of the international index of erectile function (IIEF). In a subsequent uncontrolled study on ED patients treated with DHEA, 50 mg daily for 6 months, a significant increase was observed in the scores of questions 3 and 4 of the IIEF (ability to initiate vaginal penetration and to maintain erections) in the patients with mild or no organic etiology, but not in those with diabetes or neurological disorders (Reiter et al. 2001). Recently, there has been no evidence-based data to show involvement of DHEA in ED, although promising data supporting the possibility of DHEA-specific receptors availability on vascular endothelial and smooth muscle cells (SMCs) allows the assumption of its possible involvement in the vascular mechanisms of erection (Williams et al. 2002).

Furthermore, it is interesting that DHT prevents the erectile failure seen in adrenalectomized and castrated rats essentially to the same extent as corticosteroids, but does not normalize it completely. This agrees with the dual dependence of the rat erectile mechanism on both the adrenal and gonads. In contrast, DHEA does not have any restorative effect on the erectile response in adrenalectomized or castrated animals (Sachs and Liu 1991; Shealy 1995). This could be because DHEA is only a weak androgen by itself but is putatively involved in conditions associated with aging, immune suppression, and major diseases (Sachs and Liu 1991; Shealy 1995). In addition, a nice study has demonstrated that DHEA enhances the feeling of well-being, but not libido, in older men (Sachs and Liu 1991). The effect of DHEA on the maintenance of the erectile mechanism supports the view that adrenal corticoids are involved in this process. Furthermore, the effects of DHEA on female sexuality are still under debate. A few clinical studies showed that oral DHEA treatment increased total serum testosterone levels, libido, sexual activity, and sexual satisfaction in postmenopausal women (Munarriz et al. 2002).

TESTOSTERONE AND SEXUAL FUNCTION

EFFECT ON SEXUAL MOTIVATION

Several studies had demonstrated that sexual interest was androgen dependent. Furthermore, the studies provided evidence that circulating testosterone has a relationship with various types of erections (Bancroft 1984). Spontaneous erections, particularly those that occur during sleep, and probably fantasy-induced erections were believed to be androgen dependent, whereas erections in response to visual or tactile stimuli were less androgen dependent (Bancroft and Wu 1983; Bancroft 1984). The influence on the penile erection was believed to be indirect, via the effects on libido, rather than direct on penile tissues. Consequently, it was assumed that androgens were not very useful therapeutically when men complained of erectile difficulties while their sexual desire was not impaired. Contradictory to that notion, our understanding of the role of androgen on erectile function has recently changed with more appreciation of its effect on anatomical and physiological components of erectile capacity.

REFINED ROLE OF ANDROGEN ON ERECTILE FUNCTION

Over the last 30 years, the age-related decline in circulating testosterone in men has received increasing attention, not only in relation to sexual functioning but in a wider context of male health. Moreover, experimental research has presented convincing evidence that testosterone has remarkable effects on the mechanism of erection and that testosterone deficiency impairs the anatomical and physiological substrate of erectile capacity. Several recent studies demonstrated that restoring normal level of testosterone is a prerequisite for the process of erection. It has also been shown that the complete therapeutic potential of PDE5 inhibitors will only become evident in eugonadal state. Those results changed the earlier concept that the effects of testosterone are primarily on libido and not directly on erectile tissue.

TESTOSTERONE AND AGING

A decrease in testicular function with a consequent decrease in testosterone is recognized as a common occurrence in older men (Morales 2003). We, among others, have demonstrated the lack of association between hypogonadism and ED severity during the assessment of hypogonadism prevalence. However, in a longitudinal follow-up study we have demonstrated a steady decline in testosterone levels throughout a 4-year follow-up period in patients with ED (El-Sakka and Hassoba 2006). Histological studies in men showed a significant reduction in the number and volume of Leydig cells in aging men. In the Massachusetts Male Aging Study, a prevalence of 20% hypogonadism in men older than 55 years was observed when total testosterone levels less than the normal rate for young, healthy subjects were considered (Feldman et al. 1994). Furthermore, the mechanism of erection could be more influenced by molecular and ultrastructural alterations associated with androgen depletion than the simple biochemical evidence of low testosterone levels. Therefore, it would be better to consider those prebiochemical consequences of androgen alteration in the assessment and follow-up of aging patients with ED.

By measuring total serum testosterone levels, androgenic milieu can be determined. This method is the simplest, least expensive, and most readily accessible to rule out an abnormality in serum testosterone. However, the results may be misleading, particularly in elderly men, because this subset of patients commonly exhibits increased levels of sex hormone–binding globulin, which prevents testosterone from becoming metabolically active (Gooren 1996). Free testosterone levels may provide a better estimation of testicular endocrine function, but the care required in the technique results in interlaboratory and intralaboratory variability, casting reservations on the accuracy of the test. Measuring bioavailable testosterone may be the most reliable measurement because it determines the amount of testosterone available in target tissues. Unfortunately, determinations of bioavailable testosterone are not widely performed and are more costly.

Despite the fact that the decrease in androgens as well as the increase in ED prevalence are age-related events, a direct cause and effect relationship is not confirmed (Rhoden et al. 2002). Earlier studies had demonstrated that aromatase inhibition increases serum bioavailability and total testosterone levels to the youthful normal range in older men with mild hypogonadism (Leder et al. 2004). Furthermore, it has not been proved whether any of the ED or androgen depletion could be reversed by targeting each other. More importantly, it has not been settled yet how much the two conditions influence each other, and whether the normal aging process can be considered the main factor in the pathophysiology of ED and hypogonadism. Many studies have investigated the association between androgens and other aspects of sexual function. A common conclusion was that testosterone was positively correlated with sexual desire and nocturnal erection, but not necessarily with erectile function. Furthermore, studies that evaluated the effects of androgen replacement on sexual and erectile function are not encouraging, and little or no benefit can be expected from administration of androgens to eugonadal men. However, there is no doubt that testosterone is important in maintaining sexual function. Several abnormalities of testosterone metabolism have been demonstrated in hyperprolactinemia. It seems that the hypogonadism in hyperprolactinemia is central and peripheral in origin (Earle and Stuckey 2003).

Earlier studies have shown that total testosterone, but not free testosterone index, tended to decrease with increasing body mass index (BMI) (Harman et al. 2001). Hypertensive men treated with antihypertensive agents often experience sexual dysfunction and have a mild reduction in serum testosterone levels (Rosen, Kostis, and Jekelis 1988). Ischemic heart disease (IHD) and/or coronary risk factors such as type 2 diabetes had been shown to be associated with decreased testosterone levels or a greater rate of decrease in serum testosterone levels over time (Zmuda et al. 1997 and Corona et al. 2006). Contrary to these results, however, other studies have shown that serum testosterone levels had little or no impact on IHD (Contoreggi et al. 1990). Although the effect of testosterone derangement in men with sexual dysfunction has been repeatedly reported, tremendous efforts are underway to elucidate the prevalence and clarify the role of testosterone depletion in the process of erection.

Effects of Testosterone on Ultrastructural Composition of Erectile Tissue

Thoughtful studies provide convincing evidence that there is a significant effect of testosterone on the anatomical and physiological substrate of penile erection (Gooren and Saad 2006). Furthermore, it has become much more evident that circulating levels of testosterone are closely related to the manifestations of other etiological factors in ED, such as atherosclerotic disease and diabetes mellitus (DM), and ultimately the metabolic syndrome, which is correlated with lower-than-normal testosterone levels (Shabsigh et al. 2005; Traish and Guay 2006; El-Sakka et al. 2008).

The proportion of men with ED and low testosterone levels varies among studies (2.1%–25%). Newer insights into the action of testosterone indicated that there was a significant difference in level of blood testosterone and its biological effects between subjects. This discrepancy between the two factors could be in part due to the differences in androgen sensitivity that accounted for androgen receptor properties. In a large population-based cohort of older men, no association among total testosterone, bioavailable testosterone, sex hormone–binding globulin, and ED could be established. Testosterone levels were associated with a decrease in risk of ED only in men with increased luteinizing hormone levels (Kupelian et al. 2006).

Surgical or medical castration results in androgen deprivation with a consequence of significant reduction in trabecular smooth muscle content and a marked increase in connective tissue deposition

(Traish et al. 2003). An ultrastructural study demonstrated that the cavernosal smooth muscle in castrated animals appears to be disorganized with a large number of cytoplasmic vacuoles, whereas in the intact animals, the SMCs exhibit normal morphology and are arranged in clusters (Rogers et al. 2003). Several studies have demonstrated the potential role of androgens in maintaining the structure and function of many pelvic ganglion neurons (Keast et al. 2002). Giuliano et al. suggested that testosterone acting peripherally to the spinal cord enhanced the erectile response of the cavernous nerve and castration altered the dorsal nerve ultrastructure in the rat, concomitant with loss of erectile function (Giuliano et al. 1993; Rogers et al. 2003).

In addition to the alterations in smooth muscle and connective tissue, fat-containing cells have been observed in the subtunical region of penile tissue sections from orchiectomized animals (Traish et al. 2005). The outcome of this effect could lead to venous leakage, which has been observed in a subset of hypogonadal patients with ED, who improved upon testosterone administration (Yassin, Saad, and Traish 2006).

There is a remarkable interest in understanding the mechanisms by which androgens regulate growth and differentiation of vascular SMCs. An elegant study demonstrated that androgens promote the commitment of pluripotent stem cells into a muscle lineage and inhibit their differentiation into an adipocyte lineage (Singh et al. 2006). The total number of circulating vascular progenitor cells may also be dependent on testosterone levels (Foresta et al. 2006). Regulation of progenitor cell differentiation is a complex process, dependent on numerous hormones, growth factors, and specific activation of a cascade of gene expression (Rosen et al. 2002; Bélanger et al. 2002).

Traish, Goldstein, and Kim (2007) noted marked structural changes in the cavernosal nerves in castrated animals. These structural alterations may be responsible in part for the significant reduction in the intracavernosal pressure (Armagan et al. 2006). Supporting studies imply the role for androgens in regulating the expression and activity of NOS isoforms in the corpus cavernosum in animal models (Zvara et al. 1995; Park et al. 1999). Testosterone or 5α-DHT administration restored the erectile response and NOS expression in the penis of castrated animals (Marin et al. 1999; Baba et al. 2000; Armagan et al. 2006).

DHEA AS A TREATMENT FOR ERECTILE DYSFUNCTION

Several studies have reported decreased serum DHEAS levels in patients with ED. Morales, Heaton, and Carson (2000) have cautioned that decreased secretions of DHEA and DHEAS are important risk factors for ED in aging men. DHEA levels reach their peak in the third decade of age in men (Tekdogan et al. 2006). The serum DHEAS levels were significantly lower in the younger patients with ED than in patients without ED. Interestingly, serum DHEAS levels were reported to significantly increase after treatment with sildenafil citrate in the patients with ED (especially in those younger than 50 years old). Researchers speculated that decreased serum DHEAS levels, especially in young men with ED, may either be an etiologic factor for ED or a negative consequence of it (Tekdogan et al. 2006). Another study had demonstrated that patients treated with DHEA had a statistically significant increase in all domains of the IIEF in contrast to the placebo group. DHEA treatment demonstrated the first clinical effects after 8 weeks and an impressive improvement in maintaining the erection after 16 weeks. There was no impact of DHEA therapy on prostate volume, postvoid residual, or serum levels of prolactin and prostate-specific antigen (Reiter et al. 1999).

Utilization of testosterone and DHEA has been advocated for men with sexual dysfunction associated with documented testosterone or DHEA deficiency. Earlier studies on DHEA demonstrated its usefulness in treating sexual dysfunction (Reiter et al. 1999). However, new evidence has questioned the validity of those findings when treatment relied exclusively on DHEA supplementation (Nair et al. 2006). Despite the controversy, both testosterone and DHEA are considered central neurosteroids (Saad et al. 2005) and have been found to have direct effects on endothelial function, which are fundamental events in the mechanisms involving libido and penile erections (Bernini et al. 2006). DHEA is convertible into testosterone (Baulieu et al. 2000) and administration of DHEA results in biosynthesis of active androgens by tissue targets without representation in the peripheral circulation (Labrie et al. 1997). This mechanism could better define the role of DHEA in sexual performance.

Although a positive effect of DHEA in sexual domains had been reported (Reiter et al. 1999), most of the literature had not confirmed the role of DHEA in the field of sexual medicines (Nair et al. 2006; Løvas et al. 2003). Other studies were also opposing the concept of DHEA as a neurosteroid with a central putative effect on libido (Hunt et al. 2000), although the hormone had been found to have an effect on self-esteem and mood in younger individuals with Addison's disease (Acherman and Silverman 2001). Although DHEA has been documented to have direct genomic and nongenomic effects on the vascular endothelium (Bernini et al. 2006), its clinical effect as a modulator of endothelial function with a role as a facilitator of penile erections is not evident. An experimental study supports the protective role of DHEA and DHEAS in some pathological processes. DHEA reduced the development of vascular stenosis in heterotopic heart transplants (Eich et al. 1993). Because there is also some experimental evidence for direct effects on the brain, the possibility of a central effect of DHEA should be considered.

ANDROGENS AND DIABETES

The prevalence of DM is increasing, with noninsulin-dependent (type 2) DM, accounting for 90%–95% of the patients diagnosed with diabetes (Rendell et al. 1999). However, in the U.S. population from 1980 to 2006, the increases in the crude and age-adjusted prevalence of diagnosed DM were similar. This means that the increase may be better explained by lifestyle changes rather than aging (http://www.cdc.gov/diabetes/statistics/prev/national/figage.htm).

Aging is well known to be associated with a higher prevalence of DM, and it exerts a consistent effect on androgenic status in healthy men. Several studies had addressed whether androgenic status could influence chronic diseases. In this regard, type 2 DM is of special interest because earlier studies had reported that lower serum testosterone concentration is more prevalent in patients with DM than in men without DM (Defay et al. 1998). Furthermore, low levels of testosterone had predicted insulin resistance and the development of type 2 DM in older adults (Oh et al. 2002). Therefore, androgen alteration may have deleterious effects on glycemic control and, ultimately, therapeutic strategy and prognosis of type 2 DM. A recent study has demonstrated that testosterone levels were decreased in a relatively large number of Japanese men with type 2 DM compared with healthy men in each decade of life between 40 and 69 years (Fukui et al. 2007).

The finding of increased adrenal DHEA concentrations under hypoglycemic conditions indicates a strong activation of the hypothalamic–pituitary–adrenal axis on hypoglycemic stress. The Endocrine Society demonstrated the high association between type 2 DM and low testosterone levels and its clinical practice guidelines recommended measurement of testosterone concentration in those patients (Bhasin et al. 2006). Hyperinsulinemia and increased glucose concentration are both negatively correlated with total and free testosterone levels in men (Harris et al. 1987). Although it has been suggested that a low level of testosterone plays a role in the development of insulin resistance and type 2 DM (Harman et al. 2001), the cause of the decreased testosterone in type 2 DM remains obscure. In one of our recent studies, we have demonstrated that there were significant associations between good control of DM, decreased fasting blood sugar, and achievement of normal levels of testosterone at 3- and 6-month follow-up visits. No significant associations were detected between controls of DM or decreased fasting blood sugar and change in levels of DHEAS and insulin (El-Sakka, Sayed, and Tayeb 2009). By contrast, administration of testosterone to hypogonadal men improves insulin sensitivity and glucose homeostasis (Boyanov, Boneva, and Christov 2003). In a double-blind, placebo-controlled cross-over study, Kapoor et al. (2006) reported that testosterone replacement therapy in hypogonadal men with type 2 DM reduces insulin resistance and improves glycemic control and waist circumference. Furthermore, Corona et al. (2008) demonstrated that DM-associated central obesity and insulin resistance, rather than DM per se, play the most important role in DM-associated hypogonadism.

In our recent investigation, we demonstrated that there were significant associations between low levels of total testosterone or DHEAS and poor control of DM (El-Sakka, Sayed, and Tayeb 2008). Consistent with these results, earlier studies reported no significant linear correlations between total or free testosterone with fasting plasma glucose; however, total testosterone was negatively correlated with glycosylated hemoglobin levels (Dhindsa et al. 2004). Furthermore, type 2 DM had been shown to be associated with diminished testosterone levels or a greater rate of decline in serum testosterone levels over time (Corona et al. 2006).

Among others, we expanded current knowledge in the sense that, as occurs in men without DM, the percentage of patients with type 2 DM with subnormal levels of total testosterone is much higher than that observed with subnormal levels of total testosterone in patients without DM. A result that agrees with previous reports indicated that the existence of chronic diseases subtracts 10%–15% from the values of androgens found in men without chronic diseases (Feldman et al. 2002).

The mechanisms of androgen alteration in men with DM have not been completely revealed. It has been reported that androgen deficiency in type 2 DM is commonly associated with hypogonadotropic hypogonadism (Dhindsa et al. 2004). The inverse correlation between testosterone levels and BMI was confirmed in earlier studies. However, hypogonadotropic hypogonadism in type 2 DM is only partly explained by obesity. It has been demonstrated that insulin receptors in the central nervous system play a physiological role in the regulation of male reproductive endocrine function (Bruning et al. 2000). On the contrary, a cross-sectional study on healthy men without DM failed to detect any significant association of hypogonadism with decreased insulin sensitivity after adjustment for body fat mass (Tsai et al. 2004). Insulin-induced hypoglycemia is known to suppress the pulsatile secretion of luteinizing hormone as well as the pulse generator frequency of hypothalamic GnRH in experimental investigations. Interestingly, 6 months of androgen supplementation in symptomatic hypogonadal patients with DM has not improved glycemic control (Dhindsa et al. 2004). On the contrary, another study reported that testosterone replacement therapy in hypogonadal men with type 2 DM reduces insulin resistance and improves glycemic control. Furthermore, low serum testosterone levels were reported to be associated with an adverse metabolic profile. Low testosterone levels and impaired mitochondrial function promote insulin resistance in men (Pitteloud et al. 2005). Hypogonadotropic hypogonadism in male subjects with type 2 DM is associated with a lower hematocrit and high C-reactive protein (CRP) concentration. Such patients may also have a high risk of atherosclerotic cardiovascular events in view of their markedly elevated CRP concentrations (Bhatia et al. 2006), which suggests that inflammation may play an important role in the pathogenesis of this syndrome (Dandona et al. 2008). Obstructive sleep apnea was reported to be associated with hypogonadotropic hypogonadism, with improvements in testosterone levels after intervention with continuous positive airway pressure (Luboshitzky et al. 2003).

NEW INSIGHTS ON ANDROGEN AND SEXUAL FUNCTION

Androgens regulate trabecular smooth muscle growth and connective tissue protein synthesis. Further, androgens may stimulate differentiation of progenitor cells into SMCs and inhibit their differentiation into adipocytes. Testosterone can restore apomorphine-induced erections in castrated rats through a central mechanism. Peripherally, it can mediate neurotransmission, accentuate NOS activity, and increase intracavernous pressure. Certain changes in sexual activity are precipitated following castration. The ability to ejaculate disappears early, followed by disappearance of intromission and mounting. Hormonal replacement restores all aspects of mating in these animals. The amount of testosterone required to regain such functions is much less than the normal range. In the animal model, androgen deprivation produces penile tissue atrophy, alterations in dorsal nerve structure, alterations in endothelial morphology, reductions in trabecular smooth muscle content, increases in deposition of the extra cellular matrix (ECM), and accumulation of fat-containing cells (adipocytes) in the subtunical region of the corpus cavernosum.

CONCLUSIONS

The role of circulating androgens in erectile physiology and sexual behavior remains unclear. Testosterone is necessary for normal libido, ejaculation, and spontaneous erections. There is a threshold below which sexual function is impaired. Normal androgen levels are a prerequisite for a PDE5 inhibitor to work appropriately. The androgen-dependent loss of erectile response is restored by androgen administration but not by PDE5 inhibitors alone. Androgen deprivation leads to structural alterations in the corpus cavernosum, resulting in failure of the veno-occlusive mechanisms. Testosterone levels in men are a dominant determinant of improvement in libido. Effects of castration on sexual function in humans have been shown to range widely from complete loss of libido and erection to continued normal sexual activity in about 20% of patients. Restoration of T level to about 50%–60% of the physiological range was adequate for normal sexual functioning. Assessment of androgen pattern, particularly testosterone, in all patients with sexual dysfunction could help determine the etiology and, certainly, the treatment of patients with sexual dysfunction-associated hypogonadism. However, incorporation of androgen into the armamentarium of diagnosis and treatment of sexual dysfunction necessitates further investigation.

REFERENCES

Acherman, J. C., and B. L. Silverman. 2001. Commentary: Dehydroepiandrosterone replacement for patients with adrenal insufficiency. *Lancet* 357:1381.

Armagan, A., N. N. Kim, I. Goldstein, and A. M. Traish. 2006. Dose-response relationship between testosterone and erectile function: Evidence for the existence of a critical threshold. *J Androl* 27:517–26.

Baba, K., M. Yajima, S. Carrier et al. 2000. Effect of testosterone on the number of NADPH diaphorase-stained nerve fibers in the rat corpus cavernosum and dorsal nerve. *Urology* 56:533–8.

Bancroft, J. 1984. Hormones and human sexual behavior. *J Sex Marital Ther* 10(1):3–21.

Bancroft, J., and F. C. Wu. 1983. Changes in erectile responsiveness during androgen replacement therapy. *Arch Sex Behav* 12(1):59–66.

Baulieu, E. E. 1996. Dehydroepiandrosterone (dhea): A fountain of youth? *J Clin Endocrinol Metab* 81:3147–51.

Baulieu, E. E., G. Thomas, S. Legrain et al. 2000. Dehydroepiandrosterone (DHEA), DHEA sulfate and aging: Contribution of the DHEAge Study to a sociobiomedical issue. *PNAS* 97:4279–90.

Bélanger, C., V. Luu-The, P. Dupont, and A. Tchernof. 2002. Adipose tissue intracrinology: Potential importance of local androgen/estrogen metabolism in the regulation of adiposity. *Horm Metab Res* 34:737–45.

Bernini, G., D. Versari, A. Moretti et al. 2006. Vascular reactivity in congenital hypogonadal men before and after testosterone replacement therapy. *J Clin Endocrinol Metab* 91:1691–6.

Bhasin, S., G. R. Cunningham, F. J. Hayes et al. 2006. Testosterone therapy in adult men with androgen deficiency syndromes: An Endocrine Society clinical practice guideline. *J Clin Endocrinol Metab* 91:1995–2010.

Bhatia, V., A. Chaudhuri, R. Tomar et al. 2006. Low testosterone and high C-reactive protein concentrations predict low hematocrit in type 2 diabetes. *Diabet Care* 29:2289–94.

Boyanov, M. A., Z. Boneva, and V. G. Christov. 2003. Testosterone supplementation in men with type 2 diabetes, visceral obesity and partial androgen deficiency. *Aging Male* 6:1–7.

Bruning, J. C., D. Gautam, D. J. Burks et al. 2000. Role of brain insulin receptor in control of body weight and reproduction. *Science* 289:2122–5.

Chen, H., A. S. Lin, Y. Li, C. E. Reiter, M. R. Ver, and M. J. Quon. 2008. DHEA stimulates phosphorylation of FoxO1 in vascular endothelial cells via PI3-kinase- and PKA-dependent signaling pathways to regulate ET-1 synthesis and secretion. *J Biol Chem* 283:29228–38.

Contoreggi, C. S., M. R. Blackman, R. Andres et al. 1990. Plasma levels of estradiol, testosterone and DHEAS do not predict risk of coronary artery disease in men. *J Androl* 11:460.

Corona, G., E. Mannucci, A. D. Fisher et al. 2008. Low levels of androgens in men with erectile dysfunction and obesity. *J Sex Med* 5:2454–63.

Corona, G., E. Mannucci, L. Petrone et al. 2006. Association of hypogonadism and type II diabetes in men attending an outpatient erectile dysfunction clinic. *Int J Impot Res* 18:190–7.

Crude and age-adjusted percentage of civilian, noninstitutionalized population with diagnosed diabetes, United States, 1980–2006; Data Source. Centers for Disease Control and Prevention, National Center for Health Statistics. http://www.cdc.gov/diabetes/statistics/prev/national/figage.htm.

Dandona, P., S. Dhindsa, A. Chaudhuri et al. 2008. Hypogonadotrophic hypogonadism in type 2 diabetes. *Aging Male* 11:107–17.

Davison, S. L., R. Bell, S. Donath, J. G. Montalto, and S. R. Davis. 2005. Androgen levels in adult females: Changes with age, menopause, and oophorectomy. *J Clin Endocrinol Metab* 90:3847–53.

Defay, R., L. Papoz, S. Barny et al. 1998. Hormonal status and NIDDM in the European and Melanesian populations of New Caledonia: A case-control study. *Int J Obes Relat Metab Disord* 22:927–34.

Dhindsa, S., S. Prabhakar, M. Sethi et al. 2004. Frequent occurrence of hypogonadotropic hypogonadism in type 2 diabetes. *J Clin Endocrinol Metab* 89:5462–8.

Earle, C. M., and B. G. Stuckey. 2003. Biochemical screening in the assessment of erectile dysfunction: What tests decide future therapy? *Urology* 62:727.

Eich, D. M., J. E. Nestier, D. E. Johnson et al. 1993. Inhibition of accelerated coronary atherosclerosis with dehydroepiandrosterone in the heterotopic rabbit model of cardiac transplantation. *Circulation* 87:261–9.

El-Sakka, A. I., and H. M. Hassoba. 2006. Age-related testosterone depletion in patients with erectile dysfunction. *J Urol* 176:2589–93.

El-Sakka, A. I., H. M. Sayed, and K. A. Tayeb. 2008. Diabetes-associated androgen alteration in patients with erectile dysfunction. *Int J Androl* 31:602–8.

El-Sakka, A. I., H. M. Sayed, and K. A. Tayeb. 2009. Androgen pattern in patients with type 2 diabetes associated erectile dysfunction: Impact of metabolic control. *Urology* 74:552–60.

Farrukh, I. S., W. Peng, U. Orlinska, and J. R. Hoidal. 1998. Effect of dehydroepiandrosterone on hypoxic pulmonary vasoconstriction: A Ca(2+)-activated K(+)-channel opener. *Am J Physiol* 274:L186–95.

Feldman, H. A., I. Goldstein, D. G. Hatzichristou, R. J. Krane, and J. B. McKinlay. 1994. Impotence and its medical and psychosocial correlates: Results of the Massachusetts Male Aging Study. *J Urol* 151:54–61.

Feldman, H. A., C. Longcope, C. A. Derby et al. 2002. Age trends in the level of serum testosterone and other hormones in middle-aged men: Longitudinal results from the Massachusetts Male Aging Study. *J Clin Endocrinol Metab* 87:589–98.

Foresta, C., N. Caretta, A. Lana et al. 2006. Reduced number of circulating endothelial progenitor cells in hypogonadal men. *J Clin Endocrinol Metab* 91:4599–602.

Fukui, M., J. Soh, M. Tanaka et al. 2007. Low serum testosterone concentration in middle-aged men with type 2 diabetes. *Endocr J* 54:871–7.

Giuliano, F., O. Rampin, A. Schirar, A. Jardin, and J. P. Rousseau. 1993. Autonomic control of penile erection: Modulation by testosterone in the rat. *J Neuroendocrinol* 5:677–83.

Gooren, L. J. 1996. The age-related decline of androgen levels in men: Clinically significant? *Br J Urol* 78:763.

Gooren, L. J., and F. Saad. 2006. Recent insights into androgen action on the anatomical and physiological substrate of penile erection. *Asian J Androl* 8(1):3–9.

Gurnell, E. M., and V. K. Chatterjee. 2001. Dehydroepiandrosterone replacement therapy. *Eur J Endocrinol* 145:103–6.

Harman, S. M., E. J. Metter, J. D. Tobin et al. 2001. Baltimore longitudinal study of aging: Longitudinal effects of aging on serum total and free testosterone levels in healthy men. *J Clin Endocrinol Metab* 86:724–31.

Harris, M. I., W. C. Hadden, W. C. Knowler et al. 1987. Prevalence of diabetes and impaired glucose tolerance and plasma glucose levels in US population aged 20–74 years. *Diabetes* 36:523–34.

Hunt, P. J., E. M. Gurnell, F. A. Huppert et al. 2000. Improvement in mood and fatigue after dehidroepiandrosterone replacement in Addison's disease in a randomized, double blind trial. *J Clin Endocrinol Metab* 85:4650–5.

Johnson, M. D., R. A. Bebb, and S. M. Sirrs. 2002. Uses of DHEA in aging and other disease states. *Ageing Res Rev* 1:29–41.

Kapoor, D., E. Goodwin, K. S. Channer et al. 2006. Testosterone replacement therapy improves insulin resistance, glycaemic control, visceral adiposity and hypercholesterolaemia in hypogonadal men with type 2 diabetes. *Eur J Endocrinol* 154:899–906.

Kawano, H., H. Yasue, A. Kitagawa et al. 2003. Dehydroepiandrosterone supplementation improves endothelial function and insulin sensitivity in men. *J Clin Endocrinol Metab* 88:3190–5.

Keast, J. R., R. J. Gleeson, A. Shulkes, and M. J. Morris. 2002. Maturational and maintenance effects of testosterone on terminal axon density and neuropeptide expression in the rat vas deferens. *Neuroscience* 112:391–8.

Kupelian, V., R. Shabsigh et al. 2006. Is there a relationship between sex hormones and erectile dysfunction? Results from the Massachusetts Male Aging Study. *J Urol* 176(6 Pt 1):2584–8.

Labrie, F., A. Belanger, L. Cusan, and B. Candas. 1997. Physiological changes in dehydroepiandrosterone are not reflected by serum levels of active androgens and estrogens but of their metabolites: Intracrinology. *J Clin Endocrinol Metab* 82:2403–9.

Labrie, F., A. Belanger, V. Luu-The et al. 1998. DHEA and the intracrine formation of androgens and estrogens in peripheral target tissues: Its role during aging. *Steroids* 63:322–8.

Labrie F., V. Luu-The, C. Labrie, and J. Simard. 2001. DHEA and its transformation into androgens and estrogens in peripheral target tissues: Intracrinology. *Front Neuroendocrinol* 22:185–212.

Leder, B. Z., J. L. Rohrer, S. D. Rubin, J. Gallo, and C. Longcope. 2004. Effects of aromatase inhibition in elderly men with low or borderline-low serum testosterone levels. *J Clin Endocrinol Metab* 89:1174.

Lee, S. Y., S. C. Myung, M. Y. Lee et al. 2009. The effects of dehydroepiandrosterone (DHEA)/DHEA-sulfate (DHEAS) on the contraction responses of the clitoral cavernous smooth muscle from female rabbits. *J Sex Med* 6:2653–60.

Liu, D., and J. S. Dillon. 2002. Dehydroepiandrosterone activates endothelial cell nitric-oxide synthase by a specific plasma membrane receptor coupled to Galpha(i2,3). *Biol Chem* 277:21379–88.

Liu, D., and J. S. Dillon. 2004. Dehydroepiandrosterone stimulates nitric oxide release in vascular endothelial cells: Evidence for a cell surface receptor. *Steroids* 69:279–89.

Liu, D., M. Iruthayanathan, L. L. Homan et al. 2008. Dehydroepiandrosterone stimulates endothelial proliferation and angiogenesis through extracellular signal-regulated kinase 1/2-mediated mechanisms. *Endocrinology* 149:889–98.

Liu, D., H. Si, K. A. Reynolds, W. Zhen, Z. Jia, and J. S. Dillon. 2007. Dehydroepiandrosterone protects vascular endothelial cells against apoptosis through a Galphai protein-dependent activation of phosphatidylinositol 3-kinase/Akt and regulation of antiapoptotic Bcl-2 expression. *Endocrinology* 148:3068–76.

Løvas, K., G. Gebre-Medhin, T. S. Trovik et al. 2003. Replacement of dehidroepiandrosterone in adrenal failure: No benefit for subjective health status and sexuality in a 9-month, randomized, parallel group clinical trial. *J Clin Endocrinol Metab* 88:1112–8.

Luboshitzky, R., L. Lavie, Z. Shen-Orr et al. 2003. Pituitary-gonadal function in men with obstructive sleep apnea. The effect of continuous positive airways pressure treatment. *Neuro Endocrinol Lett* 24:463–7.

Lugg, J. A., J. Rajfer, and N. F. Gonzalez-Cadavid. 1995. The role of nitric oxide in erectile function. *J Androl* 16:2–5.

Marin, R., A. Escrig, P. Abreu, and M. Mas. 1999. Androgen-dependent nitric oxide release in rat penis correlates with levels of constitutive nitric oxide synthase isoenzymes. *Biol Reprod* 61:1012–6.

Morales, A. 2003. An integral view of the neuroendocrine aspects of male sexual dysfunction and aging. *Can J Urol* 10:1777.

Morales, A., J. P. Heaton, and C. C. Carson III. 2000. Andropause: A misnomer for a true clinical entity. *J Urol* 163:705–12.

Morales, A. J., J. J. Nolan, J. C. Nelson, and S. S. C. Yen. 1994. Effects of replacement dose of dehydroepiandrosterone in men and women of advancing age. *J Clin Endocrinol Metab* 78:1360–7.

Munarriz, R., L. Talakoub, E. Flaherty et al. 2002. Androgen replacement therapy with dehydroepiandrosterone for androgen insufficiency and female sexual dysfunction: Androgen and questionnaire results. *J Sex Marital Ther* 28(1 suppl):165–73.

Nair, K. S., A. R. Rizza, P. O'Brien et al. 2006. DHEA in elderly women and DHEA or testosterone in elderly men. *N Engl J Med* 355:1647–59.

Oh, J. Y., E. Barrett-Connor, N. M. Wedick et al. 2002. Endogenous sex hormones and the development of type 2 diabetes in older men and women: The Rancho Bernardo study. *Diabetes Care* 25:55–60.

Oka, M., V. Karoor, N. Homma et al. 2007. Dehydroepiandrosterone upregulates soluble guanylate cyclase and inhibits hypoxic pulmonary hypertension. *Cardiovasc Res* 74:377–87.

Orentreich, N., J. L. Brind, R. L. Rizer, and J. H. Vogelman. 1984. Age changes and sex differences in serum dehydroepiandrosterone sulfate concentrations throughout adulthood. *J Clin Endocrinol Metab* 59:551–5.

Park, K. H., S. W. Kim, K. D. Kim, and J. S. Paick. 1999. Effects of androgens on the expression of nitric oxide synthase mRNAs in rat corpus cavernosum. *BJU Int* 83:327–33.

Penson, D. F., Ch. Ng, L. Cai, J. Rajfer, and N. F. Gonzalez-Cadavid. 1996. Androgen and pituitary control of penile nitric oxide synthase and erectile function in the rat. *Biol Reprod* 55:567–74.

Penson, D. F., Ch. Ng, J. Rajfer, and N. F. Gonzalez-Cadavid. 1997. Adrenal control of erectile function and nitric oxide synthase in the rat penis. *Endocrinology* 138:3925–32.

Pitteloud, N., V. K. Mootha, A. A. Dwyer et al. 2005. Relationship between testosterone levels, insulin sensitivity, and mitochondrial function in men. *Diabet Care* 28:1636–42.

Rand, M. J., and C. G. Li. 1995. Nitric oxide as a neurotransmitter in peripheral nerves: Nature of transmitter and mechanism of transmission. *Annu Rev Physiol* 57:659–82.

Reiter, W. J., A. Pycha, G. Schatzel et al. 1999. Dehydroepiandrosterone in the treatment of erectile dysfunction: A prospective, double-blind, randomized, placebo-controlled study. *Urology* 53:590–5.

Reiter, W. J., A. Pycha, G. Schatzl et al. 2000. Serum dehydroepiandrosterone sulfate concentrations in men with erectile dysfunction. *Urology* 55:755–8.

Reiter, W. J., G. Schatzl, I. Mark, A. Zeiner, A. Pycha, and M. Marberger. 2001. Dehydroepiandrosterone in the treatment of erectile dysfunction in patients with different organic etiologies. *Urol Res* 29:278–81.
Rendell, M. S., J. Rajfer, P. A. Wicker et al.; for the Sildenafil Diabetes Study Group. 1999. Sildenafil for treatment of erectile dysfunction in men with diabetes. *JAMA* 281:421–6.
Rhoden, E. L., C. Teloken, R. Mafessoni, and C. A. Souto. 2002. Is there any relation between serum levels of total testosterone and the severity of erectile dysfunction? *Int J Impot Res* 14:167.
Rogers, R. S., T. M. Graziottin, C. S. Lin, Y. W. Kan, and T. F. Lue. 2003. Intracavernosal vascular endothelial growth factor (VEGF) injection and adeno-assoicated virus-mediated VEGF gene therapy prevent and reverse venogenic erectile dysfunction in rats. *Int J Impot Res* 15:26–37.
Rosen, E. D., C. H. Hsu, X. Wang et al. 2002. C/EBPalpha induces adipogenesis through PPARgamma: A unified pathway. *Genes Dev* 16:22–6.
Rosen, R. C., J. B. Kostis, and A. W. Jekelis. 1988. Beta-blocker effects on sexual function in normal males. *Arch Sex Behav* 17:241.
Saad, F., C. E. Hoesl, M. Oettel, J. D. Fauteck, and A. Römmeler. 2005. Dehydroepiandrosterone treatment in the aging male—what should the urologist know? *Eur Urol* 48:724–33.
Sachs, B. D., Y.-C. Liu. 1991. Maintenance of erection of penile glans, but not penile body, after transection of rat cavernous nerves. *J Urol* 146:900–5.
Shabsigh, R., M. A. Perelman et al. 2005. Health issues of men: Prevalence and correlates of erectile dysfunction. *J Urol* 174(2):662–7.
Shealy, C. N. 1995. A review of dehydroepiandrosterone. *Integr Physiol Behav Sci* 30:308–13.
Simoncini, T., P. Mannella, L. Fornari, G. Varone, A. Caruso, and A. R. Genazzani. 2003. Dehydroepiandrosterone modulates endothelial nitric oxide synthesis via direct genomic and nongenomic mechanisms. *Endocrinology* 144:3449–55.
Singh, R., J. N. Artaza, W. E. Taylor et al. 2006. Testosterone inhibits adipogenic differentiation in 3T3-L1 cells: Nuclear translocation of androgen receptor complex with betacatenin and T-cell factor 4 may bypass canonical Wnt signaling to down-regulate adipogenic transcription factors. *Endocrinology* 147:141–54.
Tekdogan, U., A. Tuncel, D. Tuglu, M. Basar, and A. Atan. 2006. Effect of sildenafil citrate treatment on serum dehydroepiandrosterone sulfate levels in patients with erectile dysfunction. *Urology* 68:626–30.
Tilvis, R. S., M. Kähö nen, and M. Härkö nen. 1999. Dehydroepiandrosterone sulphate, diseases and mortality in a general aged population. *Aging (Milano)* 11:30–4.
Traish, A. M., I. Goldstein, and N. N. Kim. 2007. Testosterone and erectile function: From basic research to a new clinical paradigm for managing men with androgen insufficiency and erectile dysfunction. *Eur Urol* 52(1):54–70.
Traish, A. M., and A. T. Guay. 2006. Are androgens critical for penile erections in humans? Examining the clinical and preclinical evidence. *J Sex Med* 3(3):382–404; discussion 404–7.
Traish, A. M., R. Munarriz, L. O'Connell et al. 2003. Effects of medical or surgical castration on erectile function in an animal model. *J Androl* 24:381–7.
Traish, A. M., P. Toselli, S. J. Jeong, and N. N. Kim. 2005. Adipocyte accumulation in penile corpus cavernosum of the orchiectomized rabbit: A potential mechanism for venoocclusive dysfunction in androgen deficiency. *J Androl* 26:242–8.
Trivedi, D. P., and K. T. Khaw. 2001. Dehydroepiandrosterone sulfate and mortality in elderly men and women. *J Clin Endocrinol Metab* 86:4171–7.
Tsai, E. C., A. M. Matsumoto, W. Y. Fujimoto et al. 2004. Association of bioavailable, free, and total testosterone with insulin resistance: Influence of sex hormone-binding globulin and body fat. *Diabet Care* 27:861–8.
Williams, M. R., T. Dawood, S. Ling et al. 2004. Dehydroepiandrosterone increases endothelial cell proliferation in vitro and improves endothelial function in vivo by mechanisms independent of androgen and estrogen receptors. *J Clin Endocrinol Metab* 89:4708–15.
Williams, M. R. I., S. Ling, T. Dawood et al. 2002. Dehydroepiandrosterone inhibits human vascular smooth muscle cell proliferation independent of ARs and ERs. *J Clin Endocrinol Metab* 87:176–81.
Yassin, A., F. Saad, and A. M. Traish. 2006. Testosterone undecanoate restores erectile function in a subset of patients with venous leakage: A series of case reports. *J Sex Med* 3:727–35.
Zmuda, J. M., J. A. Cauley, A. Kriska, N. W. Glynn, J. P. Gutai, and L. H. Kuller. 1997. Longitudinal relation between endogenous testosterone and cardiovascular disease risk factors in middle-aged men. A 13-year follow-up of former Multiple Risk Factor Intervention Trial participants. *Am J Epidemiol* 146:609.
Zvara, P., R. Sioufi, H. M. Schipper, L. R. Begin, and G. B. Brock. 1995. Nitric oxide mediated erectile activity is a testosterone dependent event: A rat erection model. *Int J Impot Res* 7:209–19.

30 DHEA Metabolism, Supplementation, and Decline with Age
Role on Prostate Health

Julia T. Arnold

CONTENTS

USE OF DHEA IN THE UNITED STATES

The grocery checkout order includes a loaf of bread, one dozen eggs, and a bottle of 50-mg DHEA tablets. It is that easy to purchase dehydroepiandrosterone (DHEA). Because many websites claim that DHEA is "all natural" and the "hormone of youth," it is okay for anyone to take it, right? What if the buyer does not check the fine print on the side of the bottle that says "Do not take if you are pregnant or have prostate or breast cancer"? What if he does not suspect he has cancer? This chapter attempts to define how various factors impact DHEA effects on the prostate.

The use and regulation of DHEA are topics of considerable interest. In the early 1980s, DHEA was sold in the United States as a nonprescription drug for its alleged antiaging, anticancer, antiobesity, and other properties. In 1985, the U.S. Food and Drug Administration reclassified DHEA as a prescription drug based on unknown potential long-term risks, and in response to it being banned by the International Olympic Committee. In October 1994, the U.S. Congress enacted the Dietary Supplement Health and Education Act (DSHEA), which made DHEA available over the counter as a dietary supplement. In 2005 and 2007, bills were submitted to Congress that proposed to include DHEA as an anabolic steroid. This was in response to the potential use and abuse of DHEA as an anabolic steroid, and follows the Anabolic Steroid Control Act of 2004. The act added multiple anabolic steroid precursors (not including DHEA) to the list of anabolic steroids that are classified as controlled substances. Now, DHEA is still available as a dietary supplement.

DHEA is the most abundant endogenous adrenal steroid produced in men and women. The supplemental form of DHEA is synthesized from diosgenin, which is extracted from the roots of wild yam (not to be confused with the sweet potato yam). It is increasingly consumed as a dietary supplement with unsubstantiated claims of beneficial effects on body composition, and on cardiometabolic, immune, and neurobiological functions (Valenti et al. 2006). Its long-term safety (Alesci, Manoli, and Blackman 2005) and its effects on prostate tissues remain uncertain. The serum levels of DHEA and its sulfated form, dehydroepiandrosterone sulfate (DHEAS), in adult men and women can be 100–500 times higher than those of testosterone (T), and 1,000–10,000 times higher than those of estradiol. Levels of DHEA and DHEAS peak in men and women at ages 20–30 and are reduced by up to 80% by ages 60–80 (Orentreich et al. 1992; Belanger et al. 1994; Labrie et al. 1997). DHEA also circulates in the brain and functions as a neurosteroid (Rupprecht and Holsboer 1999). So, in a sense, humans are "bathed" in DHEA and DHEAS. Interestingly, humans and other primates are unique among animal species in that their adrenal glands secrete large amounts of DHEA and DHEAS (Belanger et al. 1989; Labrie et al. 1998; Labrie et al. 2001).

DHEA IS METABOLIZED TO ANDROGENS AND ESTROGENS

In older adults, the use of DHEA as a dietary supplement is of potential concern in that its androgenic or estrogenic metabolites may have adverse effects on cancer cells within the steroid hormone-responsive tissues such as prostate or breast. Cancerous lesions may lie dormant as in the prostate, where they are often found during autopsy for other diseases but had not advanced toward symptomatic prostate cancer (Breslow et al. 1977; Haas et al. 2008; Konety et al. 2005). Prostate cancer is now the most commonly diagnosed cancer for men, with 157 per 100,000 men diagnosed from 2003 to 2007. 16% of those patients died of their cancer in that same time frame (Altekruse et al. 2010). Receptors for DHEA or DHEAS have not been definitively isolated (Widstrom and Dillon 2004). But DHEA can act as a direct ligand for the androgen receptor (AR) or the estrogen receptor-β (ER-β) in the prostatic epithelium or for the AR or estrogen receptor-α (ER-α) in the prostatic stroma (Tan et al. 1997; Chen et al. 2005; Arnold and Blackman 2005; Arnold et al. 2007). DHEA was shown to activate the mutated AR in prostate cancer epithelial cells (LNCaP) and induce weak androgenic effects, as shown by its stimulation of LNCaP cell proliferation and modulation of cellular prostatic-specific antigen (PSA), AR, ER-β, and insulin-like growth factor (IGF) axis gene and protein expression (Arnold et al. 2005).

In addition to directly activating the steroid receptors AR or ER, the potential effects of DHEA on the prostate must include its ability to be metabolized to both androgenic and estrogenic ligands (Labrie et al. 2001). DHEA is metabolized to androgenic ligands such as androstenedione, T, and dihydrotestosterone (DHT), and to estrogenic ligands including 7-OH-DHEA, 3β-adiol, or 17β-estradiol (Martin et al. 2004). DHEAS is present in high levels in the prostate, as is the sulfatase that converts DHEAS to DHEA (Klein, Molwitz, and Bartsch 1989). Prostate cells can metabolize DHEA to more active androgenic and/or estrogenic steroids and control the level of intracellular active sex steroids using the catalyzing enzymes 17β-hydroxysteroid dehydrogenase (17β-HSD), 3β-HSD, 5α-reductase, and aromatase (Gingras and Simard 1999; Klein et al. 1988; Voigt and Bartsch 1986; Labrie et al. 1998; Ellem and Risbridger 2006; Ho et al. 2008). This is termed "intracrinology," in which steroid synthesis and metabolism can occur in the cells of peripheral target tissues (Labrie 1991; Labrie et al. 1993). DHEA is an important source of androgens, which, when metabolized by the prostate cells, contribute up to one-sixth of DHT present in the prostate (Geller 1985). Although this very large pool of DHEAS/DHEA is at its peak when men are young, we do not see high rates of prostate cancer in young men.

ER-β is an important mediator of estrogen action in the prostate (Weihua, Warner, and Gustafsson 2002). Activation of ER-β by endogenous and exogenous estrogens as well as phytoestrogens may play a role in modulating androgen activity. ER-β knockout mice exhibit increased epithelial proliferation compared with wild-type mice (Weihua et al. 2001), suggesting that ER-β may inhibit prostate growth. DHEA has been shown to exert direct agonist effects on ER-β in competitive receptor-binding assays, with ER-β being the preferred target for the transcriptional effects of DHEA (Chen et al. 2005). DHEA metabolites may also activate ER-β in the prostate, such as metabolism to 7α-hydroxy-DHEA (7HD), a known ligand for ER-β (Martin et al. 2004). Activation of ER-β can antagonize androgenic effects on prostate cancer growth and thus may modulate androgen effects on the prostate. Alternatively, DHEA or its estrogen metabolites induced growth in androgen-independent prostate cancer cells without a functional AR. DHEA effects were shown to act via the ER-β, which associated with Dishevelled2 (Dvl2), a Wnt signaling protein, and activated the β-catenin/TCF signaling pathway (Liu, Arnold, and Blackman 2010). Finally, DHEA can act via nonsteroid-receptor, nongenomic pathways, as shown in vascular endothelial cells (Formoso et al. 2006).

What regulates the metabolism of DHEA in the prostate is not understood. Questions include: Under what conditions is DHEA metabolized toward the androgenic metabolic pathway? When does DHEA metabolism result in estrogenic metabolites in the prostate? Under what circumstances does it remain a prohormone and not become metabolized, but rather circulate at increased levels over other steroids, potentially providing a "hormonal buffer" (Regelson, Loria, and Kalimi 1988) against endogenous androgen or estrogen levels? What regulates the orchestration of steroid-metabolizing enzymes, HSDs? What environmental or endogenous factors favor androgen versus estrogen metabolism?

In the prostate, the androgen–estrogen balance is crucial to prostate function. As a man ages, the androgen levels decrease while the estrogen levels increase. It is very important to understand how DHEA metabolism is regulated within the prostate tissue. This chapter presents the hypothesis that the surrounding tissue microenvironment plays a very important role in the direction and course of DHEA metabolism in the prostate.

DHEA SUPPLEMENTATION—CANCER PREVENTIVE OR PROMOTING IN THE PROSTATE?

The question of whether supplemental DHEA can promote or protect against prostate cancer growth has not been answered (Arnold 2009). Oral DHEA administration has been shown to increase serum DHEA levels from 8.5 to 14.7 nM in men and 7.2 to 16.1 nM in women with unchanged serum T and DHT levels (Morales et al. 1994). Another investigation, using a 12-week intervention of a daily dose of a 20% DHEA solution absorbed through the skin, found increased DHEA levels in men from 10 to 30 nM and in women from 9 to 30 nM. DHEAS levels increased from 1525 to 3000 nM in men and 750 to ~2000 nM in women. Serum levels of DHT and T remained unchanged in men and were increased 50% in women from approximately 1.3 to 2.3 nM (Labrie 1997).

Yet in some studies, higher levels of endogenous DHEA did not appear to be a risk factor for prostate cancer. An analysis of prostatic pathological samples from patients with insufficient adrenal and testicular androgen levels concluded that adrenal androgens alone did not promote prostatic growth (Oesterling, Epstein, and Walsh 1986). In an epidemiological analysis, levels of DHEA and DHEAS were decreased by 11% and 12%, respectively, in serum collected from patients who later developed prostate cancer. This study concluded that it was unlikely that serum levels of DHEA or DHEAS were important risk factors for prostate cancer (Comstock, Gordon, and Hsing 1993). Also, DHEA levels in prostate cancer patients were found to be significantly lower, but total and free T levels were significantly higher than those in an age-matched control group (Stahl et al. 1992). A study involving DHEA supplements to patients for erectile dysfunction suggested that there was no impact of DHEA treatment on the mean serum levels of PSA, T, or the mean prostate volume (Reiter et al. 1999). A comprehensive review is available summarizing studies determining steroid levels in serum of DHEA-treated patients (Kroboth et al. 1999).

Although androgens are targeted in prostate cancer therapy, the role of serum androgen levels (including androgenic DHEA metabolites) in the risk of prostate cancer or urinary symptoms is increasingly unclear (Miwa, Kaneda, and Yokoyama 2008; Trifiro et al. 2010). There are even arguments that *declining* androgen levels associated with aging may contribute to prostatic carcinogenesis. This may be due to increased prostatic atrophy that results from a decline in androgen levels, and the resultant compensatory hyperplasia that increases the risk of cancer formation (Prehn 1999). The bottom line is that while existing clinical evidence supports that DHEA would not be detrimental to normal prostate, these studies do not include measurements of DHEA levels and metabolites within the prostate tissue. Also, there is no data concerning DHEA effects on preneoplastic cells or prostate cancer cells, which may be more complex. This supports the need for more basic and clinical research on DHEA effects on prostate and its role in human prostate cancer.

PRECLINICAL STUDIES OF DHEA IN PROSTATE MODELS

Negative health effects of DHEA have been reported in preclinical experimental models, including in induction of lipid peroxidation (Swierczynski and Mayer 1996) and in hepatocarcinogenesis in rats (Rao et al. 1992). DHEA and DHEAS were stimulatory to growth of an androgen-sensitive (mammary) cancer cell line acting via AR (Begin, Luthy, and Labrie 1988). Additionally, in the presence of 3β- and 17β-HSD, prostate cells can metabolize circulating DHEA into DHT (Labrie et al. 1993).

Yet in other rodent prostate cancer models, DHEA was protective against prostate cancer induction by androgens and carcinogens (Green et al. 2001; Rao et al. 1999; Lubet et al. 1998; Perkins et al. 1997; Ciolino et al. 2003). Administration of DHEA to mice and rats inhibited tumor development in the breast, lung, skin, liver, colon, and lymphatic tissue (Schwartz and Pashko 1995a). The antiproliferative and tumor-protective effects of DHEA are thought to be due to its inhibitory effects on the glucose 6-phosphate dehydrogenase and pentose phosphate pathways (Schwartz and Pashko 1995b; Gordon, Shantz, and Talalay 1987). These pathways contribute to the generation of oxygen free radicals that may act as second messengers in stimulating hyperplasia (Singh and Lippman 1998). DHEA has also been shown to inhibit carcinogen-metabolizing enzymes, thus decreasing carcinogenic insult (Ciolino et al. 2003). The relevance of these studies to human biology is uncertain, however, as the amounts of DHEA and DHEAS are much lower in rodents than in humans (Luu-The, Pelletier, and Labrie 2005) and the physiological importance of these adrenal steroids is unknown in rodents. Mice and rats lack the activity of 17-hydroxylation in their adrenal glands and cannot make DHEA (Brock and Waterman 1999). Also, in transplanted human prostate cancer xenografts, DHEA did not stimulate growth of PC-82 tumor tissue, while androgen was capable of inducing PC-82 tumor growth (van Weerden et al. 1992).

Although epidemiological evidence does not support the role of DHEA in prostate cancer progression, the final mechanisms have not been delineated. What happens with age that may make the prostate more vulnerable? Could DHEA or DHEAS play a protective role in normal prostate, but contribute to prostate cancer progression in the context of a reactive or senescent stromal microenvironment or tumor environment that may be present in prostatic tissues at advanced ages? When one considers the latency period for cancers, the early high exposure to DHEA in young men, hypothetically, could be confounded by risk factors such as smoking, inflammation, or diet—providing a tissue microenvironment that alters DHEA metabolism, and leading to an altered androgen–estrogen balance (Carruba 2006). Similar hypotheses of early hormonal exposure are proposed for increasing breast cancer risk, resulting from early menarche or late full-term pregnancy (Harper et al. 1974; Martin and Boyd 2008). To understand DHEA effects in the prostate, it is important to understand how the prostate changes with age and the role of the tissue microenvironment.

WHY THE TISSUE MICROENVIRONMENT IS IMPORTANT TO DHEA'S EFFECTS ON THE PROSTATE

There are many factors involved in the regulation of the prostate tissue, including the epithelial cells, stromal cells, steroid hormones, and extracellular matrix. Each plays a role in normal functioning of the tissue. The prostatic epithelial glands are the "workhorse" of the prostate, producing secretory factors to be included in the prostatic ejaculate. The glands are surrounded and held together by the stromal cells and extracellular matrix. The stroma includes fibroblasts, smooth muscle cells, endothelial cells and blood vessels, neuroendocrine cells, collagen and/or smooth muscle actin, and other extracellular matrix components. The importance of stromal cells in regulating epithelial function, especially in cancer, has been extensively reviewed (Cunha 1994; Donjacour and Cunha 1991; Risbridger et al. 2001; Taylor and Risbridger 2008). In many tissues, hormonally induced stromal cells provide secondary factors, mediating the hormonal effect to epithelial cells (Cunha, Cooke, and Kurita 2004). The interconnections between these components are discussed in a review (Arnold and Isaacs 2002) that explores the theory that prostate cancer "is not only the epithelial cells' fault." The main points here are that cancer progression in the glandular epithelial cells is due not only to factors in the epithelial cells themselves but to modulation by stimulatory or inhibitory factors stemming from the local microenvironment, including the hormonal milieu. The dysfunction of any of these factors may contribute to the progression of cancer, along with outside factors such as immune-induced oxidative damage and environmental carcinogens.

An interesting perspective of cancer promotion comes from a phrase recently used in the context of both obesity and autism, but that works here as well: "Genes load the gun, environment pulls the trigger" (Peeke 2000; Neimark 2007). The epithelial cells may have inherent or acquired mutations in oncogenes or tumor-suppressor genes and may remain dormant for many years ("loading the gun"). But the expression of these mutations or epigenetic abnormalities may be influenced by the tissue microenvironment and the external environment, for example, by acute aggravation such as inflammation or environmental stressors, or induction of a reactive stromal phenotype ("pulling the trigger"). During the time that DHEA and other systemic endogenous steroid hormone levels are decreasing with age, the local tissue is senescing. With tissue aging and in tumorigenesis, the stromal microenvironment is altered as seen in many types of cancer (Schedin and Elias 2004). Stromal cells could become "activated" or reactive in response to environmental changes. For prostate stromal cells, this means that there is an increase in myofibroblastic characteristics, including increased levels of smooth muscle α-actin (Tuxhorn, Ayala, and Rowley 2001) and secreted factors (Untergasser et al. 2005). TGF-β1 and other proinflammatory cytokines are known to promote a wound-repair-type reactive myofibroblast phenotype in prostate cancer (Peehl and Sellers 1997; Tuxhorn, Ayala, and Rowley 2001; Tuxhorn et al. 2002; Rowley 1998; Deutsch et al. 2004; Kalluri and Zeisberg 2006). The stromal microenvironment in localized prostate cancer also includes a decrease in stromal AR expression and function but maintenance of stromal ER-α activity, which can adversely affect the adjacent epithelial cells (Ellem et al. 2009; Ricke et al. 2008).

The tissue microenvironment is also altered by chronic infection and inflammation in the tissues, conditions that exist in about 20% of human cancers (De Marzo et al. 2007). In the prostate, some of the earliest precancerous lesions have been characterized as proliferative inflammatory atrophy (PIA) and are associated with unusually high proliferation (De Marzo et al. 1999). In this inflammatory microenvironment, there are also macrophages possessing sulfatase activity that convert DHEAS to DHEA (Hennebold and Daynes 1994), and the cytokines IL-4 and IL-13 that can induce 3β-HSD activity in normal prostate epithelial cells, thus promoting conversion of DHEA into T and estrogen metabolites (Simard and Gingras 2001).

These factors present in inflammation may contribute to the localized alterations in steroid (DHEA) hormone metabolism. The key to understanding DHEA effects on the human prostate epithelial cells and cancer cells may be through its secondary effects as mediated by stromal cells within the prostate tissue microenvironment. One hypothesis is that secondary stromal factors mediated by DHEA are similar to growth factors produced in androgen-treated stromal cells. In many tissues, hormonally induced stromal cells provide secondary factors, mediating the hormonal effect to the epithelial cells. In addition to growth factors, these mediators may include steroid-metabolizing enzymes.

PRECLINICAL STUDIES IN THE PROSTATE STROMAL ENVIRONMENT AND INFLAMMATION AND DHEA METABOLISM

To better decipher cellular and molecular effects of DHEA on human prostate tissue, it was necessary to develop an appropriate experimental model. DHEA effects on prostate epithelial cells were mediated by prostate stromal cells in an *in vitro* cell culture model that included human epithelial cancer cells (LAPC-4 cells) harboring a normal AR and minimally responsive to DHEA (Arnold et al. 2008), and human prostatic primary stromal cells ("6S"). The stromal cells responded to androgens such as DHT by increasing secretory IGF-1, whereas DHEA did not produce the same effect (Le et al. 2005). Compared with the minimal response to DHEA in the separate cultures, when the cells were combined in coculture, an important stromal regulatory mechanism was revealed concerning DHEA effects in the prostate. In the presence of 6S stromal cells in coculture, DHEA stimulated LAPC-4 protein and gene expression of PSA to levels approaching induction by DHT. DHEA-treated 6S stromal cells produced increased T, suggesting that stromal cells can metabolize DHEA to T and possibly DHT, and thus induce increased PSA expression in the cocultured epithelial cells (Arnold et al. 2008).

The next series of studies mimicked a reactive stromal microenvironment by adding the proinflammatory cytokine TGF-β1 to the coculture model of 6S prostate stromal plus LAPC-4 prostate cancer epithelial cells. This model represents the increased levels of cytokines and characteristics of reactive stroma, similar to those present in PIA, prostatic intraepithelial neoplasia (PIN), and prostate cancer. In such studies, LAPC-4 cells were grown in coculture with prostate stromal cells (6S) and treated with DHEA +/– TGF-β1 (Gray et al. 2009). PSA expression and T secretion in LAPC-4/6S cocultures were compared with those in monocultured epithelial and stromal cells using real-time polymerase chain reaction and/or enzyme-linked immunosorbent assay. Combined administration of TGF-β1 + DHEA increased coculture production of T over DHEA treatment alone, and cocultures increased PSA protein secretion two to four times and PSA gene expression up to 50-fold. TGF-β1 greatly increased stromal-mediated DHEA effects on T production and epithelial cell PSA production.

TGF-β1 is known to influence steroid-metabolizing enzymes, as by decreasing 3β-HSD in adrenocortical cells (Rainey, Naville, and Mason 1991) and 17β-HSD in breast cancer cells (Ee et al. 1999). TGF-β1 can decrease activity of CYP7B, which metabolizes DHEA to 7α-OH-DHEA, a ligand for ER-β (Dulos and Boots 2006), as measured in inflammatory tissues. Furthermore, mechanistic studies are in progress to characterize dose and time responses of TGF-β1 modulation of the DHEA metabolism to androstenedione and T, and they have found effects on HSD metabolic enzyme gene expression in the prostate stromal cells, leading to proandrogenic metabolism and unique mechanisms of associations of TGF-β1 receptors with HSD enzymes (Liu and Arnold, manuscript submitted). Intraprostatic factors that influence DHEA metabolism may alter the balance of androgenic and estrogenic ligands affecting growth and function of the prostate. To summarize the work of this laboratory on DHEA effects on human prostate cells, we have found that DHEA has minimal effects on prostate stromal or epithelial cells when cultured alone, but a principal mechanism of DHEA effects is found by coculture of these cells. Stromal cell paracrine and intracrine activities play an important role in DHEA effects on epithelial prostate cells. TGF-β1 induces a reactive stromal phenotype and results in increased androgenic metabolism of DHEA. DHEA effects in human prostate tissues may also be influenced by these same endocrine–immune–paracrine interactions. Extrapolation of *in vitro* work to clinical significance is needed. Ongoing studies are underway to establish the connection between increased inflammation, reactive stroma, and increased androgen metabolites in human prostate tissues.

CAN DHEA HAVE ESTROGENIC EFFECTS ON THE PROSTATE?

All of this makes for a very complicated story. The complexity of the balance between androgenic and estrogenic effects of DHEA on the prostate, whether as a direct ligand or through its metabolites, is matched by the complexity of estrogen action through the ER-α versus ER-β in the prostate. The intratissular balance of androgen and estrogen levels is very important for prostatic development and differentiation. Likewise, the balance between ER-α and ER-β, including the temporal and spatial expression, determines the various responses of prostate to estrogen and is crucial for prostate health (McPherson, Ellem, and Risbridger 2008). Estrogens have been long used in prostate cancer therapy. There is substantial evidence from *in vitro*, animal, and epidemiological studies showing that estrogen can play a crucial role in human prostate carcinogenesis and tumor progression (for reviews, see Carruba 2007; Risbridger, Ellem, and McPherson 2007).

Although men have very low levels of endogenous estrogens, additional estrogen may be produced as metabolic byproducts of DHEA or aromatized from testosterone. Paracrine functions become important when considering that stromal cells possess aromatase, allowing conversion of androgens to estrogens (Ellem and Risbridger 2006; Ho et al. 2008). Estrogens may have beneficial effects that support normal growth of the prostate but can also be detrimental to prostate growth and differentiation (McPherson et al. 2007). Estrogens acting through the prostate stromal ER-α may be growth promoting, while those acting through the epithelial ER-β may be antagonistic to ER-α- or AR-activated pathways (Chang and Prins 1999; Signoretti and Loda 2001). Excess estrogen induces squamous metaplasia and can act synergistically with androgens to induce glandular hyperplasia (Isaacs 1984). On the contrary, estrogens can inhibit prostate cancer xenograft growth in female intact and ovariectomized mice in the absence of androgens (Corey et al. 2002). These inhibitory effects were postulated to occur by direct actions via the ER or by E_2 effects on other cells secreting secondary factors, which influence cancer cell growth.

Phytoestrogens may play a role in reversing metabolism of DHEA to androgens. Prostate stromal-epithelial cocultures described above were treated with red clover isoflavones including genistein, daidzein, biochanin A, and formononetin. The effect of the red clover isoflavones was to diminish the TGF-β-induced metabolism of DHEA to T and diminish the expression of PSA in the cocultured epithelial cells (Gray et al. 2009). This is an active area of research aimed at understanding mechanisms of natural products and hormonal supplements that may affect prostate health.

CONCLUSION

In the inflammatory prostate tissue microenvironment, as in PIA or PIN with associated reactive stroma, there could potentially be altered metabolism of endogenous DHEA, either to androgens or estrogens. The microenvironment of the prostate may dictate the ultimate fate of DHEA metabolism toward androgenic or estrogenic ligands. This is important, not only in consideration of DHEA used as a dietary supplement, but also in determination of its physiological role in prostate.

We still do not understand all of the endogenous or exogenous factors that promote androgenic versus estrogenic metabolism of DHEA, nor which homeostatic mechanisms regulate levels and activity of the intraprostatic steroid metabolic enzymes 3β- or 17β-HSDs. The complexities of intratissular levels and balance of androgens and estrogens, along with the contribution of DHEA to prostate function, are becoming increasingly appreciated as the role of serum androgen levels in risk of prostate cancer, it is becoming less clear (Hsing, Chu, and Stanczyk 2008). The question of whether DHEA can be cancer promoting or cancer preventive is contextual and continues to be debated (Arnold and Blackman 2005; Arnold 2009). These studies highlight the need for further rigorous, well-designed laboratory, translational, and clinical investigations of the mechanisms of action, efficacy, and safety of DHEA, so that questions regarding its potential for improving or compromising human health can finally be answered.

One final question is why endogenous DHEA and DHEAS levels are so high in humans and primates but not in other species. No one seems to have an answer. Is there an evolutionary advantage of high DHEA levels for humans? Is it based on DHEA effects as a neurosteroid? Is it because of the conversion from DHEAS in the fetal adrenal gland to large amounts of estriol during pregnancy, which is very important in uterine blood flow (Walsh and Siiteri 1975)? Does DHEA promote health or contribute to possible complications in hormone-related cancers? The combination of research studies presented in this book will help enlighten us to DHEA's place in human health.

ACKNOWLEDGMENTS

This work was supported by the Intramural Research Program, National Center for Complementary and Alternative Medicine, National Institutes of Health, and U.S. Department of Health and Human Services, Bethesda, Maryland. The author thanks Dr. Xunxian Liu and Dr. Yun-shang Piao for their critical review of this manuscript.

REFERENCES

Alesci, S., I. Manoli, and M. R. Blackman. 2005. Dehydrodepiandrosterone (DHEA). In *Encyclopedia of Dietary Supplements*, 1st ed., ed. P. Coates, M. R. Blackman, G. Cragg, M. Levine, J. Moss, and J. White, 167–76. New York: Marcel Dekker, Inc.

Altekruse, S. F., C. L. Kosary, M. Krapcho, N. Neyman, R. Aminou, W. Waldron, J. Ruhl et al. eds. 2010. *SEER Cancer Statistics Review, 1975–2007*. Bethesda, MD: National Cancer Institute.

Arnold, J. T. 2009. DHEA metabolism in prostate: For better or worse? *Mol Cell Endocrinol* 301(1–2):83–8.

Arnold, J. T., and M. R. Blackman. 2005. Does DHEA exert direct effects on androgen and estrogen receptors, and does it promote or prevent prostate cancer? *Endocrinology* 146(11):4565–7.

Arnold, J. T., N. E. Gray, K. Jacobowitz, L. Viswanathan, P. W. Cheung, K. K. McFann, H. D. Le, and M. R. Blackman. 2008. Human prostate stromal cells stimulate increased PSA production in DHEA-treated prostate cancer epithelial cells. *J Steroid Biochem Mol Biol* 111(3–5):240–6.

Arnold, J. T., and J. T. Isaacs. 2002. Mechanisms involved in the progression of androgen-independent prostate cancers: It is not only the cancer cell's fault. *Endocr Relat Cancer* 9(1):61–73.

Arnold, J. T., H. Le, K. K. McFann, and M. R. Blackman. 2005. Comparative effects of DHEA vs. testosterone, dihydrotestosterone, and estradiol on proliferation and gene expression in human LNCaP prostate cancer cells. *Am J Physiol Endocrinol Metab* 288(3):E573–84.

Arnold, J. T., X. Liu, J. D. Allen, H. Le, K. K. McFann, and M. R. Blackman. 2007. Androgen receptor or estrogen receptor-beta blockade alters DHEA-, DHT-, and E2-induced proliferation and PSA production in human prostate cancer cells. *Prostate* 67(11):1152–62.

Begin, D., I. A. Luthy, and F. Labrie. 1988. Adrenal precursor C19 steroids are potent stimulators of growth of androgen-sensitive mouse mammary carcinoma Shionogi cells in vitro. *Mol Cell Endocrinol* 58(2–3):213–9.

Belanger, B., A. Belanger, F. Labrie, A. Dupont, L. Cusan, and G. Monfette. 1989. Comparison of residual C-19 steroids in plasma and prostatic tissue of human, rat and guinea pig after castration: Unique importance of extratesticular androgens in men. *J Steroid Biochem* 32(5):695–8.

Belanger, A., B. Candas, A. Dupont, L. Cusan, P. Diamond, J. L. Gomez, and F. Labrie. 1994. Changes in serum concentrations of conjugated and unconjugated steroids in 40- to 80-year-old men. *J Clin Endocrinol Metab* 79(4):1086–90.

Breslow, N., C. W. Chan, G. Dhom, R. A. Drury, L. M. Franks, B. Gellei, Y. S. Lee et al. 1977. Latent carcinoma of prostate at autopsy in seven areas. The International Agency for Research on Cancer, Lyons, France. *Int J Canc* 20(5):680–8.

Brock, B. J., and M. R. Waterman. 1999. Biochemical differences between rat and human cytochrome P450c17 support the different steroidogenic needs of these two species. *Biochemistry* 38(5):1598–606.

Carruba, G. 2006. Estrogens and mechanisms of prostate cancer progression. *Ann N Y Acad Sci* 1089:201–17.

Carruba, G. 2007. Estrogen and prostate cancer: An eclipsed truth in an androgen-dominated scenario. *J Cell Biochem* 102(4):899–911.

Chang, W. Y., and G. S. Prins. 1999. Estrogen receptor-beta: Implications for the prostate gland. *Prostate* 40(2):115–24.

Chen, F., K. Knecht, E. Birzin, J. Fisher, H. Wilkinson, M. Mojena, C. T. Moreno et al. 2005. Direct agonist/antagonist functions of dehydroepiandrosterone. *Endocrinology* 146(11):4568–76.

Ciolino, H., C. MacDonald, O. Memon, M. Dankwah, and G. C. Yeh. 2003. Dehydroepiandrosterone inhibits the expression of carcinogen-activating enzymes in vivo. *Int J Canc* 105(3):321–5.

Comstock, G. W., G. B. Gordon, and A. W. Hsing. 1993. The relationship of serum dehydroepiandrosterone and its sulfate to subsequent cancer of the prostate. *Cancer Epidemiol Biomarkers Prev* 2(3):219–21.

Corey, E., J. E. Quinn, M. J. Emond, K. R. Buhler, L. G. Brown, and R. L. Vessella. 2002. Inhibition of androgen-independent growth of prostate cancer xenografts by 17beta-estradiol. *Clin Cancer Res* 8(4):1003–7.

Cunha, G. R. 1994. Role of mesenchymal-epithelial interactions in normal and abnormal development of the mammary gland and prostate. *Cancer* 74(3 Suppl):1030–44.

Cunha, G. R., P. S. Cooke, and T. Kurita. 2004. Role of stromal-epithelial interactions in hormonal responses. *Arch Histol Cytol* 67(5):417–34.

De Marzo, A. M., V. L. Marchi, J. I. Epstein, and W. G. Nelson. 1999. Proliferative inflammatory atrophy of the prostate: Implications for prostatic carcinogenesis. *Am J Pathol* 155(6):1985–92.

De Marzo, A. M., E. A. Platz, S. Sutcliffe, J. Xu, H. Gronberg, C. G. Drake, Y. Nakai, W. B. Isaacs, and W. G. Nelson. 2007. Inflammation in prostate carcinogenesis. *Nat Rev Cancer* 7(4):256–69.

Deutsch, E., L. Maggiorella, P. Eschwege, J. Bourhis, J. C. Soria, and B. Abdulkarim. 2004. Environmental, genetic, and molecular features of prostate cancer. *Lancet Oncol* 5(5):303–13.

Donjacour, A. A., and G. R. Cunha. 1991. Stromal regulation of epithelial function. *Cancer Treat Res* 53:335–64.

Dulos, J., and A. H. Boots. 2006. DHEA metabolism in arthritis: A role for the p450 enzyme Cyp7b at the immune-endocrine crossroad. *Ann N Y Acad Sci* 1069:401–13.

Ee, Y. S., L. C. Lai, K. Reimann, and P. K. Lim. 1999. Effect of transforming growth factor-beta1 on oestrogen metabolism in MCF-7 and MDA-MB-231 breast cancer cell lines. *Oncol Rep* 6(4):843–6.

Ellem, S. J., and G. P. Risbridger. 2006. Aromatase and prostate cancer. *Minerva Endocrinol* 31(1):1–12.

Ellem, S. J., H. Wang, M. Poutanen, and G. P. Risbridger. 2009. Increased endogenous estrogen synthesis leads to the sequential induction of prostatic inflammation (prostatitis) and prostatic pre-malignancy. *Am J Pathol* 175(3):1187–99.

Formoso, G., H. Chen, J. A. Kim, M. Montagnani, A. Consoli, and M. J. Quon. 2006. Dehydroepiandrosterone mimics acute actions of insulin to stimulate production of both nitric oxide and endothelin 1 via distinct phosphatidylinositol 3-kinase- and mitogen-activated protein kinase-dependent pathways in vascular endothelium. *Mol Endocrinol* 20(5):1153–63.

Geller, J. 1985. Rationale for blockade of adrenal as well as testicular androgens in the treatment of advanced prostate cancer. *Semin Oncol* 12(1 Suppl. 1):28–35.

Gingras, S., and J. Simard. 1999. Induction of 3beta-hydroxysteroid dehydrogenase/isomerase type 1 expression by interleukin-4 in human normal prostate epithelial cells, immortalized keratinocytes, colon, and cervix cancer cell lines. *Endocrinology* 140(10):4573–84.

Gordon, G. B., L. M. Shantz, and P. Talalay. 1987. Modulation of growth, differentiation and carcinogenesis by dehydroepiandrosterone. *Adv Enzyme Regul* 26:355–82.

Gray, N. E., X. Liu, R. Choi, M. R. Blackman, and J. T. Arnold. 2009. Endocrine-immune-paracrine interactions in prostate cells as targeted by phytomedicines. *Cancer Prev Res (Phila)* 2(2):134–42.

Green, J. E., M. A. Shibata, E. Shibata, R. C. Moon, M. R. Anver, G. Kelloff, and R. Lubet. 2001. 2-difluoromethylornithine and dehydroepiandrosterone inhibit mammary tumor progression but not mammary or prostate tumor initiation in C3(1)/SV40 T/t-antigen transgenic mice. *Cancer Res* 61(20):7449–55.

Haas, G. P., N. Delongchamps, O. W. Brawley, C. Y. Wang, and G. de la Roza. 2008. The worldwide epidemiology of prostate cancer: Perspectives from autopsy studies. *Can J Urol* 15(1):3866–71.

Harper, M. E., A. Pike, W. B. Peeling, and K. Griffiths. 1974. Steroids of adrenal origin metabolized by human prostatic tissue both in vivo and in vitro. *J Endocrinol* 60(1):117–25.

Hennebold, J. D., and R. A. Daynes. 1994. Regulation of macrophage dehydroepiandrosterone sulfate metabolism by inflammatory cytokines. *Endocrinology* 135(1):67–75.

Ho, C. K., J. Nanda, K. E. Chapman, and F. K. Habib. 2008. Oestrogen and benign prostatic hyperplasia: Effects on stromal cell proliferation and local formation from androgen. *J Endocrinol* 197(3):483–91.

Hsing, A. W., L. W. Chu, and F. Z. Stanczyk. 2008. Androgen and prostate cancer: Is the hypothesis dead? *Cancer Epidemiol Biomarkers Prev* 17(10):2525–30.

Isaacs, J. T. 1984. The aging ACI/Seg versus Copenhagen male rat as a model system for the study of prostatic carcinogenesis. *Cancer Res* 44(12 Pt 1):5785–96.

Kalluri, R., and M. Zeisberg. 2006. Fibroblasts in cancer. *Nat Rev Cancer* 6(5):392–401.

Klein, H., M. Bressel, H. Kastendieck, and K. D. Voigt. 1988. Quantitative assessment of endogenous testicular and adrenal sex steroids and of steroid metabolizing enzymes in untreated human prostatic cancerous tissue. *J Steroid Biochem* 30(1–6):119–30.

Klein, H., T. Molwitz, and W. Bartsch. 1989. Steroid sulfate sulfatase in human benign prostatic hyperplasia: Characterization and quantification of the enzyme in epithelium and stroma. *J Steroid Biochem* 33(2):195–200.

Konety, B. R., V. Y. Bird, S. Deorah, and L. Dahmoush. 2005. Comparison of the incidence of latent prostate cancer detected at autopsy before and after the prostate specific antigen era. *J Urol* 174(5):1785–8; discussion 1788.

Kroboth, P. D., F. S. Salek, A. L. Pittenger, T. J. Fabian, and R. F. Frye. 1999. DHEA and DHEA-S: A review. *J Clin Pharmacol* 39(4):327–48.

Labrie, F. 1991. Intracrinology. *Mol Cell Endocrinol* 78(3):C113–8.

Labrie, F., A. Bélanger, L. Cusan, and B. Candas. 1997. Physiological changes in dehydroepiandrosterone are not reflected by serum levels of active androgens and estrogens but of their metabolites: Intracrinology. *J Clin Endocrinol Metab* 82(8):2403–9.

Labrie, F., A. Belanger, V. Luu-The, C. Labrie, J. Simard, L. Cusan, J. L. Gomez, and B. Candas. 1998. DHEA and the intracrine formation of androgens and estrogens in peripheral target tissues: Its role during aging. *Steroids* 63(5–6):322–8.

Labrie, F., A. Dupont, J. Simard, V. Luu-The, and A. Belanger. 1993. Intracrinology: The basis for the rational design of endocrine therapy at all stages of prostate cancer. *Eur Urol* (24 Suppl. 2):94–105.

Labrie, F., V. Luu-The, C. Labrie, and J. Simard. 2001. DHEA and its transformation into androgens and estrogens in peripheral target tissues: Intracrinology. *Front Neuroendocrinol* 22(3):185–212.

Le, H., J. T. Arnold, K. K. McFann, and M. R. Blackman. 2006. Dihydrotestosterone and testosterone, but not DHEA or estradiol, differentially modulate IGF-I, IGFBP-2 and IGFBP-3 gene and protein expression in primary cultures of human prostatic stromal cells. *Am J Physiol Endocrinol Metab* 290(5):952–60.

Liu, X., J. T. Arnold, and M. R. Blackman. 2010. Dehydroepiandrosterone administration or G{alpha}q overexpression induces {beta}-catenin/T-Cell factor signaling and growth via increasing association of estrogen receptor-{beta}/Dishevelled2 in androgen-independent prostate cancer cells. *Endocrinology* 151(4):1428–40.

Lubet, R. A., G. B. Gordon, R. A. Prough, X. D. Lei, M. You, Y. Wang, C. J. Grubbs et al. 1998. Modulation of methylnitrosourea-induced breast cancer in Sprague Dawley rats by dehydroepiandrosterone: Dose-dependent inhibition, effects of limited exposure, effects on peroxisomal enzymes, and lack of effects on levels of Ha-Ras mutations. *Cancer Res* 58(5):921–6.

Luu-The, V., G. Pelletier, and F. Labrie. 2005. Quantitative appreciation of steroidogenic gene expression in mouse tissues: New roles for type 2 5alpha-reductase, 20alpha-hydroxysteroid dehydrogenase and estrogen sulfotransferase. *J Steroid Biochem Mol Biol* 93(2–5):269–76.

Martin, L. J., and N. F. Boyd. 2008. Mammographic density. Potential mechanisms of breast cancer risk associated with mammographic density: Hypotheses based on epidemiological evidence. *Breast Cancer Res* 10(1):201.

Martin, C., M. Ross, K. E. Chapman, R. Andrew, P. Bollina, J. R. Seckl, and F. K. Habib. 2004. CYP7B generates a selective estrogen receptor beta agonist in human prostate. *J Clin Endocrinol Metab* 89(6):2928–35.

McPherson, S. J., S. J. Ellem, and G. P. Risbridger. 2008. Estrogen-regulated development and differentiation of the prostate. *Differentiation* 76(6):660–70.

McPherson, S. J., S. J. Ellem, E. R. Simpson, V. Patchev, K. H. Fritzemeier, and G. P. Risbridger. 2007. Essential role for estrogen receptor beta in stromal-epithelial regulation of prostatic hyperplasia. *Endocrinology* 148(2):566–74.

Miwa, Y., T. Kaneda, and O. Yokoyama. 2008. Association between lower urinary tract symptoms and serum levels of sex hormones in men. *Urology* 72(3):552–5.

Morales, A. J., J. J. Nolan, J. C. Nelson, and S. S. Yen. 1994. Effects of replacement dose of dehydroepiandrosterone in men and women of advancing age. *J Clin Endocrinol Metab* 78:1360–7.

Neimark, J. 2007. Autism: It's not just in the head. *Discover Magazine*. http://discovermagazine.com/2007/apr/autism-it2019s-not-just-in-the-head.

Oesterling, J. E., J. I. Epstein, and P. C. Walsh. 1986. The inability of adrenal androgens to stimulate the adult human prostate: An autopsy evaluation of men with hypogonadotropic hypogonadism and panhypopituitarism. *J Urol* 136(5):1030–4.

Orentreich, N., J. L. Brind, J. H. Vogelman, R. Andres, and H. Baldwin. 1992. Long-term longitudinal measurements of plasma dehydroepiandrosterone sulfate in normal men. *J Clin Endocrinol Metab* 75(4):1002–4.

Peehl, D. M., and R. G. Sellers. 1997. Induction of smooth muscle cell phenotype in cultured human prostatic stromal cells. *Exp Cell Res* 232(2):208–15.
Peeke, P. 2000. *Fight Fat after Forty*. New York: Penguin Group.
Perkins, S. N., S. D. Hursting, D. C. Haines, S. J. James, B. J. Miller, and J. M. Phang. 1997. Chemoprevention of spontaneous tumorigenesis in nullizygous p53-deficient mice by dehydroepiandrosterone and its analog 16alpha-fluoro-5-androsten-17-one. *Carcinogenesis* 18(5):989–94.
Prehn, R. T. 1999. On the prevention and therapy of prostate cancer by androgen administration. *Cancer Res* 59(17):4161–4.
Rainey, W. E., D. Naville, and J. I. Mason. 1991. Regulation of 3 beta-hydroxysteroid dehydrogenase in adrenocortical cells: Effects of angiotensin-II and transforming growth factor beta. *Endocr Res* 17(1–2):281–96.
Rao, K. V., W. D. Johnson, M. C. Bosland, R. A. Lubet, V. E. Steele, G. J. Kelloff, and D. L. McCormick. 1999. Chemoprevention of rat prostate carcinogenesis by early and delayed administration of dehydroepiandrosterone. *Cancer Res* 59(13):3084–9.
Rao, M. S., V. Subbarao, A. V. Yeldandi, and J. K. Reddy. 1992. Hepatocarcinogenicity of dehydroepiandrosterone in the rat. *Cancer Res* 52(10):2977–9.
Regelson, W., R. Loria, and M. Kalimi. 1988. Hormonal intervention: "Buffer hormones" or "state dependency". The role of dehydroepiandrosterone (DHEA), thyroid hormone, estrogen and hypophysectomy in aging. *Ann N Y Acad Sci* 521:260–73.
Reiter, W. J., A. Pycha, G. Schatzl, A. Pokorny, D. M. Gruber, J. C. Huber, and M. Marberger. 1999. Dehydroepiandrosterone in the treatment of erectile dysfunction: A prospective, double-blind, randomized, placebo-controlled study. *Urology* 53(3):590–4; discussion 594–5.
Ricke, W. A., S. J. McPherson, J. J. Bianco, G. R. Cunha, Y. Wang, and G. P. Risbridger. 2008. Prostatic hormonal carcinogenesis is mediated by in situ estrogen production and estrogen receptor alpha signaling. *Faseb J* 22(5):1512–20.
Risbridger, G. P., S. J. Ellem, and S. J. McPherson. 2007. Estrogen action on the prostate gland: A critical mix of endocrine and paracrine signaling. *J Mol Endocrinol* 39(3):183–8.
Risbridger, G., H. Wang, P. Young, T. Kurita, Y. Z. Wang, D. Lubahn, J. A. Gustafsson, and G. Cunha. 2001. Evidence that epithelial and mesenchymal estrogen receptor-alpha mediates effects of estrogen on prostatic epithelium. *Dev Biol* 229(2):432–42.
Rowley, D. R. 1998. What might a stromal response mean to prostate cancer progression? *Cancer Metastasis Rev* 17(4):411–9.
Rupprecht, R., and F. Holsboer. 1999. Neuroactive steroids: Mechanisms of action and neuropsychopharmacological perspectives. *Trends Neurosci* 22(9):410–6.
Schedin, P., and A. Elias. 2004. Multistep tumorigenesis and the mircroenvironment. *Breast Cancer Res* 6(2):93–101.
Schwartz, A. G., and L. L. Pashko. 1995a. Cancer prevention with dehydroepiandrosterone and non-androgenic structural analogs. *J Cell Biochem Suppl* 22:210–7.
Schwartz, A. G., and L. L. Pashko. 1995b. Mechanism of cancer preventive action of DHEA. Role of glucose-6-phosphate dehydrogenase. *Ann N Y Acad Sci* 774:180–6.
Signoretti, S., and M. Loda. 2001. Estrogen receptor beta in prostate cancer: Brake pedal or accelerator? *Am J Pathol* 159(1):13–6.
Simard, J., and S. Gingras. 2001. Crucial role of cytokines in sex steroid formation in normal and tumoral tissues. *Mol Cell Endocrinol* 171(1–2):25–40.
Singh, D. K., and S. M. Lippman. 1998. Cancer chemoprevention. Part 1: Retinoids and carotenoids and other classic antioxidants. *Oncology (Williston Park)* 12(11):1643–53, 1657–8; discussion 1659–60.
Stahl, F., D. Schnorr, C. Pilz, and G. Dörner. 1992. Dehydroepiandrosterone (DHEA) levels in patients with prostatic cancer, heart diseases and under surgery stress. *Exp Clin Endocrinol* 99(2):68–70.
Swierczynski, J., and D. Mayer. 1996. Dehydroepiandrosterone-induced lipid peroxidation in rat liver mitochondria. *J Steroid Biochem Mol Biol* 58(5–6):599–603.
Tan, J., Y. Sharief, K. G. Hamil, C. W. Gregory, D. Y. Zang, M. Sar, P. H. Gumerlock et al. 1997. Dehydroepiandrosterone activates mutant androgen receptors expressed in the androgen-dependent human prostate cancer xenograft CWR22 and LNCaP cells. *Mol Endocrinol* 11(4):450–9.
Taylor, R. A., and G. P. Risbridger. 2008. Prostatic tumor stroma: A key player in cancer progression. *Curr Cancer Drug Targets* 8(6):490–7.
Trifiro, M. D., J. K. Parsons, K. Palazzi-Churas, J. Bergstrom, C. Lakin, and E. Barrett-Connor. 2010. Serum sex hormones and the 20-year risk of lower urinary tract symptoms in community-dwelling older men. *BJU Int* 105(11):1554–9.

Tuxhorn, J. A., G. E. Ayala, and D. R. Rowley. 2001. Reactive stroma in prostate cancer progression. *J Urol* 166(6):2472–83.

Tuxhorn, J. A., G. E. Ayala, M. J. Smith, V. C. Smith, T. D. Dang, and D. R. Rowley. 2002. Reactive stroma in human prostate cancer: Induction of myofibroblast phenotype and extracellular matrix remodeling. *Clin Cancer Res* 8(9):2912–23.

Untergasser, G., R. Gander, C. Lilg, G. Lepperdinger, E. Plas, and P. Berger. 2005. Profiling molecular targets of TGF-beta1 in prostate fibroblast-to-myofibroblast transdifferentiation. *Mech Ageing Dev* 126(1):59–69.

Valenti, G., L. Denti, M. Sacco, G. Ceresini, S. Bossoni, A. Giustina, D. Maugeri et al. 2006. Consensus document on substitution therapy with DHEA in the elderly. *Aging Clin Exp Res* 18(4):277–300.

van Weerden, W. M., A. van Kreuningen, N. M. Elissen, F. H. de Jong, G. J. van Steenbrugge, and F. H. Schroder. 1992. Effects of adrenal androgens on the transplantable human prostate tumor PC-82. *Endocrinology* 131(6):2909–13.

Voigt, K. D., and W. Bartsch. 1986. Intratissular androgens in benign prostatic hyperplasia and prostatic cancer. *J Steroid Biochem* 25(5B):749–57.

Walsh, P. C., and P. K. Siiteri. 1975. Suppression of plasma androgens by spironolactone in castrated men with carcinoma of the prostate. *J Urol* 114(2):254–6.

Weihua, Z., S. Makela, L. C. Andersson, S. Salmi, S. Saji, J. I. Webster, E. V. Jensen et al. 2001. A role for estrogen receptor beta in the regulation of growth of the ventral prostate. *Proc Natl Acad Sci U S A* 98(11):6330–5.

Weihua, Z., M. Warner, and J. A. Gustafsson. 2002. Estrogen receptor beta in the prostate. *Mol Cell Endocrinol* 193(1–2):1–5.

Widstrom, R. L., and J. S. Dillon. 2004. Is there a receptor for dehydroepiandrosterone or dehydroepiandrosterone sulfate? *Semin Reprod Med* 22(4):289–98.

31 DHEA and Vascular Function

Sam Rice and Aled Rees

CONTENTS

INTRODUCTION

Dehydroepiandrosterone (DHEA) and its sulfate ester, dehydroepiandrosterone sulfate (DHEAS), are circulating steroids produced in significant quantities by the adrenal glands such that peak serum concentrations, which are achieved at 20–30 years of age, exceed those of cortisol (Orentreich et al. 1984). Thereafter, concentrations of circulating DHEA/DHEAS fall progressively with age, such that at age 70, the levels are only 20%–30% of those seen in young adults (Orentreich et al. 1984).

DHEA exerts its actions either indirectly, in peripheral tissues following conversion to androgens, estrogens, or both (Labrie et al. 1995) or directly via an unidentified cell surface receptor (Liu and Dillon 2002; Webb et al. 2006). DHEAS is the main circulating form, but it is hydrophilic, and only lipophilic DHEA can be converted intracellularly to androgens and estrogens; DHEAS can be desulfated to DHEA in many tissues (Webb et al. 2006).

The marked age-related decline in DHEAS concentrations has prompted speculation that DHEA supplementation might have potential benefits in many disease processes associated with aging. A number of small clinical trials have demonstrated some benefits of DHEA therapy on well-being (Morales et al. 1994), bone turnover (Labrie et al. 1997), and lean body mass (Morales et al. 1998), although the doses used have occasionally been supraphysiological (Morales et al. 1998). Physiological DHEA replacement has also been shown to have beneficial effects on mood, self-esteem, fatiguability, and sexual function in patients with Addison's disease (Arlt, Callies et al. 1999; Hunt et al. 2000) or hypopituitarism (Johannsson et al. 2002; Brooke et al. 2006), in whom DHEAS deficiency is usually striking. In contrast, there have been comparatively few studies examining the actions of DHEA on vascular function despite much epidemiological and *in vitro* evidence to suggest a potential beneficial role. Here, we review the regulation of vascular function in humans and critically examine the evidence for a role of DHEA in vasculoprotection.

REGULATION OF VASCULAR FUNCTION

Vascular function and its dysfunction relate to the ability of the blood vessel wall to respond to environmental changes and in doing so maintain blood flow. The healthy blood vessel will respond readily to changes in pulse volume and heart rate, and in disease, this dynamism is impaired. Stiffer, less-responsive vessels occur as a result of alterations in the structural makeup of the vascular wall. Such alterations typically occur with aging but are also influenced by factors such as sustained hypertension and other disease states such as diabetes. More recently, it has become clear that vascular function is also controlled by a number of vasoactive substances generated within the vascular endothelium itself in response to various stimuli. Modern techniques not only allow investigators to determine the relative stiffness of blood vessels but also the reactivity of the vascular endothelium.

THE VASCULAR ENDOTHELIUM

The vascular endothelium is a dynamic structure with the ability to modify blood vessel diameter in response to a variety of stimuli (posture, stress, exercise) through the release of, and in response to, a number of vasoactive substances (Figure 31.1). In health, vascular tone is maintained through the action of vasodilators and vasoconstrictors. Nearly all substances that induce vasodilation do so through nitric oxide (NO). NO is produced within the endothelium from the amino acid L-arginine via the activity of endothelial nitric oxide synthase (eNOS; Loscalzo and Welch 1995). Shear stress is the most important stimulating factor for NO production (Cooke and Tsao 2001); however, acetylcholine acting on muscarinic receptors (M1 and M3) and bradykinin acting on bradykinin receptors (BR2; Moreau et al. 2005) can also activate calcium channels in endothelial cells and generate NO

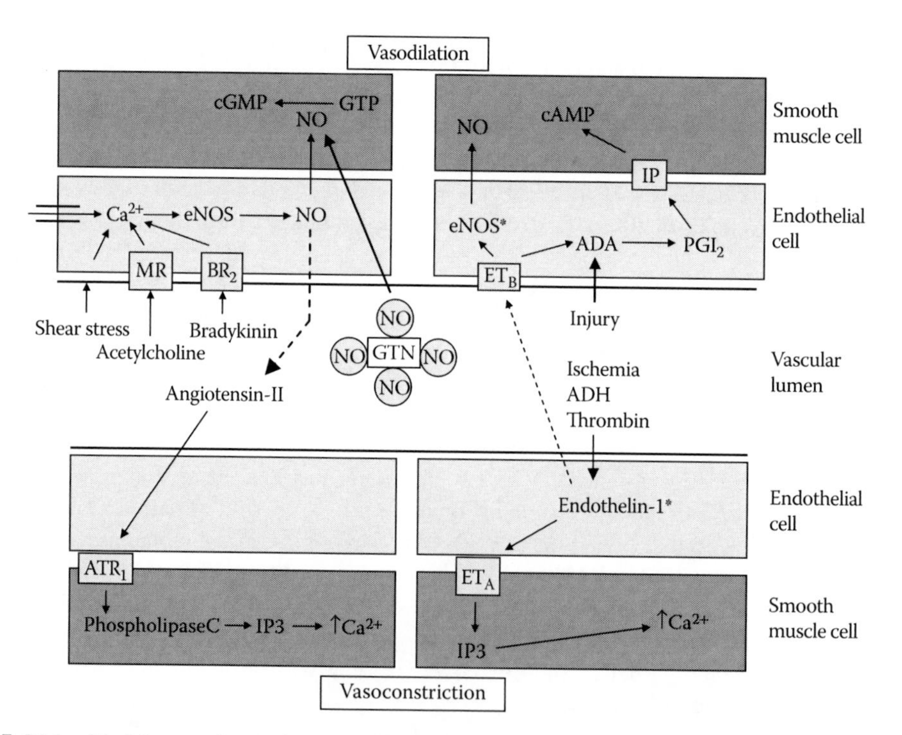

FIGURE 31.1 Modulators of vascular tone. *Possible site of DHEA action. NO = nitric oxide; eNOS = endothelial nitric oxide synthase; MR = muscarinic receptor; BR = bradykinin receptor; GTP = guanosine triphosphate; cGMP = cyclic guanosine monophosphate; ET = endothelin receptor; cAMP = cyclic adenyl monophosphate; ADA = arachidonic acid; PGI_2 = prostaglandin I_2; IP = prostaglandin receptor; ADH = antidiuretic hormone; ATR = angiotensin II receptor; IP3 = inositol triphosphate; GTN = glyceryl trinitrate.

via eNOS. NO is a lipophilic gas that can cross cell membranes. In the vascular endothelium, NO crosses the intima to reach the bundles of smooth muscle within the vessel wall. Here, NO causes the degradation of GTP, releasing cGMP, which in turn modulates cytosolic calcium and causes relaxation of smooth muscle fibers and, therefore, vessel vasodilation (Loscalzo and Welch 1995). NO can also directly inhibit the actions of vasoconstrictors such as angiotensin-II. The NO system and eNOS have been identified as potential targets for DHEA through the activity of a plasma membrane G-protein ($G\alpha_{i2,3}$) coupled receptor (Liu and Dillon 2002; Liu and Dillon 2004).

Endothelin-1 is a potent vasoconstrictor and may also be a target for DHEA (Chen et al. 2008). It is generated within the endothelium from precursor molecules (preproendothelin and big endothelin). Endothelin-1 acts on the endothelin A receptor on the vascular smooth muscle wall in response to ischemia to increase IP3 and intracellular calcium, in turn, stimulating contraction (Highsmith, Blackburn, and Schmidt 1992). Interestingly, endothelin-1 can also cause vasodilation through action on the endothelin B receptor. This triggers smooth muscle relaxation and vasodilation through prostaglandin I_2 (PGI_2) production from arachidonic acid and by stimulating eNOS.

A fine balance exists between the various vasodilating and vasoconstricting substances, such that the combination of these substances along with the intraluminal microenvironment will govern vascular tonicity. An inability to respond to these stimuli or dysregulation of these mechanisms is known as endothelial dysfunction, a condition that precedes the development of atherosclerotic plaques, contributing to lesion development and, subsequently, clinical complications (Ross 1993).

Arterial Stiffness

Increased large artery stiffness is a function of aging and relates to a reduced ability of blood vessels to expand and recoil in response to pulsed cardiac output. This dynamic ability facilitates transduction of the intermittent cardiac pulse into steady blood flow and moves away from the traditional concept of blood vessels acting solely as conduits. The proximal aorta will dilate in response to increased volume generated in systole with an associated higher pressure (systolic pressure) and will then recoil during diastole, maintaining a lower pressure (diastolic pressure) and thus flow. Therefore, the pulse pressure (difference between systolic and diastolic pressures) is both a reproducible and easily measured surrogate marker of arterial stiffness. Stiffened vessels will have a reduced ability to accommodate and distend in response to the systolic pulse; hence, higher systolic pressures will be generated in the vessel along with lower diastolic pressures secondary to impaired elastic recoil.

A number of factors are now known to regulate arterial stiffness (Figure 31.2). Traditionally, it was thought to develop solely as a result of alterations in structural components within the vascular wall such as elastin and collagen, both of which are located at the intimal medial layer.

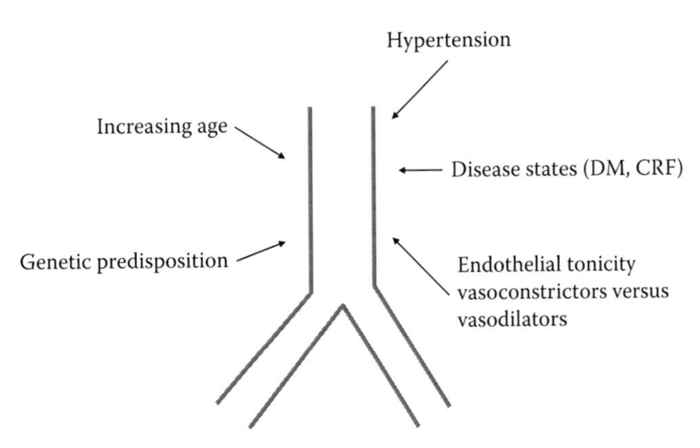

FIGURE 31.2 Regulation of arterial stiffness. DM = diabetes mellitus; CRF = chronic renal failure.

At higher pressures collagen primarily maintains vessel stiffness (e.g., in systole), whereas elastin predominantly functions at lower pressures. The ratio of elastin to collagen can therefore affect vessel stiffness, whereby low elastin-to-collagen ratios are consistent with increased stiffness. Smooth muscle hypertrophy will also contribute to increasing vessel stiffness; hence, it is intriguing to note that DHEA can inhibit vascular smooth muscle cell proliferation via a mechanism involving the mitogen-activated protein kinase signaling pathway (Yoneyama et al. 1997). However, it is now apparent that structural changes are not the only modulators of arterial stiffness, because it is intimately linked to the vascular endothelium and the balance between vasodilating and vasoconstricting substances (Cox 1976) that can themselves be potentially modulated by DHEA.

Arterial stiffness is increasingly recognized as an important cardiovascular risk factor in its own right, with evidence that it predicts a higher risk of coronary atherosclerosis independently of traditional cardiovascular risk factors (Weber et al. 2004). Furthermore, arterial stiffness, as measured by the technique of pulse wave velocity (PWV), is independently predictive of cardiovascular mortality in a number of disease states including renal failure, hypertension, and diabetes (Blacher et al. 1999; Laurent et al. 2001; Cruickshank et al. 2002).

DHEA AND VASCULAR FUNCTION: *IN VITRO* AND ANIMAL STUDIES

A number of studies have been undertaken using animal models to explore the effects of DHEA supplementation on the vasculature. Studies in rodents and rabbits have shown the potential for antiatherosclerotic actions of DHEA (Yorek et al. 2002; Gordon, Bush, and Weisman 1998; Hayashi et al. 2000; Eich et al. 1993; Cheng, Hu, and Ruan 2009), although the relevance of these findings to human disease is questionable in view of the often-used supraphysiological DHEA doses and the low-circulating DHEA(S) levels present in lower order mammals.

Several *in vitro* studies have also explored the direct and indirect actions of DHEA on human vascular endothelium and smooth muscle (Williams et al. 2004; Yoneyama et al. 1997; Williams et al. 2002; Liu and Dillon 2002; Simoncini et al. 2003; Formoso et al. 2006; Liu et al. 2007; Liu et al. 2008; Chen et al. 2008). These studies collectively support a beneficial action of DHEA on the endothelium, which appears to be both a direct and nongenomically-mediated effect, because activation of endothelial eNOS occurs both rapidly and independently of downstream conversion to testosterone or estradiol (Williams et al. 2004; Simoncini et al. 2003; Liu and Dillon 2002). More recent studies have also shown a beneficial effect of DHEA on vascular endothelial cell survival (Liu et al. 2007), proliferation and angiogenesis (Liu et al. 2008), and activation, including an ability to regulate endothelin-1 at the transcriptional level (Chen et al. 2008). DHEAS has also been shown to inhibit tumor necrosis factor-α (TNF-α)-induced vascular inflammation in endothelial cells (Altman et al. 2008), an effect involving the peroxisome proliferator-activated receptor-α (PPAR-α). These *in vitro* findings, thus, provide a biological rationale for testing the hypothesis that DHEA might reduce vascular risk in humans.

EPIDEMIOLOGICAL ASSOCIATIONS OF DHEA(S) WITH VASCULAR RISK, MORBIDITY, AND MORTALITY

Because circulating DHEA(S) levels decline markedly with age, a number of cohort studies have examined the association between DHEA(S) levels and many diseases associated with aging, including cardiovascular morbidity and mortality (Barrett-Connor, Khaw, and Yen 1986; Slowinska-Srzednicka et al. 1989; Contoreggi et al. 1990; Slowinska-Srzednicka et al. 1991; LaCroix, Yano, and Reed 1992; Mitchell et al. 1994; Newcomer et al. 1994; Barrett-Connor and Goodman-Gruen 1995; Legrain et al. 1995; Slowinska-Srzednicka et al. 1995; Berr et al. 1996; Haffner et al. 1996; Tilvis, Kähönen, and Härkönen 1999; Kähönen et al. 2000; Feldman et al. 2001; Trivedi and Khaw 2001; Arnlöv et al. 2006; Maggio et al. 2007; Ohlsson et al. 2010). Several, but not all, of these studies have demonstrated an inverse association between DHEAS levels and all-cause (Barrett-Connor,

Khaw, and Yen 1986; Berr et al. 1996; Trivedi and Khaw 2001; Mazat et al. 2001; Maggio et al. 2007; Ohlsson et al. 2010) and cardiovascular (Barrett-Connor, Khaw, and Yen 1986; Trivedi and Khaw 2001; Ohlsson et al. 2010) mortality in men. This relationship appears to be gender-specific because the association is not found in women (Trivedi and Khaw 2001). In contrast, other studies have not confirmed these findings, even in men (Barrett-Connor and Goodman-Gruen 1995; Legrain et al. 1995; Tilvis, Kähönen, and Härkönen 1999; Kähönen et al. 2000; Feldman et al. 2001). The sexually dimorphic nature of this relationship might be explained, at least in part, by sex-specific differences in the bioconversion of DHEA, whereby DHEA administration increases androgens in women (Arlt et al. 1998) and estrogens in men (Arlt, Haas et al. 1999). However, it is possible that this association of low DHEAS levels with mortality is simply an epiphenomenon of unrecognized subclinical disease because chronic disease induces a change in adrenal steroid synthesis to favor cortisol over DHEA production (Parker et al. 1985), and DHEA(S) levels decline rapidly during critical illness (Marx et al. 2003).

In the most recent study to address this area, Ohlsson and colleagues reported their analyses of the relationship between DHEA and DHEAS levels and mortality from a large cohort of elderly (69–81 years) Swedish men taking part in the prospective population-based MrOS (Multicenter Osteoporotic Fractures in Men) study (Ohlsson et al. 2010). Their study is noteworthy for a number of reasons. First, this is the largest study (2644 subjects) to examine this relationship. Second, the authors analyzed both DHEA and DHEAS levels by the "gold standard" techniques of gas chromatography- or liquid chromatography-mass spectrometry, respectively. Third, numerous covariates were available for adjustment, including smoking status, body mass index (BMI), diabetes, and history of cardiovascular disease and cancer. Fourth, there was minimal loss of follow-up data. Finally, an analysis was undertaken to minimize the potential for confounding by the presence of occult disease through the exclusion of men who died in the first 3 years of the study. Their findings confirmed an inverse association between both DHEA and DHEAS levels and mortality, with the men in the lowest quartile of DHEAS demonstrating hazard ratios for all-cause, cardiovascular disease, and ischemic heart disease mortality of 1.54–1.67. The association was nonlinear because the analyses demonstrated an apparent threshold effect for mortality, which emerged in men with the lowest DHEAS levels only (<0.37 μg/mL). This association remained largely unchanged after adjustment not only for traditional cardiovascular risk factors (age, smoking status, BMI, diabetes, hypertension, and ApoB:ApoA1 ratio) but also serum C-reactive protein levels and testosterone/estradiol status, suggesting that this relationship was neither accounted for by increased inflammation nor by peripheral conversion to androgens or estrogens. Importantly, their analyses also showed that the inverse relationship between DHEAS and mortality persisted even after the exclusion of subjects who died within the first 3 years of follow-up, suggesting that unrecognized illness did not account for the observed associations. Furthermore, as reported previously (Barrett-Connor, Khaw, and Yen 1986), DHEA levels were not predictive of death from cancer, which supports the view that DHEA levels are not simply a marker of age-related comorbidity. This large study thus provides strong support for an association between DHEA(S) levels and cardiovascular mortality in men.

In addition to the many studies examining the association between DHEA(S) levels and death from cardiovascular disease, a number of studies have also examined this relationship with cardiovascular morbidity and surrogate markers of vascular risk. As with the mortality studies, inverse associations have also been shown between DHEA/DHEAS concentrations and cardiac graft stenosis, as well as atherosclerosis extent observed at angioplasty (Herrington et al. 1990; Herrington et al. 1996). More recently, Fukui et al. (2007) demonstrated a significant inverse association between serum-DHEAS concentrations and brachial PWV, a marker of arterial stiffness, in men with type 2 diabetes. However, a similar inverse association with PWV was also found for free testosterone, although both DHEAS and free testosterone levels were found to be independent determinants of PWV on multiple regression analysis. In contrast, Davoodi et al. (2007) did not identify any association between serum-DHEAS concentration and the presence or severity of coronary angiography–identified cardiovascular disease in their study of 502 Iranian men. Serum-DHEAS levels were, however, found to be inversely associated with carotid intima media thickness, a measure of atherosclerosis, in postmenopausal women with type 2 diabetes (Kanazawa et al. 2008), a relationship that remained significant after adjustment for traditional cardiovascular risk factors including age, diabetes duration, systolic blood pressure, glycated hemoglobin (HbA1C), low density lipoprotein (LDL) cholesterol, creatinine, and smoking status. DHEAS levels have also been shown to weakly associate with flow-mediated dilatation (FMD), a measure of endothelial function, in 115 postmenopausal women with cardiovascular risk factors (Akishita et al. 2008). This association was independent of age, BMI, hypertension, hyperlipidemia, smoking status and diabetes, and supports the contention that DHEA may have a protective role on the endothelium. Several studies have also examined the relationship between DHEAS levels and cardiovascular morbidity, notably, in relation to ischemic heart disease and myocardial infarction. Although data from the Massachusetts Male Aging Study did not support an association between low DHEAS levels and mortality from ischemic heart disease in 40- to 70-year-old men, an association was found between low DHEAS levels and combined fatal and nonfatal ischemic heart disease events (Feldman et al. 2001), in agreement with LaCroix, Yano, and Reed (1992), who reported lower DHEAS levels in fatal ischemic heart disease cases. Other studies have shown an inverse association of DHEAS with risk of myocardial infarction (Slowinska-Srzednicka et al. 1989; Mitchell et al. 1994), but this is not a consistent finding (Contoreggi et al. 1990; Newcomer et al. 1994; Haffner et al. 1996; Arnlöv et al. 2006), and one study has even demonstrated a modest positive association between DHEA and DHEAS levels with myocardial infarction risk in postmenopausal women (Page et al. 2008). These inconsistent findings make it difficult to draw firm conclusions on the relationship between DHEA(S) and morbidity from vascular disease, which may relate, in part, to differences in study design (retrospective case-control, prospective case-control, population-based), sample size, and study subject characteristics (age, gender, disease status). Furthermore, all such studies are limited by their observational design, which neither necessarily implies causation nor tells us about the directionality of effects. Hence, interventional studies in humans of DHEA therapy are required in order to fully explore any potential benefits (or harm) of this hormone on the vasculature.

INTERVENTIONAL STUDIES OF DHEA AND VASCULAR FUNCTION

In contrast to the large body of studies examining the association between DHEA(S) levels and vascular morbidity and mortality, comparatively few clinical trials of DHEA therapy on vascular function have been undertaken in humans, and only a couple of these have been performed in patients with pathological DHEA(S) deficiency, in whom any potential effects of DHEA on vascular function might be expected to be greatest. Kawano and colleagues prospectively examined the short-term (12 weeks) effects of DHEA administration on endothelial function in a cohort of 24 middle-aged men with hypercholesterolemia unselected for DHEA(S) deficiency (Kawano et al. 2003). Endothelial dysfunction contributes to the pathogenesis of atherosclerosis and cardiovascular disease and precedes the development of insulin resistance (Pinkney et al. 1997). Using a randomized, double-blind, placebo-controlled crossover design, they were able to demonstrate that DHEA supplementation improved FMD of the brachial artery, an endothelium-dependent process. The authors were also able to show a reduction in plasminogen activator inhibitor type 1 (PAI-1), a suppressor of fibrinolysis with a pathogenic role in coronary artery disease (Beer et al. 1996), and steady-state glucose concentrations without alteration in fasting insulin, demonstrating improved insulin sensitivity. This study used a low dose of DHEA (25 mg), which was sufficient to restore DHEAS levels to those seen in young adults. The study had the limitation that it was conducted in hypercholesterolemic males only, in whom endothelial dysfunction is well established. Hence, other studies in men without hypercholesterolemia would be needed in order to confirm or refute these findings. The authors were also unable to determine whether these effects of DHEA were due to direct actions on the vasculature or mediated by downstream conversion to androgens or estrogens. Although total testosterone levels did not change in response to DHEA therapy, an expected finding because testicular Leydig cells are the major source of circulating androgens in adult males, estradiol levels were not measured in this study. Because DHEA may be converted *in vivo* to estradiol (Labrie et al. 1995; Arlt, Haas et al. 1999) and estradiol may itself increase eNOS activity and NO production within the endothelium (Mendelsohn and Karas 1999), their observations do not exclude an indirect action of DHEA on the vasculature mediated by peripheral conversion to estrogens. The critical importance of estrogens to endothelial health in males is highlighted by data from studies in an individual with an estrogen receptor mutation and men with aromatase deficiency. The patient with a mutation in the estrogen receptor gene, and thus estrogen insensitivity, was shown to have evidence of premature coronary artery disease (Sudhir et al. 1997). Computed tomography scanning showed premature calcification of a coronary artery, and brachial artery studies demonstrated absence of FMD in response to ischemic cuff occlusion despite preserved nitroglycerin and estradiol-induced vasodilation (Sudhir et al. 1997). Furthermore, males with aromatase deficiency display elevated LDL cholesterol, triglycerides, and insulin levels, indicative of insulin resistance (Morishima et al. 1995; Bilezikian et al. 1998). Although there are no studies of vascular function included in these latter reports, there is evidence of impaired endothelium-dependent vasodilation in the aorta of aromatase knockout mice, suggesting a role for endogenous estrogens in the regulation of endothelial function in this model (Kimura et al. 2003). Furthermore, a randomized, double-blind, placebo-controlled study in healthy young men treated with the aromatase inhibitor anastrozole showed an improvement in endothelial function independent of changes in well-established cardiovascular risk factors such as high-sensitivity CRP, lipoprotein(a), or homocysteine (Lew et al. 2003), supporting the hypothesis that physiological levels of estrogen play an important role in the male cardiovascular system.

Williams et al. (2004) substantiated the findings of Kawano et al. (2003) by demonstrating a beneficial effect of DHEA supplementation on endothelial function in their study of 36 healthy postmenopausal women free of known cardiovascular disease, albeit using a high DHEA dose of 100 mg/day. Using a randomized, double-blind, placebo-controlled, parallel group design, they showed a significant increase in FMD of the brachial artery at 12 weeks in response to DHEA, which was accompanied by a reduction in total cholesterol concentrations; no changes were observed in the placebo group. They were also able to show a beneficial effect of DHEA on endothelium-mediated vascular reactivity in the microvasculature using laser Doppler velocimetry. In contrast to the normal baseline values of DHEAS in the middle-aged men included in Kawano and colleagues' study (11.7 μg/mL), the postmenopausal women in this study had DHEAS levels that were significantly reduced at baseline and doubled in response to treatment. The authors speculated that the beneficial effects they observed with DHEA on endothelial function were likely to be a direct action of the hormone, in line with their accompanying *in vitro* findings of a direct DHEA action on endothelial cell proliferation. However, total testosterone and estrone levels also increased in response to DHEA therapy; hence, an indirect effect by peripheral androgenic or estrogenic conversion can again not be excluded. Testosterone is increasingly recognized to have important effects on vascular function, not only in men with angina, in whom it can increase ischemic threshold (Malkin et al. 2004) and may protect against increased arterial stiffness (Smith et al. 2001), but also in women, in whom the addition of testosterone to estrogen therapy has been shown to reduce vascular inflammation (Kocoska-Maras et al. 2009). Furthermore, testosterone can induce rapid vasorelaxation of vascular smooth muscle through nongenomic mechanisms at physiologically relevant concentrations (Perusquía and Stallone 2010), and it can attenuate fatty streak formation in a mouse model of androgen insensitivity (Nettleship et al. 2007), an effect that is also mediated independent of the classic nuclear androgen receptor.

Addison's disease and hypopituitarism represent excellent disease models for studying any potential benefits of DHEA replacement on vascular function because DHEA(S) deficiency is usually marked (Young et al. 1997; Miller et al. 2001) and present in a young population in which circulating levels would normally be at their highest (Orentreich et al. 1984). In a randomized, placebo-controlled, crossover study, Christiansen et al. (2007) studied the effects of 6 months' physiological

DHEA replacement (50 mg daily) on a variety of cardiovascular measures in 10 women with Addison's disease or isolated adrenocorticotropic hormone deficiency. Two women withdrew during the course of the study leaving only eight subjects available for the final analysis. Their study measures included a comprehensive assessment of several aspects of cardiovascular function, including echocardiography, cardiac output as determined by magnetic resonance imaging, endothelial function by FMD, lipids, and 24-hour blood pressure. They did not find any effect of DHEA on these measures despite restoration of DHEA and DHEAS levels to the normal range. However, the small sample size may have left their study underpowered to detect any minor changes.

In our own clinical trial of DHEA on vascular function in 40 patients with Addison's disease or hypopituitarism (Rice et al. 2009), we also chose a randomized, placebo-controlled, double-blind crossover design because a substantial within-subject correlation was expected, such that smaller sample sizes might be required to detect significant differences. We chose a washout period of 8 weeks between 3-month treatments to ensure no carryover effect, and chose a treatment dose of DHEA of 50 mg in the expectation that this would restore the levels of DHEAS to the young adult range. Although we used augmentation index to assess endothelial function, which may offer reduced precision compared with FMD (Donald et al. 2006), we expected this limitation to be outweighed by the within-subject correlation. Furthermore, our a priori sample-size calculations indicated that we had in excess of 96% power to detect a clinically meaningful change in endothelial function if present. Despite restoration of DHEAS and androstenedione levels to concentrations marginally above the physiological range for young adults in both sexes, we did not observe any changes in the primary vascular outcome measures of arterial stiffness, endothelial function, or blood pressure. Although a 12-week treatment period may not necessarily have been sufficient to precipitate changes in measures of arterial stiffness that require conformational change to the blood vessel wall, we would expect changes in endothelial responsiveness over this time frame should DHEA have had an effect on this parameter. Recognizing that any potential benefits of DHEA on vascular function might be confined to subgroups, we repeated our analyses with the data stratified by gender or disease (Addison's disease vs. hypopituitarism) group. We again failed to observe any significant effects of DHEA on vascular function other than a small reduction in central blood pressure in subjects with hypopituitarism, of borderline significance. Because our study was not powered for these subsidiary analyses, we cannot entirely exclude a small effect on vascular function in subgroups, although the absence of even a trend toward significance makes it unlikely that we were observing a type 2 error. We also examined a number of anthropometric and metabolic variables in our study but found no change in weight, BMI, fat percentage, waist circumference, or lipids following DHEA therapy, other than a small reduction in high density lipoprotein (HDL) cholesterol, which we speculate might relate to androgenic conversion (Casson et al. 1998). Finally, we assessed steady-state insulin sensitivity by the homeostasis model assessment (HOMA-IR) method but failed to detect any changes in this parameter with DHEA treatment, unlike some other reports that have observed improvements in insulin action following DHEA, using more sensitive methodologies for assessment of insulin sensitivity (Dhatariya, Bigelow, and Nair 2005). We concluded from our study that short-term DHEA therapy does not have a major influence on arterial stiffness or endothelial function in patients with hypopituitarism or Addison's disease.

CONCLUSION

There is compelling evidence from a number of *in vitro* studies that DHEA may have vasculoprotective effects, centered on its ability to activate endothelial NO production through, at least in part, direct and nongenomic means. Furthermore, many epidemiological studies demonstrate an inverse association between DHEA(S) levels and cardiovascular mortality. However, short-term clinical trials of DHEA supplementation in humans have shown conflicting results on vascular function, with both improvements and no effects reported. Interpretation of these studies is difficult because of heterogeneity in study design, including differences in DHEA dose, DHEA(S) status of subjects at baseline, treatment duration, and underlying conditions; hence, further clinical trials are required to clarify whether DHEA has a role in regulating vascular function in humans.

SUMMARY

- Vascular dysfunction is a precursor to overt cardiovascular disease.
- *In vitro* studies have demonstrated potential mechanisms through which DHEA treatment might improve vascular function and modulate cardiovascular disease risk.
- Many epidemiological studies have shown that low-circulating DHEA(S) levels are associated with premature cardiovascular mortality in men.
- Interventional studies in humans have either shown benefits or no effect of short-term DHEA therapy on vascular function.
- To date, there is insufficient evidence to support a role for DHEA therapy in the prevention or treatment of human cardiovascular disease.

REFERENCES

Akishita, M., M. Hashimoto, Y. Ohike, S. Ogawa, K. Iijima, M. Eto, and Y. Ouchi. 2008. Association between plasma dehyroepiandrosterone-sulphate levels with endothelial function in postmenopausal women with coronary risk factors. *Hypertens Res* 31:68–74.

Altman, R., D. D. Motton, R. S. Kota, and J. C. Rutledge. 2008. Inhibition of vascular inflammation by dehydroepiandrosterone sulfate in human aortic endothelial cells: Roles of PPARα and NF-κB. *Vascul Pharmacol* 48:76–84.

Arlt, W., F. Callies, J. C. van Vlijmen, I. Koehler, M. Reincke, M. Bidlingmaier, D. Huebler et al. 1999. Dehydroepiandrosterone replacement in women with adrenal insufficiency. *N Engl J Med* 341:1013–20.

Arlt, W., J. Haas, F. Callies, M. Reincke, D. Hubler, M. Oettel, M. Ernst, H. M. Schulte, and B. Allolio. 1999. Biotransformation of oral dehydroepiandrosterone in elderly men: Significant increase in circulating estrogens. *J Clin Endocrinol Metab* 84:2170–6.

Arlt, W., H. G. Justl, F. Callies, M. Reincke, D. Hubler, M. Oettel, M. Ernst, H. M. Schulte, and B. Allolio. 1998. Oral dehydroepiandrosterone for adrenal androgen replacement: Pharmacokinetics and peripheral conversion to androgens and estrogens in young healthy females after dexamethasone suppression. *J Clin Endocrinol Metab* 83:1928–34.

Arnlöv, J., M. J. Pencina, S. Amin, B. H. Nam, E. J. Benjamin, J. M. Murabito, T. J. Wang, P. E. Knapp et al. 2006. Endogenous sex hormones and cardiovascular disease incidence in men. *Ann Intern Med* 145:176–84.

Barrett-Connor, E., and D. Goodman-Gruen. 1995. The epidemiology of DHEAS and cardiovascular disease. *Ann N Y Acad Sci* 774:259–70.

Barrett-Connor, E., K. T. Khaw, and S. S. Yen. 1986. A prospective study of dehydroepiandrosterone sulphate, mortality, and cardiovascular disease. *N Engl J Med* 315:1519–24.

Beer, N., D. J. Jakubowicz, D. W. Matt, R. M. Beer, and J. E. Nestler. 1996. Dehydroepiandrosterone reduces plasma plasminogen activator inhibitor type 1 and tissue plasminogen activator antigen in men. *Am J Med Sci* 311:205–10.

Berr, C., S. Lafont, B. Debuire, J. F. Dartigues, and E. E. Baulieu. 1996. Relationships of dehydroepiandrosterone sulfate in the elderly with functional, psychological, and mental status, and short-term mortality: A French community-based study. *Proc Natl Acad Sci U S A* 93:13410–5.

Bilezikian, J. P., A. Morishima, J. Bell, and M. M. Grumbach. 1998. Increased bone mass as a result of estrogen therapy in a man with aromatase deficiency. *N Engl J Med* 339:599–603.

Blacher, J., A. P. Guerin, B. Pannier, S. J. Marchais, M. Safar, and G. London. 1999. Impact of aortic stiffness on survival in end-stage renal disease. *Circulation* 99:2434–9.

Brooke, A. M., L. A. Kalingag, F. Miraki-Moud, C. Camacho-Hübner, K. T. Maher, D. M. Walker, J. P. Hinson, and J. P. Monson. 2006. Dehydroepiandrosterone improves psychological well-being in male and female hypopituitary on maintenance growth hormone replacement. *J Clin Endocrinol Metab* 91:3773–9.

Casson, P. R., N. Santoro, K. Elkind-Hirsch, S. A. Carson, P. J. Hornsby, G. Abraham, and J. E. Buster. 1998. Postmenopausal dehydroepiandrosterone administration increases free insulin-like growth factor-I and decreases high density lipoprotein: A six-month trial. *Fertil Steril* 70:107–10.

Chen, H., A. S. Lin, Y. Li, C. E. Reiter, M. R. Ver, and M. J. Quon. 2008. DHEA stimulates phosphorylation of FoxO1 in vascular endothelial cells via PI3-kinase- and PKA-dependent signaling pathways to regulate ET-1 synthesis and secretion. *J Biol Chem* 283:29228–38.
Cheng, H., X. J. Hu, and Q. R. Ruan. 2009. Dehydroepiandrosterone anti-atherogenesis effect is not via its conversion to estrogen. *Acta Pharmacol Sin* 30:42–53.
Christiansen, J., N. H. Anderson, K. E. Sørensen, E. M. Pedersen, P. Bennets, M. Anderson, J. S. Christansen, J. O. L. Jørgensen, and C. H. G. Gravholt. 2007. Dehydroepiandrosterone substitution in female adrenal failure: No impact on endothelial function and cardiovascular parameters despite normalization of androgen status. *Clin Endocrinol* 66:426–33.
Contoreggi, C. S., M. R. Blackman, R. Andres, D. C. Muller, E. G. Lakatta, J. L. Fleg, and S. M. Harman. 1990. Plasma levels of estradiol, testosterone, and DHEAS do not predict risk of coronary artery disease in men. *J Androl* 11:870–5.
Cooke, J., and P. S. Tsao. 2001. Go with the flow. *Circulation* 103:2773–5.
Cox, R. 1976. Mechanics of canine iliac artery smooth muscle in vitro. *Am J Physiol* 230:462–70.
Cruickshank, K., L. Riste, S. G. Anderson, J. S. Wright, G. Dunn, and R. G. Gosling. 2002. Aortic pulse-wave velocity and its relationship to mortality in diabetes and glucose intolerance: An integrated index of vascular function? *Circulation* 106:2085–90.
Davoodi, G., A. Amirezadegan, M. Borumand, M. Dehkordi, A. Kazemisaeid, and A. Yaminisharif. 2007. The relationship between level of androgenic hormones and coronary artery disease in men. *Cardiovasc J Afr* 18:362–6.
Dhatariya, K., M. L. Bigelow, and K. S. Nair. 2005. Effect of dehydroepiandrosterone replacement on insulin sensitivity and lipids in hypoadrenal women. *Diabetes* 54:765–9.
Donald, A. E., M. Charakida, T. J. Cole, P. Friberg, P. J. Chowienczyk, S. C. Millasseau, J. E. Deanfield, and J. P. Halcox. 2006. Non-invasive assessment of endothelial function: Which technique? *J Am Coll Cardiol* 48:1846–50.
Eich, D., J. E. Nestler, D. E. Johnson, G. H. Dworkin, D. Ko, A. S. Wechsler, and M. L. Hess. 1993. Inhibition of accelerated coronary atherosclerosis with dehydroepiandrosterone in the heterotopic rabbit model of cardiac transplantation. *Circulation* 87:261–9.
Feldman, H., C. B. Johannes, A. B. Araujo, B. A. Mohr, C. Longcope, and J. B. Mckinlay. 2001. Low dehydroepiandrosterone and ischaemic heart disease in middle-aged men: Prospective results from the Massachusetts Male Aging Study. *Am J Epidemiol* 153:79–89.
Formoso, G., H. Chen, J. A. Kim, M. Montagnani, A. Consoli, and M. J. Quon. 2006. Dehydroepiandrosterone mimics acute actions of insulin to stimulate production of both nitric oxide and endothelin 1 via distinct phosphatidylinositol 3-kinase- and mitogen-activated protein kinase-dependent pathways in vascular endothelium. *Mol Endocrinol* 20:1153–63.
Fukui, M., H. Ose, Y. Kitagawa, M. Yamazaki, G. Hasegawa, T. Yoshikawa, and N. Nakamura. 2007. Relationship between low serum endogenous androgen concentations and arterial stiffness in men with type 2 diabetes. *Metab Clin Experimental* 56:1161–73.
Gordon, G., D. E. Bush, and H. F. Weisman. 1998. Reduction of atherosclerosis by administration of dehydroepiandrosterone. A study in the hypercholesterolemic New Zealand white rabbit with aortic intimal injury. *J Clin Investig* 82:712–20.
Haffner, S. M., S. E. Moss, B. E. Klein, and R. Klein. 1996. Sex hormones and DHEA-SO_4 in relation to ischemic heart disease mortality in diabetic subjects. The Wisconsin Epidemiologic Study of Diabetic Retinopathy. *Diabetes Care* 19:1045–50.
Hayashi, T., T. Esaki, E. Muto, H. Kano, Y. Asai, N. K. Thakur, D. Sumi, M. Jayachandran, and A. Iguchi. 2000. Dehydroepiandrosterone retards atherosclerosis formation through its conversion to estrogen. *Arterioscler Thromb Vasc Biol* 20:782–92.
Herrington, D., G. B. Gordon, S. C. Achuff, J. F. Trejo, H. F. Weisman, P. O. Kwiterovich, and T. A. Pearson. 1990. Plasma dehydroepiandrosterone and dehydroepiandrosterone sulphate in patients undergoing diagnostic coronary angiography. *J Am Coll Cardiol* 16:862–70.
Herrington, D., N. Nanjee, S. C. Achuff, D. E. Cameron, B. Dobbs, and K. L. Baughman. 1996. Dehydroepiandrosterone and cardiac allograft vasculopathy. *J Heart Lung Transplant* 15:88–93.
Highsmith, R., K. Blackburn, and D. J. Schmidt. 1992. Endothelin and calcium dynamics in vascular smooth muscle. *Annu Rev Physiol* 54:257–77.
Hunt, P. J., E. M. Gurnell, F. A. Huppert, C. Richards, A. T. Prevost, J. A. Wass, J. Herbert, and V. K. Chatterjee. 2000. Improvement in mood and fatigue after dehydroepiandrosterone replacement in Addison's disease in a randomized, double blind trial. *J Clin Endocrinol Metab* 85:4650–6.

Johannsson, G., P. Burman, L. Wiren, B. E. Engström, A. G. Nilsson, M. Ottosson, B. Jonsson, B. A. Bengtsson, and F. A. Karlsson. 2002. Low dose dehydroepiandrosterone affects behavior in hypopituitary androgen-deficient women: A placebo-controlled trial. *J Clin Endocrinol Metab* 87:2046–52.
Kähönen, M. H., R. S. Tilvis, J. Jolkkonen, K. Pitkälä, and M. Härkönen. 2000. Predictors and clinical significance of declining plasma dehydroepiandrosterone sulfate in old age. *Aging* 12:308–14.
Kanazawa, I., T. Yamaguchi, M. Yamamoto, M. Yamauchi, S. Kurioka, S. Yano, and T. Sugimoto. 2008. Serum DHEA-S is associated with the presence of atherosclerosis in postmenopausal women with type 2 diabetes mellitus. *Endocr J* 55:667–75.
Kawano, H., H. Yasue, A. Kitagawa, N. Hirai, T. Yoshida, H. Soejima, S. Miyamoto, M. Nakano, and H. Ogawa. 2003. Dehydroepiandrosterone supplementation improves endothelial function and insulin sensitivity in men. *J Clin Endocrinol Metab* 88:3190–5.
Kimura, M., K. Sudhir, M. Jones, E. Simpson, A. M. Jefferis, and J. P. F. Chin-Dusting. 2003. Impaired acetylcholine-induced release of nitric oxide in the aorta of male aromatase-knockout: Regulation of nitric oxide production by endogenous sex hormones in males. *Circ Res* 93:1267–71.
Kocoska-Maras, L., A. L. Hirschberg, B. Byström, B. V. Schoultz, and A. F. Rådestad. 2009. Testosterone addition to estrogen therapy - effects on inflammatory markers for cardiovascular disease. *Gynecol Endocrinol* 25:823–7.
Labrie, F., A. Belanger, J. Simard, V. Luu-The, and C. Labrie. 1995. DHEA and peripheral androgen and estrogen formation: Intracrinology. *Ann N Y Acad Sci* 774:16–28.
Labrie, F., P. Diamond, L. Cusan, J. L. Gomez, A. Bélanger, and B. Candas. 1997. Effect of 12-month dehydroepiandrosterone replacement therapy on bone, vagina and endometrium in postmenopausal women. *J Clin Endocrinol Metab* 82:3498–505.
Lacroix, A. Z., K. Yano, and D. M. Reed. 1992. Dehydroepiandrosterone sulfate, incidence of myocardial infarction, and extent of atherosclerosis in men. *Circulation* 86:1529–35.
Laurent, S., P. Boutouyrie, R. Asmar, I. Gautier, B. Laloux, L. Guize, P. Ducimetiere, and A. Benetos. 2001. Aortic stiffness is an independent predictor of all-cause and cardiovascular mortality in hypertensive patients. *Hypertension* 37:1236–41.
Legrain, S., C. Berr, N. Frenoy, V. Gourlet, B. Debuire, and E. E. Baulieu. 1995. Dehydroepiandrosterone sulfate in a long-term care aged population. *Gerontology* 41:343–51.
Lew, R., P. Komesaroff, M. Williams, T. Dawood, and K. Sudhir. 2003. Endogenous estrogens influence endothelial function in young men *Circ Res* 93:1127–33.
Liu, D., and J. S. Dillon. 2002. Dehydroepiandrosterone activates endothelial cell nitric-oxide synthase by a specific plasma membrane receptor coupled to $G\alpha_{i\,2,3}$. *J Biol Chem* 277:21379–88.
Liu, D., and J. S. Dillon. 2004. Dehydroepiandrosterone stimulates nitric oxide release in vascular endothelial cells: Evidence for a cell surface receptor. *Steroids* 69:279–89.
Liu, D., M. Iruthayanathan, L. L. Homan, Y. Wang, L. Yang, Y. Wang, and J. S. Dillon. 2008. Dehydroepiandrosterone stimulates endothelial proliferation and angiogenesis through extracellular signal-regulated kinase 1/2-mediated mechanisms. *Endocrinology* 149:889–98.
Liu, D., H. Si, K. A. Reynolds, W. Zhen, Z. Jia, and J. S. Dillon. 2007. Dehydroepiandrosterone protects vascular endothelial cells against apoptosis through a Gα i protein-dependent activation of phosphatidylinosirol 3-kinase/Akt and regulation of antiapoptotic Bcl-2 expression. *Endocrinology* 148:3068–76.
Loscalzo, J., and G. Welch. 1995. Nitric oxide and its role in the cardiovascular system. *Prog Cardiovasc Nurs* 38:87–104.
Maggio, M., F. Lauretani, G. P. Ceda, S. Bandinelli, S. M. Ling, E. J. Metter, A. Artoni et al. 2007. Relationship between low levels of anabolic hormones and 6-year mortality in older men: The aging in Chianti Area (InCHIANTI) study. *Arch Intern Med* 167:2249–54.
Malkin, C., P. J. Pugh, P. D. Morris, K. E. Kerry, R. D. Jones, T. H. Jones, and K. S. Channer. 2004. Testosterone replacement in hypogonadal men with angina improves ishcaemic threshold and quality of life. *Heart* 90:871–6.
Marx, C., S. Petros, S. R. Bornstein, M. Weise, M. Wendt, M. Menschikowski, L. Engelman, and G. Höffken. 2003. Adrenocortical hormones in survivors and nonsurvivors of severe sepsis: Diverse time course of dehydroepiandrosterone, dehydroepiandrosterone-sulfate, and cortisol. *Crit Care Med* 31:1382–8.
Mazat, L., S. Lafont, C. Berr, B. Debuire, J. F. Tessier, J. F. Dartigues, and E. E. Baulieu. 2001. Prospective measurements of dehydroepiandrosterone sulfate in a cohort of elderly subjects: Relationship to gender, subjective health, smoking habits, and 10-year mortality. *Proc Natl Acad Sci U S A* 98:8145–50.
Mendelsohn, M. E., and R. H. Karas. 1999. The protective effects of estrogen on the cardiovascular system. *N Engl J Med* 340:1801–11.

Miller, K. K., G. Sesmilo, A. Schiller, D. Schoenfeld, S. Burton, and A. Klibanski. 2001. Androgen deficiency in women with hypopituitarism. *J Clin Endocrinol Metab* 86:561–7.

Mitchell, L. E., D. L. Sprecher, I. B. Borecki, T. Rice, P. M. Laskarzewski, and D. C. Rao. 1994. Evidence for an association between dehydroepiandrosterone sulfate and nonfatal, premature myocardial infarction in males. *Circulation* 89:89–93.

Morales, A. J., R. H. Haubrich, J. Y. Hwang, H. Asakura, and S. S. Yen. 1998. The effects of six months treatment with a 100 mg daily dose of dehydroepiandrosterone (DHEA) on circulating sex steroids, body composition, and muscle strength in age-advanced men and women. *Clin Endocrinol* 49:421–32.

Morales, A. J., J. J. Nolan, J. C. Nelson, and S. S. C. Yen. 1994. Effects of replacement dose of dehydroepiandrosterone in men and women of advancing age. *J Clin Endocrinol Metab* 78:1360–7.

Moreau, M., N. Garbacki, G. Molinavo, N. J. Brown, F. Marceau, and A. Adam. 2005. The kallikrein-kinin system: Current and future pharmacological targets *J Pharmacol Sci* 99:6–38.

Morishima, A., M. M. Grumbach, E. R. Simpson, C. Fisher, and K. Qin. 1995. Aromatase deficiency in male and female siblings caused by a novel mutation and the physiological role of estrogens. *J Clin Endocrinol Metab* 80:3689–98.

Nettleship, J., T. H. Jones, K. S. Channer, and R. D. Jones. 2007. Physiological testosterone replacement therapy attenuates fatty streak formation and improves high-density lipoprotein cholesterol in Tfm mouse: An effect that is independent of the classical androgen receptor. *Circulation* 116:2427–34.

Newcomer, L. M., J. E. Manson, R. L. Barbieri, C. H. Hennekens, and M. J. Stampfer. 1994. Dehydroepiandrosterone sulfate and the risk of myocardial infarction in U.S. male physicians: A prospective study. *Am J Epidemiol* 140:870–5.

Ohlsson, C., F. Labrie, E. Barrett-Connor, M. K. Karlsson, O. Ljunggren, L. Vandenput, D. Mellström, and A. Tivesten. 2010. Low serum levels of dehydroepiandrosterone sulfate predict all-cause and cardiovascular mortality in elderly Swedish men. *J Clin Endocrinol Metab* 95:4406–14.

Orentreich, N., J. L. Brind, R. L. Rixer, and J. H. Vogelman. 1984. Age changes and sex differences in serum dehydroepiandrosterone sulfate concentrations throughout adulthood. *J Clin Endocrinol Metab* 59:551–5.

Page, J., J. Ma, K. Rexrode, N. Rifai, J. Manson, and S. Hankinson. 2008. Plasma dehyroepiandrosterone and risk of myocardial infarction in women. *Clin Chem* 54:1190–6.

Parker, L., J. Eugene, D. Farber, E. Lifrak, M. Lai, and G. Juler. 1985. Dissociation of adrenal androgen and cortisol levels in acute stress. *Horm Metab Res* 17:209–12.

Perusquía, M., and J. N. Stallone. 2010. Do androgens play a beneficial role in the regulation of vascular tone? Nongenomic vascular effects of testosterone metabolites. *Am J Physiol Heart Circ Physiol* 298:H1301–7.

Pinkney, J. H., C. D. A. Stehouwer, S. W. Coppak, and J. S. Yudkin. 1997. Endothelial dysfunction: Case of the insulin resistance syndrome. *Diabetes* 46:S9–13.

Rice, S. P. L., N. Agarwal, H. Bolusani, R. Newcombe, M. F. Scanlon, M. Ludgate, and D. A. Rees. 2009. Effects of dehydroepiandrosterone replacement on vascular function in primary and secondary adrenal insufficiency. *J Clin Endocrinol Metab* 94:1966–72.

Ross, R. 1993. The pathogenesis of atherosclerosis: A perspective for the 1990s. *Nature* 362:801–9.

Simoncini, T., P. Mannella, L. Fornari, G. Varone, A. Caruso, and A. R. Genazzani. 2003. Dehydroepiandrosterone modulates endothelial nitric oxide synthesis via direct genomic and nongenomic mechanisms. *Endocrinology* 144:3449–55.

Slowinska-Srzednicka, J., B. Malczewska, E. Chotkowska, A. Brzezinska, W. Zgliczynski, M. Ossowski, W. Jeske, S. Zgliczynski, and Z. Sadowski. 1995. Hyperinsulinemia and decreased plasma levels of dehydroepiandrosterone sulfate in premenopausal women with coronary heart disease. *J Int Med* 237:465–72.

Slowinska-Srzednicka, J., S. Zgliczynski, M. Ciswicka-Sznajderman, M. Srzednicki, P. Soszynski, M. Biernacka, M. Woroszylska, W. Ruzyllo, and Z. Sadowski. 1989. Decreased plasma dehydroepiandrosterone sulfate and dihydrotestosterone concentrations in young men after myocardial infarction. *Atherosclerosis* 79:197–203.

Slowinska-Srzednicka, J., S. Zgliczynski, P. Soszynski, A. Makowska, W. Zgliczynski, M. Srzednicki, M. Bednarska, E. Chotkowska, M. Woroszylska, and W. Ruzyllo. 1991. Decreased plasma levels of dehydroepiandrosterone sulphate (DHEA-S) in normolipidaemic and hyperlipoproteinemic young men with coronary artery disease. *J Int Med* 230:551–3.

Smith, J., S. Bennet, L. M. Evans, H. G. Kynaston, M. Parmer, M. D. Mason, J. R. Cockcroft, M. F. Scanlon, and J. S. Davies. 2001. The effects of induced hypogonadism on arterial stiffness, body composition, and metabolic parameters in males with prostate cancer. *J Clin Endocrinol Metab* 86:4261–7.

Sudhir, K., T. M. Chou, K. Chatterjee, E. P. Smith, T. C. Williams, J. P. Kane, M. J. Malloy, K. S. Korach, and G. M. Rubanyi. 1997. Premature coronary artery disease associated with a disruptive mutation in the estrogen receptor gene in a man. *Circulation* 96:3774–7.

Tilvis, R., M. Kähönen, and M. Härkönen. 1999. Dehydroepiandrosterone sulphate, diseases and mortality in a general aged population. *Aging Clin Exp Res* 11:30–4.

Trivedi, D., and K. T. Khaw. 2001. Dehydroepiandrosterone sulphate and mortality in elderly men and women. *J Clin Endocrinol Metab* 86:4171–7.

Webb, S. J., T. E. Geoghegan, R. A. Prough, and K. K. Michael Miller, 2006. The biological actions of dehydroepiandrosterone involves multiple receptors. *Drug Metab Rev* 38:89–116.

Weber, T., J. Auer, M. F. O'Rourke, E. Kvas, E. Lassnig, R. Berent, and B. Eber. 2004. Arterial stiffness, wave reflections, and the risk of coronary artery disease. *Circulation* 109:184–9.

Williams, M., T. Dawood, S. Ling, A. Dai, R. Lew, K. Myles, J. W. Funder, S. Krishnankutty, P. A. Komersaroff. 2004. Dehydroepiandrosterone increases endothelial cell proliferation in vitro and improves endothelial function in vivo by mechanisms independent of androgen and estrogen receptors. *J Clin Endocrinol Metab* 89:4708–15.

Williams, M., S. Ling, T. Dawood, K. Hashimura, A. Dai, H. Li, J.-P. Lui, J. W. Funder, S. Krishnankutty, and P. A. Komersaroff. 2002. Dehydroepiandrosterone inhibits human vascular smooth muscle cell proliferation independent of ARs and ERs. *J Clin Endocrinol Metab* 87:176–81.

Yoneyama, A., Y. Kamiya, M. Kawaguchi, and T. Fujinami. 1997. Effects of dehydroepiandrosterone on proliferation of human aortic smooth muscle cells. *Life Sci* 60:833–8.

Young, J., B. Couzinet, K. Nahoul, S. Braiiilly, P. Chanson, E. E. Baulieu, and G. Schaison. 1997. Panhypopituitarism as a model to study the metabolism of dehydroepiandrosterone (DHEA) in humans. *J Clin Endocrinol Metab* 82:2578–85.

Yorek, M., L. J. Coppey, J. S. Gellett, E. P. Davidson, X. Bing, D. D. Lund, and J. S. Dillon. 2002. Effect of treatment of diabetic rats with dehydroepiandrosterone on vascular and neural function. *Am J Physiol—Endocrinol Metab* 283:E1067–75.

32 DHEAS and Coping Capability

Chia-Hua Kuo

CONTENTS

SUMMARY

Our living environment is continuously changing, and failure to cope with environmental fluctuations such as mental challenge, muscular work, temperature shift, and tissue hypoxia is the primary cause of illness and death. Successful coping and adaptation against unpredictable daily challenges ensure continuation of our well-being and survival. Baseline plasma dehydroepiandrosterone sulfate (DHEAS) concentration has been found to closely associate with human longevity and fitness and is known to decline with advancing age. In this view, we performed a series of research on humans to examine the role of DHEAS in postchallenged recovery. Our recent findings suggest that DHEAS is consumed during recovery against external challenge.

BACKGROUND

Dehydroepiandrosterone (DHEA), or its sulfate derivative (DHEAS), is the most abundant steroid in humans, and it is mainly produced in the adrenal gland (Baulieu and Robel 1998; Krug, Ziegler, and Bornstein 2008; Zwain and Yen 1999a; Zwain and Yen 1999b). It is also the primary precursor in biosynthesis of many steroid hormones such as androgens, estrogens, and some other steroid hormones. DHEAS has been considered a neurosteroid by many researchers because of its high concentration in brain and its role in cognitive function (Paul and Etienne-Emile 1995).

In the animal kingdom, plasma DHEAS concentration is proportional to their life span across species. For example, DHEAS cannot be detected in rats, which typically have an average life span of only 3–4 years (Cutler et al. 1978). A wide deviation in life span was found in primates. The gray mouse lemurs with life spans of 10–13 years have plasma DHEAS ranges from 20 to 200 ng/mL (Perret and Aujard 2005), whereas the plasma DHEAS of Rhesus monkeys with life spans of 35–45 years ranges from 30 to 300 ng/mL (Pelletier et al. 2008). Humans have very high DHEAS levels (500–5000 ng/mL) with wide variation compared with all other reported species (Mika et al. 2008; Villareal and Holloszy 2006). This fact leads to the hypothesis that the acquisition of adrenarche is an evolutionary event in increasing survival. Currently, the plasma level of DHEAS has been demonstrated as an important biomarker to predict longevity in humans (Roth et al. 2002).

One way to increase DHEAS level is by oral administration of DHEA (Yang et al. 2005). This will lead to a wide array of health benefits in rats, whereas human experiment receives less positive results. Thus, human subjects would be more realistic models in elucidating the physiological role of DHEAS in human health.

Progressively decreased circulating DHEAS levels with age have long been documented in humans, with peak levels occurring between the ages of 20 and 25 and decreasing progressively thereafter by 90% after the age of 80 (Lamberts, van den Beld, and van der Lely 1997). This age-dependent decline has been linked to the gradually increasing prevalence of a wide array of metabolic disorders in the elderly (Abbasi et al. 1998; Barrett-Conner, Khaw, and Yen 1986; Herranz et al. 1995; Valenti et al. 2004; Wellman, Shane-McWhorter, and Jennings 1999).

In this chapter, we summarize evidence from our earlier studies to elucidate the role of DHEAS in physiological and psychological coping and adaptation against external challenges (Huang et al. 2006; Lee et al. 2006; Tsai et al. 2006; Wang et al. 2009; Yang et al. 2005). Since daily coping (adaptation) against environmental challenges is directly linked with survival and a state of well-being (Kirkwood 2005), we hypothesized that DHEAS could be a biomaterial consumed during recovery from external stress.

COPING AGAINST MENTAL STRESS

The physiological role of steroids has been extensively studied in the brain and peripheral nervous system after traumatic challenge (Eftal et al. 2002; Herbert 1998; Milman et al. 2008). These experiments have demonstrated the role of steroids in brain coping against damage caused by external factors, and indicate that DHEAS is involved with the recovery. Changes in mood state are, undoubtedly, involved with alteration in brain activity and have been suggested to play an adaptive role in mobilizing internal resources to cope with the external demands of a mental task (Doyle and Parfitt 1999; Keller and Nesse 2005; Lane and Terry 2000; Saito, Ikeda, and Seki 2007). We previously conducted a study in elite golfers who participated in a high-level golf tournament (Wang et al. 2009). After baseline measurement, these golfers were divided into two groups according to their competition outcomes: those who made the cut ($n = 8$), and those who failed to make the cut ($n = 6$). We found a prolonged decrease in circulating DHEAS for 5 days in the golfers who received negative competition outcome, whereas the overall mood state (measured by profile of mood states (POMS) inventory) remained stable. Because these participants are well-experienced golf players, we are not surprised that they are capable of coping with such challenging conditions. Depressed mood has been suggested as a stimulus for readjustment in mood state (Lane 2007). The depression level in the golfers recovered quickly, demonstrating that the psychological readjustment was successful and is linked with the DHEAS decline in the elite golfers. In contrast, the mood state and DHEAS were not changed for the "made the cut" group throughout the entire observation period. Mood has been proposed to be a less intense but more prolonged psychological experience that may play a role for adaptation in mobilizing personal resource (Keller and Nesse 2005; Lane and Terry 2000). Thus, the prolonged decline in DHEAS might be related to increased consumption of the steroid for brain coping against this negative psychological event. This idea is also supported by clinical studies that have shown decreases in DHEAS concentration with concurrent improvement in mood state and functioning (Fabian et al. 2001). Increasing DHEAS levels can reduce depressive symptoms in patients (Rabkin et al. 2006; Rasmusson et al. 2004; Southwick, Vythilingam, and Charney 2005; Wolkowitz et al. 1997). In this study, the possibility that muscle damage from golf competition contributes to DHEAS decline is precluded because the DHEAS decline was concurrent with a comparable decrease in circulating level of muscle creatine kinase (a muscle damage marker) after competition for both groups.

This golfer study provides additional knowledge on how athletes cope with psychological challenge in cost of endogenous DHEAS. Coping is defined as internal efforts to manage environmental demands and internal readjustment, which has unavoidably required consuming internal resources

(Lane and Terry 2000). Mental coping, generally related to situations that cannot be changed, involves managing the distressed mood evoked by adverse psychological events. Successful coping capability is reported to relate with maintenance of mental well-being (Drossman et al. 2000; Vaughn and Roesch 2003). In a real-life environment, although sometimes we could not clearly distinguish the reason and the cause, unpredictable situations encountered during our daily life can be considered as challenges that cause consumption of internal resource. Shortage in this internal resource will result in delay of recovery and development of mental illness. DHEAS appears to be the internal resource for recovery and is required for the maintenance of our good mood within normal range in our daily life. The involvement of DHEAS in the psychological coping mechanism against stress provides an additional support for the link of DHEAS to human well-being (Panjari and Davis 2007; Roth et al. 2002).

COPING AGAINST ACUTE EXERCISE

Muscular challenge against varied physical work occurs in our daily life. Fast recovery from physical work ensures an active life. We found a persistent reduction in DHEAS level with concurrent improvement in insulin sensitivity during the recovery period after resistance exercise (Tsai et al. 2006). Resistance exercise is known to generate considerable muscle damage and thereafter induces a protracted repair process. DHEAS decline phenomenon appears to be related to increased consumption for tissue repair and recovery in damaged skeletal muscle. This steroid has been documented as having a buffering action against various stresses (Cruess et al. 2000; Regelson, Loria, and Kalimi 1988), and it was found to reduce in trauma and disease (Eftal et al. 2002). The increased demand on DHEAS following resistance exercise could account for consumption of this steroid during recovery, based on the evidence that exogenous DHEA administration following various conditions causing tissue damage enhances functional recovery (Herbert 1998; Malik et al. 2003).

A significant improvement in the whole-body insulin sensitivity can be observed in middle-aged subjects receiving oral DHEA administration (increased circulating DHEAS levels by about threefold; Yang et al. 2005). Insulin is an anabolic hormone responsible for muscle protein synthesis under postprandial condition. Plasma-DHEAS concentration is closely related to the whole-body insulin sensitivity (Mika et al. 2008; Roth et al. 2002; Villareal and Holloszy 2004). This may partly account for better recovery after a catabolic challenge.

EXERCISE TRAINING EFFECTS AND DHEAS

Although exercise is a physical challenge to the human body, the benefit of exercise training on health and life expectancy are well documented (Byberg et al. 2009). This is largely because energy depletion and tissue damage can stimulate the anabolic process, such as increased insulin sensitivity and recruitment of stem cell for cell turnover (replacement of aged or unhealthy cells). Based on this fact, exercise training is generally recommended for the elderly to maintain good health status. However, it was also reported that the training effect on improving glycemic control for the elderly is not as effective as for young individuals (Short, Vittone, and Bigelow 2003). It is known that DHEAS declines with age for both males and females. We have previously investigated the effect of a 4-month exercise training program on glucose tolerance and insulin sensitivity in a group of elderly individuals aged greater than 80 years, in relation to their baseline DHEAS levels. Our result shows that the 4-month exercise training effect on improving insulin sensitivity was absent in the low-DHEAS subjects. This study was the first report demonstrating that elderly individuals with low DHEAS levels can be poor responders to exercise training intervention (Huang et al. 2006).

Insulin resistance characterized by increased glucose and insulin levels, gradually emerges with advancing age and has been suggested as an origin of many metabolic disorders, such as hypertension, vascular complications, type 2 diabetes, and cancer (Facchini et al. 2001). The related metabolic parameters, such as blood pressure and fasting glucose, have also been demonstrated as significant predictors for longevity (Mika et al. 2008). Reduction in diastolic BP by exercise training can be achieved in the oldest subjects with high-baseline DHEAS, but not for those with low-DHEAS counterparts (Huang et al. 2006). High BP is a known risk factor leading to stroke and heart failure. According to our data (Huang et al. 2006; Yang et al. 2005), individual variations in baseline DHEAS level can be one possibility that accounts for the discrepancy among many earlier studies presenting inconsistent results in BP-lowering effects of exercise training.

The aged subjects with greater DHEAS levels also exhibited greater enhancement in motor performance by training. This result could be related to the improvements in neuromuscular components secondary to the improvement in insulin sensitivity. Increasing insulin action could result in better capability to store glycogen and reduced rate of muscle protein degradation. Additionally, the sugar (ribose) for DNA synthesis during cell turnover after physical challenge is primarily from intracellular glycogen and glucose transported from circulation. Thus, high insulin sensitivity is essential for preserving greater glycogen synthesis, normal contractile property of skeletal muscle, and replacement of damaged muscle cells after acute physical challenge. The elderly are usually faced with the problem of poor insulin sensitivity and a high risk of type 2 diabetes, which can have negative impacts on nutrient delivery to peripheral motor neurons because of microvascular complication. DHEAS has been found to exert a neuroprotective effect on motor neurons, evidenced by the fact that treating DHEA in diet for more than 5 weeks prevents the diabetes-induced development of neural dysfunction (Yorek et al. 2002).

Improvement in carbohydrate metabolism by exercise training may be functionally relevant to survival and longevity (Mika et al. 2008). Environmental stress that demands muscular work persistently occurs throughout the entire lifetime and may sometimes threaten human survival. Age-dependent reduction in insulin-stimulated glycogen storage may be linked with insufficient adaptation against daily stress and explains the decrease in cumulative survival in human. Under acute stress conditions, demands on ATP increase within a short period of time. Carbohydrate has the advantage of fast degradation rate in muscle cell and can occur without oxygen supply for rapid ATP resynthesis. With advancing age, the metabolism of the human body is characterized by less reliance on carbohydrates (Rizzo et al. 2005). In young individuals, exercise is an acute stress condition that leads to rapid consumption of muscle glycogen. Afterward, the whole-body glucose tolerance and the rate of muscle glycogen storage enhance during recovery and lead to glycogen supercompensation. Exercise adaptation in improving insulin sensitivity ensures reservation of more carbohydrate fuel for coping against the recurrence of similar challenge. Normal amounts of DHEAS appear to be essential for such adaptive response.

COPING AGAINST HIGH-ALTITUDE ENVIRONMENT

High-altitude hiking represents a major stress condition to the human body because oxygen demand increases during physical activity, while oxygen availability reduces at high altitudes. The role of DHEAS on altitude adaptation has been evaluated in 12 young males who participated in a 25-day mountaineering activity, and who were evenly divided into lower and upper halves according to their sea-level DHEAS levels: low-DHEAS group ($N = 6$, 1480 ± 247 ng/mL) and high-DHEAS group ($N = 6$, 3588 ± 590 ng/mL). We found that serum-DHEAS level was reduced on day 3 of altitude activity in the young subjects with high-DHEAS level (Lee et al. 2006), whereas DHEAS level in the low-DHEAS subjects remained unchanged. To further determine whether the DHEAS reduction in the upper half was due to decreased adrenal function in response to this altitude stress, we also measured the levels of another adrenal hormone, cortisol. Our data did not provide strong support for the possibility of reduced adrenal function in the altitude activity in the high-DHEAS subjects because serum cortisol was increased at high altitudes for those DHEAS-declined subjects. Therefore, the DHEAS-depletion phenomenon is more likely due to an increased consumption for altitude adaptation. This hypothesis is strengthened by the fact that only the high-DHEAS group exhibited significant adaptive responses in improved glucose tolerance and increased erythropoiesis (red blood cell [RBC] increase) compared with the low-DHEAS counterparts.

Increase in RBC concentration is a classic adaptive response against altitude exposure. We found that only the subjects with greater DHEAS level have the normal altitude response in erythropoiesis. Intriguingly, the low-DHEAS group, without increase in RBC, exhibited about twofold greater serum erythropoietin (EPO) concentration, at both sea level and altitude, compared with the high-DHEAS subjects. This result suggests that EPO sensitivity of hematopoietic stem cells for stimulating erythropoiesis in bone marrow was lower in the subjects with low-DHEAS level.

Altitude exposure is already known to improve glucose tolerance in humans. Whether subjects can be successfully induced with an improvement in glucose tolerance and insulin sensitivity by altitude activity is also associated with their serum DHEAS concentration. Glucose is a major anaerobic fuel for most of the tissue in the human body and is particularly essential for maintaining normal ATP resynthesis as the fatty acid oxidation is hindered in hypoxic environment. Therefore, improvement in glucose tolerance could be functionally meaningful in ensuring better survival in altitude.

Increasing anaerobic fuel utilization and enhancing oxygen-carrying capability are the two fundamental physiologic strategies for survival during prolonged altitude activity. We observed that circulating DHEAS level was significantly reduced during the altitude activity in the subjects exhibiting greater baseline DHEAS concentration. Furthermore, we demonstrated that the magnitude of altitude activity effects on improving glucose tolerance, insulin sensitivity, and increasing RBC concentration were varied with the serum-DHEAS level.

COPING WITH HEAT STRESS

Coping with heat stress is one of the most important survival mechanisms for the human body because increasing body temperature can cause permanent change in protein conformation, leading to death. In our living environment, the ambient temperature is continuously changing, and this factor has been reported to affect human mortality (Laaidi, Laaidi, and Besancenot 2006). In humans, sweat evaporation from skin surface is the most important way to dissipate excessive body heat. Although the hot spring bath has been frequently used to alleviate stress, this model exerts a high magnitude of heat stress compared with dry, hot weather because skin is surrounded by hot water, which blocks sweat dissipation. A state of insulin resistance, evidenced by increased area under the curve of glucose and HOMA-IR (insulin resistance indicator), was induced in 16 healthy males with a 30-minute hot spring immersion at 41°C. A wide range of individual variation for the heat-induced stress response was identified. All subjects were then evenly divided into lower and upper halves according to their baseline DHEAS concentrations. We found that changes in insulin-resistance measures after heat stress are associated with baseline DHEAS level. In particular, hot spring immersion-induced insulin resistance state was restricted only to the low-DHEAS group. This result, together with the evidence from a number of studies from our group and others investigating various types of acute stress (Grillon et al. 2005; Lee et al. 2006; Regelson, Loria, and Kalimi 1988; Wang et al. 2009), implicates again that DHEAS plays a role in physiologic coping against acute stress.

To maintain the set point of body temperature, the human body must increase skin blood flow, which thus results in an acute reduction in diastolic blood pressure. Under this condition, the heart rate must increase, aiming to prevent an acute reduction of blood supplies (oxygen and substrates) to the rest of the peripheral tissues, other than skin. In the heat stress study, we found that the low-DHEAS group exhibits slower heart rate recovery after hot spring immersion, suggesting a lower-coping capability against the heat stress.

During hot spring immersion, circulating growth hormone and cortisol are increasing, together with the onset of insulin resistance (Møller et al. 1989). Cortisol and growth hormone are potent antagonists of insulin action (Holmang and Bjorntorp 1992; Rosenfeld et al. 1982). Increases in growth hormone might not contribute to the insulin resistance induced after hot spring immersion because the levels of growth hormone in both groups exhibited similar changes. Thus, the result of this heat-stress study suggests that the insulin resistance induced by hot spring immersion in the low-DHEAS group was associated with the increased cortisol. This hormone functions to increase glucose output from the liver and reduce glucose uptake by skeletal muscle (Kirschbaum et al. 1997; Rizza, Mandarino, and Gerich 1982), which could elevate plasma glucose availability for the brain, aimed to prevent hypoglycemia during heat stress.

In the heat stress study, DHEAS was also reduced after hot spring immersion. Together with our earlier altitude study, this result points to a possibility that DHEAS is associated with universal stress coping but convergent on a lack of oxygen (Lee et al. 2006). In fact, heat stress has been demonstrated to cause a hypoxia-mimetic effect (Fan et al. 2008; Katschinski et al. 2002; Maloyan et al. 2005; Shein et al. 2005). Hypoxia-induced factor-1α (HIF-1α), a marker of hypoxia stress, has been reported to be induced by a mild-heat environment (Maloyan et al. 2005; Shein et al. 2005). It is possible that increasing blood distribution to skin can cause a short-term reduction in oxygen delivery to the rest of the tissues in the human body. Oxygen is the final acceptor of electrons transported from reducing substrates for ATP production in virtually all cells and tissues. Mismatch in oxygen demand and supply to tissues represents coping failure against acute heat stress. Evidence from this study suggests a possibility that DHEAS is an essential coping resource for regulating physiological homeostasis against environmental challenges, leading to oxygen redistribution among tissues.

Our observation on lower baseline growth hormone in the low-DHEAS subjects suggests that DHEAS may affect the normal production of growth hormone. Growth hormone is functioned to increase lipolysis (Lucidi et al. 2000) and protein synthesis (Lucidi et al. 2000). Its biological action under heat stress could be important to reverse protein denature and degradation at the cost of fat consumption. Increasing DHEAS level by DHEA administration (Yang et al. 2005) has been reported to increase the levels of growth hormone and insulin-like growth factor-1 (IGF-1; Alessandro et al. 2001) and lower the growth hormone dose requirement for hypopituitary patients (Brooke et al. 2006). Conversely, growth hormone treatment has no effect on DHEAS levels (Boonstra et al. 2004). This evidence highlights the causal relationship between DHEAS and growth hormone production.

CONCLUSION

Successful coping with the changing environment ensures well-being in humans. Evidence from our studies and others suggests that DHEAS is the common coping resource against both physiological and psychological challenges (Fabian et al. 2001; Lee et al. 2006; Morgan et al. 2004). Mental distress significantly lowers circulating DHEAS level for young athletes against negative competition outcome, while mood can be well maintained at this cost. Muscle-damaging exercise causes a similar effect in young people. We previously found that low-DHEAS elderly individuals (more than 80 years old) received less benefit from exercise training in metabolic improvements, whereas the high-DHEAS counterparts displayed normal training adaptation. For the young subjects (~30 years old) with low DHEAS, normal altitude adaptation in erythropoiesis is less effective, and erythropoietin level is doubled compared with high-DHEAS counterparts. Under heat exposure, the baseline stress hormone levels were raised to greater extent and cardiovascular adaptations were less efficient for those low-DHEAS young individuals compared with high-DHEAS age-matched individuals. DHEA supplementation is one way to enhance circulating DHEAS level, which can significantly suppress the postexercise muscle-damage marker (unpublished data). Because survival and well-being are directly linked with successful adaptation against the changing environment, our results suggest that (1) DHEAS is consumed after external stress, which may be involved with recovery from external challenges; and (2) humans with low DHEAS have low adaptability because of poor recovery against challenges, which implicates the lower life span in the low-DHEAS cohorts reported previously.

REFERENCES

Abbasi, A., E. H. Duthie, L. Sheldahl, C. Wilson, E. Sasse, I. Rudman, and D. E. Mattson. 1998. Association of dehydroepiandrosterone sulfate, body composition, and physical fitness in independent community dwelling older men and women. *J Am Geriatr Soc* 46:263–73.

Alessandro, D. G., S. Massimo, S. Claudia, P. Simone, L. Stefano, and R. G. Andrea. 2001. Oral dehydroepiandrosterone supplementation modulates spontaneous and growth hormone-releasing hormone-induced growth hormone and insulin-like growth factor-1 secretion in early and late postmenopausal women. *Fertil Steril* 76:241–8.

Barrett-Conner, E., K. Khaw, and S. Yen. 1986. A prospective study of dehydroepi-androsterone sulfate, mortality and cardiovascular disease. *N Engl J Med* 315:1519–24.

Baulieu, E.-E., and P. Robel. 1998. Dehydroepiandrosterone (DHEA) and dehydroepiandrosterone sulfate (DHEAS) as neuroactive neurosteroids. *Proc Natl Acad Sci U S A* 95:4089–91.

Boonstra, V. H., P. G. H. Mulder, F. H. de Jong, and A. C. S. Hokken-Koelega. 2004. Serum dehydroepiandrosterone sulfate levels and pubarche in short children born small for gestational age before and during growth hormone treatment. *J Clin Endocrinol Metab* 89:712–7.

Brooke, A., L. Kalingag, F. Miraki-Moud, C. Camacho-Hubner, K. Maher, D. Walker, J. Hinson, and J. Monson. 2006. Dehydroepiandrosterone (DHEA) replacement reduces growth hormone (GH) dose requirement in female hypopituitary patients on GH replacement. *Clin Endocrinol* 65:673–80.

Byberg, L., H. Melhus, R. Gedeborg, J. Sundstrom, A. Ahlbom, B. Zethelius, L. G. Berglund, A. Wolk, and K. Michaelsson. 2009. Total mortality after changes in leisure time physical activity in 50 year old men: 35 year follow-up of population based cohort. *BMJ* 338:b688.

Cruess, D. G., M. H. Antoni, M. Kumar, and N. Schneiderman. 2000. Reductions in salivary cortisol are associated with mood improvement during relaxation training among HIV-seropositive men. *J Behav Med* 23:107–22.

Cutler, G. B. J., M. Glenn, M. Bush, G. D. Hodgen, C. E. Graham, and D. L. Loriaux. 1978. Adrenarche: A survey of rodents, domestic animals, and primates. *Endocrinology* 103:2112–8.

Doyle, J., and G. Parfitt. 1999. The effect of induced mood states on performance profile areas of perceived need. *J Sports Sci* 17:115–27.

Drossman, D. A., J. Leserman, Z. Li, F. Keefe, Y. J. B. Hu, and T. C. Toomey. 2000. Effects of coping on health outcome among women with gastrointestinal disorders. *Psychosom Med* 62:309–17.

Eftal, G., O. Kagan, C. Brian, S. Krzysztof, B. Earl, and S. Maria. 2002. Dehydroepiandrosterone as an enhancer of functional recovery following crush injury to rat sciatic nerve. *Microsurgery* 22:234–41.

Fabian, T. J., M. A. Dew, B. G. Pollock, C. F. Reynolds, B. H. Mulsant, M. A. Butters, M. D. Zmuda, A. M. Linares, M. Trottini, and P. D. Kroboth. 2001. Endogenous concentrations of DHEA and DHEA-S decrease with remission of depression in older adults. *Biol Psychiatry* 50:767–74.

Facchini, F. S., N. Hua, F. Abbasi, and G. M. Reaven. 2001. Insulin resistance as a predictor of age-related diseases. *J Clin Endocrinol Metab* 86:3574–8.

Fan, J. L., J. D. Cotter, R. A. I. Lucas, K. Thomas, L. Wilson, and P. N. Ainslie. 2008. Human cardiorespiratory and cerebrovascular function during severe passive hyperthermia: Effects of mild hypohydration. *J Appl Physiol* 105:433–45.

Grillon, C., D. S. Pine, J. M. Baas, M. Lawley, V. Ellis, and D. S. Charney. 2005. Cortisol and DHEA-S are associated with startle potentiation during aversive conditioning in humans. *Psychopharmacology* 29:1–8.

Herbert, J. 1998. Neurosteroids, brain damage, and mental illness. *Exp Gerontol* 33:713–27.

Herranz, L., A. Megia, C. Grande, P. Gonzalez-Gancedo, and F. Pallardo. 1995. Dehydroepiandrosterone sulfate, body fat distribution and insulin in obese men. *Int J Obes* 19:57–60.

Holmang, A., and P. Bjorntorp. 1992. The effects of cortisol on insulin sensitivity in muscle. *Acta Physiol Scand* 144:425–31.

Huang, Y., M. Chen, C. Fang, W. Lee, S. Yang, and C. Kuo. 2006. A possible link between exercise-training adaptation and dehydroepiandrosterone sulfate- an oldest-old female study. *Int J Med Sci* 3:141–7.

Katschinski, D. M., L. Le, D. Heinrich, K. F. Wagner, T. Hofer, S. G. Schindler, and R. H. Wenger. 2002. Heat induction of the unphosphorylated form of hypoxia-inducible factor-1alpha is dependent on heat shock protein-90 activity. *J Biol Chem* 277:9262–7.

Keller, M. C., and R. M. Nesse. 2005. Is low mood an adaptation? Evidence for subtypes with symptoms that match precipitants. *J Affect Disord* 86:27–35.

Kirkwood, T. B. 2005. Time of our lives. What controls the length of life? *EMBO Rep* 6:S4–8.

Kirschbaum, C., E. G. Bono, N. Rohleder, C. Gessner, K. M. Pirke, A. Salvador, and D. H. Hellhammer. 1997. Effects of fasting and glucose load on free cortisol responses to stress and nicotine. *J Clin Endocrinol Metab* 82:1101–5.

Krug, A. W., C. G. Ziegler, and S. R. Bornstein. 2008. DHEA and DHEA-S, and their functions in the brain and adrenal medulla. In Weizman, A. (Ed.). Neuroactive steroids in brain function, behavior, and neuropsychiatric disorders (p. 227–39). New York: Springer.

Laaidi, M., K. Laaidi, and J. P. Besancenot. 2006. Temperature-related mortality in France, a comparison between regions with different climates from the perspective of global warming. *Int J Biometeorol* 51:145–53.

Lamberts, S. W. J., A. W. van den Beld, and A. J. van der Lely. 1997. The endocrinology of aging. *Science* 278:419–24.

Lane, A. M. 2007. *Mood and Human Performance: Conceptual, Measurement, and Applied Issues.* Hauppauge, NY: Nova Science Publishers.

Lane, A. M., and P. C. Terry. 2000. The nature of mood: Development of a conceptual model with a focus on depression. *J Appl Sport Psychol* 12:16–33.

Lee, W. C., S. M. Chen, M. C. Wu, C. W. Hou, Y. C. Lai, Y. H. Laio, C. H. Lin, and C. H. Kuo. 2006. The role of dehydroepiandrosterone levels on physiologic acclimatization to chronic mountaineering activity. *High Alt Med Biol* 7:228–36.

Lucidi, P., S. Laureti, S. Santoni, M. Lauteri, N. Busciantella-Ricci, G. Angeletti, F. Santeusanio, and P. De Feo. 2000. Administration of recombinant human growth hormone on alternate days is sufficient to increase whole body protein synthesis and lipolysis in growth hormone deficient adults. *Clin Endocrinol* 52:173–9.

Malik, A. S., R. K. Narayan, W. W. Wendling, R. W. Cole, L. L. Pashko, A. G. Schwartz, and K. I. Strauss. 2003. A novel dehydroepiandrosterone analog improves functional recovery in a rat traumatic brain injury model. *J Neurotrauma* 20:463–76.

Maloyan, A., L. Eli-Berchoer, G. L. Semenza, G. Gerstenblith, M. D. Stern, and M. Horowitz. 2005. HIF-1 alpha-targeted pathways are activated by heat acclimation and contribute to acclimation-ischemic cross-tolerance in the heart. *Physiol Genomics* 23:79–88.

Mika, E., A. Hisashi, F. Ako, F. Kumiko, S. Akira, O. Maki, K. Shun-ichi, N. Yasuki, S. Yoshiyuki, and I. Tsutomu. 2008. Serum dehydroepiandrosterone sulfate levels predict longevity in men: 27-year follow-up study in a community-based cohort (Tanushimaru Study). *J Am Geriatr Soc* 56:994–8.

Milman, A., O. Zohar, R. Maayan, R. Weizman, and C. G. Pick. 2008. DHEAS repeated treatment improves cognitive and behavioral deficits after mild traumatic brain injury. *Eur Neuropsychopharmacol* 18:181–7.

Møller, N., R. Beckwith, P. C. Butler, N. J. Christensen, H. Orskov, and K. G. Alberti. 1989. Metabolic and hormonal responses to exogenous hyperthermia in man. *Clin Endocrinol* 30:651–60.

Morgan III, C. A., S. Southwick, G. Hazlett, A. Rasmusson, G. Hoyt, Z. Zimolo, and D. Charney. 2004. Relationships among plasma dehydroepiandrosterone sulfate and cortisol levels, symptoms of dissociation, and objective performance in humans exposed to acute stress. *Arch Gen Psychiatry* 61:819–25.

Panjari, M., and S. R. Davis. 2007. DHEA therapy for women: Effect on sexual function and wellbeing. *Hum Reprod Update* 13:239–48.

Paul, R., and B. Etienne-Emile. 1995. Dehydroepiandrosterone (DHEA) is a neuroactive neurosteroid. *Ann N Y Acad Sci* 774:82–110.

Pelletier, G., C. Labrie, C. Martel, and F. Labrie. 2008. Chronic administration of dehydroepiandrosterone (DHEA) to female monkey and rat has no effect on mammary gland histology. *J Steroid Biochem Mol Biol* 108:102–8.

Perret, M., and F. Aujard. 2005. Aging and season affect plasma dehydroepiandrosterone sulfate (DHEA-S) levels in a primate. *Exp Gerontol* 40:582–7.

Rabkin, J. G., M. C. McElhiney, R. Rabkin, P. J. McGrath, and S. J. Ferrando. 2006. Placebo-controlled trial of dehydroepiandrosterone (DHEA) for treatment of nonmajor depression in patients with HIV/AIDS. *Am J Psychiatry* 163:59–66.

Rasmusson, A. M., J. Vasek, D. S. Lipschitz, D. Vojvoda, M. E. Mustone, Q. Shi, G. Gudmundsen, C. A. Morgan, J. Wolfe, and D. S. Charney. 2004. An increased capacity for adrenal DHEA release is associated with decreased avoidance and negative mood symptoms in women with PTSD. *Neuropsychopharmacology* 29:1546–57.

Regelson, W., R. Loria, and M. Kalimi. 1988. Hormonal intervention: "buffer hormones" or "state dependency." The role of dehydroepiandrosterone (DHEA), thyroid hormone, estrogen and hypophysectomy in aging. *Ann N Y Acad Sci* 521:260–73.

Rizza, R. A., L. J. Mandarino, and J. E. Gerich. 1982. Cortisol-induced insulin resistance in man: Impaired suppression of glucose production and stimulation of glucose utilization due to a postreceptor detect of insulin action. *J Clin Endocrinol Metab* 54:131–8.

Rizzo, M. R., D. Mari, M. Barbieri, E. Ragno, R. Grella, R. Provenzano, I. Villa, K. Esposito, D. Giugliano, and G. Paolisso. 2005. Resting metabolic rate and respiratory quotient in human longevity. *J Clin Endocrinol Metab* 90:409–13.

Rosenfeld, R. G., D. M. Wilson, L. A. Dollar, A. Bennett, and R. L. Hintz. 1982. Both human pituitary growth hormone and recombinant DNA-derived human growth hormone cause insulin resistance at a postreceptor site. *J Clin Endocrinol Metab* 54:1033–8.

Roth, G. S., M. A. Lane, D. K. Ingram, J. A. Mattison, D. Elahi, J. D. Tobin, D. Muller, and E. J. Metter. 2002. Biomarkers of caloric restriction may predict longevity in humans. *Science* 297:811.

Saito, M., M. Ikeda, and H. Seki. 2007. A study on adaptive action development-interference of mood states with perceived performance and perceived efficacy. *Int J Ind Ergon* 5:48–51.

Shein, N. A., M. Horowitz, A. G. Alexandrovich, J. Tsenter, and E. Shohami. 2005. Heat acclimation increases hypoxia-inducible factor 1 alpha and erythropoietin receptor expression: Implication for neuroprotection after closed head injury in mice. *J Cereb Blood Flow Metab* 25:1456–65.

Short, K. R., J. L. Vittone, and M. L. Bigelow. 2003. Impact of aerobic exercise training on age-related changes in insulin sensitivity and muscle oxidative capacity. *Diabetes* 52:1888–96.

Southwick, S. M., M. Vythilingam, and D. S. Charney. 2005. The psychobiology of depression and resilience to stress: Implications for prevention and treatment. *Annu Rev Clin Psychol* 1:255–91.

Tsai, Y. M., S. W. Chou, Y. C. Lin, C. W. Hou, K. C. Hung, H. W. Kung, T. W. Lin, S. M. Chen, C. Y. Lin, and C. H. Kuo. 2006. Effect of resistance exercise on dehydroepiandrosterone sulfate concentrations during a 72-h recovery: Relation to glucose tolerance and insulin response. *Life Sci* 79:1281–6.

Valenti, G., L. Denti, M. Maggio, G. Ceda, S. Volpato, S. Bandinelli, G. Ceresini, A. Cappola, J. M. Guralnik, and L. Ferrucci. 2004. Effect of DHEAS on skeletal muscle over the life span: The InCHIANTI Study. *J Gerontol A Biol Sci Med Sci* 59:M466–72.

Vaughn, A. A., and S. C. Roesch. 2003. Psychological and physical health correlates of coping in minority adolescents. *J Health Psychol* 8:671–83.

Villareal, D. T., and J. O. Holloszy. 2004. Effect of DHEA on abdominal fat and insulin action in elderly women and men: A randomized controlled trial. *JAMA* 10:2243–8.

Villareal, D. T., and J. O. Holloszy. 2006. DHEA enhances effects of weight training on muscle mass and strength in elderly women and men. *Am J Physiol* 291:E1003–8.

Wang, H. T., S. M. Chen, S. D. Lee, M. C. Hsu, K. N. Chen, Y. F. Liou, and C. H. Kuo. 2009. The role of DHEA-S in the mood adjustment against negative competition outcome in golfers. *J Sports Sci* 27:291–7.

Wang, J. S., S. M. Chen, S. P. Lee, S. D. Lee, C. Y. Huang, C. C. Hsieh, and C. H. Kuo. 2009. Dehydroepiandrosterone sulfate linked to physiologic response against hot spring immersion. *Steroids* 74:945–9.

Wellman, M., L. Shane-McWhorter, and J. P. Jennings. 1999. The role of dehydroepiandrosterone in diabetes mellitus. *Pharmacotherapy* 19:582–91.

Wolkowitz, O. M., V. I. Reus, E. Roberts, F. Manfredi, T. Chan, W. J. Raum, S. Ormiston et al. 1997. Dehydroepiandrosterone (DHEA) treatment of depression. *Biol Psychiatry* 41:311–8.

Yang, S. C., C. Y. Chen, Y. H. Liao, F. C. Lin, W. C. Lee, Y. M. Cho, M. T. Chen, C. H. Chou, and C. H. Kuo. 2005. Interactive effect of an acute bout of resistance exercise and dehydroepiandrosterone administration on glucose tolerance and serum lipids in middle-aged women. *Chin J Physiol* 48:23–9.

Yorek, M. A., L. J. Coppey, J. S. Gellett, E. P. Davidson, X. Bing, D. D. Lund, and J. S. Dillon. 2002. Effect of treatment of diabetic rats with dehydroepiandrosterone on vascular and neural function. *Am J Physiol* 283:E1067–75.

Zwain, I. H., and S. S. C. Yen. 1999a. Dehydroepiandrosterone: Biosynthesis and betabolism in the brain. *Endocrinology* 140:880–7.

Zwain, I. H., and S. S. C. Yen. 1999b. Neurosteroidogenesis in astrocytes, oligodendrocytes, and neurons of cerebral cortex of rat brain. *Endocrinology* 140:3843–52.

33 DHEA and Memory

Elizabeth Sujkovic, Radmila Mileusnic, and Jonathan P. Fry

CONTENTS

INTRODUCTION

As its sulfate ester, dehydroepiandrosterone (DHEA) is the major circulating steroid hormone in the young adult primate and is secreted from the cortex of the adrenal gland. This circulating dehydroepiandrosterone sulfate (DHEAS) serves as a pool of the free steroid DHEA, following uptake then desulfation by the steroid sulfatase (Sts) enzyme in various tissues (Reed et al. 2005; Miller 2009; Hobkirk 1985). Secretion of DHEAS by the adrenal gland shows a characteristic pattern over life, suggestive of functional roles, although few have been fully elucidated. In this chapter, we will focus on the possible role of DHEA in learning and memory, either as the free steroid or as DHEAS.

Endocrine production of DHEAS begins with a high output from the fetal adrenal gland, and in association with placental uptake, desulfation, and further metabolism, is the major contributor of maternal estrogens during late pregnancy. However, adrenal DHEAS production declines after birth, and plasma concentrations of DHEAS then remain low in humans until the age of 6–9 years, when they rise at the onset of adrenarche. Circulating DHEAS concentrations continuously increase during puberty, reaching a peak concentration of around 5 μM during the second decade of life, although there is variability between individuals and concentrations are generally higher in males than in females (Smith et al. 1975; Orentreich et al. 1984; Campbell 2006). A slow decline in the production of DHEAS then follows through adulthood, culminating in noticeably low plasma concentrations in the elderly (Orentreich et al. 1984).

With the above age-related changes in adrenal DHEAS production, there has long been an interest in the influence of DHEA on learning and memory. This was reinforced nearly 30 years ago by the discovery that DHEAS and smaller amounts of the free steroid DHEA could be detected in adult male rat brains at concentrations that did not change upon removal of the adrenal glands and testes (Corpéchot et al. 1981). This led to the concept of neurosteroids: steroids produced within the nervous system. In addition to rodents, the presence of DHEA and DHEAS has been reported in the brains of humans (Lacroix et al. 1987; Lanthier and Patwardhan 1986), other primates (Robel et al. 1987), amphibians (reviewed in Mensah-Nyagan et al. 2001), and avians (Migues, Johnston, and Rose 2002; Tsutsui and Yamazaki 1995). Nevertheless, and, ironically, in view of the fact that characterization of DHEA in rat brain led to the concept of neurosteroids, the synthesis of DHEA in the mammalian central nervous system remains a controversy. In endocrine glands, the CYP17 (17α-hydroxylase/17,20-lyase) enzyme catalyzes the production of DHEA from pregnenolone via the intermediate 17α-hydroxypregnenolone. This enzyme shows little or no activity in brain tissue (Mellon 2007) and the adult male rat brain, which contains both pregnenolone and DHEA, lacks detectable concentrations of the above intermediate (Ebner et al. 2006). Thus, the possibility remains that as in other tissues, circulating DHEAS is imported into the brain and desulfated to DHEA. Consideration of the role of DHEA and DHEAS in learning and memory is further complicated by two factors: (1) no specific receptor sites have been identified for either of these steroids in the brain, and (2) apart from desulfation of DHEAS to DHEA, little is known at present of their downstream metabolism in the brain to potential neuroactive steroids. In this chapter, we will start by reviewing the possible sites of action for DHEA and DHEAS in the brain, and then outline the processes thought to underlie learning and memory before discussing how these might be influenced by DHEA(S).

MODES OF ACTION OF DHEA(S) IN THE BRAIN

Steroid molecules classically bind to intracellular nuclear or cytoplasmic receptors to regulate gene expression. As mentioned in the previous section, DHEA appears to have no such receptor, and so effects on transcription would appear to depend on metabolism of other steroids, which in peripheral tissues are known to include both androgens and estrogens. However, little is known of DHEA(S) metabolism by brain other than 17-keto reduction to produce free and sulfated androstenediol *in vivo* (Kishimoto and Hoshi 1972). Such metabolism could underlie the androgenic actions of DHEA in mouse brain, although some such activity can be detected even after blocking metabolism to the more potent androgens, testosterone and dihydrotestosterone (see Mo et al. 2009). In addition, there are well-documented antiglucocorticoid actions of DHEA, which occur also in the brain (see Maninger et al. 2009) and could involve interactions at the receptors for glucocorticoids and/or competition with the metabolic activation of these stress hormones (Muller, Hennebert, and Morfin 2006). Whatever the mechanism, such actions of DHEA are likely to be of significance for consideration of the effects of stress on learning and memory, particularly with age.

Besides actions mediated probably indirectly through nuclear steroid receptors in the brain, DHEA(S) has been shown to have direct effects on neuronal membranes, especially at receptors for neurotransmitters. The latter includes antagonism at type A receptors for the inhibitory amino acid transmitter γ-aminobutyric acid ($GABA_A$ receptors; Majewska et al. 1990; Spivak 1994) and enhancement of N-methyl-D-aspartate (NMDA) type receptors for the excitatory amino acid glutamate (Compagnone and Mellon 1998; Lhullier et al. 2004). DHEA(S) also shows agonism at type 1 sigma (σ_1) receptors (Urani, Privat, and Maurice 1998), which underlie further interactions at NMDA and acetylcholine receptors in the brain (Monnet et al. 1995; Matsuno et al. 1994; also see review by Monnet and Maurice 2006). Overall, the above actions on the neuronal membrane could be considered as excitatory. However, their relevance to any physiological influences of DHEA(S) on learning and memory must be approached with caution as such actions are reported at concentrations of these steroids higher than those observed in brain tissue *in vivo*.

Drugs such as barbiturates and benzodiazepines, which are agonists and/or enhancers of the action of GABA at $GABA_A$ receptors, are associated with impairment of memory acquisition. Thus, it might be expected that the $GABA_A$ antagonistic action of DHEA(S) would enhance learning (Chapouthier and Venault 2002; Castellano and McGaugh 1990). Likewise, the ability of DHEA(S) to enhance postsynaptic NMDA receptor function will increase the influx of Ca^{2+} ions, a form of signalling associated with the phenomenon of long-term potentiation of synaptic transmission and thought to underlie memory formation. We will return to the above putative sites of action for DHEA(S) later in this chapter, but we will first give a brief overview of current concepts and experimental approaches to learning and memory.

LEARNING AND MEMORY

Learning is typically understood as the process by which new information or knowledge is acquired, while memory is the process by which organisms store, retain, and recall information. Memory formation is processed in three stages: acquisition (the initial phase in the formation of a memory trace), consolidation (the phase during which the stabilization of the memory trace takes place), and retrieval (the "actualisation" of the memory trace). Once formed, memory can last from minutes to years, suggesting the existence of complex mechanisms for altering patterns of neuronal connectivity involving changes in gene expression, protein synthesis, and ultimately cellular structure.

Memory is often classified according to its endurance into short-term and long-term memory. Short-term memory has a limited capacity and allows recall for a period of several seconds to a minute without rehearsal. In contrast, long-term memory can store much larger quantities of information for a very long time, sometimes for a whole life span. However, research on different animal models and on humans indicates that long-term memory could be divided into at least two major types of memory systems: memory of "how," also called procedural (or implicit) memory; and memory of "that," also called declarative (or explicit) memory. Thus, we learn in two fundamentally different ways: we learn about the world around us by acquiring knowledge of people and places and we learn how to do things by acquiring different skills. Procedural memory is not based on conscious recall of information, in contrast to declarative memory, which requires conscious recall (Squire 1986; Squire 2004; Tulving 1985). According to information type carried, each one of these categories could be further subdivided into different subtypes, as shown in Figure 33.1. Why do we need model systems and different tasks to study memory? Because only by using different models and different tasks might we ultimately learn how different aspects of behavior emerge from biological correlates of neural cell functions.

Standard laboratory tasks used in most memory laboratories today may be aversive or appetitive, single or multiple trial. They include passive avoidance and fear conditioning (both single trial) and versions of the Morris water maze to test spatial memory (multiple trials). All of them might be used to study short-term and long-term memory, declarative as well as nondeclarative (procedural). The merit of one-trial tasks is that they are sharply timed; the brevity of the training trial allows for a separation of events surrounding the training experience from the processes that occur during

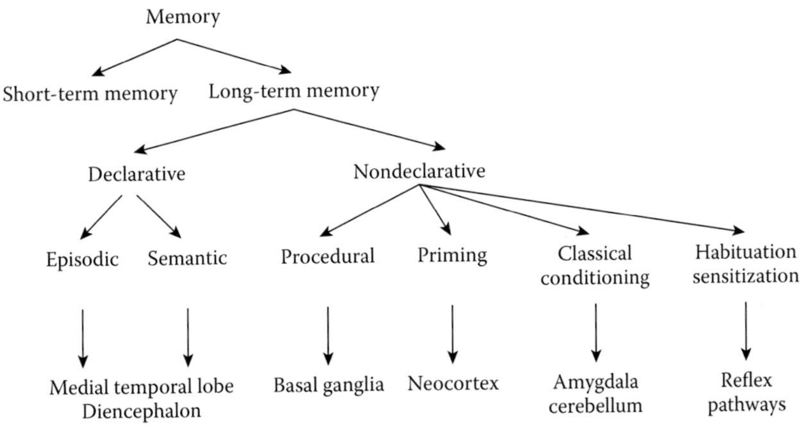

FIGURE 33.1 Taxonomy of memory: the multiplicity of memory systems and brain regions involved.

memory consolidation. However, during passive avoidance training and fear conditioning, animals could acquire both declarative memories and procedural memories. Because procedural memories take many more trials to acquire than declarative memories, measuring retention latencies after one-trial passive avoidance training most likely reflects declarative fear memories. Nevertheless, single-trial learning is not typical of learning, in general, because many instances of animal and human learning are based on the acquisition of experience in a number of repeated trials, involving processes such as generalization, categorization, and discrimination.

Our laboratory has a long-standing experience in using the chick as a model system to study memory. The training task that we use is a one-trial passive avoidance task, in which chicks learn to avoid pecking at a small bead coated with the bitter, distasteful, but nontoxic, methylanthranilate (MeA). The task has the merits of being rapid and sharply timed (for a review see Rose 2000). Chicks that are trained to peck a bead typically coated with 100% MeA (strong training) show a disgust reaction (backing away, shaking their heads, and wiping their bills) and will avoid a similar but dry bead for at least 48 hours. Another version of the task exists, where the bead is coated in 10% MeA (weak training) and chicks show avoidance for up to 8 hours, subsequently. Because the pecking response requires a positive, accurate act by the bird, it also controls for effects on attentional, visual, and motor processes. In its strong form, the task can be used to identify the molecular cascade involved in memory formation and the interventions that impair consolidation; in its weak form, the task can be used to explore potential memory-enhancing agents.

Because of their closer evolutionary relatedness to humans, mammals are usually preferred to birds as model systems. However, the bird brain is not a primitive form of the mammalian brain, and it has striking homologies and analogies with that of the mammal. Interestingly, birds have cognitive capacities that were thought to be characteristic of primates. For example, sensory learning in songbirds shares important features with forms of sensory imprinting described in mammals and, similarly to mammals, results in a long-lasting and perhaps permanent memory (for review see Bolhuis and Gahrb 2004; Reiner et al. 2004).

The biological mechanisms that underlie learning, memory storage, and, eventually, changes in behavior involve a cascade of molecular events such as inter- and intra-cellular communication systems, cell adhesion molecules, growth factors, and gene expression. In chicks, the combination of interventive and correlative studies has revealed a cascade of molecular processes occurring in defined brain regions, notably the left *intermediate medial mesopallium* (IMMP) and *medial striatum*. Briefly, within minutes of pecking at the bitter bead, there is (1) enhanced glutamate release, (2) upregulation of NMDA-sensitive glutamate receptors, and (3) the opening of N-type conotoxin-sensitive calcium channels. These transient synaptic responses result in the activation of protein kinases and the expression of immediate early genes such as *c-fos* and *c-jun*, and subsequently, the family of late genes coding for glycoproteins, which, inserted into the pre- and postsynaptic membranes, alter synaptic structure and connectivity. In other words, the early phase of memory, or short-term memory, depends on transient synaptic changes. Without activating transcription of new genes in the nucleus, short-term memories will quickly fade. The progression from short-term to long-term memory, a stable form of memory that can last from hours to a lifetime, requires the mobilization of molecular processes that go beyond the synapse, to the neuronal nucleus, with the activation of gene expression and protein synthesis to modify permanently the synaptic structure (Rose 2000).

ACTIONS OF DHEA(S) ON LEARNING AND MEMORY

Animal Studies

The role of the steroid DHEA and its sulfate ester on learning and memory has been studied in animal models by using various learning paradigms. The first observations to suggest that DHEA and its sulfate ester may have memory-enhancing properties in experimental animals were reported by Flood and colleagues. In a study using the T-maze footshock active avoidance test, immediate

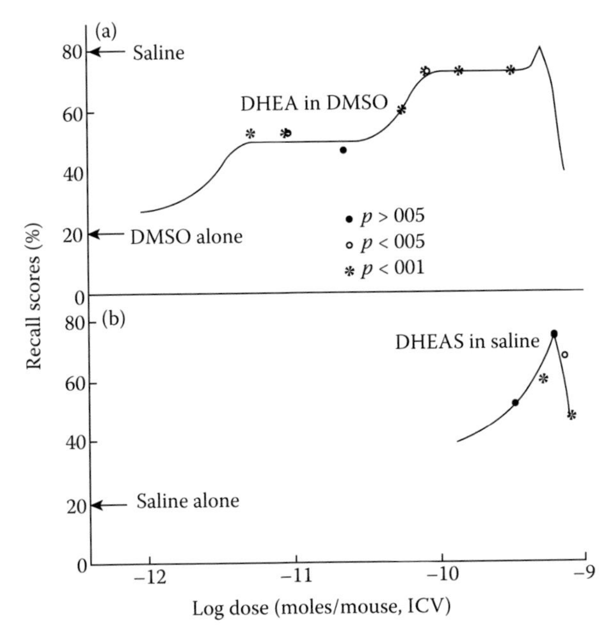

FIGURE 33.2 Effects of intracerebroventricularly injected DHEA and DHEAS on memory retention in mice. (a) DHEA in DMSO in well-trained animals: five training trials in the T maze; the buzzer, loud; intertrial interval, 45 seconds; footshock level, 0.35 mA. (b) DHEAS in saline in poorly trained animals; four training trials; buzzer, muffled; intertrial interval, 30 seconds; footshock level, 0.30 mA. Injections of test solutions were made within 3 minutes after training. Retention was tested 1 week after the training trials. (Reprinted from Roberts, E., L. Bologa, J. F. Flood, and G. E. Smith. 1987. *Brain Res* 406(1–2):357–62. Copyright (1987), with permission from Elsevier.)

post-training intracerebroventricular (i.c.v.) injections of DHEA or DHEAS were found to facilitate memory retention 1 week later in adult male mice (Roberts et al. 1987; see Figure 33.2). This improved memory retention was also seen after addition of DHEAS to the drinking water (Flood, Smith, and Roberts 1988). A great deal of research has followed focusing on the influence of DHEA and DHEAS in learning and memory processes and investigating their mechanisms of action. To examine their role during the acquisition, consolidation, and retrieval stages of memory processing, these steroids have been administered at various times throughout the learning process, as outlined in the following section.

DHEA(S) Pretraining

Subcutaneous (s.c.) administration of DHEAS to male mice 60 minutes before training on a passive avoidance task enhanced memory retention tested 24 hours later with a maximal effect at a dose of 1 mg/kg (Reddy and Kulkarni 1998a; Figure 33.3a).

Additionally, DHEAS has been administered to both male and female adult rats trained on the elevated plus maze test, in which decreased transfer latency (the time taken by the mouse to move into one of the enclosed arms) between training and 24 hour-delay retention testing was used as a measure of learning and memory. DHEAS was shown to have a memory-enhancing effect when administered 30 minutes prior to training at 5 mg/kg s.c., with testing done 24 hours later. Interestingly, the memory-facilitating effects in this study were observed in male but not in female rats, demonstrating a sex-specific effect (Reddy and Kulkarni 1999). Another study has reported memory enhancement following DHEA administration (28 mg/kg, intraperitoneal [i.p.]) to male rats 30 minutes before training on the step-down inhibitory avoidance task, with testing carried out at both 1.5 hours and 24 hours later (Lhullier et al. 2004).

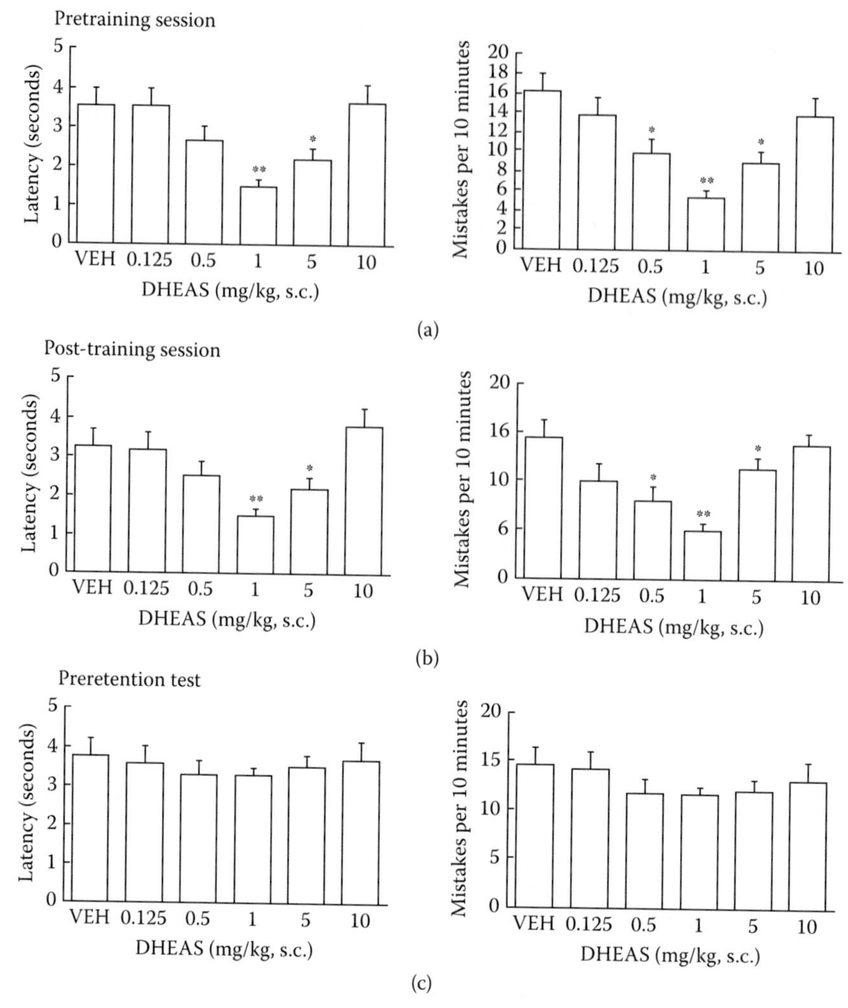

FIGURE 33.3 The effect of dehydroepiandrosterone sulfate (DHEAS; 0.125–10 mg/kg) administered subcutaneously to mice (a) 60 minutes before (pretraining), (b) immediately after (post-training) the training session, or (c) 60 minutes before the retention test. The training was for passive avoidance of an electric shock to the feet, as tested by the step-down latency (left panel) and the number of mistakes (right panel). The retention test was performed 24 hours after the training session. The results are shown as mean ± standard error of the mean with six to eight animals per group. Values marked * or ** were significantly different from those obtained with mice injected with the solvent vehicle (VEH, 0.1% Tween 80 in saline) alone. (Reprinted from Reddy, D. S., and S. K. Kulkarni. 1998a. *Brain Res* 791(1–2):108–16. Copyright (1998), with permission from Elsevier.)

The memory-enhancing effects of DHEA and DHEAS have also been examined in 1-day-old chicks using the one-trial passive avoidance-learning paradigm, which, as mentioned above, has the advantage of a sharply timed training event. Following direct injection 15 minutes before training into the IMMP, a brain region known to be specifically involved in the early stages of memory formation in these animals (see Rose 2000), both DHEA and DHEAS facilitated memory retention in male and female chicks tested 24 hours later. The DHEAS appeared more potent than DHEA, with significant effects seen at minimum doses of 0.04 and 0.3 ng per chick, respectively. With the microinjection procedure used, such doses would have resulted in concentrations at the site of injection of 22 μM for DHEAS and 208 μM for DHEA, before diffusion into the brain tissue (Migues, Johnston, and Rose 2002). Also in chicks, a similar enhancement of memory retention tested 24 hours later was seen when

DHEAS was administered at 20 mg/kg i.p. 30 minutes before training (Sujkovic et al. 2007; Figure 33.4a). Use of radioactively labeled DHEAS allowed an estimate of the amount of the i.p. dose reaching the brain and assuming a uniform distribution with no metabolism of the label, gave an estimated local concentration of about 0.5 μM. The latter study also found the memory-enhancing effect of DHEAS only to be significant in males and not females, as reported above for rats.

DHEA(S) Post-Training

Several studies have examined the effects of DHEA and DHEAS on the consolidation of memory by administering these steroids after training.

As already mentioned, an earlier study, using the T-maze footshock active avoidance paradigm in adult male mice, showed immediate post-training i.c.v. injections of DHEA or DHEAS facilitating memory retention 1 week later (Roberts et al. 1987; Flood, Smith, and Roberts 1988). The same authors showed that DHEAS also enhances memory retention, when administered i.c.v. at 30 and 60 minutes post-training, but not at 90 or 120 minutes. The most effective doses of DHEA and DHEAS would have resulted in local i.c.v. concentrations of around 20 mM, although these are likely to have quickly diluted on diffusion into the brain tissue. Enhancement of memory retention was also seen following systemic administration of DHEAS to mice trained on the same task; the most effective doses being 20 mg/kg s.c. immediately after training or 43 mg/kg/day for 1 week in the drinking water (Flood, Smith, and Roberts 1988). Other authors have shown DHEAS to facilitate memory retention for a passive avoidance task in male mice when injected s.c. 60 minutes post-training and tested 24 hours later, with the most effective dose being 1 mg/kg (Reddy and Kulkarni 1998a; Figure 33.3b).

In 1-day-old chicks trained on the one-trial passive avoidance memory task, post-training administration of DHEA (3 ng) or DHEAS (4 ng) into the IMMP could enhance memory retention when administered at 30 and 60 minutes post-training but not at 180 minutes (Migues, Johnston, and Rose 2002). Similarly, i.p. administration of DHEAS 20 mg/kg has been shown to enhance memory retention at 30 minutes and 4.5 hours post-training in male 1-day-old chicks trained on the above taste avoidance learning paradigm (Sujkovic et al. 2007; see Figure 33.4b and c). In the latter study, as for DHEAS administered pretraining, effects were only significant in male and not in female chicks.

DHEAS Pretesting

Effects of DHEA and DHEAS on the recall of learning have been addressed by administering these steroids prior to recall of a learnt experience and have consistently failed to show significant effects on this aspect of memory function. Thus, DHEAS administered at 0.125–10 mg/kg, s.c. at 1-hour preretention test for the step-down passive avoidance task in mice had no effects on recall (Reddy and Kulkarni 1998a, see Figure 33.3c). Likewise, Maurice, Su, and Privat (1998) have reported that administration of DHEAS (20 mg/kg, s.c.) to mice 30 minutes before testing on the step-down type of passive avoidance task could not facilitate memory retention 14 days post-training. We have also been unable to detect changes in memory performance following administration of DHEAS 20 mg/kg i.p. 30 minutes before testing to 1-day-old chicks trained on the one-trial passive avoidance task (Sujkovic et al. 2007; see Figure 33.4d).

Antiamnesic Effects

A number of studies have examined whether DHEA(S) could prevent and/or reverse pharmacological impairment of memory. Again, these show effects of DHEA(S) on acquisition and consolidation, rather than recall of memory.

Using the T-maze footshock active avoidance paradigm, Flood, Smith, and Roberts (1988) reported antiamnesic effects of DHEAS. In this study, 15 minutes before training, mice were administered with anisomycin (ANI; 20 mg/kg, s.c.), an inhibitor of protein synthesis, followed by immediate post-training injections of DHEAS (162 ng, i.c.v.) and an additional ANI injection 1.75 hours later.

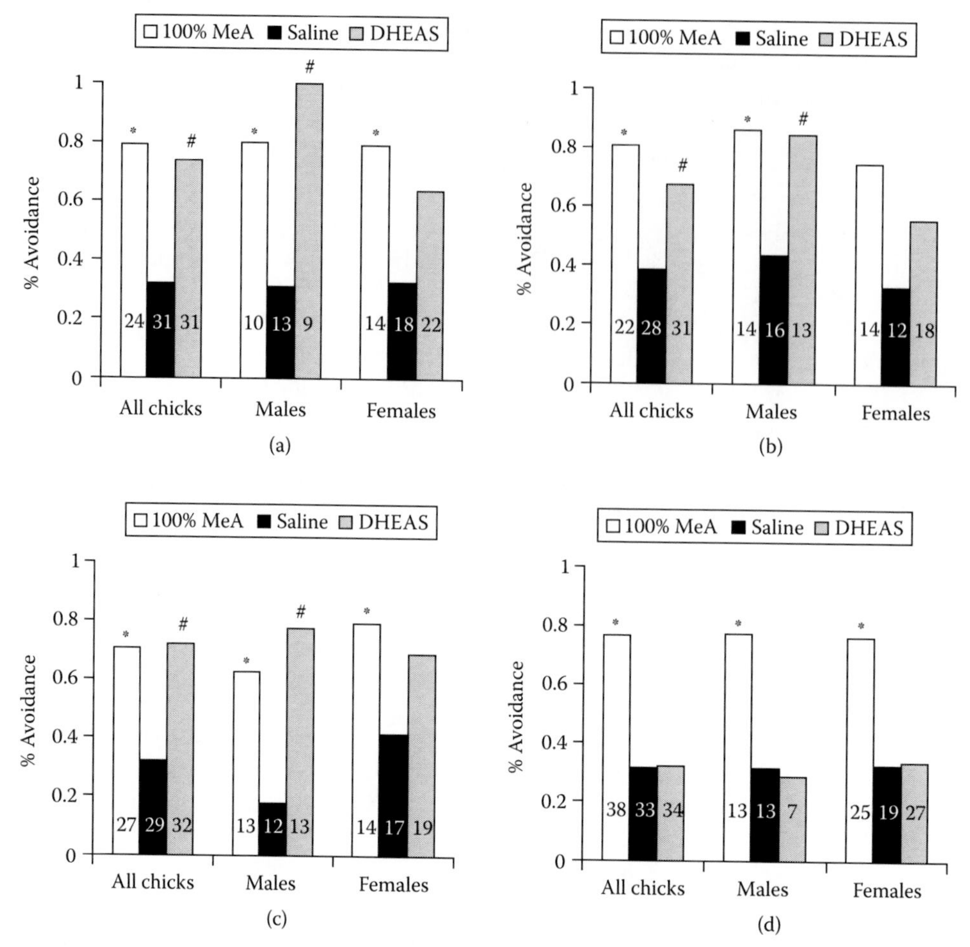

FIGURE 33.4 The effect of DHEAS (20 mg/kg) or saline administered i.p. either (a) 30 minutes before training, (b) post-training at 30 minutes, or (c) 4.5 hours or (d) 30 minutes pretesting on recall for the weak aversive stimulus (WS) of 10% methylanthranilate (MeA). Noninjected controls were trained for with the strong stimulus (SS) of 100% MeA. Recall was tested 24 hours after training and the results presented as percent avoidance. A statistically significant difference was observed when percent avoidance was compared between (*) SS group and saline-injected WS chicks or between (#) DHEAS-injected WS group and saline alone, also before WS. Numbers of animals in each group are presented in the relevant bars. (Reprinted from Sujkovic, E., R. Mileusnic, J. P. Fry, and S. P. Rose. 2007. *Neuroscience* 148(2):375–84. Copyright (2007), with permission from Elsevier.)

Here, DHEAS was shown to reverse the amnesic effects of ANI when tested at 1 week post-training. Therefore, it appears that DHEAS may influence the process of protein synthesis that accompanies consolidation of memory. This is consistent with our observation that DHEAS enhances memory in the 1-day-old chick if administered at 30 minutes or 4.5 hours post-training (Sujkovic et al. 2007). These time points have been shown previously to be accompanied by protein synthesis, which is crucial for the formation of memory in the chick model system (reviewed by Rose 2000).

Using the same behavioral paradigm, Flood, Smith, and Roberts (1988) have also studied whether DHEAS could reverse memory impairment induced by the muscarinic cholinergic receptor antagonist, scopolamine (SCO). Here, the learning deficit induced by SCO (1 mg/kg, s.c.) injected immediately after training could be reversed by subsequent injection with DHEAS (162 ng, i.c.v.) at 45 minutes post-training. By contrast, DHEA (0.300, 1.350, 6.075 μg/kg) has failed to reverse a SCO (1 mg/kg)-induced amnesia when administered to mice trained on a spontaneous alternation task, a measure of short-term memory formation (Bazin et al. 2009).

Both DHEA and DHEAS could block ethanol-induced memory impairment in male mice trained on win-shift foraging paradigm. In this study, steroids were injected at the dose of 0.05 mg/kg, i.p. 30 minutes before the testing trial, while ethanol was injected at 0.5 g/kg, i.p. at 10 minutes pretesting (Melchior and Ritzmann 1996). Given that ethanol-induced memory impairment can be potentiated by $GABA_A$ agonists such as muscimol but inhibited by its antagonists such as picrotoxin and bicuculline (Castellano and Pavone 1988), the observed effects of steroids in this study are consistent with their interactions with the $GABA_A$ receptor complex.

In another study, DHEAS at the doses of 5 and 10 mg/kg s.c. injected at 30 minutes before training could prevent the amnesic effects of the NMDA receptor antagonist dizocilpine (0.1 mg/kg, i.p.) administered at 15 minutes pretraining to mice trained on both the step-down passive avoidance and elevated plus maze tasks (Reddy and Kulkarni 1998b).

The possible interactions of DHEA and DHEAS with σ_1 receptors are further supported by two studies (Maurice, Su, and Privat 1998; Reddy and Kulkarni 1998a), which showed the memory-enhancing effects of both these steroids given s.c. to be blocked in mice when coadministered with the σ_1-receptor antagonist, haloperidol. Indeed, the σ_1 receptor may be a necessary target for anti-amnesic compounds because the reversal of dizocilpine-induced amnesia by PRE-084 (a selective σ_1 receptor agonist) and by DHEAS in mice trained on spontaneous alternation and passive avoidance task could be blocked by a 16-mer oligodeoxynucleotide antisense to the σ_1-receptor cDNA, which induced downregulation of σ_1-receptor expression in the brain (Maurice et al. 2001). The abilities of DHEA and DHEAS to reverse pharmacologically induced amnesia are fully consistent with the sites of action for these steroids on neurotransmitter receptors described in the section, "Modes of Action of DHEA(S) in the Brain."

Steroids and Aging

Other animal studies have addressed the possibility that DHEAS can also ameliorate age-related impairments in learning and memory. Flood and Roberts (1988) tested the effects of immediate post-training injection of DHEAS (20 mg/kg, s.c.) to middle-aged (18-month-old) and old (24-month-old) mice trained on the aversive T-maze footshock active avoidance task. Single treatment with DHEAS in this study was found to result in memory enhancement in both old and middle-aged animals when tested 1 week or 1 month after training, respectively.

Additionally, the effects of DHEAS have been examined on the impaired memory of 16-month-old as compared with 3-month-old mice. When given before training on the step-down passive avoidance and elevated plus maze tasks, a dose of 10 mg/kg s.c. DHEAS significantly attenuated the age impairment of memory (Reddy and Kulkarni 1998b).

In another study, the effects of DHEAS on working memory as measured on the win-shift water escape task have been studied in 18- to 20-month-old male and female mice. Here, post-training administration of DHEAS orally for 1 week, in the drinking water at the dose of 1.5 mg/mouse/day resulted in memory enhancement, compared with mice given water alone, in both males and females (Markowski et al. 2001).

Using aged SAMP8 mice (a model to study amyloid beta toxicity, which is thought to be linked with Alzheimer's disease), one study has reported improved learning and memory when DHEAS was administered in drinking water (0.3 mg/mL) for 8 weeks to animals trained on the T-maze footshock avoidance paradigm (Farr et al. 2004).

Conclusions from Animal Studies

From the evidence of animal studies reviewed here, both DHEA and DHEAS are more likely to facilitate memory when administered pretraining and post-training, but not pretesting. Thus, both these steroids appear to enhance the acquisition and consolidation but not the retrieval stages of memory. The sulfated steroid, DHEAS, is more potent in this respect than the free steroid DHEA, at least when a reliable comparison of their potencies can be made by direct injection into the chick brain (Migues, Johnston, and Rose 2002). As reported by Sujkovic, Mileusnic, and Fry (2009), one possibility is that this higher potency of DHEAS arises, at least in part, from its slower clearance than DHEA from chick brain. The latter study showed desulfation of a small proportion of the DHEAS injected into the brain to yield free DHEA during the period of memory formation and consolidation in the chick. Similar comparisons of the potencies of DHEAS and DHEA following i.c.v. injections in mice cannot be made as the studies reviewed here used different solvent vehicles and assessed the memory-enhancing potencies of the free and sulfated steroids against different strengths of training. Nevertheless, several studies in rats and mice have shown inhibition of the Sts enzyme to increase the potency of DHEAS to enhance memory and reverse drug-induced amnesia (see Johnson et al. 2000). Such effects should be interpreted with caution because these Sts inhibitors have been given peripherally, where they will inhibit desulfation of DHEAS in the liver (see Johnson et al. 2000) and are also sulfate esters themselves, competing with circulating DHEAS for access into the brain (Nicolas and Fry 2007).

As for whether or not DHEA and DHEAS require further metabolism in order to induce memory-enhancing effects, the evidence from reversal of pharmacologically induced amnesia in animals is consistent with the actions of these native steroids at the neurotransmitter receptor sites identified *in vitro*. Also, our study of the metabolism of DHEA and DHEAS after direct injection into chick brain showed no detectable conversion to other steroids (see Figure 33.5), at least during the time taken for the early stages of memory formation (Sujkovic, Mileusnic, and Fry 2009). However, the animal studies described here have used intracerebral doses of DHEA(S) likely to produce localized concentrations of these steroids in at least the micromolar range and over 1000-fold higher than the endogenous brain concentrations (see Migues, Johnston, and Rose 2002; Sujkovic et al. 2007; Ebner et al. 2006). Moreover, the other studies in which animals were administered with DHEA(S) peripherally frequently showed bell-shaped dose–response curves, suggesting an optimum dose beyond which additional actions become detrimental to learning and memory. More laboratory investigations are required into the actions and brain metabolism of DHEA(S) at physiological concentrations. Inevitably, these will be limited by the low adrenal

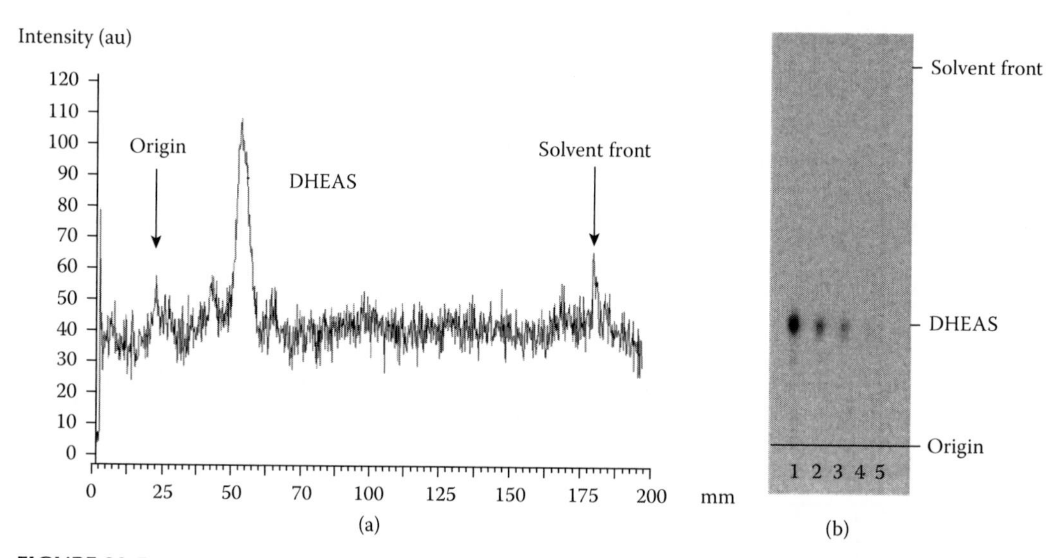

FIGURE 33.5 Typical chromatographic profile (a) from thin layer chromatography (TLC) of steroid sulfate fraction extracted from chick brain at 5 minutes after an intracranial injection of 3H-dehydroepiandrosterone sulfate (DHEAS) to the intermediate medial mesopallium and (b) phosphorimage of TLC for steroid sulfate fractions at: (1) 5 minutes, (2) 10 minutes, (3) 30 minutes, (4) 1 hour, and (5) 5 hours after injection of this label. The peak for DHEAS is labeled as 1. (Reprinted from Sujkovic, E., R. Mileusnic, and J. P. Fry. 2009. *J Neurochem* 109(2):348–59. Copyright (2009), with permission from John Wiley and Sons.)

production of DHEA(S) in rodents and other laboratory animals in comparison with primates (see Cutler et al. 1978).

Human Studies

Unsurprisingly, given the above evidence for the enhancement of learning and memory in animals by DHEA(S) and the well-known increase in the circulating concentrations of these steroids before puberty in humans, followed by a decline through adulthood, DHEA(S) have received attention not only as markers for healthy cognitive development and aging but also as potential treatments to enhance performance. Although some associations have been reported between plasma DHEAS and general mental health and performance in the elderly, there appear to be no significant correlations with more precise cognitive measures, and DHEAS has no predictive value for future performance (see Wolf and Kirschbaum 1999; Vallée et al. 2001). Rigorous assessment of the clinical potential of DHEA(S) to enhance cognition requires controlled investigations of DHEA supplementation of middle-aged or elderly subjects without dementia. A Cochrane Collaboration review of the literature to March 2008 could include only five such studies, with no evidence of a beneficial effect of DHEA supplementation on cognition (Grimley Evans et al. 2006). However, one study found that oral DHEA supplementation of elderly men and women for 2 weeks, sufficient to elevate their plasma DHEAS concentrations to young adult levels, protected against stress-induced deterioration in a test of attention (selecting target shapes on a sheet of paper; Wolf et al. 1998; see Figure 33.6). This is interesting in view of the antiglucocorticoid effects of DHEA mentioned in the section "Modes of Action of DHEA(S) in the Brain," although the same study found DHEA supplementation to exacerbate the effects of stress on declarative memory (visual–verbal memory), possibly by enhancing cortisol release. Unlike DHEA(S), adrenal production of cortisol does not show marked age-related changes. This means that compared with healthy young adults with peak DHEA(S) production, children and the elderly have cortisol/DHEA(S) ratios four to five times higher in both blood and cerebrospinal fluid and so would be expected to be especially predisposed to the deleterious effects of stress on learning and memory (see Herbert 1998). Supplementation of elderly subjects for

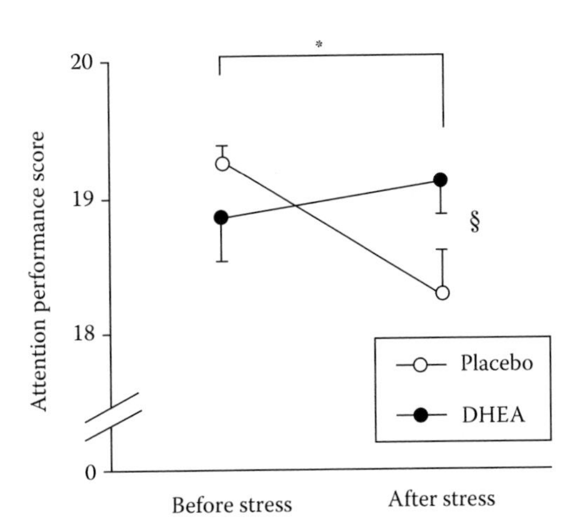

FIGURE 33.6 Effects of DHEA and stress on human performance in an attention test. Performance (using parallel versions) was tested before and after stress. The symbol * indicates a significant difference between pre- and poststress performance for subjects given placebo and the symbol § a significant difference between DHEA and placebo. (Reprinted from Wolf, O. T., B. M. Kudielka, D. H. Hellhammer, J. Hellhammer, and C. Kirschbaum. 1998. *Psychoneuroendocrinology* 23(6):617–29. Copyright (1998), with permission from Elsevier.)

3 months with DHEA sufficient to reduce the cortisol/DHEA ratio (as assessed in saliva) did indeed reduce confusion and anxiety but had no significant effect on cognition, as measured in tests of speed, attention, and episodic memory (van Niekerk, Huppert, and Herbert 2001).

CONCLUDING REMARKS

Studies in laboratory animals have consistently shown an enhancement of the acquisition and consolidation, rather than the recall, of memory by DHEA(S). Perhaps this gives some clue to the lack of clear effects in elderly human subjects, for whom recall of life skills and events might be expected to assume a greater importance. Nevertheless, controlled studies of learning and memory (of new information) in elderly human subjects have shown no significant effects of DHEA(S) supplementation. The ability of DHEA(S) to enhance the acquisition and consolidation of memory in laboratory animals is greater in males than in females. This may reflect actions mediated after conversion of the DHEA(S) to estrogens, which should already be higher in the females with functioning ovarian cycles. By the same logic, and reverting to human subjects, DHEA(S) supplementation would be expected to have clearer effects in elderly, postmenopausal women than men. We suggest this because although both groups will have declining adrenal DHEA(S) production, men should still receive a contribution of these steroids from the testes (see Labrie 2010). Supplementation of healthy elderly subjects with DHEA was indeed reported to improve mood in women rather than men (Wolf et al. 1997) but, as in other studies, there were no convincing improvements in cognitive performance.

So why have clear actions of DHEA(S) on learning and memory in laboratory animals not been followed by equally convincing studies in humans? There are probably several answers to this question including at least the following: (1) the age-related changes in DHEA(S) will vary between individuals; (2) circulating DHEA(S) will fall more rapidly with age in women than men; (3) supplementation may occur too late in the decline of DHEA(S) with aging; (4) supplementation is usually with oral DHEA, which can elevate plasma DHEAS but is likely also to produce other steroid metabolites; (5) the effects of DHEA(S) on cognition will vary according to the levels of the stress hormone cortisol; and (6) there could be more than one mode of action for DHEA(S) and/or its metabolites.

In our view, more animal studies are required to elucidate the metabolism and mode of action of DHEA(S) in the brain. Such studies can be expected to inform clinical investigations of human subjects, which need to be sufficiently powered to uncover significant effects through the variables listed in the previous paragraph.

SUMMARY POINTS

- DHEA(S) enhances the acquisition and consolidation stages of memory in animal models of memory.
- DHEA(S) does not appear to enhance recall in animal models of memory.
- DHEA(S) can reverse pharmacologically induced amnesia in animal models.
- DHEA(S) may play a role in age-dependent cognitive decline.
- DHEA(S) supplementation has not yet been convincingly shown to enhance learning and memory in normal human aging.

REFERENCES

Bazin, M. A., L. El Kihel, M. Boulouard, V. Bouët, and S. Rault. 2009. The effects of DHEA, 3beta-hydroxy-5alpha-androstane-6,17-dione, and 7-amino-DHEA analogues on short term and long term memory in the mouse. *Steroids* 74:931–7.

Bolhuis, J. J., and M. Gahrb. 2004. Neuronal mechanisms of birdsong memory. *Nat Rev Neurosci* 7:347–57. Review.

Campbell, B. 2006. Adrenarche and the evolution of human life history. *Am J Hum Biol* 18:569–89. Review.

Castellano, C., and J. L. McGaugh. 1990. Effects of post-training bicuculline and muscimol on retention: Lack of state dependency. *Behav Neural Biol* 54:156–64.
Castellano, C., and F. Pavone. 1988. Effects of ethanol on passive avoidance behavior in the mouse: Involvement of GABAergic mechanisms. *Pharmacol Biochem Behav* 29:321–4.
Chapouthier, G., and P. Venault. 2002. GABA-A receptor complex and memory processes. *Curr Top Med Chem* 2:841–51.
Compagnone, N. A., and S. H. Mellon. 1998. Dehydroepiandrosterone: A potential signalling molecule for neocortical organization during development. *Proc Natl Acad Sci USA* 95:4678–83.
Corpéchot, C., P. Robel, M. Axelson, J. Sjovall, and E. E. Baulieu. 1981. Characterization and measurement of dehydroepiandrosterone sulfate in rat brain. *Proc Natl Acad Sci USA* 78:4704–7.
Cutler Jr., G. B., M. Glenn, M. Bush, G. D. Hodgen, C. E. Graham, and D. L. Loriaux. 1978. Adrenarche: A survey of rodents, domestic animals, and primates. *Endocrinology* 103:2112–8.
Ebner, M. J., D. I. Corol, H. Havlíková, J. W. Honour, and J. P. Fry. 2006. Identification of neuroactive steroids and their precursors and metabolites in adult male rat brain. *Endocrinology* 147:179–90.
Farr, S. A., W. A. Banks, K. Uezu, F. S. Gaskin, and J. E. Morley. 2004. DHEAS improves learning and memory in aged SAMP8 mice but not in diabetic mice. *Life Sci* 75:2775–85.
Flood, J. F., and E. Roberts. 1988. Dehydroepiandrosterone sulfate improves memory in aging mice. *Brain Res* 448:178–81.
Flood, J. F., G. E. Smith, and E. Roberts. 1988. Dehydroepiandrosterone and its sulfate enhance memory retention in mice. *Brain Res* 447:269–78.
Grimley Evans, J., R. Malouf, F. A. H. Huppert, and J. K. Van Niekerk. 2006. Dehydroepiandrosterone (DHEA) supplementation for cognitive function in healthy elderly people. *Cochrane Database Syst Rev* Issue 4 Art. No.: CD006221.
Herbert, J. 1998. Neurosteroids, brain damage, and mental illness. *Exp Gerontol* 33:713–27.
Hobkirk, R. 1985. Steroid sulfotransferases and steroid sulfate sulfatases: Characteristics and biological roles. *Can J Biochem Cell Biol* 63:1127–44.
Johnson, D. A., T. Wu, P. Li, and T. J. Maher. 2000. The effect of steroid sulfatase inhibition on learning and spatial memory. *Brain Res* 865:286–90.
Kishimoto, Y., and M. Hoshi. 1972. Dehydroepiandrosterone sulphate in rat brain: Incorporation from blood and metabolism in vivo. *J Neurochem* 19:2207–15.
Labrie, F. 2010. DHEA, important source of sex steroids in men and even more in women. *Prog Brain Res* 182:97–148.
Lacroix, C., J. Fiet, J. P. Benais, B. Gueux, R. Bonete, J. M. Villette, B. Gourmel, and C. Dreux. 1987. Simultaneous radioimmunoassay of progesterone, androst-4-enedione, pregnenolone, dehydroepiandrosterone and 17-hydroxyprogesterone in specific regions of human brain. *J Steroid Biochem* 28:317–25.
Lanthier, A., and V. V. Patwardhan. 1986. Sex steroids and 5-en-3 beta-hydroxysteroids in specific regions of the human brain and cranial nerves. *J Steroid Biochem* 25:445–9.
Lhullier, F. L., R. Nicolaidis, N. G. Riera, F. Cipriani, D. Junqueira, K. C. Dahm, A. M. Brusque, and D. O. Souza. 2004. Dehydroepiandrosterone increases synaptosomal glutamate release and improves the performance in inhibitory avoidance task. *Pharmacol Biochem Behav* 77:601–6.
Majewska, M. D., S. Demirgoren, C. E. Spivak, and E. D. London. 1990. The neurosteroid dehydroepiandrosterone sulfate is an allosteric antagonist of the GABAA receptor. *Brain Res* 526:143–6.
Maninger, N., O. M. Wolkowitz, V. I. Reus, E. S. Epel, and S. H. Mellon. 2009. Neurobiological and neuropsychiatric effects of dehydroepiandrosterone (DHEA) and DHEA sulfate (DHEAS). *Front Neuroendocrinol* 30:65–91. Review.
Markowski, M., M. Ungeheuer, D. Bitran, and C. Locurto. 2001. Memory-enhancing effects of DHEAS in aged mice on a win-shift water escape task. *Physiol Behav* 72:521–5.
Matsuno, K., T. Senda, K. Matsunaga, and S., Mita. 1994. Ameliorating effects of sigma receptor ligands on the impairment of passive avoidance tasks in mice: Involvement in the central acetylcholinergic system. *Eur J Pharmacol* 261:43–51.
Maurice, T., V. L. Phan, A. Urani, and I. Guillemain. 2001. Differential involvement of the sigma(1) (sigma(1)) receptor in the anti-amnesic effect of neuroactive steroids, as demonstrated using an in vivo antisense strategy in the mouse. *Br J Pharmacol* 134:1731–41.
Maurice, T., T. P. Su, and A. Privat. 1998. Sigma1 (sigma 1) receptor agonists and neurosteroids attenuate B25-35-amyloid peptide-induced amnesia in mice through a common mechanism. *Neuroscience* 83:413–28.
Melchior, C. L., and R. F. Ritzmann. 1996. Neurosteroids block the memory-impairing effects of ethanol in mice. *Pharmacol Biochem Behav* 53:51–6.

Mellon, S. H. 2007. Neurosteroid regulation of central nervous system development. *Pharmacol Ther* 116:107–24. Review.

Mensah-Nyagan, A. G., D. Beaujean, V. Luu-The, G. Pelletier, H. Vaudry. 2001. Anatomical and biochemical evidence for the synthesis of unconjugated and sulfated neurosteroids in amphibians. *Brain Res Brain Res Rev* 37:13–24. Review.

Migues, P. V., A. N. Johnston, and S. P. Rose. 2002. Dehydroepiandosterone and its sulphate enhance memory retention in day-old chicks. *Neuroscience* 109:243–51.

Miller, W. L. 2009. Androgen synthesis in adrenarche. *Rev Endocr Metab Disord* 10:3–17. Review.

Mo, Q., S. Lu, C. Garippa, M. J. Brownstein, and N. G. Simon. 2009. Genome-wide analysis of DHEA- and DHT-induced gene expression in mouse hypothalamus and hippocampus. *J Steroid Biochem Mol Biol* 114:135–43.

Monnet, F. P., V. Mahe, P. Robel, and E. E. Baulieu. 1995. Neurosteroids, via sigma receptors, modulate the [3H]norepinephrine release evoked by N-methyl-D-aspartate in the rat hippocampus. *Proc Natl Acad Sci USA* 92:3774–8.

Monnet, F. P., and T. Maurice. 2006. The sigma1 protein as a target for the non-genomic effects of neuro(active) steroids: Molecular, physiological, and behavioral aspects. *J Pharmacol Sci* 100:93–118. Review.

Muller, C., O. Hennebert, and R. Morfin. 2006. The native anti-glucocorticoid paradigm. *J Steroid Biochem Mol Biol* 100:95–105. Review.

Nicolas, L. B., and J. P. Fry. 2007. The steroid sulfatase inhibitor COUMATE attenuates rather than enhances access of dehydroepiandrosterone sulfate to the brain in the mouse. *Brain Res* 1174:92–6.

Orentreich, N., J. L. Brind, R. L. Rizer, and J. H. Vogelman. 1984. Age changes and sex differences in serum dehydroepiandrosterone sulfate concentrations throughout adulthood. *J Clin Endocrinol Metab* 59:551–5.

Reddy, D. S., and S. K. Kulkarni. 1998a. The effects of neurosteroids on acquisition and retention of a modified passive-avoidance learning task in mice. *Brain Res* 791:108–16.

Reddy, D. S., and S. K. Kulkarni. 1998b. Possible role of nitric oxide in the nootropic and antiamnesic effects of neurosteroids on aging- and dizocilpine-induced learning impairment. *Brain Res* 799:215–29.

Reddy, D. S., and S. K. Kulkarni. 1999. Sex and estrous cycle-dependent changes in neurosteroid and benzodiazepine effects on food consumption and plus-maze learning behaviors in rats. *Pharmacol Biochem Behav* 62:53–60.

Reed, M. J., A. Purohit, L. W. Woo, S. P. Newman, and B. V. Potter. 2005. Steroid sulfatase: Molecular biology, regulation, and inhibition. *Endocr Rev* 26:171–202. Review.

Reiner, A., D. J. Perkel, L. L. Bruce et al. 2004. Revised nomenclature for avian telencephalon and some related brainstem nuclei. *J Comp Neurol* 473:377–414.

Robel, P., E. Bourreau, C. Corpéchot et al. 1987. Neuro-steroids: 3 beta-hydroxy-delta 5-derivatives in rat and monkey brain. *J Steroid Biochem* 27:649–55.

Roberts, E., L. Bologa, J. F. Flood, and G. E. Smith. 1987. Effects of dehydroepiandrosterone and its sulfate on brain tissue in culture and on memory in mice. *Brain Res* 406:357–62.

Rose, S. P. 2000. God's organism? The chick as a model system for memory studies. *Learn Mem* 7:1–17.

Smith, M. R., B. T. Rudd, A. Shirley, P. H. Rayner, J. W. Williams, N. M. Duignan, and P. V. Bertrand. 1975. A radioimmunoassay for the estimation of serum dehydroepiandrosterone sulphate in normal and pathological sera. *Clin Chim Acta* 65:5–13.

Spivak, C. E. 1994. Desensitization and noncompetitive blockade of GABAA receptors in ventral midbrain neurons by a neurosteroid dehydroepiandrosterone sulfate. *Synapse* 16:113–22.

Squire, L. R. 1986. Mechanisms of memory. *Science* 232:1612–9.

Squire, L. R. 2004. Memory systems of the brain: A brief history and current perspective. *Neurobiol Learn Mem* 82:171–7. Review.

Sujkovic, E., R. Mileusnic, and J. P. Fry. 2009. Metabolism of neuroactive steroids in day-old chick brain. *J Neurochem* 109:348–59.

Sujkovic, E., R. Mileusnic, J. P. Fry, and S. P. Rose. 2007. Temporal effects of dehydroepiandrosterone sulfate on memory formation in day-old chicks. *Neuroscience* 148:375–84.

Tsutsui, K., and T. Yamazaki. 1995. Avian neurosteroids. I. Pregnenolone biosynthesis in the quail brain. *Brain Res* 678:1–9.

Tulving, E. 1985. How many memory systems are there? *Am Psychol* 40:385–98.

Urani, A., A. Privat, and T. Maurice. 1998. The modulation by neurosteroids of the scopolamine-induced learning impairment in mice involves an interaction with sigma1 (sigma1) receptors. *Brain Res* 799:64–77.

Vallée, M., W. Mayo, G. F. Koob, and M. Le Moal. 2001. Neurosteroids in learning and memory processes. *Int Rev Neurobiol* 46:273–320.

van Niekerk, J. K., F. A. Huppert, and J. Herbert. 2001. Salivary cortisol and DHEA: Association with measures of cognition and well-being in normal older men, and effects of three months of DHEA supplementation. *Psychoneuroendocrinology* 26:591–612.

Wolf, O. T., and C. Kirschbaum. 1999. Actions of dehydroepiandrosterone and its sulfate in the central nervous system: Effects on cognition and emotion in animals and humans. *Brain Res Brain Res Rev* 30:264–88.

Wolf, O. T., B. M. Kudielka, D. H. Hellhammer, J. Hellhammer, and C. Kirschbaum. 1998. Opposing effects of DHEA replacement in elderly subjects on declarative memory and attention after exposure to a laboratory stressor. *Psychoneuroendocrinology* 23:617–29.

Wolf, O. T., O. Neumann, D. H. Hellhammer, A. C. Geiben, C. J. Strasburger, R. A. Dressendörfer, K. M. Pirke, and C. Kirschbaum. 1997. Effects of a two-week physiological dehydroepiandrosterone substitution on cognitive performance and well-being in healthy elderly women and men. *J Clin Endocrinol Metab* 82:2363–7.

34 DHEA and Aggression

Gregory E. Demas, Kiran K. Soma, and H. Elliott Albers

CONTENTS

INTRODUCTION

Aggression is one of the most important social behaviors, is displayed by virtually all animals, and serves a wide range of adaptive functions. Nonetheless, the precise mechanisms that regulate aggression remain largely unknown. Aggressive behavior occurs whenever the interests of two or more individuals are in conflict, typically over limited resources (e.g., food, territories, and mates). Aggression is difficult to define and has been defined in a variety of ways. Aggression has traditionally been defined as overt behavior with the intention of inflicting physical damage on another individual (Moyer 1971). Moyer (1971) divided aggression into specific subtypes, including predatory aggression, intermale aggression, fear-induced aggression, irritable aggression, maternal aggression, territorial defense, and instrumental aggression, based on differences in the social conditions in which the behavior was elicited. A simplified classification of aggressive behavior has been suggested in which aggression is divided into offensive and defensive aggression (Blanchard and Blanchard 1988). Offensive aggression involves behaviors used in an attack, whereas defensive behaviors do not involve an active approach to the opponent but, rather, they serve as a defense against an attack. This latter classification system provides a more functional framework with which to identify and characterize aggressive behavior across a range of vertebrate taxa. A key aspect of both classifications is that, although different forms of aggression share specific behavioral features, the environmental factors eliciting these responses, as well as the biological substrates underlying their manifestation, may differ markedly.

Aggressive behavior has received extensive study under a wide range of environmental settings and experimental conditions. The experimental models used, the types of aggression measured, and the species tested, however, can vary considerably from study to study. Thus, it is often difficult to compare results across a range of studies. Although a relatively large number of experimental paradigms have been developed to evaluate aggression in animal models, one of the most prevalent models of assessing offensive aggression has been the resident-intruder model. This model simulates rodent territorial aggression and involves introducing a group-housed, nonexperimental "intruder" into the home cage of an experimental animal, and the amount and duration of aggressive behavior (e.g., chases, attacks, bites) are subsequently recorded in a timed test. Another less-commonly employed but useful model is the neutral arena model, which involves placing two animals in a novel "neutral" cage and recording the amount of aggression directed toward each animal. In addition, this latter model has the added benefit of assessing the formation of dominance relationships among animals because "territories" have not been established at the time of testing.

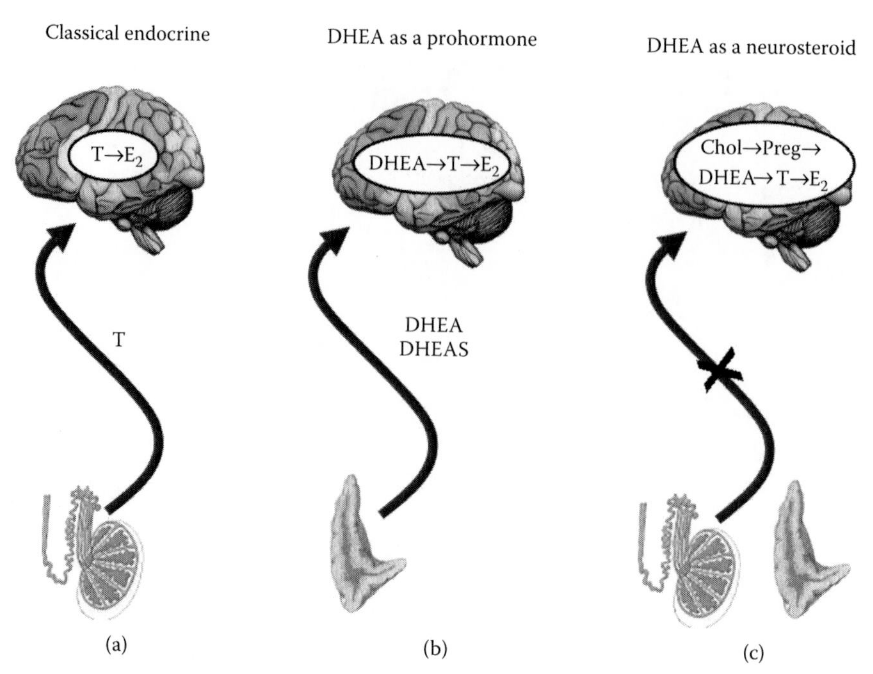

FIGURE 34.1 Steroid hormones can reach the brain to affect aggression via several possible mechanisms. (a) The classical endocrine pathway involves gonadal testosterone acting directly, or indirectly via local conversion to estradiol, on the brain. (b) Adrenal DHEA can act as a prohormone by being converted locally to T or E_2. (c) DHEA can act as a neurosteroid by be produced locally in the brain in the absence of gonadal and adrenal steroid production.

Much of the early research on the neuroendocrine mechanisms of aggression focused on the role of gonadal steroid hormones, and predominantly testosterone (T), as the primary factor regulating aggression. In fact, most behaviorally oriented textbooks discuss the role of T in aggression almost exclusively, with much less emphasis on the role of other factors (e.g., dehydroepiandrosterone [DHEA]) in regulating this behavior. One of the goals of this chapter is to suggest that the idea that a single hormone mediates aggression is overly simplistic. For example, recent evidence indicates that T can be converted to 17β-estradiol (E_2) within the brain and that E_2 may mediate aggressive behavior, at least in some species and contexts (Soma, Tramontin, and Wingfield 2000). Alternatively, androgen precursors such as DHEA may be produced in extragonadal peripheral tissues (e.g., adrenals) or de novo within the brain. In fact, it is suggested that both adrenal and neurally derived androgens, called neurosteroids (Baulieu 1991), may play an important role in regulating aggressive responses (Simon 2002). Collectively, these important findings have helped elucidate several possible pathways by which androgens can act on the brain, either directly or indirectly, to affect aggression (Figure 34.1). While numerous endocrine and neural factors are now known to affect aggressive behaviors, this chapter will focus on emerging evidence suggesting a role of DHEA in the regulation of aggression across vertebrate taxa.

DHEA AND RODENT AGGRESSION

Despite the traditional research focus on the role of T in the regulation of aggression in vertebrates, it is becoming increasingly clear that steroid hormones other than T (e.g., DHEA) play important roles in the regulation of aggressive behavior, either by acting independent of or in conjunction with T (Demas et al. 2007; Soma et al. 2008). For example, DHEA is shown to be a potent inhibitor of female typical aggression in mice (Bayart et al. 1989; Haug et al. 1989, 1992; Perché et al. 2000; Young et al. 1996). Attack behavior in ovariectomized female mice, intact female mice, or castrated male mice toward lactating female mice can be significantly reduced following treatment with exogenous DHEA (Bayart et al. 1989; Brain and Haug 1992). Because this effect requires prolonged (~15 days) DHEA treatment, the results are consistent with a genomic mechanism of action (Lu et al. 2003).

It has been shown that prenatal T treatment enhances DHEA-induced inhibition of aggression in female offspring (Perché et al. 2001). It is important to note, however, that this testing paradigm (i.e., aggression toward a lactating intruder) is employed far less than more traditional models of aggression (e.g., intermale aggression); thus, the role of DHEA in mediating other forms of aggression has received very little experimental attention. One finding, however, suggests that increases in aggressive behavior can be induced by a single injection of the sulfated form of DHEA (i.e., dehydroepiandrosterone sulfate [DHEAS]) immediately prior to behavioral testing in mice tested in a neutral arena (Nicolas et al. 2001). Furthermore, treatment with COUMATE, a drug that inhibits the steroid sulfatase enzyme that converts DHEAS to DHEA, also increased aggression in male mice (Nicolas et al. 2001). Despite these intriguing results, more studies are needed to fully elucidate the role of DHEA and DHEAS in intermale aggression.

The effects of DHEA on aggression have led to an attempt at determining the physiological mechanisms underlying this behavioral response, with much of the work focused on interaction of this steroid with GABA neurotransmission (reviewed in a study by Simon 2003). It has been demonstrated that DHEA alters brain levels of pregnenolone sulfate (PREG-S), a neurosteroid that inhibits GABAergic actions via the $GABA_A$ receptor. Specifically, DHEA-induced changes in GABA activity appear to be responsible for the effects of DHEA on aggression. These results are consistent with previous findings demonstrating an inhibitory effect of GABA on aggression (Miczek et al. 1994, 1997). Although the precise mechanisms of action are still unknown, DHEA can modulate the actions of both GABA and glutamatergic NMDA receptors (Labrie 1998; Mellon and Vaudry 2001). Alternatively, DHEA may act by further conversion in the brain to T and E_2 via the enzymes 3β-HSD and aromatase (Soma et al. 2004).

More recent findings suggest that an additional mechanism for the effects of DHEA on aggression may exist (Simon 2003). Although it has long been assumed that DHEA is incapable of directly interacting with the androgen receptor (AR), recent data suggest that DHEA upregulates AR in the mouse brain (Lu et al. 2003; Mo et al. 2004). Intact and gonadectomized, T-treated male mice do not display aggression toward female mice, and prolonged treatment with DHEA is necessary to elicit antiaggressive effects. Thus, it has been speculated that androgens such as DHEA play an inhibitory role, presumably through a genomic action by binding directly to AR and altering AR transcription and translation (Simon 2003). For example, DHEA can compete for recombinant AR binding, upregulate neural AR protein levels in mouse brains and immortalized GT1-7 hypothalamic cells, and induce transcription through AR in CV-1 cells, suggesting direct actions of DHEA on AR (Simon 2003). Although this idea is intriguing, the affinity of DHEA for AR is low relative to T, so high local levels of DHEA would be necessary for this mechanism to be plausible.

Several studies have examined the role of androgens including T and DHEA in aggression in a seasonal context. Many nontropical rodent species are seasonal breeders, maintaining reproductive function during the summer and curtailing breeding during the winter. Ambient day length (photoperiod) is the proximal environmental cue used by individuals within these species to coordinate their reproduction to the appropriate season (Goldman 2001). For example, reproductive activity (with high levels of circulating androgens) is maintained during long "summer-like" days (e.g., >12.5 hours of light/day), whereas reproductive regression, including virtual collapse of the gonads and marked decreases in T, occur during the short "winter-like" days (e.g. <12.5 hours of light/day; Goldman 2001). Interestingly, although short-day exposure produces a substantial reduction in circulating T levels in males, it also significantly increases circulating levels of DHEA (Caldwell, Smith, and Albers 2008). Surprisingly, male Syrian hamsters (*Mesocricetus auratus*) maintained in short days actually display *increased* resident-intruder aggression compared with long-day animals despite gonadal regression and relatively low T levels (Garrett and Campbell 1980). Specifically, adult male Syrian hamsters housed in short days for 9 weeks display approximately twice the amount of aggression in a resident-intruder test compared with long-day controls when tested 4 hours before dark, despite gonadal regression (Garrett and Campbell 1980). After prolonged maintenance in short days (>15 weeks), hamsters typically undergo spontaneous gonadal recrudescence (i.e., increased testicular mass and circulating T), despite continued maintenance in short days. The short-day increases in aggressive behavior largely disappear in animals undergoing spontaneous recrudescence, returning to long-day levels of aggression by 21 weeks (Garrett and Campbell 1980). Short-day increases in aggression in male Syrian hamsters have been confirmed (Jasnow et al. 2002; Caldwell and Albers 2004). For example, Syrian hamsters housed in short days (light:dark [LD] 10:14) for 10 weeks displayed a significantly greater number of attacks and a longer duration of attacks than did long-day hamsters when tested using a resident-intruder test (Jasnow et al. 2002). It has been shown that short-day gonadal regression does not affect the frequency of a form of social communication in Syrian hamsters called "flank marking" (Caldwell and Albers 2003). Flank marking, which is an androgen-dependent behavior in hamsters housed in long days, continues to be displayed at high levels during social encounters and in response to conspecific odors in male hamsters exposed to short days despite low circulating levels of gonadal steroids (Albers et al. 2002; Caldwell and Albers 2003; Gutzler et al. 2011). These findings suggest that neuroendocrine factors other than T may mediate flank marking as well as overt aggression in this species.

In Syrian hamsters, unlike most rodent species, females are more aggressive than males (Ciacco, Lisk, and Rueter 1979; Marques and Valenstein 1977). Not surprisingly, photoperiodic changes in aggression have been demonstrated in female Syrian hamsters (Badura and Nunez 1989; Fleming et al. 1988). Female hamsters were housed in long (LD 14:10) or short days (LD 6:18) for 12 weeks, and then both offensive and defensive aggression were tested (Fleming et al. 1988). Female hamsters maintained in short days displayed significantly less defensive aggression compared with long-day animals and thus had a higher ratio of offensive to defensive aggression than long-day animals (Fleming et al. 1988). To further examine the physiological mechanisms mediating short-day aggression in females of this species, the effects of short-day exposure on circulating levels of adrenal steroids were assessed (Gutzler et al. 2009). Specifically, animals were housed in either long or short days, and circulating concentrations of cortisol, DHEA, and DHEAS were determined (Figure 34.2). While exposure to short days decreased cortisol and DHEA levels (Gutzler et al. 2009), DHEAS concentrations were significantly higher in short-day exposed hamsters (Figure 34.3). Furthermore, short days increased aggression, regardless of the endocrine state of the animals. Exogenous E_2, however, reduced aggression in long-day hamsters but not in short-day hamsters, suggesting that exposure to short photoperiod renders females less sensitive to E_2-induced decreases in aggressive behavior, at least in females of this species (Gutzler et al. 2009). The precise role of the decrease in DHEA and the increase in DHEAS in short-day hamsters and its relationship with increased aggression, however, remain unknown.

Unlike Syrian hamsters, male Siberian hamsters (*Phodopus sungorus*) display significantly more aggression than females. It has previously been demonstrated that short-day male Siberian hamsters are significantly more aggressive than long-day animals (Jasnow et al. 2000; Jasnow et al. 2002), consistent with previous studies in Syrian hamsters. Specifically, male Siberian hamsters housed in short days (LD 8:16) for 10 weeks display a greater number of attacks during a resident-intruder test and have a lower latency to initial attacks, relative to long-day (LD 16:8) animals. As previously reported for many rodent species, prolonged maintenance on short days (i.e., 20 weeks) resulted in spontaneous reproductive recrudescence in which the gonads, and thus T, returned to normal long-day levels (Jasnow et al. 2000). Interestingly, gonadally recrudesced hamsters displayed less aggression than gonadally regressed animals even though both groups experienced the same photoperiod and melatonin signal; levels of aggression in recrudesced hamsters were generally indistinguishable

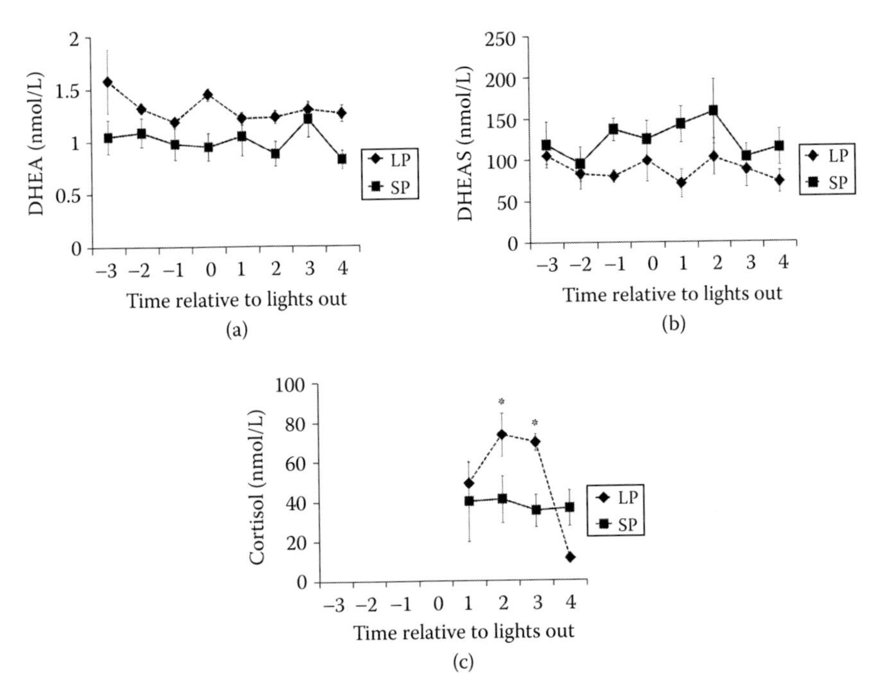

FIGURE 34.2 Effects of photoperiod on circulating DHEA, DHEAS, and cortisol concentrations across time in female Syrian hamsters. (a) Photoperiod altered DHEA concentrations, but time did not affect circulating hormone levels. (b) Photoperiod altered DHEAS concentrations, but as with DHEA, there was no effect of time on DHEAS. (c) Cortisol concentrations were not significantly different as a result of photoperiod, but a significant effect of time on hormone levels was found, with significantly higher cortisol concentrations in long photoperiod (LP) females than in short photoperiod (SP) females 2 and 3 hours after the onset of the dark phase. (Adapted from Gutzler, S. J., M. Karom, D. W. Erwin, and H. E. Albers. 2011. *Behav Brain Res.*)

from long-day hamsters (Jasnow et al. 2000). These results support previous findings in male Syrian hamsters (Garrett and Campbell 1980). When short-day Siberian hamsters were implanted with Silastic capsules containing T (to achieve long day-like levels), aggression actually *decreased* compared with short-day control animals (Jasnow et al. 2000), suggesting that short-day increases in aggression may be *inversely* related to serum T concentrations.

Preliminary studies in male Siberian hamsters indicate that endogenous serum DHEA levels are elevated under short days, when aggression is relatively high (Demas and Jasnow 2004). Exogenous DHEA does not increase aggression in either long- or short-day housed Siberian hamsters (Scotti et al. 2008). Specifically, long- and short-day housed hamsters were implanted with Silastic capsules containing DHEA or no hormone (control). If short-day increases in DHEA occur in hamsters and regulate photoperiodic changes in aggression in this species, then it is predicted that exogenous DHEA mimicking long-day levels will increase aggression in long-day hamsters to levels comparable to those in short-day control animals. Further, short-day hamsters receiving DHEA would show levels of aggression significantly higher than all other experimental groups. While short days did increase aggression as predicted, exogenous DHEA had no effect on aggression in either long- or short-day hamsters (Scotti et al. 2009). Giving exogenous DHEA, however, may not affect the rate of conversion of this prohormone to biologically active steroids (e.g., T, E_2). Thus, elevated DHEA may be necessary but not sufficient to elicit increased aggression in this and other species.

To further test the hypothesis that changes in circulating DHEA levels or DHEA metabolism mediate aggression in Siberian hamsters, daily fluctuations in DHEA levels were assessed prior to and in response to aggressive interactions (Scotti et al. 2009). Specifically, circulating DHEA and T concentrations were measured during the day (noon) and night (midnight). Although there were

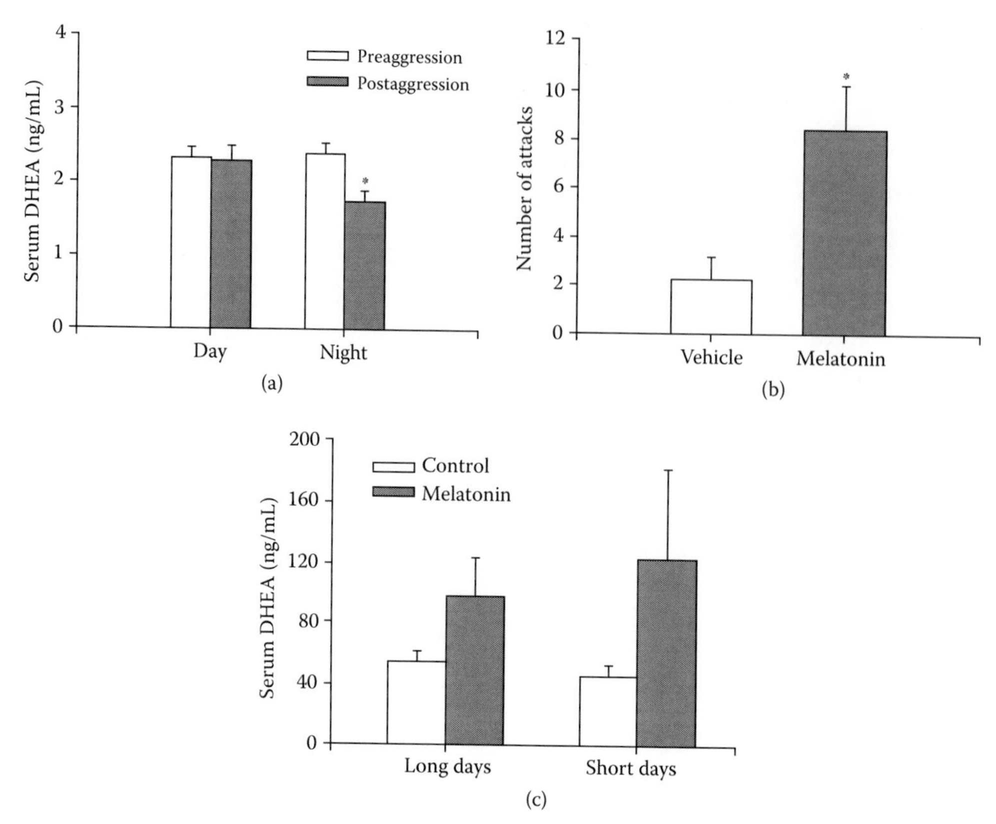

FIGURE 34.3 (a) Mean (± SEM) pre- and postaggression serum DHEA in Siberian hamsters. Hamsters were tested during the day and night, when circulating melatonin levels were relatively low and high, respectively. (Adapted from Scotti, M. L., A. E. M. Newman, K. L. Schmidt, T. N. Bonu, K. K. Soma, and G. E. Demas. 2009. *Horm Behav* 56:376–81.) (b) Resident-intruder aggression, as measured by the number of attacks during a 5-minute test in hamsters treated with exogenous melatonin or no melatonin (control). (Adapted from Demas et al. 2004.) (c) *In vitro* adrenal DHEA in cultured adrenal glands from hamsters housed in either long or short days. Hamsters were housed in their respective photoperiods for 10 weeks prior to testing. Adrenal DHEA release was stimulated by application of adrenocorticotropic hormone alone or in combination with melatonin. (Demas, Scotti and Soma, unpublished data.)

no significant differences in serum DHEA concentrations at these times, there was a trend towards reduced circulating levels of DHEA at midnight (Scotti et al. 2009). In contrast, in male Syrian hamsters, there were robust diurnal changes in circulating DHEA levels, and DHEA levels peaked 30 minutes prior to lights off and remained elevated during the night (Pieper and Lobocki 2000). Next, the effects of a brief (5 minutes), aggressive encounter on serum androgens were examined. Postaggression DHEA levels were significantly lower than preaggression DHEA levels in animals tested during the night but not during the day (Figure 34.3a). An opposite pattern of results was found for serum T levels; postaggression T levels were significantly higher than preaggression T levels during the night but not during the day. Consistent with these results, resident-intruder aggression was significantly greater during the night, when circulating levels of melatonin were at their relative peak. These data suggest that circulating DHEA may be converted to active sex steroids within the brain to influence aggressive behavior. The enzyme 3β-hydroxysteroid dehydrogenase/isomerase (3β-HSD) catalyzes the conversion of DHEA to androstenedione (AE), which can then be converted to T by 17β-HSD. Aggressive encounters at night may cause rapid increases in 3β-HSD activity in the brain or periphery. Increased 3β-HSD activity would be consistent with the pattern of results in animals tested at nighttime (rapid decrease in DHEA and increase in T levels in serum).

A recent study has examined circulating DHEA levels in territorial red squirrels (Boonstra et al. 2008). Both males and females of this species are highly aggressive during both the breeding and nonbreeding seasons due to the defense of food stores in their territories. Plasma DHEA concentrations in both males and females were considerably higher than levels typically seen in laboratory rodents (e.g., rats and mice; Boonstra et al. 2008). Furthermore, circulating DHEA levels displayed both seasonal and yearly variation that was negatively correlated with testis mass and positively correlated with population density (Boonstra et al. 2008). Plasma DHEA increased following ACTH challenges, suggesting that DHEA was predominantly of adrenal origin. Although aggression is not specifically addressed, these findings are consistent with the idea that DHEA is elevated at times of high territorial aggression in the field.

In virtually all mammals, photoperiodic responses are mediated by changes in the pineal indolamine melatonin. Melatonin is secreted predominantly during darkness, whereas daylight inhibits pineal melatonin secretion (Goldman 2001). Thus, changes in ambient day length result in changes in the pattern of secretion of melatonin. In this manner, it is the precise pattern of melatonin secretion, and not the amount of hormone per se, that provides the biochemical "code" for day length (Goldman 2001). Pinealectomy, which eliminates melatonin secretion and renders animals physiologically "blind" to day length, prevents the short-day increase in aggression in female Syrian hamsters, whereas treatment of long-day hamsters with exogenous short day-like melatonin increases aggression in female Syrian hamsters (Fleming et al. 1988). Ovariectomy, in contrast, has no effect on aggression. This finding suggests that photoperiodic changes in aggression are independent of changes in gonadal steroids in female Syrian hamsters (Fleming et al. 1988). A subsequent study in female Syrian hamsters confirmed these findings and provided further support for a role of pineal melatonin in mediating photoperiod changes in aggression. Specifically, a higher percentage of female hamsters housed in short days (LD 6:18) showed aggressive behavior compared with long-day housed (LD 16:8) hamsters (Badura and Nunez 1989). Consistent with previous findings, short-day aggression was attenuated by pinealectomy, but treatment with exogenous E_2 (alone or in combination with progesterone) had no effect on aggression. These results support the hypothesis that photoperiodic changes in aggression are mediated by pineal melatonin, but independent of gonadal steroids, at least in female Syrian hamsters. Furthermore, timed daily melatonin injections mimicking short-day patterns of the hormone in long-day, pineal-intact animals will produce short day-like increases in aggression. Because these injections occurred for only 10 days, gonadal mass and circulating levels of T are unaffected, supporting the hypothesis that photoperiodic changes in aggression are not mediated by changes in gonadal steroids in this species (Jasnow et al. 2002).

More recent research in male Siberian hamsters (Demas et al. 2004) confirms previous findings that treatment of long-day animals with short-day like levels of melatonin mimics photoperiodic changes in aggression; long-day hamsters given daily timed injections of melatonin 2 hours before lights out to mimic short-day levels of the hormone displayed elevated aggression in a resident-intruder test compared with control animals (Figure 34.3b). As with previous studies, these results were not likely due to changes in gonadal steroids, as serum T was unaffected by this injection protocol.

The effects of melatonin on aggression in rodents may be due to direct actions of this hormone on neural substrates mediating aggression (e.g., hypothalamus, limbic system). Alternatively, melatonin-induced aggression may be indirectly due to changes in hypothalamic-pituitary-adrenal (HPA) activity, as adrenal hormones (e.g., glucocorticoids, DHEA) have been implicated in aggressive behavior (Haller and Kruk 2003). In support of the latter hypothesis, changes in both the size and function of the adrenal gland are associated with changes in aggression (Paterson and Vickers 1981). In addition, male house mice housed in a LD 12:12 photoperiod and treated with melatonin display increased territorial aggression but decreased adrenal masses compared to saline-treated animals (Paterson and Vickers 1981). The increases in aggression displayed by melatonin-treated animals, however, can be blocked by adrenalectomy (Paterson and Vickers 1981). Experimental reductions in both adrenomedullary catecholamines and adrenocortical glucocorticoids are associated with decreased aggression in rodents (Haller and Kruk 2003; Paterson and Vickers 1981), and reductions of glucocorticoids via pharmacological blockade of adrenocorticotropic hormone (ACTH) release can attenuate melatonin-induced increases in aggression in mice (Paterson and Vickers 1981). Thus, exogenous melatonin, despite reducing adrenal mass, appears to increase aggression by stimulating adrenocortical steroid release. These results are particularly intriguing given that house mice are generally reproductively nonresponsive to photoperiod (Nelson 1990).

More recently, research has implicated changes in adrenocortical hormones in mediating melatonin-induced aggression in Siberian hamsters. Long-day hamsters treated with short-day like levels of melatonin displayed increased aggression, comparable to levels seen in short-day animals (Demas et al. 2004). Interestingly, melatonin-induced aggression could be blocked by bilateral adrenalectomy, consistent with previous results in house mice (Paterson and Vickers 1981). Adrenal demedullation, which eliminates adrenal catecholamines (i.e., epinephrine), but leaves adrenocortical steroid release (i.e., cortisol, DHEA) intact, had no effect on melatonin-induced aggression (Demas et al. 2004). Collectively, these results support the hypothesis that the effects of exogenous melatonin on aggression are mediated by the effects of this hormone on adrenocortical steroids. However, it is currently unknown which class of steroid hormones may mediate this effect, as both adrenal androgens (e.g., DHEA) and glucocorticoids (e.g., cortisol) have been implicated in aggression in rodents (Haller and Kruk 2003; Schlegel et al. 1985). In laboratory rats and mice, corticosterone is the predominant adrenal glucocorticoid, and these species secrete little to no adrenal DHEA. In contrast, in hamsters, as in humans, cortisol is the primary adrenal glucocorticoid (Albers et al. 1985), and both hamsters and humans secrete measurable amounts of DHEA and its sulfated form, DHEAS (Mellon and Vaudry 2001; Pieper et al. 2000).

Melatonin also facilitates DHEA secretion from cultured adrenal glands in both hamsters and mice (Haus et al. 1996; Scotti et al. 2009; Figure 34.3c). Specifically, incubation of cultured adrenal glands for 2 hours with a combination of ACTH and melatonin results in higher concentrations of DHEA in the culture media compared with ACTH alone. These results suggest that melatonin plays a permissive role in the regulation of adrenal DHEA release. Circulating DHEA may, in turn, be converted to active sex steroids within the brain to influence aggressive behavior. As discussed above, the enzyme 3β-HSD catalyzes the conversion of DHEA to AE, which can then be subsequently converted to T. It is possible that increased aggression when melatonin levels are elevated (e.g., short days or nighttime) may be driven by increased adrenal DHEA and subsequent rapid increases in 3β-HSD activity in the brain or periphery, thus leading to increased T. Increased 3β-HSD activity would be consistent with the pattern of results in animals tested at nighttime (rapid decrease in DHEA and increase in T levels in serum) previously reported (Scotti et al. 2008).

In addition to peripheral secretion of androgens, it is now commonly accepted that the brain is a significant source of steroid production. The idea of "neurosteroids" (i.e., brain-derived steroid hormones) was first introduced to describe the high levels of the androgen DHEA, and its sulfated form DHEAS, seen in rat brain even after castration and adrenalectomy (Baulieu 1981; Corpechot et al. 1981). It is now established that DHEA, among other steroid hormones (e.g., allopregnanolone [ALLO]), can be synthesized de novo within the central nervous system and can act locally on specific neural substrates to regulate behavior (Simon 2003). For example, a recent study has demonstrated that intermale aggression is associated with changes in brain neurosteroid synthesis (Pinna, Costa, and Guidotti 2005). Specifically, administration of T propionate (TP) to male mice decreased brain ALLO content by about 40% and was correlated with increased aggression. Increasing brain ALLO levels pharmacologically attenuated aggressive behavior in these mice. Similar changes in brain DHEA may also occur, but this remains to be examined in rodents. It remains unclear, however, how neurosteroids might be synthesized in adult rodent brains, given that a key synthetic enzyme (P450c17) is nondetectable in the rodent brains in several studies (Mellon and Vaudry 2001; but see Hojo et al. 2004).

DHEA AND AVIAN AGGRESSION

As with the rodent studies discussed above, most avian species show seasonal breeding, with distinct breeding and nonbreeding seasons. Recent investigations of avian species that exhibit year-round territorial aggression have found that nonbreeding aggression can be independent of circulating T levels. The males and females of many bird species exhibit high levels of territorial aggression throughout the year (Soma and Wingfield 1999). Often the territoriality expressed during the reproductive and nonreproductive seasons is quantitatively and qualitatively similar (Wingfield and Hahn 1994; Soma and Wingfield 2001). While aggression during the breeding season is generally regulated by gonadal steroids, aggression outside of the breeding season may be regulated by nongonadal steroids (Wingfield and Soma 2002; Soma et al. 2000; Soma 2006).

Several studies have focused on nonbreeding aggression in song sparrows (*Melospiza melodia morphna*), a common North American songbird (Soma, Sullivan, and Wingfield 1999; Soma et al. 2000; Soma and Wingfield 2001). Male song sparrows are highly territorial during the spring (breeding season) and autumn/winter (nonbreeding season; Wingfield and Hahn 1994). Note that the testes are completely regressed (≤ 1 mm in length) and plasma T levels are basal or nondetectable during the nonbreeding season (Soma et al. 2003). Furthermore, castration of nonbreeding male song sparrows has no effect on territorial aggression in autumn (Wingfield 1994). These data led to the hypothesis that aggression during the nonbreeding season is not regulated by sex steroids.

To test this hypothesis, male song sparrows were treated with the aromatase inhibitor fadrozole in the nonbreeding season (Soma et al. 2000). Fadrozole strongly decreases aggressive behavior in nonbreeding song sparrows (Soma et al. 2000). Moreover, the effects of fadrozole are rescued by E_2 replacement (Soma et al. 2000). These data suggest that sex steroids, in particular estrogens, are necessary for the expression of aggressive behavior in the nonbreeding season, even though plasma sex steroid levels are nondetectable (Soma et al. 2000). Similar results were obtained in two other field experiments (Soma, Sullivan, and Wingfield 1999; Soma et al. 2000).

The source of androgen substrate for brain aromatase in the nonbreeding season could be circulating DHEA (Soma and Wingfield 2001). Although DHEA cannot be directly aromatized, DHEA can be metabolized to AE by 3β-HSD, an aromatizable androgen. In contrast to T and E_2, DHEA is detectable and elevated in the circulation of nonbreeding birds (Soma and Wingfield 2001; Newman, Pradhan, and Soma 2008; Newman and Soma 2009; Newman et al. 2010; Pradhan et al. 2010). DHEA concentrations in the adrenals and regressed testes of nonbreeding birds are even higher than plasma levels, suggesting that both organs could secrete DHEA into the general circulation (Soma and Wingfield 2001; Newman et al. 2009). Interestingly, DHEA concentrations in the brain are also very high (Newman and Soma 2009).

Treatment of nonbreeding male song sparrows with a physiological dose of DHEA increases territorial singing and the size of HVC, a brain region involved in the production of songs (Soma, Wissman et al. 2002). DHEA treatment also regulates cell proliferation and new cell incorporation into HVC in adult song sparrows (Newman et al. 2010). These are some of the largest reported effects of DHEA on adult neuroplasticity and similar to the effects of T and E_2 on song behavior and neuroanatomy (Soma et al. 2004). However, DHEA treatment does not affect other territorial behaviors or, importantly, T-dependent secondary sexual characteristics (Soma et al. 2002). Other studies demonstrate that DHEA, unlike T, does not suppress immune function in song sparrows (Owen-Ashley et al. 2004).

Further investigations examined the metabolism of DHEA to active sex steroids within the songbird brain (Soma et al. 2004; Pradhan et al. 2008; Schlinger, Pradhan, and Soma 2008). To do so, brain tissue was incubated with ^{3}H-DHEA *in vitro*. The biochemical assay measures the conversion of ^{3}H-DHEA to ^{3}H-AE by the enzyme 3β-HSD. ^{3}H-AE can be metabolized subsequently to ^{3}H-T or ^{3}H-estrogens. In captive adult zebra finches (*Taeniopygia guttata*), brain tissue clearly metabolizes ^{3}H-DHEA to ^{3}H-AE, which is in turn aromatized to ^{3}H-E_1. Importantly, trilostane, a specific 3β-HSD inhibitor, blocks the production of ^{3}H-AE, and fadrozole, a specific aromatase inhibitor, reduces the production of ^{3}H-E_1. The song sparrow brain can also convert DHEA to androgens and

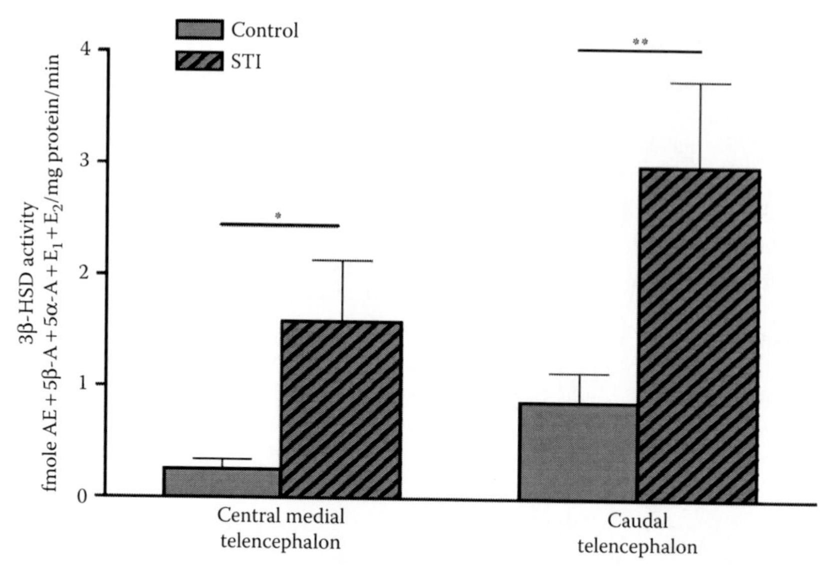

FIGURE 34.4 Effect of a simulated territorial intrusion (STI) for 30 minutes on DHEA metabolism by brain 3β-HSD in wild male song sparrows during the nonbreeding season. Exogenous NAD+ was not added. The STI rapidly increased DHEA metabolism by 3β-HSD in the central medial telencephalon ($n = 9$ control and 8 STI) and caudal telencephalon ($n = 7$ control and 8 STI). $^*p < .05$, $^{**}p < .01$. (Adapted from Pradhan, D. S., A. E. Newman, D. W. Wacker, J. C. Wingfield, B. A. Schlinger, and K. K. Soma. 2010. *Horm Behav* 57:381–9.)

estrogens (Pradhan et al. 2010). Brain 3β-HSD activity is higher during the nonbreeding season than during the breeding season, consistent with a greater role of neurosteroids in the nonbreeding season. Moreover, aggressive interactions rapidly (within 30 minutes) upregulate brain 3β-HSD activity in the nonbreeding season (Pradhan et al. 2010; Figure 34.4). Taken together, these data support the hypothesis that nonbreeding song sparrows combine peripheral DHEA synthesis with neural DHEA metabolism to maintain territorial behavior when gonadal T secretion is low. Moreover, it remains possible that the brain itself is a significant site of DHEA synthesis in nonbreeding song sparrows (Newman and Soma 2009).

Other studies have examined the spotted antbird (*Hylophylax n. naevioides*; Hau et al. 2000; Hau, Stoddard, and Soma 2004). The spotted antbird is a tropical species that exhibits year-round territorial aggression (Hau, Stoddard, and Soma 2004). Despite the presence of year-round territorial aggression, these birds generally have low or nondetectable levels of plasma T, even during the breeding season (Hau et al. 2000). Nonetheless, experiments have shown that T or its estrogenic metabolites do play a role in male territoriality (Hau et al. 2000). Results of a recent study that examined male and female antbirds during the nonbreeding season indicate that both sexes exhibit robust aggressive behavior during the nonbreeding season, and in both sexes, plasma DHEA levels are detectable and higher than plasma T and E_2 levels (Hau, Stoddard, and Soma 2004). In addition, plasma DHEA levels in males are positively correlated with aggressive vocalizations. Plasma DHEA, therefore, might serve as a precursor for synthesis of sex steroids in the brain throughout the year in these birds, in both males and females (Hau, Stoddard, and Soma 2004).

DHEA AND HUMAN AGGRESSION

Although the study of DHEA as a regulator of aggression in primates, including humans, has received some limited experimental attention, much less is known compared with nonprimates. Alterations in DHEA and DHEAS, however, have been implicated in a range of psychiatric disorders in humans (Maninger et al. 2009). Circulating levels of DHEAS are generally 1000-fold higher than DHEA in humans. Studies from rodents and birds suggest that androgens such as DHEA may regulate aggression in situations where aggression seems otherwise T-independent. Some studies in nonhuman primates also suggest that aggressive behavior can be unrelated to fecal T levels (Lynch, Ziegler, and Strier 2002; Ostner, Kappeler, and Heistermann 2002). Although the zona reticularis of the primate adrenal glands can secrete high levels of DHEA and DHEAS (Rehman and Carr 2004), the role of adrenal DHEA and DHEAS in primate aggression is largely unknown. In some primate studies, measures of T may be confounded by adrenal DHEA and DHEAS, causing difficulty in interpretation. For example, metabolites of T and DHEA may cross-react with T antibodies in urine and fecal assays in macaques (Mohle et al. 2002). Thus, potential correlations between aggression and fecal or urinary androgens are confounded, making it difficult to determine with certainty the identity and origin of the relevant steroid. In addition, a lack of correlation between aggression and excreted T does not necessarily mean that aggression is androgen independent, as other androgens such as DHEA may regulate behavior. One study has assessed circulating DHEAS levels in a population of wild baboons (Sapolsky et al. 1993). DHEAS concentrations were high in both male and female baboons and showed marked age-related decreases in both sexes; however, circulating levels of DHEAS were not compared with aggression (Sapolsky et al. 1993).

Adrenal DHEA appears to play a role in human aggression, as indicated by studies on "conduct disorder," which is typically defined as a collection of symptoms including aggression directed toward humans or animals, destruction of property, theft, and serious violations of rules. Prepubertal boys with conduct disorder were found to have higher levels of plasma DHEAS, but not T, than normal control boys (van Goozen et al. 1998). Also, DHEAS concentrations were correlated with the intensity of aggression as rated by parents and teachers. In another study, plasma DHEAS concentrations were found to be higher in boys with conduct disorder than in boys with attention-deficit hyperactivity disorder (ADHD) or normal controls (van Goozen et al. 2000). Finally, a recent study examined plasma levels of cortisol, DHEA, and DHEAS in delinquent adolescent boys diagnosed with conduct disorder compared with healthy controls. Hormone levels were correlated with aggression as determined by the Child Behavior Checklist and the Overt Aggression Scale (Golubchik et al. 2009). Delinquent boys tended to have higher DHEAS levels than control boys, but did not show any differences in either DHEA or cortisol (Golubchik et al. 2009). Collectively, these data suggest an important relationship between changes in DHEAS and aggression, at least in male adolescents diagnosed with clinically relevant psychiatric conditions.

These studies suggest that adrenal androgen precursors play an important role in the onset of aggression in adolescent boys. Adrenal androgen precursors may also contribute to the regulation of aggression in women. Adolescent and adult women with congenital adrenal hyperplasia, who were exposed to high levels of adrenal androgen precursors in the prenatal and early postnatal periods, were found to have greater self-reported aggression ratings than were control women (Berenbaum and Resnick 1997). Adrenal androgen precursor levels have been determined in adolescent girls diagnosed with conduct disorder (Pajer et al. 2006). Specifically, blood samples were drawn from adolescent girls with either conduct disorder or no psychiatric disorder, and samples were assessed for DHEA, DHEAS, cortisol, as well as for gonadal androgens and estrogens. Girls with conduct disorder scored higher on a clinical aggression scale and demonstrated significantly lower cortisol to DHEA ratio but did not differ from control girls on any other hormone measurement (Pajer et al. 2006). Furthermore, girls diagnosed with aggressive conduct disorder had lower cortisol to DHEA ratios than those with nonaggressive conduct disorder.

DHEA can play a role in adult aggression as well. In a study of alcohol withdrawal, serum levels of DHEAS and cortisol, as well as DHEAS response to treatment with exogenous dexamethasone (i.e., dexamethasone suppression test), were determined in adult alcohol-dependent or healthy control men (Ozsoy and Esel 2008). Alcohol-dependent subjects displayed reduced basal as well as dexamethasone-induced levels of DHEAS compared to control subjects in late alcohol withdrawal. When alcohol-dependent subjects were separated into high and low aggression groups, lower basal DHEAS levels were seen only during early alcohol withdrawal in high aggression individuals, whereas DHEA was lower only during late withdrawal in low aggression individuals relative to control subjects. In contrast, dexamethasone-induced decreases in DHEAS were observed during both early and late alcohol withdrawals whereas lower DHEA levels were only seen during early withdrawal (Ozsoy and Esel 2008). Although it is not entirely clear what these results mean, they do indicate that there are significant changes in HPA axis activity, suggesting an important link between DHEA and aggressive behavior, at least under conditions of drug withdrawal. Whether a link between DHEA, DHEAS, and aggressive behavior exists in healthy adult and adolescent men and women remains to be determined. Regardless, it is clear that considerably more research on DHEA and human aggression is needed.

CONCLUSIONS

In many vertebrate species, including humans, rates of aggression in males are often associated with circulating levels of T. This relationship has been extensively demonstrated in rodents, predominantly rats and mice, where castration has been shown to reduce aggression and exogenous T treatment restores aggressive behavior (Edwards 1969, 1970; Barfield 1971). In birds, a positive correlation between plasma T and rates of aggression has been repeatedly documented during periods of social challenge, and particularly during mate competition. It is important to remember, however, that in many cases, aggression is independent of or even inversely related to circulating levels of T. Seasonal aggression provides a valuable paradigm in which to study the neuroendocrine mechanisms of aggression. In many species (e.g., song sparrows, Siberian and Syrian hamsters), high rates of aggression in males occur outside the period of high circulating T levels (Soma et al. 2000; Garrett and Campbell 1980; Lynch, Ziegler, and Strier 2002, respectively). The neuroendocrine mechanisms regulating aggression under these conditions are poorly understood, but recent data provide some interesting possibilities. In birds, the metabolism of T to E_2 may mediate aggression during the mating season. During the nonmating season when circulating T is basal, steroid metabolism may still mediate aggression, albeit via a different endocrine mechanism (Soma et al. 2000). Despite low levels of circulating T, the androgen precursor DHEA remains elevated in blood, and conversion of DHEA to E_2 within the brain appears to play an important role in nonbreeding season aggression in some species (Soma et al. 2000; Soma and Wingfield 2001). In rodents, the aromatization of androgens such as DHEA also appears to regulate, at least in part, intermale aggression. Furthermore, it seems likely that melatonin might regulate aggression under these conditions by affecting DHEA secretion or metabolism (Haus et al. 1996).

Human and nonhuman primates secrete relatively high levels of DHEA and DHEAS from the adrenal glands, but the role of DHEA in mediating primate aggression has received little attention. Recent evidence, however, suggests that plasma DHEAS levels are associated with conduct disorder, and, in particular, with the aggressive symptoms of conduct disorder in prepubertal boys and adolescent girls (van Goozen et al. 1998; Pajer et al. 2006), as well as in alcohol withdrawal–induced aggression in adults (Ozsoy and Esel 2008).

Traditionally, a nearly exclusive focus on T in the regulation of aggressive behavior has limited our understanding of additional neuroendocrine mechanism regulating aggression and strengthens the misperception that aggression is simply a function of changes in this one steroid. More recently, however, the idea that circulating T is not always associated with physical aggression and instead is more often associated with competition in specific situations has been suggested in humans (Mazur and Booth 1998) and nonhumans (Demas et al. 2007; Soma et al. 2008). Among several factors recently identified as potential regulators or modulators of aggression, DHEA has received considerable attention. While much of this initial work has been correlational and thus a causal role for this hormone has not been identified, continued research on the role of DHEA and its metabolites in the regulation of aggression in human and nonhumans animals, via its actions as both an adrenal hormone and as a neurosteroid, will improve our understanding of normal and abnormal aggression and violence.

ACKNOWLEDGMENTS

This work was supported in part by an Eli Lilly Endowment METACyt grant, Indiana University Faculty Research Support Program and NSF IOB 0543798 to G.E.D, NSERC, Canadian Institutes of Health Research, Michael Smith Foundation for Health Research, and Canada Foundation for Innovation to K.K.S. and NSF IBN 9876754 and NSF IOS 0923301 to H.E.A.

REFERENCES

Albers, H. E., and M. Bamshad. 1998. Role of vasopressin and oxytocin in the control of social behavior in Syrian hamsters (*Mesocricetus auratus*). *Prog Brain Res* 119:395–408.

Albers, H. E., K. L. Huhman, and R. L. Meisel. 2002. Hormonal basis of social conflict and communication. In *Hormones, Brain and Behavior*, ed. D. W. Pfaff, A. P. Arnold, A. M. Etgen, S. E. Fahrbach, and R. T. Rubin, 393–433. New York: Academic Press.

Albers, H. E., L. Yogev, R. B. Todd, and B. D. Goldman. 1985. Adrenal corticoids in hamsters: Role in circadian timing. *Am J Physiol* 248:R434–8.

Badura, L. L., and A. A. Nunez. 1989. Photoperiodic modulation of sexual and aggressive behavior in female golden hamsters (*Mesocricetus auratus*): Role of the pineal gland. *Horm Behav* 23:27–42.

Barfield, R. J. 1971. Activation of sexual and aggressive behavior by androgen implanted into male ring dove brain. *Endocrinology* 89:1470–7.

Baulieu, E. E. 1981. Steroid hormones in the brain: Several mechanisms? In *Steroid Hormone Regulation of the Brain*, ed. K. Fuxe, and J. A. Gustafsson, 3–14. Oxford: Pergamon Press.

Baulieu, E. E. 1991. Neurosteroids: A new function in the brain. *Biol Cell* 71:3–10.

Bayart, F., J. P. Spetz, I. Citton, and M. Haug. 1989. The role of gender and hormonal state on aggression during encounters between residential and intruder mice. *Med Sci Res* 17:517–19

Berenbaum, S. A., and S. M. Resnick. 1997. Early androgen effects on aggression in children and adults with congenital adrenal hyperplasia. *Psychoneuroendocrinology* 22:505–15.

Blanchard, D. C., and R. J. Blanchard. 1988. Ethoexperimental approaches to the biology of emotion. *Annu Rev Psychol* 39:43–68.

Blanchard, D. C., R. L. Spencer, S. M. Weiss, R. J. Blanchard, B. McEwen, and R. R. Sakai. 1995. Visible burrow system as a model of chronic social stress: Behavioral and neuroendocrine correlates. *Psychoneuroendocrinology* 20:117–34.

Boonstra, R., J. E. Lane, S. Boutin, A. Bradley, L. Desantis, A. E. Newman, and K. K. Soma. 2008. Plasma DHEA levels in wild, territorial red squirrels: Seasonal variation and effect of ACTH. *Gen Comp Endocrinol* 158:61–7.

Brain, P. F., and M. Haug. 1992. Hormonal and neurochemical correlates of various forms of animal "aggression." *Psychoneuroendocrinology* 17:537–51.

Brain, P. F., and A. E. Poole. 1974. Some studies on the use of "standard opponents" in intermale aggression testing in TT albino mice. *Behaviour* 50:100–10.

Caldwell, H. K., and H. E. Albers. 2003. Short-photoperiod exposure reduces vasopressin (V1a) receptor binding but not arginine-vasopressin-induced flank marking in male Syrian hamsters. *J Neuroendocrinol* 10:971–7.

Caldwell H. K., and H. E. Albers. 2003. Short-photoperiod exposure reduces vasopressin (V1a) receptor binding but not arginine-vasopressin-induced flank marking in male Syrian hamsters. *J Neuroendocrinol* 15:971–7.

Caldwell, H. K., and H. E. Albers. 2004. Effect of photoperiod on vasopressin-induced aggression in Syrian hamsters. *Horm Behav* 46:444–9.

Caldwell, G. S., S. E. Glickman, and E. R. Smith. 1984. Seasonal aggression independent of seasonal testosterone in wood rats. *Proc Natl Acad Sci USA* 81:5255–7.

Caldwell, H. K., D. A. Smith, and H. E. Albers. 2008. Photoperiodic mechanisms controlling scent marking: Interactions of vasopressin and gonadal steroids. *Eur J Neurosci* 27:1189–96.

Ciacco, L. A., R. D. Lisk, and L. A. Rueter. 1979. Prelordotic behavior in the hamster: A hormonally modulated transition from aggression to sexual receptivity. *J Comp Physiol Psychol* 93:771–80.

Corpechot, C., P. Robel, M. Axelson, J. Sjovall, and E. E. Baulieu. 1981. Characterization and measurement of dehydroepiandrosterone sulfate in rat brain. *Proc Natl Acad Sci* 78:4704–7.

Demas, G. E., H. E. Albers, M. Cooper, and K. K. Soma. 2007. Novel mechanisms underlying neuroendocrine regulation of aggression: A synthesis of bird, rodent and primate studies. In *Behavioral Neurochemistry, Neuroendocrinology and Molecular Neurobiology*, ed. J. D. Blaustein, 337–72. New York: Springer.

Demas, G. E., K. M. Polacek, A. Durazzo, and A. M. Jasnow. 2004. Adrenal hormones mediate melatonin-induced increases in aggression in male Siberian hamsters (*Phodopus sungorus*). *Horm Behav* 46:582–91.

Edwards, D. A. 1969. Early androgen stimulation and aggressive behavior in male and female mice. *Physiol Behav* 4:333–8.

Edwards, D. A. 1970. Post-neonatal androgenization and adult aggressive behavior in female mice. *Physiol Behav* 5:465–67.

Fleming, A. S., A. Phillips, A. Rydall, and L. Levesque. 1988. Effects of photoperiod, the pineal gland and the gonads on agonistic behavior in female golden hamsters (*Mesocricetus auratus*). *Physiol Behav* 44:227–34.

Garrett, J. W., and C. S. Campbell. 1980. Changes in social behavior of the male golden hamster accompanying photoperiodic changes in reproduction. *Horm Behav* 14:303–19.

Goldman, B. D. 2001. Mammalian photoperiodic system: Formal properties and neuroendocrine mechanisms of photoperiodic time measurement. *J Biol Rhythms* 16:283–301.

Golubchik, P., T. Mozes, R. Maayan, and A. Weizman. 2009. Neurosteroid blood levels in delinquent adolescent boys with conduct disorder. *Eur Neuropsychopharmacol* 19:49–52.

Gutzler, S. J., M. Karom, W. D. Erwin, and H. E. Albers. 2011. Seasonal regulation of social communication by photoperiod and testosterone: Effects of arginine-vasopressin, serotonin and galanin in the medial preoptic area—anterior hypothalamus. *Behav Brain Res*. 216:214–9.

Gutzler, S. J., M. Karom, W. D. Erwin, and H. E. Albers. 2009. Photoperiodic regulation of adrenal hormone secretion and aggression in female Syrian hamsters. *Horm Behav* 56:481–9.

Haller, J., and M. R. Kruk. 2003. Neuroendocrine stress responses and aggression. In *Neurobiology of Aggression*, ed. M. P. Mattson, 93–118. Totawa, NJ: Humana Press.

Haller, J., G. B. Makara, and M. R. Kruk. 1998. Catecholaminergic involvement in the control of aggression: Hormones, the peripheral sympathetic, add central noradrenergic systems. *Neurosci Biobehav Rev* 22:85–97.

Hau, M., S. T. Stoddard, and K. K. Soma. 2004. Territorial aggression and hormones during the non-breeding season in a tropical bird. *Horm Behav* 45:40–9.

Hau, M., M. Wikelski, K. K. Soma, and J. C. Wingfield. 2000. Testosterone and year-round territorial aggression in a tropical bird. *Gen Comp Endocrinol* 117:20–33.

Haug, M., P. F. Brain, and A. B. Kamis. 1986. A brief review comparing the effects of sex steroids on two forms of aggression in laboratory mice. *Neurosci Biobehav Rev* 10:463–8.

Haug, M., F. Johnson, and P. E. Brain. 1992. Biological corrlelates of attack on lactating intruders by female mice: A topical review. In *Of Mice and Women: Aspects of Female Agression*, ed. K. Bjorkvist and P. Niemala, 381–93. San Diego: Academic Press.

Haug, M., M. L. Ouss-Schlegel, J. F. Spetz, P. F. Brain, V. Simon, E. E. Baulieu, and P. Robel. 1989. Suppressive effects of dehydroepiandrosterone and 3-beta-methylandrost-5-en-17-one on attack towards lactating female intruders by castrated male mice. *Physiol Behav* 46:955–9.

Haug, M., J. Young, P. Robel, and E. E. Baulieu. 1991. Inhibition by dehydroepiandrosterone of aggression responses of spayed female mice towards lactating intruders is potentiated by neonatal androgenization. *C R Acad Sci III* 312:511–6.

Haus, E., G. Y. Nicolau, E. Ghinea, L. Dumitriu, E. Petrescu, and L. Sackett-Lundeen. 1996. Stimulation of the secretion of dehydroepiandrosterone by melatonin in mouse adrenals in vitro. *Life Sci* 58:PL263–7.

Hojo, Y., T. A. Hattori, T. Enami, A. Furukawa, K. Suzuki, H. T. Ishii et al. 2004. Adult male rat hippocampus synthesizes estradiol from pregnenolone by cytochromes P45017alpha and P450 aromatase localized in neurons. *Proc Natl Acad Sci U S A* 101:865–70.

Jasnow, A. M., K. L. Huhman, T. J. Bartness, and G. E. Demas. 2000. Short-day increases in aggression are inversely related to circulating testosterone concentrations in male Siberian hamsters (*Phodopus sungorus*). *Horm Behav* 38:102–10.

Jasnow, A. M., K. L. Huhman, T. J. Bartness, and G. E. Demas. 2002. Short days and exogenous melatonin increase aggression in male Syrian hamsters (*Mesocricetus auratus*). *Horm Behav* 42:13–20.

Labrie, F. 2003. Extragonadal synthesis of sex steroids: Intracrinology. *Ann Endocrinol (Paris)* 64:95–107.

Labrie, F., A. Bélanger, J. Simard, V. Luu-The, and C. Labrie. 1995. DHEA and peripheral androgen and estrogen formation: Intracrinology. *Ann NY Acad Sci* 774:16–28.

Labrie F., A. Bélanger, V. Luu-The, C. Labrie, J. Simard, L. Cusan, J. L. Gomez, and B. Candas. 1998. DHEA and the intracrine formation of androgens and estrogens in peripheral target tissues: Its role during aging. *Steroids* 63:322–8.

Labrie, F., V. Luu-The, C. Labrie, and J. Simard. 2001. DHEA and its transformation into androgens and estrogens in peripheral target tissues: Intracrinology. *Front Neuroendocrinol* 22:185–212.

Lu, S. F., Q. Mo, S. Hum, C. Garippam, and N. G. Simon. 2003. Dehydroepiandrosterone upregulates neural androgen receptor level and transcriptional activity. *J Neurobiol* 57:163–71.

Lynch, J. W., T. E. Ziegler, and K. B. Strier. 2002. Individual and seasonal variation in fecal testosterone and cortisol levels of wild male tufted capuchin monkeys, *Cebus apella nigritus*. *Horm Behav* 41:275–87.

Maninger, N., O. M. Wolkowitz, V. I. Reus, E. S. Epel, and S. H. Mellon. 2009. Neurobiological and neuropsychiatric effects of dehydroepiandrosterone (DHEA) and DHEA sulfate (DHEAS). *Front Neuroendocrinol* 30:65–91.

Marques, D. M., and E. S. Valenstein. 1977. Individual differences in aggressiveness of female hamsters: Response to intact and castrated males and to females. *Anim Behav* 25:131–9.

Mazur, A., and A. Booth. 1998. Testosterone and dominance in men. *Behav Brain Sci* 21:353–97.

Mellon, S. H., and H. Vaudry. 2001. Biosynthesis of neurosteroids and regulation of their synthesis. *Int Rev Neurobiol* 46:33–78.

Miczek, K. A., J. F. DeBold, A. M. van Erp, and W. Tornatzky. 1997. Alcohol, GABAA-benzodiazepine receptor complex, and aggression. *Recent Dev Alcohol* 13:139–71.

Miczek, K. A., E. Weerts, M. Haney, and J. Tidey. 1994. Neurobiological mechanisms controlling aggression: Preclinical developments for pharmacotherapeutic interventions. *Neurosci Biobehav Rev* 18:97–110.

Mo, Q., S. F. Lu, S. Hu, and N. G. Simon. 2004. DHEA and DHEA sulfate differentially regulate neural androgen receptor and its transcriptional activity. *Brain Res Mol Brain Res* 126:165–72.

Mohle, U., M. Heistermann, R. Palme, and J. K. Hodges. 2002. Characterization of urinary and fecal metabolites of testosterone and their measurement for assessing gonadal endocrine function in male nonhuman primates. *Gen Comp Endocrinol* 129:135–45.

Moyer, K. E. 1971. *The Physiology of Hostility*. Chicago: Markham.

Nelson, R. J. 1990. Photoperiodic responsiveness in house mice. *Physiol Behav* 48:403–8.

Newman, A. E., S. A. MacDougall-Shackleton, Y. S. An, B. Kriengwatana, and K. K. Soma. 2010. Corticosterone and dehydroepiandrosterone have opposing effects on adult neuroplasticity in the avian song control system. *J Comp Neurol* 518:3662–78.

Newman, A. E., D. S. Pradhan, and K. K. Soma. 2008. Dehydroepiandrosterone and corticosterone are regulated by season and acute stress in a wild songbird: Jugular versus brachial plasma. *Endocrinology* 149:2537–45.

Newman, A. E., and K. K. Soma. 2009. Corticosterone and dehydroepiandrosterone in songbird plasma and brain: Effects of season and acute stress. *Eur J Neurosci* 29:1905–14.

Nicolas, L. B., W. Pinoteau, S. Papot, S. Routierm, G. Guillaumetm, and S. Mortaud. 2001. Aggressive behavior induced by the steroid sulfatase inhibitor COUMATE and by DHEAS in CBA/H mice. *Brain Res* 922:216–22.

Ostner, T., P. M. Kappeler, and M. Heistermann. 2002. Seasonal variation and social correlates of androgen excretion in male redfronted lemurs (*Eulemur fulvus rufus*). *Behav Ecol Sociobiol* 52:485–95.

Owen-Ashley, N. T., D. Hasselquist, and J. C. Wingfield. 2004. Androgens and the immunocompetence handicap hypothesis: Unraveling direct and indirect pathways of immunosuppression in song sparrows. *Am Nat* 164:490–505.

Ozsoy, S., and E. Esel. 2008. Hypothalamic-pituitary-adrenal axis activity, dehydroepiandrosterone sulphate and their relationships with aggression in early and late alcohol withdrawal. *Prog Neuropsychopharmacol Biol Psychiatry* 32:340–7.

Pajer, K., R. Tabbah, W. Gardner, R. T. Rubin, R. K. Czambel, and Y. Wang. 2006. Adrenal androgen and gonadal hormone levels in adolescent girls with conduct disorder. *Psychoneuroendocrinology* 31:1245–56.

Paterson, A. T., and C. Vickers. 1981. Melatonin and the adrenal cortex: Relationship to territorial aggression in mice. *Physiol Behav* 27:983–7.

Perché, F., J. Young, P. Robel, N. G. Simon, and M. Haug. 2000. Prenatal testosterone treatment potentates the aggression-suppressive effect of the neurosteroid dehydroandrosterone in female mice. *Agg Behav* 27:130–8.

Perché, F., J. Young, P. Robel, N. G. Simon, and M. Haug. 2001. Prenatal testosterone treatment potentates the aggression-suppressive effect of the neurosteroid dehydroandrosterone in female mice. *Agg Behav* 27: 130–8.

Pieper, D. R., and C. A. Lobocki. 2000. Characterization of serum dehydroepiandrosterone secretion in golden hamsters. *Proc Soc Exp Biol Med* 224:278–84.

Pieper, D. R., C. A. Lobocki, E. M. Lichten, and J. Malaczynski. 1999. Dehydroepiandrosterone and exercise in golden hamsters. *Physiol Behav* 67:607–10.

Pinna, G., E. Costa, and A. Guidotti. 2005. Changes in brain testosterone and allopregnanolone biosynthesis elicit aggressive behavior. *Proc Natl Acad Sci U S A* 102:2135–40.

Pinxten, R., E. De Ridder, J. Balthazart, L. Berghman, and M. Eens. 2000. The effect of castration on aggression in the nonbreeding season is age-dependent in male European starlings. *Behaviour* 137:647–61.

Pradhan, D. S., A. E. Newman, D. W. Wacker, J. C. Wingfield, B. A. Schlinger, and K. K. Soma. 2010. Aggressive interactions rapidly increase androgen synthesis in the brain during the non-breeding season. *Horm Behav* 57:381–9.

Pradhan D. S., Y. Yu, and K. K. Soma. 2008. Rapid estrogen regulation of DHEA metabolism in the male and female songbird brain. *J Neuro* 104: 244–53.

Rehman, K. S., and B. R. Carr. 2004. Sex differences in adrenal androgens. *Semin Reprod Med* 22:349–60.

Sapolsky, R. M., J. H. Vogelman, N. Orentreich, and J. Altmann. 1993. Senescent decline in serum dehydroepiandrosterone sulfate concentrations in a population of wild baboons. *J Gerontol* 48:B196–200.

Schlegel, M. L., J. F. Spetz, P. Robel, and M. Haug. 1985. Studies on the effects of dehydroepiandrosterone and its metabolites on attack by castrated mice on lactating intruders. *Physiol Behav* 34:867–70.

Schlinger, B. A., D. S. Pradhan, and K. K. Soma. 2008. 3beta-HSD activates DHEA in the songbird brain. *Neurochem Int* 52:611–20.

Scotti, M.-A., J. Belén, G. E. Jackson, and G. E. Demas. 2008. The role of testosterone and DHEA in the mediation of seasonal territorial aggression in male Siberian hamsters (*Phodopus sungorus*). *Physiol Behav* 95:633–40.

Scotti, M. L., A. E. M. Newman, K. L. Schmidt, T. N. Bonu, K. K. Soma, and G. E. Demas. 2009. Aggressive encounters differentially affect circulating androgen levels in male Siberian hamsters (*Phodopus sungorus*). *Horm Behav* 56:376–81.

Scotti M. A., K. L. Schmidt, A. E. Newman, T. Bonu, K. K. Soma, G. E. Demas. 2009. Aggressive encounters differentially affect serum dehydroepiandrosterone and testosterone concentrations in male Siberian hamsters (Phodopus sungorus). *Horm Behav* 56:376–81.

Simon, N. G. 2002. Hormonal processes in the development and expression of aggressive behavior. In *Hormones, Brain and Behavior*, ed. D. W. Pfaff, A. P. Arnold, A. M. Etgen, S. E. Fahrbach, and R. T. Rubin, 339–2. New York: Academic Press.

Simon, N. G., J. R. Kaplan, S. Hu, T. C. Register, and M. R. Adams. 2004. Increased aggressive behavior and decreased affiliative behavior in adult male monkeys after long-term consumption of diets rich in soy protein and isoflavones. *Horm Behav* 45:278–84.

Soma, K. K. 2006. Testosterone and aggression: Berthold, birds, and beyond. *J Neuroendocrinol* 18:543–51.

Soma, K. K., N. A. Alday, M. Hau, and B. A. Schlinger. 2004. Dehydroepiandrosterone metabolism by 3β-hydroxysteroid dehydrogenase/Δ5-Δ4 isomerase in adult zebra finch brain: Sex difference and rapid effect of stress. *Endocrinology* 145:1668–77.

Soma, K. K., R. K. Bindra, J. Gee, J. C. Wingfield, and B. A. Schlinger. 1999. Androgen-metabolizing enzymes show region-specific changes across the breeding season in the brain of a wild songbird. *J Neurobiol* 41:176–88.

Soma, K. K., V. N. Hartman, J. C. Wingfield, and E. A. Brenowitz. 1999. Seasonal changes in androgen receptor immunoreactivity in the song nucleus HVc of a wild bird. *J Comp Neurol* 409:224–36.

Soma, K. K., B. A. Schlinger, J. C. Wingfield, and C. J. Saldanha. 2003. Brain aromatase, 5α-reductase and 5β-reductase change seasonally in wild male song sparrows: Relationship to sexual and aggressive behaviors. *J Neurobiol* 56:209–21.

Soma, K. K., M.-A. Scotti, A. E. Newman, and G. E. Demas. 2008. Novel mechanisms of steroid regulation of aggression. *Front Neuroendocrinol* 29:476–89.

Soma, K. K., K. A. Sullivan, A. D. Tramontin, C. J. Saldanha, B. A. Schlinger, and J. C. Wingfield. 2000. Acute and chronic effects of an aromatase inhibitor on territorial aggression in breeding and nonbreeding male song sparrows. *J Comp Physiol A* 186:759–69.

Soma, K. K., K. Sullivan, and J. Wingfield. 1999. Combined aromatase inhibitor and antiandrogen treatment decreases territorial aggression in a wild songbird during the nonbreeding season. *Gen Comp Endocrinol* 115:442–53.

Soma, K. K., A. D. Tramontin, and J. C. Wingfield. 2000. Oestrogen regulates male aggression in the nonbreeding season. *Proc Roy Soc London B* 267:1089–96.

Soma, K. K., and J. C. Wingfield. 1999. Endocrinology of aggression in the nonbreeding season. In *22nd International Ornithological Congress*, ed. N. Adams, and R. Slotow, 1606–20. Durban, South Africa: University of Natal.

Soma, K. K., and J. C. Wingfield. 2001. Dehydroepiandrosterone in songbird plasma: Seasonal regulation and relationship to territorial aggression. *Gen Comp Endocrinol* 123:144–55.

Soma, K. K., A. M. Wissman, E. A. Brenowitz, and J. C. Wingfield. 2002. Dehydroepiandrosterone (DHEA) increases territorial song and the size of an associated brain region in a male songbird. *Horm Behav* 41:203–12.

van Goozen, S. H. M., W. Matthys, P. T. Cohen-Kettenis, J. H. H. Thijssen, and H. van Engeland. 1998. Adrenal androgens and aggression in conduct disorder prepubertal boys and normal controls. *Biol Psychiat* 43:156–8.

van Goozen, S. H., E. van den Ban, W. Matthys, P. T. Cohen-Kettenis, J. H. Thijssen, and H. van Engeland. 2000. Increased adrenal androgen functioning in children with oppositional defiant disorder: A comparison with psychiatric and normal controls. *J Am Acad Child Adolesc Psychiat* 39:1446–51.

Wingfield, J. C. 1994. Control of territorial aggression in a changing environment. *Psychoneuroendocrinology* 19:709–21.

Wingfield, J. C., and T. Hahn. 1994. Testosterone and territorial behaviour in sedentary and migratory sparrows. *Anim Behav* 47:77–89.

Wingfield, J. C., and K. K. Soma. 2002. Spring and autumn territoriality in song sparrows: Same behavior, different mechanisms? *Integr Comp Biol* 42:11–20.

Young, J., C. Corpéchot, F. Perché, B. Eychenne, M. Haug, E. E. Baulieu, and P. Robel. 1996. Neurosteroids in the mouse brain: Behavioral and pharmacological effects of a 3 beta-hydroxysteroid dehydrogenase inhibitor. *Steroids* 61:144–9.

35 DHEA and Alzheimer's Disease

Laïla El Kihel

CONTENTS

BACKGROUND

Alzheimer's disease (AD) is a neurodegenerative disorder characterized by a selective neuronal loss and the presence of abnormal extracellular accumulation of β-amyloid peptide (AβP) in the sensitive structures of the brain. AβP derives from a type I integral membrane amyloid precursor protein (APP), by two consecutive proteolytic cleavages operated by β- and γ-secretase (Selkoe 2001). After the β-secretase enzyme cleaves the APP extracellular domain, γ-secretase cleaves the remaining segment to release the AβP.

These AβPs are the principal constituents of senile plaques, which are composed of 37–43 amino acids (Wiltfang et al. 2002; Verdier, Zarándi, and Penke 2004; Kaminsky et al. 2010). AβPs can vary in length, and the 40-residue peptide Aβ(1–40) represents the predomiant Aβ species in normal and AD brains, followed by the 42-residue peptide Aβ(1–42) (Finder and Glockshuber 2007). Aβ(1–42) differs from Aβ(1–40) by two additional hydrophobic residues at the C-terminus; it is considered the main amyloidogenic species and is most likely responsible for neuropathology in AD, despite its 10-fold lower *in vivo* concentrations as compared to Aβ1–40 (Finder et al. 2010).

Gradual deposition of β-amyloid protein in the form of neurotic plaques, apparition of neurofibrillary tangles (NFTs), and progressive cognitive deficits accompany the emergence of AD. Soluble AβPs dramatically derange synaptic plasticity, dendritic spine number and motility, and pre- and postsynaptic composition without overt neuronal loss (Walsh et al. 2002; Cleary et al. 2005; Calabrese et al. 2007b). Genetic studies of early-onset cases of familial AD indicated that APP mutations and AβP metabolism are associated with the disease (Bertram and Tanzi 2004; De Kimpe and Scheper 2010).

The mechanism by which Aβ expresses its neurotoxic effects may involve the induction of reactive oxygen species (ROS) and the elevation of intracellular free calcium levels (Liu and Schubert 2009; Reddy 2009). AD is one among the neurodegenerative diseases that involve a dysregulation in the brain's glutamatergic system (Thorns et al. 1997). The rapid effects involve the activation of the *N*-methyl-D-aspartate receptor (NMDAR) that leads to a large Ca^{2+} influx that may be detrimental to cell viability (Gardoni and Di Luca 2006).

Although intracellular Ca^{2+} ($[Ca^{2+}]i$) is necessary for a number of physiological processes, excessive amounts may lead to neuronal dysfunction and cell death (Sattler and Tymianski 2000). Neuronal increases in Ca^{2+} can activate a number of enzymes, such as phospholipases (Chalbot et al. 2009), proteases, endonucleases, and nitric oxide synthase (NOS). Increase in the activity of these enzymes is associated with neuronal cell death (De Strooper, Vassar, and Golde 2010). Excessive glutamatergic stimulation is also associated with an increase in $[Ca^{2+}]i$ required for neuronal NOS (nNOS) activation and nitric oxide (NO) production within the neuron, and this can result in increased cell death (Calabrese et al. 2007a; Provias and Jeynes 2008). Therefore, the maintenance of proper Ca^{2+} homeostasis may be effective in preventing the progression of glutamate associated neuronal degeneration (Praticò 2008).

The relationship between AD and circulating levels of DHEA has also been studied extensively (Rasmuson et al. 2002; Hoskin et al. 2004).

Numerous studies investigating the relationship between DHEA and AD have shown the involvement of DHEA in various mechanisms, such as the modulating effect on GABAA, NMDAR, and σ1-receptor. Studies in both central nervous system (CNS) and peripheral target cells suggest that DHEA may also exert genomic effects via the androgen receptor (AR).

About 10% of the causes of AD are believed to have a genetic background (Hampl and Bicíková 2010). Therefore, in this chapter, we will focus only on the (patho) biochemical mechanisms, with an interest in the involvement of DHEA. After a brief description of AD, we will summarize the recent reports of DHEA distribution related to AD. We will then examine the metabolism of DHEA in the aging brain of AD patients, and the effect of DHEA on the biochemical pathways affected in AD. The chapter will conclude with a discussion of possible innovative therapeutic strategies.

DHEA DISTRIBUTION AND ALZHEIMER'S DISEASE

Circulating DHEA and Alzheimer's Disease

DHEA and DHEA sulfate (DHEAS) concentrations have been found to decline with progressing age in both men and women. The relationship between AD and circulating levels of DHEA has been studied extensively. Most studies have not found differences in DHEAS levels between AD patients and nondemented patients (Legrain et al. 1995), whereas other studies have reported decreased levels of DHEAS in AD patients; for example, Cho et al. (2006) explored the quantification of DHEAS by liquid chromatography–mass spectrometry (LC-MS) in the plasma of patients with AD and compared it with that in normal subjects. The level of DHEAS was significantly decreased in the plasma of patients with AD. Other authors have studied the correlation between AD (in men and women) and circulating levels of DHEA and testosterone. These results do not show the evident implication of these two hormones in AD in men. However, women with AD had significantly higher levels of DHEA than in controls, suggesting that DHEA may be involved in the progression of the disease (Rasmuson et al. 2002). Bo et al. (2006) have investigated the association of serum DHEAS levels with dementia of Alzheimer's type (DAT) patients and impairment in selected cognitive domains (memory, language, attention, and working memory). No significant association between the presence of DAT or impairment of cognitive domains and DHEAS levels was found. Baseline DHEAS levels were not associated with cumulative mortality in patients and controls.

Following these results, we can conclude that causal relationships between the DHEA levels and impaired cerebral functions in elderly or AD patients are still difficult to assert, and mixed results have been observed with DHEA replacement therapy in the elderly (Weill-Engerer et al. 2003).

Analysis of DHEA and DHEAS in Individual Brain Regions of Patients with Alzheimer's Disease and Aged Nondemented Patients by Gas Chromatography–Mass Spectrometry

In a recent study (Weill-Engerer et al. 2002), the concentrations of DHEA, DHEAS, and other neurosteroids were measured by gas chromatography–mass spectrometry (GC-MS), in individual brain regions of AD patients and aged nondemented controls, including six brain regions: hippocampus, amygdala, frontal cortex, striatum, cerebellum, and hypothalamus.

These results showed declined levels of all steroids in all AD patients' brain regions when compared with controls. The steroids found at the highest concentrations were, in decreasing order, pregnenolone > DHEA> progesterone > pregnenolone sulfate > DHEAS>> allopregnanolone, in all regions of both groups. DHEAS was significantly lower in the AD group compared with controls in the striatum, cerebellum, and hypothalamus.

However, a significant overall difference between regions was observed in the control group for each steroid measured, which are, in decreasing order, hypothalamus > striatum > frontal cortex > cerebellum > amygdala ~ hippocampus.

Weill-Engerer et al. also studied the relationship between the levels of brain steroids and specific structural brain abnormalities related to AD, including the extracellular senile plaques composed of AβP and the intracellular NFTs containing intraneuronal bundles of paired helical filaments (PHFs) that result from the aggregation of pathogenic tau proteins (PHF-tau). These results show that high levels of key proteins implicated in the formation of plaques and NFTs were correlated with decreased brain levels of pregnenolone sulfate and DHEAS, suggesting a possible neuroprotective role of these neurosteroids in AD.

Another study reported elevated DHEA levels in the frontal cortex, hippocampus, and hypothalamus of AD patients when compared with controls, using high-performance liquid chromatography (HPLC) purification and GC-MS (Brown et al. 2003).

It can be noted that these data are contrary to the previous hypotheses that speculated that DHEA may be important in memory and AD. The results presented here demonstrate that, contrary to the previous hypotheses, levels of DHEA are much higher in the AD brain than in the normal brain.

Cerebrospinal Fluid DHEA Levels and Alzheimer's Disease

Brown et al. (2003) also reported that DHEA levels in cerebrospinal fluid (CSF) of AD patients were significantly higher than age-matched controls. These authors suggest that DHEA is formed in the AD brain by the oxidative stress metabolism of a precursor, and DHEA levels in the CSF in response to $FeSO_4$ may serve as an indicator of AD pathology.

A second report also demonstrated higher DHEA levels in the CSF of patients with AD and vascular dementia (VD) using GC-MS preceded by HPLC (Kim et al. 2003). It is known that the human brain transforms DHEA into DHEAS, 7α-hydroxy-DHEA, 7β-hydroxy-DHEA, and 16α-hydroxy-DHEA (Figure 35.1). DHEA accumulation in the brain may result from a decreased production of such metabolites. Therefore, these authors have measured and compared the CSF levels of DHEA, DHEAS, 7α-hydroxy-DHEA, 7β-hydroxy-DHEA, and 16α-hydroxy-DHEA in patients with AD, controls, and patients with another common dementia such as VD. The neurosteroids found in the CSF of patients with AD and VD, compared with that in normal subjects, were DHEA precursors (pregnenolone and its sulfate), DHEA metabolites (7α-, 7β- and 16α-hydroxy-DHEA), DHEA, and DHEAS. CSF levels of DHEA metabolites were reported for the first time by Kim et al., who showed that CSF levels of DHEA were higher in AD and VD patients when compared with controls. However, CSF levels of DHEAS were higher in controls than in AD and VD patients. Taken together, these differences could help discriminate between controls and AD or VD patients. No significant difference was found between controls and AD and VD patients with regard to CSF levels of DHEA precursors and DHEA metabolites.

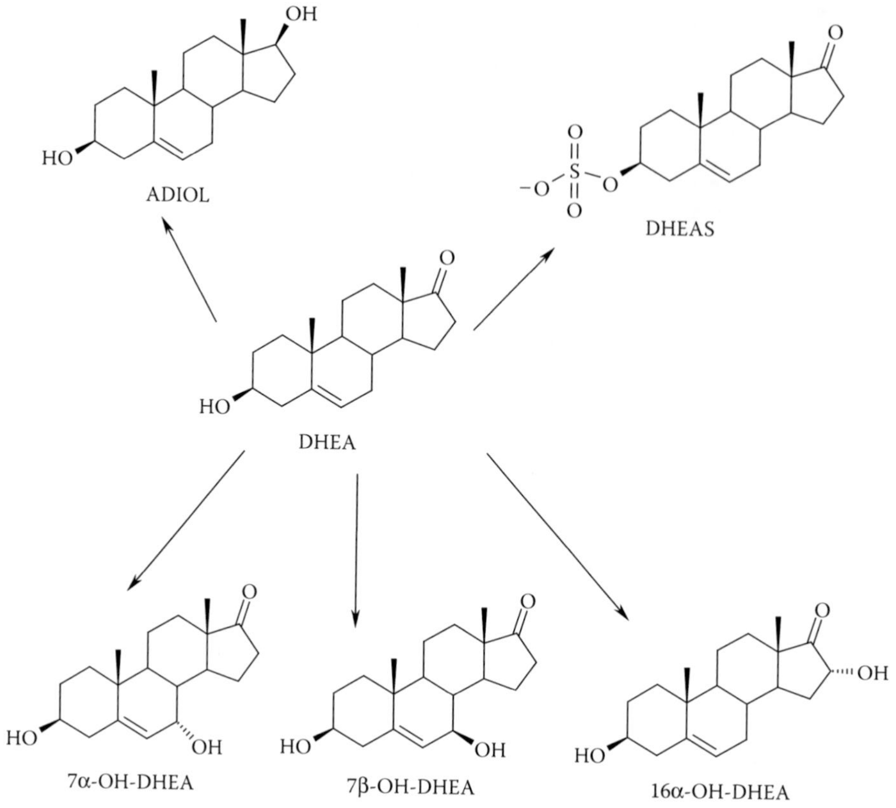

FIGURE 35.1 DHEA and its metabolites.

In contrast, the DHEA/(7α-hydroxy-DHEA + 7β-hydroxy- DHEA) ratio was significantly higher in AD and VD patients than in controls, and the 7β-hydroxy-DHEA/DHEA ratio was significantly higher in controls than in AD and VD patients. Specific differences between AD and VD patients were found by the 7α-hydroxy-DHEA/7β-hydroxy-DHEA ratio, which was significantly higher in AD patients and controls than in VD patients. However, the 7α-hydroxy-DHEA/DHEA ratio was found to be significantly higher in AD patients than in VD patients and controls. Taken together, these CSF steroid ratios help us discriminate between controls and AD or VD patients.

On the other hand, significant differences were found with the ratios of 7α-hydroxy-DHEA/DHEA, DHEAS/DHEA, and 16α-hydroxy-DHEA/DHEA, which were significantly higher in controls than in AD and VD patients respectively. These CSF steroid ratios could be used for differentiating AD and VD patients from controls.

Following their results, Kim et al. suggested that in AD and VD, the increased DHEA levels do not result in increased sulfatation or increased hydroxylation at the 7α, 7β, and 16α positions. In addition, the sulfotransferase and the P450s responsible for 7α- and 16α-hydroxylations of DHEA are either present in lower levels or transformed through natural polymorphism into less-efficient enzymes.

More recently, Naylor et al. (2008) attempted to establish the correlation between the neurosteroid levels of CSF and those of the brain in patients with AD. DHEA and pregnenolone levels were measured by GC-MS preceded by HPLC in CSF and temporal cortex brain and compared with controls. These findings showed increased CSF DHEA levels in AD patients, which is consistent with the findings of Kim et al. (2003) and Brown et al. (2003). In addition, these results indicate that CSF DHEA levels are correlated with temporal cortex brain DHEA levels, which are elevated in AD patients, and may serve as an indicator for the pathophysiology of AD.

In conclusion, it is possible that an increase in DHEA in AD may be reflective of an adaptive or compensatory mechanism resulting from β-amyloid deposition because β-amyloid administration increases DHEA formation in oligodendrocytes. Oxidative stress also plays an important role in the pathophysiology of AD, and DHEA has neuroprotective effects against various factors that result in oxidative stress. Previous studies also provide evidence that changes in the levels of the sulfated derivative of DHEA (DHEAS) may be important in AD pathophysiology. Many roles have previously been attributed to DHEAS, including the prevention and reduction of the neurotoxic effects.

It may be noteworthy that the reduced levels of both DHEA and DHEAS in AD patients may negatively impact several neuroprotective processes. Investigating both DHEA and DHEAS, as well as the ratio of DHEA to DHEAS, will therefore be important in future studies focusing on the clinical relevance of these neurosteroids to the pathophysiology of AD.

Metabolism of DHEA in the Aging Brain of AD Patients and Nondemented Patients

In the human brain, the metabolism of DHEA, when compared to other metabolites, is still poorly understood, particularly in aging people and AD patients. Weill-Engerer et al. (2003) described the *in vitro* transformation of DHEA into 7α-hydroxy-DHEA (7α-OH-DHEA) and androst-5-ene (ADIOL) for the first time in the aging brain of AD patients and controls (Figure 35.1). DHEA metabolites were purified by HPLC and identified by GC-MS in the frontal cortex, hippocampus, amygdala, cerebellum, and striatum of both AD patients and controls.

Regional differences were observed in the two groups, with significantly higher formation of 7α-OH-DHEA in the frontal cortex compared with other brain regions and of ADIOL in cerebellum and striatum compared to frontal cortex. Interestingly, a trend toward a negative correlation was found between the density of cortical β-amyloid deposits and the amount of 7α-OH-DHEA formed in the frontal cortex and that of ADIOL formed in the hippocampus.

Currently, the significance of the production of 7α-OH-DHEA and ADIOL from DHEA in the aged human brain, including in AD, is still unclear. Weill-Engerer et al. suggested that it is likely to regulate cerebral DHEA availability, and therefore biologically active androgens/estrogens derived from DHEA, and may contribute to the control of DHEA activities in the CNS.

EFFECT OF DHEA ON THE BIOCHEMICAL PATHWAYS AFFECTED IN ALZHEIMER'S DISEASE

Numerous biochemical pathways are affected in AD, the specific cause of which remains undetermined. Although genetic susceptibility factors have been identified, aging remains the predominant risk factor.

AD is a multifactoral pathology, characterized not only by an increase in cerebral deposition of AβP, the major constituent of senile plaques that can potentially cause cognitive impairments, but also by neuroinflammation, oxidative damage, and neurodegeneration in critical brain regions (hippocampus, frontal cortex) also involved in memory and cognition.

DHEA and Oxidative Stress in Alzheimer's Disease Pathology

ROS and reactive nitrogen species (RNS), also called free radicals, are highly reactive substances formed by the incomplete reduction of oxygen. They include superperoxide ($O_2.^-$), peroxide (H_2O_2), hydroxyl radical (.OH), and peroxynitrite anions ($ONOO^-$; Taupin 2010).

Oxidative stress due to excessive generation of ROS has been implicated in many neurodegenerative diseases, and evidence of increased oxidative stress has been shown in the AD brain.

It is also established that oxidative stress contributes to the formation of amyloid plaques and NFTs. Some authors suggest that the source of oxidative stress in the AD brain is caused by AβP, which is formed by the two sequential cleavages of β-APP operated by β-secretase (β-amyloid-converting enzyme [BACE]) and γ-secretase at the N- and C-termini of Aβ. There are two major forms of β-secretase enzyme: BACE1 (501 amino acids) and BACE2 (518 amino acids; Ahmed et al. 2010). It is reported that the levels of BACE1 are increased in the vulnerable regions of the AD brain, but the underlying mechanism is unknown (Tamagno et al. 2008).

Superperoxide ($O_2.^-$), Peroxide (H_2O_2), and Hydroxyl Radical (.OH)

ROS that are commonly formed include superperoxide ($O_2.^-$) produced by the mitochondrial respiratory chain, peroxide (H_2O_2) generated from the conversion of superoxide by the enzyme superoxide dismutase, and hydroxyl radical (.OH) produced by the reduction of hydrogen peroxide. Metals are also a source of redox-generated free radicals. Under normal physiological conditions, these species are detoxified by several mechanisms; however, when the ROS and/or RNS are overproduced, as occurs in many diseases involving chronic inflammation, these reactive species can cause oxidative damage to cellular proteins, DNA, and membranes. Elevation of lipid oxidation (LPO), protein oxidation (PO), and endogenous antioxidant defense systems has been observed in postmortem brain tissue from AD patients (Wang, Markesbery, and Lovell 2006; Keller et al. 2005).

Bastianetto et al. (1999) found that DHEA protects rat primary hippocampal cells and human hippocampal sections in AD patients from oxidative stress–mediated toxicity. Moreover, it has also been shown that DHEA protects the frontal cortex of AD patients from oxidative stress–mediated injury (Ramassamy et al. 1999).

Brown, Cascio, and Papadopoulos (2000) examined the ability of the cell lines from human brains to make DHEA via an alternative pathway induced by the treatment of intracellular free radicals with $FeSO_4$ inducing ROS and oxidative stress. Oligodendrocytes and astrocytes make DHEA via this pathway, but neurons do not. In searching for a natural regulator of DHEA formation, treatment of oligodendrocytes, steroid-synthesizing cells of the human CNS, with β-amyloid, increases both ROS and DHEA formation. The effects of β-amyloid were blocked by vitamin E (antioxidant). These results show that the human brain makes steroids in a cell-specific manner and suggest that DHEA synthesis can be regulated by intracellular free radicals.

To determine if this pathway exists in the human brain, Brown et al. (2003) have measured the levels of DHEA in hippocampus, hypothalamus, and frontal cortex in AD patients and age-matched controls. DHEA levels are significantly increased in the AD brain in all three areas examined and are maximal in AD hippocampus. This may be a reflection of increased oxidative stress in the AD brain, potentially due to the actions of Aβ. Another study shows that DHEA reduces the expression and activity of BACE in NT2 neurons exposed to oxidative stress (Tamagno et al. 2003). These and other authors have shown that the expression and activity of BACE1 are increased by oxidants and by the lipid peroxidation products and that there is a significant correlation of BACE1 activity with oxidative markers in sporadic AD brain tissue (Tong et al. 2005; Borghi et al. 2007). How DHEA exerts its neuroprotection function has not been fully elucidated; another possible mechanism is via its active metabolites, 7α-OH-DHEA and 7β-OH-DHEA (Li and Bigelow 2010).

Nitric Oxide

Oxidative stress could also stimulate damage via the overexpression of inductible and neuronal-specific NOS in the pathology of a number of neurodegenerative diseases such as AD. The RNS $ONOO^-$ is formed when O_2^- combines with NO. Nitric oxide is a signaling molecule produced by neurons and endothelial cells in the brain. Three isoforms of NOS have been described: neuronal (nNOS), endothelial (eNOS), and inducible (iNOS). NO can be scavenged with superoxide (O_2^-) to generate peroxynitrite ($ONOO^-$). $ONOO^-$ is a potent oxidant and the primary component of nitroxidative stress. $ONOO^-$ can undergo homolytic or heterolytic cleavage at high concentrations to produce NO_2^+, NO_2, and OH, highly reactive oxidative species and secondary components of nitroxidative stress (Malinski 2007). In AD brain and CSF, increased levels of nitrated proteins have been found, implying a role for RNS in AD pathology (Castegna et al. 2003). In addition, vascular

NO activity appears to be a major contributor to this pathology before any overexpression of NOS isoforms is observed in the neuron, glia, and microglia of the brain tree, where the overexpression of the NOS isoforms causes the formation of a large amount of NO (Aliev et al. 2009). Increased NOS activity has been demonstrated previously after a single Aβ-related peptide administration into the rat brain, as monitored by ex vivo measurement of the enzyme activity (Rosales-Corral et al. 2004). Loss of nNOS-containing neurons has been reported after the intrahippocampal injection of Aβ(1–40) (Li et al. 2004). Some studies showed that DHEA is able to increase NOS activity. DHEA (100 nM) increases endothelial NOS (eNOS) protein in cultured endothelial cells after 16 h of incubation with the steroid (Williams et al. 2004). This effect is not likely to be dependent on DHEA metabolites and is due to the decreased eNOS protein turnover (Simoncini et al. 2003). DHEAS does not increase NOS activity in endothelial cells (Liu and Dillon 2004), but its effect on other cell types (especially neurons) remains to be determined. According to the literature and to the best of our knowledge, no study has shown the effect of DHEA on NO synthase in AD, and it is a field that remains to be explored.

Monoamine Oxidase

A prominent feature that accompanies aging is an increase in monoamine oxidase (MAO), which is an important source of oxidative stress. MAO is a mitochondria-bound isoenzyme, which catalyzes the oxidative deamination of dietary amines and monoamine neurotransmitters. The byproducts of these reactions include a number of potentially neurotoxic species, such as hydrogen peroxide (H_2O_2). Two different types of MAO, named A and B, have been characterized. MAO-A is present in catecholaminergic neurons, and MAO-B is present in astrocytes, interneurons, and serotonergic neurons. Several authors reported high brain MAO-B activity in neurodegenerative diseases such as Parkinson's and AD without any changes in MAO-A enzyme activity (Bortolato, Chen, and Shih 2008). Previous studies suggest a modulatory effect of DHEA on MAO. Recently, one study investigated the effects of exogenous administration of DHEA on the following age-related parameters such as monoamine oxidase activity, lipid peroxidation, and lipofuscin accumulation in brain regions of aging rats (Kumar et al. 2008). The results showed that chronic DHEA treatment reduced the aging-induced (14 and 24 months) increase of total MAO activity in the whole brain hemispheres, without altering enzyme activity in adult (4-month-old) rats. DHEA inhibited MAO activity, but it remains to be determined if that effect involves one or both MAO isoforms as well as if it occurs acutely in young animals. A more recent study evaluated the acute effect of DHEA on MAO activity in the corpus striatum (CS) and the nucleus accumbens (NAc) *in vivo* and *in vitro* (Pérez-Neri, Montes, and Ríos 2009). It found an acute inhibitory effect of DHEA (120 mg/kg) on the total MAO activity in the NAc, but not in the CS. No significant difference was observed when MAO-A and MAO-B activities were independently analyzed. When assayed *in vitro*, total MAO, MAO-A, and MAO-B activities were found to be reduced by DHEA in the NAc and in the CS, respectively (IC_{50}, 4.7–56.1 μM). Another investigation determined platelet MAO-B activity in patients with AD subdivided according to the severity of dementia into groups of patients in the early, middle, and late phase of AD (Muck-Seler et al. 2009). This is the first report of the significantly reduced platelet MAO-B activity in patients in the late phase of AD compared to patients in the early phase of AD. The reason for the altered platelet MAO-B activity during the progress of AD is at present unknown. There are several factors that might influence platelet MAO-B activity such as aging, sex, alcohol abuse, smoking, different medication, and ethnicity (Oreland 2004).

DHEA and Calcium Flux in Alzheimer's Disease

Calcium signaling is utilized by neurons to control a variety of functions, including membrane excitability, neurotransmitter release, gene expression, free radical species formation, synaptic plasticity, and learning and memory, as well as pathophysiology, including necrosis, apoptosis, and degeneration (Bezprozvanny 2009). Intracellular levels are maintained by receptor-operated, voltage-gated, or store-operated calcium channels in the plasma membrane and by ER-resident channels. Disruption of cellular Ca^{2+} homeostasis in neurons of AD patients was observed for many years. The correlations between the pathological hallmarks of AD (amyloid plaques and NFTs) and perturbed cellular Ca^{2+} homeostasis have been established in the studies of patients and in animal and cell culture models of AD. Numerous studies have specifically implicated that increased levels of AβP induce neurotoxic factors including ROS and cytokines, which impair cellular Ca^{2+} homeostasis and render neurons vulnerable to apoptosis and excitotoxicity (Yamamoto et al. 2007). It is also important to note that aging is the principal risk factor in AD, and calcium dysregulation in aged brains is one of the molecular hypotheses of aging-dependent brain impairment (Thibault, Gant, and Landfield 2007). Recent approaches have emphasized the importance of AβP oligomerization in the pathogenesis of AD, which causes synaptic degeneration and neuronal loss. Recent findings have demonstrated that amyloid β oligomers induce calcium dysregulation and neuronal death through the activation of ionotropic glutamate receptors, which cause mitochondrial dysfunction as indicated by mitochondrial Ca^{2+} overload, oxidative stress, and mitochondrial membrane depolarization (Alberdi et al. 2010).

The precise contribution of calcium dysregulation to the pathogenesis of this disease remains unclear, but studies indicate that systemic calcium changes accompany almost the whole brain pathology process that is observed in AD (Yu, Chang, and Tan 2009).

It has been shown that DHEA and DHEAS inhibit depolarization, which induces an increase in intracellular calcium only at relatively high concentrations above 10 (for DHEA) or 60 (for DHEAS) μM in cultured hippocampal neurons (Kurata et al. 2001), but to the best of our knowledge, no study concerning the effect of DHEA on Ca channel in the case of AD has been reported.

DHEA and *N*-Methyl-d-Aspartate Receptors in Alzheimer's Disease

Among the underlying mechanisms leading to neurodegeneration is the excessive activation of glutamate receptors by excitatory amino acids including NMDARs. Overactivation of the NMDA subtype of glutamate receptor is known to trigger excessive calcium influx, contributing to neurodegenerative conditions. Such dysregulation of calcium signaling results in the generation of excessive free radicals, including ROS and RNS. The NMDAR is used as a target for clinical AD treatment, which implies that the protein network of NMDAR is involved in synaptic dysfunction in AD (Robinson and Keating 2006).

DHEA (but not its sulfate) is a potent positive modulator of NMDARs, causing an increase in Ca^{2+} flux (Compagnone and Mellon 2000). Both *in vivo* and *in vitro* studies have shown that DHEA functions as a neurotrophic or neuroprotective factor to prevent NMDA-induced neurotoxicity (Kurata et al. 2004). The results of these authors showed that DHEA (1–60 μM) has significant neuroprotective effects against NMDA-induced neurotoxicity, whereas 1–30 μM DHEA did not inhibit the NMDA-induced $[Ca2^{+}]i$ increases. They also demonstrated that 10 μM DHEA inhibited NMDA-induced NOS activity and NO production. However, the exact role of NMDAR activation in AD is far from being elucidated (Gardoni and Di Luca 2006).

DHEA and σ-Receptors

σ1-Receptor has received considerable attention in the regulation of cognitive function. It is classified into at least two subtypes, namely, σ1 and σ2. The σ1-receptor is a unique intraneuronal protein that modulates intracellular Ca^{2+} mobilization and extracellular Ca^{2+} influx, leading to a wide spectrum of neuromodulatory activities (Maurice 2002). It has also been well documented that σ1-receptors regulate the activity of NMDAR channels (Martina et al. 2007).

σ1-Receptors in patients with AD were studied by Sultana et al. (2006). The density of σ1-receptors in the cerebellum was significantly lower in AD than in controls, although K1 in AD was comparable with that in controls. Although the cerebellum was formerly thought to be unaffected in AD, many studies have revealed cerebellar changes in AD patients.

A more recent study investigated the mapping of σ1-receptors in AD using [11C]SA4503 positron emission tomography (PET; Mishina et al. 2008). These results showed that the density of cerebral and cerebellar σ1-receptors is reduced in early AD. Although an endogenous ligand for the σ-receptors remains unclear, some studies have reported that steroid hormones such as progesterone and testosterone might interact with σ-receptors. The σ1-receptor agonists are also expected as drugs for improving the cognitive deficits of AD (Maurice 2002). On the one hand, antiamnesic potencies of σ-receptor agonists and DHEAS have also been evaluated in an AD-type amnesia model created by βA_{25-35} (Maurice, Su, and Privat 1998); on the other hand, DHEA is an endogenous σ1-agonist (Maurice 2002). In addition, selective σ1-agonists, as well as DHEA, showed marked neuroprotective activity *in vitro* against oxidative stress-related damages. Acting chronically through the σ1-receptor may indeed offer a new way to alleviate the cognitive disturbances observed in AD and promote long-term improvements (Maurice 2002). However, the exact molecular mechanism of the cellular protective action of the σ1-receptor remains elusive (Hayashi and Su 2008). To date, however, little has been reported concerning the role of DHEA(S)/σ1-receptor interaction in AD.

CONCLUSION

AD is the most common form of neurodegenerative disease. One of the major problems with AD diagnostics and treatment is the inability of clinicians and biochemists to determine the onset of the disease. Numerous biochemical pathways are affected in AD, the specific cause of which remains undetermined. However, a common key feature in AD and the aging brain is that it is particularly vulnerable to oxidative damage induced by increased oxidative stress, but there is much debate as to whether this is a cause or a consequence of AD (Praticò 2008; Sayre, Perry, and Smith 2008).

From a biochemical point of view, the typical feature of AD is the general impairment of oxidative metabolism in brain tissues, and the biochemical mechanisms behind these events are in most instances known or are being intensively studied (Moreira et al. 2010). This chapter demonstrates the complexity of the biochemical mechanisms in AD, which are often interrelated and often involve oxidative stress. A series of recent studies has begun to focus on oxidative stress, suggesting that it is a primary event in AD, and the possibility that oxidative stress is a possible primary event in AD indicates that antioxidant-based therapies are perhaps the most promising weapons against this devastating neurodegenerative disorder (Moreira et al. 2006; Bonda et al. 2010).

DHEA is a multifunctional steroid that is known to be involved in a variety of functional activities in the CNS. DHEA(S) and its metabolites are among the important factors that are involved in the pathogenesis of AD. The recent research of AD biochemistry focuses on molecular signaling and the role of steroids such as DHEA. Although a large number of studies have been carried out on DHEA, the mechanisms of neuroprotection remain unclear.

A series of various studies and observations in human trials (Von Mühlen et al. 2007; Yamada et al. 2007), animal models, and *in vitro* findings (Pérez-Neri, Montes, and Ríos 2009) support the potential utility of DHEA as a therapeutic intervention. A significant clinical interest in DHEA is based on many observations, including an important decline of production since early adulthood, literature evidence showing changes in the steroid levels associated with multiple pathologies, and a pronounced replacement therapy with DHEA, which may alleviate age-associated declines in a range of functions. Since great controversy in this subject still remains and no sufficient data are available in the literature to support its secure recommendation, it is indispensable to better understand DHEA's role in oxidative stress and peripheral blood.

Current therapeutic approaches suggest that drugs acting at a single target may be insufficient for the treatment of multifactorial neurodegenerative diseases such as AD, which is characterized by the coexistence of multiple etiopathologies (Weinreb et al. 2009). In this way, DHEA(s) are still intriguing hormones, but it is not enough evidence to recommend routine treatment with DHEA. However, few DHEA analogs have been developed, and this approach must give more attention to designing other similar molecules for the development of novel pharmaceutical drugs (Matsuya et al. 2009; Bazin et al. 2009). The study and exploration of the biochemical mechanisms of steroid hormone biosynthesis could be promising ways to synthesize new analogs of DHEA as new promising drugs.

REFERENCES

Ahmed, R. R., C. J. Holler, R. L. Webb et al. 2010. BACE1 and BACE2 enzymatic activities in Alzheimer's disease. *J Neurochem* 112(4):1045–53.

Alberdi, E., M. V. Sánchez-Gómez, F. Cavaliere et al. 2010. Amyloid beta oligomers induce Ca^{2+} dysregulation and neuronal death through activation of ionotropic glutamate receptors. *Cell Calcium* 47(3):264–72.

Aliev, G., H. H. Palacios, A. E. Lipsitt et al. 2009. Nitric oxide as an initiator of brain lesions during the development of Alzheimer disease. *Neurotox Res* 16(3):293–305.

Bastianetto, S., C. Ramassamy, J. Poirier et al. 1999. Dehydroepiandrosterone (DHEA) protects hippocampal cells from oxidative stress-induced damage. *Brain Res Mol Brain Res* 66:35–41.

Bazin, M. A., L. El Kihel, M. Boulouard et al. 2009. The effects of DHEA, 3beta-hydroxy-5alpha-androstane-6,17-dione, and 7-amino-DHEA analogues on short term and long term memory in the mouse. *Steroids* 74(12):931–7.

Bertram, L., and R. E. Tanzi. 2004. The current status of Alzheimer's disease genetics: What do we tell the patients? *Pharmacol Res* 50:385–96.

Bezprozvanny, I. 2009. Calcium signaling and neurodegenerative diseases. *Trends Mol Med* 15(3):89–100.

Bo, M., M. Massaia, P. Zannella et al. 2006. Dehydroepiandrosterone sulfate (DHEA-S) and Alzheimer's dementia in older subjects. *Int J Geriatr Psychiatry* 21(11):1065–70.

Bonda, D. J., X. Wang, G. Perry et al. 2010. Oxidative stress in Alzheimer disease: A possibility for prevention. *Neuropharmacology.* 59(4–5):290–4.

Borghi, R., S. Patriarca, N. Traverso et al. 2007. The increased activity of BACE1 correlates with oxidative stress in Alzheimer's disease. *Neurobiol Aging* 28:1009–14.

Bortolato, M., K. Chen, and J. C. Shih. 2008. Monoamine oxidase inactivation: From pathophysiology to therapeutics. *Adv Drug Deliv Rev* 60:1527–33.

Brown, R. C., C. Cascio, and V. Papadopoulos. 2000. Pathways of neurosteroid biosynthesis in cell lines from human brain: Regulation of dehydroepiandrosterone formation by oxidative stress and beta-amyloid peptide. *J Neurochem* 74(2):847–59.

Brown, R. C., Z. Han, C. Cascio, and V. Papadopoulos. 2003. Oxidative stress-mediated DHEA formation in Alzheimer's disease pathology. *Neurobiol Aging* 24:57–65.

Calabrese, V., C. Mancuso, M. Calvani et al. 2007a. Nitric oxide in the central nervous system: Neuroprotection versus neurotoxicity. *Nat Rev Neurosci* 8:766–75.

Calabrese, B., G. M. Shaked, I. V. Tabarean et al. 2007b. Rapid, concurrent alterations in pre- and postsynaptic structure induced by naturally-secreted amyloid-beta protein. *Mol Cell Neurosci* 35:183–93.

Castegna, A., V. Thongboonkerd, J. B. Klein et al. 2003. Proteomic identification of nitrated proteins in Alzheimer's disease brain. *J Neurochem* 85:1394–401.

Chalbot, S., H. Zetterberg, K. Blennow et al. 2009. Cerebrospinal fluid secretory Ca^{2+}-dependent phospholipase A_2 activity is increased in Alzheimer disease. *Clin Chem* 55(12):2171–9.

Cho, S. H., B. H. Jung, W. Y. Lee et al. 2006. Rapid column-switching liquid chromatography/mass spectrometric assay for DHEA-sulfate in the plasma of patients with Alzheimer's disease. *Biomed Chromatogr* 20(10):1093–7.

Cleary, J. P., D. M. Walsh, J. J. Hofmeister et al. 2005. Natural oligomers of the amyloid-beta protein specifically disrupt cognitive function. *Nat Neurosci* 8:79–84.

Compagnone, N. A., and S. H. Mellon. 2000. Neurosteroids: Biosynthesis and function of these novel neuromodulators, *Front Neuroendocrinol* 21:1–56.

De Kimpe, L., and W. Scheper. 2010. From alpha to omega with Abeta: Targeting the multiple molecular appearances of the pathogenic peptide in Alzheimer's disease. *Curr Med Chem* 17(3):198–212.

De Strooper, B., R. Vassar, and T. Golde. 2010. The secretases: Enzymes with therapeutic potential in Alzheimer disease. *Nat Rev Neurol* 6(2):99–107.

Finder, V. H., and R. Glockshuber. 2007. Amyloid-beta aggregation. *Neurodegener Dis* 4(1):13–27.

Finder, V. H., I. Vodopivec, R. M. Nitsch et al. 2010. The recombinant amyloid-β peptide Aβ1–42 aggregates faster and is more neurotoxic than synthetic Aβ1–42. *J Mol Biol* 396:9–18.

Gardoni, F., and M. Di Luca. 2006. New targets for pharmacological intervention in the glutamatergic synapse. *Eur J Pharmacol* 545(1):2–10.

Hampl, R., and M. Bicíková, 2010. Neuroimmunomodulatory steroids in Alzheimer dementia. *J Steroid Biochem Mol Biol* 119(3–5):97–104.

Hayashi, T., and T. P. Su. 2008. An update on the development of drugs for neuropsychiatric disorders: Focusing on the σ 1 receptor ligand. *Expert Opin Ther Targets* 12(1):45–58.

Hoskin, E. K., M. X. Tang, J. J. Manly et al. 2004. Elevated sex-hormone binding globulin in elderly women with Alzheimer's disease. *Neurobiol Aging* 25(2):141–7.

Kaminsky, Y. G., M. W. Marlatt, M. A. Smith et al. 2010. Subcellular and metabolic examination of amyloid-beta peptides in Alzheimer disease pathogenesis: Evidence for Abeta(25–35). *Exp Neurol* 221(1):26–37.

Keller, J. N., F. A. Schmitt, S. W. Scheff et al. 2005. Evidence of increased oxidative damage in subjects with mild cognitive impairment. *Neurology* 64(7):1152–6.

Kim, S. B., M. Hill, Y. Kwak et al. 2003. Neurosteroids: Cerebrospinal fluid levels for Alzheimer's disease and vascular dementia diagnostics. *J Clin Endocrinol Metab* 88:5199–206.

Kumar, P., A. Taha, D. Sharma et al. 2008. Effect of dehydroepiandrosterone (DHEA) on monoamine oxidase activity, lipid peroxidation and lipofuscin accumulation in aging rat brain regions. *Biogerontology* 9(4):235–46.

Kurata, K., M. Takebayashi, A. Kagaya et al. 2001. Effect of β–estradiol on voltage-gated Ca2+ channels in rat hippocampal neurons: A comparison with dehydroepiandrosterone. *Eur J Pharmacol* 416:203–12.

Kurata, K., M. Takebayashi, S. Morinobu et al. 2004. Beta-estradiol, dehydroepiandrosterone, and dehydroepiandrosterone sulfate protect against N-methyl-D-aspartate-induced neurotoxicity in rat hippocampal neurons by different mechanisms. *J Pharmacol Exp Ther* 311(1):237–45.

Legrain, S., C. Berr, N. Frenoy et al. 1995. Dehydroepiandrosterone sulfate in a long-term care aged population. *Gerontology* 41:343–51.

Li, A., and J. C. Bigelow. 2010. The 7-hydroxylation of dehydroepiandrosterone in rat brain. *Steroids* 75(6):404–10.

Li, L., J. Dai, L. Ru et al. 2004. Effects of transhinone on neuropathological changes induced by amyloid β-peptide$_{1-40}$ injection in rat hippocampus. *Acta Pharmacol Sin* 25(7):861–8.

Liu, D., and J. S. Dillon. 2004. Dehydroepiandrosterone stimulates nitric oxide release in vascular endothelial cells: Evidence for a cell surface receptor. *Steroids* 69:279–89.

Liu, Y., and D. R. Schubert. 2009. The specificity of neuroprotection by antioxidants. *J Biomed Sci* 16:98–112.

Malinski, T. 2007. Nitric oxide and nitroxidative stress in Alzheimer's disease. *J Alzheimer's Dis* 11(2):207–18.

Martina, M., M. E. Turcotte, S. Halman et al. 2007. The sigma-1 receptor modulates NMDA receptor synaptic transmission and plasticity via SK channels in rat hippocampus. *J Physiol* 578(1):143–57.

Matsuya, Y., Y. Yamakawa, C. Tohda et al. 2009. Synthesis of sominone and its derivatives based on an RCM strategy: Discovery of a novel anti-Alzheimer's disease medicine candidate "denosomin." *Org Lett* 11(17):3970–3.

Maurice, T. 2002. Improving Alzheimer's disease-related cognitive deficits with sigma1 receptor agonists. *Drug News Perspect* 15:617–25.

Maurice, T., T. P. Su, and A. Privat. 1998. Sigma (σ) receptor agonists and neurosteroids attenuate β-amyloid peptide-induced amnesia in 25–35 mice through a common mechanism. *Neuroscience* 83:413–28.

Mishina, M., M. Ohyama, K. Ishii et al. 2008. Low density of sigma1 receptors in early Alzheimer's disease. *Ann Nucl Med* 22(3):151–6.

Moreira, P. I., L. M. Sayre, X. Zhu et al. 2010. Detection and localization of markers of oxidative stress by in situ methods: Application in the study of Alzheimer disease. *Methods Mol Biol* 610:419–34.

Moreira, P. I., X. Zhu, A. Nunomura et al. 2006. Therapeutic options in Alzheimer's disease. *Expert Rev Neurother* 6(6):897–910.

Muck-Seler, D., P. Presecki, N. Mimica et al. 2009. Platelet serotonin concentration and monoamine oxidase type B activity in female patients in early, middle and late phase of Alzheimer's disease. *Prog Neuropsychopharmacol Biol Psychiatry* 33(7):1226–31.

Naylor, J. C., C. M. Hulette, D. C. Steffens et al. 2008. Cerebrospinal fluid dehydroepiandrosterone levels are correlated with brain dehydroepiandrosterone levels, elevated in Alzheimer's disease, and related to neuropathological disease stage. *J Clin Endocrinol Metab* 93(8):3173–8.

Oreland, L. 2004. Platelet monoamine oxidase, personality and alcoholism: The rise, fall and resurrection. *Neurotoxicology* 25:79–89.

Pérez-Neri, I., S. Montes, and C. Ríos. 2009. Inhibitory effect of dehydroepiandrosterone on brain monoamine oxidase activity: In vivo and in vitro studies. *Life Sci* 85(17–18):652–6.

Praticò, D. 2008. Evidence of oxidative stress in Alzheimer's disease brain and antioxidant therapy: Lights and shadows. *Ann N Y Acad Sci* 1147:70–8.

Provias, J., and B. Jeynes. 2008. Neurofibrillary tangles and senile plaques in Alzheimer's brains are associated with reduced capillary expression of vascular endothelial growth factor and endothelial nitric oxide synthase. *Curr Neurovasc Res* 5:199–205.

Ramassamy, C., D. Averill, U. Beffert et al. 1999. Oxidative damage and protection by antioxidants in the frontal cortex of Alzheimer's disease is related to the apolipoprotein E genotype. *Free Radic Biol Med* 27:544–53.

Rasmuson, S., B. Nasman, K. Carlstrom et al. 2002. Increased levels of adrenocortical and gonadal hormones in mild to moderate Alzheimer's disease. *Dement Geriatr Cogn Disord* 13(2):74–9.

Reddy, P. H. 2009. Role of mitochondria in neurodegenerative diseases: Mitochondria as a therapeutic target in Alzheimer's disease. *CNS Spectr* 14(8 Suppl 7):8–13.

Robinson, D. M., and G. M. Keating. 2006. Memantine: A review of its use in Alzheimer's disease. *Drugs* 66(11):1515–34.

Rosales-Corral, S., D.-X. Tan, R. J. Reiter et al. 2004. Kinetics of the neuroinflammation-oxidative stress correlation in rat brain following the injection of fibrillar amyloid-β onto the hippocampus in vivo. *J Neuroimmunol* 150:20–8.

Sattler, R., and M. Tymianski. 2000. Molecular mechanisms of calcium-dependent excitotoxicity. *J Mol Med* 78:3–13.

Sayre, L. M., G. Perry, and M. A. Smith. 2008. Oxidative stress and neurotoxicity. *Chem Res Toxicol* 21:172–88.

Selkoe, D. J. 2001. Alzheimer's disease: Genes, proteins, and therapy. *Physiol Rev* 81:741–66.

Simoncini, T., P. Mannella, L. Fornari et al. 2003. Dehydroepiandrosterone modulates endothelial nitric oxide synthesis via direct genomic and nongenomic mechanisms. *Endocrinology* 144:3449–55.

Sultana, R., D. Boyd-Kimball, H. F. Poon et al. 2006. Redox proteomics identification of oxidized proteins in Alzheimer's disease hippocampus and cerebellum: An approach to understand pathological and biochemical alterations in AD. *Neurobiol Aging* 27:1564–76.

Tamagno, E., M. Guglielmotto, M. Aragno et al. 2008. Oxidative stress activates a positive feedback between the gamma- and beta-secretase cleavages of the beta-amyloid precursor protein. *J Neurochem* 104(3):683–95.

Tamagno, E., M. Guglielmotto, P. Bardini et al. 2003. Dehydroepiandrosterone reduces expression and activity of BACE in NT2 neurons exposed to oxidative stress. *Neurobiol Dis* 14(2):291–301.

Taupin, P. 2010. A dual activity of ROS and oxidative stress on adult neurogenesis and Alzheimer's disease. *Cent Nerv Syst Agents Med Chem* 10:16–21.

Thibault, O., J. C. Gant, and P. W. Landfield. 2007. Expansion of the calcium hypothesis of brain aging and Alzheimer's disease: Minding the store. *Aging Cell* 6:307–17.

Thorns, V., M. Mallory, L. Hansen et al. 1997. Alterations in glutamate receptor 2/3 subunits and amyloid precursor protein expression during the course of Alzheimer's disease and Lewy body variant. *Acta Neuropathol* 94(6):539–48.

Tong, Y., W. Zhou, V. Fung et al. 2005. Oxidative stress potentiates BACE1 gene expression and Abeta generation. *Neural Transm* 112:455–69.

Verdier, Y., M. Zarándi, and B. Penke. 2004. Amyloid beta-peptide interactions with neuronal and glial cell plasma membrane: Binding sites and implications for Alzheimer's disease. *J Pept Sci* 10(5):229–48.

Von Mühlen, D., G. A. Laughlin, D. Kritz-Silverstein et al. 2007. The Dehydroepiandrosterone and wellness (DAWN) study: Research design and methods. *Contemp Clin Trials* 28:153–68.

Walsh, D. M., I. Klyubin, J. V. Fadeeva et al. 2002. Naturally secreted oligomers of amyloid beta protein potently inhibit hippocampal long-term potentiation in vivo. *Nature* 416:535–9.

Wang, J., W. R. Markesbery, and M. A. Lovell. 2006. Increased oxidative damage in nuclear and mitochondrial DNA in mild cognitive impairment. *J Neurochem* 96:825–32.

Weill-Engerer, S., J. P. David, V. Sazdovitch et al. 2002. Neurosteroid quantification in human brain regions: Comparison between Alzheimer's and non-demented patients. *J Clin Endocrinol Metab* 87:5138–43.

Weill-Engerer, S., J. P. David, V. Sazdovitch et al. 2003. In vitro metabolism of dehydroepiandrosterone (DHEA) to 7alpha-hydroxy-DHEA and Delta5-androstene-3beta, 17beta-diol in specific regions of the aging brain from Alzheimer's and non-demented patients. *Brain Res* 969(1–2):117–25.

Weinreb, O., S. Mandel, O. Bar-Am et al. 2009. Multifunctional neuroprotective derivatives of rasagiline as anti-Alzheimer's disease drugs. *Neurotherapeutics* 6(1):163–74.

Williams, M. R. I., T. Dawood, S. Ling et al. 2004. Dehydroepiandrosterone increases endothelial cell proliferation in vitro and improves endothelial function in vivo by mechanisms independent of androgen and estrogen receptors. *J Clin Endocrinol Metab* 89:4708–15.

Wiltfang, J., H. Esselmann, M. Bibl et al. 2002. Highly conserved and disease-specific patterns of carboxy-terminally truncated Abeta peptides 1-37/38/39 in addition to 1–40/42 in Alzheimer's disease and in patients with chronic neuroinflammation. *J Neurochem* 81:481–96.

Yamada, Y., H. Sekihara, M. Omura et al. 2007. Changes in serum sex hormone profiles after short-term low-dose administration of DHEA to young and elderly persons. *Endocr J* 54:153–62.

Yamamoto, S., T. Wajima, Y. Hara et al. 2007. Transient receptor potential channels in Alzheimer disease. *Biochim Biophys Acta* 1772:958–67.

Yu, J. T., R. C. Chang, and L. Tan. 2009. Calcium dysregulation in Alzheimer's disease: From mechanisms to therapeutic opportunities. *Prog Neurobiol* 89(3):240–55.

Index

C

D

E

F

G

H

I

J

K

L

M

T

U

V

W